LANDOLT-BÖRNSTEIN

Numerical Data and Functional Relationships in Science and Technology

New Series

Editors in Chief: K.-H. Hellwege · O. Madelung

Group V: Geophysics and Space Research

Volume 3

Oceanography

H. G. Gierloff-Emden · N. K. Højerslev · G. Krause
H. Peters · G. Siedler · G. Weichart · P. Wille

Editor: J. Sündermann

Subvolume a

Springer-Verlag Berlin Heidelberg New York
London Paris Tokyo

LANDOLT-BÖRNSTEIN

Zahlenwerte und Funktionen aus Naturwissenschaften und Technik

Neue Serie

Gesamtherausgabe: K.-H. Hellwege · O. Madelung

Gruppe V: Geophysik und Weltraumforschung

Band 3

Ozeanographie

H. G. Gierloff-Emden · N. K. Højerslev · G. Krause
H. Peters · G. Siedler · G. Weichart · P. Wille

Herausgeber: J. Sündermann

Teilband a

Springer-Verlag Berlin Heidelberg New York
London Paris Tokyo

ISBN 3-540-15092-7 Springer-Verlag Berlin Heidelberg New York
ISBN 0-387-15092-7 Springer-Verlag New York Heidelberg Berlin

CIP-Kurztitelaufnahme der Deutschen Bibliothek
Zahlenwerte und Funktionen aus Naturwissenschaften und Technik/Landolt-Börnstein. – Berlin; Heidelberg; New York; London; Paris; Tokyo: Springer. Parallelt.: Numerical data and functional relationships in science and technology. – Teilw. mit d. Erscheinungsorten Berlin, Heidelberg, New York. Teilw. mit d. Erscheinungsorten Berlin, Heidelberg, New York, Tokyo. – N.S. teilw. Gesamthrsg.: O. Madelung
NE: Landolt, Hans [Begr.]; PT – N.S. Gesamthrsg.: K.-H. Hellwege; O. Madelung. Gruppe 5, Geophysik und Weltraumforschung. Bd. 3. Ozeanographie. Teilbd. a. H. G. Gierloff-Emden ... Hrsg.: J. Sündermann. – 1986.
ISBN 3-540-15092-7 (Berlin ...)
ISBN 0-387-15092-7 (New York ...)
NE: Hellwege, Karl-Heinz [Hrsg.]; Madelung, Otfried [Hrsg.]; Gierloff-Emden, Hans-Günter [Mitverf.]; Sündermann, Jürgen [Hrsg.]

Printed in Germany

Typesetting: Brühlsche Universitätsdruckerei, Giessen
Printing: Druckhaus Langenscheidt KG, Berlin
Bookbinding: Lüderitz & Bauer-GmbH, Berlin

2163/3020–543210

Editor

J. Sündermann
Institut für Meereskunde der Universität Hamburg
Heimhuder Str. 71, D-2000 Hamburg 13

Contributors

H. G. Gierloff-Emden
Institut für Geographie der Universität München
Luisenstr. 37, D-8000 München 2

N. K. Højerslev
Institut for Fysisk Oceanografi, Københavns Universitet
Haraldsgade 6, DK-2200 N København

G. Krause
Alfred-Wegener-Institut für Polar- und Meeresforschung
Columbusstraße, D-2850 Bremerhaven

H. Peters
Institut für Meereskunde an der Universität Kiel
Düsternbrooker Weg 20, D-2300 Kiel 1

G. Siedler
Institut für Meereskunde an der Universität Kiel
Düsternbrooker Weg 20, D-2300 Kiel 1

G. Weichart
Deutsches Hydrographisches Institut, Labor Sülldorf,
Wüstland, D-2000 Hamburg 55

P. Wille
Forschungsanstalt der Bundeswehr für Wasserschall- und Geophysik
Klausdorfer Weg 2–24, D-2300 Kiel 14

Vorwort

In der 6. Auflage des Landolt-Börnstein war das Sachgebiet Ozeanographie in den Band Astronomie und Geophysik (1952) integriert und umfaßte 116 Seiten. Hiermit wird jetzt im Rahmen der Neuen Serie des Landolt-Börnstein ein eigener Band (V/3) mit dem Titel "Oceanography" vorgelegt, der einen Umfang von ca. 1200 Seiten aufweist. Dieser Band wird in drei Teilbänden (V/3a, b, c) erscheinen. Schon daran lassen sich die stürmische Zunahme unserer Kenntnisse und die wachsende Bedeutung der Ozeanographie in einem Zeitraum von 34 Jahren ermessen. Ich bin dem Springer-Verlag dankbar, daß er diese Entwicklung erkannt und der Ozeanographie entsprechenden Raum in der neuen Serie bereitgestellt hat.

Es kann dennoch nicht behauptet werden, daß die Ozeanographie bislang den Reifegrad und die Abgeschlossenheit der klassischen Physik erreicht hätte. Viele, auch grundlegende, Fragen befinden sich noch in der Bearbeitung oder sind gar noch unerforscht. Entsprechend schwer ist es gewesen, in der Darstellung der Ozeanographie den traditionellen Stil des Landolt-Börnstein, der ja Figuren und Tabellen liefern soll, zu treffen. Ich habe mich sehr bemüht, im Zusammenwirken mit den Autoren hier eine gute Mischung aus grundlegender Information und quantitativen Angaben zu finden. Es hat sich indessen nicht vermeiden lassen, daß auch mehr erläuternder Text und mehr Formeln in den Band eingegangen sind als in diejenigen Bände, die den besser erforschten Teilen der Physik gewidmet sind. Ich danke hier vor allem dem Gesamtherausgeber für sein Verständnis. Ich glaube, daß die schließlich gefundene Form einen guten Kompromiß darstellt und für die ozeanographische Fachwelt interessant und nützlich ist.

An dieser Stelle möchte ich den Begriff Ozeanographie für dieses Werk präzisieren, ohne einen Anspruch auf Allgemeingültigkeit zu stellen. Gemeint ist *physikalische* Ozeanographie, ein Teilgebiet der Geophysik und damit der Physik. Die Kapitel "Topography" und "Coastal Oceanography" stellen unter diesem Gesichtspunkt Randbereiche dar, da sie mehr geographisch orientiert sind. Andererseits sind sie für die Physik des Ozeans von großer Bedeutung.

Damit bin ich bei der generellen Gliederung des Bandes. Obwohl diese Gliederung zweifellos eine größere Systematik und Vollständigkeit aufweist als die Gliederung in der 6. Auflage des Landolt-Börnstein, ist sie sicher immer noch etwas subjektiv und lückenhaft. Hier mußte ich mich nach dem aktuellen Bedarf der Ozeanographie richten, aber auch nach den Spezialgebieten und den Interessen der beitragenden Fachkollegen. Es ist daher nicht auszuschließen, daß – vom Standpunkt einer lehrbuchartigen Gesamtdarstellung – manches Kapitel zu speziell erscheint, während vielleicht ein anderes zu allgemein ist. Es kann auch sein, daß ein wünschenswertes Kapitel ganz fehlt. Ich habe mich aber nach Kräften bemüht, in Zusammenarbeit mit den Autoren das Gesamtgebiet der physikalischen Ozeanographie einigermaßen zu überdecken, Lücken zu schließen und Beiträge aufeinander abzustimmen. Mehr erscheint bei einem solchen Gemeinschaftswerk, das ja die speziellen Kenntnisse eines jeden Autors nutzen soll, nicht geboten.

Ich bin den Autoren, die sich alle sehr schnell und engagiert bereiterklärt haben mitzuarbeiten, für ihren Enthusiasmus, ihre Kompromißbereitschaft und ihre Geduld zu größtem Dank verpflichtet. Es gehört einiger Idealismus dazu, in einer Zeit, in der kurze Originalveröffentlichungen mehr gefragt sind als längere systematische Beiträge, dieses

Arbeitspensum aufzubringen. Es bedarf auch eines gewissen Mutes, auf einem sich schnell entwickelnden Fachgebiet Wissen zu einem bestimmten Zeitpunkt festzuschreiben. Dies kann nur unter der Prämisse geschehen, daß es sich eben um den gegenwärtigen Erkenntnisstand handelt, der künftig gewiß Änderungen unterworfen sein wird. Es gehört schließlich einige Einsicht dazu, im Interesse einer Homogenisierung des Gesamtwerkes und im Interesse seiner Anpassung an den Stil des Landolt-Börnstein, die Einwände und Korrekturwünsche von Herausgeber, Gesamtherausgeber und Redaktionsstab (Dr. W. Finger, Frau G. Burfeindt und Frau H. Hämmer) zu akzeptieren. Alles dies ist in weitgehender Harmonie geschehen.

Leider hat es sich bei den unterschiedlichen Belastungen der Autoren, bei ihren vielseitigen anderweitigen Verpflichtungen, nicht vermeiden lassen, daß Verzögerungen bei einzelnen Beiträgen aufgetreten sind, die den Abschluß des Gesamtwerkes doch erheblich beeinträchtigt haben. Um nicht weitere Zeit zu verlieren, wurde deswegen Teilband V/3c (Kapitel 6 bis 9) zuerst publiziert. Als nächstes schließt sich der vorliegende Teilband V/3a an, der die Kapitel 1 bis 3 umfaßt. Teilband V/3b (Kapitel 4 und 5) wird möglichst bald folgen.

Dieser Band wird – wie alle Bände des Landolt-Börnstein – ohne finanzielle Unterstützung durch andere Stellen gedruckt.

Hamburg, August 1986 **Der Herausgeber**

Preface

In the 6th edition of Landolt-Börnstein, the field of oceanography was incorporated into the volume on Astronomy and Geophysics (1952) and covered 116 pages. In this new edition a whole volume (V/3) of the New Series of Landolt-Börnstein is devoted to oceanography with a total of, approximately, 1200 pages, to be published in three subvolumes (V/3a, b, c). This demonstrates the tremendous increase in knowledge and the growing significance of oceanography over the past 34 years. I am grateful to Springer-Verlag for recognizing this development and providing the appropriate space in the New Series for this subject.

In spite of such an impressive accumulation of new knowledge, it can hardly be claimed that oceanography has reached the maturity and conclusiveness of classical physics. Many questions, even fundamental issues, are still in early stages of development or have not yet been explored. Accordingly, it has been rather difficult to represent oceanography in terms of the traditional Landolt-Börnstein style by means of figures and tables.

Editor and authors have made a considerable effort to find the proper combination of basic information and data. However, it has not been possible to avoid including more explanatory text and more formulas than have been necessary in the volumes dealing with more developed areas of physical sciences. I am particularly grateful to the editor-in-chief for his understanding. I believe the final form represents a good compromise and will be interesting and helpful to the oceanographic community.

Let me clarify here that under our general topic we are dealing with *physical* oceanography, a part of geophysics and physics. From this point of view the chapters "Topography" and "Coastal Oceanography" represent peripherical areas, since they are more geographically oriented. On the other hand, they are very pertinent to the physics of the ocean.

This brings me to the general organization of this volume. Although it is undoubtedly more systematic and comprehensive than in the previous (6th edition) Landolt-Börnstein volume, it is certainly still somewhat subjective and incomplete. Here it was necessary to consider not only the present needs of oceanography but also the specific fields and interests of the contributors. It is therefore possible that, in comparison to a textbook, some chapters might appear to be too specialized or too general. It is also possible that a desirable chapter is entirely missing. I have tried my best, in cooperation with the authors, to cover more or less completely the entire field of physical oceanography, to close gaps and to balance the individual contributions. The remaining inhomogeneity is surely more than compensated by the advantages of letting the authors demonstrate their individual expertise.

I am much obliged to the authors who all promptly committed their cooperation, for their enthusiasm, patience and willingness to compromise. It requires a certain amount of idealism to take on the extra work involved in writing long and systematic articles in a time when short, original publications are the order of the day. It requires courage to draw the line at a particular point in time and write about factual knowledge in such a rapidly developing field; this must be done under the premise that this transient state of the art will be revised in the future. Subsequently, it requires mutual sense to accept the reservations and editorial modifications of the editor-in-chief, the editor and the editorial staff (Dr. W. Finger, Frau G. Burfeindt, and Frau H. Hämmer) in the interests of a homogeneous whole in the Landolt-Börnstein style. This all proved to be an agreeable harmonious process.

Unfortunately, the heavy work load of some of the authors prevented them from submitting their contributions punctually, which delayed completion of the subvolumes. To avoid further delay, subvolume V/3c (Chapters 6 to 9) has been published first. It is now being followed by the present subvolume V/3a, covering Chapters 1 to 3. Subvolume V/3b will come out as soon as possible.

Like all other volumes of Landolt-Börnstein, this volume is published without any outside financial support.

Hamburg, August 1986 **The Editor**

Contents

1 Topography of the ocean floor

1.0 List of abbreviations

ACMRR	Advisory Committee on Marine Resources Research
ACUF	Advisory Committee on Undersea Features
ASDIC	Allied Submarine Devices Investigation Committee includes SONAR
BBL	Benthic Boundary Layer
BLOB	Plateau of unknown origin
CCD	Calcite Compensation Depth or Carbonate Compensation Depth
CLIMAP	Climate Long Range Investigation Mapping and Prediction
CNEXO	Centre National pour l'Exploitation des Océans (France)
CNRI	Continental Rise and Small Ocean Basins
CONS	Continental Shelf and Slope
CZCS	Coastal Zone Color Scanner
DHI	Deutsches Hydrographisches Institut, Hamburg
DOD	Deutsches Ozeanographisches Datenzentrum (in DHI)
DSDP	Deep Sea Drilling Project
DSL	Deep Scattering Layer
FAMOUS	French American Ocean Undersea Study 1973/74
GEBCO	General Bathymetric Chart of the Oceans
GLORIA	Geologic Long Range inclined ASDIC
HEBBLE	High Energetic Benthic Boundary Layer Experiments
IAPSO	International Association for the Physical Sciences of the Ocean
ICES	International Council for the Exploration of the Sea
ICSU	International Council of Scientific Unions
IDOE	International Decade of Ocean Exploration 1971–1980
IGCP	International Geologic Correlation Project No. 148 for Evaluation and Development of Quantitative Stratigraphic Correlation Techniques
IHB	International Hydrographic Bureau (Monaco)
IHO	International Hydrographic Organization (Monaco)
IIOE	International Indian Ocean Expedition 1960–1965
IOC	Intergovernmental Oceanographic Commission
IODE	International Oceanographic Data Exchange
IPOD	International Program of Ocean Drilling
ISSC	International Subcommission on Stratigraphic Classification
IUGS	International Union of Geophysical Science
JOIDES	Joint Oceanographic Institution for Deep Earth Sampling
LIBEC	Light behind the Camera
MORB	Mid-Oceanic Ridges Basalt
MPS	Marine Polymetallic Sulfid
NASA	National Aeronautics and Space Administration (USA)
NAVSAT	Navigation Satellite
NAVSTAR	Navigation System Time and Ranging (Satellite Navigation System)
NGS	National Geographic Society
NGSDC	National Geophysical and Solar-terrestrial Data Center
NNSS	Navy Navigation Satellite System
NOAA	National Oceanographic and Atmospheric Administration (USA)
OCBN	Ocean Basin with Abyssal Hills and Plains
ODP	Ocean Drilling Program (from 1985 onward)
RADAR	Radio Detecting and Ranging
RIDG	Ridge
RISE	Oceanic Rise and Ridge
RV	Research Vessel
SCAR	Scientific Committee on Antarctic Research
SCOR	Scientific Committee on Oceanic Research of IGU

Gierloff-Emden

For navigation, velocity is counted in knots [kn], 1 kn = 1 nm/h ≙ 0.514 m/s, while for hydrology, [$m\,s^{-1}$] or [$cm\,s^{-1}$] is used.

Depths are related to a datum level, the sea level (for sea level, mean sea level (msl), and chart datum, see ch. 9), and are given in [m] or [km]. On nautical and bathymetrical charts depths are also given in fathoms [fm] and feet [ft]. A transcription is not always possible since graphs and contour lines represented in fathoms and feet do not fit in equivalent metric decadic contour line intervals.

Areas are given in [km^2], or as map squares of a certain latitude. Areas are also indicated as "Marsden squares", i.e. squares of 10° latitude and 10° longitude on a Mercator projection map. The Marsden squares are subdivided into 100 one-degree subsquares, numbered from 00 to 99 to give the position to the nearest degree. The system is still used for geologic data acquisition in IODE.

Table 1. Nautical measurement units of nonmetric scale, definition and conversion to metric units. See also Tables 2···4 in ch. 9.

	Nonmetric unit	Abbre-viation	Definition	Conversion to metric unit
Length	inch	in		0.0254 m
	foot	ft	12 in	0.3048 m
	yard	yd	3 ft	0.9144 m
	fathom	fm	2 yd	1.8288 m
	nautical mile	nm		1.8520 km [1])
	league		3 nm	5.5560 km
Velocity	knot	kn	nm/h	1.8520 km/h
	feet/second		ft/s	0.3048 m/s

[1]) By definition for International nautical mile (nm).

1.1.4 The boundaries of the ocean

The interface, hydrosphere of the ocean – lithosphere of the continents – atmosphere, is a triple interface [80 G]. The hydrosphere, consisting of the total free water of the earth (whether as gas, liquid, or solid), is largely represented by the world ocean. The marginal boundaries of the ocean, i.e. the boundaries between ocean and continents (or islands), depend on sea level and intervals of time: they change periodically, annually (summer beach, winter beach), day by day (tides), or episodically (storm tides).

The (geologic) boundary continent – ocean is mostly defined seaward off the coast, often in coincidence with the continental slope. See Fig. 1. [82 S 1].

benthic: pertaining to submarine bottom terrain regardless of water depth

Terms related to the distance from the coast:

neritic: close to the coast, shallow sea; refers to marine sediments and organisms
pelagic: far from the coast, deep sea, refers to marine sediments and organisms
offshore: shelf area outside the intertidal zone, refers to engineering and exploitation and coastal dynamics

Terms of areas related to the depth of ocean areas:

supralittoral: coastal area above High Water (spray zone)
littoral: coastal area of intertidal zone
sublittoral: submarine coastal area below Low Water on the shelf
bathyal: deep area from shelf edge to the deep of the continental rise
abyssal: very deep area of ocean basins
hadal: deepest areas of deep sea trenches

The coastal boundary is defined as the intersection of the ocean surface and the land at a particular datum level [78 E]. This boundary is significant for all rules of the law of a nation, and the law of the sea, see ch. 9. [84 L 1].

The length of the ocean coasts exceeds by far the length of the earth's circumference. Depending on the accuracy of the measurements and on taking into account medium to small scale configurations of the coast, the lengths differ between 260000 km and 440000 km [33 K, 78 L, 70 M, 76 G 2, 73 N, 80 G, 82 K].

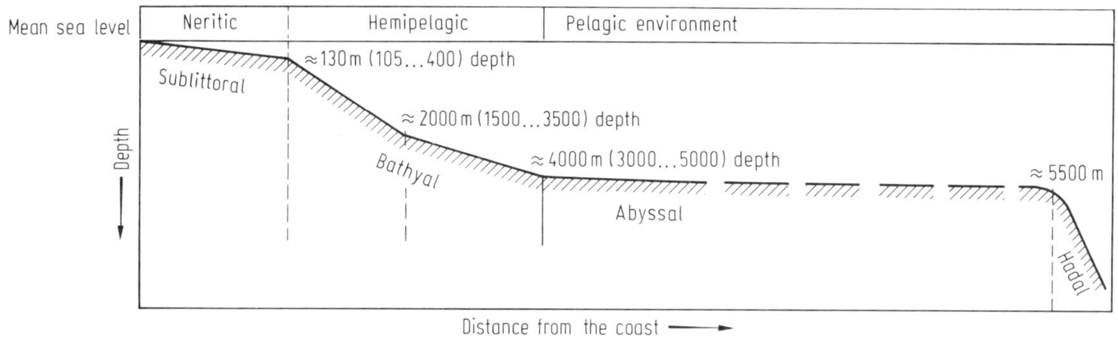

Fig. 1. Depth zones of the sea floor. Diagram defining the most common terms used in connection with sea floor depth and distance from land. Diagram not true to scale. Profiles are strongly exaggerated: in reality slopes are gentle. Based on [82 S 1] modified. See also ch. 9.

1.1.5 The spatial extension of the ocean

1.1.5.1 The horizontal distribution of the world ocean

Antipodal relationship of land and sea:

Antarctica is antipodal to the Arctic Ocean, Australia antipodal to the North Atlantic, North America to the Indian Ocean, Europe and Africa to the Pacific, and the South Atlantic Ocean is not far from antipodal to eastern Asia. The equator separates the northern and the southern hemisphere with the maximum of ocean distribution in the southern hemisphere corresponding remarkably well to the maximum of land distribution in the northern hemisphere [63 C]. Since the oceans cover approximately twice the area of the land masses, 95% of land is antipodal to oceans. This relationship is significant only of the present geologic age and stage. See Figs. 2, 3.

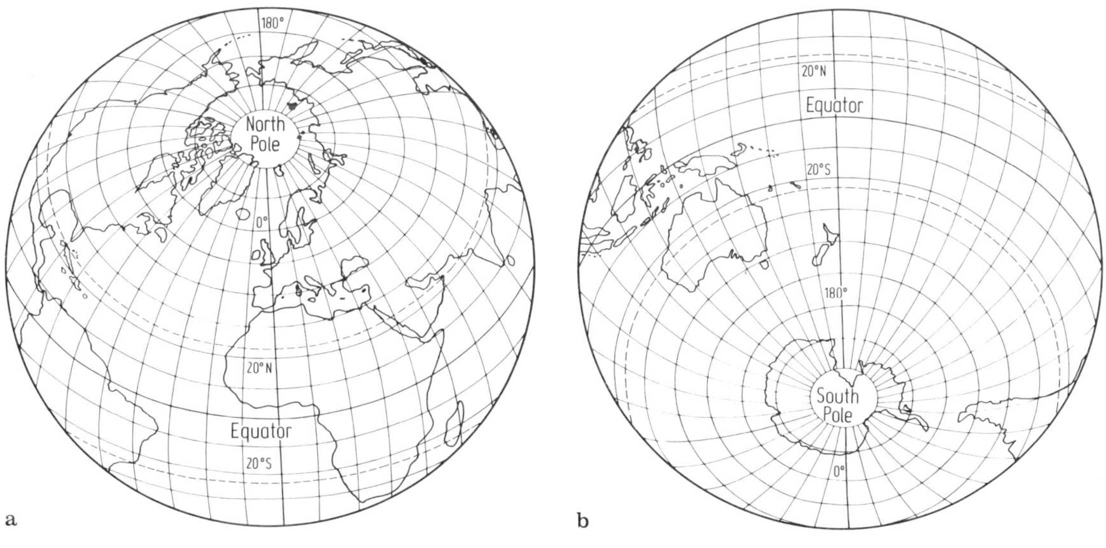

Fig. 2. Distribution of the sea and the land on hemispheres [71 W]. (a) Surface of land hemisphere: 53.5% sea (central point at 47°15′ N, 2°30′ W near the coast of France). (b) Surface of sea hemisphere: 88.5% sea (central point at 47°15′ S, 177°30′ E near the coast of New Zealand). Antipodal island position on 50° S, 180° E.

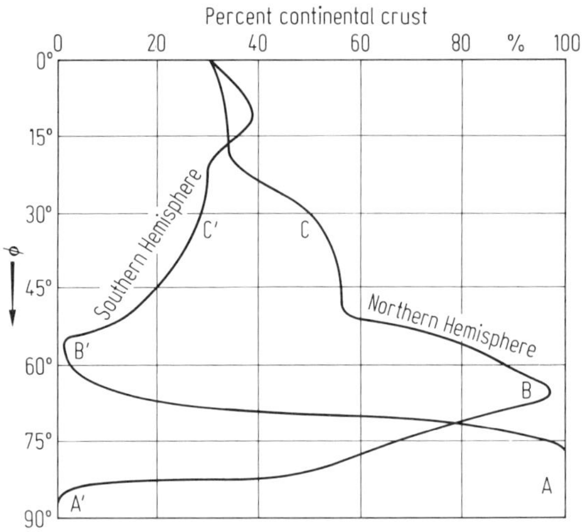

Fig. 3. Antipodal relationship of land and sea. Percentage of continental crust plotted against latitude [63 C]. Seas shallower than 2000 m are plotted as continental. AA′: South polar continent, North Polar Ocean, BB′: Northern land girdle, southern ocean girdle, CC′: Lands taper south, oceans taper north.

Ocean boundaries by convention:

Atlantic Ocean–Pacific Ocean at 67°16′W, Atlantic Ocean–Indian Ocean at 20° E, and Indian Ocean–Pacific Ocean at 146°55′E. [53 I]. Fig. 6.

The surface of the earth is made up of six continents: Eurasia, Africa, North America, South America, Australia, and Antarctica, which cover about 29% of the earth's surface, and three principal divisions of the world ocean: Atlantic Ocean (with Arctic Ocean), Indian Ocean, and Pacific Ocean, which cover the remaining 71%. See Fig. 7.

The planisphere is showing the continuity of the world ocean. Between the latitudes of about 55° S and 65° S essentially no land is present and the ocean encircles the entire globe. In general, the southern hemisphere can be considered as a sea hemisphere because most of the land lies on the northern hemisphere. See Figs. 4, 5.

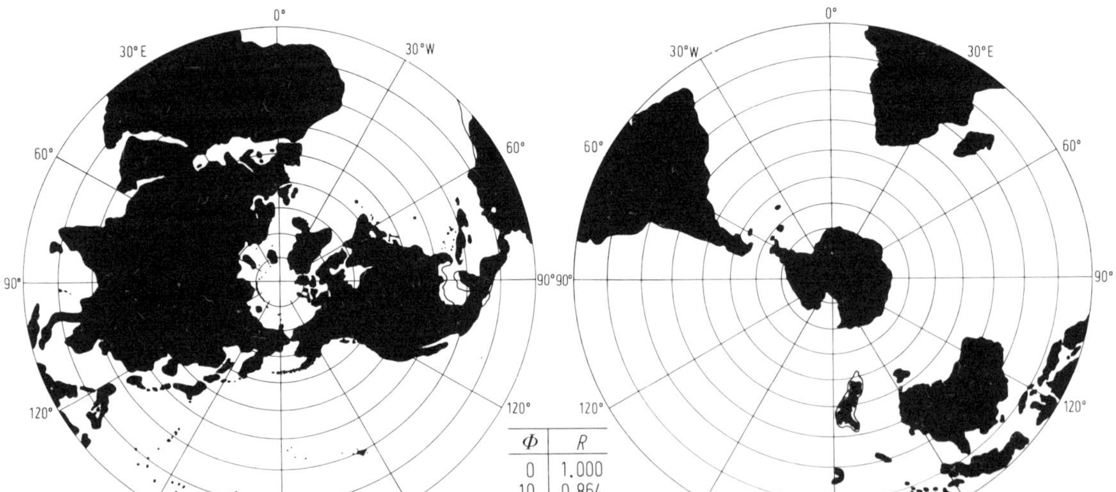

Φ	R
0	1.000
10	0.864
20	0.735
30	0.615
40	0.498
50	0.392
60	0.288
70	0.190
80	0.095
90	0.000

Fig. 4. Hemispheric projections of land and sea (azimuthal equal area projection). [62 A]. Spread of sea and land on the northern and southern hemisphere. R lists the radii of circles of latitude ϕ relative to the earth's radius.

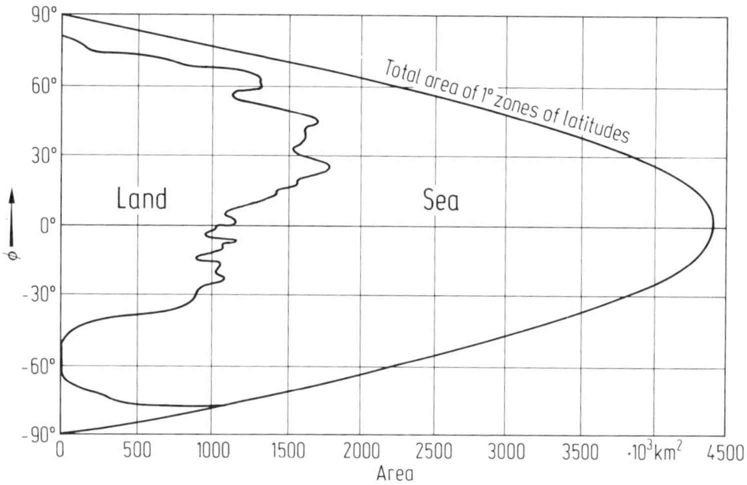

Fig. 5. Distribution of areas of land and sea on the northern and the southern hemisphere [71 W]; northern hemisphere 155 · 10^6 km^2 (sea ≈61%, land 39%), southern hemisphere 207 · 10^6 km^2 (sea ≈81%, land 19%). The ocean covers about 71% of the globe.

Table 2. Distribution of sea and land between parallels of latitudes [21 K]. *)

ϕ	Northern hemisphere				Southern hemisphere			
	Sea	Land	Sea	Land	Sea	Land	Sea	Land
	10^6 km^2		%		10^6 km^2		%	
90°···85°	0.979		100.0			0.978		100.0
85°···80°	2.545	0.384	85.2	12.8		2.929		100.0
80°···75°	3.742	1.112	77.1	22.9	0.522	4.332	10.7	89.3
75°···70°	4.414	2.326	65.5	34.5	2.604	4.136	38.6	61.4
70°···65°	2.456	6.116	28.7	71.3	6.816	1.756	79.5	20.5
65°···60°	3.123	7.210	31.2	69.8	10.301	0.032	99.7	0.3
60°···55°	5.399	6.613	45.0	55.0	12.006	0.006	99.9	0.1
55°···50°	5.529	8.066	40.7	59.3	13.388	0.207	98.5	1.5
50°···45°	6.612	8.458	43.8	56.2	14.693	0.377	97.5	2.5
45°···40°	8.411	8.016	51.2	48.8	15.833	0.594	96.4	3.6
40°···35°	10.029	7.627	56.8	43.2	16.483	1.173	93.4	6.6
35°···30°	10.806	7.943	57.7	42.3	15.782	2.967	84.2	15.8
30°···25°	11.747	7.952	59.6	40.4	15.438	4.261	78.4	21.6
25°···20°	13.354	7.145	65.2	34.8	15.450	5.049	75.4	24.6
20°···15°	14.981	6.164	70.8	29.2	16.147	4.998	76.4	23.6
15°···10°	16.553	5.080	76.5	23.5	17.211	4.422	79.6	20.4
10°··· 5°	16.628	5.332	75.7	24.3	16.898	5.062	76.9	23.1
5°··· 0°	17.387	4.737	78.6	21.4	16.792	5.332	75.9	24.1
90°··· 0°	154.695	100.281	60.7	39.3	206.364	48.611	80.9	19.1
All oceans and seas						361.059 × 10^6 km^2 ≙ 70.8%		
All land						148.892 × 10^6 km^2 ≙ 29.2%		

*) The total is not in coincidence with Table 3 but for the distribution according to parallels of latitudes no other updating of this table is available.

1.1.5.2 The horizontal extension of the oceans and the continents

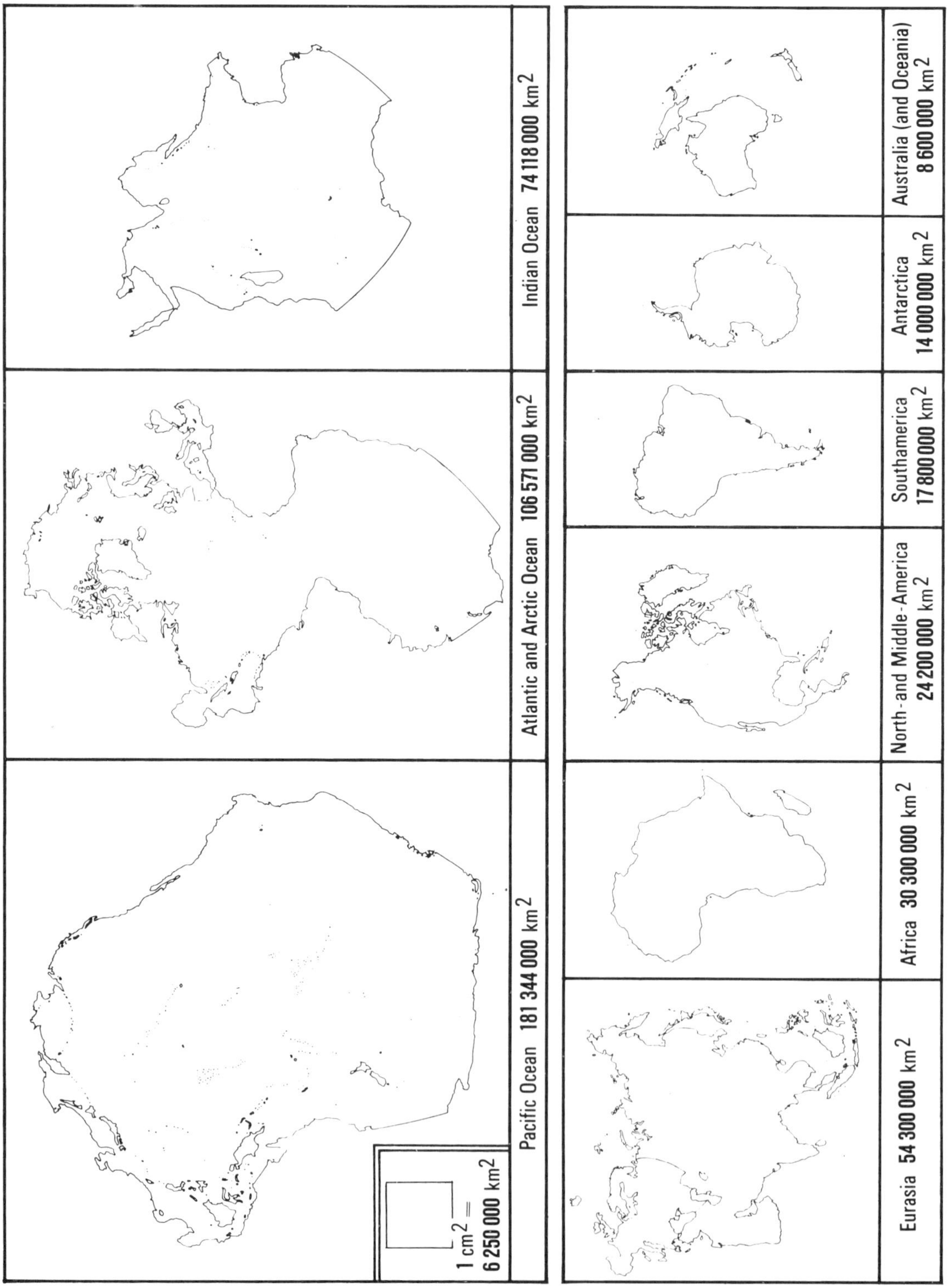

1.1.5.3 The horizontal classification of the oceans

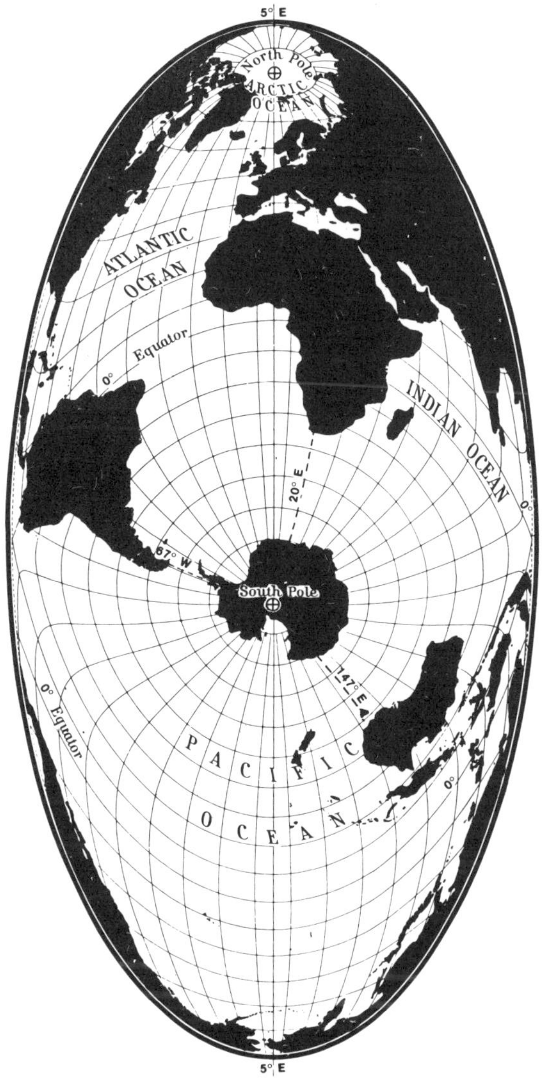

Fig. 7. Planisphere of the earth with oceans and continents in equal area projection (Mollweide projection). Geographical grid: latitudes each 10°, longitudes each 10°, NS-axis through 5° East of Greenwich. North America and Asia partly encompass the margin of the under part of the planisphere of this projection. Dashed lines represent the boundaries between the oceans. For a display of the oceans on regular icosahedron world maps projection, see [77 N, 80 G].

←

Fig. 6. The oceans and the continents: areas on equal area projection. Areas: Oceans [66 M 1], continents [80 A], graphics [71 W]. The addition of both of these data is 511 200 000 km^2. This sum exceeds the (standard) area of the earth, counted with 510 100 000 km^2, see Table 1 with a surplus of 1 100 000 km^2. The area for the Atlantic Ocean is 94 314 · 10^6 km^2, for the Arctic Ocean 12 257 · 10^6 km^2. The noncoincidence is caused by the new standards of the oceanic area [66 M 1] in comparison with older accounts: the areas of islands to oceanic areas or to continental areas are not in coincidence with the new data [66 M 1]. See Table 3.

Gierloff-Emden

1.1.5.4 Limits and nomenclature of oceans and seas

The limits given in Fig. 8 of the Artic, Atlantic, Pacific, and Indian Oceans exclude the seas lying within each of them. The boundary line between the North and South Atlantic Oceans and the North and South Pacific Oceans is the equator. The southern boundary line between the South Atlantic and South Pacific Oceans is the meridian of Cape Horn (67°16′W).

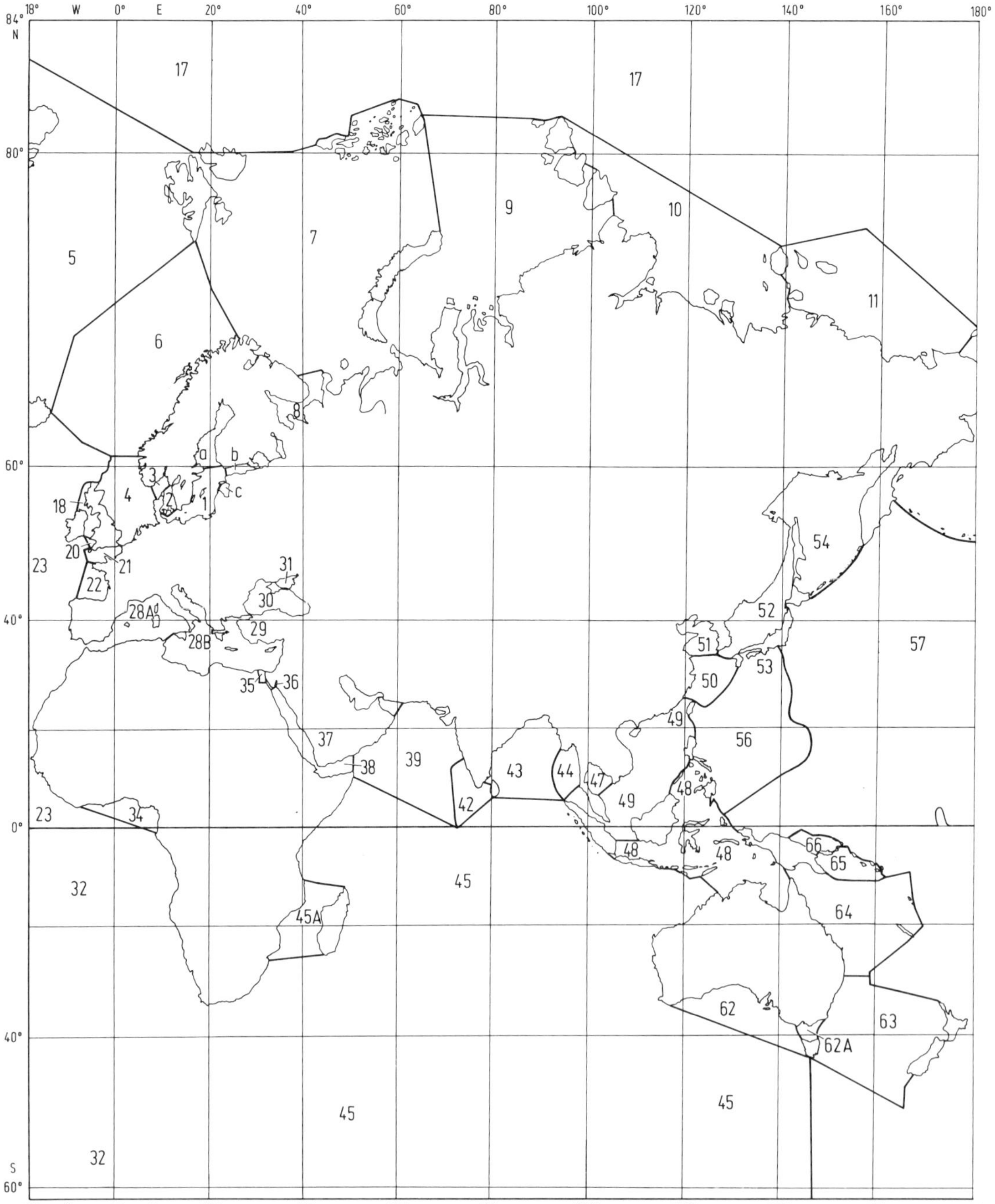

For caption to Fig. 8, see p. 12.

The Antarctic or Southern Ocean has been omitted. There exists no real justification for applying the term "Ocean" to this body of water, the northern limits of which are difficult to lay down owing to their seasonal change. The limits of the Atlantic, Pacific and Indian Oceans have therefore been extended south to the Antarctic Continent. [53 I].

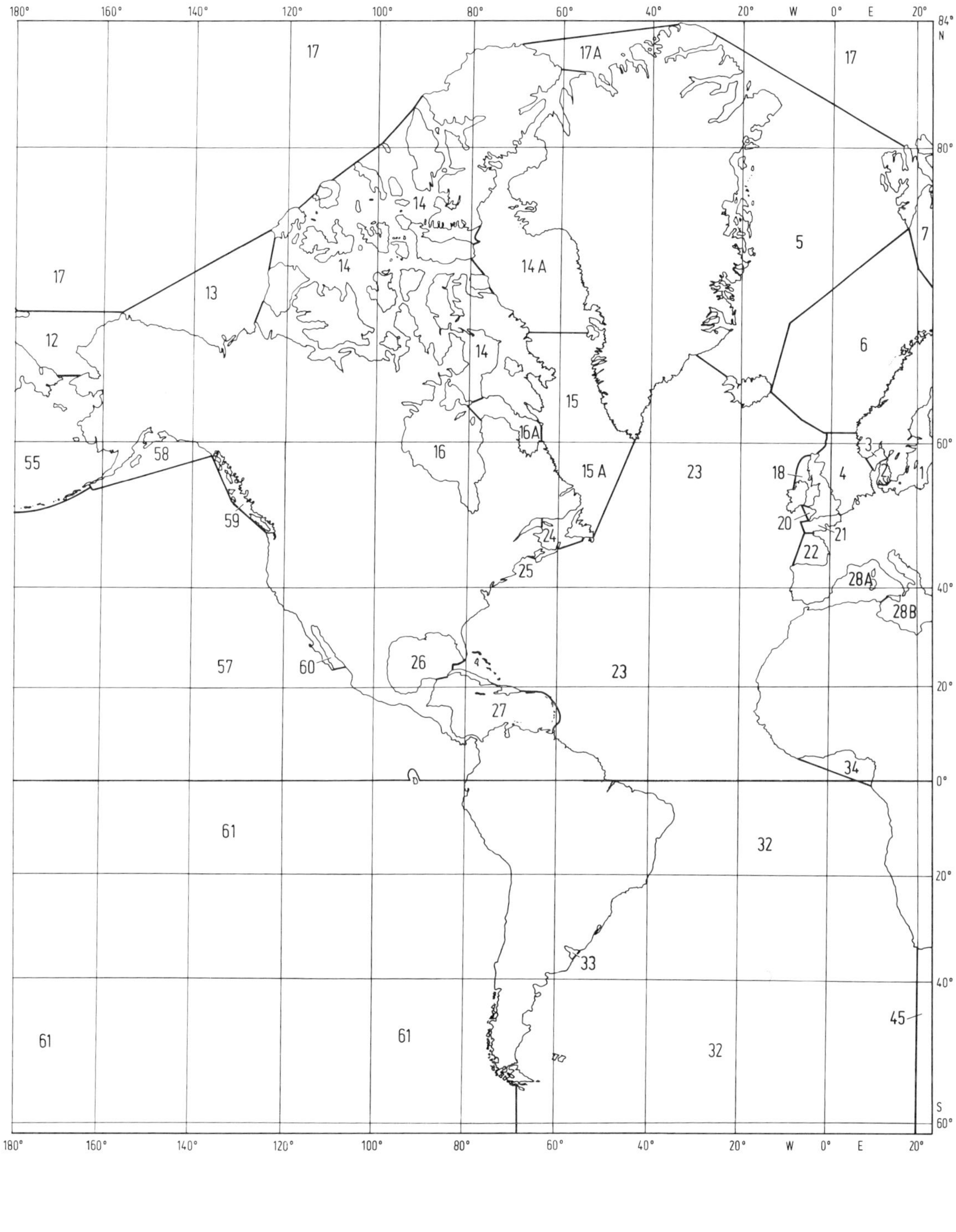

Gierloff-Emden

←

Fig. 8. Limits of oceans and seas [53 I].

Atlantic Ocean
North Atlantic Ocean
1 Baltic Sea (sub-divisions: a, b, c)
a Gulf of Bothnia
b Gulf of Finland
c Gulf of Riga
2 Kattegat, Sound and Belts
3 Skagerrak
4 North Sea
5 Greenland Sea
6 Norwegian Sea
7 Barents Sea
8 White Sea
9 Kara Sea
10 Laptev (or Nordenskjôld) Sea
11 East Siberian Sea
12 Chuckchi Sea
13 Beaufort Sea
14 The North Western Passages
14A Baffin Bay
15 Davis Strait
15A Labrador Sea
16 Hudson Bay
16A Hudson Strait
17 Arctic Ocean
17A Lincoln Sea
18 Inner Seas off the west coast of Scotland
19 Irish Sea and St. George's Channel
20 Bristol Channel
21 English Channel
22 Bay of Biscay
23 North Atlantic Ocean
24 Gulf of St. Lawrence
25 Bay of Fundy
26 Gulf of Mexico
27 Caribbean Sea
28 Mediterranean Sea
A Western Basin
B Eastern Basin
29 Sea of Marmara
30 Black Sea
31 Sea of Azov
South Atlantic Ocean
32 South Atlantic Ocean
33 Rio de La Plata
34 Gulf of Guinea

Indian Ocean
35 Gulf of Suez
36 Gulf of Aqaba
37 Red Sea
38 Gulf of Aden
39 Arabian Sea
40 Gulf of Oman
41 Gulf of Iran (Persian – Arabian Gulf)
42 Laccadive Sea
43 Bay of Bengal
44 Andaman or Burma Sea
45 Indian Ocean
45A Mozambique Channel
62 Great Australian Bight

Pacific Ocean
North Pacific Ocean
46 Malacca and Singapore Straits
47 Gulf of Thailand (Siam)
48 East Indian Archipelago (Indonesia)
49 South China Sea (Nan Hai)
50 Eastern China Sea (Tung Hai)
51 Yellow Sea (Hwang Hai)
52 Japan Sea
53 Seto Naikai or Inland Sea
54 Sea of Okhotsk
55 Bering Sea
56 Philippine Sea
57 North Pacific Ocean
58 Gulf of Alaska
59 The Coastal Waters of E. Alaska and British Columbia
60 Gulf of California
South Pacific Ocean
61 South Pacific Ocean
62A Bass Strait
63 Tasman Sea
64 Coral Sea
65 Solomon Sea
66 Bismarck Sea

Published by the International Hydrographic Bureau – Monaco, 3rd edition, 1953 (IHB, later renamed IHO). [53 I].

1.1.5.5 Spatial dimensions of the oceans and adjacent seas

Volume of the ocean:

Murray (1888) calculated the volume of the ocean at 323 722 150 cubic miles, which equals about $1325 \cdot 10^9\,km^3$, Kossinna (1921) obtained $1370 \cdot 10^9\,km^3$, and Smith (1966) obtained $1350 \cdot 10^9\,km^3$. It appears unlikely that this value is in error by more than a few percent. The method of calculation is essentially the same as that of Murray and Kossinna. The midpoint value of a depth interval is multiplied by the area of that interval, and the volumes of the intervals are summed [66 M 1]. [1894 K, 33 K]. See Tables 3, 4.

Table 3. Area, volume, and depth of the oceans [66 M 1, 21 K].

Oceans and adjacent seas	Area 10^6 km^2	Volume 10^6 km^3	Depth [m] mean		maximum[1])
			[66 M 1]	[21 K]	
Pacific	166.241	696.189	4188	4282	
Asiatic Mediterranean [2])	9.082	11.366	1252	1182	4536
Bering Sea	2.261	3.373	1492	1437	4150
Sea of Okhotsk	1.392	1.354	973	838	3374
Yellow and East China Seas	1.202	0.327	272	188	2717
Sea of Japan	1.013	1.690	1667	1350	3695
Gulf of California	0.153	0.111	724	813	2600
Pacific and adjacent seas, total	181.344	714.410	3940	4013	
Atlantic	86.557	323.369	3736	3805	
American Mediterranean	4.357	9.427	2164	2216	7680
Mediterranean	2.510	3.771	1502	1487	5015
Black Sea	0.508	0.605	1191	1115	2212
Baltic Sea	0.382	0.038	101	55	459
Atlantic and adjacent seas, total	94.314	337.210	3575	3641	
Indian	73.427	284.340	3872	3963	
Red Sea	0.453	0.244	538	491	2920
Persian Gulf	0.238	0.024	100	25	73
Indian and adjacent seas, total	74.118	284.608	3840	3929	
Arctic	9.485	12.615	1330	1240	4506
Arctic Mediterranean	2.772	1.087	392	277	
Arctic and adjacent seas, total	12.257	13.702	1117	1020	
Totals and mean depths	362.033	1349.929	3729	3814	

[1]) Different sources.
[2]) Also called "South China Sea".

Table 4. Depth distribution of the world ocean [66 M 1, 21 K].

Depth km	Area 10^6 km^2		Area %		Cumulative area (shallower than deeper limit of depth interval) 10^6 km^2	Cumulative area %	
	[66 M 1]	[21 K]	[66 M 1]	[21 K]	[66 M 1]	[66 M 1]	[21 K]
0 ··· 0.2	27.123	27.491	7.49	7.6	27.123	7.49	7.6
0.2··· 1	16.012	15.437	4.42	4.3	43.135	11.91	11.9
1 ··· 2	15.844	15.184	4.38	4.2	58.978	16.29	16.1
2 ··· 3	30.762	24.347	8.50	6.8	89.740	24.79	22.9
3 ··· 4	75.824	70.800	20.94	19.6	165.565	45.73	42.5
4 ··· 5	114.725	119.092	31.69	33.0	280.289	77.42	75.5
5 ··· 6	76.753	84.317	21.20	23.3	357.042	98.62	98.8
6 ··· 7	4.461	3.919	1.23	1.1	361.503	99.85	99.9
7 ··· 8	0.380	0.328	0.10		361.883	99.96	
8 ··· 9	0.115	0.126	0.03		361.998	99.99	
9 ···10	0.032	0.018	0.01	0.1	362.031	100.00	100.0
10 ···11	0.002		0.00		362.033	100.00	

1.2 Hypsography

1.2.1 The Hypsographic Curve

The mean depth of the ocean is 3729 m, i.e. 1/1700 of the mean radius of the earth. The mean depth of the ocean basins is 4700 m. The water body of the oceans can be subdivided into the shallow sea (depth < 200 m), and the deep sea (depth > 200 m). The sum up of submarine depth areas in bathymetric levels, and the presentation of their statistical distribution leads to the so-called "Hypsographic Curve", i.e. the cumulative sum of hypsometric data. Each point (x, y) of the curve represents with the ordinate (y) a discrete height of the earth's surface relief, and with the abscissa (x) the amount of area of the earth's relief up to this (y) very height.

Because of the lack of more recent computations until 1966 these curves were generally drawn according to the state of knowledge about the submarine topographic data from the 1920's and 1930's [21 K, 33 K, 42 M]. Fig. 9 is based on recent data [66 M 1]. A computation of data into curves for the individual oceans, see Figs. 14, 16, 20 [66 M 1].

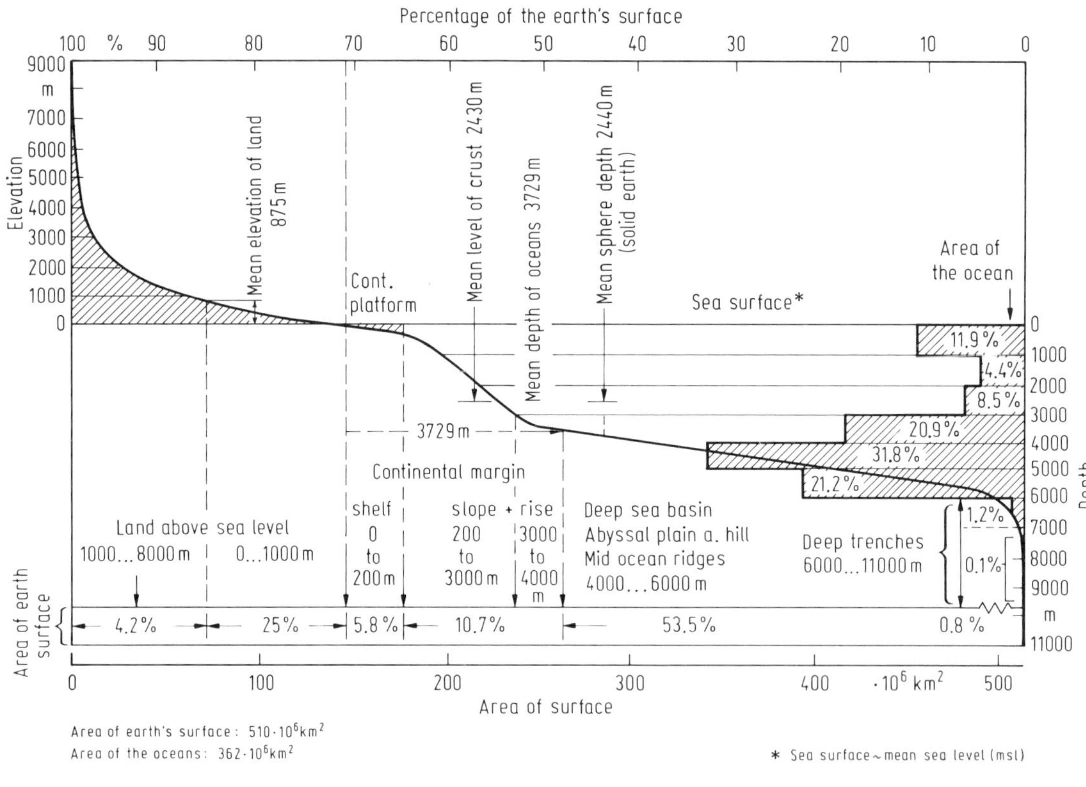

Fig. 9. The Hypsographic Curve of world oceans and continents. Hypsographic Curve showing the area of the earth's solid surface above any given level of elevation or depth. Area of surface not true to scale. (The smooth curve represents the cumulative sum of areas denoted by marked bars). The levels + 100 m and − 5000 m show maximum extensions, representing the two levels of the solid earth crust. Shelf and continental slope are significant features of global scale of the relief of the world's oceans. The mid-ocean ridges as a global feature with crests and rifts of 1500 m of relative relief average do not appear on this statistical display. [85 G].

The determination of the mean depth of the ocean depends on different methods of computation:

Profile measurement of depth of profiles (in grid areas): 1878 Krümmel. Planimetry of depth zones according bathymetric lines: 1883 Lapparent, 1888 Murray.

Areas of geographic grids mean depths: 1888 Heiderich, 1889 Supan.

Computation and formulas of this methods: [1894 K].

The statistical distribution of the heights and depths of the earth's crust shows that the frequency in no way approaches a Gaussian probability curve. On the contrary, there are two height-intervals which occur with high frequency while the other, less frequent, intervals group themselves around these two culmination points (see Fig. 10 and Table 5).

If topographical features were formed by random uplift and subsidence a random distribution centered on the mean elevation (dashed line in Fig. 10) would be expected. Actually two dominant levels are existing, one level at 0 m +100 m to −100 m), and another level at −4700 m depth (−3800 m··· −5700 m) (solid line in Fig. 10). The upper level (0 m), represents average elevation of the continental platforms (continental lowlands and continental shelves), while the lower level is that of the abyssal sea floor at −5000 m. The dominating two levels occur on the hypsographic profiles around the earth as presented on Figs. 11, 12.

The significant extended level around 5000 m depth represents, in general, the younger part of crust because this sea floor is of an age not older than approximately $200 \cdot 10^6$ years (Jurassic), i.e. this topography is in coincidence with geologic history characterized by sea floor spreading and plate tectonics. See Figs. 71, 72, and 76.

The subdivision of the sea floor according to the statistics of bathymetric data is presented in Table 5.

Palaeohypsography:

The development of hypsographic conditions since $200 \cdot 10^6$ years ago from the situation of the so-called "Pangaea" continent to the situation of the continents of present days was plotted by computer display [81 H 1, 83 J 2]. The change from the hypsography of Pangaea $200 \cdot 10^6$ years ago to that of the presently existing continents has been gradual, and affected the different continents in different ways according to their location with respect to spreading centers, subduction zones, and colliding continental fragments, so-called terranes. See Fig. 53. For a detailed representation on maps of the oceans, see [82 R 2, 83 O].

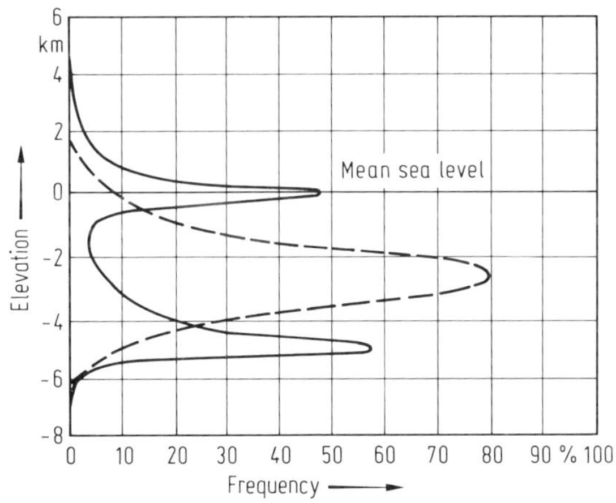

Fig. 10. Statistical distribution of levels of the solid earth surface. [75 H]. Solid line: actual distribution. Dashed line: distribution for random uplift and subsidence. It is especially noticeable that the mean crust level of −2430 m (depth interval −2000 m to −3000 m) occurs infrequently, and is indeed very near the minimum between the two maxima.

1.2.2 Global hypsographic profiles according to the Hypsographic Curve

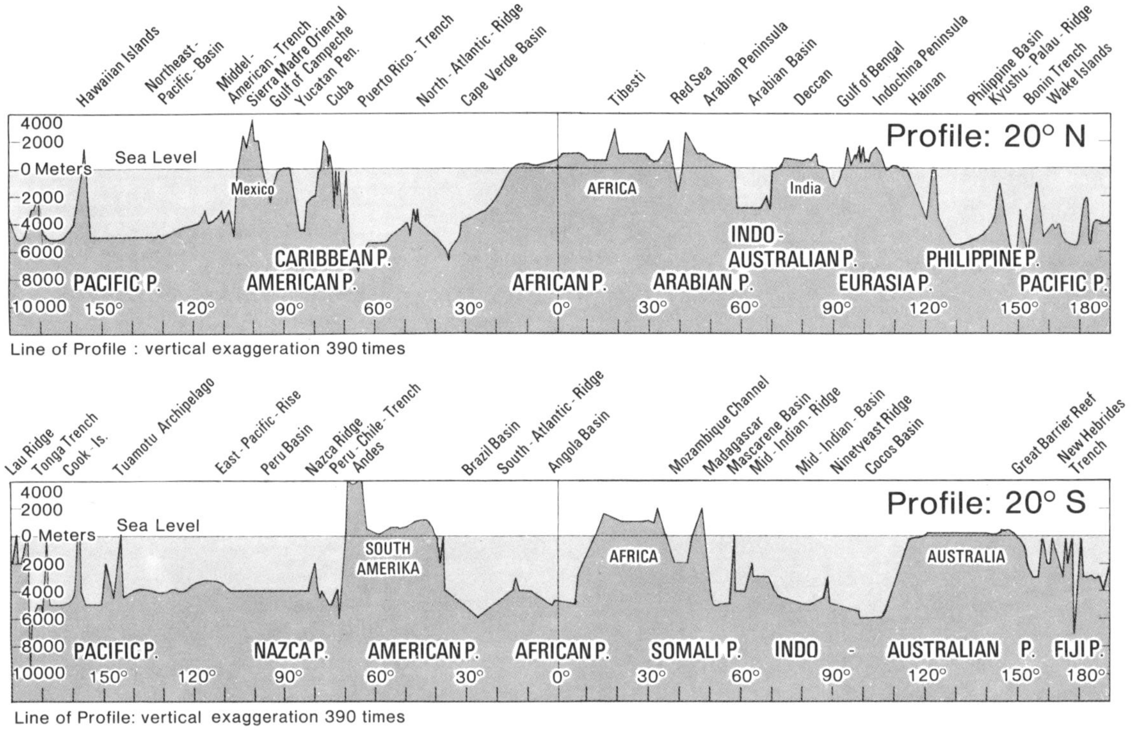

Fig. 11. Hypsographic profiles of the sea floor and of the continents around the globe, 20° N and 20° S (dashed lines in Fig. 12). [83 G 1, 83 G 2]. The two levels at −5000 m of the sea floor, and of 0···1000 m of the continents are significant.

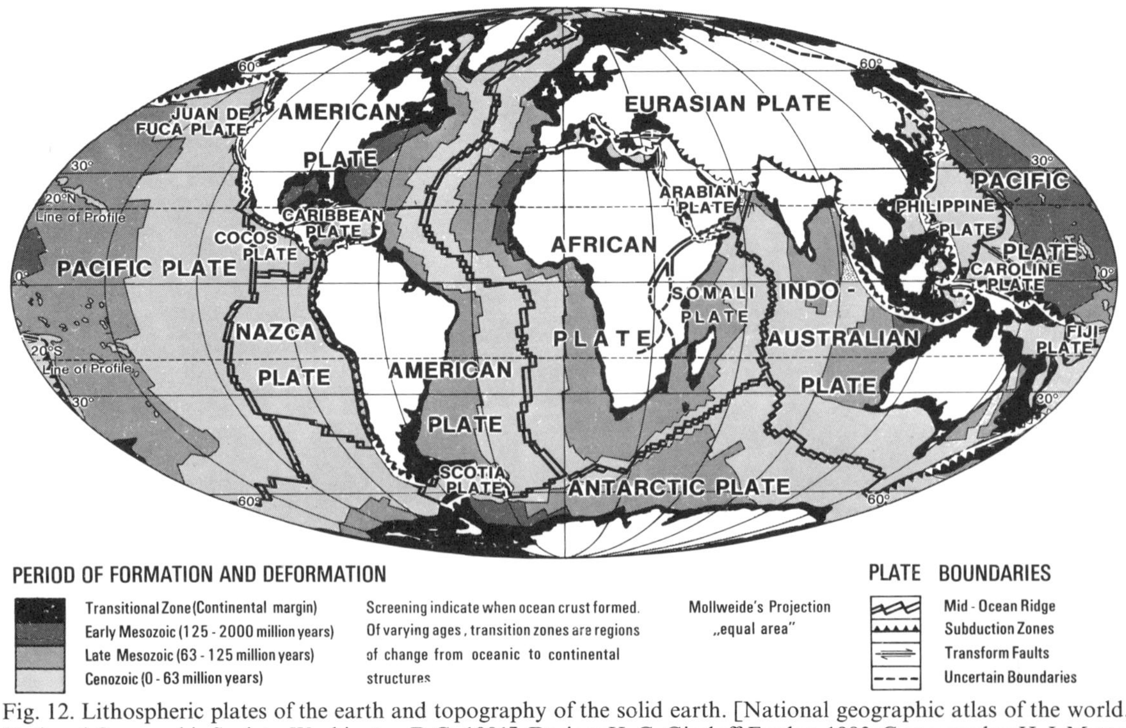

Fig. 12. Lithospheric plates of the earth and topography of the solid earth. [National geographic atlas of the world. National Geographic Society, Washington D.C., 1981]. Design: H. G. Gierloff-Emden, 1983. Cartography: H. J. Mette. The positions of the profiles depicted in Fig. 11 are indicated by dashed lines.

1.2.3 Hypsometric data of the oceans

Table 5. Depth zones in the oceans [66 M 1].

Depth interval [km]	0···0.2	0.2···1	1···2	2···3	3···4	4···5	5···6	6···7	7···8	8···9	9···10	10···11	Total area (ocean)	Volume	Mean depth
	Area [10^6 km^2]													10^6 km^3	m
Ocean															
Pacific Ocean	2.712	4.294	5.403	11.397	36.233	58.162	44.691	2.896	0.313	0.105	0.032	0.002	166.241	696.189	4188
Asiatic Mediterranean	4.715	0.841	0.948	1.104	0.608	0.707	0.149	0.007	0.005	0	0	0	9.082	11.366	1252
Bering Sea	1.050	0.135	0.172	0.234	0.670	0	0	0	0	0	0	0	2.261	3.373	1492
Sea of Okhotsk	0.368	0.549	0.311	0.047	0.115	0	0	0	0	0	0	0	1.392	1.354	973
Yellow and East China Seas	0.977	0.137	0.072	0.015	0.001	0	0	0	0	0	0	0	1.202	0.327	272
Sea of Japan	0.238	0.154	0.199	0.204	0.218	0	0	0	0	0	0	0	1.013	1.690	1667
Gulf of California	0.071	0.032	0.040	0.010	0	0	0	0	0	0	0	0	0.153	0.111	724
Atlantic Ocean	6.080	4.474	3.718	7.436	16.729	28.090	19.324	0.639	0.058	0.010	0	0	86.557	323.369	3736
American Mediterranean	1.021	0.465	0.589	0.667	0.906	0.586	0.112	0.008	0.002	0	0	0	4.357	9.427	2164
Mediterranean	0.513	0.564	0.437	0.766	0.224	0.006	0	0	0	0	0	0	2.510	3.771	1502
Black Sea	0.177	0.064	0.117	0.149	0	0	0	0	0	0	0	0	0.508	0.605	1191
Baltic Sea	0.381	0.001	0	0	0	0	0	0	0	0	0	0	0.382	0.038	101
Indian Ocean	2.622	1.971	2.628	7.364	18.547	26.906	12.476	0.911	0.001	0	0	0	73.427	284.340	3872
Red Sea	0.188	0.195	0.068	0.003	0	0	0	0	0	0	0	0	0.453	0.244	538
Persian Gulf	0.238	0	0	0	0	0	0	0	0	0	0	0	0.238	0.024	100
Arctic Ocean	3.858	1.569	0.968	1.249	1.573	0.269	0	0	0	0	0	0	9.485	12.615	1330
Arctic Mediterranean	1.913	0.567	0.174	0.118	0	0	0	0	0	0	0	0	2.772	1.087	392
Total each depth	27.123	16.012	15.844	30.762	75.824	114.725	76.753	4.461	0.380	0.115	0.032	0.002	362.033	1349.929	3729

Table 5 (continued)

Depth interval [km]	0···0.2	0.2···1	1···2	2···3	3···4	4···5	5···6	6···7	7···8	8···9	9···10	10···11	% of world ocean in each ocean
	Area in % of each ocean												
Ocean													
Pacific Ocean	1.631	2.583	3.250	6.856	21.796	34.987	26.884	1.742	0.188	0.063	0.019	0.001	45.919
Asiatic Mediterranean	51.913	9.255	10.433	12.151	6.698	7.780	1.636	0.076	0.058	0	0	0	2.509
Bering Sea	46.443	5.975	7.623	10.330	29.629	0	0	0	0	0	0	0	0.625
Sea of Okhotsk	26.475	39.479	22.383	3.403	8.260	0	0	0	0	0	0	0	0.384
Yellow and East China Seas	81.305	11.427	5.974	1.239	0.055	0	0	0	0	0	0	0	0.332
Sea of Japan	23.498	15.176	19.646	20.096	21.551	0.033	0	0	0	0	0	0	0.280
Gulf of California	46.705	20.848	25.891	6.556	0	0	0	0	0	0	0	0	0.042
Atlantic Ocean	7.025	5.169	4.295	8.590	19.327	32.452	22.326	0.738	0.067	0.012	0	0	23.909
American Mediterranean	23.443	10.674	13.518	15.313	20.796	13.440	2.572	0.193	0.051	0	0	0	1.203
Mediterranean	20.436	22.475	17.413	30.515	8.940	0.221	0	0	0	0	0	0	0.693
Black Sea	34.965	12.587	23.077	29.371	0	0	0	0	0	0	0	0	0.140
Baltic Sea	99.832	0.168	0	0	0	0	0	0	0	0	0	0	0.105
Indian Ocean	3.570	2.685	3.580	10.029	25.259	36.643	16.991	1.241	0.001	0	0	0	20.282
Red Sea	41.454	43.058	14.920	0.568	0	0	0	0	0	0	0	0	0.125
Persian Gulf	100.000	0	0	0	0	0	0	0	0	0	0	0	0.066
Arctic Ocean	40.673	16.539	10.209	13.167	16.580	2.834	0	0	0	0	0	0	2.620
Arctic Mediterranean	69.013	20.454	6.274	4.260	0	0	0	0	0	0	0	0	0.766
Percent of world ocean in each depth interval	7.492	4.423	4.376	8.497	20.944	31.689	21.201	1.232	0.105	0.032	0.009	0.001	

Table 6. Major deeps and their location and maximal depths [66 M 1, 63 H 3, 63 H 4].

Trench [c])	Max. depths m
Marianas Trench	11 034 ± 50
(specifically Challenger Deep)	10 915 ± 20 [a])
	≈ 10 915
	10 863 ± 35
	10 850 ± 20 [a])
Tonga	10 882 ± 50
	10 800 ± 100
Kuril – Kamchatka	10 542 ± 100
	9 750 ± 100 [b])
Philippine	
(vicinity of Cape	10 497 ± 100
Johnson Deep)	10 265 ± 45
	10 030 ± 10 [a])
Kermadec	≈ 10 047
Idzu – Bonin	
(includes "Ramapo Deep"	≈ 9 810
of the Japan Trench)	
(vicinity of Ramapo Depth)	≈ 9 695
Puerto Rico (Atlantic Oc., N)	9 200 ± 20
New Hebrides (North)	9 165 ± 20 [a])
North Solomons (Bougainville)	≈ 9 103
	8 940 ± 20 [a])
Yap (West Caroline)	≈ 8 527
New Britain	≈ 8 320
	8 245 ± 20 [a])
South Solomons	8 310 ± 20 [a])
South Sandwich	≈ 8 264
(Atlantic Oc., S)	
Peru – Chile	8 055 ± 10
Palau	≈ 8 054
Sunda (Indian Oc.)	≈ 7 455
Romanche Fracture Zone	≈ 7 864
(Atlantic Oc., Equator)	
Aleutian (uncorrected, taken	≈ 7 679
with nominal sounding	
velocity of 1500 m s^{-1})	
Nansei Shoto (Ryuku)	≈ 7 507
Java (Indian Oc.)	≈ 7 450
New Hebrides (South)	7 070 ± 20 [a])
Middle America	6 662 ± 10

[a]) These soundings were taken during an expedition, April, June 1962, aboard R.V. Spencer, F. Baird. A Precision Depth Recorder was employed, and the ship's track crossed over (within the limits of celestial navigation) points from which maximum depth had been reported.

[b]) This is the maximum sounding obtained in the vicinity of the Vitiaz Depth by French and Japanese vessels in connection with dives of the bathyscaph "Archimède", July 1962.

[c]) All trenches are in the Pacific Ocean (exceptions marked).

1.2.4 Hypsometry of ocean basin provinces based on a physiographic classification

The floors of the oceans can be subdivided into *physiographic provinces* according to topographic and morphologic-genetic characteristics. A statistical distribution which takes the differentiated bathymetric divisions of the provinces into account is presented in Fig. 13.

Province	Land	Continental shelf and slope	Continental rise and small ocean basins	Oceanic basins with abyssal hills and plains	Trenches associated areas	Oceanic rise and ridge	Composite volcanic ridge	Volcanic sediments and guyots	Ridges or plateaus of unknown origin
Code		CONS	CNRI	OCBN	TNCH	RISE	VRCM	VOLC	RIDG/BLOB
Example Location		Blake Plateau	N. W. Atlantic	Atlantic Composite	Marianas rise and trench	Mid Indian Rise	Tuamotu Ridge	Rehoboth Seamount	Rio Grande Rise

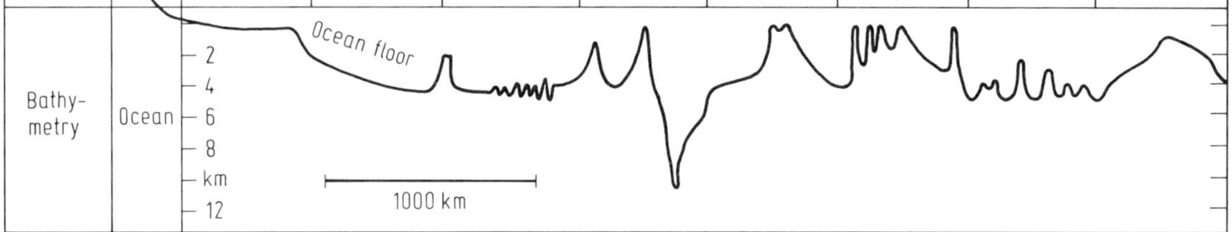

Distribution of major sediment facies	Terrigenous sediments		Pelagic sediments					
Area in 10^3 km^2 Total $362033 \cdot 10^3$ km^2	55421	19242	151522	6070	118607	2298	3435	5438
Depth range in km	0...4.0	1.5...5.0	2.0...7.0	0...10.0	0...6.0	0...7.0	0...6.0	0.2...6.0
Distribution of world oceans in%	15.3	5.3	41.8	1.7	32.7		3.2	
% Pacific	13.1	2.7	43.0	2.9	35.9		2.4	
% Atlantic	17.7	8.0	39.3	0.7	32.3		2.0	
% Indian	9.1	5.7	49.2	0.3	30.2		5.5	
% Arctic	68.2	20.8	0	0	4.2		6.8	

Fig. 13. The physiographic provinces of the oceans and seas: their dimensions, area, depth, areal percentage, and a curve of the sea floor according to the morphologic-bathymetric characteristics of the physiographic provinces. Upper three lines: submarine topography and morphology. Center: bathymetric according to morphologic criteria (not statistical). Lower 7 lines: statistics according to morphologic criteria [66 M 1, 74 C 2, 74 S 2]. Tabulation of data by 10° squares of latitude (φ) and longitude (λ) within a square, the areas between depth intervalls 0···200 m, 200···1000 m, and between 1 km contours down to 11 km were compiled for each physiographic province.

Hypsometry of ocean basin provinces

Physiographic provinces: These "provinces" are regions or groups of features that have distinctive topography and usually characteristic structures and relations to other provinces. Province boundaries are based on detailed physiographic diagrams where available. Provinces do not overlap nor are they superimposed in this study. Thus the area of a volcano rising from an ocean basin is included only in province "Volcano" and excluded from province "Ocean Basin". See Figs. 14···20, Tables 7···9.

1. Continental Shelf and Slope (CONS), the whole region from the shoreline to the base of the steep continental slope. Shelf and slope are grouped together because they are merely the top and front of the margins of continental blocks.

2. Continental Rise and partially filled sedimentary basins (CNRI). Gently sloping or almost flat, they appear to have characteristic features resulting from the accumulation of a thick fill of sediment eroded from an adjacent continent and overlying an otherwise relatively normal oceanic crust. In this respect, the Gulf of Mexico and the western Basin of the American Mediterranean differ from the continental rise off the eastern United States only because they are relatively enclosed.

3. Ocean Basin (OCBN), the remainder after removing all other provinces. Abyssal plains and abyssal hills and archipelagic aprons are common features of low relief.

4. Oceanic Rise and Ridge (RISE), commonly called "mid-ocean ridges" despite the fact that they continue across ocean margins. They form one worldwide system with many branches. Boundaries are taken in most places as the outer limit of essentially continuous slopes from the crest.

5. Ridge not known to be volcanic (RIDG), relatively long and narrow and with steep sides. Most have unknown structure and some or most may be volcanic.

6. Individual Volcano (VOLC), with a boundary defined as the base of steep side slopes.

7. Island Arc and Trench (TNCH), includes whole system of low swells and swales subparallel to trenches. Continental equivalents or extensions of island arcs, such as Japan, are excluded.

8. Composite Volcanic Ridge (VRCM), formed by overlapping volcanoes and with a boundary at the base of steep side slopes.

9. Poorly defined elevation (BLOB), with nondescript side slopes and length no more than about twice its width. Crustal structure unknown; may be thin continental type [66 M 1].

For Table 7, see p. 22.

Table 8. Distribution of provinces in oceans and adjacent seas. For abbreviations for provinces, see Fig. 13 and text [66 M 1].

Oceans and adjacent seas	RISE	OCBN	VOLC VRCM RIDG BLOB	CONS	TNCH	CNRI	% of world ocean in each ocean group
	% of each ocean						
Pacific and adjacent seas	35.9	43.0	2.5	13.1	2.9	2.7	50.1
Atlantic and adjacent seas	32.3	39.3	2.0	17.7	0.7	8.0	26.0
Indian and adjacent seas	30.2	49.2	5.4	9.1	0.3	5.7	20.5
Arctic and adjacent seas	4.2	0	6.8	68.2	0	20.8	3.4
Percent of world ocean in each province	32.7	41.8	3.1	15.3	1.7	5.3	

Table 9. Bathymetric charts used for hypsometric calculations.

Title	Scale	Projection	Ref.
Pacific Ocean	1:7270000 [1])	Lambert azimuthal equal-area	64 M 1
Indian Ocean	1:7510000 [1])	Lambert azimuthal equal-area	[2])
Antarctica	1:9667000	Polar azimuthal equal-area	[3])
Atlantic Ocean	1:10150000 [1])	Lateral projection with oval isoclines	[2])
Tectonic Chart of the Arctic	1:10000000	Polar azimuthal equal-area	[4])
Mediterranean Sea	1:2259000	Mercator	[5])
Northern Hemisphere	1:25000000	Polar azimuthal equal-area	[2])

[1]) Scale of photographic enlargement used for measuring.
[2]) Main Administration in Geodesy and Cartography of the Government Geological Committee, USSR.
[3]) American Geographical Society, New York.
[4]) Geological Institute, Academy of Science, Moscow.
[5]) Unpublished chart of the Mediterranean, modified from contours compiled by R. Nason from various sources. US Navy Hydrographic Office chart 4300 used as base.

Table 7. Province areas in each ocean and total areas of provinces and oceans. For abbreviations for provinces, see Fig. 13 and text [66 M 1].

Oceans and adjacent seas	RISE	OCBN	VOLC	CONS	TNCH	CNRI	VRCM	RIDG	BLOB	Total area of each ocean
	Area [10^6 km^2]									
Pacific Ocean	65.109	77.951	2.127	11.299	4.757	2.690	1.589	0.494	0.227	166.241
Asiatic Mediterranean	0	0	0.003	7.824	0.023	1.233	0	0	0	9.082
Bering Sea	0	0	0	1.286	0.281	0.694	0	0	0	2.261
Sea of Okhotsk	0	0	0	1.254	0.023	0.115	0	0	0	1.392
Yellow and East China Seas	0	0	0	1.119	0.082	0	0	0	0	1.202
Sea of Japan	0	0	0.005	0.798	0	0.210	0	0	0	1.013
Gulf of California	0.042	0	0	0.111	0	0	0	0	0	0.153
Atlantic Ocean	30.519	35.728	0.882	12.658	0.447	5.381	0	0.412	0.530	86.557
American Mediterranean	0	1.346	0.060	1.889	0.201	0.861	0	0	0	4.357
Mediterranean	0	0	0	1.465	0	1.046	0	0	0	2.510
Black Sea	0	0	0	0.263	0	0.245	0	0	0	0.508
Baltic Sea	0	0	0	0.382	0	0	0	0	0	0.382
Indian Ocean	22.426	36.426	0.358	6.097	0.256	4.212	0.407	2.567	0.679	73.427
Red Sea	0	0.070	0	0	0.383	0	0	0	0	0.453
Persian Gulf	0	0	0	0.238	0	0	0	0	0	0.238
Arctic Ocean	0.513	0	0	5.874	0	2.267	0.302	0	0.528	9.485
Arctic Mediterranean	0	0	0	2.483	0	0.289	0	0	0	2.772
Total area each province	118.607	151.522	3.435	55.421	6.070	19.242	2.298	3.473	1.965	362.033

1.2.5 Hypsometric Curves of single ocean basins and ocean basin provinces

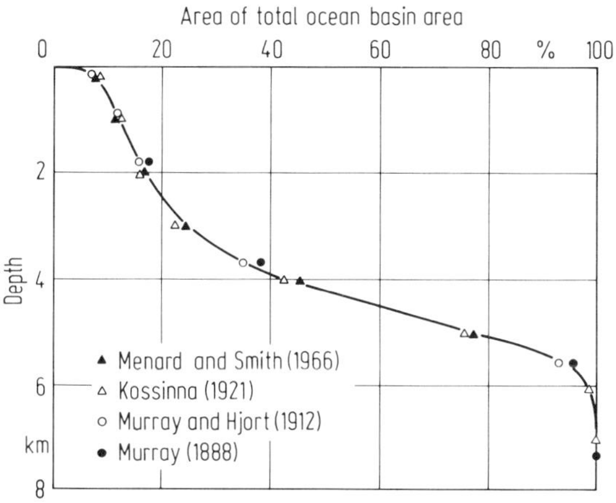

Fig. 14. Hypsometry of all ocean basins, plotted cumulatively [66 M 1].

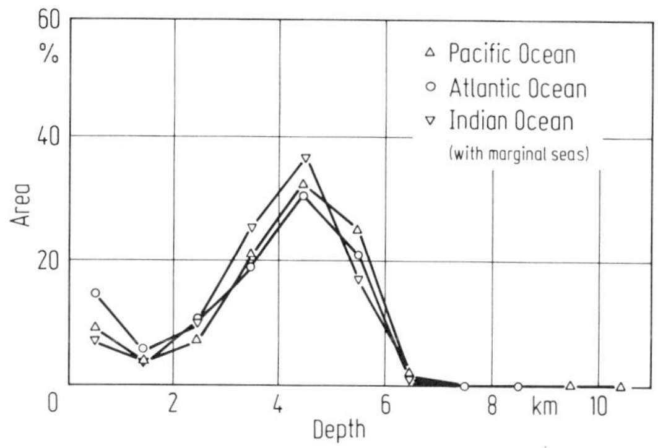

Fig. 15. Hypsometry of individual major ocean basins [66 M 1]. Area in % of individual ocean basin area.

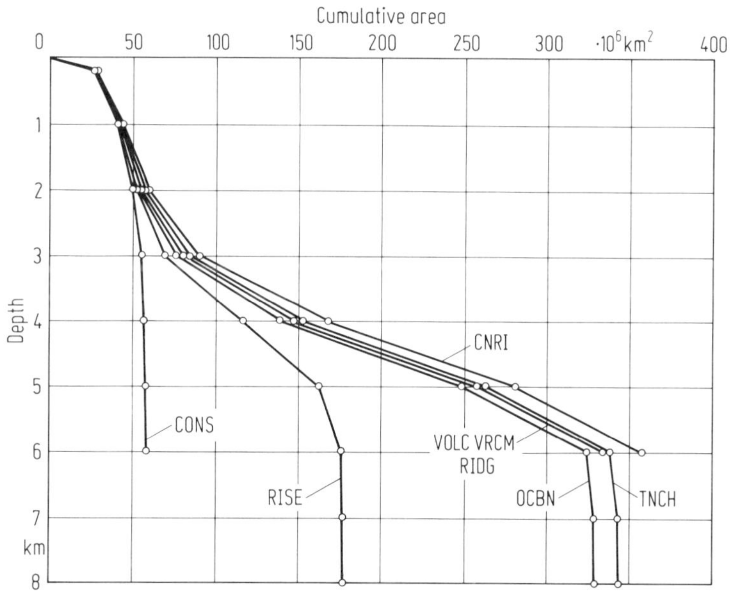

Fig. 16. Hypsometry of ocean basins, plotted cumulatively by provinces [66 M 1]. For abbreviations for provinces see Fig. 13, and text.

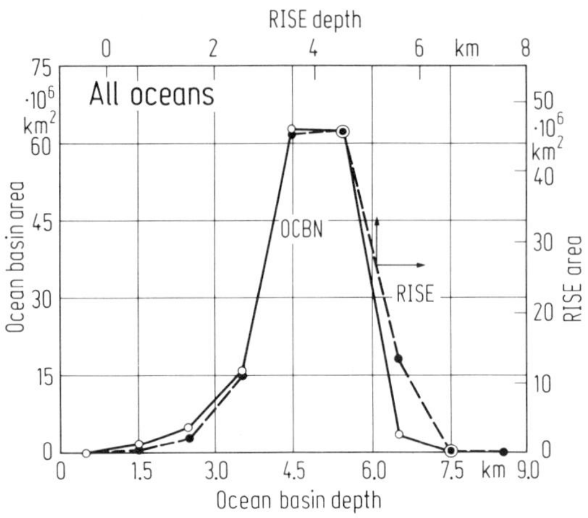

Fig. 17. Hypsometric Curve of ocean basins [66 M 1]. Hypsometric Curve of all ocean basins for rise and ridge (RISE) province normalized to curve for ocean basin (OCBN) province to show close similarity.

The symmetrical distribution curves for ocean basins (Fig. 17) indicate that a considerable area is below the most common depth interval. Very extensive regions deeper than 6000 m exist in the northwestern Pacific and eastern Indian Oceans.

Table 10. Characteristics of oceanic rises [66 M 1].

Relief gives the elevation of oceanic rises and ridges above adjacent ocean basins as determined from difference in mean depth and shifting of Hypsographic Curves respectively.

	All provinces	OCBN (1)	RISE (2)	Mean (1) -Mean (2)	Shift of distribution curves
	Mean depth [m]			Relief [m]	
World ocean	3729	4753	3970	783	1000
Pacific Ocean and marginal seas	3940	4896	3968	928	1100
Atlantic Ocean and marginal seas	3575	4670	4008	662	900
Indian Ocean and marginal seas	3840	4530	3945	585	800

The mean depths of rise and ridge have a limited range, from 3945 m in the Indian Ocean to 4008 m in the Atlantic Ocean. This uniformity seems remarkable considering the widespread and diverse evidence that oceanic rises and ridges are tectonically among the more unstable features of the surface of the earth. It is all the more remarkable because the local relief, or elevation above the adjacent ocean basin, differs substantially in different oceans. The relief of rise and ridge in an ocean basin can be estimated by subtracting the mean depth of rise and ridge from that of ocean basin or by determining the deepening required for the rise and ridge Hypsometric Curves to give a best fit of individual ocean basin Hypsometric Curves.

Comparing the means gives the relief of rise and ridge above ocean basin as 585 m in the Indian Ocean, 662 m in the Atlantic Ocean, and 928 m in the Pacific Ocean. The reliefs from matching curves are 800, 900, and 1200 m, respectively. The greater relief obtained by the curve-matching method results from ignoring the shallow tails of the depth distribution. See Figs. 14···20.

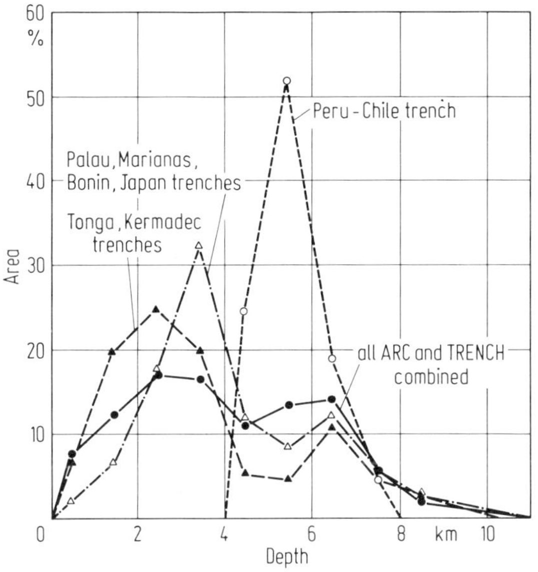

Fig. 18. Hypsometry of all arc and trench provinces and of some groups of arcs and trenches [66 M 1]. Area in % of the total area of the respective arc and trench provinces.

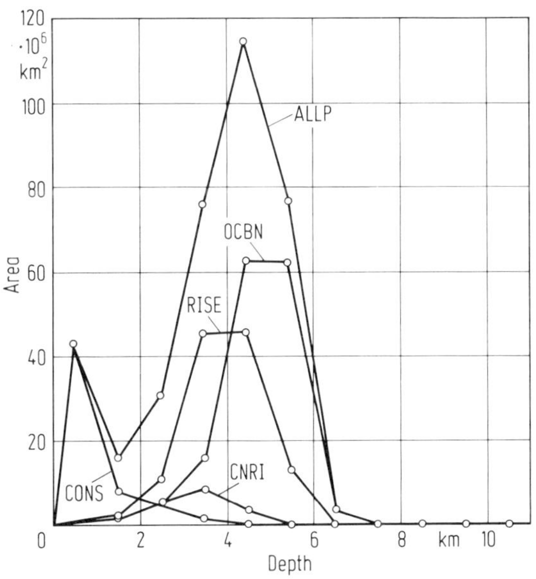

Fig. 19. Hypsometry of all ocean basins [66 M 1]. This diagram is for all provinces combined (ALLP), and for individual major provinces. For abbreviations of provinces, see Fig. 10 and text. For hypsometry of major provinces of individual ocean basins, see Fig. 20.

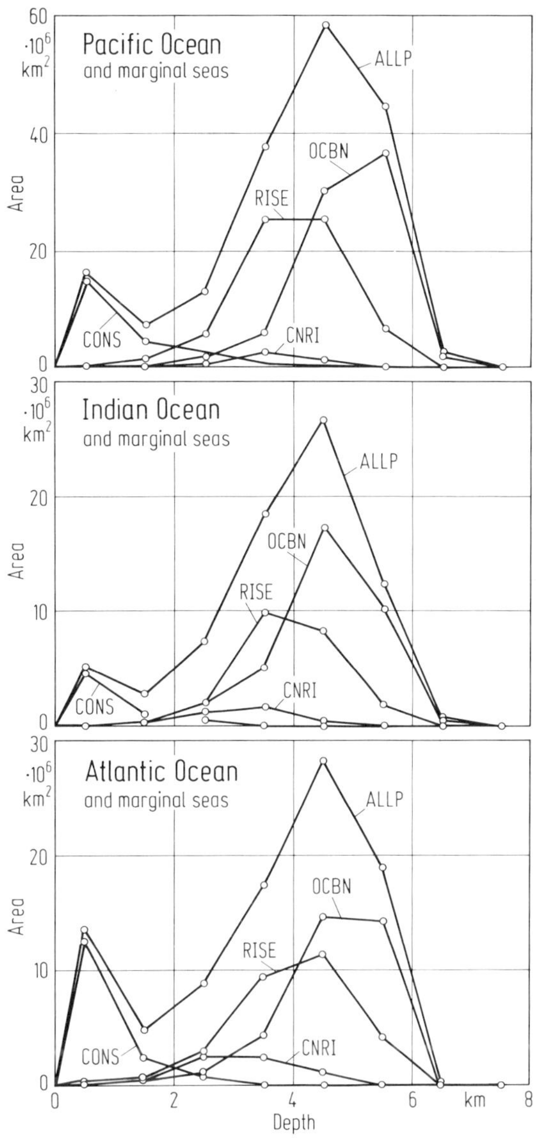

Fig. 20. Hypsometry of all provinces (ALLP) and individual provinces in major basins [66 M 1]. For abbreviations of provinces see Fig. 13, and text. The depth distribution in provinces in different ocean basins closely resembles the composite distribution in the world ocean. The sum of the depth distribution of all provinces (ALLP) in an ocean basin is double peaked, but for the individual provinces it is single peaked, and relatively symmetrical. The mean depths of all provinces in the three major ocean basins, including marginal seas, range from 3575 m for the Atlantic Ocean to 3940 m for the Pacific. The range of mean depths of the ocean basin province in each of these ocean basins is similar.

1.3. Bathymetry

1.3.1 The measurable parameters

1.3.1.1 The position of a topographic data

Our knowledge about topography of the ocean floors that is global and of comparable quality for the different regions can be represented on small scale maps (1 : 5000000 or 1 : 20000000) so far only. Distinct smaller areas are mapped on larger scale bathymetric charts of special issues. [80 G, 84 L 1]. The depth of the ocean floor is still surveyed very unequally in different regions. Coastal areas and offshore bottom charts are available in large scales as nautical charts or special charts.

The presentation of topographic data on bathymetric charts depends on the accuracy of measurement and positioning. The geographic position at sea is, to some extent, uncertain, so that disagreements in depth may well be caused by the uncertainty in position. Referring to IHB (1968), part B, section I (a), the standard deviation of a fix at sea, i.e. its precision based on repeatability, combined with the fluctuations in plotting, should not exceed 1.5 mm at the scale of the survey [77 I]. See Table 11. A fix is a relatively accurate position without reference to any former position. The terrain-accuracy standards, associated to work/tasks, are [84 L 1]: General navigation (long range) = 3000 m, fisheries research = 400 m, waste diposal = 250 m, cable laying = 100 m, deep sea drilling = 25 m.

Table 11a. Accuracy of position finding systems at sea, based on average, spatial and temporal conditions. Compare [84 L 1].

Navigation systems	Signal characteristics	Accuracy	System characteristics
Astronomical	optical	1 nm(≙ 1′)	referred to points of the celestial sphere changing with time
Radiolocation	radiofrequency	0.1 ··· 1 km	depending on the system, beacon-array and range
Satellite	microwave	300 m	NNSS, 5 satellites, since 1965, depending on Doppler-effect-signal-system
Satellite	microwave	30 m (≙ arc 1″)	NAVSTAR, 18 satellites (6 per orbit), 1990 onward, global-positioning system [82 N] (advanced program only 12 satellites), satellite navigation correlated to WGS
		< 30 m	1972 ellipsoid, compare subsect. 1.1.3 for military use, claims a precision in time measurement of about 0.1 micro second
Transponder	acoustic	1 ··· 10 m	referred to buoys, or fixed instruments at the sea floor, used to control surveying with submersibles
Range or directions			referred to targets on shore, offshore, or buoys, electronically operated
radar	microwave or	0.1 ··· 1 m	
laser	optical	0.01 ··· 0.1 m	geodetic precision
Integrated systems	combined	very accurate	depending on array, computer controlled
Inertial	acceleration	variable with updating intervals	useful for submersible missions, not yet used for surveying from sea surface

1.3.1.2 The depth of a topographic data

Charted depth is the vertical distance from the chart datum to the bottom. At present, there exists a number of analogue or digital echo sounders, side-looking and side-scan sonars, penetrating, profiling, etc., instruments with hull-mounted or towed transducers. In order to avoid the disturbing hyperbolic recording of peaks ahead, astern or abeam, a very narrow-beam echo sounder, vertically stabilized, is required. The problem is the high attenuation of the signal's energy in sea water [75 I 1, 54 L, 84 L 1, 70 K].

The detectability of sea floor topography by sounding

DSL: Deep scattering layer: False bottom. The stratified population of organisms in most oceanic waters, which scatter the sound.

Detectable is the stochastic fluctuation to which repeated depth measurements are subject. The depth is uncertain by 0.1%···0.5%, depending on the acoustic system and the seawater condition (salinity, temperature, pressure). Because no higher precision is available each data for a certain point of the ocean floor is not a measurable constant in the sense of geodetic data, but has the character of index data. Soundings are measured in units of travel time of the pulsed signal from the system carrier to the bottom and back (echo-time). The signal travel time, to be converted to distance to the bottom (depth), has to be corrected for deviation of direction, bottom configuration, sea state, and the parameters of sound transmission in sea water (temperature, salinity, pressure).

At the basis of the depth determination is the echo-sounding formula:

$$h=\frac{1}{2}ct$$

where h denotes depth, c sound speed in sea water, and t total travel time of the signal.

The actual value of the sound speed may vary from 1450 to 1550 m s^{-1} (average 1500 m s^{-1}), depending on salinity and temperature of the sea water. The sounding unit τ in modern use is the 400th part of a second. This unit is used on all US Navy charts of deep sea. USA standards: 800 fathoms/s, 4.742 feet/s at 32 °F, 35‰ salinity, 1 atmosphere pressure.

Tables for the conversion of travel time to depth (in use for soundings up to 1982): Matthews Tables [30 M]. In use for 1982 onward: The tables at present internationally agreed upon have been adopted by the XIIth International Hydrographic Conference, April 1982, at the site of the International Hydrographic Organization at Monaco. The present tables have also been prepared under the auspices of the United Kingdom Hydrographic Department, and have been issued as: "NP 139 – Echo-sounding correction tables".

Since 1980 a very important innovation of surveying of the sea floor with a simultan plain sounding, gaining large strips, has been introduced: the so-called "Sea-Beam-System", developed in Boston, USA.

Sea-beam is a bathymetric range survey system, which sounds with a fan of 16 dense sonar beams simultaneously. The swath of the sea-beam's fan is equal to 80% of the sea depth, i.e. at 5000 m depth a strip of 4000 m width can be detected. [81 S, 80 A, 82 S 1]. Compare subsect. 1.4.3.4.1 and Fig. 56. For the long range side-scan device system GLORIA, see subsect. 1.4.5, Fig. 63.

For higher resolution of the topography, near bottom echo-sounders are occasionally used thus avoiding the long water path and consequent geometrical problems. Scanning narrow-beam sounders operating in near-vertical mode allow a contoured strip of the ocean floor either side of the ship's track to be generated continuously. By contrast side-scan sonars do not give quantitative depths data, but delineate the shapes of features in plan view as much as radar does. Long range side-scan sonar can operate in ocean depths giving sonograph strips 20 km wide, so that they can be built up into a mosaic view in conjunction with bathymetric survey to aid contouring [73 L 1]. Acoustic systems that scan the sea bottom from near the sea surface, including even the modern multibeam techniques, are unable to resolve vertical relief of less than 10 m or horizontal dimensions of less than 100 m in deep water.

1.3.1.3 Ambiguity of the detected configuration of the topography of the floor of the ocean

The course of a sounding vessel and therefore the sounded profile is, in general, not perpendicular to a submarine slope which is to be detected and surveyed. Therefore calculations of the data of different sounding profiles are necessary to get the real slope inclination of the relief. Before these operations the usual geometric corrections of the sounding system have to be made (concerning sound propagation with a spheric front against the bottom) [84 L 1], Fig. 21b, c.

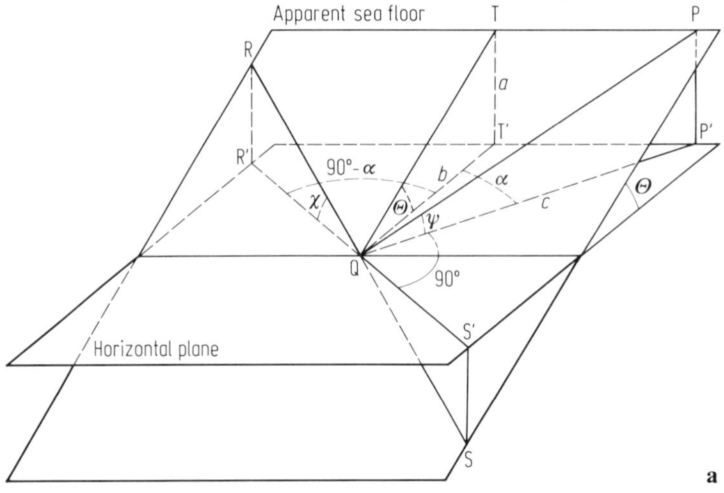

Fig. 21a. Showing in perspective two courses steered at right angles to each other, projected as P′Q and QR′ on the horizontal plane through Q. The two apparent slope angles observed during these two courses are ψ and χ, respectively, thus defining the apparent slope angle ϕ of the sea floor,

$$\tan^2\phi = \tan^2\psi + \tan^2\chi,$$

from which the real value θ of the slope angle can be calculated,

$$\sin^2\theta = \tan^2\phi.$$

Since $\tan\psi = \tan\phi\cos\alpha$, the direction α of the maximum slope relative to the first course steered follows from

$$\cos^2\alpha = \tan^2\psi/(\tan^2\psi + \tan^2\chi),$$

relating α to the two apparent slope angles observed. The sign of α follows from a sketch of the situation. [84 L 1].

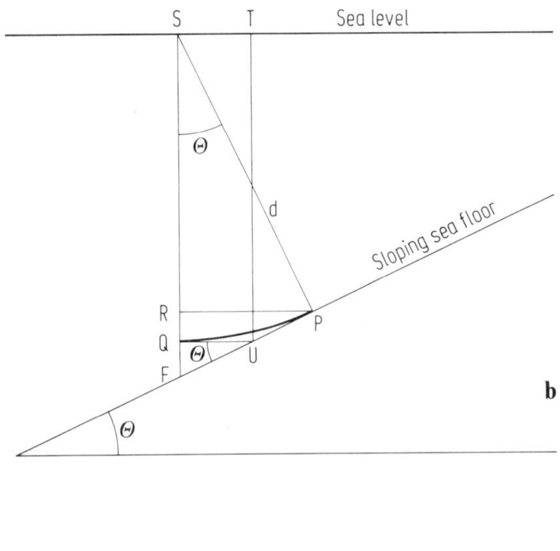

Fig. 21b. Sounding vessel at S measuring "depth" SP = SQ, but requiring real depth SF. The measured distance SP = SQ = TU is real depth at T, so that ST = QU represents position correction. [84 L 1].

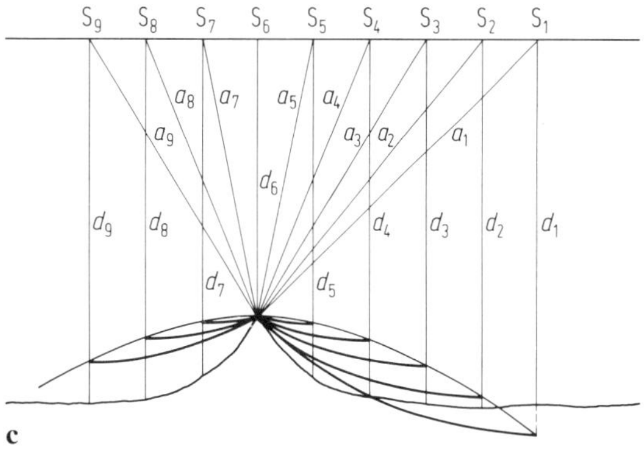

Fig. 21c. Hyperbolic presentation of a peak on the sea floor. The positions S_1 to S_9 are those of the sounding vessel in which depths d_1 to d_9 are measured while simultaneously reflections from respective distances a_1 to a_9 are received. [84 L 1].

c

Depth and position corrections over a sloping sea floor:

Conditions and computing of the error made in depth and in the positioning of the depth figure because of acoustic measurement of depth over a sloping sea floor using a wide-beam echo sounder: It is possible to determine the direction of slope of the sea floor and, therefore, the direction of the acoustic ray which traverses the shortest distance (perpendicular to the surface of the sea floor). In Fig. 21b a section is shown along the plane through the transducer, perpendicular to the sea floor and in the direction of the maximum slope.

The sounding vessel at S measuring "depth" SP = SQ finds a depth QF too small. The value of QF follows from QF = SF − SP. If for SP is written the measured distance d and θ denotes the slope, the depth correction amounts to:

$$\mathrm{QF} = d \sec\theta - d = d(\sec\theta - 1)$$

The measured distance to a sloping sea floor is always smaller than the depth vertically beneath the sounding vessel. The measured value SP = d is the correct depth at position T, i.e. the depth TU. The displacement ST = QU is found from (after [84 L 1]): $\mathrm{QU} = d(\sec\theta - 1)\cot\mathrm{an}\,\theta$, correction tables concerning QF and QU for d vs. θ see [84 L 1].

Determination of slopes of the sea floor topography:

Assuming the slope of the sea floor to be uniform over a larger area, the problem of finding the direction of maximum slope can be solved if the sounding vessel steers two courses, the second at right angles to the first. Fig. 21a.

In the case that the sounding profiles are steamed parallel in larger distance the relief may not be matched as best as possible for quantitative relief detection. This is still the reality for areas of the sea floor. The example may demonstrate the extent to which the so-called apparent sea floor and the real sea floor topography differ.

Determination of the relationship of surface echo pattern and deep sea sedimentary topography:

Large sedimentary areas of the sea floor often can be characterized by a similar type of echo return pattern, and thus presumably by a similar sedimentary morphology. This morphology provides important information on regional sedimentary processes, but the echo pattern often does not correctly display the bottom morphology. The amplitude of the sea-floor relief may be difficult to determine where side echoes cross in the trough or where hyperbolic echoes dominate.

Wide-beam surface-ship echo-sounding records (3.5 and 12 kHz) are frequently utilized to study the nature and distribution of deep sea sediments and sedimentary morphology. Wide-beam echo pattern is sensitive to water depth as well as to the amplitude and wavelength of a gently-varying, low-amplitude sea bed (e.g., a sinusoidal wave), and the structure of such features can be obscured on wide-beam records in deep water. The shape of the overlapping trough echo zone can be a sensitive indicator of sea-floor structure when it is obscured by side echoes. For a uniformly accreting undulatory surface, the overlapping echoes form a vertical band; for a migrating wave this echo band dips opposite to the migration direction; and for a filling trough its width increases with depth. Hyperbolic echoes can be formed by a number of sea-bed features, and the hyperbola pattern may preserve information on the nature of the topography. Several characteristics of surface echo profiles can aid in the description and understanding of the sea floor morphology. These include the linearity, orientation, spacing, amplitude, and structure of the sediment surface as well as the regional echo distribution pattern. Discrete hyperbolic echoes may not be observed from closely spaced point or line reflectors. After [84 L 1].

Table 11b. The acceptable standard deviation in repeated depth measurements, after application of corrections for roll, heave and pitch, for different depths. Both d and s_d expressed in meters. From [84 L 1].

d	s_d	d	s_d	d	s_d
0	0.14	50	0.46	100	0.85
5	0.17	55	0.495	200	1.70
10	0.20	60	0.53	300	2.55
15	0.23	65	0.57	400	3.40
20	0.26	70	0.61	500	4.25
25	0.29	75	0.65	600	5.10
30	0.32	80	0.69	700	5.95
35	0.355	85	0.73	800	6.80
40	0.39	90	0.77	900	7.65
45	0.425	95	0.81	1000	8.50

The above standard deviations are based on different equations for the following depth ranges:

d from 0 to 30 m: $s_d = 0.14\,\mathrm{m} + 0.006\,d$

30 to 60 m: $s_d = 0.11\,\mathrm{m} + 0.007\,d$

60 to 100 m: $s_d = 0.05\,\mathrm{m} + 0.008\,d$

100 m and over: $s_d = 0.0085\,d$

1.3.1.4 Convertion of sounded data into contourlines of bathymetric maps

The technique of continuous sounding along ship-sailed profiles and the display of sounded depths on tapes combined with integrated navigation and digitally computed bathymetric profiles of geometrically corrected distances leads to the problem: from data acquisition to data display of the depths in the mode of contour lines, the discreta of a bathymetric map.

Contour lines of bathymetric maps before the 70's were drawn by hand after visual interpretation of the data. [70 C, 80 G].

The development of future bathymetric systems which exploit data-processing technology to the maximum will reflect as great an advance over systems presently in use as does current echo-sounding over the lead-line techniques years ago. Numerical topographic mapping technique with a matrix of x, y, and z information (latitude, longitude, and depth) as collected through a survey, and stored in a computer can deliver reliable maps. Fig. 22. Automatical systems based on integrated sounding-survey systems are developed. Such system is used in the DHI by Bettac [84 B 2].

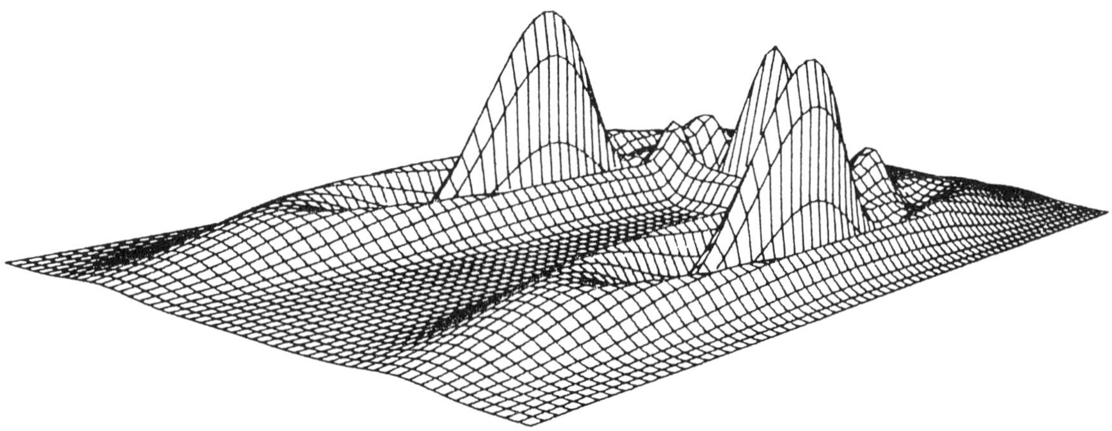

Fig. 22. Computer-generated relief drawing of three seapeaks with surrounding moats and walls. This technique can depict the relief from different perspectives, and at different vertical exaggerations. California Computer Products, Inc., Anaheim, California. From [70 C]. For "moats" compare subsect. 1.4.3.4.

1.3.2 Topographic features of the ocean floor on profiles

A bathymetric profile is a drawing presenting depths from sounding data along a surveyed line, with depth and distance as vertical and horizontal parameters, respectively.

1.3.2.1 Bathymetric profiles, geometric conditions

Profiles of "global" or oceanic size depend on the spheric geometry. The deeper parts of the profiles are decreasing in scale in comparison with the upper parts due to earth curvature, i.e. the extent of features of greater depths is exaggerated in comparison with features of lower depths on a linear profile base, where earth curvature is not taken into account. See Figs. 23, 24.

The earth curvature profile (based on spheric geometric condition)

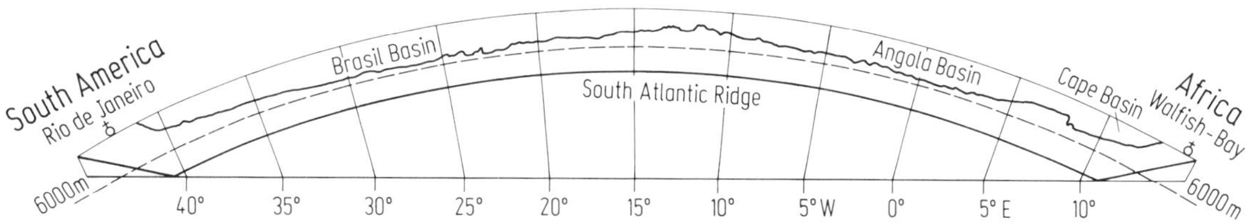

Fig. 23. Transoceanic bathymetric profile of the southern Atlantic Ocean along 23° S (average latitude of the sounding profile of the German research vessel "Meteor" in 1925) [36 S]. Exaggeration of the profile: 30 times. Length of the profile between 45° W and 15° E = 60° = $\frac{1}{6}$ of the latitude circle of 23° S ≅ 4600 km. The profile conforms to scale with the earth curvature. Compare Fig. 27.

The linear profile (based on plain geometric condition)

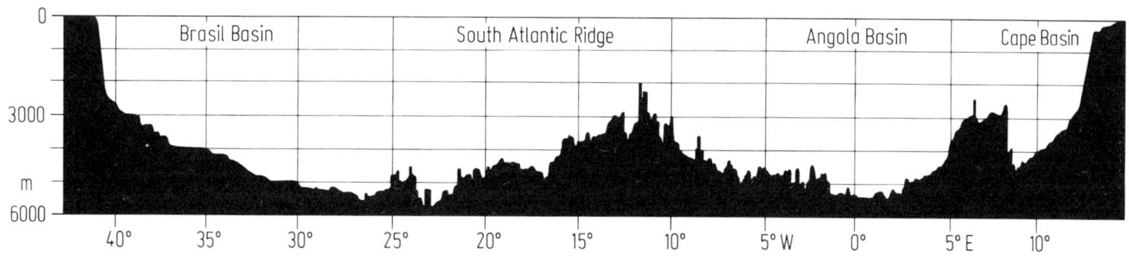

Fig. 24. Transoceanic bathymetric profile of the southern Atlantic Ocean along 23 °S (average latitude of the sounding profile of the German research vessel "Meteor" in 1925) [36 S]. Scale of the length of the profile as in Fig. 23. Exaggeration: 180 times. The profile is represented in linear configuration, where earth curvature is not taken into account. Compare Fig. 27.

The reference grid of a bathymetric profile.

Bathymetric profiles are surveyed and displayed both on the same or on separate geographical reference grid system projections. Most profiles are surveyed (sounded and plotted) along true courses, i.e. the rhumb line (loxodrome). The display may be presented also on great circles. Fig. 25.

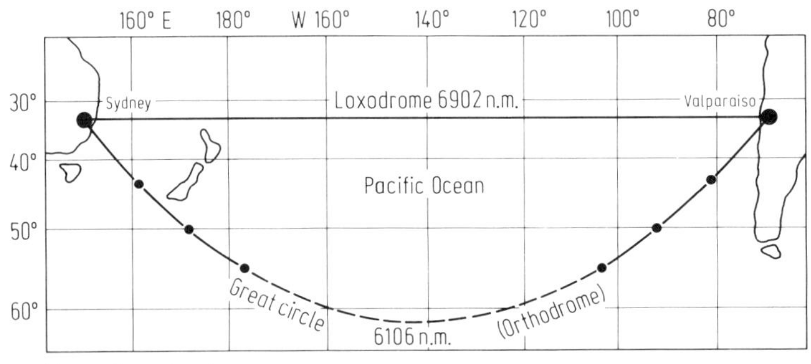

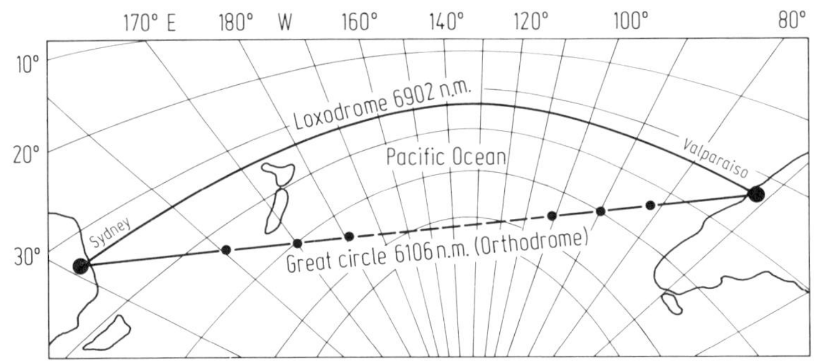

Fig. 25. Bathymetric profile along a surveyed or a displayed line [80 G]. Top: Profile on Mercator projection in which the longitudes and the latitudes appear on the cylinder projection as straight lines, intersecting each other in right angles. The loxodrome or rhumb line, which crosses all meridians at a constant angle, appears on this Mercator projection as a straight line. The loxodrome is mostly used as the surveying course. The GEBCO-chart is produced in this projection, except for the Polar region, Fig. 31.

Bottom: Profile on gnomonic projection in which a projection from the center of the globe is on a flat sheet tangential at a point. The great circle (orthodrome) appears as a straight line. The orthodrome is mostly used for long range navigation.

The slope gradient of a bathymetric profile.

The inclination of the bottom is given as the tangent of the angle between the sloping plane and the horizontal, expressed as a ratio of integers. These ratios are referred to as gradients. Slope values are expressed in arc degrees, percent grade, or height per distance. Fig. 26 shows various gradient ratios at a 40:1 vertical exaggeration.

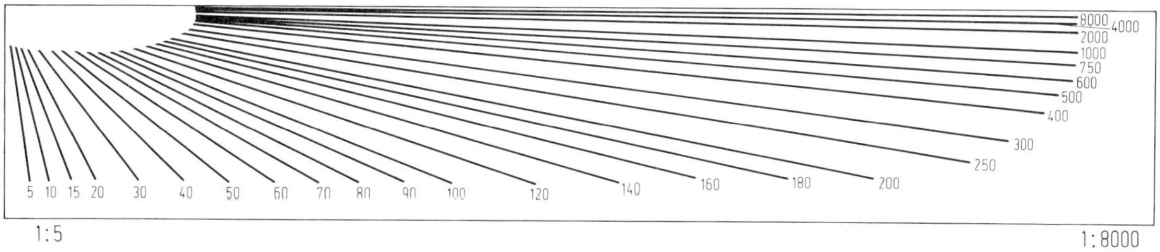

Fig. 26. Gradients from 1:5 to 1:8000 shown at a 40:1 vertical exaggeration, from [59 H].

1.3.2.2 Topographic features of particular oceans on transoceanic bathymetric profiles

The symmetric sea floor configuration as a general model from the Atlantic Ocean.

The oceans can be divided into three major provinces, the continental margins, the ocean basin floor, and the mid-ocean ridges, all are illustrated in this cross-section of the Atlantic Ocean (Fig. 27). The smooth terrain of the continental margins and the abyssal plains is well-differentiated from the rough terrain of the mid-oceanic ridge and abyssal hills. The rough terrain is generally thought to be an expression of the effects of tectonic displacements and volcanism at the surface of the ocean crust. The irregular surface of the basement is thought to have been formed initially at the crest of the mid-oceanic ridge (where lithospheric divergence and axial accretion occur), and then buried more and more by sediments with time. The smooth terrain has resulted from sedimentary deposition, and, locally, erosion. Note: the fracture zones (transform faults) interrupting the ridge-and valley terrain of the Mid-Atlantic Ridge, the symmetry of the entire sea floor about the crest of the mid-oceanic ridge, and the extreme flatness of the abyssal plains. Vertical exaggeration is 100 [70 H 1, 70 H 2].

Bathymetric profiles of the Pacific and Indian Oceans

In contrast to the Atlantic ocean which is dominated by the mid-oceanic ridge, about 80% of the Pacific Ocean is ocean basin floor, made up primarily of abyssal hills and interrupted, particularly in the western portion, by seamounts, and oceanic rises. Much of the Pacific Ocean continental margin consists of island arcs, marginal trenches, and outer ridges, rather than continental rises. Figs. 28, 46.

The profiles of the Indian Ocean taken during the International Indian Ocean Expedition (IIOE) crosses with IV the mid-ocean ridge. See Fig. 29. Compare Figs. 33, 35, 36.

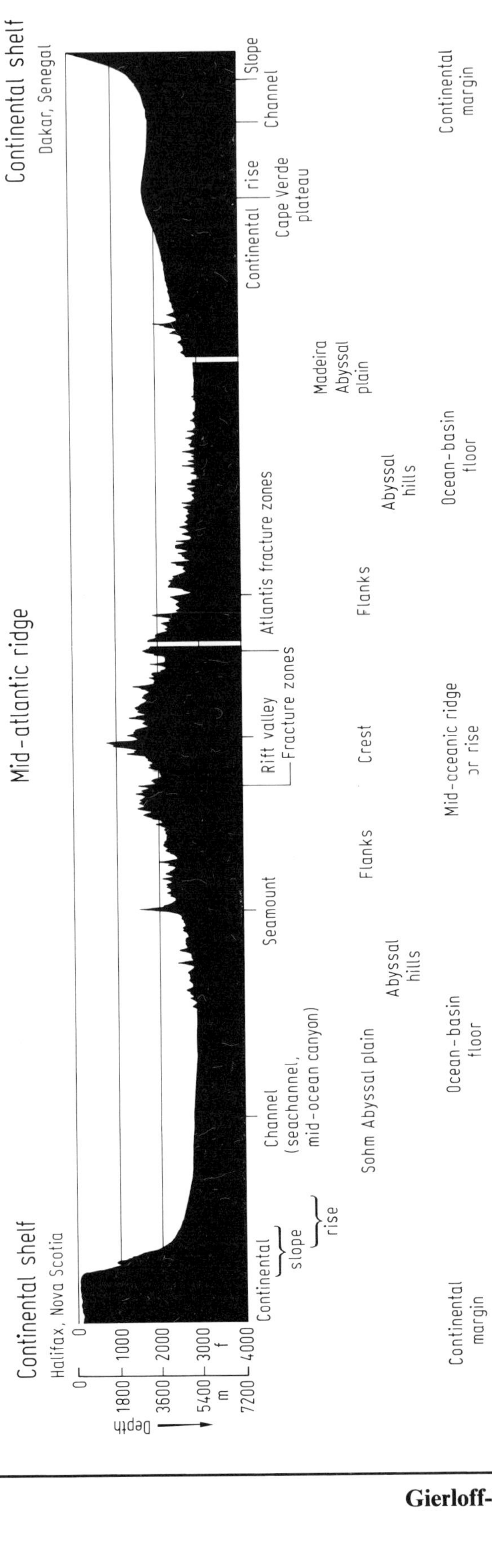

Fig. 27. Transatlantic profile between Halifax, Nova Scotia, and Dakar, Senegal [77 H 2]. Vertical exaggeration is 100. Profile not along a great circle but with three different courses. Earth curvature is not taken into account. Length ≈5500 km. Compare Figs. 23, 24.

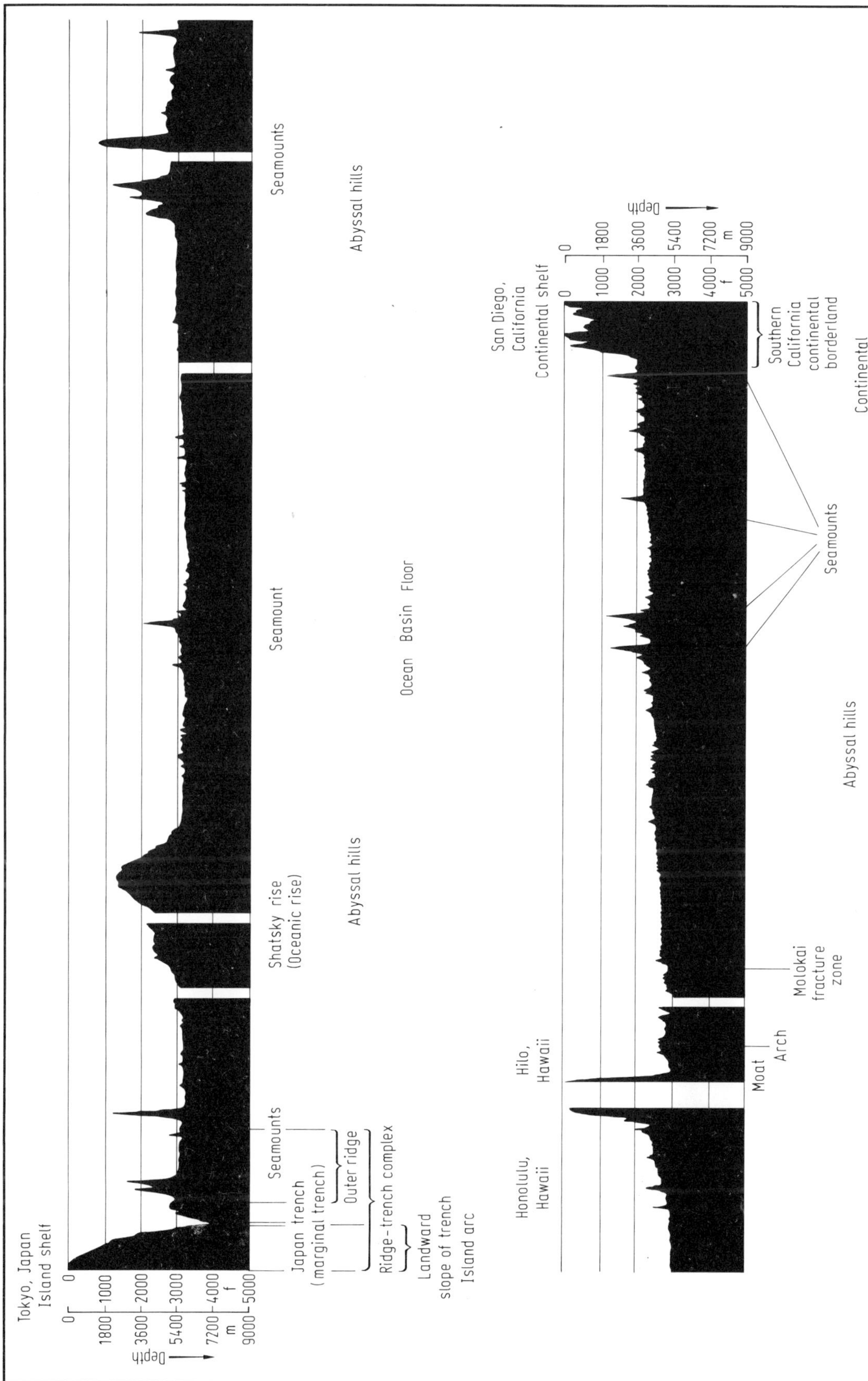

Fig. 28. Trans-Pacific profile between Tokyo, Japan, and San Diego, California. [70 H 2]. Profile along nine different courses, not along a great circle. Length ≈ 8000 km. Vertical exaggeration ≈ 80 times.

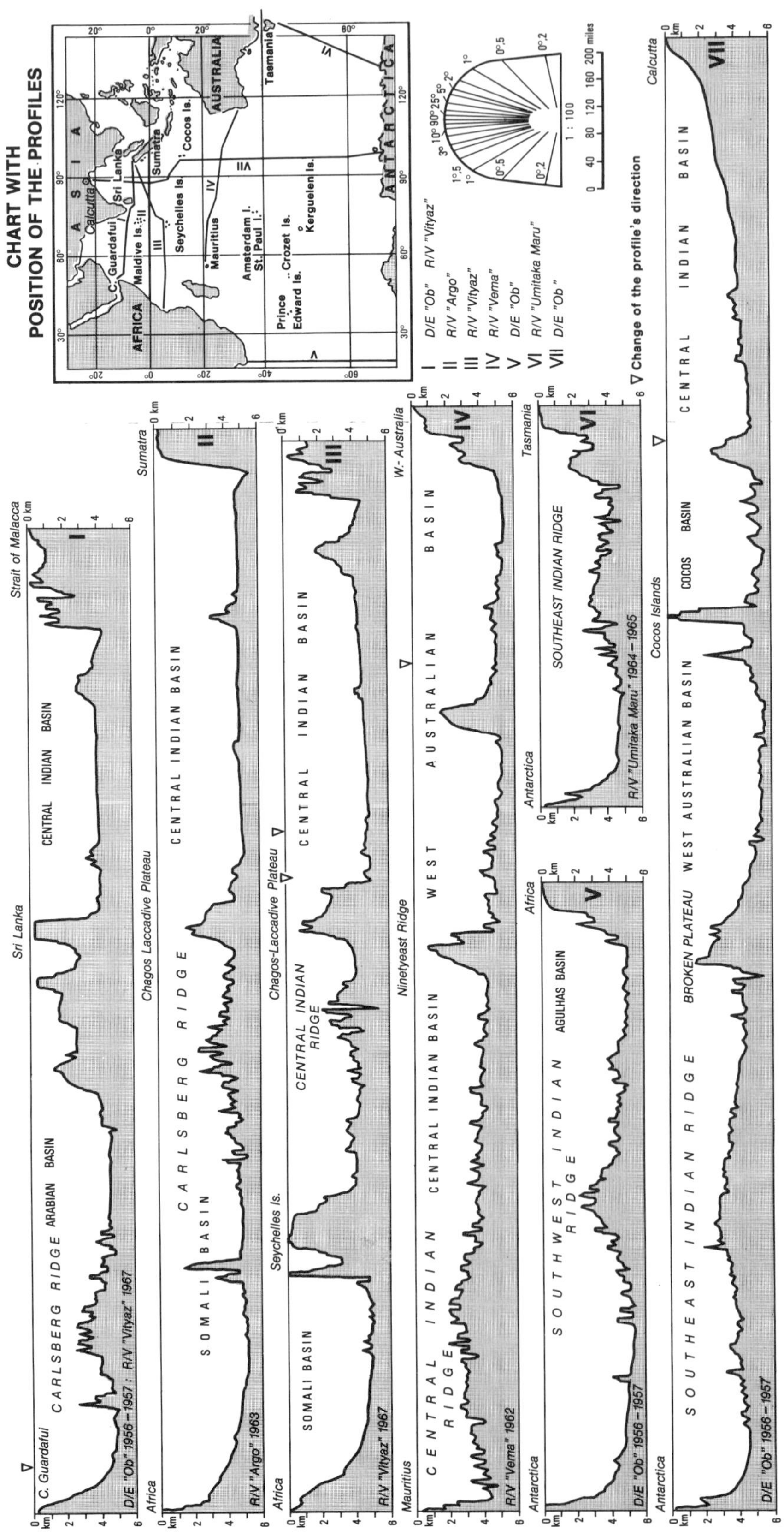

Fig. 29. Transoceanic profiles of the Indian Ocean [7511]. During the International Indian Ocean Expedition (IIOE) of 1960–1965, thousands of kilometers of bathymetric profiles were surveyed by ships of several nations.

1.3.2.3 Classification of topographic features on acoustic sounded ocean floor and subbottom stratigraphic reflection (bottom penetrating sounding) profiles

Seismic reflection profiles may illustrate the modification of ocean bottom terrain due to sediment accumulation. Fig. 30. Compare Table 14.

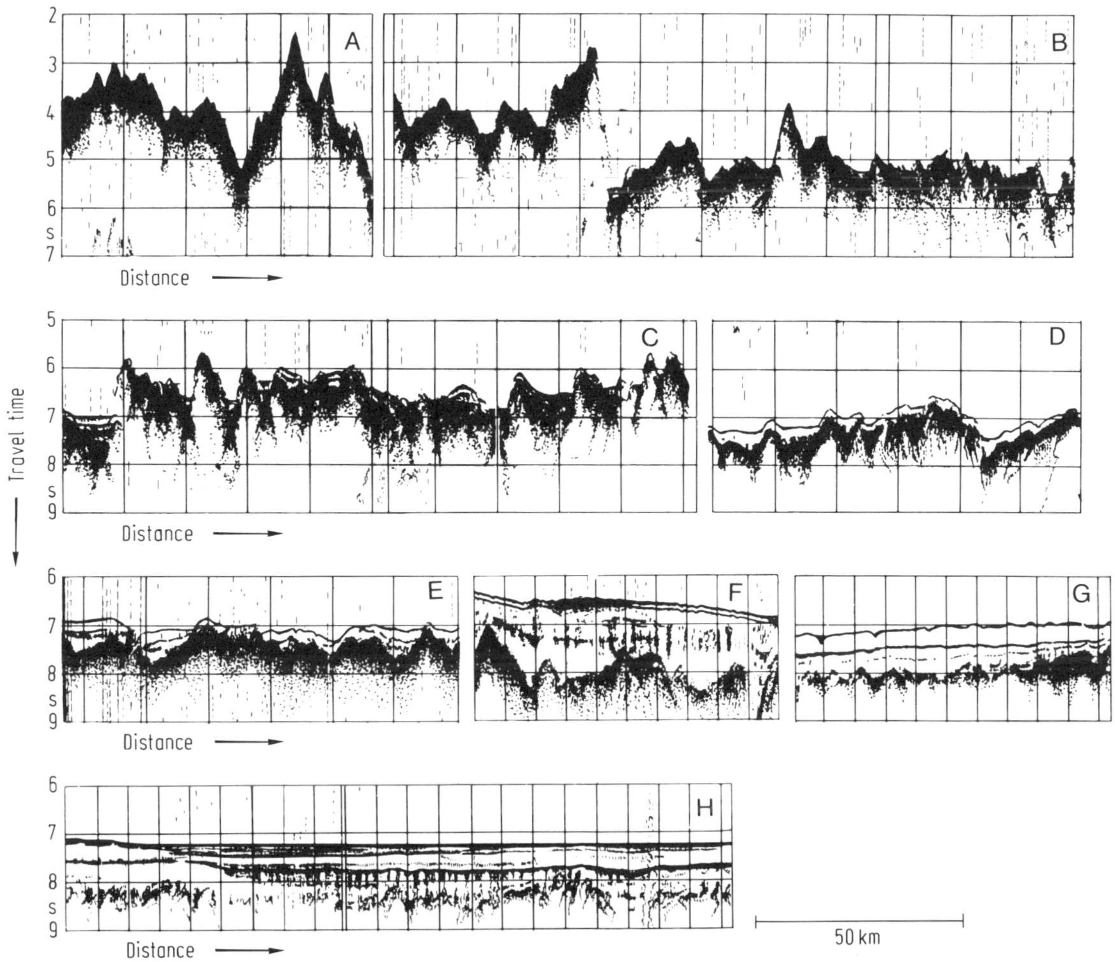

Fig. 30. Acoustic sounded sea floor and subbottom profiles for terrains of the Mid-Atlantic Ridge (A···D). Bermuda Rise (E···G), and Hatteras Abyssal Plain (H) [77 H 3]. (A) Completely exposed basement ribs are found near the crest of the Mid-Atlantic Ridge. (B) At its upper flanks accumulation of sediments occurs mainly in intermontane basins. At its middle (C) and outer (D) flanks continued sedimentation has reduced the size and the ruggedness of the original relief. (E) For the Bermuda Rise the basement is almost completely covered by sediments. At its northern (F) and western (G) fringes currents have initiated the formation of hills or dunes exhibiting regular patterns. (H) Turbidity currents transporting terrigenous sediments from the continental shelves are responsible for sediment cover formation in abyssal plains.

1.3.3 Classification of maps and charts presenting ocean floor topography

1.3.3.1 Nautical charts

Nautical charts display isobathic contour lines and particular numeric depth data. These charts are produced for national purpose by national hydrographic services, in connection with IOC and IHO. Coastal areas are well mapped in large scales (with contour lines of 2.5, 5, 10, 40···100, 200 m), the floors of the wide oceans, however, are mapped only with large contour line intervals or scattered depth numbers on very small scale charts, and not really bathymetrically, i.e. topographic features of the sea floor are represented only to some degree.

The navigator requires informations on different features. Emphasis must be laid on features that are either hazards to navigation, such as tidal areas or shoals and reefs, and the 10 m and 20 m-line, or aids to navigation, such as the 200 m-line (approaching the shelf-edge). On English charts there is often still the 100 fathom-line (180 m) used to mark the shelf-boundary. The scales of nautical charts include a huge variety. A project of an international nautical chart in the scales of 1 : 10000000 and 1 : 3500000 is in operation since 1970 (IHO). Several sheets are issued.

Moreover nautical charts have to present the most recent situation of lights, radio beams, and all floating buoys and beacons which are not subjects of interest for sea floor bathymetry.

Examples of nautical charts:

Chart No. 1: United States of America, Nautical chart symbols and abbreviations

Chart No. 5011: Great Britain, Symbols and abbreviations used on Admirality Charts

Seekarte Nr. 1: Federal Republic of Germany, DHI, Abkürzungen und Zeichen in der Seekarte (chart abbreviations).

For a list of nautical chart catalogues of the nations, see: List of chart catalogues, published by the member states of the IHO in "Extracts from IHO bulletin, Monaco, published by IHB, Monaco, periodically. [72 K, 84 L 1], and NOAA chart-catalogue, USA.

Geologic maps of the sea floor

Geologic-tectonic-geomorphologic maps of the sea floor:

Geologic-tectonic-geomorphologic features are mapped in the "Geological world atlas", Unesco, Paris-Washington, sheets from 1976 and later. [76 G 2], Antarctica (sheet 17, 1 : 10 000 000), Pacific Ocean (sheet 20, 1 : 36 000 000), Indian Ocean (sheet 21, 1 : 29 000 000), Atlantic Ocean (sheet) 22, 1 : 34 000 000). "IIOE geological-geophysical atlas", Academy of Sciences of the USSR (1975).

A program of France is [71 C 2].

NOAA-atlas 3, the central North Atlantic ocean basins and continental margins [80 R].

1.3.3.2 Bathymetric charts

A bathymetric chart is a map which delineates the shape of the bottom of a body of water, or a portion thereof, by the use of depth contours (isobaths), often called "chart-contour-imagery-line" on the ground, of which all points are at the same elevation above or below a specified datum [72 K, 80 G]. For distinct regions of the oceans, see List of chart catalogues, published by the member states of the IHO, Monaco [81 I], the International bathymetric charts of the Mediterranean (IBCM), ten sheets, 1 : 1 000 000, IHB, and the "Bathymetry of the Northeast Atlantic", charts 1 : 2 400 000, 4 sheets, 1975···1983, of the Institute of Oceanographic Sciences. A list of world series of bathymetric charts and a list of ocean series of bathymetric charts of different scales, nations and regions, updated from 1950···1970, are presented in [73 L 1]. For atlasses with general bathymetric maps, world cover, smaller scales, see [68 D, 72 K, 80 G, 84 L 1]. Figs. 31, 32.

The General Bathymetric Chart of the Oceans (GEBCO)

The bathymetry of the oceans is presented on the General Bathymetric Chart (GEBCO) by International Hydrographic Organisation (IHO). The fifth edition of the Bathymetric Chart of the Oceans (GEBCO) is the most important and the authentic presentation of the bathymetry of the oceans. It consists of 16 charts on Mercator Projection covering the world from 72° N to 72° S, on a scale of 1 : 10000000 at the equator (greater at higher latitudes!); and two polar sheets on Polar Stereographic Projection to 64° N and 64° S respectively, on a scale of 1 : 6000000 at the pole. On this scale of 1 : 10000000, a distance of 10 km on the sea floor is equal 1 mm on the map. The IHO is responsible, in conjunction with 19 volunteering Hydrographic Offices in its member states, for maintenance of 655 base master sounding sheets on a scale of 1 : 1 000 000. In 1982 the 5th edition of GEBCO was completed [70 G 2, 80 I]. Bathymetric information from GEBCO maps, supplemented by nautical charts and other maps, gives the present state of our knowledge of the topography of the ocean floor to be adequate for 16% of the oceans, sufficient for major features only for 22%, and too sparse to determine for 62%.

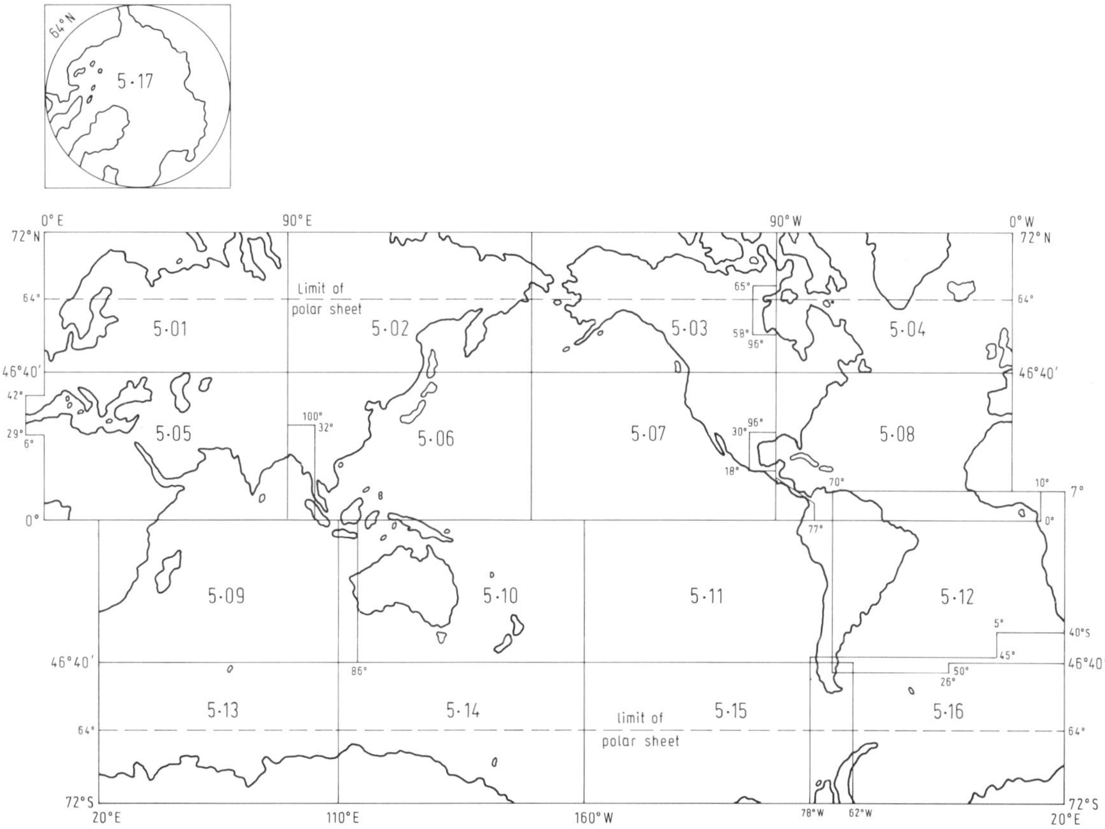

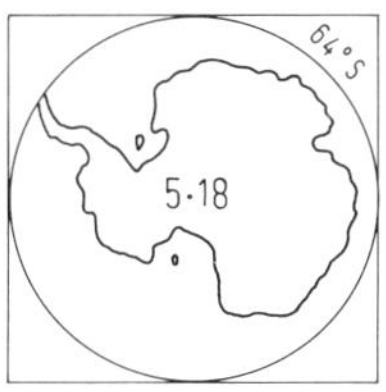

Fig. 31. General Bathymetric Chart of the Oceans (GEBCO). Assembly diagram for GEBCO sheets (5th edition), 1976 onward, by IHO, Monaco [76 G 1]. The fifth edition consists of 16 sheets covering the world from 72° N to 72° S, 1 : 10 000 000, on Mercator projection, and 2 polar sheets, 1 : 6 000 000, on polar stereographic projection, to 64 °N and S respectively. GEBCO international distribution service: Hydrographic Chart Distribution Office, Department of Fisheries and Oceans, Ottawa, Canada. Compare Fig. 32.

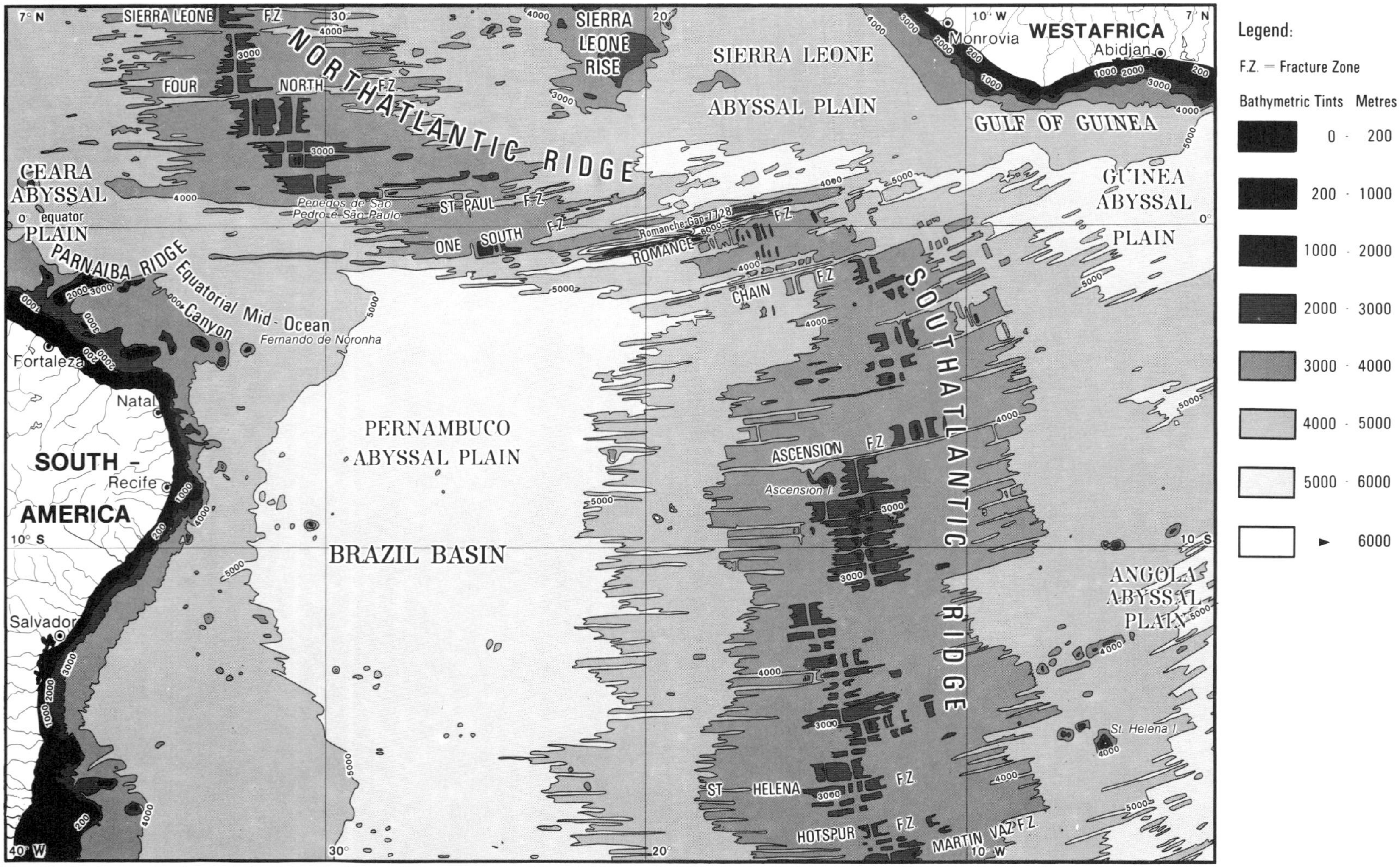

Fig. 32. Bathymetric chart of the Atlantic Ocean floor (section of the middle part) GEBCO 5–12. The contour lines do not match significant relief but are lines of decimal interval-bathymetric condition. They do, however, represent a graphic display of structural morphologic conditions of the ocean floor; the shelf (200 m), the continental slope (200···300 m), the continental rise (3000···4000 m), the mid-oceanic ridge with transcurrent troughs (at the fracture zones), and the central rift valley. Scale 1:20000000. Compare Figs. 33, 62. The use of contours implies that the fracture zones are better determined than they really are in fact: the chart shows contours representing fracture zones in abyssal plains [84 E].

1.3.3.3 Physiographic maps of the ocean floor

Physiographic maps of the oceans represent the characteristic features of the relief. They have the advantage that it is easy to visualize the three-dimensional relief which is presented in two-dimensional form. Conventional contour charts are not always the most effective way to present the sea bed morphology, i.e. the relief of the ocean floor, Fig. 32, [59 H]. Figs. 33···36, Table 12.

The relief of the sea floor is represented physiographically on the global maps "The Floor of the Oceans" which are based on bathymetric studies by B.C. Heezen and M. Tharp, supported by US Navy Office of Naval Research and The National Geographic Society, and on partial ocean maps [58 H, 59 H, 77 H 1]. The scales range from 1 : 10 000 000 for parts of the oceans to 1 : 40 000 000 for global ocean maps [81 N]. The National Geographic Society, USA, produced also a physical globe with physiographic design of ocean floors, issued in different scales.

Although these maps are generally called and titled as "physiographic maps", they suit to this term only, if they are drawn according to a legend of physiographic classification, not if they are only drawn with a hatchure system to visualize geomorphology or topography. However, these physiographic maps may present many topographic details, they necessarily are subjective. Some authors do not count this type of maps with this perspective appearance of signatures as "physiographic" [84 E], so at least the term is still ambiguous.

Table 12. Physiographic maps and diagrams of the oceans (recent editions).

Series title	Organisation responsible and country	Date of publication	Scale	Ref.
World Ocean floor	US Navy, Office of Naval Research	1977	1 : 23230300	Heezen, B.C., Tharp, M., painted Berann, H.C. (1977)
National Geographic Physical Globe	National Geographic Society (USA)	1979	1 : 31750000	Darley, R.J., Bond, W.H.
World Ocean floor	National Geographic Society (USA)	1981	1 : 42440000	Garrett (1981)
Ocean floor in Atlas of the World	National Geographic Society (USA)	1981	1 : 10000000 to 1 : 40000000	Tharp, M., Lloyd, I.T., Townsend, Cartographic Staff of N.G.S.

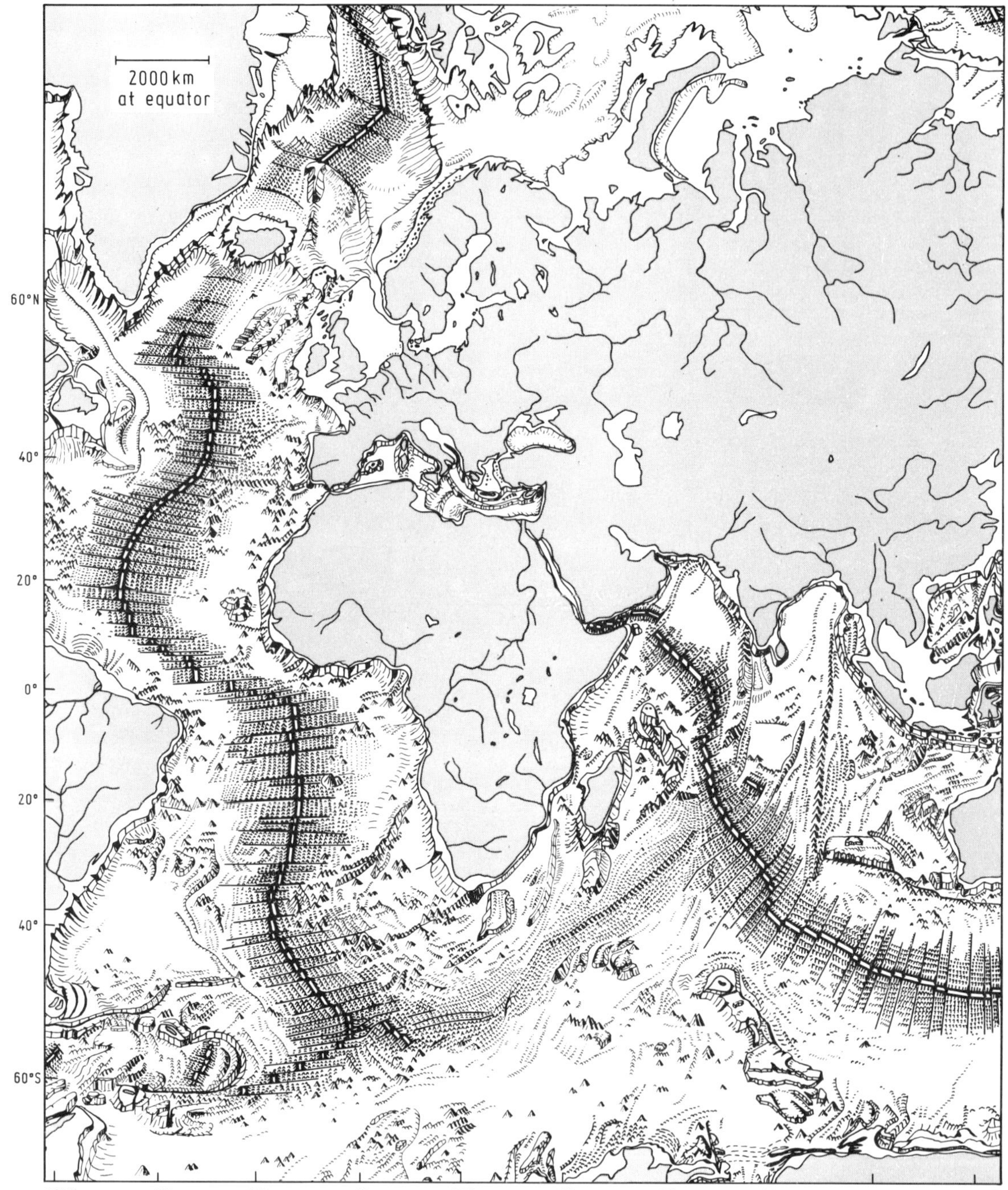

Fig. 33. Global physiographic map of the sea floor [83 G 1, 83 G 2]. Morphologic features: Mid-oceanic ridges with rift valleys, fracture zones (transform faults), continental shelves, plateaus, seamounts. Projection: Mercator, after model "World ocean floor" based on bathymetric studies by B. C. Heezen and M. Tharp, Lamont Geological Observatory, painted by H. C. Berann, compiled by C. J. Boberschmidt, issue 1977. Scale at equator: 1 : 148 000 000. Design: H. G. Gierloff-Emden, 1982. Cartography: H. J. Mette 1982. Compare Figs. 62, 72 and 76.

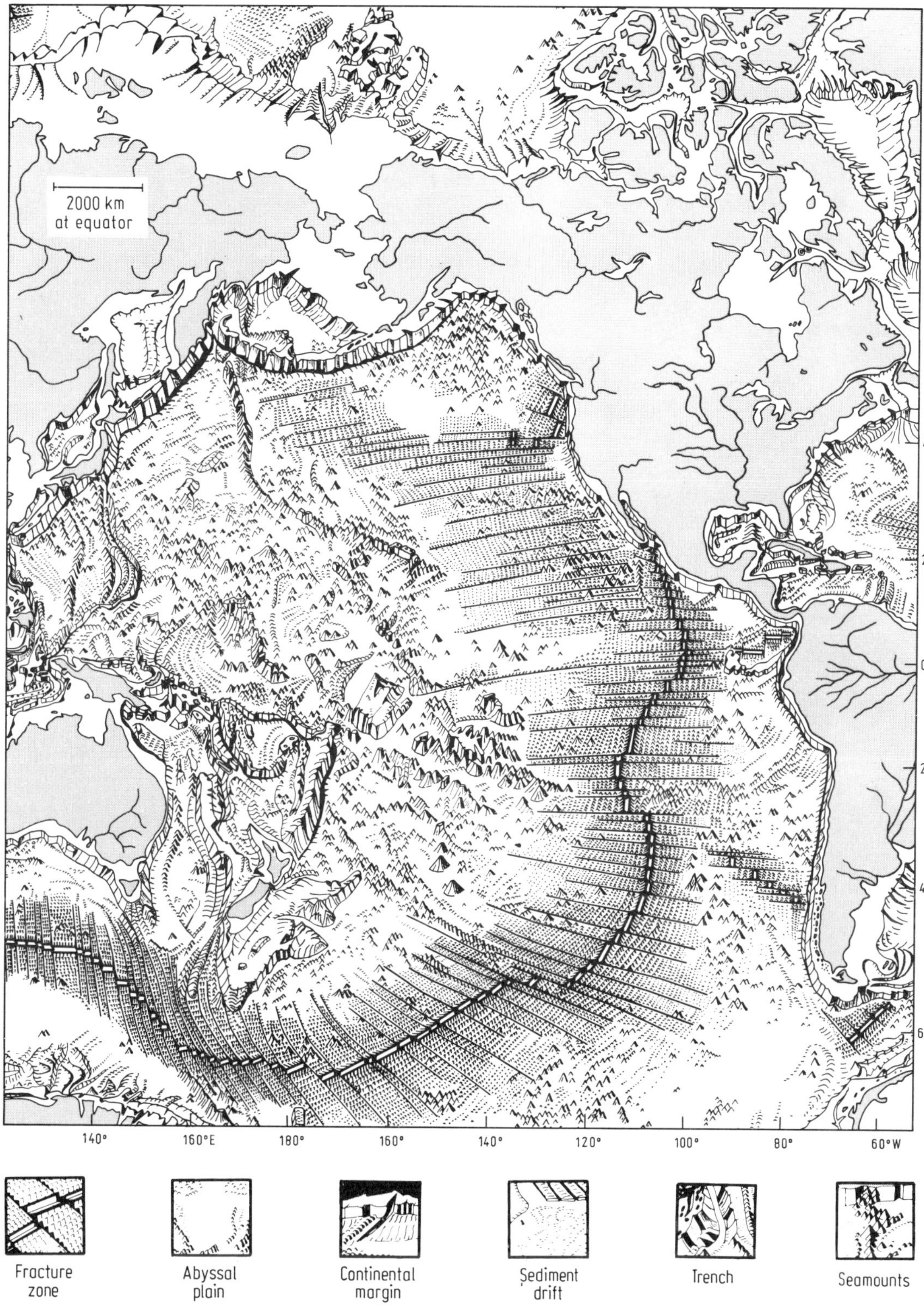
2000 km
at equator
60°N
40°
20°
0°
20°
40°
60°S
140°
160°E
180°
160°
140°
120°
100°
80°
60°W
Fracture zone
Abyssal plain
Continental margin
Sediment drift
Trench
Seamounts

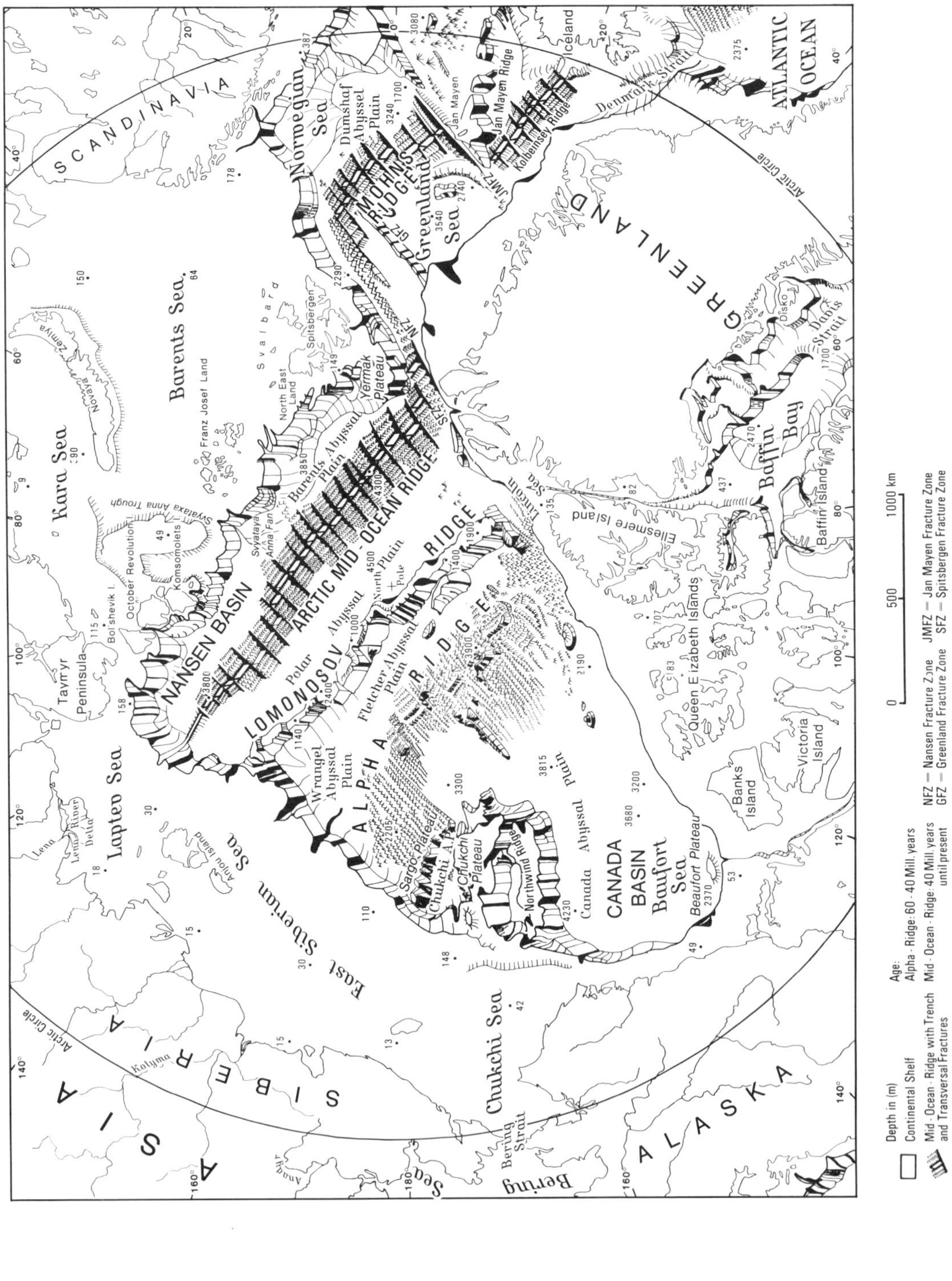

Fig. 34. Physiographic map of the Arctic Ocean floor [72 H, 74 A]. Compare Fig. 50.

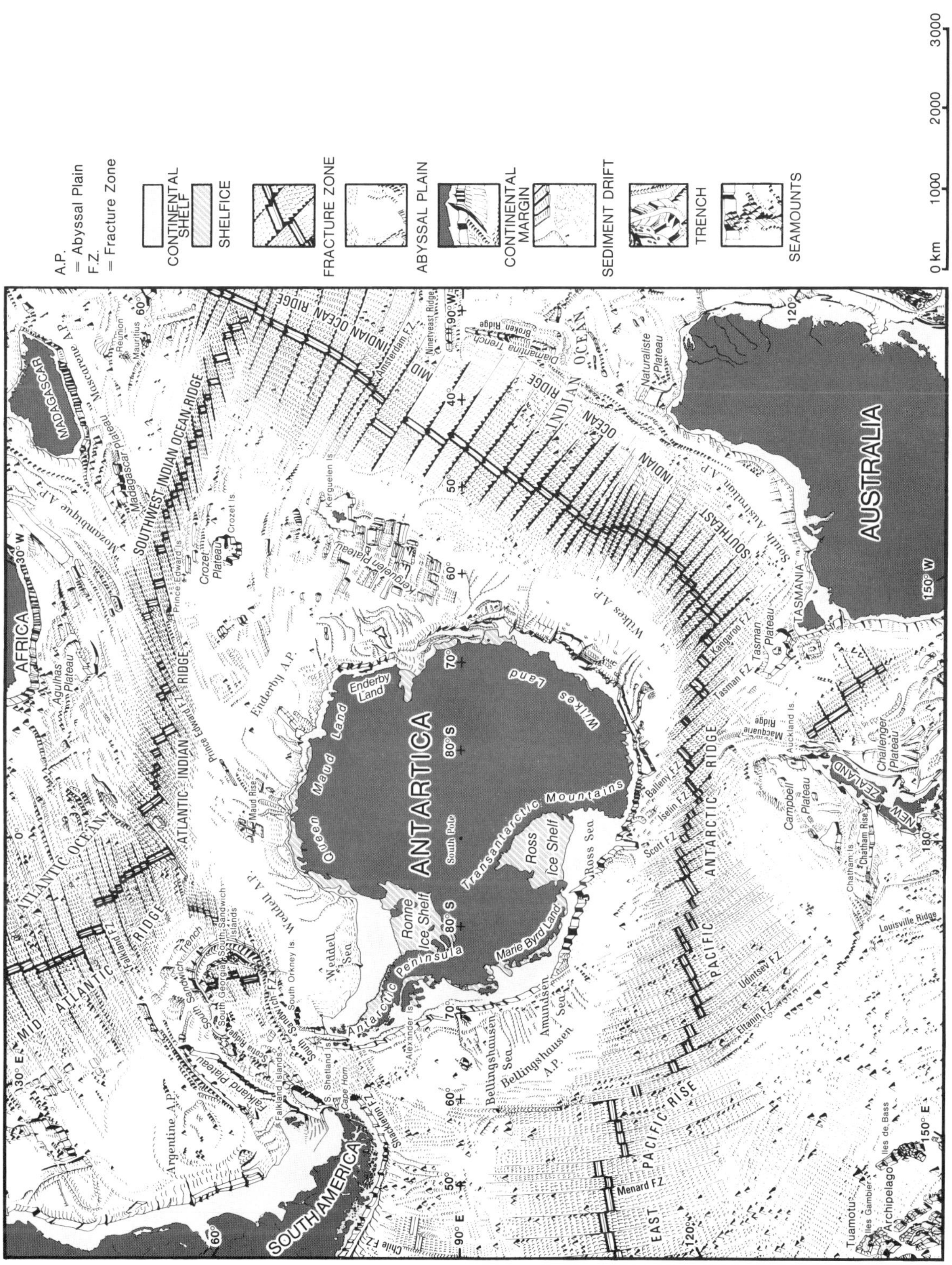

Fig. 35. Physiographic map of the floors of the oceans around Antarctica. [83 G 1, 83 G 2, 72 H, 74 A, 81 V, 81 G]. Compare [83 O] and Fig. 43.

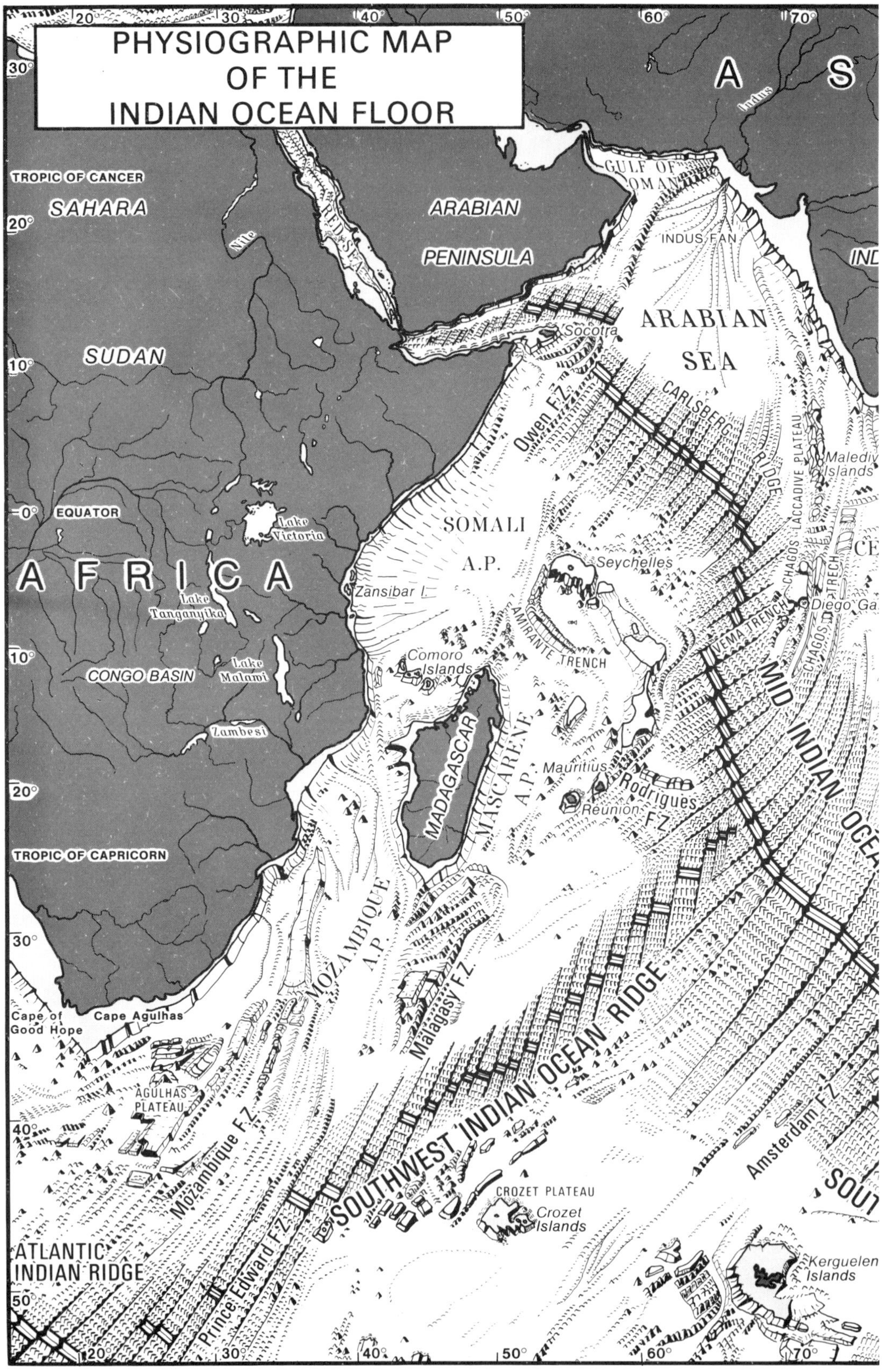
PHYSIOGRAPHIC MAP
OF THE
INDIAN OCEAN FLOOR
TROPIC OF CANCER
SAHARA
ARABIAN
PENINSULA
GULF OF OMAN
INDUS FAN
ARABIAN
SEA
SUDAN
Socotra
Owen F.Z.
CARLSBERG RIDGE
LACCADIVE PLATEAU
Maledive Islands
EQUATOR
AFRICA
Lake Victoria
SOMALI
A.P.
Seychelles
Zansibar I.
Lake Tanganyika
AMIRANTE TRENCH
CHAGOS LACCADIVE
CHAGOS TRENCH
Diego Ga
VEMA TRENCH
Comoro Islands
CONGO BASIN
Lake Malawi
Zambesi
MADAGASCAR
MASCARENE A.P.
Mauritius
Rodrigues F.Z.
Reunion
MID INDIAN OCEAN
TROPIC OF CAPRICORN
MOZAMBIQUE A.P.
Malagasy F.Z.
SOUTHWEST INDIAN OCEAN RIDGE
Cape of Good Hope
Cape Agulhas
AGULHAS PLATEAU
Mozambique F.Z.
Amsterdam F.Z.
CROZET PLATEAU
Crozet Islands
ATLANTIC INDIAN RIDGE
Prince Edward F.Z.
Kerguelen Islands
Indus
Nile
RED SEA
A S
20° 30° 40° 50° 60° 70°

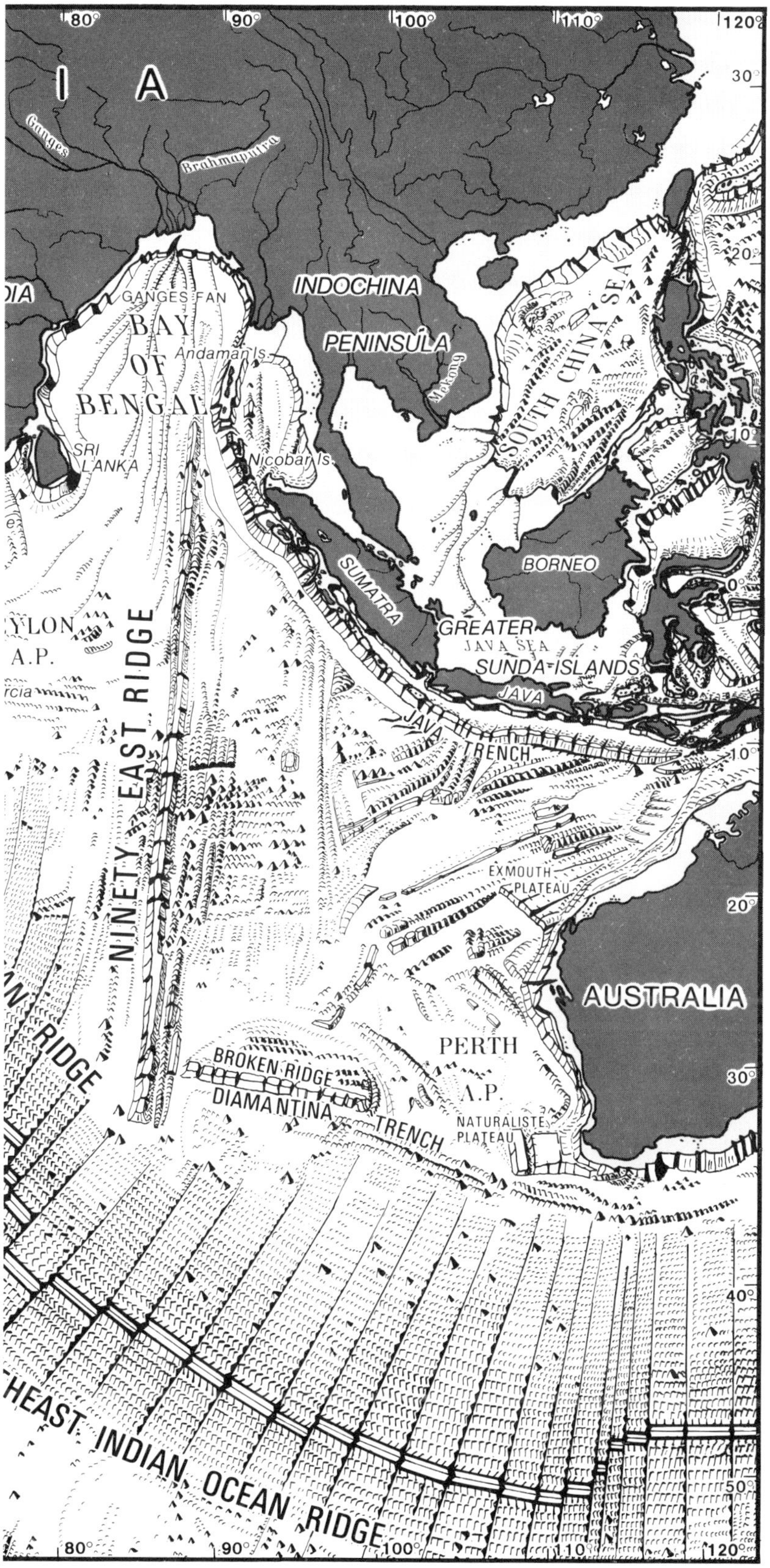

Fig. 36. For caption see next page.

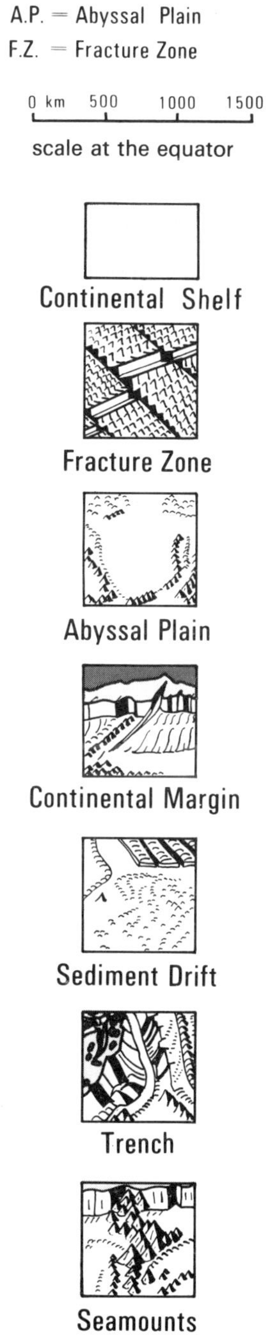

←

Fig. 36. Physiographic map of the Indian Ocean. The ocean floor of the Indian Ocean is distinctly marked by the triple junction of the mid-ocean ridges and by the two meridional spurs of the Ninetyeast Ridge and Maledivian Ridge. After sources of Fig. 33 and "National atlas of the world", Washington 1981: National Geographic Society. Design: H. G. Gierloff-Emden, 1983. Cartography: H. J. Mette, 1983.

1.4 Topography of the ocean floor: features and pattern

1.4.1 Classification of the ocean floor features

The classification of the important and large morphologic units as published by B.C. Heezen [59 H], has found its way into oceanographic literature such as "The Sea", vol. III [63 H 2]. Most of the terminology of this classification has been found applicable throughout the oceans. Fig. 37. The liberal use of descriptive adjectives such as marginal, abyssal, and mid-oceanic may be used for standard terminology on charts.

The following publication serves as an international reference work for the terminology of ocean bottom features and for the sea floor morphology:

"Ocean Bottom Features – Terminology and Nomenclature" [77 H 2], revised by the Advisory Committee on Undersea Features (ACUF) of the US Board on Geographic Names, and adapted by the Office of the Geographer, US Department of State, and US Naval Oceanographic Office, and US Board on Geographic Names, and Advisory Committee of the GEBCO, p. 23.

For nomenclature of the ocean bottom features, see IHB Series, Monaco [71 I].

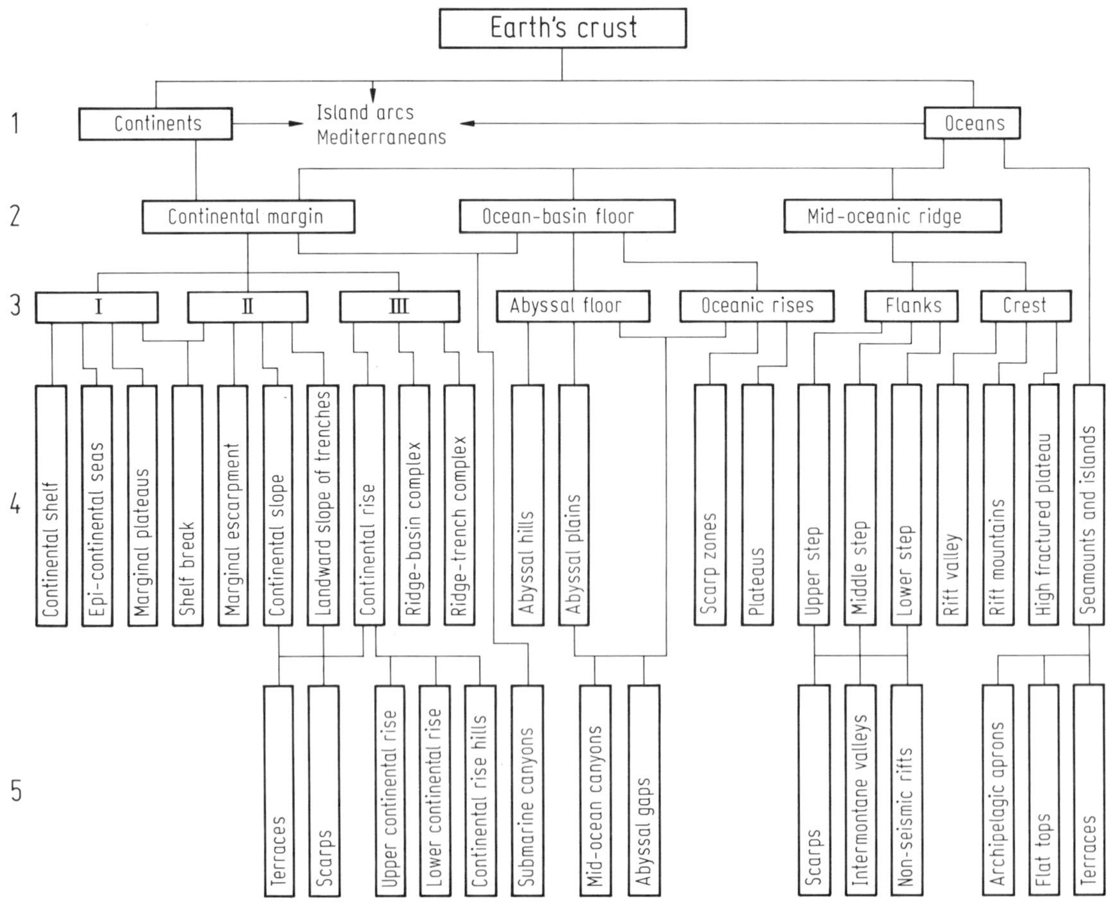

Fig. 37. Classification of submarine topography. Line 1: first-order features of the crust; line 2: major topographic features of the ocean; line 3: categories of provinces and superprovinces; line 4: provinces; line 5: subprovinces and other important features. Outline of [59 H].

1.4.1.1 Physiographic provinces of the Atlantic Ocean profile and spatial dimensions

The median symmetry of the Atlantic Ocean's topography represents a more simple model than those of the Pacific Ocean or the Indian Ocean [59 H]. Compare [82 N, 84 E], Figs. 38a, b, Table 13.

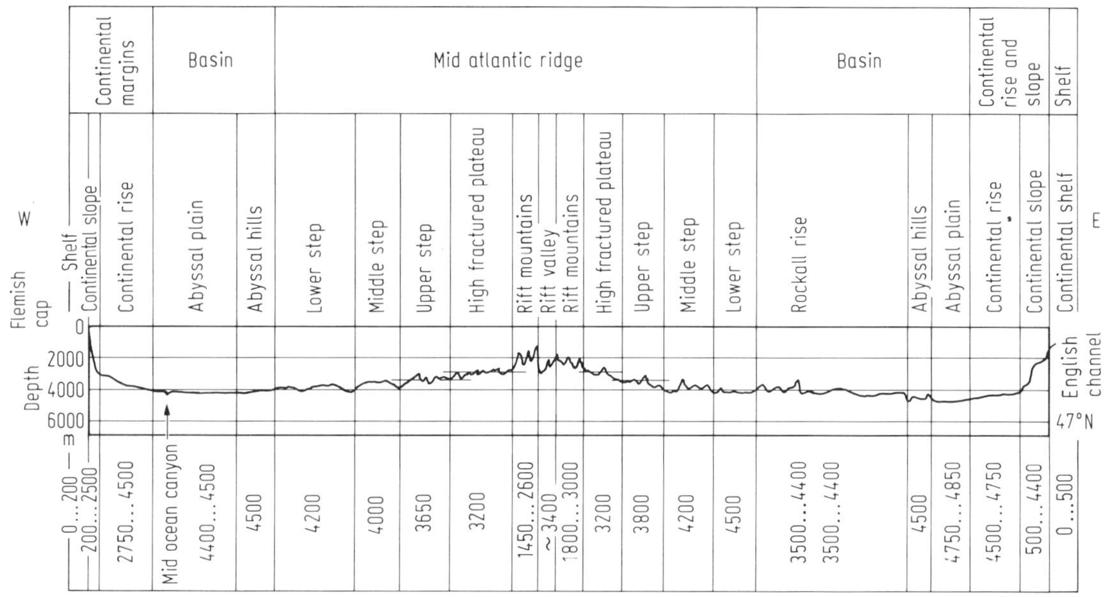

Fig. 38a. Typical cross section of the sea floor in the North Atlantic Ocean along 47° N in W–E direction. Depth rounded to decimal figures, profile exaggerated 60:1, length of the profile ≈3500 km. The profile lies to the east of the New Foundland Bank from Flemish Cap to English Channel [59 H]. Earth curvature is not taken into account. Compare Fig. 38b.

Table 13. Spatial dimensions (metric) of features of North Atlantic Ocean. After [63 U 1].

	Width from ridge to ridge	Approximate mean depth, absolute	Approximate max. depth, absolute	Approximate relative depth, from ridge to bottom
	km	m		
Rift valley	27··· 55	3400···4250	2300···5200	915···3200
Rift mountains	46··· 50	1800···2530	3600	300···1300
High fractured plateau	150	2720···3660	3730	360··· 730
Upper steps	180	3000···4200	4400	200··· 850
Middle steps	180	3920···4600	4700	200··· 400
Lower steps	200	4300···5000	5150	between
Abyssal hills	120···550	4400	4500···5000	150 and 300
Abyssal plain	120···700	4400	4500···5350	

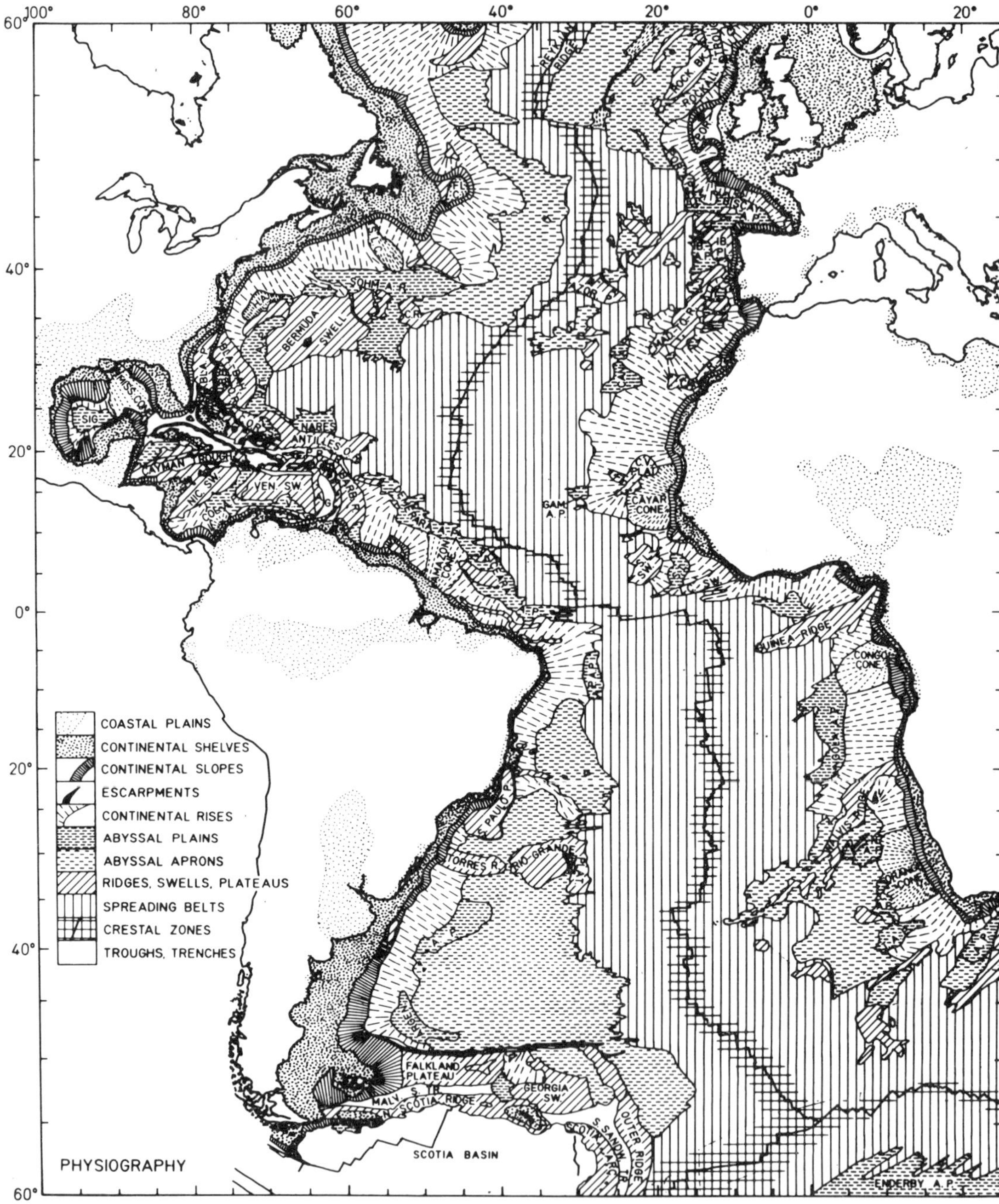

Fig. 38b. Physiographic diagram of Atlantic Ocean floor based upon interpretations of topography supplemented by echo-sounding profiles, seismic data, and publications of many authors [84 E]. This physiographic diagram presents the main features as drawn after bathymetric charts by K.O. Emery and E. Uchupi. [84 E]. For the term "physiographic" compare subsect. 1.3.3.3., see Tables 14, 17b and Figs. 32, 33, 49.

Table 14 (continued)

Feature	Typical dimensions	Comments	Example
Abyssal hills	Width: 100 m···100 km Height: few to 1500 m	Found seaward of abyssal plains, usually in belts	Western Atlantic Abyssal Hills Province
Abyssal plains	Width: few to 1000 km Depth: usually >3000 m Gradient: <1 : 1000 (to 1 : 10000) Relief: nil	Small plains occur in trench bottoms; plains are very flat, gentle	Nares and Sohm Plains, North Atlantic
Oceanic rises	Width: 300···500 km Height: up to 5000 m	Aseismic features with highly variable relief (smooth slopes – rough scarps)	Bermuda Rise
Aseismic ridges	Width: to 150 km Length: to 4000 km Height: up to 4000 m	Strongly asymmetrical	Walvis Ridge
Seamounts	Width: 2···100 km Height: >1000 m Depth: 0···2000 m (top surface)	Submerged volcanoes; often wave truncated and submerged (guyots); may have atolls	East Pacific
Oceanic plateaus	Dimensions similar to oceanic rises Gradient: <1 : 1000 on top; >1 : 40 on sides	Also termed "Microcontinents"	Mascarene Plateau Indian Ocean
Mid-oceanic ridge	Width: 2000···4000 km Length: 35000···40000 miles Height: 1000···3000 m Depth: 0···5000 m	Includes flank and crest regions. World's greatest mountain range	Mid-Atlantic Ridge
Ridge flank	Width: 500···1500 km Relief: up to 1000 m Depth: >3000 m	Includes fractured plateau, and steps on flank	Mid-Atlantic Ridge
Ridge crest	Width: 500···1000 km Relief: 2000 m Depth: 2000···4000 m	Includes rift zone, which may be 20···50 km wide	Mid-Atlantic Ridge

1.4.2 Major topographic features of the continental margin complex

The continental margins of the world may be subdivided into three classes of physiographic provinces:

1. Provinces lying at the general level of the continental shelves, the epicontinental seas, and the marginal plateaus.

2. Provinces lying between the general level of the continental platforms, and the general level of the deep ocean floor. These include the continental slopes, the marginal escarpments, and the landward slopes of trenches.

3. Provinces bordering the continental platform, and extending to the ocean basin floor. These include the continental rises, the marginal basin outer-ridge complexes, and the marginal trench outer-ridge complexes [70 H 1, 74 B 2], Fig. 39.

Continental terrace:

Together, the relatively flat continental shelf and the steeply sloping continental slope constitute a large-scale bench or terrace called the continental terrace, Fig. 39.

1.4.1.2 Spatial dimensions of the features of submarine topo

Table 14. Typical dimensions of major features of submarine topography [68 H 1]. Comp 44b, 51.

Feature	Typical dimensions	Comments	Examp
Continental shelf	Width: few to >300 km Relief: <20 m Depth: <200 m Gradient: <1:1000	Seaward boundary, or shelf break occurs where gradient changes to >1:40	Northe Stat
Marginal plateau	Similar to shelf except depth: between 200···1200 m	No seaward barrier or sill	Blake l
Epicontinental sea	Similar to shelf except depth: between 100···1500 m Relief: often >40 m	Seaward barrier or sill	Gulf of
Continental slope	Width: few to 150 km Relief: locally >2000 m Depth: drops from 100+ to 2000+ m Gradient: >1:40 (3···6°)	Upper boundary is shelf break; high relief is associated with canyons	Northe State
Marginal escarpment	Width: few km Height: 2000···4000 m Depth: begin at 1000···3000 m Gradient: >1:10	Precipitous slope	Blake E
Landward slope of trench	Height: several thousand meters Depth: drop from 500···5000 m Gradient: >1:40	Always associated with trench	North o
Continental rise	Width: few to >300 km Relief: <40 m Depth: 1500···5000 m Gradient: 1:1000 to 1:700	Seaward limit is change to <1:1000, although 1:2500 (and 1:50) can occur	Northeas States
Outer ridge	Width: >150 km Height: 200···2000 m	Lies seaward of a basin or trench	North of
Marginal basin	Variable dimensions Depth: <5000 m Gradient: often <1:1000	Defined by position *re* outer ridge, continental slope	Blake Pla
Marginal trench	Width: 30···100 km Length: 300···5000 km Depth: 3000···10000 m Relief: >2000 m	Narrow, steep sided	Puerto Ri
Continental borderland	Similar to shelf except relief: 100···1000 m Depth: up to 2000 m Gradient: 1:1000 to 1:40	Rises and depressions are common	Off South
Submarine canyon	Width: 1···15 km Relief: 20···2000 m Depth: 20···2000 m Gradient: <1:40	Most commonly cut continental rise; length to 500 km	Hudson C off New

Continental margin:
One of the first-order physiographic subdivisions of the oceans, consisting of a zone lying between the emerged portion of the continental block and the provinces of the deep ocean floor. It includes the continental shelf, continental slope, and the continental rise provinces, Fig. 39.

1.4.2.1 Terminology of continental margins and adjacent features

The terminology is developed concerning the topographic inventory of the ocean floor. The limiting gradients and depths of the "sea floor" to be used in the rules of the jurisdiction according to the United Nations Conference on the Law of the Sea (UNCLOS) are not in coincidence with the topographic terminology, and the limitation of those features [84 L 1, 74 B 2]. Compare Fig. 1 and LB, NS, vol. V/3, ch. 9, Figs. 1, 7, 10. See Fig. 39.

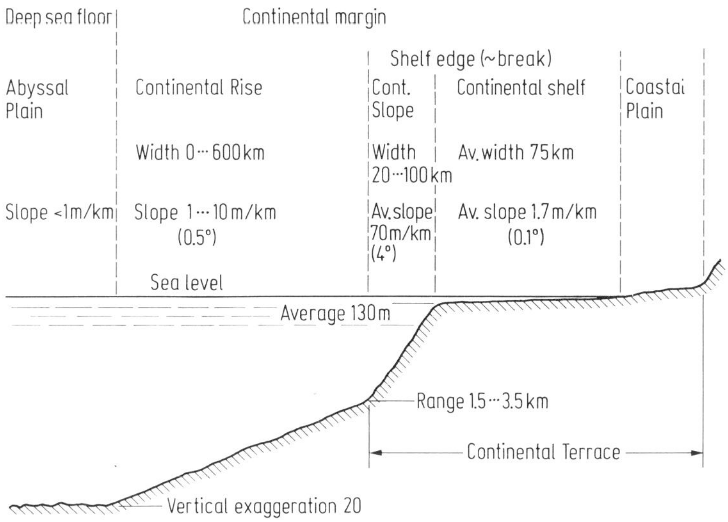

Fig. 39. Diagrammatic profile of one type of continental margin showing average or range of depth, width, and slope of component parts [66 F].

1.4.2.2 Classification of continental margins

The ocean margins, i.e. the region of transition between continent and deep ocean, differ greatly in their characteristics, depending on whether they occur in mid-plate areas (on the continent's trailing edge), or on the collision edge of a continent, or along a shear zone. The one thing most of the ocean margins of the Atlantic type (passive type) have in common is the occurrence of large masses of sediment. Ocean margins are generally referred to as continental margins [82 R 1]. See Fig. 43, Table 14.

Early classifications of continental margins by E. Suess [1885 S] were based on regional geographic distribution rather than structure. The modern terminology and classification are based on the processes of global tectonics, creation and destruction of oceanic crust. Destruction is related to subduction, the underlying cause for trench and mountain formation, seismicity and volcanism, crustal deformation and metamorphism, and opening of backarc basins. Several terms are applied to the same types of margins [82 K]. However, margins of one of these types may be of great diversity [66 D]. See Fig. 40, 43, Table 15. [74 B 2].

Most profiles of the slopes connecting the continents to the ocean basins are of one of the three types delineated in Fig. 40. Table 15.

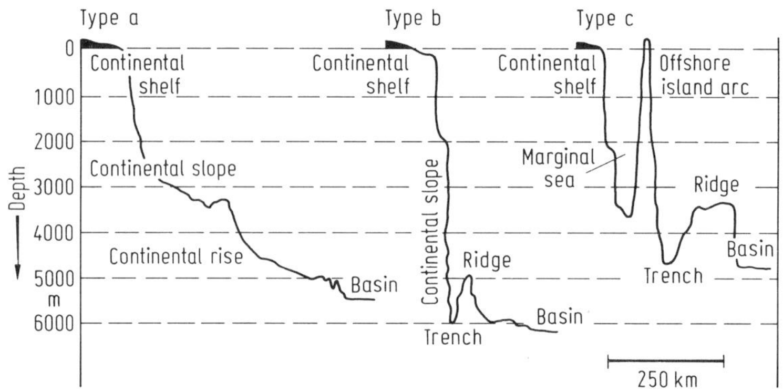

Fig. 40. Features of continental margins [60 E 2]. Type a approaches the "Atlantic type", with extended slope, thick continental sediments, and is characterized as "divergent, passive, aseismic". Type b corresponds to a profile existing in coastal regions of very bold relief (the coast of Peru for example), approaching the "Pacific type" found at ocean plate subduction zones, with steep slope and is characterized as "convergent, active, seismic". It has only very thin continental sediments. Type c is found off the coast of a chain of islands isolating a marginal sea, as off the north coast of Venezuela and is characterized by "translation, active or passive", Pacific, and Atlantic. For the special features and conditions of shelf, margin reef morphology, see [77 V, 74 B 2].

Table 15. Statistics on continental margins [66 M 1]. (World ocean without adjacent seas).

World ocean without adjacent seas	Shelf			Continental slope			C. rise	Trenches
	Area (0···200 m) 10^6 km^2	Average width km	Average slope	Area 10^6 km^2	Average width km	Average slope	Area 10^6 km^2	Area 10^6 km^2
Atlantic Ocean	6.080			6.578			5.381	0.447
(% of its area)	(7.0%)	115	0°28′	(7.6%)	260	1°19′	(6.2%)	(0.5%)
Indian Ocean	2.622			3.475			4.212	0.256
	(3.6%)	91	0°23′	(4.7%)	182	1°35′	(5.7%)	(0.3%)
Pacific Ocean	2.712			8.587			2.690	4.757
	(1.6%)	52	0°49′	(5.2%)	139	3°13′	(1.6%)	(2.9%)

1.4.2.3 The shelf

Shelf (continental shelf, island shelf) [71 H]:

A zone adjacent to a continent or an island that extends from the low water line to a depth at which there is usually a marked increase of slope to greater depth, the shelf edge.

The continental shelf consists of that part of the continental platform extending from the shore line to the edge of the continental terrace which is submerged, shallow (generally less than 200 m), very gently sloping regionally (less than 1 : 1000), and very low in local relief (less than 20 m). Typical elevations of the sea floor of the shelves are shallows, called "banks" [75 S].

Locally the shelf may be of higher relief and extend to greater depths. The shelves are significantly shaped by climatic zone-bounded agences as there are ice, waves, corals, and by effects of changing sea level. See ch. 8 and Table 16a.

The sedimentation on shelves was investigated by Emery [72 E], the specific shaping by Swift [72 S], the geographic distribution of types and their features by Vanney [77 V].

The width of the shelf generally ranges from a few kilometers to more than 400 km. A series of benches or terraces typically occurs on the continental shelf. Continental shelves are of specific importance due to their bearing of hydrocarbons and minerals for economic use. They also encompass the most important fishery regions of the oceans.

Shelf edge, shelf break:

The line along which there is a marked increase of slope at the outer margin of a shelf. Shelf edges seaward of major deltas differ considerably from those off open coasts. See Fig. 41.

Shelf terrace:

A steplike feature of the continental shelf, often occuring in a series [77 H 2].

Shelf canyon:

A submarine canyon (or channel) on the continental shelf, or that part of a submarine canyon which crosses the continental shelf.

Ravines on sediment slopes, and gullies:

Ordinarily, where slopes have been built seaward, they are creased by a series of small ravines. These are usually not more than a few meters deep, with exceptional depths of up to 30 m. They are slightly curving in trend but have few tributaries. They are discontinuous, starting at any point on the advancing slope, and stopping part way down the slope. The origin of the ravines appears to be submarine slumping [67 S, 81 C]. See Fig. 42 [85 E].

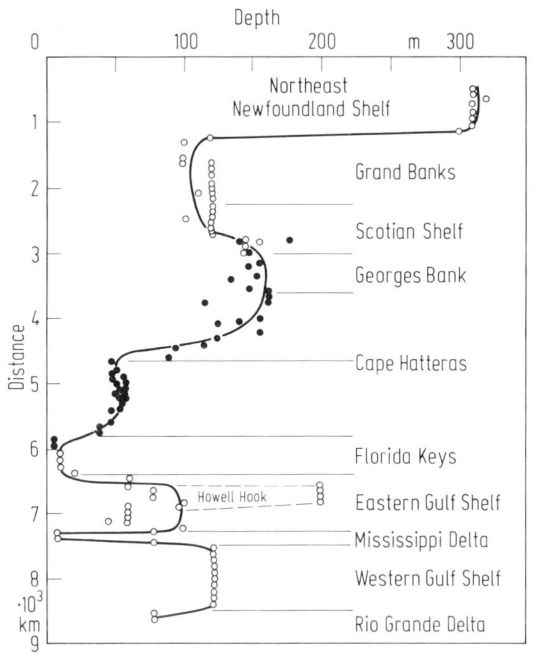

Fig. 41. Depth of shelf break between Labrador shelf and the Rio Grande delta. Based on precision-sounding profiles (solid dots) and on depth contours (open circles). Greater depths of shelf break at north are believed to be the result of glacial erosion and incomplete isostatic rebound: shallower depths at south are caused by rapid deposition of calcareous sediment and reef growth [72 E].

Table 16a. Important shelf regions: areas and prevailing depths [49 K].

Shelves	Area $10^3\,km^2$	Prevailing depth m
Northwest European	1050	<100
Norwegian	93	200···300
Barents	830	200···300
North Sibirian	1330	< 50
Island-Faröer	115	200···300
New Foundland	345	150···200
Florida–Texas	385	< 50
Campeche	170	< 50
Guyanan	485	< 50
New Scotian	370	50···100
South Brazilian	370	< 50
Patagonian	960	50···100
Guinean	210	50···100
Southwest African	140	50···200
Agulhas	75	>100
Sambesi	55	< 50
Bombay	230	50···100
Northwest Australian	590	50···100
Arafura	930	50···100
South Australian	320	50···100
Tasmanian	160	50···100
Queensland	190	<100
Birma	290	<100
Borneo–Java	1850	50···100
Tongking–Hongkong	435	<100
Tunghai	915	<100
Okhots–Sachalin	715	50···100
Bering	1120	< 50

The high-energy environment of shelves:

High-energy environments, defined as the terrain where waves and currents are strong enough to shape geomorphology, or where morphology is sufficiently unstable to be affected by water movements, most frequently occur nearshore and on the inner parts of continental shelves. Beaches, tidal inlets and channels, submarine deltas and bottom forms, barrier islands, and river mouths are examples, which, generally, combine into larger assemblies like estuaries, barrier island chains and parallel submarine shallows. The size of the material moved ranges from large pebbles to finegrained mud. Sand dunes frequently form an essential part of the sedimentary structure [75 S, 77 V]. For high-energy environments the interplay between morphology and hydraulic forces is important [84 I, 76 D 2, 76 D 3]. For the Agulhas Current at the African continent in the Indian Ocean, compare [81 M].

Significant quantities of fine-grained sediments are delivered by large river systems to continental shelf regions. Deposition rates can be large in specific areas off deltas, with the sediments participating in various mass-movement processes on the subaqueous slopes (slumps and slides, debris flows, mud flows). They give rise to rapid sediment accumulation at shelf breaks and progradation of the shelf edge [81 C, 83 C 2, 70 G 1].

The dynamics causes processes of relief forming of the ocean floor. See [71 H, 72 B 1, 78 R, 80 G].

Fig. 42. Morphology, sedimentation and instability on a shelf and continental slope in front of a delta (Mississippi Delta) [77 C], compare [83 C 2]. For the delta furrows and the benthic boundary layer compare [83 A 1]. These processes form turbidite features at the slopes, i.e., turbidity currents deposits. [52 H, 64 H 2, 83 A 2]. For mudlumps see [51 M].

Table 16b. Examples of submarine landslides on low slope angles. After [79 P 2].

Location	Slope deg	Length (*L*) km	Depth (*D*) m	*D*/*L* %
Grand Banks	3.5	110	365	0.33
Hawaii	6.0	160	2000	1.25
Kidnappers	1.0···4.0	11	50	0.45
Bay of Biscay		21	250	1.19
Rockall	2.0	160	330	0.20
Bassein	6.0	37	360	0.97
Agulhas		106	375	0.35
Copper River Delta	1.0	18	115	0.64
Albatross Bank	7.0	5.3	300	5.66
Portlock Bank	4.0	6.5	200	3.07
Kayak Trough	1.0	15.0	115	0.76
Atlantic East Coast	3.8	3.4	30	0.88
	5.7	4.8	80	1.66
	6.8	2.3	18	0.78

1.4.2.4 The continental slope and rise

Definitions after [77 H 2]. For general references, see [74 B 2]. For regional example, see Fig. 43.

Continental slope (island slope):

The descending slope seaward from the shelf edge to the beginning of a rise or the point where there is a general reduction in slope. See Figs. 40, 41, 43.

At the edge of the continental shelf, the depth generally falls off rapidly from 100···200 m to 1500···3500 m. The regional slope is quite steep (greater than 1 : 40) and the boundaries with the relatively flat continental rise provinces are generally abrupt. This zone is characterized by a steep regional slope, is generally less than 200 km wide, and is typically dissected by submarine canyons. It forms a continuous ring encircling the continents.

Continental rise:

A gentle slope with a generally smooth surface, rising toward the foot of the slope.

The rise consists of a zone generally 100···1000 km wide which is characterized by a gentle seaward gradient (generally 1 : 100 to 1 : 700) and low local relief (generally less than 40 m). Seaward boundaries are generally abrupt, occurring at the point where the regional gradient decreases to that of the abyssal plains. Submarine canyons or deep-sea channels may extend over the continental rise. Deep-sea fans and outer ridges are features associated with the continental rise. In the western North Atlantic, the continental rise is locally divisible into two terraces, an upper continental rise, and a lower continental rise [84 E, 80 M].

Marginal basin:

A basin at the foot of a continental slope or a marginal escarpment which is slightly shallower than the general level of the deep-ocean floor, and is bounded on the seaward side by an outer ridge. Its floor may be occupied in part by an abyssal plain.

Marginal escarpment:

A precipitous escarpment (gradient steeper than 1 : 10), which forms the seaward slope of a marginal plateau. They are typically 1000···3000 m deep at their upper limits and 2000···4000 m in height [84 J].

Marginal plateau:

A shelf-like feature which lies at greater depths than the continental shelf, and is separated from the continental shelf by an incipient continental slope. They are characterized by an absence of a seaward barrier or sill, a general depth of 2000···2400 m, and a surface morphology very similar to that of the continental shelf. A very special continental borderland is that of southern California with a checkerboard pattern of basins and plateaus [60 E 1].

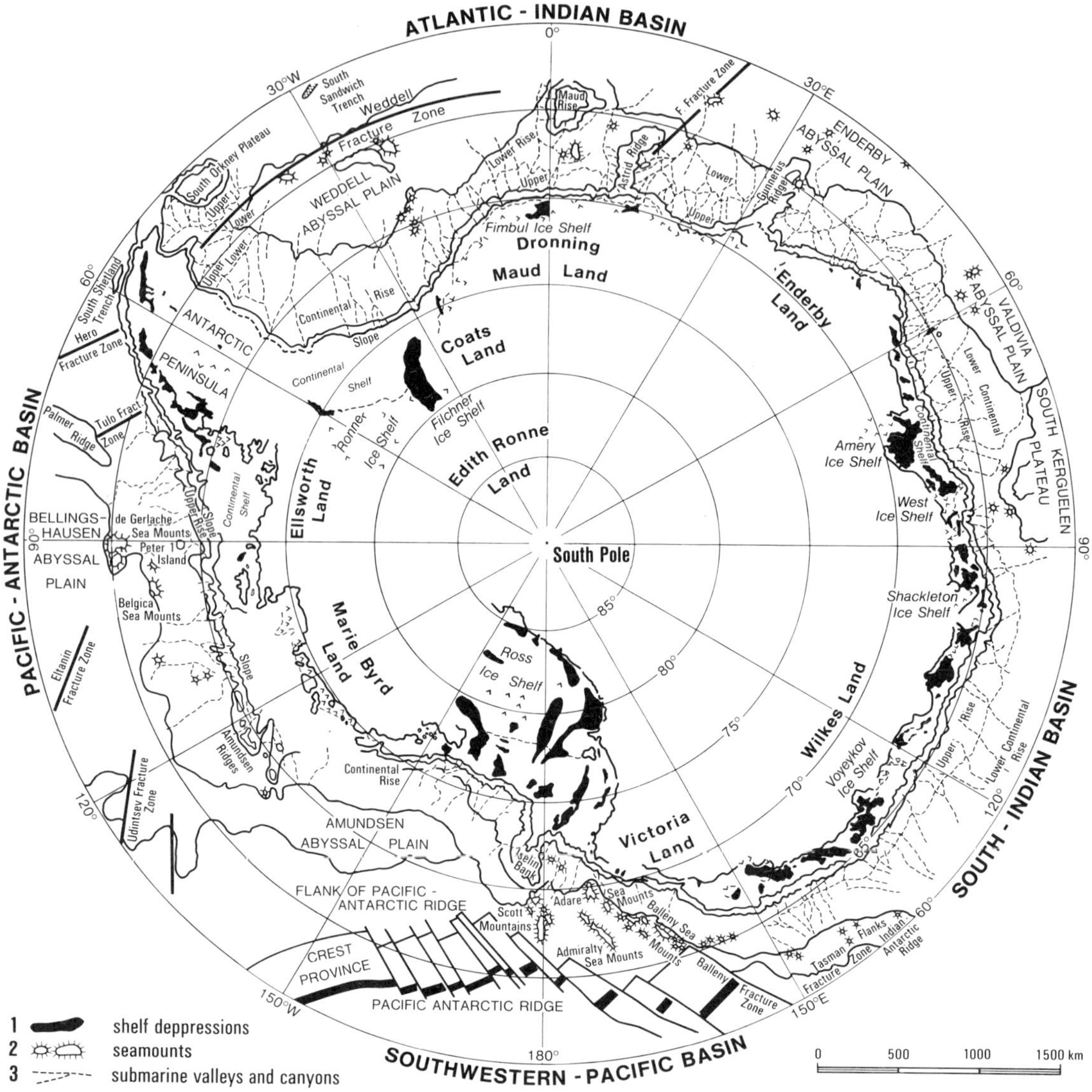

Fig. 43. Continental margin of Antarctica with major features of submarine topography: shelf, continental rise, abyssal plains, fracture zones. Compare [82 H, 81 V, 83 T]. Design: H.G. Gierloff-Emden, 1983. Cartography: H.J. Mette. After [83 T]. Compare [83 H 3, 81 V] for the Ross-Ice-Shelf, [81 J] for the Weddell Sea, and [76 V] for the Bellingshausen Sea. Compare Fig. 35.

Marginal trench:

A narrow, steep-sided trench roughly paralleling the continental margin at the seaward base of a continental platform or an island arc. They are generally characterized by maximum depths at least 2000 m below the general level of the ocean floor, and are separated from the ocean floor by a low outer ridge which rises 200···1000 m above the general level of the adjacent ocean floor. Trench plains may occur in the deepest portions of trenches. The greatest depths of the oceans occur in marginal trenches.

1.4.2.5 Canyons, abyssal cones, and fans

Definitions after [77 H 2]

Canyon (submarine canyon):

A relatively narrow, deep depression with steep sides, the bottom of which has a continuous slope. The origin is multigenetic.

Submarine canyons are a common occurrence on the continental slopes (slope canyons) of the world. Where the regional gradient is quite steep, they are well-developed and deep, ranging in size from 10···1000 m in depth, 2···20 km in width, and from a few kilometers to several hundred kilometers in length. Submarine canyons represent a great variety of scale in length, width, depth, and configuration depending on many parameters [63 S 2, 65 S, 66 S, 76 W 2]. See Table 17a. Major canyons, such as the Hudson Submarine Canyon, are actually part of a channel system which extends over the continental shelf (shelf canyon or channel), down the continental slope (slope canyon), and out onto the continental rise (seachannel). [73 R 2]. A complete summary and map of the canyons of the Atlantic Ocean is outlined in the Geology of the Atlantic Ocean [84 E].

Mid-ocean canyon:

A steep-walled, flat-floored, persistent linear depression, 2···10 km wide, and 10···200 m deep. Mid-ocean canyons are commonly flanked by rims or levees which are higher than the surrounding plain or fan; these levees are often higher on one side of the channel than on the other side. They are frequently associated with abyssal gaps.

Abyssal cone:

A large (areal dimensions generally 200···2000 km) composite cone-shaped feature of the continental rise which is made up of a series of coalescing fans, typically crossed by radiating channels.

The surface relief is typically smooth (0···50 m) with regional gradients ranging from 1 : 40 to 1 : 1000. Fine textured hills may occur locally. Shallow channels (10···100 m depth) generally dissect the surface, radiating outward from the base of the continental slope and extending across the cones in much the same sense as channels extend across a river delta. In some cases, these channels are bordered by levees. A distinction is made between cones, abyssal cones (which are larger composite features), and fans, abyssal fans, or deep-sea fans, which are smaller, independent features although deep-sea fan has been used interchangeably with abyssal cone in certain cases [66 H].

Fan (deep-sea fan):

A relatively smooth feature normally sloping away from the lower termination of a canyon. Fans resemble river deltas in their general form, even to the extent of having radiating channels crossing their surface. A distinction has been made between deep-sea fans, which are singular features of limited areal extent, and abyssal cones, which are composite features extending over large ares. The term, deep-sea fan, anyway has also been used synonymously with abyssal cone in some cases. The term "delta" has also been applied [63 H 2, 78 S 2, 77 H 2, 81 C]. See Table 17a. Compare LB, NS, vol. V/3, ch. 9.9.3. See Figs. 44a, 44b, Table 17b.

A special issue on "Deep Water Fans and Related Turbidities" encompasses the problems of this topic [83 B].

For modern fans, the available data-acquisition techniques and resulting data sets generally do not overlap in scale and resolution, see Fig. 44b. Both the sediment source area and the morphology of the system can be defined.

Acoustic definition of sedimentary layers (reflectors) using reflection-profiling systems is limited, and the minimum spacing of reflectors, which in some cases do not correspond with distinct lithologic boundaries, is commonly greater than the length of core samples recovered from modern fans; therefore, acoustic facies cannot be confidently correlated with sedimentary facies. Little is known about the internal make-up of modern submarine fans because sedimentary structures are inadequately resolved in the narrow core samples [83 N 2].

Table 17a. Dimensions of representative submarine canyon-fan systems of the world [70 R].

Name of canyon-fan system	Length [km]		Width [km]	Gradient				Depth [m]			Levees	App. No. of distributaries
	canyon	fan	fan	canyon		fan		canyon		fan		
								head	mouth	base		
1 Bering (USA, Alaska)	407	–	–	0°27′	1:125	–	–	183	3400	–	–	–
2 Zhemchug (USA, Alaska)	233	–	–	0°48′	1:71	–	–	183	3480	–	–	–
3 Ganges-Bengal (India)	183	2570	1091	0°27′	1:128	04′	1:890	46	1500	4390	yes	50
4 Congo (Africa)	222	278	185	0°34′	1:101	15′	1:227	37	2150	3840	yes	7
5 Monterey (USA, Calif.)	111	305	222	1°31′	1:38	14′	1:250	16	3050	4570	yes	7
6 Mississippi (USA, La.)	222	222	148	30′	1:117	29′	1:120	18	1280	2930	yes	7
7 Astoria (USA, Ore.)	115	166	102	57′	1:60	18′	1:190	1140	2090	2840	yes	40
Willapa (USA, Wash.)	111			58′	1:56	–	–	1166	2150	–	yes	–
8 Hudson (USA, N.Y.)	92	148	148	1°15′	1:46	23′	1:147	101	2100	3110	yes	4
9 Rhone (France)	28	166	166	3°11′	1:18	24′	1:144	183	1690	2840	yes	20
10 La Jolla (USA, Calif.)	14	30	22	2°17′	1:25	04′	1:54	15	550	1100	yes	1
11 Redondo (USA, Calif.)	15	7	11	2°12′	1:26	11′	1:48	9	586	755	yes	8

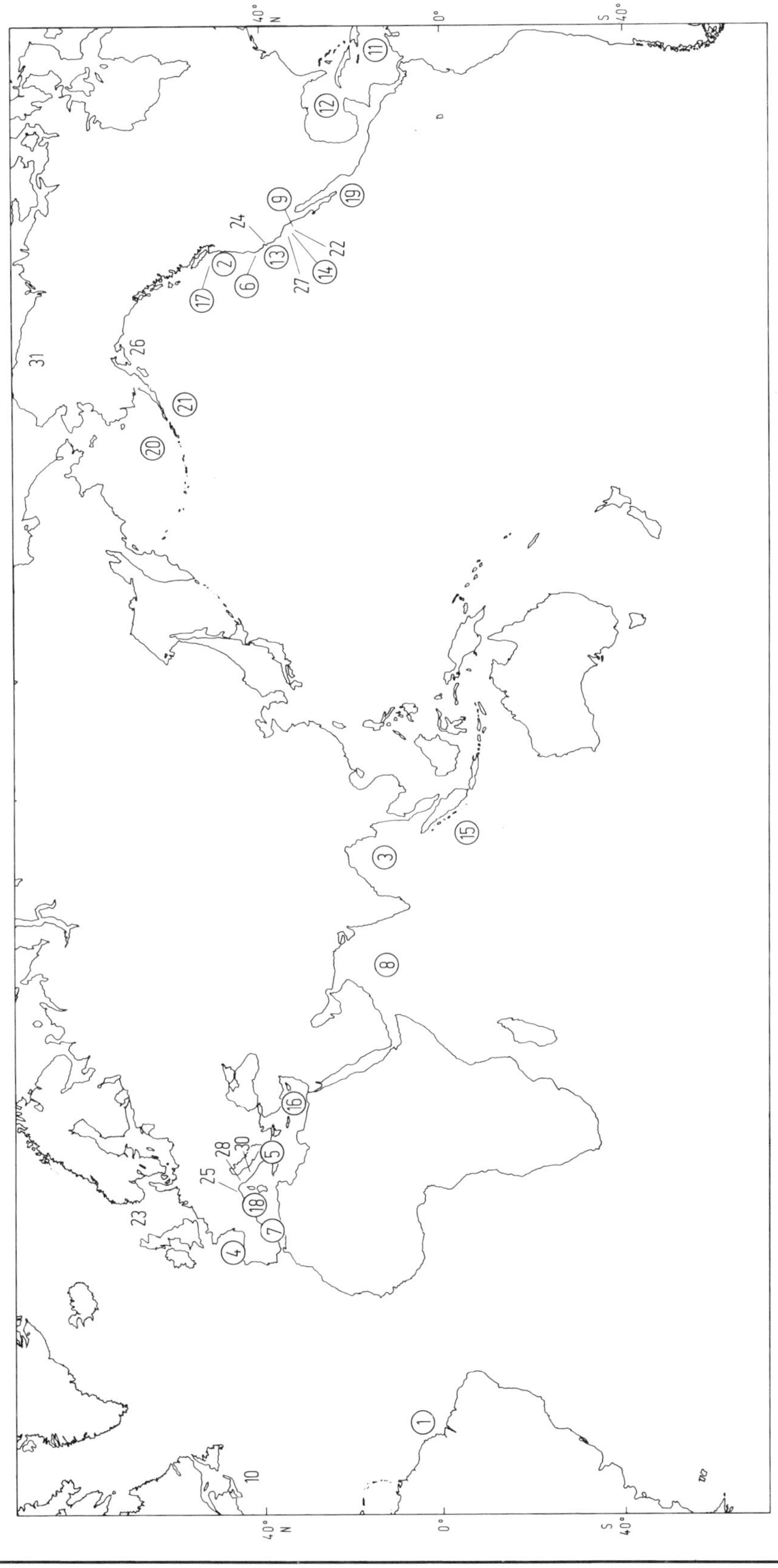

Fig. 44a. Comparison of fan outlines at a scale of 1:5000000, based on 42 references [83 N 2]. The fans are identified by numbers listed in Table 17b. The outlines of modern fans (1···21) reflect morphologic expression of the deposit. For ancient fans (22···31), paleogeographic reconstructions are used where possible, otherwise the outcrop pattern is represented. At this scale 1 cm $\hat{=}$ 50 km. Fans or cones at the African and South American coast are not taken into account. See Table 17b.

Table 17b. Diagnostic parameters for comparing modern submarine fans (1⋯21) and ancient turbidite systems (22⋯31): tabulation of key descriptive features [83 N 2]. The location of the fans is shown in Fig. 44a.

1. Age abbreviations used: Q-Quaternary, H-Holocene, Pl-Pleistocene; T-Tertiary, Pli-Pliocene, M-Miocene, O-Oligocene, E-Eocene, Pa-Paleocene; K-Cretaceous; J-Jurassic; Cz-Cenozoic; Mz-Mesozoic. Subdivisions of early, middle, and late are given as an e, m or l (e.g. Pll-late Pleistocene, Mm-Middle Miocene, Te-early Tertiary, Plml-middle-late Pleistocene).

2. Type of margin: A-accretionary; P-Passive; T-transform; S-subduction.

3. Type of crust: C-continental; O-oceanic.

4. Minimum (min) value is area of recognizable convex contours. Maximum (max) value includes the associated abyssal plain. Single values represent the best known maximum extent.

Name, area	Age fan/(basin)	Basin setting			Dimensions		
		type of margin	type of crust	synsedimentary control	length, width (radius) km	area km² max. thickness m	volume km³
	1	2	3		4		
1. Amazon, Brazilian Margin	Me-H/(K)	P	O	no	700 min, 250⋯700	$3.3 \cdot 10^5$ 4200	$7 \cdot 10^5$ min
2. Astoria, Oregon Margin, NE Pacific	Pl/(T)	A	O	yes	250 min, 130	$3.2 \cdot 10^4$ 2200	$2.7 \cdot 10^4$
3. Bengal, Bay of Bengal, NE Indian Ocean	E/(Ke)	P	O	no	2800, 1100	$3 \cdot 10^6$ >5000	$4 \cdot 10^6$
4. Cap Ferret, Bay of Biscay, NE Atlantic	Neogene/ (Mz-Cz)	P	C	yes	(75)	$1.6 \cdot 10^3$ 1800	$1.3 \cdot 10^4$
5. Crati, Gulf of Taranto, Southern Italian Margin	H/(T)	A	C	yes	16, 4⋯5	60 30	0.9
6. Delgada, Central California Margin, NE Pacific	Ml/(Ol)	T	O	no	(>350)	$4.4 \cdot 10^4$ 3000	$4 \cdot 10^4$
7. Ebro, Baleric Sea, Eastern Spain	Pl/(T)	P	C	no	100, 50	$5 \cdot 10^3$ 370	$1.7 \cdot 10^3$
8. Indus, Arabian Sea, NW Indian Ocean	O-H/(Kl)	P	O	no	1500, 960 max	$1.1 \cdot 10^6$ >3000	$1 \cdot 10^6$
9. La Jolla, Southern California Borderland, NE Pacific	Q/(M-Pli)	T	C	yes	40, 50	1200? 1600	1175?
10. Laurentian, Eastern Canadian Margin, NW Atlantic	Q/(J-K)?	P	O	no	500⋯1500, 200⋯400	$1.8 \cdot 10^5$ to $4.2 \cdot 10^5$ 2000	$1 \cdot 10^5$?
11. Magdalena, Columbian Margin, SW Caribbean	M-Pl/(Kl)	A	O	no	(230)	$5.3 \cdot 10^4$ 3000	$>4 \cdot 10^4$

5. Control on shape: D-depositional; S-structural; B-both.
6. Slash mark "/" indicates changing source type from a to b (i.e. a/b).
7. Instrumentation used for data source: A-echo-sounding profiles (surface-ship, wide beam); B-SEABEAM; C-GLORIA: D-Sea Marc I & II; E-deep-towed systems (for narrow beam sounding, side-looking sonar, and/or high resolution profiles); F-multichannel seismic profiles; G-single channel seismic profiles; H-high resolution seismic profiles: I-seismic refraction profiles, including sonobuoys; J-cores; piston and gravity; K-deep drilling: cores, cuttings, logs; L-photography; M-submersible; N-dredge and/or grab; O-nephelometer; P-well data: cuttings, logs; a-bed by bed measured sections; b-general outcrop mapping; c-Paleocurrent measurements; d-cross section mapping.
8. Degree of control: H-high, M-moderate, L-low.

Shape descriptive control	Source geometry (type)	Sediment		Type of data			
		primary lithology	grain size range (dominant)	instru-mentation	degree of control		
					morphologic	structural	sample
5	6			7		8	
elongated partial cone D	point or migrating (canyon)	terrigenous	clay to pebbles (clay)	A, C, G, H, I, J, L	H	M	L
trapezoidal D	point (canyon)	volcaniclastic and glacial	mud to gravel (silt)	A, G, J, K	M	L	L
elongated B	point or migrating (canyon)	terrigenous	mud to medium sand (mud)	A, F, G, H, I, J, K	M	M	L
elongated dissymmetric B	bipoint (canyons)	terrigenous	clay to sand (silt)	A, B, F, H, J, L, M	H	H	M
elongated B	migrating (delta)	siliclastic, fluvial	mud to medium sand (mud)	A, D, G, H, J, L, N	M	M	L
elongated bi-lobed D	bipoint (canyons)	unknown	mud to sand (mud)	A, F, G, H, J, K	L	M	L
oval S	migrating (delta)	lithicwacke, fluvial	mud to medium sand (fine sand)	A, F, G, J	L	H	M
fan B	migrating point (canyon-delta)	terrigenous	clay to sand (mud)	A, F, G, H, I, J, K, L, O	M	M	L
pear shaped B	point (canyon)	batholithic and metamorphic	mud to gravel (fine sand)	A, E, F, G, H, J, L, M	H	H	H
elongated, lobate D	point/line (canyon/ slump)	glacial	clay to gravel (medium-fine sand and mud)	A, C, F, G, H, I, J, K	M	M	L
trapezoidal fan D	migrating point (canyon-delta)	terrigenous	clay to sand (mud)	A, F, G, H, I, J, L	M	M	L

Table 17b (continued)

Name, area	Age fan/(basin)	Basin setting: type of margin	type of crust	synsedimentary control	Dimensions: length, width (radius) km	area km² max. thickness m	volume km³
	1	2	3		4		
12. Mississippi, Gulf of Mexico	Pl/(T)	P	C/O	yes	540, 570	$>3.0\cdot10^5$ 4000	$2.9\cdot10^5$
13. Monterey, Central California Margin, NE Pacific	M-H/(Ol)	T	O	no	400, 250	$7.5\cdot10^4$ 2000	$5\cdot10^4$
14. Navy, Southern California Borderland, NE Pacific	Pll-H/(Tl)	T	C	yes	(40)	560 900	75
15. Nicobar, East Central Indian Ocean	M-Pl/(M)	A/T	O	yes	(2200)	$1.5\cdot10^6$ 3000	?
16. Nile, Egyptian Margin, Eastern Mediterranean	Pl-H/(M)	P	C	yes	280, 500	$7.0\cdot10^5$ >3000	$1.4\cdot10$
17. Nitinat, Washington Margin, NE Pacific	Plml/(T-Q)	A	O	no	260, 80	$2.3\cdot10^4$ 1000	9000
18. Rhone, Gulf of Lions, Southern French Margin	Neogene/ (M)	T	C/O	yes	440, 210	$7\cdot10^4$ 1500	$1.2\cdot10^4$
19. San Lucas, Southern Baja California Margin, NE Pacific	M/(M)	P	O	no	(60)	6000 1000	3000
20. Zhemchug, Central Bering Sea	Pl/(Te)	P	O	no	>115, ?	? 2000	?
21. Zodiac, Aleutian Margin, Alaska NE Pacific Ocean	E-M/(Kl-E)	A	O	no	?	$1.1\cdot10$ 1600	$3\cdot10^5$
22. Blanca, Southern California Borderland, NE Pacific	M	T	C	?	215, 30	$4\cdot10^3$ 1000	1300
23. Brae, North Sea	Jl	P	C	yes	(5⋯10) [1]	20⋯50 600	5⋯8 each
24. Butano, Central California Coast Ranges	E	S	C	yes	80, 40	3200 3000	9600
25. Cengio, Piedmont Basin, NW Italy	O-M	A	C	yes	6.4, 4.8	30 170	3.0

[1]) overlapping fans, each (5⋯10)

Shape descriptive control	Source geometry (type)	Sediment: primary lithology	Sediment: grain size range (dominant)	Type of data: instrumentation	degree of control: morphologic	structural	sample
5	6			7		8	
conical B	migrating point (canyon)	terrigenous clastics	mud to gravel (silty mud)	A, C, D, E, F, G, H, J, L	H	H	M
quarter cone D	bipoint (canyons)	terrigenous clastics	mud to cobble (silty mud)	A, E, F, G, H, J, L	M	H	L
interconnected elongated basins D	point (canyon)	granitic and volcaniclastic	mud to gravel (sandy silt)	A, E, G, H, J, L	H	M	H
triangular B	point (canyon)	terrigenous	mud to fine sand (?)	A, F, G, H, I, J, K	L	M	L
fan (arcuate) B	migrating point (delta)	terrigenous	mud to sand (silty mud)	A, G, H, I, J, K	H	L	L
triangular B	shifting point (canyon)	volcaniclastic	mud to coarse sand (silt)	A, G, H, I, J	M	M	L
elongated B	migrating point (canyons)	terrigenous	mud to fine sand (coarse silt)	A, B, F, G, H, J	H	M	L
rectangular D	point (canyons)	granitic	mud to pebble (fine sand)	A, E, G, H, I, J, K	M	M	L
? D	point (canyon)	terrigenous	mud to sand (?)	A, F, G, H, I, J	L	L	L
partial cone B	point	plutonic and metamorphic	clay to fine sand (?)	A, G, H, I, J, K, L	L	M	L
elongated S	migrating line (fan delta)	volcaniclastic and metamorphic	mud to boulders (sand)	F, G, J, b	M	L	L
slope-apron elongated B	line	terrigenous	mud to boulders (sand and mud)	F, K, P	L	M	H
elongated ellipsoidal B	point (2 canyons)	arkosic sandstone	clay to boulders (medium sand)	a, b, c	M	M	H
arcuate S	point (fan delta)	tectonites and volcaniclastics	mud to granule (medium sand)	a, b, c, d	L	L	M

(continued)

Table 17b (continued)

Name, area	Age fan/(basin)	Basin setting: type of margin	type of crust	synsedimentary control	Dimensions: length, width (radius) km	area km^2 max. thickness m	volume km^3
	1	2	3		4		
26. Chugach, Gulf of Alaska, Aleutian Alaska Margin	Kl	S	O	yes	2000, 100	$2.0 \cdot 10^5$ 10000	$2.0 \cdot 10^6$
27. Ferrelo, Southern California Borderland, NE Pacific	E	A	O	?	340, 65	$3.1 \cdot 10^4$ 1800	$1.5 \cdot 10^4$
28. Gottero, Ligurian Apennines, NW Italy	Kl-Pa	A	O	yes	(30, 50)	3500 1500	5200
29. Hecho, Southern Pyrenees, Northern Spain	E/(Tl)	A	C	yes	175, 40···50	$8 \cdot 10^3$ 3500	$2.5 \cdot 10^4$
30. Marnoso-Arenacea, Northern Apennines, Italy	Mm-l/(E)	A	C	yes	[2])	150···7500 1000	?
31. Torok-Fortress Mountain, North Slope, Alaska	Ke/ (Neocomian)	P?	C	?	150, 300	$4.5 \cdot 10^3$ 3400	$8 \cdot 10^4$

[2]) major 25···50, 6···15
minor 20···25, 4···10

Shape descriptive control	Source geometry (type)	Sediment		Type of data			
		primary lithology	grain size range (dominant)	instru-mentation	degree of control		
					morphologic	structural	sample
5	6			7		8	
narrow prism S	point and line (canyons)	feldspatholithic sandstone	clay to cobbles (fine sand)	a, b, c	L	L	M
elongated S	point (canyon)	siliceous volcanic and granitic	mud to cobbles (sand)	F, G, J, K, b	M	M	M
semicircle D	point (canyon)	arkosic sandstone	clay to pebble (fine-medium sand)	a, b, c	M	M	M
elongated S	point (canyons)	lithic sandstone	mud to boulders (sand)	a, b, c, d	L	M	M
elongated D	migrating point (fan delta?)	terrigenous-fluvial	mud to gravel (fine-medium sand)	a, b, c, d	L	M	M
delta aprons D	line (slope gullies)	terrigenous	mud to gravel (fine sand and mud)	F, K, P	L	L	L

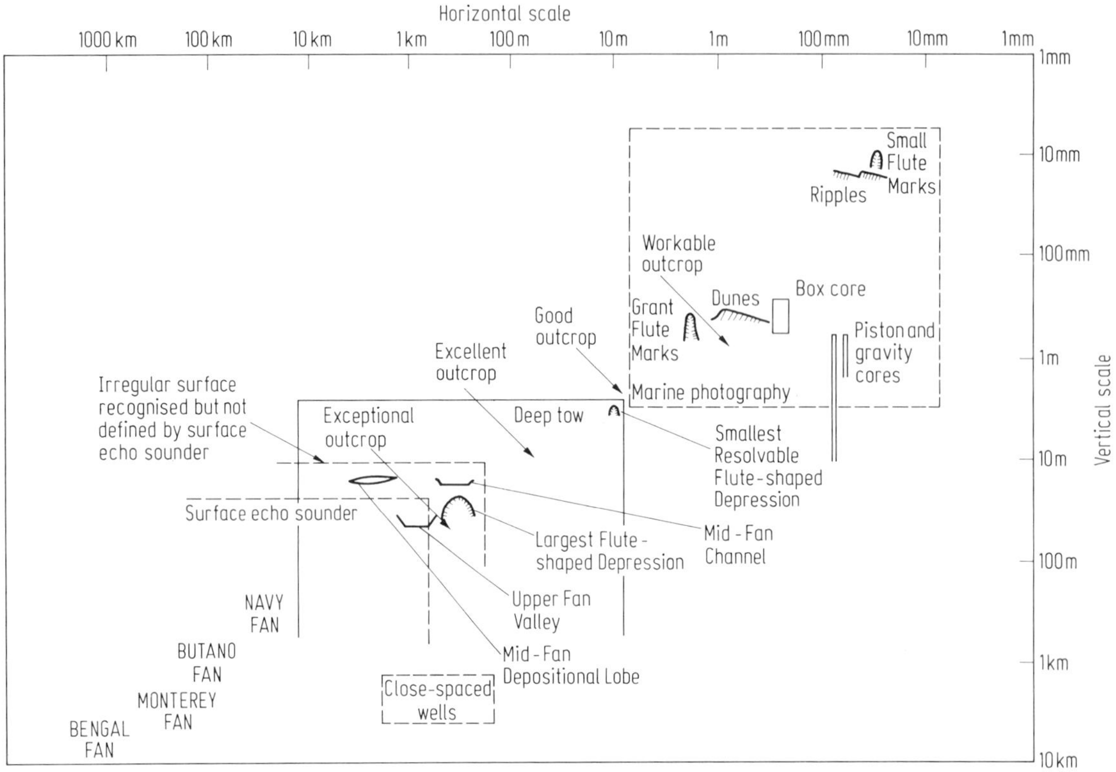

Fig. 44b. Horizontal and vertical scale comparison for modern submarine-fan data sets with sedimentologic observations based on outcrop studies of ancient turbidite deposits. Nondeep-tow systems are capable of resolving features in areas noted by dashed lines. From [83 N 2]. Compare Fig. 63, Table 17b.

1.4.2.6 Trenches

Deep-sea trenches are mostly "marginal trenches", i.e. of types A (facing thalassocraton), and B (quasi cratonic) of the classification after Fairbridge [66 F]. Type C (transverse) includes transverse fracture zones of ocean ridges, as there is the Vitiaz Trench, and is of different origin. Marginal trenches are narrow, steep-sided troughs roughly parallel to continental margins at the seaward base of a continental platform, such as off western, South America, or at an island arc, such as the Bonin-Mariana Islands, or off Middle America. See Fig. 45.

Trenches are generally separated from the ocean floor by an outer ridge, which rises 200···100 m above the level of the adjacent ocean floor. In parts of the Indonesian region trenches occur as double, parallel features [82 K].

Trenches have a characteristic asymmetrical V-shaped section with the steeper slope (8···20°) on the overriding (inward) side. The trench bottom is usually filled with a thin sediment cover which forms a small flat plain, 1···3 km wide, occasionally wider. This plain is not continuous over great distances along the trench axis; rather, small elongate plains are limited by small ridges transverse to the trench axis, with slight differences in depth from one to another adjacent plain.

The extraordinary uniformity of trench topography could be demonstrated with 35 bathymetric profiles across various trenches which have been projected perpendicularly to the consuming plate boundary. The average slopes of the flanks, the width of the topographic depression, and the distance between trench and arc are all so consistent that they suggest common physical constraints with regard to the mechanism of formation of these features. After [73 L 1].

Most of deep sea trenches are located around the Pacific Ocean. Only few of them exist in the Atlantic Ocean and in the Indian Ocean. The maximum depths of the oceans are located in the deep-sea trenches. See Tables 6 and 18 [66 F].

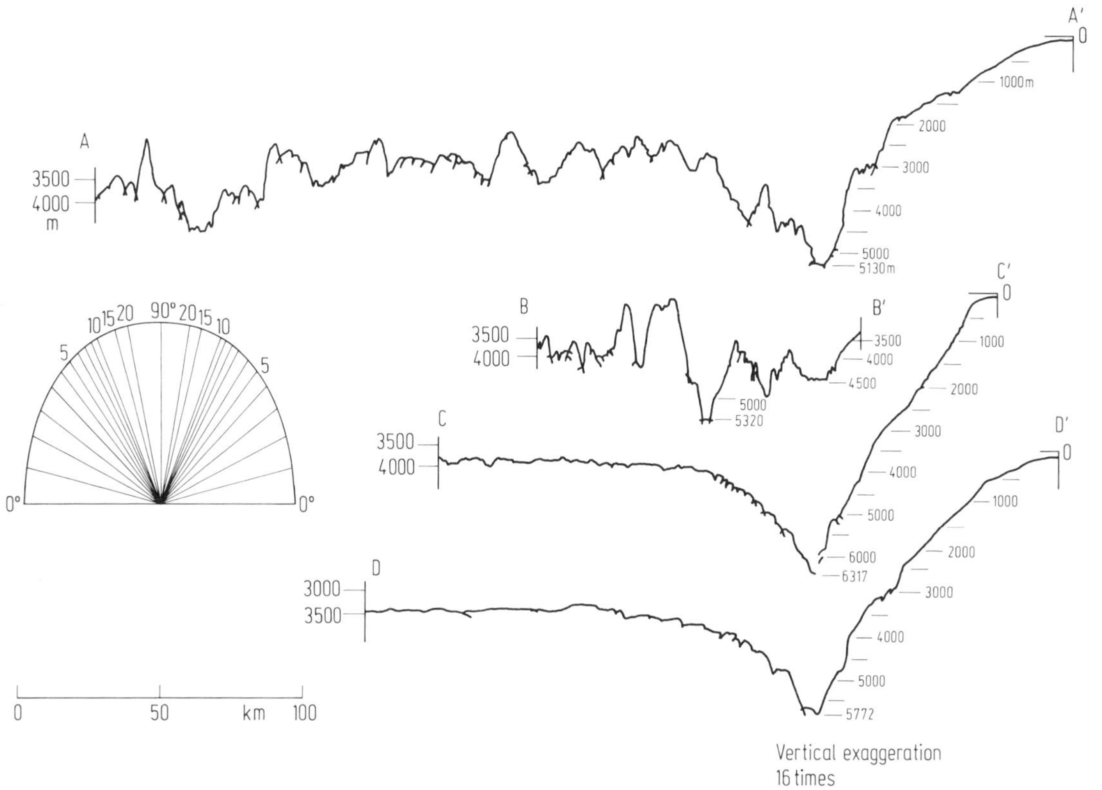

Fig. 45. Topography of the Middle America Trench: cross sections of the Middle America Trench [63 H 3, 63 H 4]. The Middle America Trench is created at the subduction zone of the Cocos Plate to the southwest margin of Middle America, which belongs to the Caribbean Plate. See Fig. 72. Dimensions of the trench: length 2800 km, mean width 40 km, depth 5000···6600 m. Profiles: horizontal scale in [km], vertical scale in [m], vertical exaggeration: 16 ×. Profile location, from left to right, i.e. from ≈225° to 45° sounding courses, A···A′: from the sea to Manzanillo, Mexico, B···B′: from the sea to Petalco Bay, Mexico, C···C′: from the sea to Champerico, Guatemala, D···D′: from the sea to Gulf of Fonseca. Interpretation: The continental slope is steep and convex to the trench; the seaward slope of the trench is less steep than the continental slope, smooth in the southern part C···C′, D···D′, and rugged in the northern part A···A′, B···B′, while the bottom of the trench is flat to V-shaped [80 A]. Compare Table 18.

The continental margin of southern Mexico north of the Tehuantepec Ridge and south of the Rivera Fracture Zone is a narrow crustal transition zone. The margin lacks a well-developed forearc basin and the active volcanic arc is subparallel to the trench and not clearly related in a normal fashion to the Benioff Zone. A geological survey of a small part of the margin 100 km south of Acapulco reveals significant variations in structure over short distances. Multi-channel seismic reflection data, sonobuoy refraction data and dredging indicate that the shelf and upper slope are underlain by continental crust. Geometry derived from the reflection data suggests that the continental crust is less than 10 km thick at its seaward termination.

Migrated reflection data reveal structures on the lower slope and within trench axis turbidite fill which are interpreted as folds and thrust faults. Structures and landward dipping reflections suggest that the base of the slope is an accreting zone with trench turbidite fill being uplifted and emplaced. Both reflection and refraction data suggest the trench inner wall contains both slope deposits and uplifted trench deposits. There is no evidence of significant amounts of oceanic crust within the slope. Volume calculations reveal that some of the sedimentary rocks are being consumed along with the oceanic crust. From [80 S 2].

Table 18. Trenches and related troughs: locations and depths [66 F]. Depths may differ 0.1 to 1% from real scale due to measurement methods and conditions.

Type A and B of the classification after Fairbridge [66 F]. Type A (facing thalassocraton), type B (quasi cratonic). Type C (transverse) includes transverse fracture zones of ocean ridges such as the Vitiaz Trench, and is of different mode of origin.

	Depth m	Length km	Mean width km	Area extent km²	Volume km³
Type A (facing thalassocraton)					
Western Pacific:					
Kurile-Kamchatka Trench	10 542	2200	120	264 000	1 320 000
Japan Trench	8 412	800	100	80 000	336 000
Idzu-Bonin Trench (including Ramapo Deep)	9 810	800	90	72 000	352 800
Marianas Trench (including Challenger Deep)	11 022	2550	70	17 850	98 200
Yap (West Caroline) Trench	8 527	700	40	28,000	72 500
Palau Trench	8 054	400	40	16 000	64 000
New Guinea Trough	5 311	440	60	26 400	70 300
West Melanesian Trench	6 534	1100	60	66 000	216 000
Southwest Pacific:					
Tonga Trench	10 800 ··· 10 882	1400	55	77 000	415 800
Kermadec Trench	10 047	1500	60	90 000	450 000
Hikurangi Trough	3 590	700	40	28 000	48 400
Eastern Pacific:					
Aleutian Trench	7 679	3700	50	185 000	673 000
Cedros Trough	6 225	400	40	16 000	49 600
Middle America Trench (including Acapulco Deep)	6 662	2800	40	96 000	316 800
Peru-Chile Trench	8 055	5900	100	590 000	2 360 000
Cape Horn Trough	4 395	1050	50	52 500	115 000
Indian Ocean					
Aru Trough	3 652	900	25	22 500	42 500
Timor Trough	3 276	800	40	32 000	51 200
Java (Indonesian) Trench	7 450	4500	80	180 000	666 000
Makran Trough	3 500	400	100	40 000	70 000
Atlantic Ocean:					
Puerto Rico Trench	8 385	1550	120	186 000	779 000
Cayman Trench	7 093	1450	70	101 500	360 000
Dominican Trench	≈ 6 200	700	30	21 000	65 000
South Sandwich Trench	8 428	1450	90	130 500	558 000
Birdwood Trough	4 758	980	40	38 200	90 800
Type B (quasi-cratonic)					
Western Pacific:					
Ryukyu (Nansei Shoto) Trench	7 507	2250	60	135 000	506 000
Philippine Trench (Cape Johnson Deep)	10 030 ··· 10 497	1400	60	84 000	420 000
Luzon Trough	4 510	600	50	30 000	67 500
Manila Trough	5 245	350	40	14 000	36 400
Moru Trough	5 842	270	50	13 500	39 400
Sangir Trough	2 867	700	70	49 000	68 600
Minahassa Trough	5 520	550	60	33 000	92 000
Flores Trough	5 123	540	40	21 600	56 160

Table 18 (continued)

	Depth m	Length km	Mean width km	Area extent km^2	Volume km^3
Southwest Pacific:					
New Britain Trench	8 320	750	40	30 000	124 800
Bougainville (North Solomons) Trench	8 310	500	50	25 000	103 800
Kiriwina Trough	5 200	700	35	24 500	63 700
San Cristobal (South Solomons) Trench	8 310	950	40	38 000	159 700
Torres (North Hebrides) Trench	9 162	750	60	45 000	206 000
New Hebrides Trench	9 165	1200	70	84 000	385 000
Papuan Trough	2 300	750	80	60 000	69 000
New Caledonia Trough	3 932	600	70	42 000	84 000
Atlantic and Mediterranean:					
Malvines Trough	3 520	1250	80	100 000	175 600
Hellenic (Crete) Trough	5 121	1500	40	60 000	153 000
Indian Ocean:					
Amirante Trench	9 074	680	30	21 400	95 000
Type C (transverse)					
Western Pacific:					
East Caroline (Mussau) Trough	6 920	400	30	12 000	41 400
Southwest Pacific:					
Vitiaz Trench	6 150	550	60	33 000	105 600
Yasawa Trough	4 810	480	20	9 600	23 000
Also: Vanikoro, Strathmore, Alexa, Hazel Holme, Rotuma, Bayonnaise, Horne and Wallis Troughs: all small examples in the Melanesian Border Plateau					
Macquarie Trough	5 851	600	60	36 000	105 000
Atlantic Ocean:					
Romanche Trench	7 856	300	20	6 000	21 900
Indian Ocean:					
Chagos Trench	5 408	2450	70	101 500	264 000
Vema Trench	6 402	700	25	17 500	56 000
Mauritius Trench	5 564	1080	30	32 400	95 400
Diamantina Trench	8 230	2160	30	64 800	265 700
Ob Trench	5 714	550	30	16 500	45 400
Type D (parallel, intermediate)					
Indonesia:					
Bali Trough	4 251	2300	90	207 000	434 700
Savu Trough	3 370	270	50	13 500	29 950
Banda (Weber) Trough	7 440	650	80	52 000	192 400
And many smaller troughs					

1.4.2.7 Island arcs and trench systems

Island arc:

A group of islands usually having a curving arch-like pattern, generally convex toward the open ocean, with a deep trench or trough on the convex side and usually enclosing a deep basin on the concave side. Seamounts, tablemounts, and banks are frequently interspersed with the islands in an island arc.

In the Pacific Ocean the arcs are typically 250···500 km in width, with their axes located 150···250 km from the central feature. The crests of the arches are generally 0.5···1.5 km above the level of the moat and the general level of the surrounding ocean floor [77 H 2, 79 D, 77 T, 80 H 2, 82 M].

Gierloff-Emden

Some arcs are single, such as the Kurile arc between Kamchatka and Japan, consisting of a row of active or recently extinct volcanoes behind a trench. Others, such as Sumatra, are double arcs, with an additional line of non-volcanic islands between the trench and the volcanic arc [79 D, 74 I]. Fig. 46.

Structural significance of the arcs:

Most of the features of island arcs and young fold mountains can be explained in terms of the plate tectonics theory. The arcs will bring together many of the aspects of earth structure, gravity, earthquakes and volcanoes. The island-arcs of the Pacific delimit one fundamental geological boundary in the earth's surface: this has been called the "andesite line" after the prevalent type of andesitic lava which occurs in the arcs and fold mountain zones landward of trenches ("andes") but is absent from the Pacific Ocean floor and islands beyond [63 H 3, 63 H 4, 71 K]. See Figs. 47 and 48. For seismicity and volcanism see the atlas [78 K].

In the circumpacific belt the "andesite" is representative of the andesite-dacite-rhyolite volcanic rocks. The andesite line has turned out to be the line where oceanic lithosphere is subducting along Benioff-zones [78 H 1, 82 S 3]. See Fig. 47.

Classification and morphologic terminology of arc regions:

For specified classifications and modes of origin, see [77 S 1, 79 D].

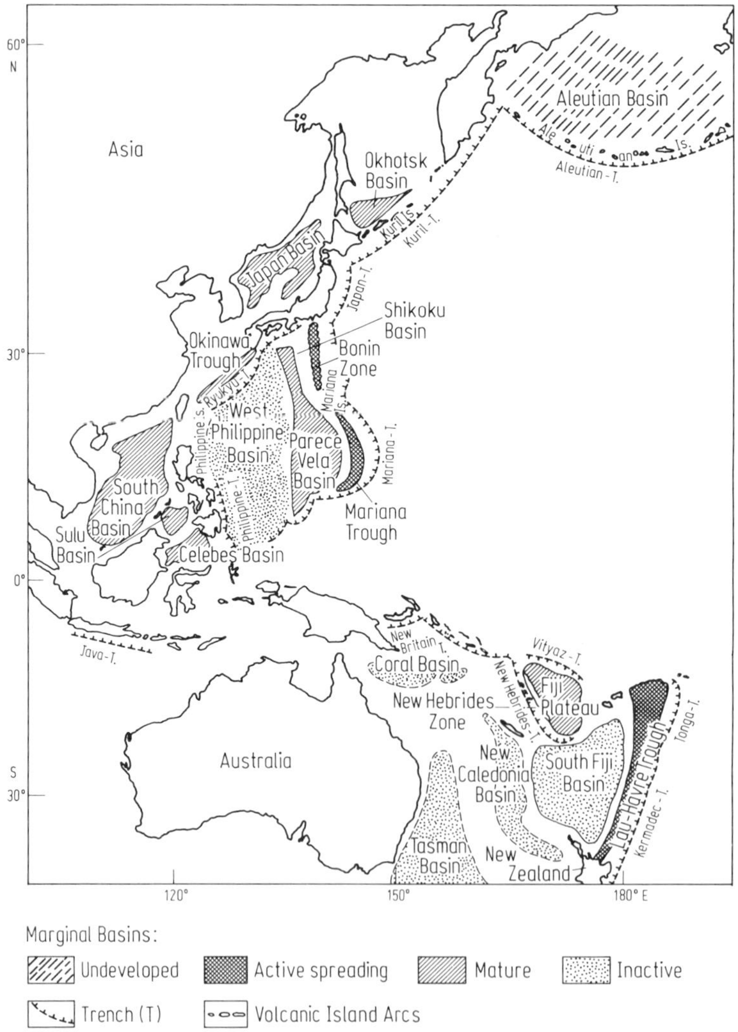

Fig. 46. Volcanic arcs, trenches, and marginal seas in the Northwest Pacific Ocean (north of the equator). The marginal seas encompass both shelf sea regions and deep basins. After [71 K]. See Fig. 55.

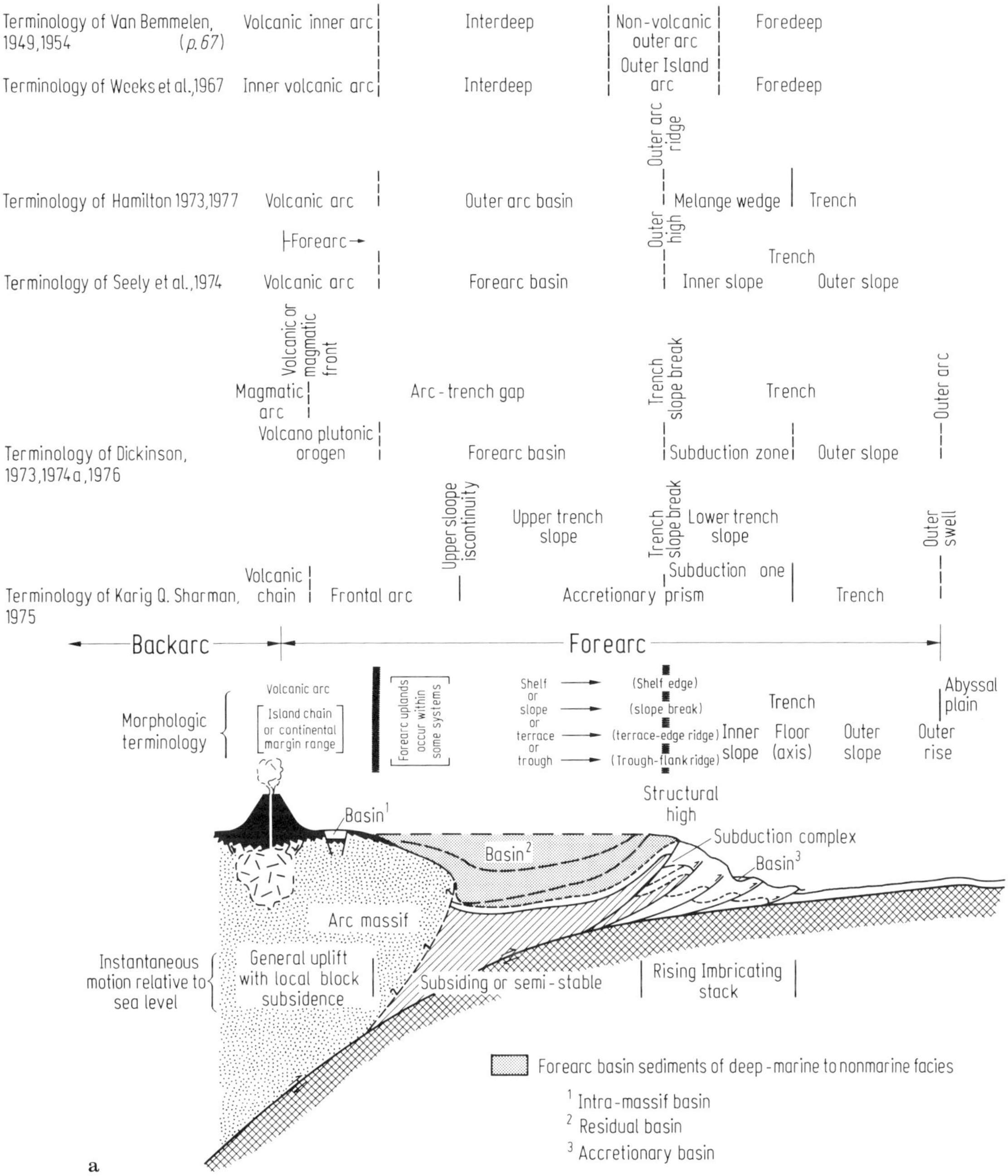

Fig. 47a. Generalized model of the forearc region showing terminology of different workers for the various structural features [77 S 1]. For specified classifications and modes of origin, see [77 S 1, 79 D].

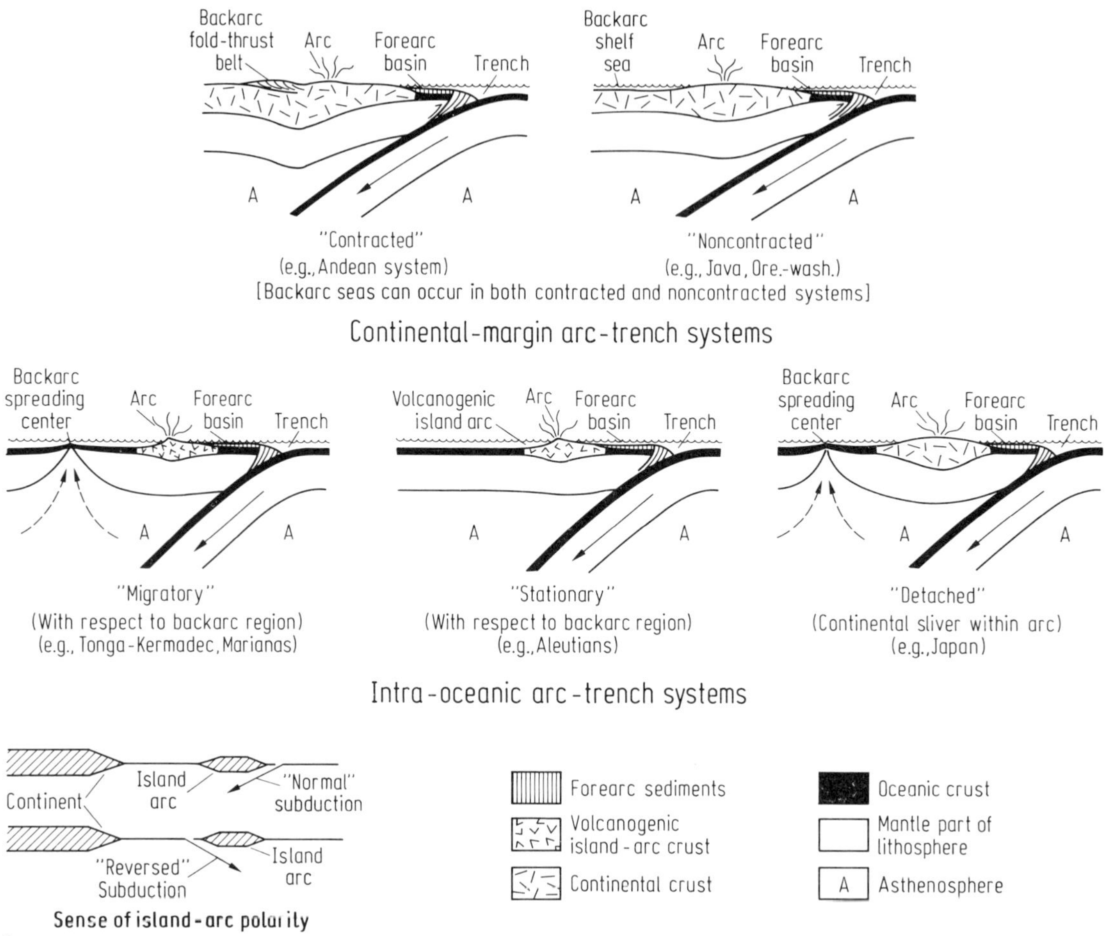

b

Fig. 47b. The subduction of the ocean's crust. Thick solid line and arrow represent the Benioff-zone where seismic activity happens. Settings of arc-trench systems as defined by backarc features [77 S 1].

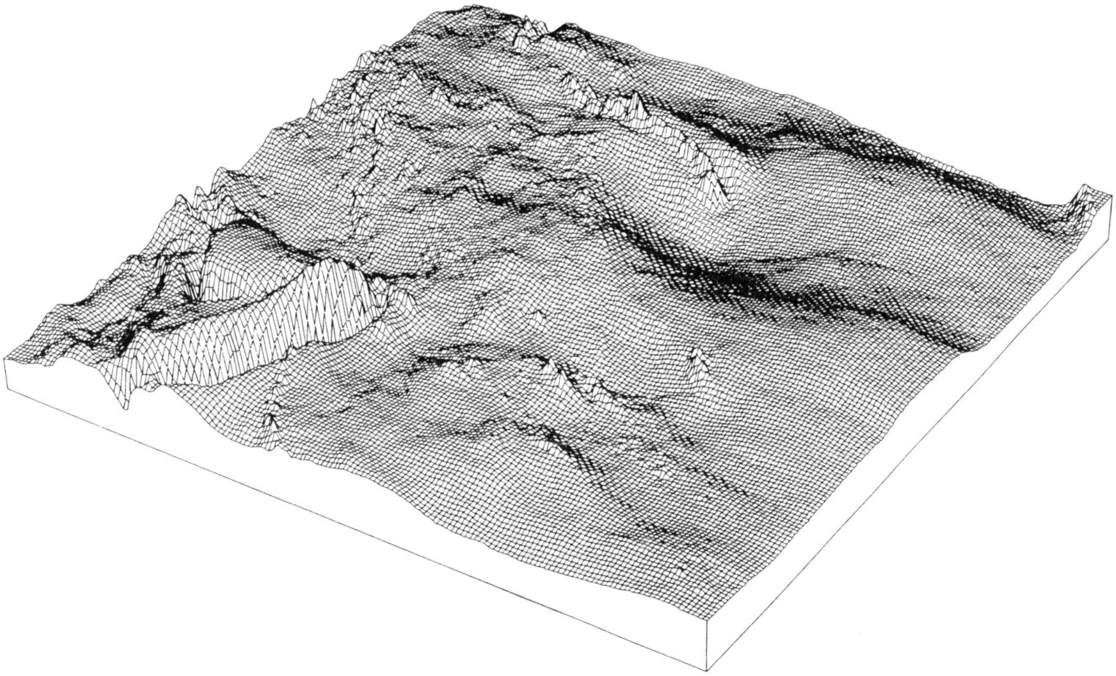

Fig. 48. The geoid in the central Pacific as determined from Seasat altimeter data. The sea surface topography is correlated with ocean floor topography due to gravity anomalies representing at the left the Tonga-Kermadec Trench, at the top the Hawaiian seamount chain. Compare Figs. 57, 58. Undulation amplitude 15···20 m. Left sector presents Lau Ridge and Tonga Islands. From [85 E].

1.4.3 The ocean-basin floor complex

Ocean-basin floor:

One of the three morphological divisions of the ocean bottom. The ocean-basin floor includes the deep-ocean floor, exclusive of the mid-oceanic ridge (rise) and the continental margin. The three major subdivisions of the ocean-basin floor are:

1. abyssal floor,
2. oceanic rise, and
3. seamounts and seamount groups.

The abyssal floor is made up of abyssal plain and abyssal hill provinces [71 H, 84 E]. See Figs. 49, 51, 52.

Oceanic rise:

A large area (measured in hundreds of kilometers) which rises to a general level of a few hundred meters above the surrounding abyssal floor. Relief is variable, ranging from gentle to extremely rugged. The term "oceanic rise" has also been applied to Pacific sectors of the mid-oceanic ridge (East Pacific Rise). [74 B 2, 84 E].

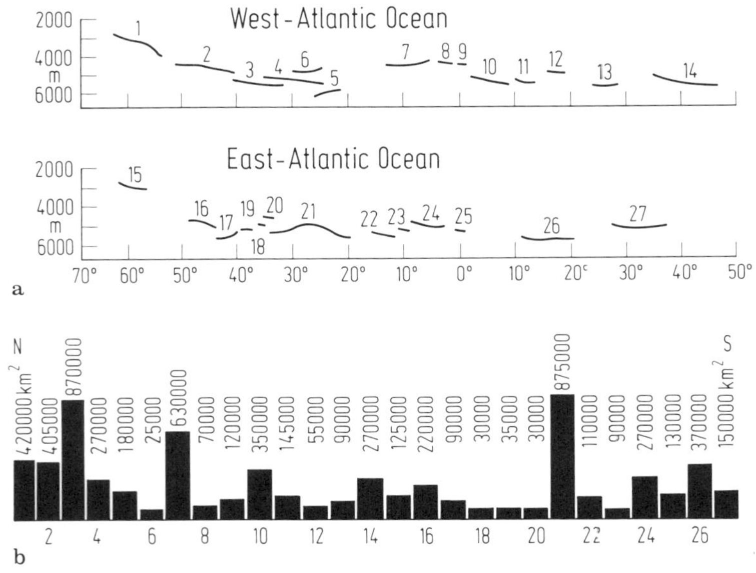

Fig. 49. Abyssal plains of the Atlantic Ocean. (a) Depths in [m], (b) area in [km²]. 1 Labrador, 2. Newfoundland, 3 Sohm, 4 Hatteras, 5 Nares, 6 Blake-Bahama, 7 Demerara, 8 Ceara 1, 9 Ceara 2, 10 Pernambuco, 11 Brazil 1, 12 Brazil 2, 13 Brazil 3, 14 Argentin., 15 Iceland, 16 Biskay, 17 Iberia, 18 Tagus, 19 Horseshoe, 20 Seine, 21 Madeira, 22 Green-Cape, 23 Gambia, 24 Sierra Leone, 25 Guinea, 26 Angola, 27 Cape [72 B 2]. See Fig. 38b.

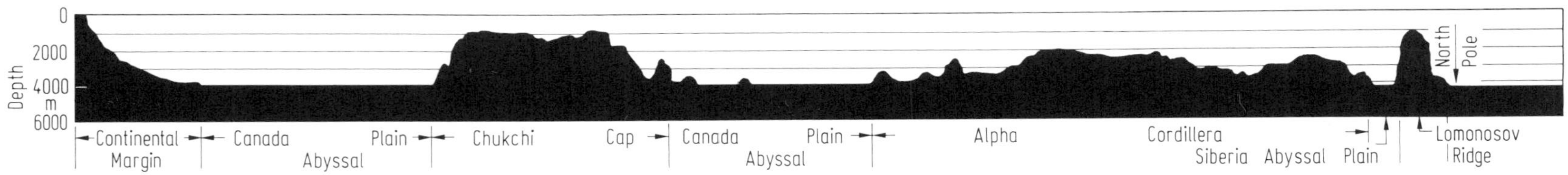

Fig. 50. Abyssal plains in the North Polar Ocean (Arctic Ocean). Bathymetric profile at meridian 155° W from Point Barrow (left) to the North Pole (right). Soundings from submersible (SSN) US "Nautilus". Vertical exaggeration: 26 times [61 D 2]. Compare Fig. 34.

1.4.3.1 Abyssal plains

An abyssal plain is an area of the ocean-basin floor in which the ocean bottom is essentially flat, and the regional gradient is less than 1 : 1000. They characteristically occur adjacent to the outer margins of the continental rise. The horizontal extent of an abyssal plain may range from less than 200 km to more than 2000 km. On some flat surfaces it has a continuous gradient in one direction, whereas other directions are characterized by broad irregularities [70 H 2]. They are widespread in the Atlantic and Indian Oceans, and occur in marginal seas such as the western Mediterranean, Gulf of Mexico, and the Caribbean. See Figs. 49 and 50. [84 D 2].

Abyssal apron:

A newly named but major physiographic feature of the ocean floor is the abyssal apron. As shown by Fig. 38b these features lie oceanward of continental rises and of most abyssal plains. During the past, many of the Atlantic representatives have been included as parts of the continental rises; others were considered aprons deposited by turbidity currents from sediment sources other than the continents.

[84 E] detailed studies of the features showed them to be deposited by geostrophic bottom currents from high latitudes, termed contour currents by Heezen and Hollister [52 H, 71 H] and others.

The total area identified as abyssal aprons in the Atlantic Ocean is about $9.60 \cdot 10^6$ km^2, or 70% of the area of true continental rises and nearly double the area of abyssal plains.

Seismic evidence indicates that abyssal plains are deposits of great thicknesses of sediments over the original, now buried, topographic irregularities. Abyssal plains occur in areas, such as the Atlantic, where abundant sediments are derived from the continents. Since relatively few rivers drain into the Pacific Ocean, and because it is rimmed by a trench system that traps sediment, the Pacific Ocean basin receives less sediment and fewer plains form [82 K]. The landward edge of the abyssal plains is generally marked by an abrupt change in slope to values from 1 : 100 to 1 : 700 – typical of the continental rise. In some cases the deep-sea fans of the rise merge with the abyssal plains, while in others the abyssal plains are cut by deep-sea channels at the seaward ends of submarine canyons.

An abyssal gap is a constricted passage connecting two abyssal plains which, in the vicinity of the abyssal gap, lie at different levels, as, for instance, the Vema Channel in the South Atlantic Ocean [84 J].

1.4.3.2 Abyssal hills

Abyssal hill:

A hill which occurs on the ocean-basin floor. It may be equidimensional or elongated in shape. Relief is generally in the 50···1000 m range. Horizontal dimensions are generally 1···10 km, but may range as high as 50 km or more. Hillside slopes generally range between 1···15°. The term is generally not used to describe singular features, but rather to refer to one of many which, together, make up an irregular bottom surface. An abyssal hill generally takes its shape from that of the underlying basement surface, although the term may be applied to hills which are of purely sedimentary origin. Though not so common in the Atlantic and Indian Oceans, they cover 80···85% of the Pacific Ocean floor [82 K]. See Figs. 51, 52.

The abyssal hills are structurally and genetically an integral part of the mid-oceanic ridge, but they are morphologically separate because of their occurrence at a lower level, away from the regional slopes of the mid-oceanic ridge, and because of their generally (though not always) lower relief, due to burial by sediments [70 H 2].

Abyssal hills province:

A province of the ocean-basin floor which is essentially entirely occupied by abyssal hills. The surface of this smooth terrain slopes radially outward from the central feature; thus, the overall shape is roughly conical. The regional gradient may range from 1 : 50 on the upper portion of the apron near the central feature, to 1 : 1000 or less on its outer reaches. Local occurrences of deep-sea channels have been interpreted. Total width of aprons is highly variable, but widths of 100···400 km seem to be typical [77 H 2].

Ocean-basin hills (swales):

Hills, or ridges and valleys, occurring on the ocean-basin floor. Ocean-basin hills are distinguished from abyssal hills, in that they are generally fine-textured, with very low relief (generally less than 100 m), being similar in morphology to continental rise hills. In certain localities, ocean-basin hills have been referred to as dunes [66 H, 84 R].

1.4.3.2.1 Morphometric classification of the submarine, elevated relief

Abyssal hills, seamounts, peaks, and other elevated areas such as plateaus, have been classified according to their extension, length, width, and height for 15 selected profiles chosen from the East-Pacific Ocean between California and the Hawaiian Islands, each covering 60 nm. Their bathymetric-statistical evaluation of bathymetric features gave the following results: The majority of these hills has a height of ca. 150 m above the sea floor, and a width of 2···8 km. Several profiles have Gaussian distribution. There is no periodic distribution along a horizontal profile. The frequency (distribution of the individual hills along the profile) is related to the elevation of the hills [64 M 1, 65 K]. See Figs. 51 and 52.

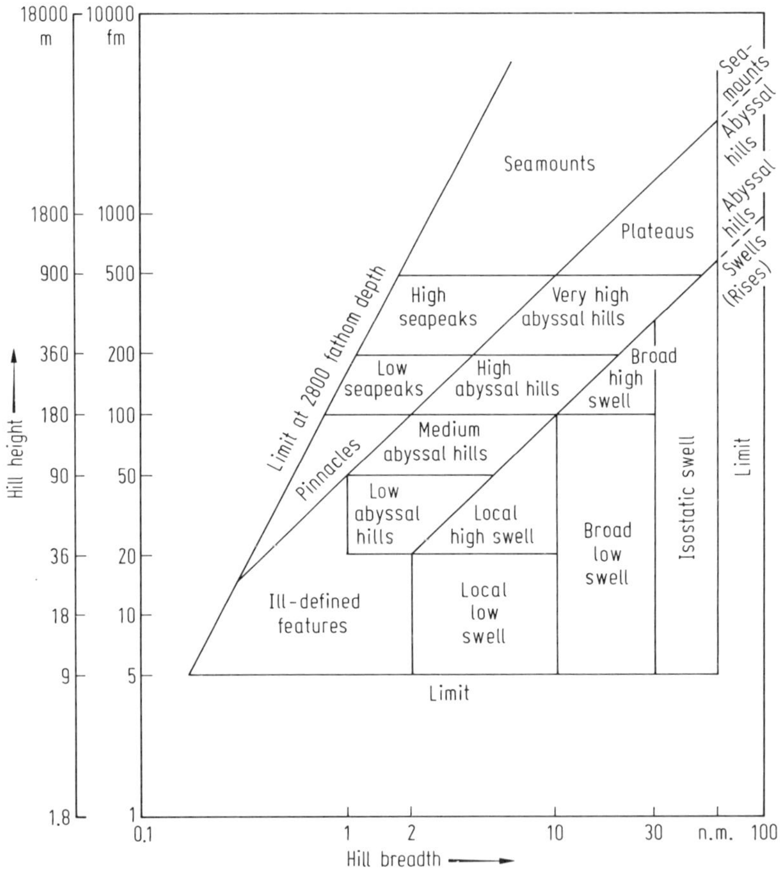

Fig. 51. Classification of elevated relief of the sea floor. (Classification of positive bathymetric features). This classification is based on criteria of the following categories: i) Metric: height (high, middle, low); width (large, middle, narrow). ii) Geomorphologic-genetic: shape (geologic-genetic form) [65 K]. Compare Fig. 44b, Table 17b.

→

Fig. 52. Bathymetric profiles divided into abyssal hills for selected profiles between San Francisco, and Hawaii [65 K]. Dashed lines connect the bottoms of the valleys.

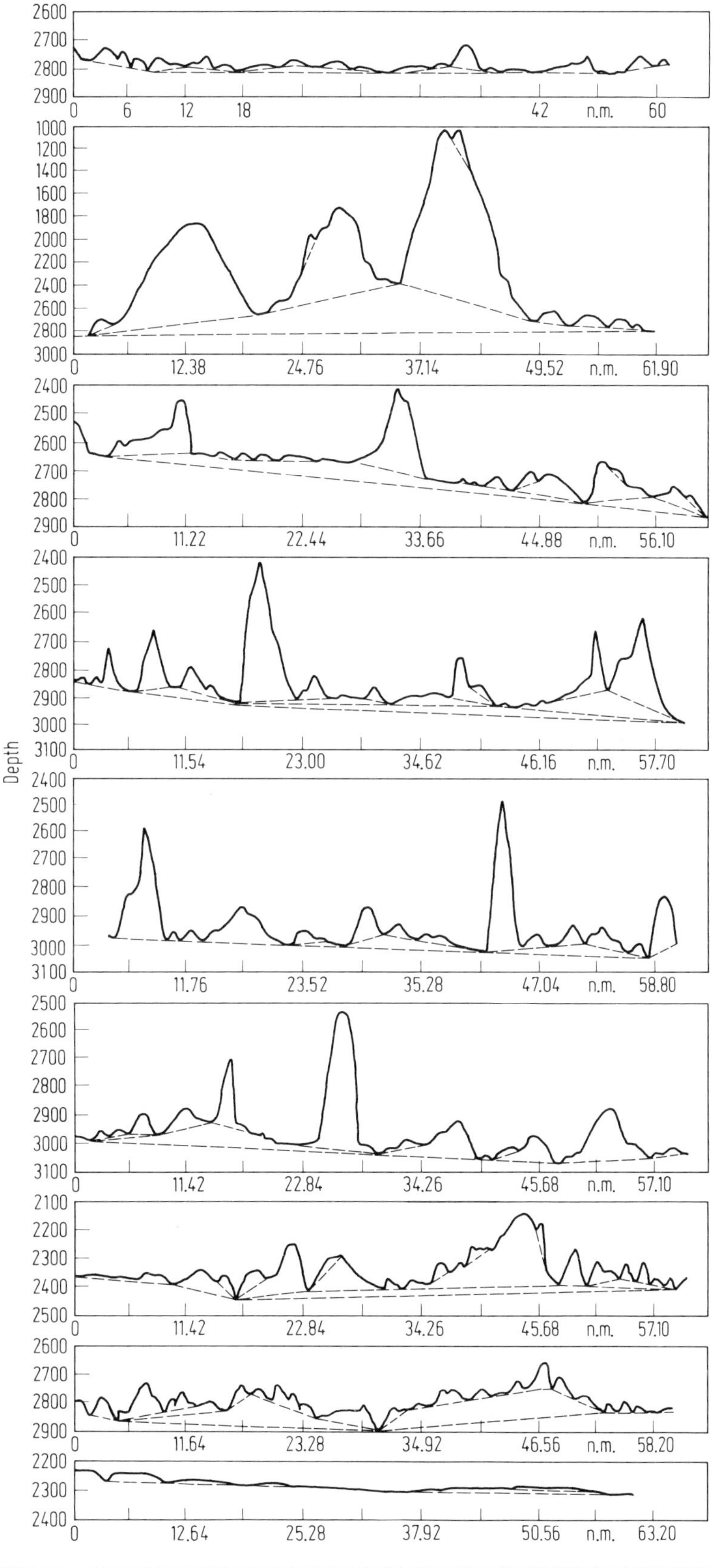
Depth
2600
2700
2800
2900
0 6 12 18 42 n.m. 60
1000
1200
1400
1600
1800
2000
2200
2400
2600
2800
3000
0 12.38 24.76 37.14 49.52 n.m. 61.90
2400
2500
2600
2700
2800
2900
0 11.22 22.44 33.66 44.88 n.m. 56.10
2400
2500
2600
2700
2800
2900
3000
3100
0 11.54 23.00 34.62 46.16 n.m. 57.70
2400
2500
2600
2700
2800
2900
3000
3100
0 11.76 23.52 35.28 47.04 n.m. 58.80
2500
2600
2700
2800
2900
3000
3100
0 11.42 22.84 34.26 45.68 n.m. 57.10
2100
2200
2300
2400
2500
0 11.42 22.84 34.26 45.68 n.m. 57.10
2600
2700
2800
2900
0 11.64 23.28 34.92 46.56 n.m. 58.20
2200
2300
2400
0 12.64 25.28 37.92 50.56 n.m. 63.20

Fig. 52

1.4.3.3 Submarine plateaus

Submarine plateau:

A comperatively flat-topped elevation of considerable extent across the summit and usually rising more than 200 m on at least one side from the surrounding regions. See Fig. 53. The importance of submarine plateaus in the sense of deviating ocean currents is evident.

Marginal plateau is a shelf-like feature which lies at greater depth than the continental shelf and is separated from the continental shelf by an incipient continental slope. The marginal plateau is characterized by absence of a seaward barrier or sill, a general depth of 200···2400 m, and a surface morphology very similar to that of the continental shelf. Marginal plateaus are closely related to continental borderlands, but while the latter are defined as lying seaward of the normal continental shelf, their surface is highly irregular [66 F].

Fig. 53 see p. 82

Oceanic plateaus are anomalously high parts of the sea floor, which comprise about 10% of the ocean floor. They are not at present parts of the continents, active volcanic arcs, or active spreading ridges. Some plateaus are probably submerged continental fragments that subsided. Other plateaus are of oceanic rather than continental origins.

In the mid-ocean area there have been identified: *aseismic ridges* which appear to consist of basic igneous oceanic crust with some lower-density volcanics, but only some of them may have formerly continental components, *oceanic rises*, the broad sea floor elevation, generally taken to be less precipitous and rugged than a ridge, true *mid-ocean plateaus*, to which Heezen has applied the genetic connotation *microcontinent*; these are isolated continental fragments.

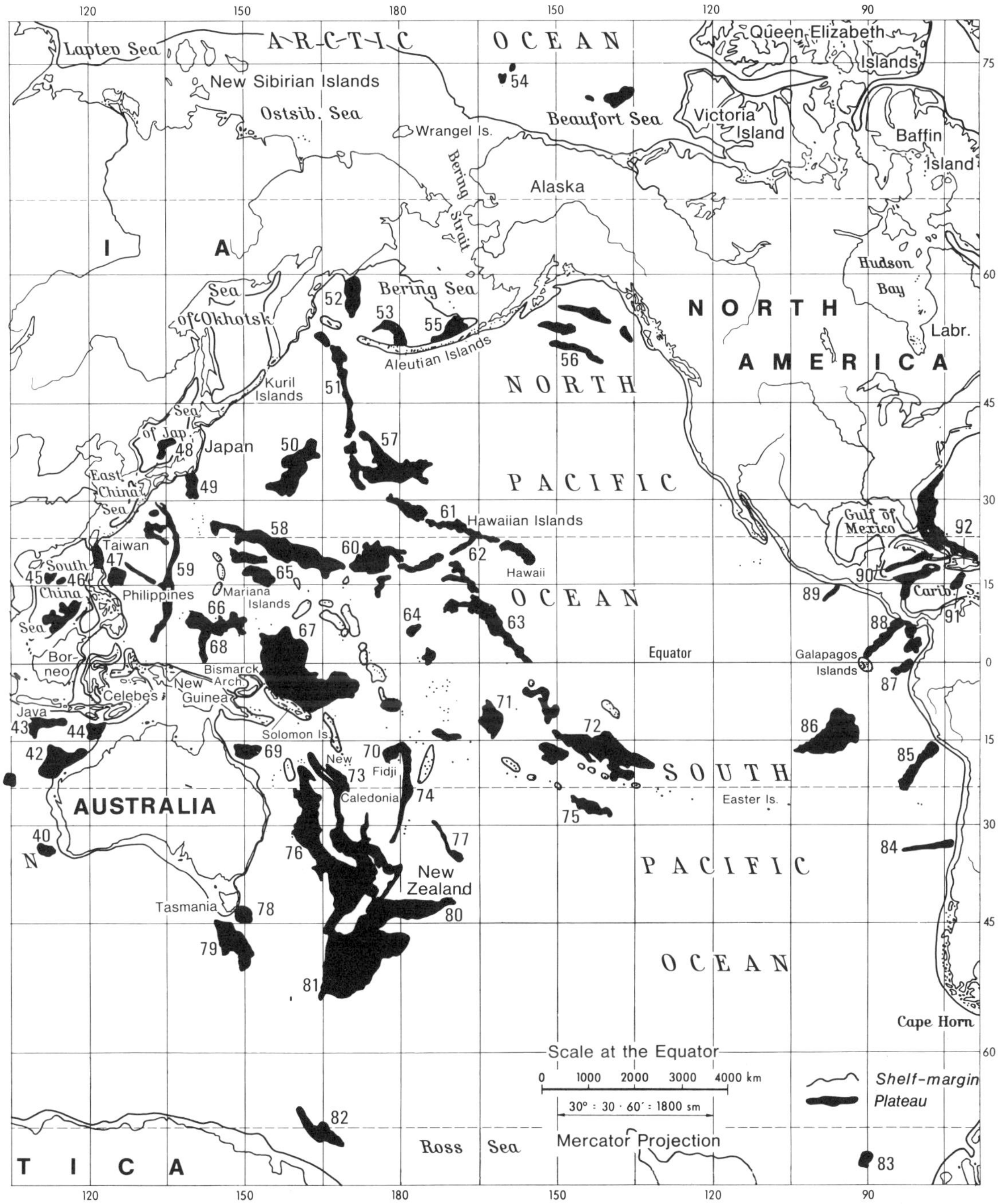

Gierloff-Emden

←

Fig. 53. Distribution of plateaus in the ocean (shaded areas). After [81 B 1] in [83 T].

Atlantic Ocean
1 Voering Plateau
2 Faroe Plateau
3 Rockall Plateau
4 Porcupine Bank
5 Orphan Knoll
6 Flemish Cap
7 Bermuda Rise
8 Galicia Bank
9 Madeira Plateau
10 East Canary Islands
11 Mallorca Balearic Rise
12 Medina Plateau
13 Eratosthenes Plateau
14 Aves Ridge
15 Barracuda Ridge
16 Barbados Ridge
17 Cape Verde Plateau
18 Curacao Ridge
19 Sao Paulo Plateau
20 Walvis Ridge
21 Rio Grande Rise
22 Cape Rise
23 Falkland Plateau
24 South Georgia Rise
25 North Scotia Ridge
26 South Scotia Ridge
27 Maud Rise
90 Cayman Trench
91 Beata Ridge
92 Blake Plateau

Indian Ocean
28 Thirty – East Spur
29 Grido Bank
30 Kerguelen – Gaussberg Ridge
31 Crozet Plateau
32 Agulhas Plateau
33 Mozambique Ridge
34 Madagascar Plateau
35 Mascarene Plateau
36 Seychelles Bank
37 Chagos Laccadive Ridge
38 Ninetyeast Ridge
39 Broken Ridge
40 Naturaliste Plateau
41 Wallaby Plateau
42 Exmouth Plateau
43 Roo Rise
44 Scott Plateau

Arctic Ocean
54 Chukchi Rise

Pacific Ocean
45 Paracel Isl.
46 Macclesfield Bank
47 Benham Seamount
48 Yamato Ridge
49 Izu Ridge
50 Shatsky Rise
51 Emperor Seamounts
52 Shirshov Ridge
53 Bowers Ridge
55 Umnak Plateau
56 Alaska Seamounts
57 Hess Rise
58 Markus Ridge
59 Palau – Kyushu Ridge
60 Mid Pacific Mts.
61 Hawaiian Ridge
62 Necker Ridge
63 Line Is. Ridge
64 Magellan Rise
65 Magellan Seamounts
66 Caroline Ridge
67 Ontong – Java Plateau
68 Eauripik Rise
69 Queensland Plateau
70 Fiji Plateau
71 Manihiki Plateau
72 Tuamotu Ridge
73 Norfolk Plateau
74 Lau Ridge
75 Tubuai Is. Ridge
76 Lord Howe Rise
77 Loisville Ridge
78 East Tasman Rise
79 South Tasman Rise
80 Chatham Rise
81 Campbell Plateau
82 Balleny Island
83 Peter Island
84 Juan Fernandez Ridge
85 Nazca Ridge
86 Galapagos Rise
87 Carnegie Ridge
88 Cocos Ridge
89 Tehuantepec Ridge

Certain areas described as "*plateaus*" from the mid-oceanic regions are preferably defined as "rises", etc., since precision depth recording shows that their surfaces are markedly irregular.

Essentially flat-topped platforms have probably a continental origin, although the mechanism of their subsidence is not known.

Significance of oceanic plateaus:

The migration of oceanic plates was verified by deep sea drilling and other methods. This could explain the migration of submarine plateaus with the plates to the continents. Some of these oceanic rises or plateaus are moving with the oceanic plates in which they are embedded, and eventually, they will be accreted to the surrounding continents [81 B 1].

If incorporated to the crust of continents, these plateaus are named "*terranes*" [83 J 3]. See Fig. 53. The scientific term is "tectono-stratigraphic terran".

1.4.3.4 Seamounts

Seamount:

An elevation rising 1000 m or more, and of limited extent across the summit. Seamounts were discovered during the expedition of the RV "Challenger", 1872···1874.

Seamounts are distributed throughout all the provinces of the deep ocean floor. They occur randomly scattered, but more typically, they are clustered in groups or linear rows. Slopes on the flanks of seamounts are typically in the 5···15° range, in the upper slope up to 30°. Seamounts rise mostly abruptly from the ocean floor; their bases have been buried by downsinking. Most are roughly conical in shape. The total number of seamounts in the Pacific Ocean has been estimated at 10000 [63 H 2]. Seamounts of the Atlantic Ocean, see Fig. 54.

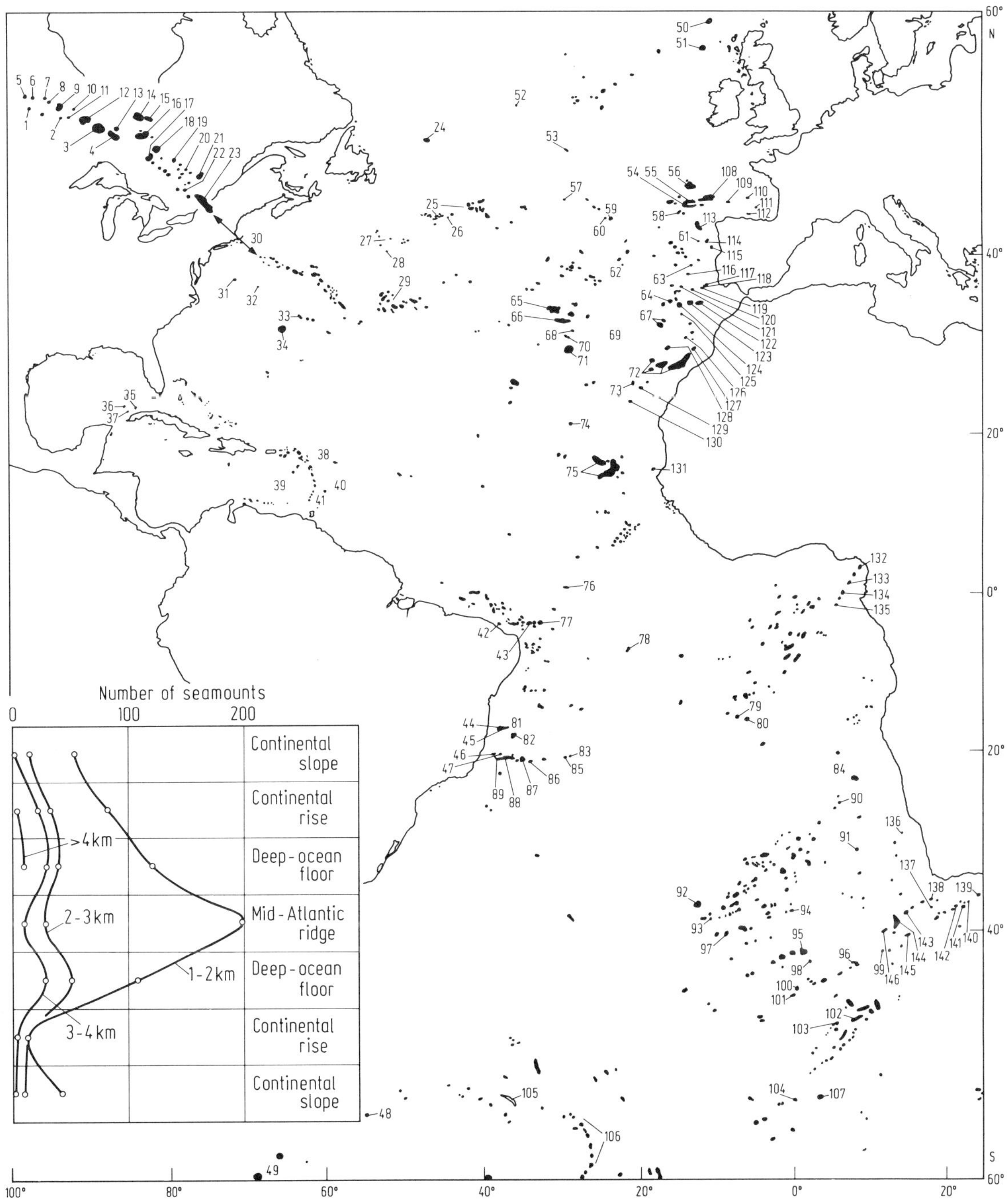

Fig. 54. Larger seamounts, banks, and islands in the Atlantic Ocean: position and statistical distribution. From [84 E]. The arrow at North America coast (upper left) shows the seamount group amplified, placed on the continental area.

1 Mytilus Smt.
2 Asterias Smt.
3 Atlantis II Smt.
4 Gosnold Smt.
5 Bear Smt.
6 Physalia Smt.
7 Retriever Smt.
8 Picket Smt.
9 Balanus Smt.
10 Kiwi Smt.
11 Panulirus Smt.
12 Kelvin Smt.
13 Sheldrake Smt.
14 Gregg Smt.
15 San Pablo Smt.
16 Manning Smt.
17 Vogel Smt.
18 Rehoboth Smt.
19 Allegheny Smt.
20 Michael Smt.
21 Gerda Smt.

(continued)

Fig. 54 caption (continued)

22 Gillis Smt.
23 Nashville Smt.
24 Orphan Kn.
25 Newfoundland Smts.
26 Shredder Smt.
27 Fogo Smts.
28 Spur Smt.
29 Corner Smts.
30 New England Smt. Chain
31 Knauss Kn.
32 Caryn Kn.
33 Muir Smt.
34 Bermuda I.
35 Jordan Kn.
36 Catoche Kn.
37 PDR Kn.
38 Leeward Is.
39 Aves I.
40 Barbados I.
41 Windward Is.
42 Ceara Plateau
43 Rocky Atoll
44 Sulfur Bk.
45 Minerva Bk.
46 Champlain Bk.
47 Eclaireur Bk.
48 Rhine Bk.
49 Sars Smt.
50 Rosemary Bk.
51 Anton-Dohrn Smt.
52 Minia Smt.
53 Faraday Bk.
54 North Charcot Smt.
55 South Charcot Smt.
56 Amorican Smt.
57 Bald Smt.
58 Coruna Smt.
59 Antialtair Smt.
60 Crumb Smt.
61 Vasco da Gama Smt.
62 Azores Is.
63 Tore Smt.
64 Lion Smt.
65 Atlantis Smt.
66 Plato Smt.
67 Madeira Is.
68 Irving Smt.
69 Cruiser Smt.
70 Hyeres Smt.
71 Great Meteor Smt.
72 Canary Is.
73 Papp Smt.
74 Kane Smt.
75 Cape Verde Is.
76 St. Peter and St. Paul Rocks
77 Fernando de Noronha I.
78 Ascension I.
79 Bonaparte Smt.
80 St. Helena I.
81 Rogers Bk.
82 Hotspur Bk.
83 Martim Vas I.
84 Ewing Smt.
85 Trinidade I.
86 Dogaressa Bk.
87 Davis Bk.
88 Montague Bk.
89 Vitoria Bk.
90 Valdivia Smt.
91 Vema Smt.
92 Tristan da Cunha I.
93 Crawford Smt.
94 Zapiola Smt.
95 Discovery Smt.
96 Andre Smt.
97 McNish Smt.
98 Shannon Smt.
99 De Jager Smt.
100 Herdman Smt.
101 Schwabenland Smt.
102 Meteor Smt.
103 Alfred Mertz Smt.
104 Spiess Ridge
105 South Georgia I.
106 South Sandwich I.
107 Bouvet I.
108 Biscay Smt.
109 Cantabria Kn.
110 Gascony Kn.
111 Jovellanos Smt.
112 Le Danois Bk.
113 Galicia Bk.
114 Vigo Smt.
115 Oporto Smt.
116 Ashton Smt.
117 Ormande Bk.
118 Gorringe Ridge
119 Gettysberg Bk.
120 Hirondelle Bk.
121 Josephine Smt.
122 Coral Patch Smt.
123 Ampere Smt.
124 Unicorn Smt.
125 Seine Bk.
126 Dacia Smt.
127 Conception Bk.
128 Salvage I.
129 Echo Bk.
130 Tropical Bk.
131 Cayar Smt.
132 Fernando Poo I.
133 Principe I.
134 São Tome I.
135 Annobon I.
136 Tripp Smt.
137 Argentina Smt.
138 Protea Smt.
139 Davie Smt.
140 Shackleton Smt.
141 Natal Smt.
142 Mallory Smt.
143 Erica Smt.
144 Schmitt Ott Smt.
145 Richardson Smt.
146 Panzarini Smt.

A seamount chain is a series of seamounts in a line, rising independently from an otherwise relatively flat sea floor. (See Fig. 58a, b, Hawaiian – Emperor seamount group). [82 S 1]. In certain linear groups, the seamounts coalesce so that the individual summits are aligned along a ridge or rise, termed a seamount range. A seamount group is a group or cluster of seamounts without definite alignment [64 M 1, 77 H 3]. Fig. 58.

A seamount, or knoll, of conical shape is interpreted as being of volcanic origin. Most of the seamounts would be included in this category, and it has been suggested that the term "seamount" may be dropped in favor of volcano. See Fig. 56. The terms "volcanic hill" and "volcanic knobs" suggest a volcanic origin for smaller topographic features.

Guyot (tablemount):

A seamount having a comparatively smooth, flat top. Tablemounts have been defined as having a minimum depth greater than 200 m (<200 m = pedestal). They occur in all oceans, but, by far, the largest number occurs in the Pacific Ocean where minimum depth generally ranges between 1000 and 2000 m. Their general shape is that of a volcano whose top had once surmounted the sea level and then been truncated. The "flat" top is usually not flat, but slopes gently outward from one or more crests near the center [63 M]. In tropical areas coral often grows, forming reefs around seamounts. After the seamount subsides, only the coral atoll remains. After deep submergence they may be flat topped, or peaks with surrounding terraces, sunken islands of the Mid-Pacific mountains [46 H, 56 H, 64 M 1]. See Fig. 57.

Moat:

An annular depression that may not be continuous located at the base of many seamounts, islands, and other isolated elevations.

Moat and arch:

Many islands, island groups, seamounts, tablemounts, or banks situated on the ocean basin floor are surrounded by a moat, which is in turn encircled by a broad swell or arch. Compare Fig. 22.

1.4.3.4.1 Seamounts and volcanic arcs and earthquake zones

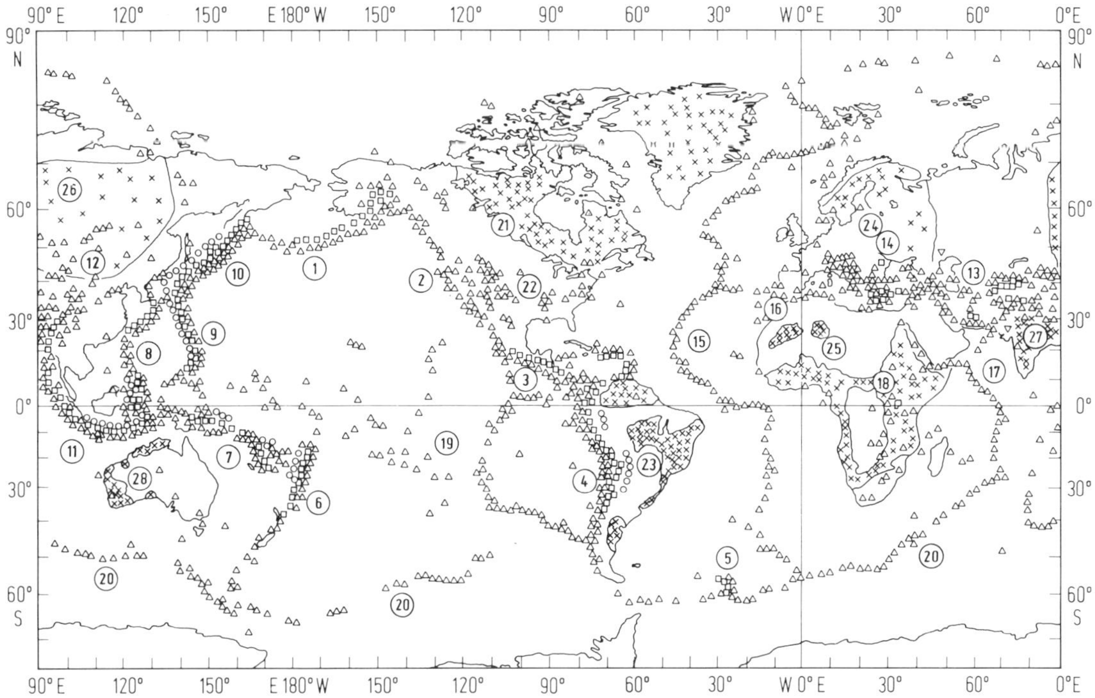

Fig. 55. Earthquake regions (compiled by [84 S2]. Triangles: $h_0 = 0 \cdots 70$ km; squares: $h_0 = 71 \cdots 300$ km; circles: $h_0 = 301 \cdots 800$ km; crosses: old shields; time interval: 1961···1969, strongly schematized. Compare Figs. 46, 54, 57.

A. Circumpacific belt
1. Aleutians, Alaska
2. Western Canada and USA
3. Central America
4. South America
5. Southern Antilles
6. New Zealand, Kermadec-, Tonga- and Fiji-Islands
7. Loyalty Islands, New Guinea
8. Celebes, Philippines, Taiwan, Ryu-Kyu-Islands
9. Carolines, Mariana Islands
10. Japan, Kamchatka

B. Eurasiatic belt
11. Indonesia, Burma
12. China, Mongolia, Southern Siberia
13. Iran, Pakistan, Central Asia
14. Mediterranean

C. Oceanic ridges- rift zones
15. Midatlantic ridge, (Eastern Siberia)
16. Azores-Gibraltar-ridge, (Northern Africa)
17. Indian Cordillera, Red Sea
18. Eastafrican rift system
19. Pacific ridges and rises
20. Circumantarctic ridge

D. Shields and other conteninental areas
21. Canada (without west coast)
22. USA (without west coast)
23. Eastern South America
24. Northern and Eastern Europe
25. Africa (without Northern Africa and the eastafrican rift system)
26. Siberia (without Southern Siberia)
27. India (without Himalaya mountain range)
28. Australia

1.4.3.4.2 Topography of a seamount, surveyed with the Sea-Beam System: the "Sonne Seamount" south of Hawaii

In 1981, the Sea-Beam System was installed on the German (FRG) research vessel "Sonne". It is capable of producing real-time contour strip charts of the ocean floor. The system was tried out in the central Pacific Ocean during the cruise SO-21 in November, 1981. For the first time it was possible to record flat deep-sea areas, and to make an almost complete survey of a bigger seamount, situated 550 km SSW of the main island Hawaii, rising from a depth of 5500 to 2180 m below sea level. The position of the highest peak is 15°07′ N, 156°40.3′ W [81 S, 82 U, 82 S 2]. See Figs. 56a, b. In 1984 the Sea-Beam System of the German (FRG) RV "Polarstern" was used to survey the Meteor Bank in the Atlantic Ocean (29°24′ N, 28°58′ W) [84 U].

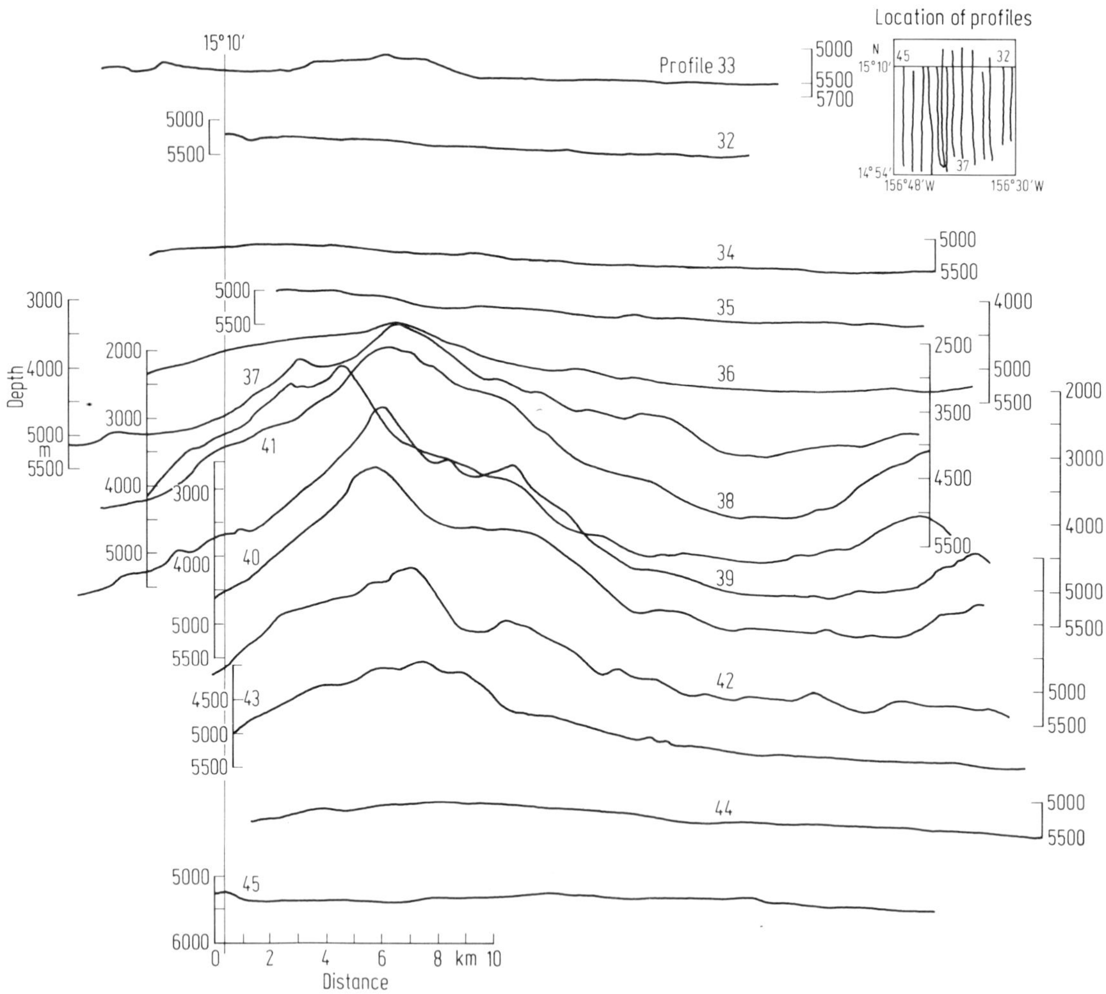

Fig. 56a. Series of sounding profiles above the Sonne Seamount evaluated from the two-dimensional sea-beam plottings (exaggeration 1 : 5). N–S-sections seen from the west. Depths in [m]. Pacific Ocean surveying: "Sonne" (1981) [82 U].

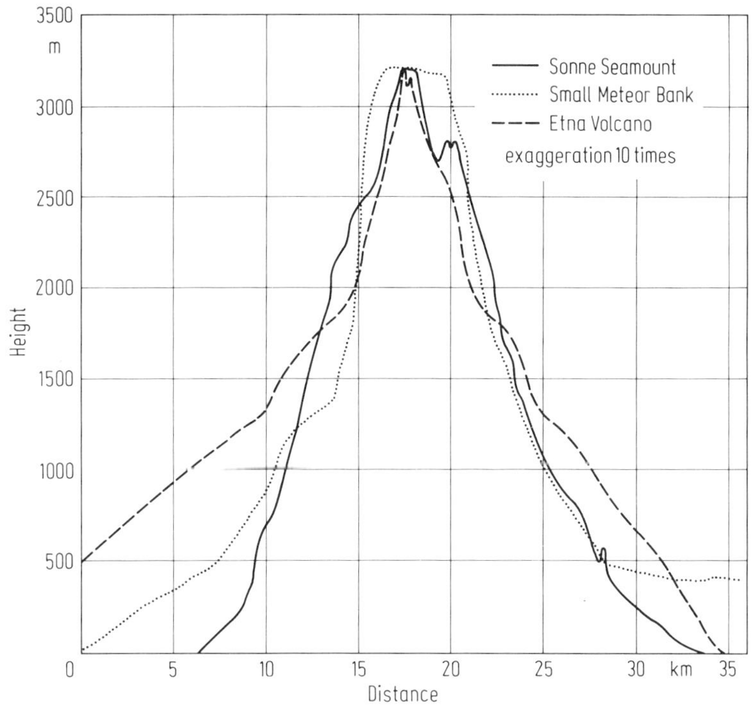

Fig. 56b. W–E-profile of the Sonne Seamount in comparison with corresponding profiles of Mt. Etna, and the "Small Meteor Bank", North Atlantic Ocean [82 U].

1.4.3.4.3 Seamounts in the Pacific Ocean and hot spots

Pacific island chains can be interpreted as tracks formed by the movement of the sea floor over stationary *hot spots*. The theory of this mechanism was created by [72 M and 73 W]. The tracks of the hot spots represent a plate external system, a so-called average hot spot frame of reference: The Hawaiian Islands grow older toward the northwest, beginning with Hawaii itself. Two other chains parallel to the Hawaiian Islands display a similar pattern and ages. They are the Austral group, which begins with the MacDonald Seamount, and the Tuamotu group, which begins with Pitcairn Island. All three chains could be generated by the same clockwise rotation of the Pacific Plate. The age sequence of the Hawaiian Islands is continued in the Emperor Seamounts, which strike northward from a bend formed $40 \cdot 10^6$ years ago. The change in direction implies that until then the rotation of the plate was centered on a different pole. Radiometric dating of volcanoes along the chain shows age progression. See Fig. 59. The average rate of migration along the Hawaiian chain (10 cm/year) compares to only about 6 cm/year along the Emperor chain. There are 18 major volcanoes per 1000 km along the Hawaiian ridge [82 K, 72 J, 80 F 2]. Seamounts also extend to the north from the Austral and Tuamotu groups, but there is no convincing evidence of their ages forming a linear sequence. See Figs. 56···59, compare ch. 1.6.1.

"A postulated "hot spot" beneath the Galapagos platform is suggested as the cause of: (1) decreased seismicity along the spreading center for a 400 km E–W distance from the islands; (2) distinctive petro-chemistry of tholeiites from the islands and adjacent oceanic crust generated by the Galapagos Rise; (3) high-amplitude magnetic anomalies in a 1000 km E–W band including and just north of the Galapagos platform; and (4) morphologic shape and the regionally elevated sea floor of the Galapagos Rise as it approaches the insular platform". From [76 J].

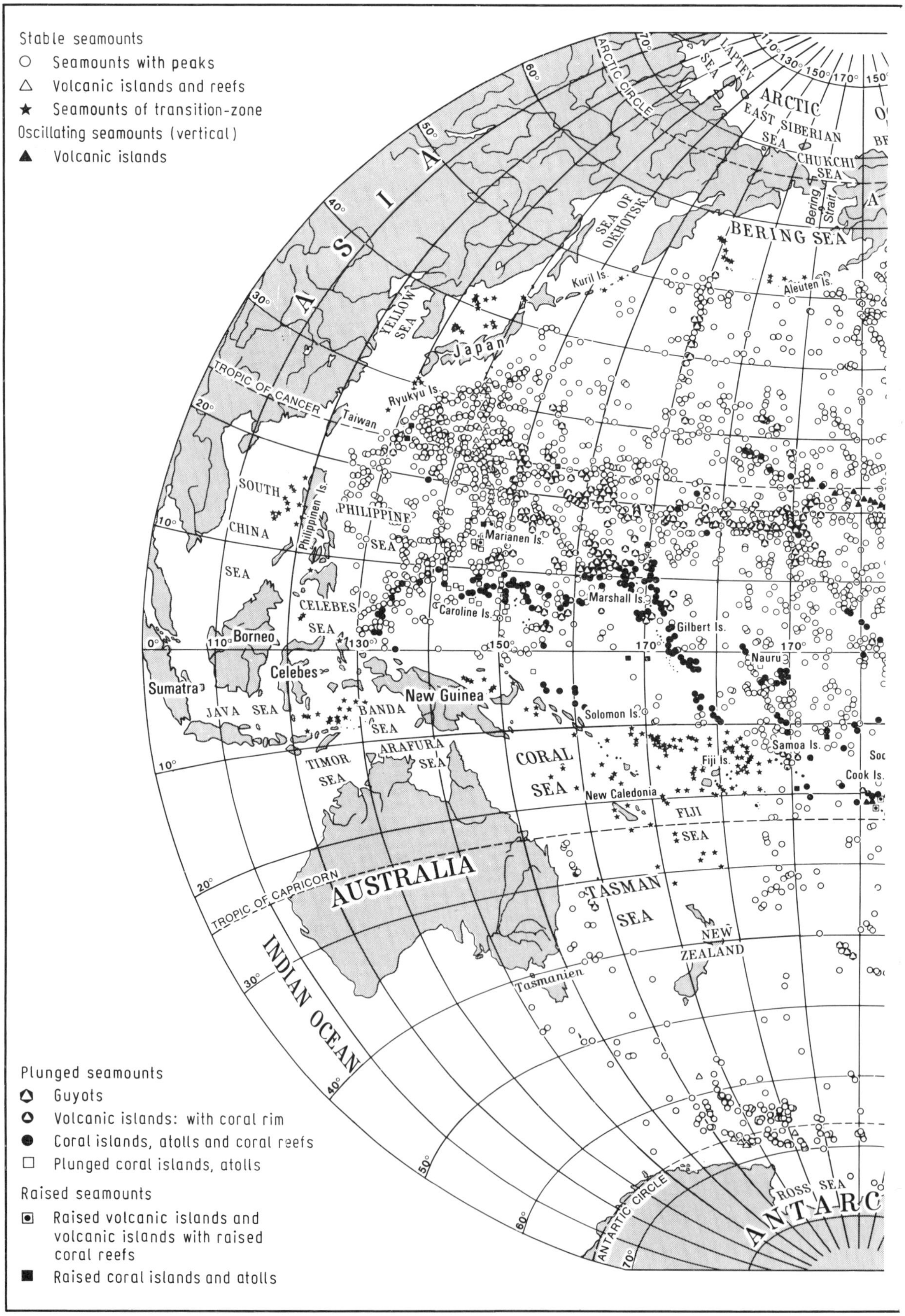
Stable seamounts
Seamounts with peaks
Volcanic islands and reefs
Seamounts of transition-zone
Oscillating seamounts (vertical)
Volcanic islands
Plunged seamounts
Guyots
Volcanic islands: with coral rim
Coral islands, atolls and coral reefs
Plunged coral islands, atolls
Raised seamounts
Raised volcanic islands and volcanic islands with raised coral reefs
Raised coral islands and atolls
ASIA
ARCTIC CIRCLE
LAPTEV SEA
ARCTIC
EAST SIBERIAN SEA
CHUKCHI SEA
Bering Strait
BERING SEA
SEA OF OKHOTSK
Kuril Is.
Aleuten Is.
YELLOW SEA
Japan
TROPIC OF CANCER
Ryukyu Is.
Taiwan
SOUTH CHINA SEA
Philippinen Is.
PHILIPPINE SEA
Marianen Is.
CELEBES SEA
Caroline Is.
Marshall Is.
Gilbert Is.
Nauru
Borneo
Celebes
Sumatra
JAVA SEA
BANDA SEA
New Guinea
Solomon Is.
Samoa Is.
Fiji Is.
Cook Is.
TIMOR SEA
ARAFURA SEA
CORAL SEA
New Caledonia
FIJI SEA
AUSTRALIA
TROPIC OF CAPRICORN
TASMAN SEA
NEW ZEALAND
Tasmanien
INDIAN OCEAN
ANTARCTIC CIRCLE
ROSS SEA
ANTARC

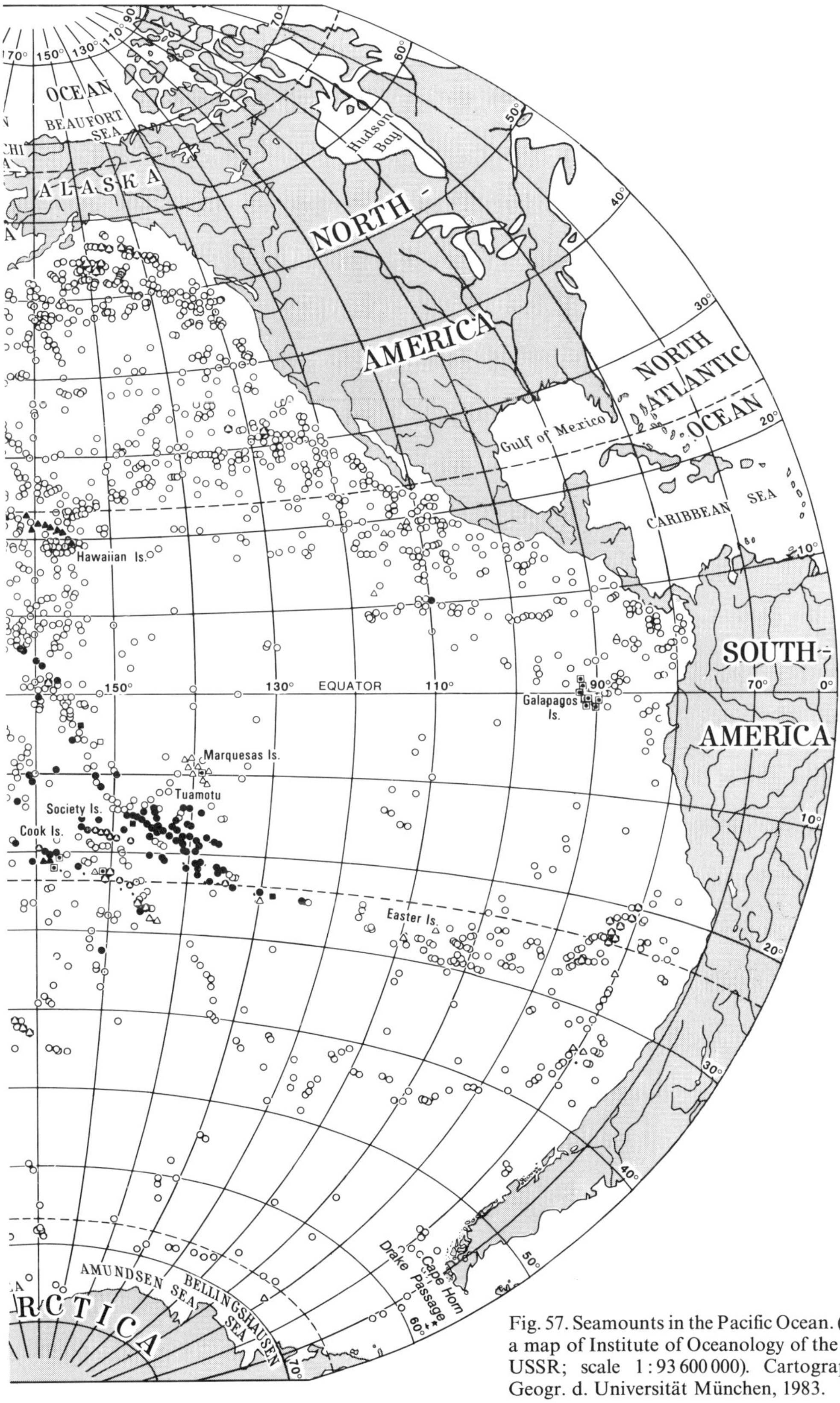

Fig. 57. Seamounts in the Pacific Ocean. (State 1968; after a map of Institute of Oceanology of the Academy of the USSR; scale 1:93600000). Cartography: Institut f. Geogr. d. Universität München, 1983.

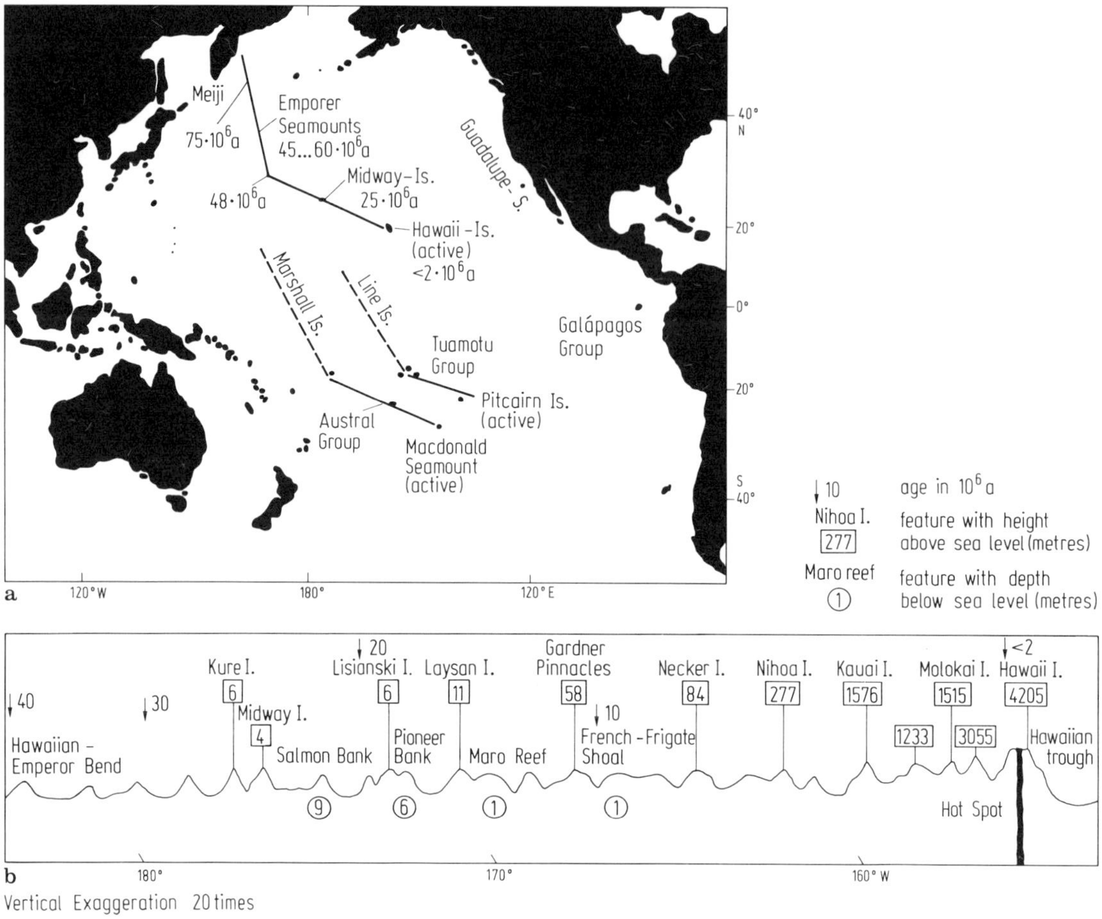

Fig. 58a. Geometric pattern of volcanic islands and seamounts in the Pacific Ocean. Long linear chains are significant topographic pattern. Age sequence is indicated for the Hawaiian Islands, Midway Islands, and the Emperor seamounts. The age of the bend of the seamount chains is obviously of critical importance for, if the hot spots were basically stationary, the bends record a change in the direction of plate motion. A date of 42···44 m/year (late Eocene) for the bend in the Hawaiian-Emperor seamount chains was estimated. For the Atlantic Ocean's hot spots, see [76 B, 85 V].

Fig. 58b. Volcanic island and sea mountain chain from Hawaii to Hawaiian-Emperor bend. The Hawaiian islands are part of an island chain formed by volcanoes moving with the crustal plate away from a hot spot. Radioisotope dating of the lava confirms the increase in age along the chain, with the active crater of Kilauea at the eastern end, and Midway Islands along the chain with lavas of $25 \cdot 10^6$ a (Tertiary). The seamounts at the north-west end of the chain have lavas of $100 \cdot 10^6$ a (Cretaceous). Compare Figs. 72, 74.

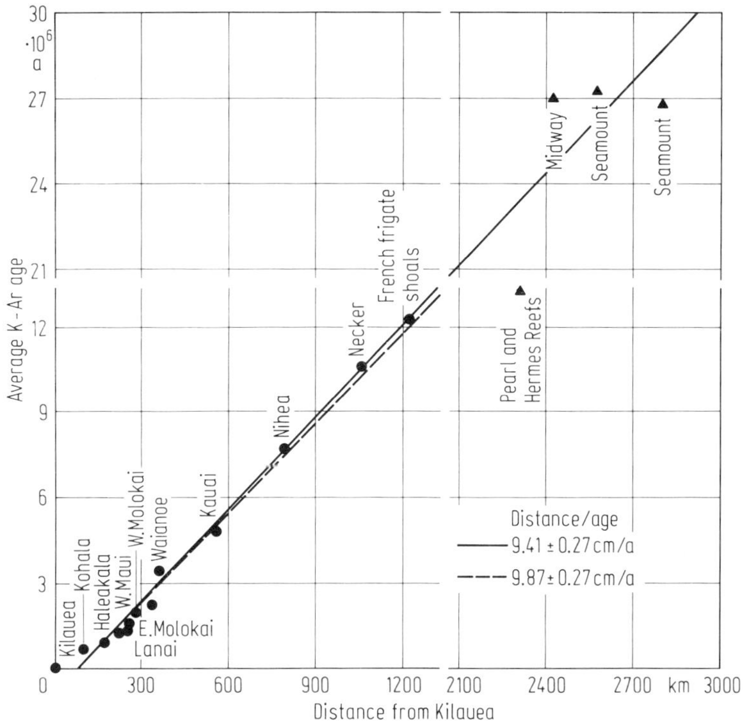

Fig. 59. Plot of average K–Ar age of shield-building volcanism for volcanoes of the Hawaiian Island chain vs. distance from Kilauea Circles indicate data from tholeiitic basalts of the main shield; triangles show results from alkalic lavas that postdate the main shield-building phase [80 M]. The full line and the dashed line are fits to the data for the two distance-age ratios indicated. Compare Fig. 72.

1.4.4 Major topographic features of the Mid-Oceanic Ridge complex

Mid-Oceanic Ridge (Rise) (definitions after [77 H 3], compare Fig. 38a):

The median elevation or swell of the ocean floor; the Mid-Oceanic Ridge or Rise is a broad cordillera or mountain range which rises to its highest elevation at the median crest or axis and slopes away on either flank. The local relief consists of ridge and valley terrain paralleling the axis, cross-cut by fracture zones; local relief ranges from 100···2000 m. The ridge crest is generally 1000···3000 m above the adjacent ocean-basin floor, while the width generally exceeds 1000 km. The Mid-Ocean Ridge lies nearly exactly in a median position only in the Atlantic Ocean. This ridge is essentially continuous through the Arctic, Atlantic, Indian, and Pacific Oceans, surrounding Antarctica, for a total distance of more than 65 000 km. It also has limbs which extend toward and into the continental blocks [66 H, 70 H 2]. This specific feature was discovered by B. Heezen [58 H], whereas ridges in the oceans were known before [42 W]. Figs. 33···36 and 60, 61, 62, 71.

Axis of Mid-Oceanic Ridge:

The center line of the Mid-Oceanic Ridge, usually defined as the central rift valley, or at the crest in regions where no rift valley occurs, about which the ridge is generally symmetrical in its regional morphology [77 H 2, 68 H 1]. Figs. 32, 38a, 38b, 61.

Crest of Mid-Oceanic Ridge (Rise):

The provinces lying at the axis of the Mid-Oceanic Ridge, including the median rift, the rift mountains, and the high fractured plateau, where present. This zone is characterized by steep slopes (5···15°), virtual absence of sediment cover, and in certain parts of the oceans, high relief. Depth to the crest varies from one area to another in the oceans, but the crest is always regionally shallower than the flank provinces on either side of the crest. In the North Atlantic, the crest is rugged, 100···300 km wide, and may be separated from the flank provinces by one or more scarps. In the South Pacific Ocean, the crest is not separated from the flank provinces by a distinct discontinuity as in the Atlantic Ocean. The axis or center line of the ridge, centrally situated along the crest, is seismically active and is frequently the locus of a high-amplitude magnetic anomaly [70 H 2].

Rift mountains:

One of the crest provinces of the Mid-Oceanic Ridge in the Atlantic, the rift mountains consist of a belt of rugged ridge-and-valley terrain lying on either side of the central rift valley. Fig. 61 [83 C 2].

Concerning the Mediterranean Basins, compare [84 R, 84 D 1].

Rift valley:

The central valley or "crack" in the earth's crust, lying at the axis of the Mid-Oceanic Ridge along most of its extent. It consists of a long, linear valley paralleling the seismically active axis of the Mid-Oceanic Ridge. Fig. 61. The rift valley of the North-Atlantic Ocean ridge crosses the island of Iceland where the rift is developed on the surface [84 E].

High fractured plateau:

One of the crest provinces of the Mid-Oceanic Ridge in the Atlantic Ocean, the high fractured plateau is not identificable in all parts of the world. Where identified, it lies on either side of the rift mountains, being the outermost of the crest provinces. It lies at a comparable level to the rift mountains, though with a lower local relief (400···800 m).

Flank of Mid-Oceanic Ridge (Rise):

That part of the Mid-Oceanic Ridge which lies on either side of its crest, between the crest and the abyssal plains and hills on the ocean-basin floor. The two flanks are generally symmetrical. They are characterized by moderate to rugged basement relief (100···1000 m), altered by variable sediment cover. The sediment cover generally thickens farther out on the flanks near the abyssal hills and abyssal plains. The basement surface generally takes the form of ridge-and-valley terrain paralleling the Mid-Oceanic Ridge axis, interrupted at intervals by fracture zones. The spacing of ridges and valleys is typically 10···60 km. Slopes of the exposed basement relief range between 5···15°, and higher. The sediment cover generally is characterized by slopes of 0···5°. The slopes along fracture zones and scarps may exceed 15°, ranging up to 30°, or more [70 H 2]. Figs. 33···36.

Classification of topography and geology of the Atlantic Ocean with oceanic ridges and margins. Fig. 60 Geologic-topographic development of the Atlantic Ocean (time scale from 1···5):

1. Late Cenozoic: terrigenous sediments developing a continental rise wedge
2. Cenozoic: shelf edge break shifted 300 km to the west, subsidence slowed
3. Early Cretaceous: sedimentation over reef dams, 300 km to the west
4. Cretaceous: rifting of the Atlantic Ocean, sea floor spreading
5. Late Jurassic: reef forming over the margin (shallow sea), rifting, dams.

The kinematics and geometry of the features of the Mid-Oceanic Ridge complex are presented in: [73 L 2, 83 O, 85 J 2, 85 B 2]. Compare Fig. 76.

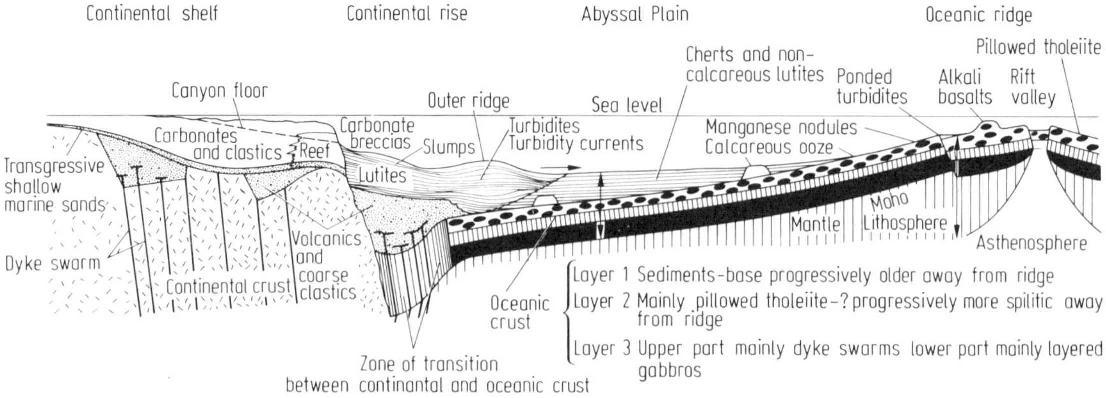

Fig. 60. Geological cross section for the western North Atlantic Ocean showing relationship between continental crust, oceanic crust, and sediment facies. Model not to scale. The extension and the displacement of the ocean floor, created by the sea-floor spreading of the mid-oceanic ridge of the Atlantic Ocean, have caused the tectonic structure of the eastern continental margin of the American continent and the westward displacement of this continent. Compare Fig. 72. The basaltic volcanic suite at the spreading ridge consists of rugged pillow lava on top and Mid-Oceanic Ridge Basalt (MORB) flows and layers. After [70 D 2].

1.4.4.1 The rift valley of the Mid-Oceanic Ridge

The central rift valley contains another, even narrower, inner-rift valley a few kilometers wide. The youngest volcanic rocks of the Mid-Oceanic Ridge have been observed and collected from this region, including closely associated central volcanic hills. On a small scale, fissures 1···5 m wide and up to 10 m deep extend parallel the axis of the rift for a few hundred meters. Basalt is the important rock type; gabbro, serpentinite, and other igneous rocks are also present. The floor of the "graben" is marked by active fissures parallel to the bounding graben faults. Tensional tectonics prevail. See Fig. 61. Some rifts show a median ridge in the valley such as the rift valley of the Red Sea.

Theories involving extension of the Mid-Oceanic Ridge require an explanation of ridge morphology in terms of relative migration of the oceanic crust in both directions away from the rift valley, with the accompanying extensional tectonic features. Addition of new crust via intrusion would be taking place primarily through the mid-oceanic rift, and possibly secondarily through fractures on either side of the rift. The surface expression of the spreadings is the rift topography [59 H, 62 H, 62 D, 72 B 1, 76 L, 76 W 3, 84 B 4]. See Fig. 61.

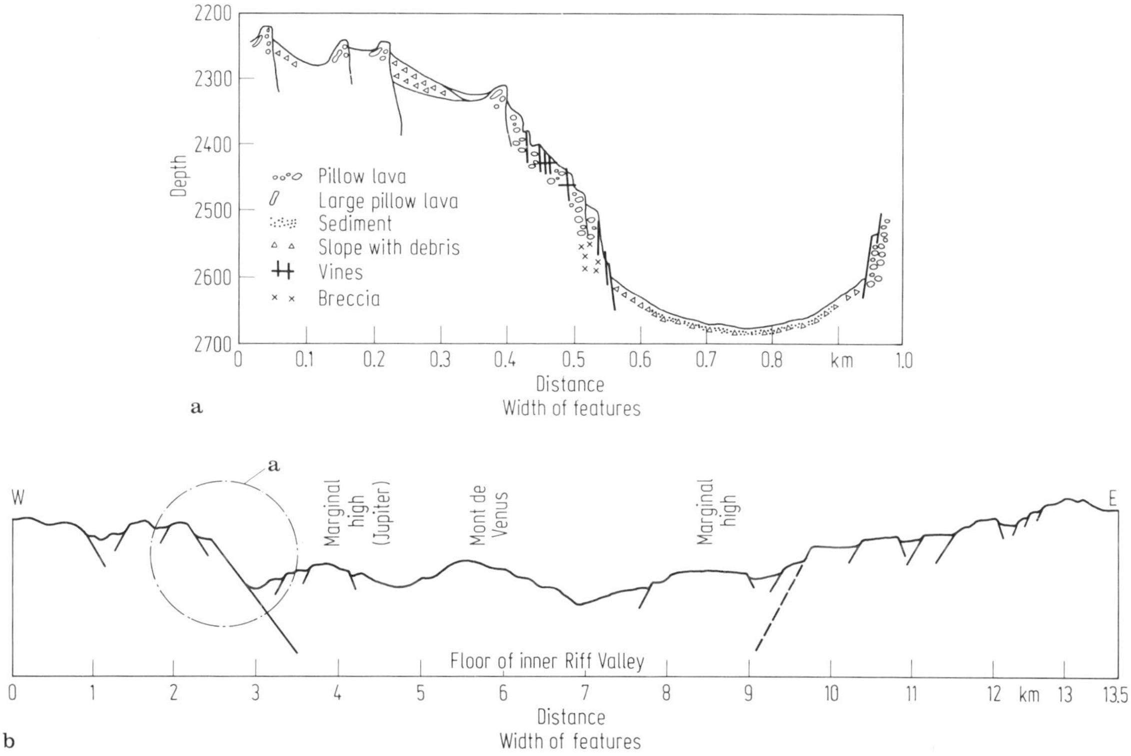

Fig. 61. Morphology of a rift valley of the Mid-Atlantic Ridge at 36°50′ N in the North-Atlantic Ocean, south of Azores Islands. The profiles are not exaggerated. Profile (a) from bathyscaph "Archimède", and diving saucer "Cyana". Profile (b) from "Entrecasteaux" (sonar), by FAMOUS (French American Undersea Study) (1973/74). CNEXO, France, branch Bretagne, "Arcyana" working group. Compare [75 R 2] and Table 17b.

Topography in an active rift system is produced both by volcanism along rifts and by normal faulting. See Photographic atlas of the Mid-Atlantic Ridge rift valley [77 B]. A sea-beam sounding of the East-Pacific Rise has presented remarkable small dome structures at the crest of the rift valley. (Hékinian and Ballard [75 A, 81 B 2, 82 B, 84 H 1].

Thermal springs on rift valleys

Submarine thermal springs are found on the bottom of rift valleys. [84 S 1, 82 F, 83 R].

Volumes of water expelled from the hot springs range from zero to probably several liters/m of length of spring deposit/s, but they are not well established from submersible observations. Another estimate, derived from summed differences between theoretical heat flow from sea-floor spreading and from measured heat flow, is 130···260 km^3/year. The volume of water expelled averages 3000···6000 m^3 per m length of spreading belt per year, 1···2 cm^3/s [84 E]. Observations from surface ships and submersibles suggest that possibly 90% of the total length of spreading belt is presently venting less than 10% of the water, so that active lengths probably average 10···20 cm^3/s, with the most active sections venting considerably faster. This range of water discharge accords with the observed range of zero to abundant metallic deposits along spreading belts of the world ocean floor [84 E, 84 R 2].

Thermal waters, hot brines and therefore sedimented metalliferous muds were found on the bottom of the Red Sea [69 R].

Submarine thermal springs on the Galapagos Rift were found during exploration with the submersible "Alvin" in 1977, which spill in form of plumes from vents, which are located predominantly along the axial ridge within the rift valley in 2500 m depth. These vents are of the scale of 1···30 m height. Similar investigations were made in 1981 with the submersibel "Cyana" and sea-beam of the RV "Charcot" in the East Pacific Rise. [79 C, 84 H 1, 84 R 2].

Table 19a. Individual Fault-block dimensions. Data have been corrected to a line perpendicular to contours [82 S 5]. The division of fault blocks into fault zones and terraces has a structural basis. [75 A, 79 T, 82 F].

	No. of Faults	Vertical Displacement		Fault Spacing		Fault Scarp Dip			Block Top Dip		
		Range	Avg.	Range	Avg.	Range	Avg. In	Avg. Out	Range	Avg. In	Avg. Out
		m				deg					
Fault zones	50	1···76	10.6	6···113	29.6	15···90	53	–	0···63	8.5	19
Terraces	21	1···19	–	6···182	48.0	7···88	46	40	0···27	5.0	–
Fault zones and terraces	74	1···76	9.5	5···182	37.0	7···90	52	47	0···63	7.5	13

1.4.4.2 Fracture zones

Fracture zone:

An extensive linear zone of unusually irregular topography of the sea floor, characterized by features, such as volcanoes, seamounts, steep-sided, or steep asymmetric ridges, troughs, or escarpments. The fracture zones were explained by Dietz and Menard [53 D, 70 M 2]. These fractur zones are caused by slide-strike motion, and then called transform faults. See Figs. 61, 62 and 73.

Fracture zones cut across the Mid-Oceanic Ridge (Rise) usually within 10 or 20 degrees of the normal to the axis, and frequently extend for long distance (500···1000 km) down the flanks of the ridge. In some cases they can be traced, either at the surface or in the subsurface, from the Mid-Oceanic Ridge (Rise) across the ocean-basin floor to the continental margins. They serve as conduits for the flow of bottom water. They generally consist of linear ridges and valleys. Frequently, the ridges are substantially higher than the surrounding ridge-and-valley terrain, and the primary valleys substantially deeper. In many areas the fracture zones separate areas of the basement which lie at different levels. Some fracture zones show median ridges such as in the Charlie Gibbs Fracture Zone, 25° N 18° W in the North-Atlantic Ocean. Total relief may range from extremely high (4000 m or more) to moderate (100 m). The crest of the Mid-Oceanic Ridge is frequently offset along fracture zones, and that part of fracture zones lying between offset limbs of the axis is generally seismically active (63 H 2). Fig. 62.

Fracture zones of the eastern North Pacific Ocean and of the South Pacific Ocean extend to 4000 km length total, changing the bearing of several segments. Maps and statistics of topographic parameters see [70 M 2].

Recent charts of the North Atlantic floor use contours rather than hatchures to present idealized topography.

These contours permit the drawing of fracture zones and short ridges parallel to the rift valley. The short ridges depict former side scarps of the rift valley left behind as new ocean floor formed within the active spreading belt. Judging from the contours, these short ridges are spaced an average of about 25 km apart, in good agreement with their spacings where actual detailed surveys are available, after [84 E, 84 B 4].

Remaining are areas of disagreement concerning the physiographic chart in reference to the contour line chart (in Figs. 32, 33, and 62). The dotted trends of the fracture zones of Fig. 62 as superimposed from other sources to this figure are showing jumps of several fracture zones diagonally across the short ridges.

The main problem for the new chart is that the user must consider these conceptual rather than factual for such third-order physiographic features as fracture zones.

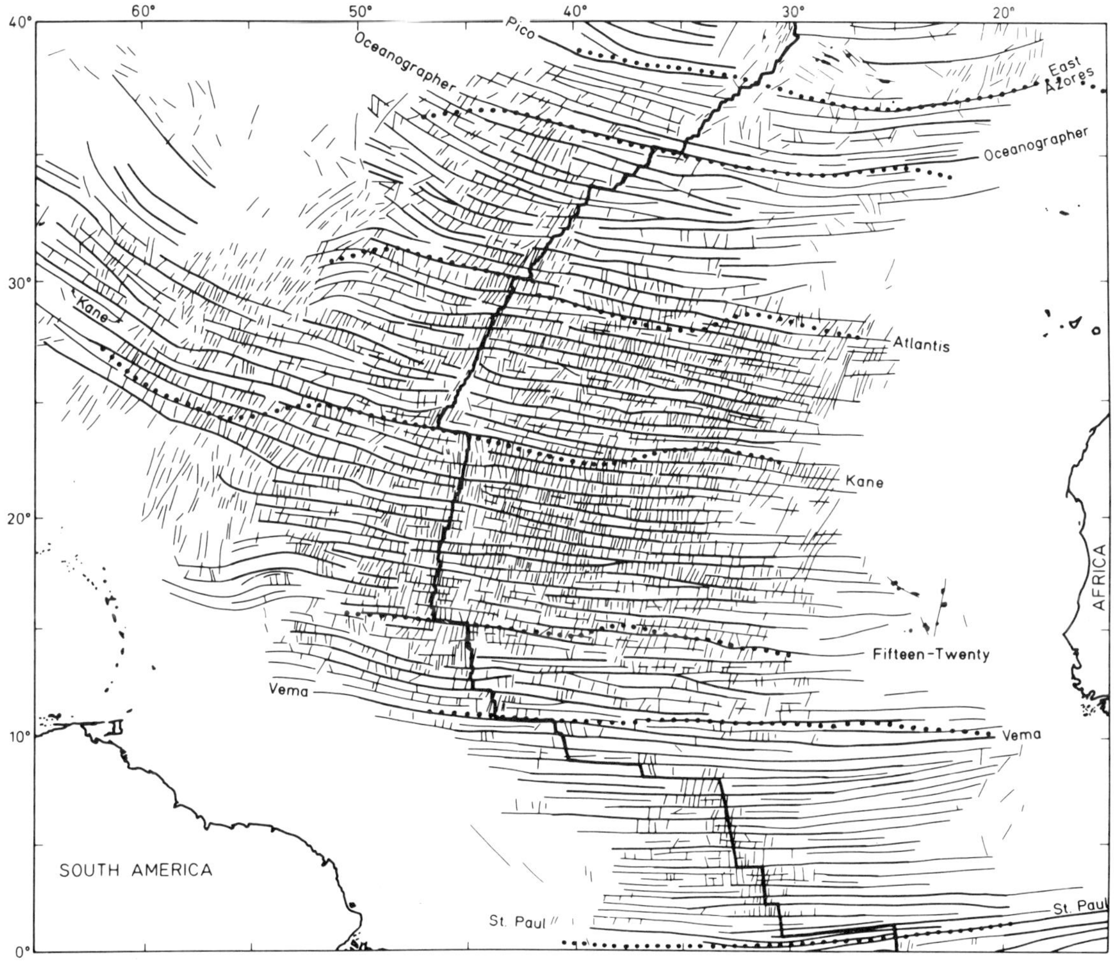

Fig. 62. Map of fracture zones and adjacent features at the North Atlantic Mid-Oceanic Ridge. Interpretation of contours on chart of US Navy Oceanographic Office (1977). Fracture zones (long smooth east-west lines) and crests of small ridges (short lines transverse to fracture zones). Dots depict major named fracture zones as mapped on other sources. Rift valley (irregular wide line) is based upon earthquake epicenters. The displacement of stripes along the transform faults causes the location of older parts of the ocean floor vis-à-vis younger parts, i.e. sunken parts vis-à-vis higher parts, herewith producing steep and high escarpments of the topography. Compare Fig. 33 and Fig. 73. [After 84 E]. Compare Fig. 32.

1.4.5 Small scale topography of the ocean floor

1.4.5.1 Classification of small scale relief and microrelief

Small scale: range of features 1···100 m in diameter, and height.

Microrelief: range of features 0.1···1 m, i.e. the roughness of the bottom.

Apart from the relief that exists on hard or consolidated rocks, there is a multitude of forms that will develop because of sedimentation. Sediment transport includes ice rafted blocks in subpolar zones, currents, waves (formation of ripple marks on the sea floor), sediment movement, and transport (as a result of sliding, turbidity currents), and chemical processes, and by organisms (bioturbation). The microrelief of the sea floor also includes manganese nodules, size 1···20 cm in diameter, which are scattered in fields of 10···100 km, mainly in the Pacific Ocean and Indian Ocean [73 R 1, 84 S 1]. Different forms of lava cover large areas of the ocean floor. A typical form is the pillow lava.

Investigations in the 70's and 80's detected specific features at the mid-ocean rift valleys, so-called "chimneys", built on blow pipes of gas exhaustion close to the areas of hot brines as, for instance, at the Galapagos Rift [81 B 1, 84 R 2].

Microforms causes by animals are presented in the classical book of B.C. Heezen and Ch. Hollister: "The face of the deep" [71 H]. Compare Fig. 63.

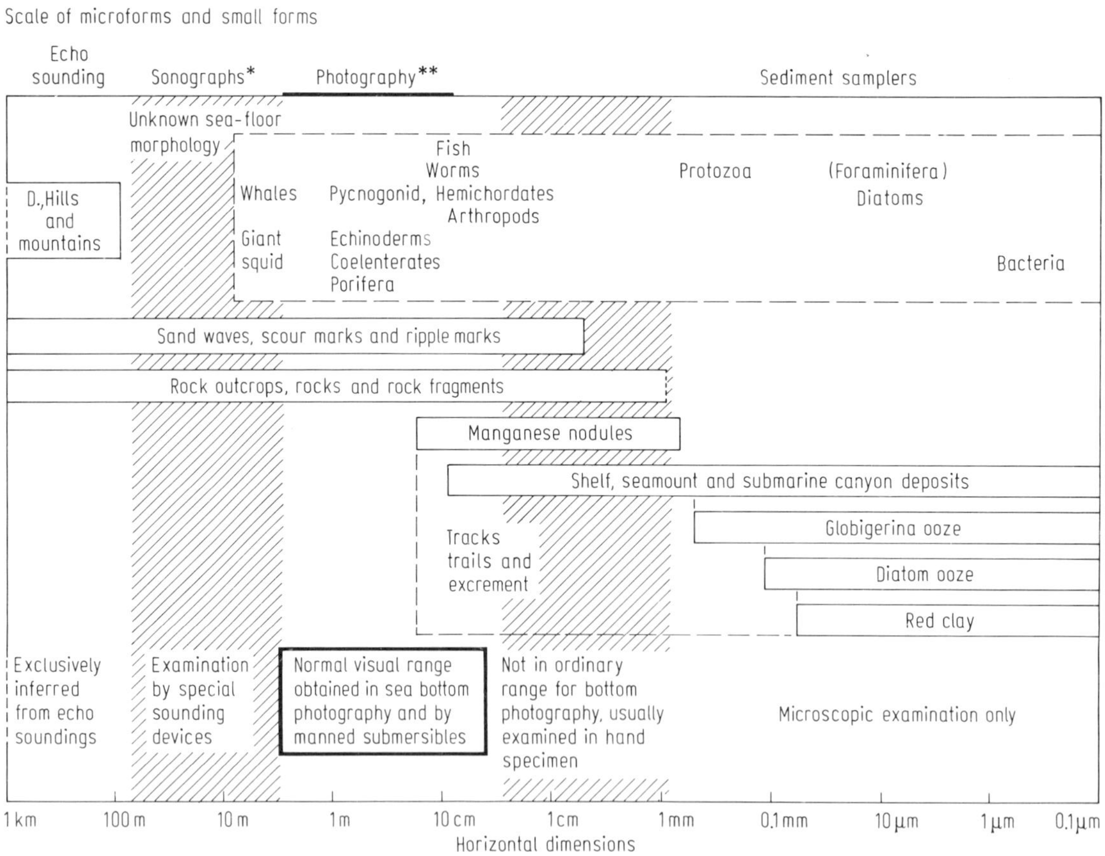

Fig. 63. Dimensions and scale classification of small- and micro-features on the ocean floor and the visual window [71 H]. * Sonographs, see [72 B 1]. ** Photography can be extended to the scale of 30 × 30 m with strobe light system, LIBEC (Light behind the Camera). D.: Diapirs. Compare Fig. 44b.

A limited range of objects can be visually observed. With the normal procedure, most bottom dwelling organisms are too small to be seen in photographs. However, rock outcrops, ripple marks, scour marks, large organisms such as starfish, holothurians, sea urchins, and tracks and trails of large animals are satisfactorily photographed with the normal combination of film size, fim grain, and lens – subject distance. See Fig. 63, Table 19.

The development of the sonar technic in the 1970's led to sonographic display of features from 1 to 100 m size: "Sonographs of the sea floor, a picture atlas" [72 B 1] and "Photographic atlas of the Mid-Atlantic Ridge rift valley" [77 B].

A gap of detection of small forms was filled with the introduction of the new deep-sea long-range sonar, called GLORIA, by A.S. Laughton, UK, on board of BRS "Discovery" in 1975.

GLORIA is a long range side-scan device developed at the Institute of Oceanographic Sciences, Wormley, England. In deep ocean and under ideal conditions (good weather that permits smooth towing, good signal-to-noise ratio, a minimum of sound wave refraction, etc.) a maximum range of up to 27 km is possible. At that distance it takes the sound pulse 36 s for a round trip from ship to target and back. An array of transducers and control mechanisms is towed astern the ship at a depth of 100 m. The towed vehicle weighs 6.7 tons in air (3.5 tons in water) and is one of the cumbersome of oceanographic instruments. The superb pictures of the ocean floor justify the effort required to use GLORIA [76 W 3].

Table 19b. Small scale features of the ocean floor, selection. See figures, tables and references of this article.

Feature	Characteristic of geomorphology	Characteristic of time scale	Presented in	Ref.
Rolling surface	hummocky surface, also of consolidated rocks	permanent		73 S
Dunes	sediment ridges, sandwaves	continuously changing	Table 20 Fig. 65	68 A, 73 R 1, 77 V
Knobs	conical summits	permanent	Figs. 51, 52, 63	
Knolls	single elevations	permanent	Fig. 64	
Diapirs	domes, protruding from deeper layers, lava at rift valleys, salt at continental margins, or bottom of semi-enclosed seas, (can be at the scale of 0.1···1 km)	permanent	Fig. 64	72 E
Slumps	slides or turbidites	episodically	Fig. 42a	67 S, 77 S 2, 77 C 2, 72 E, 73 R 1, 71 H, 83 C 2
Mudlumps	single elevations, relative flat	permanent over some years		51 M
Erosionic features	high energy benthic boundary layer, occurring due to tracks of eddies (HEBBLE: High Energetic Benthic Larger Experiment)	episodically to quasi-periodically monthly	Fig. 70a, b, c	84 H 2, 84 S 4
Erosionic and sedimentary	ice dynamics, grounding and scouring, with scattered ice rafted blocks and boulders	episodically permanent		71 H, 72 B 1, 78 R, 80 G
Fault-block terraces	structural framework of rift valleys	permanent	Table 19a	82 S 5

1.4.5.2 Diapirs

Diapirs are topographic features often found on continental margins and on the sea floor. Many diapirs are caused by salt domes which intruded upper layers and produced domes on the surrounding submarine sea floor. The caps of the diapirs consist of different rocks as there are algae reefs in the region of the Gulf of Mexico [63 S 1, 72 E]. See Fig. 64. Diapirs may be of high interest for offshore oil and gas exploration.

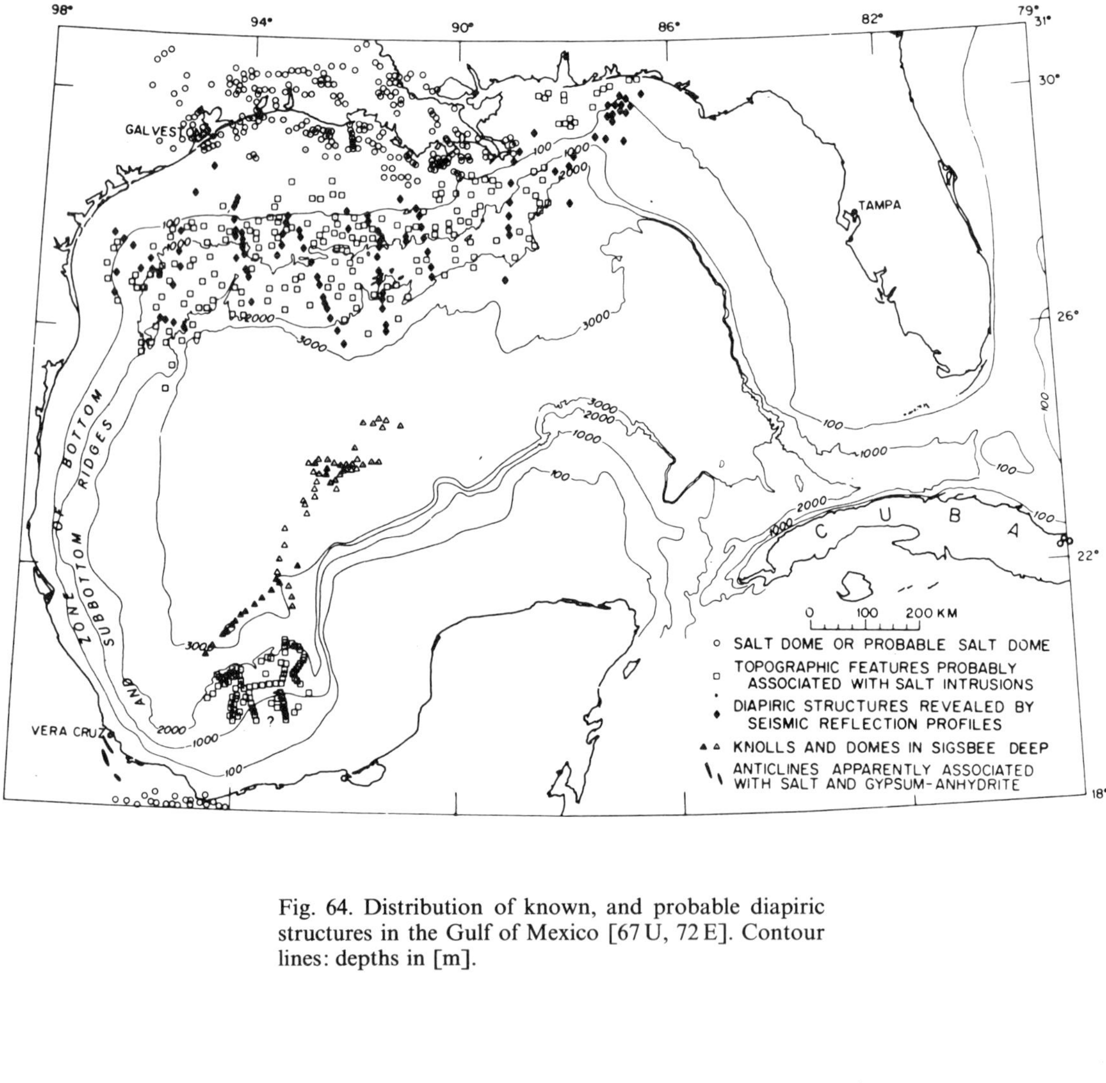

Fig. 64. Distribution of known, and probable diapiric structures in the Gulf of Mexico [67 U, 72 E]. Contour lines: depths in [m].

1.4.5.3 Sandwaves, dunes, classification

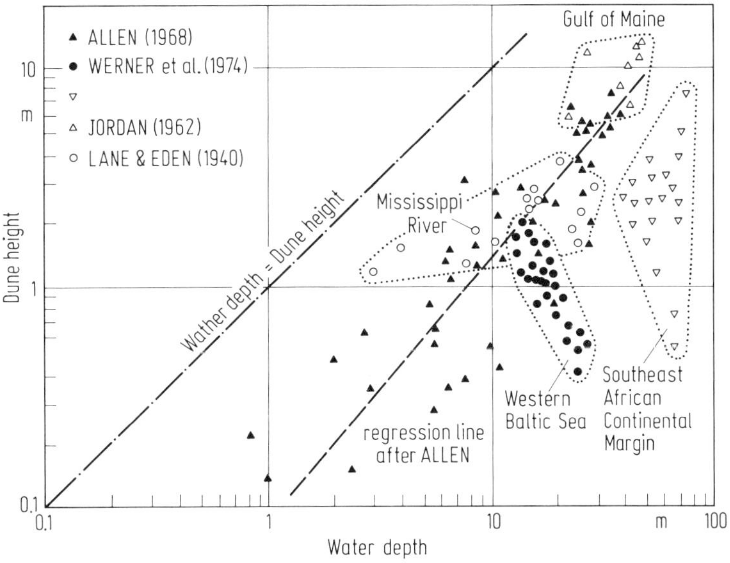

Fig. 65. The relationship between dune height and water depth. Note the disagreement between the predicted trend of Allen [68 A, 83 A] and the trend of some open-sea regions. [78 F]. For the precise sounding of submarine dunes and ripples with modern 150 kHz-fan sounder, see [80 U].

Table 20. Occurrence of ripples in various depositional environments. [73 R 1].

	Megacurrent ripples	Smallcurrent ripples	Wave ripples	Climbing ripple lamination	Rhombohedral ripples	Antidunes
Lagoon	–	0	+	0 *)	–	–
Intertidal flat	+	+ +	+ +	0	0	0
Tidal channel and inlet	+ +	+ +	–	0	–	–
Backshore and foreshore	+	+	+ +	0	+	+
Upper shoreface	+	+	+	–	–	–
Lower shoreface	–	0	+	–	–	–
Transition zone	–	0	+	–	–	–
Muddy shelf	–	–	0	–	–	–
Sandy shelf	+	+	0	–	–	–
Continental slope and rise	–	0	–	–	–	–
Deep sea	–	–	–	–	–	–
Sandy deep sea	–	+	?	–	–	–
Turbidite	0	+	–	+	–	0
Seamounts	–	+	0	–	–	–

*) Climbing ripple lamination of only wave ripple origin.

+ + abundant + common 0 rare – absent

Gierloff-Emden

Table 20a. Longitudinal dunes (chains), i.e. crest in direction of causing current. [77 V], compare [73 R 1]. L = length, W = width, D = depth, ε = interval distance, λ = length of wave, H = amplitude, v = velocity of currents.

Region (site, bank)	Author (year)	Dimension			Ripples			v
		$L \cdot W$ km · km	D_{max} m	D_{min} m	ε km	λ m	H m	$m\,s^{-1}$
North Sea (Norfolk Banks)	Caston and Stride (1970), Caston (1972)	12.2···51.9 · 0.7···2.2	42	0.4	4···8	100	2···6	1.2
North Sea (Hinder Banks)	Veenstra (1964)	27 · 8	30	5	4···5	130···170	2···7	0.9
North Sea (The Falls)	Houbolt (1968)	65 · 2	7	4.2	10···20	–	6	2.0
North Sea, south	Caston and Stride (1973)	12 · 2.4	33	24	4···5	400···1000	2···3	*1.5 **0.7 ***0.4
Thames Bay	Langhorne (1973)	20···30 · 3···5	24	1···5△	3···8	50···400	8	*2.0
Strait of Dover-Calais	van Veen (1936)	10···12 · 0.75	27	4.6	5···10	80···140	3···7	*1.5 **1.0 ***0.4
Celtic Sea	Bouysse et al. (1976)	60 · 7.5	155	111	16	500	11	*0.75···1.0
New England (Vineyard Sound: Middle Ground, Lucas Shoal)	Smith (1969)	7···11 · 6···1	18	2	1	60···120	2···4	*1.2 ***0.45
Gulf of Maine (Georges Bank)	Stewart and Jordan (1964)	20···60 · 2···4	44	4	7	300···500	7···15	*0.5···1.0
Middle Atlantic Bight (False Cape Ridge System)	Swift et al. (1972, 1973)	4···10 · 0.9···1.3	21	7	1.4···6.1	–	–	*0.5
White Sea	Chakhotin et al. (1972)	40 · 5	30	△	10···30	15···20	3···5	*2.0···3.0
Gulf of Corea	Off (1963)	60···90 · 2···3	31	5	6	–	–	*3.00 ***2.0
Strait of Malacca	Keller and Richards (1967)	48 · 3···5	30	10	3···10	–	–	***0.75···1.0

* surface
** minimum depth
*** near to bottom
△ peak raised to Low Water

Table 20b. Transversal dunes (waves), i.e. crest perpendicular to causing current. [77 V], compare [73 R 1]. L = length, W = width, D = depth, λ = length of wave, H = amplitude, v = velocity of currents, av. = average value.

Region (site, bank)	Author (year)	Dimension			Crest		v
		$L \cdot W$ km · km	D_{max} m	D_{min} m	λ m	H m	m s^{-1}
North Sea	Dingle (1965)	60 · 40	55	30	150···450	5···15	*1
North Sea, south (Flamish Bay)	Terwindt (1971)	100 · 100	30	17	125···1250	3···15	***0.8
North Sea, south-east (Flamish-Bay)	McCave (1971)	200 · 75	40	18	200···495	2···10	**0.6
Channel west (Banc de la Schôle)	Hommeril (1967)	7.5 · 3.5	6	40	100···300	5···10	*1.8
Irish Sea, north (Warts Bank)	Jones et al. (1965)	3.3 · 1.4	40	10.5	50···330 av. = 125	2···12 av. = 5.8	***0.5 ***0.5
Irish Sea, south	Harvey (1966)	4 · 1	170	140	200	10···24 av. = 15	*1
Bay of Cardigan (Tremadoc Bay)	Caston (1965)	5 · 5	11	3	650···950	0.30···5.2	*0.5
Celtic Sea (Bank of La Chapelle)	Cartwright and Stride (1958), Carruthers (1963)	22 · 10	165	151	850	5···12 av. = 7.6	*1 ***0.20···0.25
Irish Sea, south	Hinschberger (1970)	4 · 1.5	90	45	100	5···10	**1.3 ***0.7
Belle Isle Strait	Vanney (1969)	4 · 4	23	8.3	500···1000	3···12	*1···1.5
New Scotland (Brown Bank)	Drapeau (1970)	–	48	40	90	4	*1.00 ***0.20
Gulf of Maine (middle part of Georges Bank)	Jordan (1962)	30 · 20	50	7	240···1000 av. = 760	10···15 av. = 12	*0.5···1
Strait of Malacca	Keller and Richards (1967)	50 · 50	70	28	241···900	4.6···15.3	> 1
Bering Strait (Clarence Port)	Jordan (1962)	18 · 18	27	16	440···830 av. = 570	3···6 av. = 4	–

* surface
** minimum depth
*** near to bottom

1.4.5.4 Microtopography of the ocean floor: rock fragments, nodules, ripples, biogenic features

Table 21a. Pattern of region of glacial marine environment for the Bellingshausen Sea, South Pacific Ocean, Antarctic region. Summary list of lithofacies occurring in each regime. Occurrences are grouped into the sedimentation zones, subglacial = G, ice-proximal = P, ice shelf = S, iceberg = B, proximal = BP, distal = BD, and their frequency of occurrence, common (large, thick) = c, uncommon (small, thin) = u, rare = r. From [84 P].

Facies	Glacimarine regimes 1	2	3	4	5	6	7	8
Subglacial till	Gc	Gu-r	Gc	Gr	Gc	Gu-r	Gc	Gr
Dropstone diamicton	Sc	–	Sc-u	–	BPc	–	BPc	–
Stratified dropstone diamicton	Sc	–	Su	–	BPc-u	–	–	–
Dropstone mud	Sc, BDc	Su, Bu	–	–	BPu, BDc	BPr	Bc	–
Stratified dropstone mud	–	–	Su-r	–	BDu-r	–	BPc	–
Grounding line sediment and morainal bank	Pc	Pu-r	Pc	Pr	Pc	Pu-r	Pc	Pr
Subaqueous outwash fan	Pu	–	Pu	–	Pc	–	Pc	–
Interlaminated sand/silt/clay	Pu	–	Pu-r	–	BPc-u	–	–	–
Subaqueous sediment gravity flow deposits	Pc, Sc, Bc	Pu, BPr *)	Pc, Sc, Bc	BPr *)	Pc, Bc	BPr *)	Pc, Bc	BPr*)
Interstratified gravity flow deposits and dropstone diamicton	Pc *)	–	Pc *)	–	–	–	–	–
Interstratified gravity flow deposits and (dropstone) mud	–	–	–	–	Pc *)	–	–	–
Interstratified gravity flow deposits and dropstone diamicton and mud	–	–	–	–	–	–	BPc *)	–
Palimpsest lag	–	Bc-u	Bc-u	Sc, Bc	–	–	–	Bc
Biogenic bank and/or debris	BDu	Br	BDu	–	BPc-u	–	BDr	–
Deep-sea biogenic mud (ooze)	–	–	BDu	BDu-r	–	BDu-r	–	BDr
Dropstone ooze	BDc	BDr	–	–	–	–	–	–
Interstratified dropstone mud and ooze	–	–	–	–	BDc	–	BDc	–

*) Facies occur when the grounding line is at the edge of the continental shelf.

Table 21b. Average late Quaternary input of sand-sized debris to high latitude oceans; the "non-volcanic" column excludes ice rafted or other volcanic input originating predominantly from ocean island sources (after Ruddiman, 1977, Table 4). From [84 P].

Ocean	Mean input rate across area ($mg\,cm^{-2}\,ka^{-1}$)	
	all debris	nonvolcanic
North Atlantic	170	153
Norwegian Sea	100	90
Arctic	20	18
North Pacific	80	15
Antarctic	30	5

A comprehensive book on the microtopographic features is [71 H], on glacial marine sedimentation [83 M 1].

Three variables control the type of glacial debris source and where and when debris is introduced to an ocean:

(1) type of glacial ice source (ice sheet or valley glacier);

(2) condition of a grounded glacier (melting/freezing or frozen base and cold or warm internal ice); and

(3) type of glacier front (ice shelf or tidewater).

Using these variables in different combinations eight glacimarine regimes are postulated to exist. By integrating information from Holocene sediment and Pleistocene glacimarine sequences with the eight inductive regimes, lithofacies relationships are predicted and modelled. The models show a complexity of glacimarine lithofacies. From [84 P]. For the Bellingshausen Sea compare [67 H]. Table 21a, b.

1.4.5.5 Manganese nodules, micro-features of the ocean floor

Manganese nodules are probably valuable resources of the deep-sea floor (also called iron-manganese or ferromanganese deposits). These deposits can occur as round spheres from about 0.01···0.2 m in diameter, as coatings on rocks and other objects, or as long slabs, commonly called manganese pavements. Regardless of their form, they are mainly composed of hydrated oxides of iron and manganese. They generally form around a nucleus, such as a small shell, a rock, or a shark's tooth, and slowly grow outward in concentric rings. Ferromanganese deposits are found throughout the deep-sea marine environment as well as in lakes or other shallow-water bodies. The distribution pattern is irregular over the ocean floor and the other regions, respectively [64 M 2, 78 R]. The generation of most types of metal deposits can be meaningfully related to specific plate tectonic environments, and tectonic settings provide a viable framework within ore-generating systems can be studied [84 S 1]. See Fig. 66.

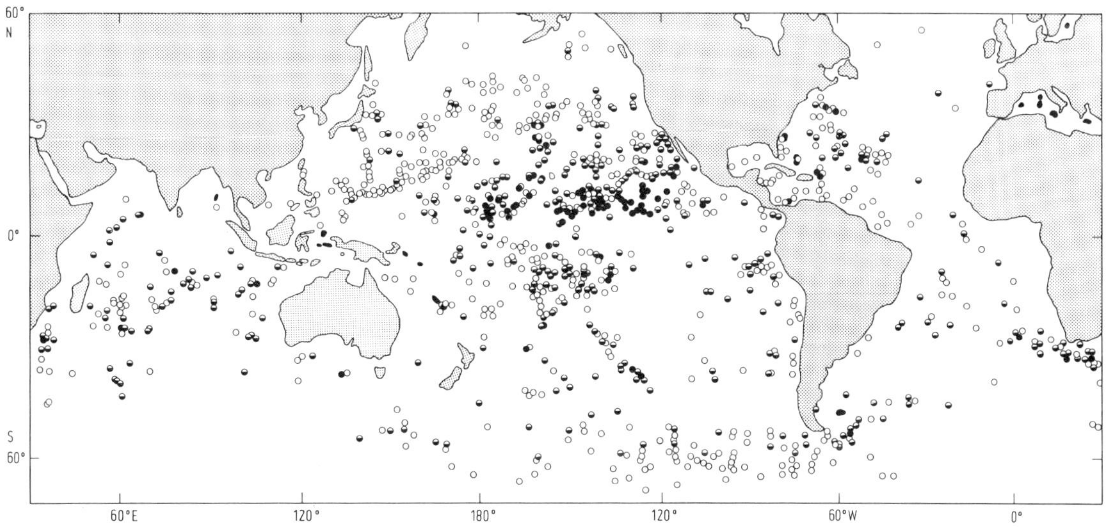

Fig. 66. Global distribution of manganese nodules. Filled circles are analyzed nodules containing more than 1% copper. Half-filled circles are analyzed nodules containing less than 1% copper. Open circles are nodules for which analyses are not available [78 H 2]. The world ocean highest concentration spreads around 5° to 13° N of the East Pacific Ocean. Maps differentiated after mineral types compare [80 R 2].

For handbooks, see [79 U 2] and [84 S 1]: "Metal deposits in relation to plate tectonics".

Ferro-manganese nodules are especially abundant in areas of low sedimentation rate, such as abyssal plains. Nodules form at an extremely slow rate that may be of the order of 1 mm in a million years, and so they would be easily buried where the sedimentation rate is high. The number of nodules on the sea floor can be immense: one of the highest concentrations observed (as determined by bottom photographs) was 100 kg/m^2 or about $120 \cdot 10^3$ t/km^2. A mining operation would need about $12 \cdot 10^3 \cdots 30 \cdot 10^3$ t/km^2. Estimates are that as much as 25% of the sea floor is covered by nodules and that over $1.5 \cdot 10^{12}$ t are located in the Pacific Ocean. See Tables 22a, b. For energy and chemistry of the "dynamic environment of the sea floor" see [82 F].

Table 22a. Average composition of ferro-manganese nodules from the different oceans in % weight [80 R 2].

Minerals	% nodule weight South Pacific	North Pacific	West Indian	East Indian	Atlantic Blake Plateau	Atlantic
Manganese	16.61	12.29	13.56	15.83	15	16.18
Iron	13.92	12.00	15.75	11.31	11	21.82
Nickel	0.433	0.422	0.322	0.512	0.4	0.297
Cobalt	0.595	0.144	0.358	0.153	0.3	0.309
Copper	0.185	0.294	0.102	0.330	0.1	0.109
Lead	0.073	0.015	0.061	0.034	–	–
Molybdenum	0.035	0.018	0.029	0.031	0.03	–

Table 22b. Metals in ferro-manganese deposits. From [77 C 2, 75 K].

	Depth m	Weight per cent Mn	Fe	Ni	Co	Cu	Weight ratios Fe/Mn	Ni/Mn	Co/Mn	Cu/Mn
Blake Plateau	650	18.4	13.5	0.623	0.345	0.092	0.74	0.034	0.019	0.005
Seamounts	2400	14.9	23.6	0.281	0.339	0.074	1.59	0.019	0.023	0.005
Mid-Atlantic Ridge	2650	15.3	20.5	0.242	0.414	0.076	1.33	0.016	0.027	0.005
Abyssal Atlantic	4150	15.5	23.0	0.308	0.234	0.141	1.49	0.020	0.015	0.009
Abyssal Pacific	4300	19.3	11.8	0.846	0.290	0.706	0.61	0.044	0.015	0.037

1.5 Sedimentation of the ocean bottom

1.5.1 The sediment coverage of the ocean floor

Sediment stratigraphy is the archiv of time resolution of ocean evolution. Extended areas of the ocean floor are covered by sediments which constitute the surface of the bottom. Because of these sediments, individual morphological features, and the former relief are buried. See Fig. 67. The sediment cover of the sea floor is not very thick (with the exception near passive continental margins (1···15 km); it is thin near the mid-ocean ridges (10···100 m). Typical basins of the Atlantic Ocean have sediments of 500 m thickness,the Pacific Ocean sediments of 300 m thickness. Maps with isolines of the sediment thickness are called isopaches. For the North-Atlantic Ocean those are presented by [84 E], for the northern part see [78 H 2]. The thicknesses of sediments mentioned above are an indication that the deep-sea floor of the earth is not very old in geological history, and that the oldest and deepest parts exist far away from the mid-ocean ridges [82 S 1, 82 K]. See Figs. 60, 67.

For the development and variation of sedimentation on the ocean floor, see the CLIMAP project [76 C].

For the development of marine sediments, see [80 B 2] and [75 R 2].

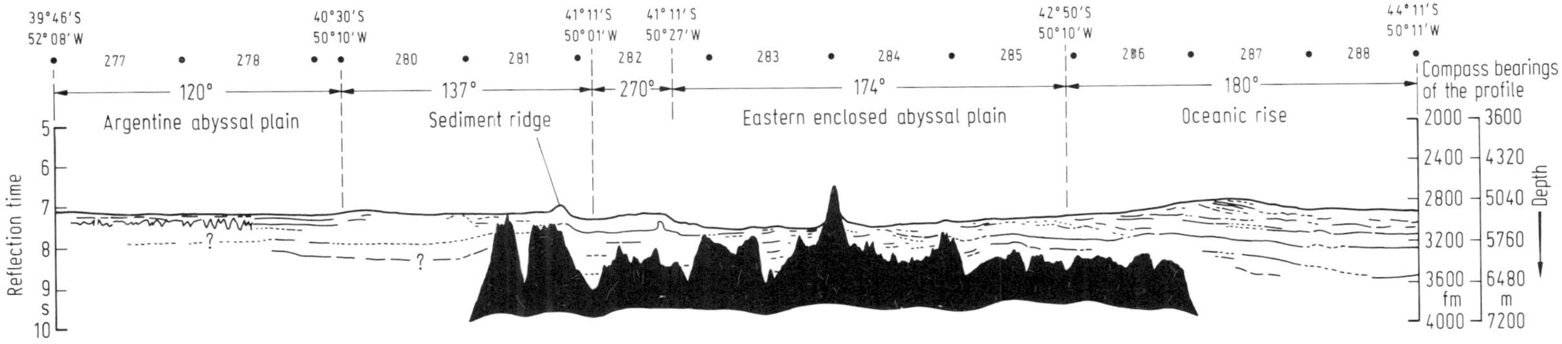

Fig. 67. Profile of the topography of the sea floor and buried features. Cross section of the Argentine Abyssal Plain, South Atlantic Ocean [71 L]. Black areas: solid rocks. Full and dashed lines: reflection elements in sedimentary rocks. Question marks: uncertain reflection elements. For buried sea floor features at the North Atlantic Ocean see [82 R 1], and compare Table 24.

Table 22c. Sedimentation accumulation rates, Legs I and II Glomar Challenger cruise. From [75 K].

Site 9 (32°37′ N, 59°10′ W)	
Pleistocene-recent	4.3 cm/10^3 a
Basal Pliocene to Pleistocene	3.6 cm/10^3 a
Eocene to basal Pliocene	1.2 cm/10^3 a
Upper Cretaceous to basal Eocene	0.2 cm/10^3 a
Pliocene/Pleistocene	78.0 m/ 1.8 Ma
Miocene/Pliocene	213.0 m/ 6.0 Ma
Eocene	683.2 m/44.0 Ma
Senonian	764.9 m/85.0 Ma

Site 10 (32°37′ N, 52°20′ W)	
Mio/Pliocene to Recent	0.80 cm/10^3 a
Eocene/Oligocene to Miocene/Pliocene	0.12 cm/10^3 a
Mid-Eocene to Eocene/Oligocene	0.25 cm/10^3 a
Mid to lower Eocene	1.43 cm/10^3 a
Lower Eocene to upper Maastrichtian	0.78 cm/10^3 a
Upper Maastrichtian to lower Campigian	1.20 cm/10^3 a
Miocene/Pliocene	40 m/ 6 Ma
Eocene/Oligocene	80 m/38 Ma
Upper Mid-Eocene	99 m/45 Ma
Lower Eocene	185 m/51 Ma
Upper Maastrichtian	295 m/65 Ma
Lower Campanian	475 m/80 Ma

Site 11 (29°29′ N, 45°07′ W)	
Lower Pleistocene	1.4 cm/10^3 a
	20.1 m/1.4 Ma
	1.6 cm/10^3 a
	267.1 m/16.5 Ma
Site 12 (19°40′ N, 26°01′ W)	
Mostly Pleistocene	0.93 cm/10^3 a
Lower Eocene	0.36 cm/10^3 a
Pleistocene	41.8 m/ 4.5 Ma
Lower Eocene	218.0 m/51.0 Ma

1.5.2 The mass supply of sediments

Sediments are transported to the oceans by rivers, ice, and wind. The largest amount is provided by fluvial transport, i.e. material transports on the bottom, and by suspension loads. The bottom transport of rivers amounts to 1/10 of the suspended load (average) [76 D 2]. See Table 23. Compare LB, NS, V/3, subsects. 9.9.1···3.

A rough calculation may be based on an estimated sediment supply from rivers of 12 km^3/a. The distribution of this amount of sediments on the area of $362 \cdot 10^6\,km^2$ of the sea floor, a sedimentation rate of about 30 mm/1000 a and a corresponding erosion rate of 60 mm/1000 a for the continents can be calculated [82 S 1]. Calculations based on other estimates give twice as large sedimentation rates [82 H]. For wind transport of dust, see [80 G]. See Table 24, and Table 22c on previous page. For ice transported sediments see [71 H, 72 B 1, 78 R, 80 G].

Local accumulation rates can be as high as 1 m/year, such as in front of active delta distributaries. Typical long-term accumulation rates on carbonate or clastic shelves are from 10···40 mm/year. Pelagic ooze sedimentation will normally not exceed 30 mm/1000 years, although under upwelling areas it may reach 0.1 m/year. Resedimentation of material to deeper water results in accumulation rates on modern deep-sea fans from 0.1···2 m/1000 years and up to 10 m/1000 years in small tectonically active basins. Turbidity currents are one process by which resedimentation occurs, and estimates of their frequency mostly range from 1 per 500···10 000 years for deep-sea clastics and from 1 per 20 000···100 000 years for carbonates. Muddy turbidity currents off some active river-delta systems may occur with a frequency as high as one per year. [83 B].

For palaeoceanographic conditions, see [83 H 1]. For energy and chemistry see [82 F].

Organic sedimentation is correlated with carbonate metabolism and carbonate chemistry in the oceans.

High temperature and a low CO_2 content in the water increase the product $[Ca^{2+}][CO_3^{2-}]$ by increasing the concentration of carbonate ion. The effect of CO_2 is readily seen from the following equations:

$$CO_2 + H_2O = H^+ + HCO_3^-$$

$$HCO_3^- = H^+ + CO_3^{2-} \quad [82\,S\,1].$$

The *calcite compensation depth* (CCD), called also carbonate line, a "Lysocline", is the particular depth level in the ocean where the solution tends to drop the carbonate value toward zero. In areas where the ocean reaches great depth, the recent carbonate sedimentation rate from shells or skeletons of micro-organism is zero or very low. Fig. 68.

Table 23. Selected rivers of the world ranked by sediment yield. World rivers with drainage areas exceeding 100 000 km^2, grouped by continent and ranked by drainage area, discharge, and suspended load, after [68 H1].

River (country)	Drainage area $10^3\,km^2$	Rank	Average discharge $10^3\,m^3/s$	Rank	Suspended load (10^6 tons/a)	Rank
North America:						
Mississippi (USA)	3200	3	17.0	7	750	6
Mackenzie (Canada)	1800	11	7.9	18	15	29
St. Lawrence (USA and Canada)	1300	14	14.0	11	3.6	35
Nelson (Canada)	1070	16	2.3	37		
Yukon (Canada and USA)	932	21	5.1	25	88	18
Columbia (USA)	670	29	7.2	20	36	25
Colorado (USA)	640	31	0.6	52	160	12
Rio Grande (USA)	461	34	0.14	55	19	28
Fraser (Canada)	238	44	3.2	32		
Brazos-Colorado (USA) Texas	232	45				
Mobile (USA)	110	55	1.7	44		

Table 23 (continued)

River (country)	Drainage area $10^3\,km^2$	Rank	Average discharge $10^3\,m^3/s$	Rank	Suspended load (10^6 tons/a)	Rank
South America:						
Amazon (Brazil)	5800	1	175.0	1	1000	3
Parana (Argentina)	2200	8	15.0	10	97	17
Tocantins (Brazil)	910	23	10.0	15		
Orinoco (Venezuela)	880	24	17.0	8	86···45	19
São Francisco (Brazil)	675	27	2.8	34		
Magdalena (Colombia)	240	43	7.5	19	250	11
Uruguay (Argentina and Uruguay)	230	46	3.9	26	14	32
Africa:						
Congo (Congo)	4000	2	40.0	2	68	21
Nile (Egypt)	3000	4	2.8	33	120···88	15
Zambezi (Mozambique)	1305	13	7.1	21	100	16
Niger (Nigeria)	1100	15	6.1	23	67···40	22
Orange (South Africa)	1040	19	0.29	54	150	13
Australia:						
Murray (Australia)	1065	17			32	25
Europe:						
Volga (USSR)	1330	12	8.1	17	19	27
Danube (Romania)	820	25	6.2	22	83	20
Dnieper (USSR)	500	33	1.7	42		
Dvina (USSR)	360	37	3.5	30		
Pechora (USSR)	282	40	2.6	35		
Ural (USSR)	220	47	0.4	53		
Vistula (Poland)	200	49	1.1	49		
Rhine (Netherlands)	145	52	2.2	38	0.5	36
Rhone (France)	96	56	1.7	43	45	24
Ebro (Spain)	83	57				
Po (Italy)	70	58	1.5	47		
Asia:						
Yenisei (USSR)	2600	5	17.0	6	10	34
Lena (USSR)	2510	6	15.5	9		
Ob (USSR)	2500	7	12.5	13	14	31
Yangtze Kiang (China)	1900	9	22.0	3	500···275	8
Amur (USSR)	1805	10	9.6	16	52	23
Ganges (India)	1060	18	19.0	5	1800···1500	2
Brahmaputra (E. Pakistan)	935	20	20.0	4	730	7
Indus (W. Pakistan)	930	22	5.6	24	400	9
Mekong (S. Vietnam)	800	26	11.0	14	1000	4
Hwang Ho (China)	672	28	3.3	31	1850···630	1
Koluma (USSR)	642	30	3.8	27		
Shatt-al-Arab (Iraq)	540	32	1.4	48	1000	5
Irrawaddy (Burma)	430	35	13.6	12	350	10
Indigirka (USSR)	365	36	1.8	41		
Krishna (India)	310	38	2.0	40		
Godavari (India)	300	39	3.6	28		
Salween (Burma)	280	41	1.5	46		
Yana (USSR)	250	42	1.0	50		
Anadyr (USSR)	205	48	1.6	45		
Pyasina (USSR)	190	50	2.6	36		
Chao-Phraya (Thailand)	160	51	0.9	51	11···5	33
Sankai (Si) (China)	120	53	3.6	29		
Red (N. Vietnam)	119	54	2.2	39	130	14

Table 24. Rates of accumulation of recent and subrecent sediments [82 S 1, p. 185]. After [74 B 1] in [74 B 2].

Facies	Area	mm/1000 a
Terrigenous mud	California Borderland	50···2000
	Ceara Abyssal Plain	≈200
Calcareous ooze	North Atlantic (40···50° N)	35···60
	North Atlantic (5···20° N)	40···14
	Equatorial Atlantic	20···40
	Caribbean	≈ 28
	Equatorial Pacific	5···18
	Eastern Equatorial Pacific	≈ 30
	East Pacific Rise (0···20° S)	20···40
	East Pacific Rise (≈30° S)	3···10
	East Pacific Rise (40···50° S)	10···60
Siliceous ooze	Equatorial Pacific	2···5
	Antarctic (Indian Ocean)	2···10
	North and Equatorial Atlantic	2···7
Red clay	South Atlantic	2···3
	Northern North Pacific (muddy)	10···15
	Central North Pacific	1···2
	Tropical North Pacific	0···1

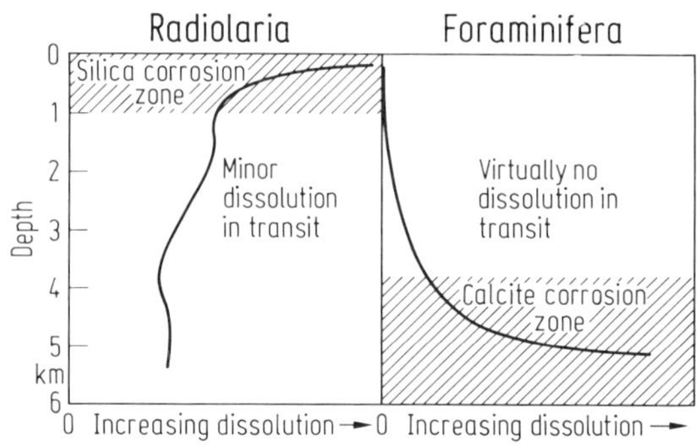

Fig. 68. Comparison of dissolution profiles of radiolaria and planktonic foraminifera based on field experiments. Most dissolution of radiolaria (and diatoms) occurs in shallow waters. In contrast, most dissolution of carbonate microfossils occurs on the sea floor at depths greater than 3.5 km. After [75 B]. Compare [75 K].

1.5.3 Sediment types

Pelagic clays are extremely fine-grained lithogenous and volcanogenic deposits. *Oozes* consist of biogenous minerals: shells of planktonic foraminifera, radiolarians, coccolithophores, and diatoms. Hemipelagic deposits are the same as clays and oozes except with large admixtures of shelf-derived sediment, and of continental material. For sedimentation of hot brines, see [82 B, 69 R]. See Table 25.

Table 25. Classification of deep-sea sediments [74 B 1, 82 S 1].

I. (Eu-)pelagic deposits (oozes and clays)
   $<25\%$ of fraction $>5\ \mu m$ is of terrigenic, volcanogenic, or neritic origin
   Median grain size $<5\ \mu m$ (except authigenic minerals and pelagic fossils):
   A. Pelagic clays, $CaCO_3$ and siliceous fossils $<30\%$
      1. $CaCO_3$: 1···10%; (slightly) calcareous clay
      2. $CaCO_3$: 10···30%; very calcareous (or marl) clay
      3. Siliceous fossils: 1···10%; (slightly) siliceous clay
      4. Siliceous fossils: 10···30%; very siliceous clay
   B. Oozes. $CaCO_3$ or siliceous fossils $>30\%$
      1. $CaCO_3$: $>30\%$; $<\frac{2}{3}$ $CaCO_3$: marl ooze; $>\frac{2}{3}$ $CaCO_3$: chalk ooze
      2. $CaCO_3$: $<30\%$; $>30\%$ siliceous fossils: diatom or radiolarian ooze

Table 25 (continued)

II. Hemipelagic deposits (muds).
$>25\%$ of fraction $>5\,\mu m$ is of terrigenic, volcanogenic, or neritic origin
Median grain size $>5\,\mu m$ (except authigenic minerals and pelagic fossils):
A. Calcareous muds. $CaCO_3 > 30\%$
1. $<\frac{2}{3}$ $CaCO_3$: marl mud; $>\frac{2}{3}$ $CaCO_3$: chalk mud
2. Skeletal $CaCO_3$: $>30\%$: foram; nanno; coquina
B. Terrigenous mud. $CaCO_3 < 30\%$. Quartz, feldspar, mica dominant
Prefixes: quartzose, arkosic, micaceous
C. Volcanogenic muds. $CaCO_3 < 30\%$. Ash, palagonite, etc., dominant

III. Pelagic or hemipelagic deposits
1. Dolomite-sapropelite cycles
2. Black (carbonaceous) clay and mud: sapropelites
3. Silicified claystones and mudstones: chert
4. Limestone

1.5.4 Sediment distribution in the oceans

Mechanisms for deep-sea sedimentation are:
1. Settling from the water column
2. Bottom transportation by gravity flows, including turbidity currents [52 H, 64 H 2], debris flows, grain flows, nepheloid suspension zones, and slumping [77 C 2, 83 C 2].
3. Transportation by geostrophic bottom currents, including contour currents [66 H 2, 84 H 2].
4. Chemical and biochemical precipitation on the ocean floor [82 K, 82 F].

The distribution patterns of marine sediments are shown in Fig. 69. This shows the dominance of topographically controlled calcareous ooze and clays throughout the basins. Siliceous oozes are concentrated at high latitudes, in the equatorial Pacific Ocean and Indian Ocean, and beneath some coastal upwellings, such as off western South America. Glacial-marine sediments are dominant at high latitudes. See Fig. 69a.

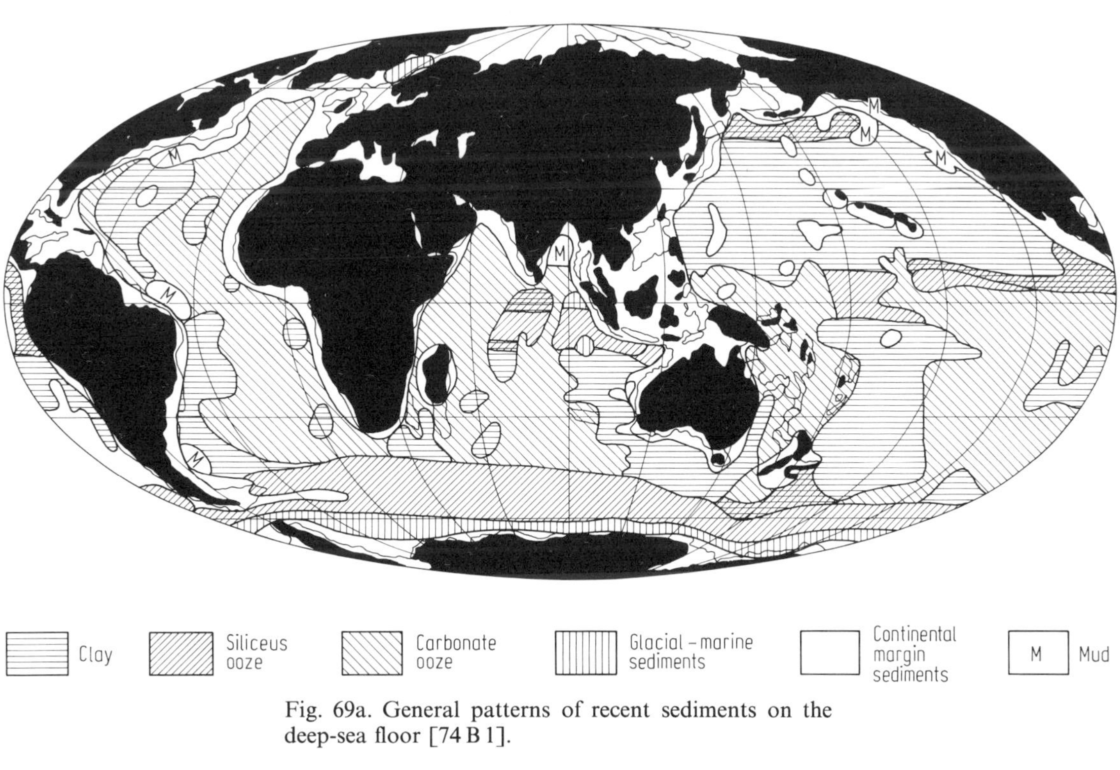

Fig. 69a. General patterns of recent sediments on the deep-sea floor [74 B 1].

1.5.5 Ocean bottom currents and sediment transport causing micro-topography

Current velocities implied by deep-sea sediment structures: See Table 26, Fig. 69b.

Current cross-stratifications in sands and silts have been observed in cores from deep ocean-basins. An estimate can be made of the minimum current velocities required to form craig and tail, mound-and-tails, ribbons, ripples, scour marks, furrows, cross-bedding, mudwaves, and other features observed in deep-sea photographs and cores if the critical traction transport velocities are known for deep-sea sediment particles. Current effects are frequently seen in globigerin ooze, a material of about 0.4 mm median grain size. Such features suggest velocities of at least 15 cm s^{-1}. The ripples seen in many photographs of mineral sands require higher velocities. The scour and rippling of finer sediments are more difficult to interpret but may imply much higher velocities [68 A, 73 R 1, 82 F, 79 E, 84 L 2]. Although there is considerably uncertainty as to the critical

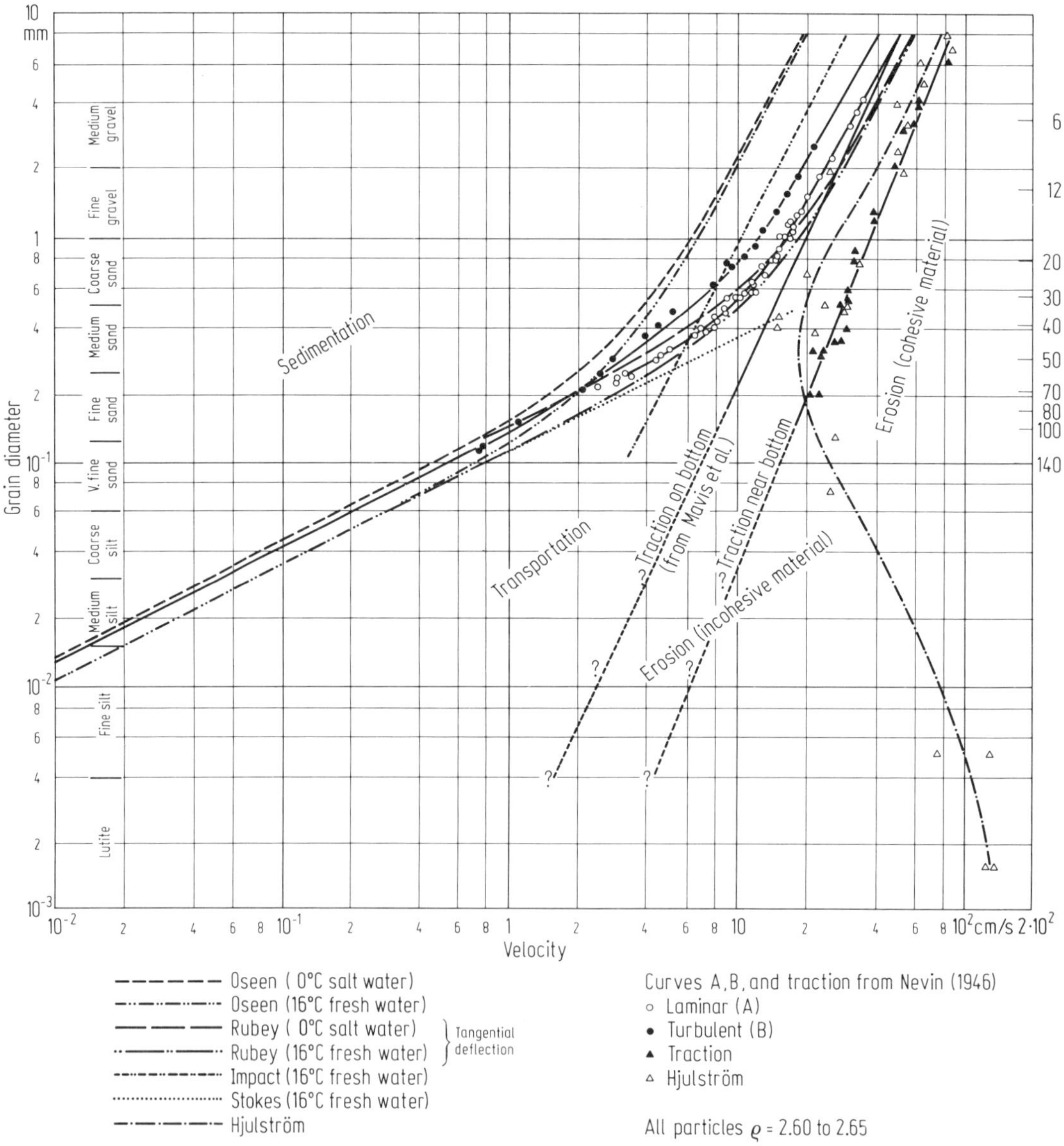

Fig. 69b. Current velocities required for erosion transportation and deposition of stream-bed materials. The traction (near bottom) curve was determined by Nevin (1946) and extrapolated to bed conditions by applying a factor of 0.5. The resulting curve coincides with that of Mavis et al. (1935). Considerable uncertainty exists for sizes smaller than 0.2 mm. After [64 H 2]. Compare [83 A]. For sea bed mechanics see [84 S 3], and for the dynamic environment of the sea floor see [82 F].

Table 26. Abyssal sediment dynamics: Recent near-bottom, deep-sea, direct current measurements, and maximum size of particles that could be eroded, and transported by these currents [66 H, 66 F, 64 H 2]. Compare Fig. 67.

Location	Water depth m	Depth of measurement m	Current velocity $cm\,s^{-1}$	Maximum size of particles that could be		
				eroded [mm]		transported [mm]
				Mavis et al. (1935), Nevin (1946)	Hjulstrom (1935)	Rubey (1933)
Continental slope (33°07′ N 75°42′ W	3230	3230	11.5 ± 1.5	0.30 (medium sand)	none	0.60 (coarse sand)
(33°08′ N 75°41′ W)	3300 (approx.)	2800	6.1	0.07 (very fine sand)	none	0.35 (medium sand)
Bermuda – Plantagenet Bank	3760	3755	6.7	0.10 (very fine sand)	none	0.40 (medium sand)
200 miles west of Bermuda	4600	4000	42	2.5···6 (medium gravel)	2 (medium gravel)	6.00 (medium gravel)
Southwest of Cape Farewell (59° N)	2750	2245	10	0.25 (fine sand)	none	0.55 (coarse sand)
Northeast of Newfoundland (51.5° N)	2800	2700	9 (approx.)	0.18 (fine sand)	none	0.51 (coarse sand)
Northeast of Newfoundland (51.5° N)	2500	2300	9 (approx.)	0.18 (fine sand)	none	0.51 (coarse sand)
50 miles south of Bermuda	4400	4400	7.6	0.15 (fine sand)	none	0.43 (medium sand)
North wall Puerto Rico Trench	6600	6599	14	0.49 (medium sand)	none	0.88 (coarse sand)
"Thresher" search area (42° N–65° W)	2300	2300	15	0.57 (medium sand)	none	0.91 (coarse sand)

traction velocity required, the minimum velocity for transport (settling velocity) is well determined. Once put into motion, particles of 0.02 mm size would not come to rest until current velocities decreased to 0.03 cm s^{-1}, and particles of 0.2 mm size would not stop until velocities fell below 2 cm s^{-1} [66 F]. See Table 21b, Fig. 69b. For the influence of abyssal circulation on sedimentary accumulation in space and time, compare [84 H 1].

Deep-sea currents and sea floor topography

A deep sea current system is world-wide developed due to the density differences of oceanic water masses caused by temperature and salinity data [34 W].

The deep sea currents are moving permanently along the ocean basins under the coriolis effect, trending towards the right hand side of the continental margins, causing sea floor sediment mass movement [74 H 2].

"Deep sea storms" are catastrophic, nonperiodical events, derived from large eddies, which were developed at the sea surface and sunken down to the sea floor. The distribution pattern is correlated to sea surface eddies.

The velocity of deep sea storm water movement is 20 times higher than the velocity of permanent deep sea currents, and reaches 0.6 m/s. The time scale is 6 days for the event, the nonperiodic repition is 1···3 months. The space scale is 200 · 1000 km. Deep sea storms take of large amounts of sediment masses of sediment particles of high diameters as deep sea currents causing erosion and sedimentation of the sea floor [84 H 2]. Recent investigations were made with HEBBLE. Figs. 70a, b, c.

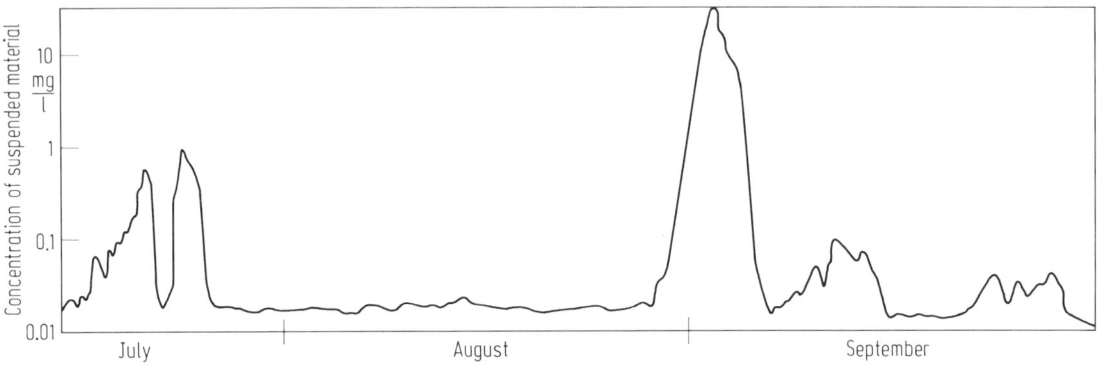

Fig. 70a. The concentration of suspended material at the ocean floor. Variation during one week was measured as 1 to 1000. Maxima were correlated with the event of heavy storms of the continental rise in June and September. After [84 H 2].

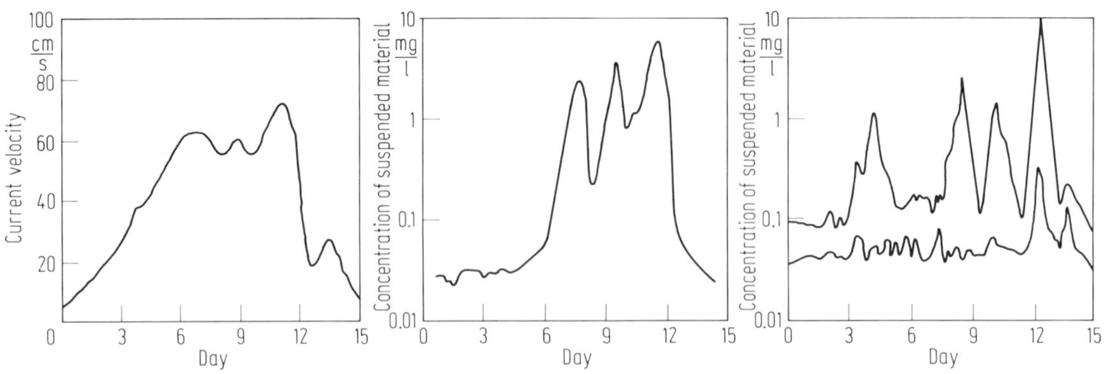

Fig. 70b. Variation of current velocity with the concentration of suspended material at the deep sea floor during the event of "deep sea storms", i.e. eddies. Left: current velocity, middle: concentration of suspended material, right: concentration of suspended material in different height above the sea floor, 0.5 m over bottom and 2 m above the sea floor, respectively. From [84 H 2].

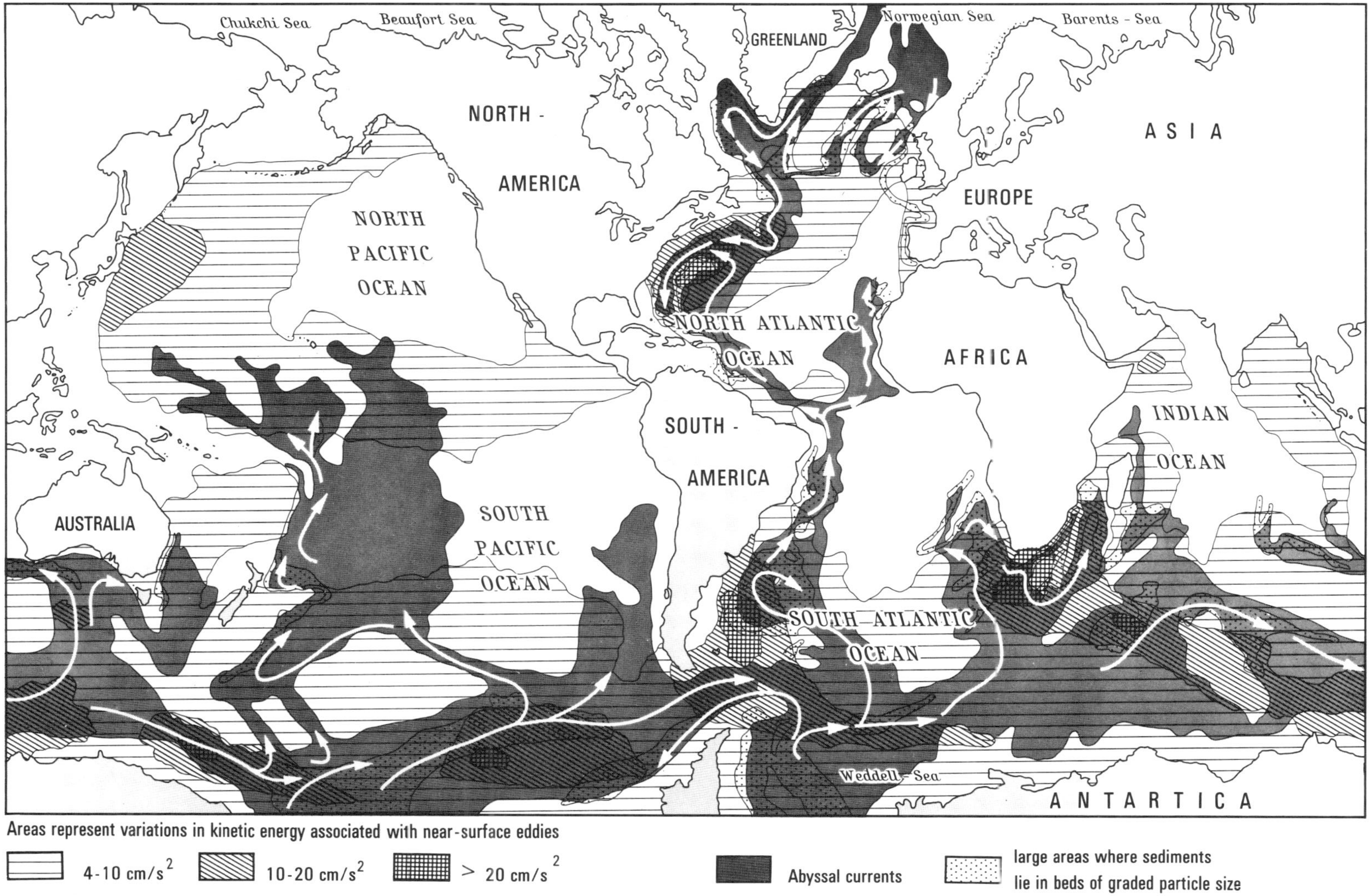

Fig. 70c. The global pattern of deep sea circulation. Drawn after [84 H 2]. Cartography: Institut f. Geographie, LMU München, 1985. The areas represent variations in kinetic energy associated with near-surface eddies. Abyssal currents marked after 3 velocities. For sea bec mechanics, see [84 S 3].

1.6 Global geological structure and topography

1.6.1 Sea floor spreading and global topography

The global topography is significantly caused by the *sea-floor spreading*. Virtually all of the present area of the oceans has been created by sea-floor spreading, and displaced during the past $200 \cdot 10^6$ years, since the beginning of the jurassic period during the last 5% of the recorded geologic history of the earth. [66 V, 69 H 2, 60 H 2, 61 D 1, 83 O]. A standard handbook is edited by C. Emiliani [81 E]. The event of a breakup of a former consolidation of the North American, South American, African, and Eurasian blocks into a *Pangaea* [70 D 2] developed with the process of the sea-floor spreading. The theory of sea-floor spreading was developed in the late fifties and the beginning sixties by H.H. Hess, B.C. Heezen, R. Dietz [61 D 1, 62 D] and Morgan and T. Wilson. The geomagnetic anomalies systematic patterns were discovered by F.J. Vine, W.C. Dittmann, and developed by LePichon (1966). [68 L, 73 H 1, 82 K, 61 D 1, 65 B 2, 85 B 2, 85 S]. Fig. 71.

The world encircling belt of mid-oceanic ridges can be traced from the Siberian continental shelf near the Lena River Delta, across the Arctic Ocean (represented by the Mid-Arctic Ridge), see Fig. 34, and through the Norwegian Sea to Iceland, where it continuous through North Atlantic Ocean [80 L, 84 M, 84 E], the Mid- and South Atlantic Ocean with jagged edges of the continental margins, splitting into a *triple junction* and continuing around the south of Africa and into the Indian Ocean [84 E]. Between Madagascar and India, it splits into a triple junction. [85 J 2]. Figs. 35, 36.

One branch runs northwest into the Gulf of Aden. Here it divides again, one branch continuing up the Red Sea, the other passing into the African rift valley system. The other Indian Ocean branch extends to the south of Australia and New Zealand [77 W 1], across the South Pacific, forming another triple junction, then running northward into the Gulf of California, and appears particularly at the Canadian coast north of Cape Mendocino encompassing the "Gorda Plate" on the continental side [82 K]. See Fig. 71, compare Fig. 33···36. The ridge encompasses Antarctica from the junction onward.

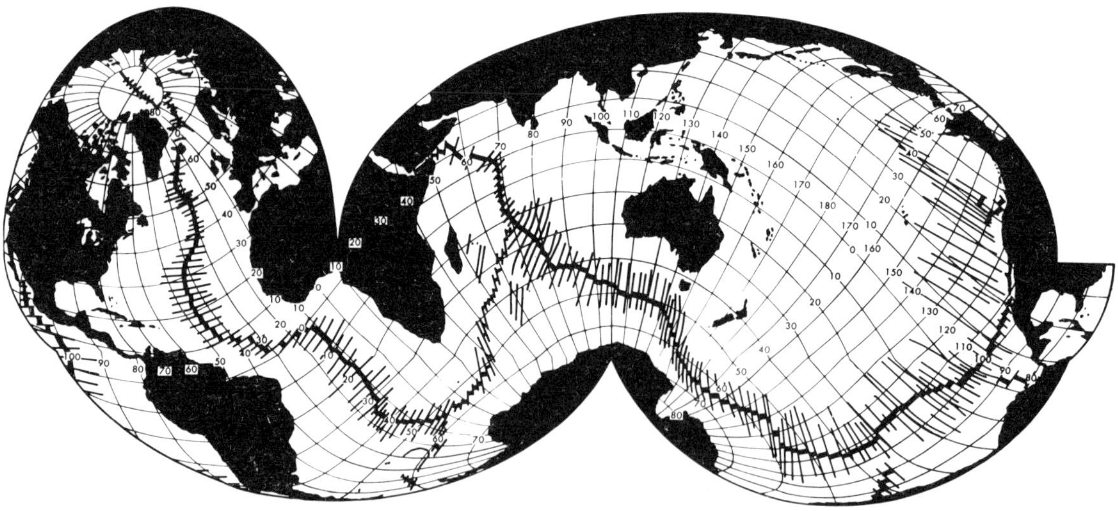

Fig. 71. Map of the mid-ocean rdiges as the sea-floor spreading zones: a global structural topographic feature. Mollweide projection as modified by F.C. Fuglister in [62 A] (oblique projection with meridian at the equator as sine wave within the elliptical outline; equatorial projection cut from the poles to the equator; the oceans appear uninterrupted; southern oceans broken into segments [61 D 1]). The mid-ocean ridges with a length of approximately 65 000 km encompass the earth along the continuing world ocean but the spreading sections between offsets by fracture zones total only 45 000 km. The link around the Antarctic continent is split due to the projection. A link from the Arctic Ocean to the Northeast-Pacific is hypothetical. Design: H.G. Gierloff-Emden, 84. Cartography: P. Schade, Institut f. Geographie der LMU München.

The Mid-Oceanic Ridges are characterized by basic rocks. The age of these rocks was determined from drill cores and dredged samples (IPOD, JOIDES) by radiometric methods, and from the lineaments of magnetic anomalies of the reversals of geomagnetic polarity which there show a symmetric pattern of alternating strips. (See LB, NS vol. V/2). Figs. 76, 79.

A world map of the magnetic sea-floor-spreading anomalies in the scale 1 : 46 460 000, Mercator projection, was published by H.A. Roeser and M. Rilat in 1982, which presents crustal ages depending on reversal time scales. [82 R 2]. Compare "Atlas of continental displacement, 200 million years to the present" by H.G. Owen [83 O] and [85 B 2, 85 S].

Slower spreading ocean ridges form new sea floor from volcanic activity at a rate of (1···5)cm/a. Fast-spreading ridges form new sea floor at rates up to 10 times as fast in the Pacific Ocean. Approximately 2/3 of ocean ridges are slower-spreading. When the Atlantic Ocean grows, the Pacific Ocean ought to shrink.

Hot spots: Although an absolute coordinate system is lacking, the "hot spots" with their tracks, i.e. chains of expired volcanoes are subjects to define a plate-external system, as so-called average hot spot frame of reference. [85 V]. The track of 14 hot spots during the last 150 million years over the spreading Atlantic Ocean as a reference system is presented by [85 V]. Compare Fig. 72.

Terranes is the term for small crustal blocks joining continents by "micro plate tectonics". Terranes are partly former submarine plateaus or island arcs (see Figs. 47 and 53), once transported by plate movements to continents and added in complex processes. Because of the genetic process, the topography of present submarine plateaus have become recently of high interest [83 J 3]. See Fig. 53.

The theory of A. Wegener of the continent-drifting [28 W] is herewith renewed but with entirely different causes and interpretation [76 D 1, 70 D 2, 82 W, 65 B 2, 82 V, 83 C 3, 85 S, 63 W, 83 F, 83 O].

The sediment accumulation on the sea floor is transported to the trenches by plate tectonic movement, and then subducted into the mantle of the earth, a process of permanent reneval and destruction. [79 S 1, 81 H 1].

The *turnover time* of the upper mantle is estimated at $300 \cdot 10^6$ years. [82 K]. The sea-floor spreading produces $2\,km^2$ new ocean bottom on an average during $200 \cdot 10^6$ years.

The map of "The bedrock geology of the world" from R.L. Larson and W.C. Pitmann III [85 L] depicts the age of the ocean basins from the Jurassic time $\sim 200 \cdot 10^6$ years age up to the present, and the age of continental bedrocks from older geologic times and more in details, and presents the age intervals of the ocean floor produced due to sea-floor-spreading, in the scale of the map of 1:23 230 000.

1.6.2 Plate tectonics and ocean floor topography

The theory of plate tectonics, developed after 1963 by J.T. Wilson, explains that the crust of the earth is formed by some large and some smaller plates of crust and lithosphere, together a mosaic with continuous movement depending on sea floor spreading by newly created crust at the rift valley region [63 W], and by subduction at the plate margins in collision zones [68 M, 73 C, 73 L 2, 83 O, 85 J 2]. Fig. 73.

The western boundary of the Pacific Plate is investigated with respect to the palaeontologic aspect [82 S 3]. The "Proposed Satellite Laser Ranging" and "Very Long Baseline Interferometry Sites for Crustal Dynamics" investigations are planned to get quantitative data on plate tectonics [79 L 1]. The program "Crustal Dynamics Research" as proposed by "Space Science and Applications Note" [85 N] will include a global network of 20 fixed laser stations and several mobile stations (most of the North American and the Western European Plate) to measure plate motions by detection of geodetical satellites.

The maps of "A geophysical atlas for interpretation of satellite-derived data" show the global plate tectonic situation [79 L 2].

The maps of the "Atlas of continental displacement" show the development of "200 million years to the present" [83 O].

The map of the ocean floor topography, derived from the 1978 SEASAT-satellite altimetry and gravity anomaly, as produced by W.F. Haxby, Lamont, and as presented by J. Francheteau [83 F], shows the structure of the world ocean floor with the whole Mid-Oceanic-Ridge configuration and other details. Compare [85 E, 85 N, 85 R].

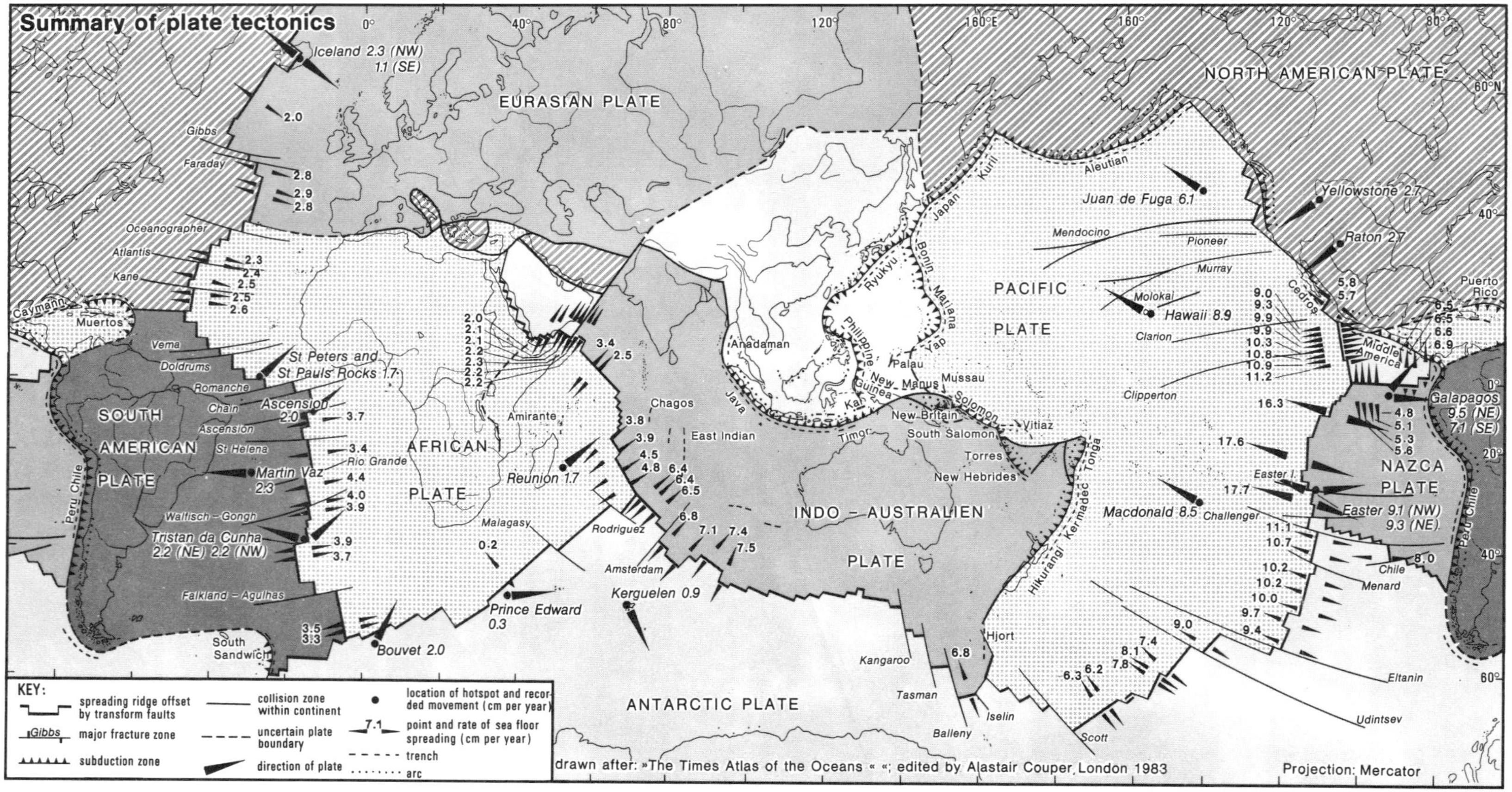

Fig. 72. Classifications of the global distribution of lithospheric plates have so far ranged from 6 to 26 plates covering Earth's surface. Several plate boundaries are conjectural, for example that around Scotia Arc in the South Atlantic, while some are uncertain, for example, that across China, within the Eurasian plate. Most classifications acknowledge 12 plates, although some of these have imprecise boundaries. Of the other possibilities, a Somalian plate, between the Carlsberg Ridge and the East African Rift system, is often recognised, but its southern boundary is very uncertain. A few small plates are vestiges of older plates which have been consumed in the continual process of plate evolution. For example, the Gorda plate, west of Seattle, USA, is possibly a remnant of the so-called Farallon plate which existed within the past 30 million years west of North America. Compare Figs. 33, 62. For long-extended fracture zone see ch. 1.4.4.2. Compare [85 J 2]. Redrawn after [83 O] "The Times atlas of the oceans", p. 32. Cartography: Institut für Geographie LMU München.

1.6.3 Topography and geometry of sea floor spreading and plate tectonics

For the specific geometric conditions, see LB, NS, vol. V/2, especially [85 J 2], and J.T. Wilson [76 W 2], and J.F. Dewey "Plate tectonics" [72 D].

The active part of the fracture zone is the ridge-ridge *transform fault*. Beyond this active part the fracture zone is the frozen trace of the fault. The fracture zones provide traces for the latitudinal circles around the pole of rotation (which need no coincidence with that of the rotation of the earth). See Fig. 73a, b, c.

Two lithospheric plates may be in contact along a common boundary on which one plate merely slides past the other with no motion that would cause the plates either to separate or to converge. The plane along which motion occurs is a nearly vertical fracture extending down through the entire lithosphere; it is called a transform fault. A fault is a plane of rock fracture along which there is motion of the rock mass on one side with respect to that on the other.

A triple junction is a point where three plates come together.

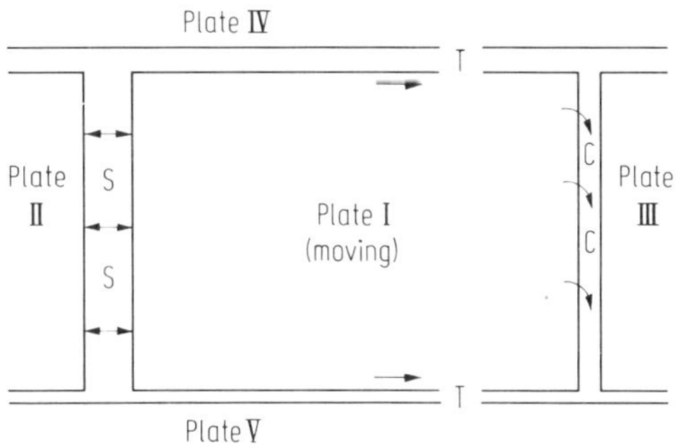

Fig. 73a. The three major kinds of active plate boundaries: Plate I is moving.

Spreading boundaries (S): New lithosphere is being formed by accretion (plate boundary II–I).

Converging boundaries (C): Subduction is in progress, lithosphere being consumed (plate boundary I–III).

Transform boundaries (T): Plates are gliding past each other on a transform fault (plate boundaries I–IV and I–V). Compare LB V/2, subvolume b, ch. 5.3.

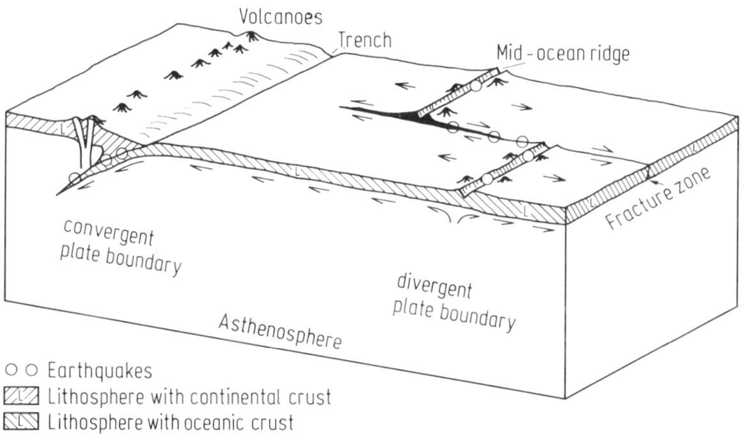

Fig. 73b. Relationship of plate motions to plate boundaries and earthquakes. Divergent plate boundary: mid-ocean ridge, shallow earthquakes. Convergent plate boundary: trench, shallow, intermediate, and deep earthquakes. Lateral boundary: fracture zone, shallow earthquakes on active part only. Horizontal motions are of the order of 1···10 cm/a. Based on a diagram by [68 I 2]. Model not to scale. The displacements of stripes along the transform faults cause the location of older, sunken parts vis-à-vis younger, higher parts, herewith steep and high escarpment along the transform faults. This is very distinctly marked in the northern Pacific Ocean where escarpments extend over several 1000 km. Compare Fig. 33. Compare LB V/2, subvolume b, ch. 5.3.

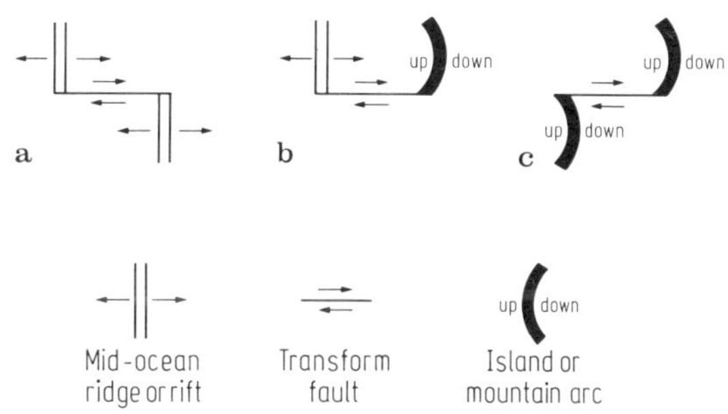

Fig. 73c. Different types of transform faults [70 W]. (a) ridge-ridge, (b) ridge-trench, (c) trench-trench. Model not to scale. Compare LB V/2, subvolume b, ch. 5.3.

1.6.4 Variation of ocean floor topography due to sea floor spreading in time-scale

As the sea floor spreads away from the crest, the lithosphere cools and sinks about 1000 m during the first 10 million years. The next 1000 m of sinking takes about 26 million years. See Fig. 78. With the sinking of the sea floor, the topography which sinks down the flanks of the ridges is smoothened by the sediment cover. The typical remaining relief is the "abyssal hill morphology". See Figs. 74 and 75. [85 N].

Subsidence is variable and also episodic in nature. The greatest rates, up to 12 m/1000 years occur locally in small basins within transform margins and in arc and trench basins (up to 5 m/1000 years) along convergent margins. Average rates outside of these tectonically active regions rarely exceed 1 m/1000 years and are significantly lower (10···40 mm/1000 years) along passive margins and adjacent oceanic basins. By comparison with vertical motion, horizontal plate movements may be an order of magnitude greater, from 10 to 100 m/1000 years. Spreading rates across oceanic ridges and displacement rates within transform zones are often of these magnitudes.

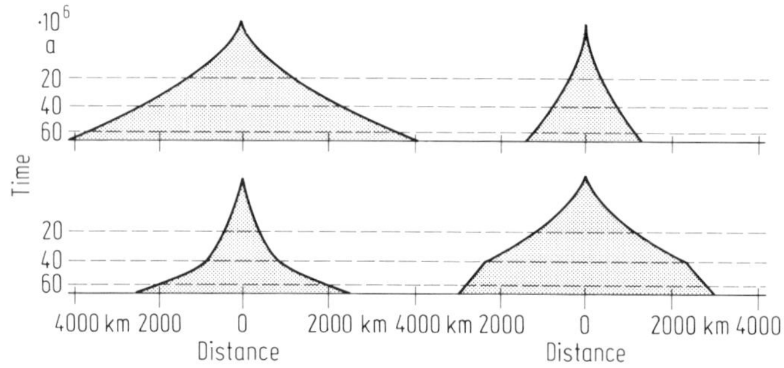

Fig. 74. Relationship between volume of mid-oceanic ridge and spreading rate. Upper cross sections of ridges spreading at 6 cm/a (left) and at 2 cm/a (right). Lower cross sections of the same ridges 40 million years after a change from 6 cm/a to 2 cm/a (left) and a change from 2 cm/a to 6 cm/a (right). The changes in ridge volume must produce corresponding changes in sea level [79 P]. Compare [81 E, 84 W], wherein an evidence of the correlation of the curves of the changes of the sea level and the mid-oceanic ridge volume is presented concerning the development during the past $700 \cdot 10^6$ years.

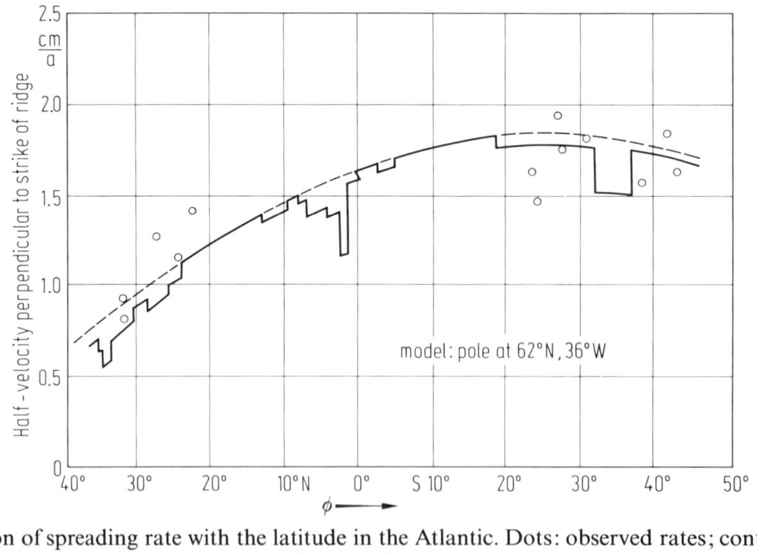

Fig. 75. Variation of spreading rate with the latitude in the Atlantic. Dots: observed rates; continuous line: predicted rates at right-angles to the ridge axis; dashed line; predicted rate parallel to the spreading direction [68 M].

→

Fig. 76. Age of the ocean floor in 10^6 a. The North-American Plate did partly override the older parts of the northern parts of the Pacific Plate. Drawn after "The Times atlas of the oceans," ed. A. Couper, London, 1983, p. 40···41. Cartography: Institut f. Geographie, LMU München, 1985. The dots with numbers mark the legs of the DSDP with the vessel "Glomar Challenger". Compare "Atlas of continental displacement", H.G. Owen [83 O]. Compare LB V/2, subvolume b, subsect. 4.2.1., Fig. 16; sect. 5.3, Figs. 7, 8 and Table 9. [85 B 2]. Compare "The new global map of the age of the ocean basins [85 L].

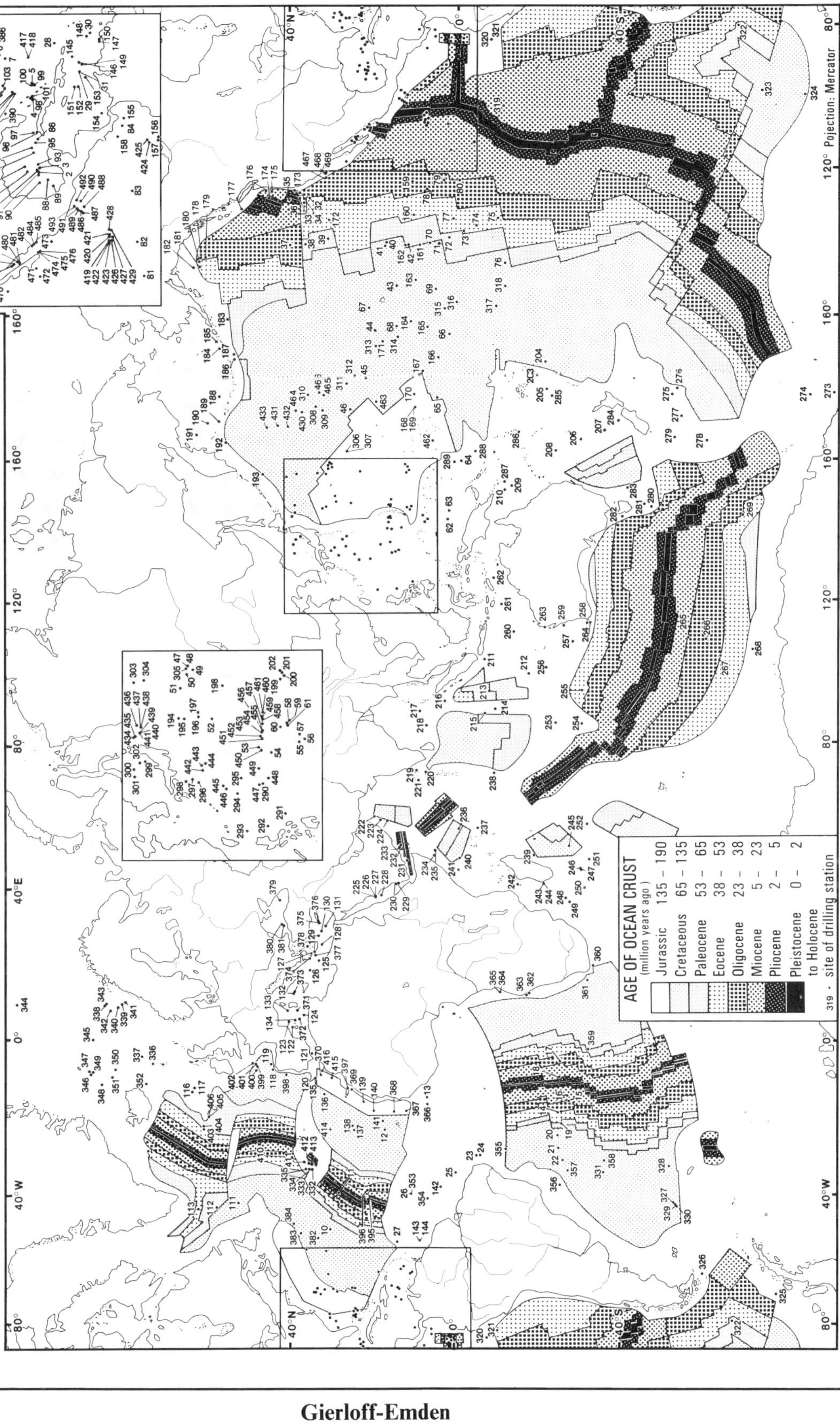

Age of ocean floor drawn after: „The Times Atlas of the Oceans," edited by Alastair Couper; London 1983, pp. 40–41. Cartography: Institut für Geographie d. LMU – München, 5/1985

1.6.5 Variation of ocean floor topography due to subsidence in time-scale

It is evident that the depth of the ocean floor is increasing from areas of central mid-ocean ridges to the more distant areas, which are sinking for a longer time. The depth of the ocean floor increases from (2700 ± 300) m at a spreading center to (5800 ± 300) m in the oceanic crust of early cretaceous age (in general). Regional example: The axis of the Mid-Atlantic Ridge is 2500 m below sea level. The 3000-m-isobath (contour of equal depth) lies on oceanic crust which is $2 \cdot 10^6$ years old, the 4000-m-isobath on crust which is $20 \cdot 10^6$ years old, and the 5000-m-isobath on a crust which is $50 \cdot 10^6$ years old [82 K, 77 S 2]. The correlation of the depth and of the age of regions of the sea floor was calculated and proved with the age determination of sediments and drilled cores. These methods both endorse the theory of the genetics of the sea floor by sea-floor spreading, the creation, the elevation, and the subsidence of certain parts of the ocean floor [62 D, 77 P 1]. From [84 S 2]. Fig. 77, Table 27.

The East Pacific Rise in the North Pacific shows a uniform decrease from a crustal elevation of about 2700 m to a depth of 5500 m for crust that is 75 million years old. The other mid-ocean ridges show the same uniform increase of depth with age. These ridges have a shallower crustal elevation than the East Pacific Rise. The similarity between the ridges of the Pacific, Atlantic, and Indian Oceans is evidence of a fundamental principle underlying the relation between depth and age of the oceanic crust.

Topographic method

Sclater [71 S, 79 S 1] examined topographic and magnetic profiles in regions of easily identified magnetic lineations and relatively uniform spreading in the Pacific, Atlantic and Indian Ocean. Regions of thick sediment deposits and large fracture zones were avoided. "Anomalous" regions were also avoided: on the southern mid-Atlantic Ridge between Walvis and Rio Grande ridges, and on the northern mid-Atlantic Ridge near the Azores and north of 45° N. This is the major drawback of the topographic method: there are known exceptions to the empirical relationship between depth and age which cannot yet be predicted on any basis.

Sclater et al. have demonstrated that, to a first approximation, there is a unique relationship between depth and age, $D = f(A)$.

It is apparent that the relation between depth and age is compatible with plate theory, and, with certain limitations, can be used within this framework to predict the age of the oceanic crust [82 K]. See Fig. 77.

J. G. Sclater developed the *palaeobathymetry* [71 S, 79 S 1, 81 H 1, 80 S 3, 83 J 2].

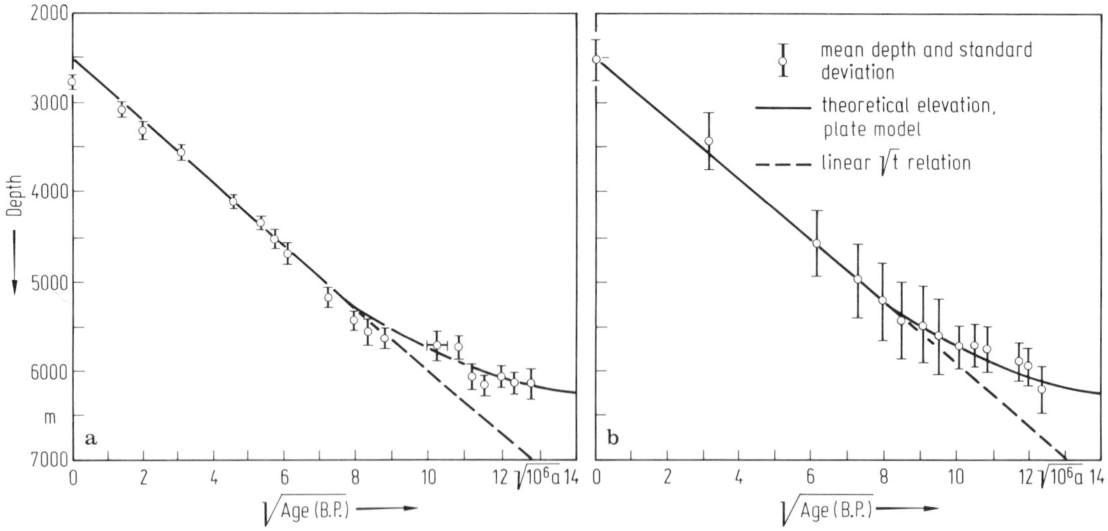

Fig. 77. Plots of mean depths and standard deviations of oceanic crust vs. the square root of its age for (a) the North Pacific and (b) the North Atlantic [77 P 1].

Table 27. Mean depth of ocean floor as a function of age (corrected meters) *). From [73 L 2].

Age 10^6 years	1	2	3	4	5	6	7	8	9	Average
0	2783	2596	2712	2700	2606	2606	2518	2683	2370	2619
2	3091	2840	2982	3003	2870		2788			2929
4	3306	3068	3191	3270	3096		2989	3291		3173
6		3196	3381	3460	3221		3247			3303
8		3353	3539	3642	3400		3472			3481
10	3554	3546	3730	3889	3581	3571	3571	3847	3149	3604
15					4103	3848				3975
21	4092	3969	4089		4184	4093	4046	4111		4083
25						4241				4241
29	4343	4261	4338							4314
32										
33.5	4507									4507
35		4636								4636
38	4695	4565					4612			4624
46							4911			4911
53	5179	4983					5136			5099
60							5227			5227
63		5060		5389						5224
64	5342									5342
67		5270								5270
70.5	5562						5591			5576
77.5	5638	5397								5517

*) After Sclater et al. [71 S].

1 = North Pacific; 2 = South Pacific < 3 cm/year; 3 = South Pacific > 3 cm/year; 4 = southeast Indian Ocean; 5 = central Indian Ocean; 6 = Carlsberg Ridge; 7 = South Atlantic; 8 = North Atlantic < 40° N; 9 = North Atlantic > 40° N.

1.7 Chronology of sea floor topography

1.7.1 The time scale as a geologic factor

The idea of *geologic time*, which is fundamental to geology, starts with James Hutton (1726–1797); its protagonists were Charles Lyell (1797–1875), and Charles Darwin (1809–1882). [1842 D].

The relative time scale depends on palaeontologic classification, the absolute time scale on radiometric data [64 A], and the geomagnetic time scale on palaeomagnetism. See Fig. 79 and LB, NS, vol. V/2, Physics of the interior of the earth.

Before the understanding of sea-floor spreading and plate tectonics, the oceans were supposed to be a permanent, stable receptacle of continental and volcanic debris. As a treatise on older theories, see Ph.H. Kuenen in "Marine Geology" [50 K]. M. Ewing found that the sedimentary columns for typical basins in the Atlantic are only about 500 m thick, and in the Pacific, a mere 300 m. It became clear that the old view of the deep ocean floor as a sedimentary memory reaching back for billions of years was wrong. For *palaeotopography* of the oceans, see J. P. Kenneth, E. Seibold and W. H. Berger [82 K, 82 S 1, and 65 B 2] and M. Sarntheim, "CLIMAP program" [82 S 1]. See also [80 S 3].

The geologic age of the existing sea floor topography. See Fig. 76; compare [85 L].

There is no record in the abyss which reaches back as far as $200 \cdot 10^6$ years (pre-jurassic), and either there was no abyss then or the relics of these ancient seas have been completely destroyed. The deposits of earlier seas are found exclusively on the continents. Only some submarine plateaus are of older age as terranes. See Fig. 53.

This understanding developed from knowledge of measured data of the sea floor gained by drilling of sediment layers, deep-sea drilling (DSDP, IPOD), seismic records, acoustic seismic measurements, radiometric technique, gravity, and heat flow.

The drilling vessel, the "Glomar Challenger" sailed for the first leg from Orange, Texas, on July 20, 1968. By the end of Leg 78 (April 2, 1981), the "Glomar Challenger" had:

Drilled 910 holes at 543 sites for a total of 208235 m

Retrieved 16164 cores totaling a length of 72832 m

Traveled a total of 591962 km.

Among its technological records are:

Deepest penetration: 1741 m below the sea floor (site 398, Leg 47, in the North Atlantic Ocean)

Longest drill string suspended: 7060 m (site 461A, Leg 60, in the Pacific Ocean)

Deepest penetration in basalt: 623 m (site 448A, Leg 59, in the Pacific Ocean)

Oldest sediments cored: Upper Jurassic (site 534, Leg 76, Blake-Bahama Basin) [81 E].

Quantitative stratigraphic correlation

IGCP Project No. 148 "Evaluation and Development of Quantitative Stratigraphic Correlation Techniques" was initiated in 1976 for the purpose of developing computer-based mathematical theory and analysis of geological information which can be applied to automated correlation techniques in stratigraphy. Specific problems were solved by establishing regional standards of ordered stratigraphic events and performing correlations on the basis of these standards.

Publications emanating from the project, including computer program, have been listed annually in "Geological Correlation". The IGCP catalogues 1973–1979 and 1980–1982 contain further publications [83 A]. See Fig. 76. The new "Ocean Drilling Program" (ODP) started 1985 with the vessel "Joides Resolution".

1.7.2 The geomagnetic time scale

The geomagnetic time scale is based on the sequence of polarity reversals of the earth's magnetic field. This sequence, determined from profiles of magnetic intensity, run across the width of the expanding oceans, and, dated by correlations with vertical sequences in sediments and lava rocks of the oceanic plates, has provided an important tool in the unraveling of ocean floor history. The width of the expansion rate of geomagnetic patterns of the reversal induced bands varies regionally with the ocean floor, so that geomagnetic bands of the same age in different parts of the world differ in width. There are differences between the data of the 3 larger oceans. Figs. 78, 76. Compare [85 B 2, 85 S].

Nomenclatural convention in magnetic polarity scale studies:

Periods of one polarity of the order of 10^6 years are called epochs, and those lasting about $50 \cdot 10^3 \cdots 100 \cdot 10^3$ years are termed events. The four epochs recognized for the Pleistocene and Pliocene (Brunhes, Matuyama, Gauss, Gilbert), are named after past workers in terrestrial magnetism, while palaeomagnetic events have been named after the locations where they were first recognized. Older palaeomagnetic epochs have been numbered. See Fig. 79.

"Magnetostratigraphic polarity units", a supplementary chapter of the ISSC (International Stratigraphic G), was published by the IUGS (International Subcommission on Stratigraphic Classification) and the IUGS/IAGA Subcommission on a magnetic polarity time scale.

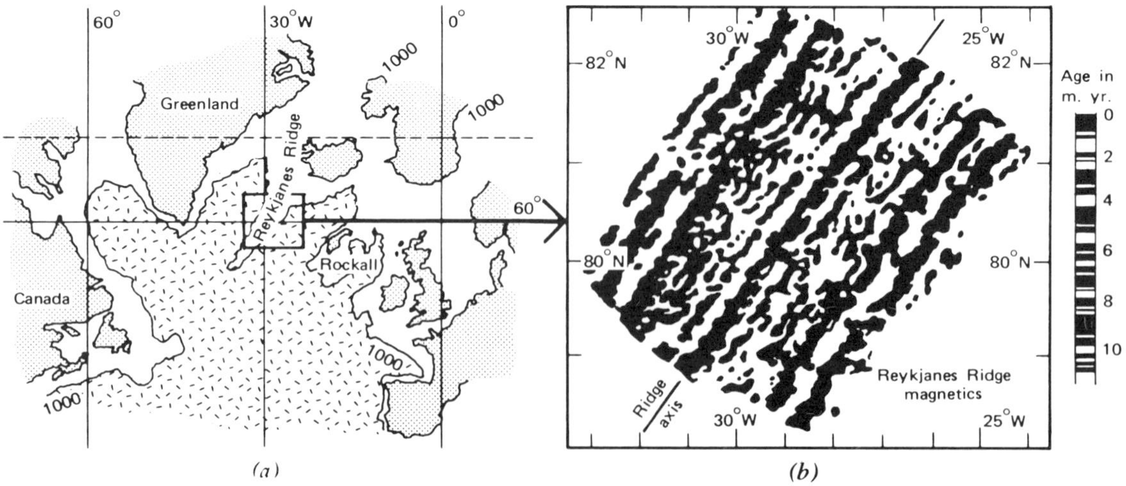

Fig. 78. (a) The location of the region enlarged in (b) along the Reykjanes Ridge in the northern Atlantic. The contour is at 1000 fathoms. (b) The magnetic anomaly patterns on both sides of the ridge axis. Black areas are regions of positive magnetic anomaly (in the direction of the earth's present field), and negative anomalies are the white regions in between. From F. J. Vine [69 V]. Compare LB V/2, subvolume b, ch. 4.2.1, Fig. 15.

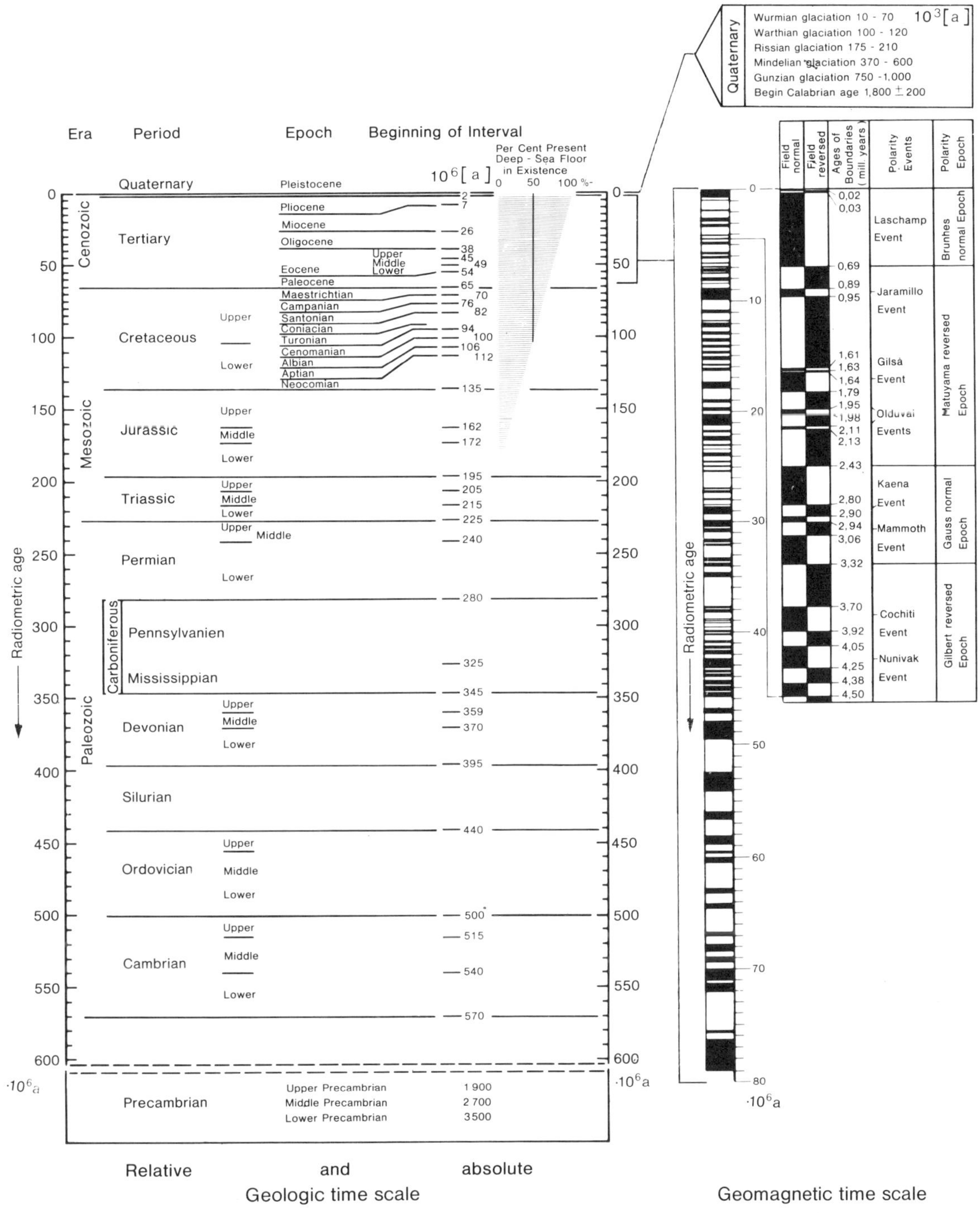

Fig. 79. Geologic and geomagnetic time scales. Compilation after Heezen and Hollister [71 H] and other sources. Gierloff-Emden 1983. Left: relative geologic time scale after paleontologic features in the stratigraphic classification and absolute radiometric age. Top: subclassification of the Quaternary. Right: geomagnetic time scale for the last $80 \cdot 10^6$ a and $4.5 \cdot 10^6$ a, respectively [69 C]. Black: normal polarity. White: reversed polarity. Event: short period of reversed polarity. The age of the present sea floor reaches back to $200 \cdot 10^6$ a, i.e. the upper Jurassic [77 H 1]. Additional references [69 W] for the period $0 \cdots 5 \cdot 10^6$ a are [82 K] and for $(60 \cdots 100) \cdot 10^6$ a [82 K]. First crust, oldest rocks $4000 \cdot 10^6$ a. Used by the IUGS Commission on Geochronology as a base for discussion of a revised geological time scale. Compare [69 W] and [85 B 2, 85 S]. Compare LB V/2, subvolume b, ch. 4.3.3, Figs. 24⋯28.

1.8 References for 1

1725 M Marsili, L.F.: Histoire physique de la mer, Amsterdam **1725**.
1842 D Darwin, Ch.: The structure and distribution of coral reefs, London: **1842**.
1855 M Maury, M.F.: The physical geography of the sea, London: T. Nelson & Son, **1855** and **1860**, p. 493, maps. Reprint: Cambridge, Mass.: Harvard Univ. Press **1963**.
1885 S Suess, E.: Das Antlitz der Erde. 3 Vols., Wien, Leipzig: **1885–1909**.
1888 M Murray, J.: On the height of the land and the depth of the ocean. Scot. Geogr. Mag. **4** (1888) lpp.
1894 K Karstens, K.: Eine neue Berechnung der mittleren Tiefen der Ozeane, Kiel, Leipzig: Lipsius & Tischer, **1894**, p. 32, Tab. 27.
1899 I International Geographical Congress: Transactions of the Seventh International Congress of 1899, Berlin. Part 2. **1901**, 370–392.
07 K Krümmel, O.: Handbuch der Ozeanographie. Vol. 1, 2nd ed., Stuttgart: Engelkorn **1907**, p. 526.
12 M Murray, J., v. Hjort, J.: The depths of the ocean, London: Macmillan **1912**, p. 821.
21 K Kossinna, E.: Veröff. Inst. Meereskunde, N.F., R.A, Berlin **2**, 9 (1921) 69.
28 W Wegener, A.: Die Entstehung der Kontinente und Ozeane, Braunschweig: Vieweg & Sohn **1928**, p. 231.
30 M Matthews, D.J.: Tables of the velocity of sound in pure water and sea water for use in echo-sounding and sound-ranging, Brit. Admiralty Hydrogr. Dep. Pub. H.D. 282, 2nd ed. **1930**.
33 K Kossinna, E., in: Handbuch der Geophysik, 2, (Gutenberg, B., ed.), Berlin: Borntraeger **1933**, p. 869–954.
33 M Maurer, H., Stocks, Th.: Wiss. Ergebn. Deut. Atlant. Exped. „Meteor", 1925–27, **2**, 1 (1933).
34 W Wüst, G.: Naturwissenschaften **16** (1934).
35 S Schott, G.: Geographie des Indischen und Stillen Ozeans, Hamburg: Boysen **1935**, p. 435.
36 F Farquharson, W.J.: The John Murray Expedition. Vol. 3, Chart I (Gulf of Aden) (echo-sounding of "F.G. Mabahiss" 1933/34), London: British Museum, **1936**.
36 S Stocks, Th.: Geogr. Zschr. **42**, 5 (1936) 170.
36 W Wüst, G.: Pet. Geogr. Mitt. **2** (1936) 33. Engl. transl.: Hydrogr. Rev. Monaco, **13**, 2, **1936**, 46.
39 S Stocks, Th.: Meteorwerk III, 1, 2, Berlin **1939**.
42 M Meinardus, W.: Ann. Hydrogr. Maritim. Met., Berlin **70** (1942) 225.
42 S Sverdrup, H.U., Johnson, M.W., Fleming, R.H.: The oceans, their physics, chemistry and general biology, Englewood Cliffs, N.J.: Prentice Hall, **1942** (new ed. 1970), p. 1087.
42 W Wißmann, H.V., Rathjens, C., Kossmat, F.: Geol. Rdsch. **33**, 4–6 (1942) 221.
44 S Schott, G.: Geographie des Atlantischen Ozeans, 3rd ed., Hamburg: Boysen **1944**, p. 438.
46 H Hess, H.H.: Am. J. Sci. **244** (1946) 722.
49 K Kalle, K.: Die waagerechte Gliederung und die Tiefen des Meeres. Handbuch der Seefischerei Nordeuropas, Stuttgart: E. Schweizerbart'sche Verlags-Buchhandlung **1949**.
50 K Kuenen, P.H.: Marine Geology, New York: J. Wiley & Sons **1950**, p. 568.
51 M Morgan, J.P.: La. Geol. Surv. Geol. Bull. **35** (1951) 208.
52 H Heezen, B.C., Ewing, M.: Am. J. Sci. **250** (1952) 849.
53 D Dietz, R.S., Menard, jr., H.W.: J. Geol., Chicago **61** (1953) 99.
53 I IHO.: Limits of oceans and seas. Int. Hydrographic Organisation, Spec. Publ. 28, Monaco **1953**, p. 38.
54 L Luskin, B., Heezen, B.C., Ewing, M.: Deep Sea Res. **1** (1954) 131.
55 W Wiseman, J.D.H., Ovey, C.D.: Deep-Sea Res. **2** (1955) 93.
56 H Hamilton, E.L.: Geol. Soc. Am. Mem. **64** (1956) 97.
57 W Wüst, G., in: Wissenschaftliche Ergebnisse der Deutschen Atlantischen Expedition auf dem Forschungs- und Vermessungsschiff Meteor 1925–27 (Defant, A., ed.), Berlin: W. de Gruyter **1957**, 201–420.
58 H Heezen, B.C., Tharp, M.: Physiographic diagrams of the oceans. Descriptive sheets to accompany physiographic diagrams. Geol. Soc. Am., New York **1958**.
58 M Menard, H.W.: Bull. Geol. Soc. Am. **69**, 9 (1958) 1179.
59 H Heezen, B.C., Tharp, M., Ewing, E.: The floors of the ocean. Geol. Soc. Am. Spec. Pap. 65, **1959**, p. 122.
60 E 1 Emery, K.O.: The sea off southern California, a modern habitat of petroleum, New York: J. Wiley & Sons **1960**, p. 366.
60 E 2 Ewing, J.I., Worzel, J.L., Ewing, M.: J. Geophys. Res. **67**, 6 (1960) 2509.
60 H 1 Heezen, B.C.: Turbidity currents. Encyclopedia of technology, New York: McGraw Hill **1960**.
60 H 2 Heezen, B.C.: Sci. Am. **10** (1960) 1.
60 U Udintsev, G.B.: Bottom relief of the western part of the Pacific Ocean. Oceanological Researches **2**, 5 (1960).
61 D 1 Dietz, R.S.: Nature, London **190** (1961) 845.

61 D 2 Dietz, R.S., Shumway, G.: Bull. Geol. Soc. Am. **72** (1961) 1319.
61 P Pfannenstiel, M.: Geogr. Rdsch. **13** (1961) 87.
62 A Arx, W.S.: An introduction to physical oceanography, Reading, Mass.: Addison Wesley Comp. **1962**, p. 422.
62 D Dietz, R.S., in: Continental drift (Runcorn, S.K., ed.), New York: Academic Press **1962**, p. 33.
62 H Hess, H.H., in: Petrological studies, a volume in honor of A.F. Buddington (Engel, A.E.J., et al., eds.), New York: Geological Society of America, **1962**, 599–620.
63 C Carey, S.W.: Aust. J. Sci. **25** (1963) 369; 479.
63 D Dietz, R.S.: J. Geol. **71** (1963) 314.
63 F Fisher, R.L., Hess, H.H.: Trenches, in: The Sea, Vol. 3, 3rd ed. (Hill, M.N., ed.), New York, London: Wiley Interscience **1963**, 411–436.
63 G Guilcher, A., in: The Sea, Vol. 3 (Hill, M.N., ed.), New York, London: Wiley Interscience **1963**, 281–311.
63 H 1 Heezen, B.C., Ewing, M., in: The Sea, Vol. 3, (Hill, M.N., ed.), New York, London: Wiley Interscience **1963**, 388–410.
63 H 2 Heezen, B.C., Menard, H.W., in: The Sea, Vol. 3 (Hill, M.N., ed.), New York, London: Wiley Interscience **1963**, 233 280.
63 H 3 Hill, M.N. (ed.): Topography of the Middle America Trench, in: The Sea, Vol. 3, New York, London: Wiley Interscience **1963**.
63 H 4 Hill, M.N. (ed.): The Sea. Ideas and observations on progress in the study of the sea, Vol. 3, New York, London: Wiley Interscience **1963**, p. 963.
63 M Menard, H.W., Ladd, H.S., in: The Sea. Vol. 3 (Hill, M.N., ed.), New York, London: Wiley Interscience **1963**, p. 365–387.
63 S 1 Shepard, F.P.: Submarine geology. 2nd. ed., New York: Harper & Row (with chapters by Inman, D.L. and Goldberg, E.D.) **1963**, p. 557.
63 S 2 Shepard, F.P.: Submarine canyons, in: The Sea, Vol. 3 (Hill, M.N., ed.), New York, London: Wiley Interscience **1963**.
63 U 1 Ulrich, J.: Erdkunde **17** (1963) 190.
63 U 2 Ulrich, J.: Geogr. Rdsch., Braunschweig **15** (1963) 163.
63 V Vine, F.J., Matthews, D.H.: Nature, London **199** (1963) 947.
63 W Wilson, J.T.: Sci. Am. **208** (1963) 85.
64 A Afanassyev, G.D., et al.: The project of a revised geological time-scale in absolute chronolgy. Contr. Geol. Sovietiques Congr. Geol. Intern. 22ᵉ Sess. India **1964**, p. 287–324.
64 D Dietz, R.S.: Am. Sci. **52**, 1 (1964) 50.
64 H 1 Heezen, B.C., Gerard, R.D., Tharp, M.: J. Geophys. Res. **69**, 4 (1964) 733.
64 H 2 Heezen, B.C., Hollister, C.D.: Mar. Geol. **1** (1964) 141.
64 M 1 Menard, H.W.: Marine Geology of the Pacific, New York: McGraw-Hill **1964**, p. 271.
64 M 2 Mero, J.L.: The mineral resources of the sea. Elsevier Oceanographic Series 1, Amsterdam: Elsevier Scientific Publ. Comp. **1964**, p. 312.
65 B 1 Blackett, P.M.S., Bullard, E., Runcorn, S.K. (eds.): A symposium on continental drift, London **1965**, p. 323.
65 B 2 Bullard, E., Everett, J.E., Smith, A.G.: Philos. Trans. R. Soc. London, Ser. A **258** (1965) 41.
65 E Emery, K.O.: Bull. Am. Assoc. Petrol. Geolog. **49** (1965) 1379.
65 H Hydrography of the Navy. Admiralty manual of hydrographic surveying, Vol. I., London: Her Majesty's Stationary Office **1965**, p. 671.
65 K Krause, D.C., Menard, H.W.: Mar. Geol. **3** (1965) 169.
65 L LePichon, X., Houtz, R.E., Drake, C.L., Nafe, J.E.: J. Geophys. Res. **70**, 2 (1965) 319.
65 S Shepard, F.P.: Bull. Am. Assoc. Petr. Geol. **49** (1965) 304.
65 W Wilson, J.T.: Nature London **207** (1965) 343.
66 D Dietz, R.S.: Am. J. Sci. **264** (1966) 177.
66 F Fairbridge, R.W. (ed.): The encyclopedia of oceanography. Encyclopedia of earth science series, Vol. 1, New York: Reinhold **1966**, p. 1021.
66 H 1 Heezen, B.C., Tharp, M.: Philos. Trans. R. Soc. London, Ser. A, **259** (1966) 137.
66 H 2 Heezen, B.C., Hollister, C.D., Ruddiman, W.F.: Science **152** (1966) 502.
66 M 1 Menard, H.W., Smith, S.M.: J. Geophys. Res. **71**, 18 (1966) 4305.
66 M 2 Meteor-Forschungsergebnisse, Reihe C, Geologie u. Geophysik (Deutsche Forschungsgemeinschaft, ed.), Berlin: Borntraeger **1966** onward.

66 S Shepard, F.P., Dill, R.F.: Submarine canyons and other sea valleys, Chicago: Rand McNally **1966**, p. 381.
66 V Vine, F.J.: Science, Washington **154** (1966) 1405.
67 C Cox, A., Dalrymple, G.B., Doell, R.R.: Sci. Am. **216**, 2 (1967) 44.
67 H Hollister, Ch.D., Heezen, B.C.: The floor of the Bellingshausen Sea. J. Hopkins oceanographic studies **3** (1967) 178, Fig. 17-1.
67 S Shepard, F.P.: Submarine geology, New York: Harper & Row **1967**, p. 557.
67 U Uchupi, E.: Gulf Coast Ass. Geol. Soc. Trans. **17** (1967) 161.
67 V Vacquier, V., Uyeda, S.: Bull. Earthquake Res. Inst., Tokyo Univ. **45** (1967).
68 A Allen, J.R.: Current ripples, their relation to patterns of water and sediment motion, Amsterdam: North Holland Publ. **1968**, p. 433.
68 B Beaglehole, J.C. (ed.): James Cook, the complete journals, 4 Vols., 1 Atlas. London: **1961–67**.
68 C Colter, Ch.H.: A history of nautical astronomy, London: Hollis-Carter **1968**, p. 387.
68 D Dietrich, G., Ulrich, J.: Atlas zur Ozeanographie. Meyers Großer Physischer Weltatlas, BI-Hochschulatlanten, Vol. 7, Bibliogr. Inst. Mannheim **1968**, p. 76.
68 H 1 Heezen, B.C., Wilson, L., in: Encyclopedia of geomorphology (Fairbridge, R.W., ed.), Vol. 3, New York: Reinhold, **1968**, p. 1079–1097.
68 H 2 Heirtzler, J.R.: Sci. Am. **219**, 6 (1968) 60.
68 H 3 Holeman, J.H.: Water Resources Research **4**, 4 (1968) 737.
68 I 1 Initial Reports of the Deep Sea Drilling Project, **1968** onward.
68 I 2 Isacks, B., Oliver, J., Sykes, L.R.: J. Geophys. Res. **73** (1968) 5855.
68 L LePichon, X.: J. Geophys. Res. **73** (1968) 3661.
68 M Morgan, W.J.: J. Geophys. Res. **73** (1968) 1959.
68 V Vine, F.J.: The history of the earth's crust, in: Phinney, R.A., Princeton Univ. Press **1968**, 73–89.
69 C Cox, A.: Science **163** (1969) 237.
69 D Dewey, J.F.: Earth Planet. Sci. Lett. **6** (1969) 189.
69 E 1 Emery, K.O.: Sci. Am. **221**, 3 (1969) 32.
69 E 2 Emiliani, C., Rona, E.: Science **166** (1969) 1551.
69 H 1 Heezen, B.C., Johnson, G.L.: Bull. Inst. Oceanogr. Monaco **67** (1969) 1.
69 H 2 Heirtzler, J.R.: Naturw., Berlin **56**, 7 (1969) 341.
69 L Lowrie, A., Escowitz, E.C.: Global Ocean Floor Analysis and Res. Data Ser. **1** (1969) 975pp.
69 R Ross, D.A., Degens, E.T.: Hot brines and recent heavy metal deposit in the Red Sea, Berlin, Heidelberg, New York: Springer **1969**, p. 600.
69 V Vine, F.J.: J. Geol. Educ. **17** (1969) 56.
69 W Wedepol: Handbook of Geochemistry, Vol. 1, Berlin, Heidelberg, New York: Springer **1969**.
70 A Admiralty manual of navigations. Vol. 1 and 2, London: Her Majesty's Stationary Office **1970**, p. 510 and 328.
70 C Cohen, P.M.: Bathymetric navigation and charting, Annapolis: US Naval Inst. **1970**, 1380.
70 D 1 Dewey, J.F., Bird, J.M.: J. Geophys. Res. **75** (1970) 2632.
70 D 2 Dietz, R.S., Holden, J.C.: Sci. Am. **223**, 4 (1970) 30.
70 G 1 GEBCO: Regulations for the bathymetric charts of the oceans, Monaco: IHB **1970**, p. 18.
70 G 2 Gierloff-Emden, H.G., Schroeder-Lanz, H., Wieneke, F.: Beiträge zur Morphologie des Schelfes und der Küste bei Kap Sines (Portugal), Meteor Forsch. Erg., R. C. 3, Berlin-Stuttgart: Boerntraeger **1970**, 65–84.
70 H 1 Heezen, B.C., Tharp, M.: Major topographic divisions of the continental margins, Washington: Office of the Geographer, Dept. of the State **1970**.
70 H 2 Holcombe, T.L., Heezen, B.C.: Patterns of relative relief, slope and topographic texture in the North Atlantic Ocean, Techn. Rep. No. 8, Cu-8-70 Naval ships systems, Lamont-Doherty Geol. Observatory of Columbia University, Palisades, New York, **1970**, p.116.
70 K Krause, D.C., Kanaev, V.: Int. Hydr. Rev. **47** (1970) 23.
70 L Luvendyk, B.P.: Geol. Soc. Am. Bull. **81** (1970) 2237.
70 M 1 Maxwell, A.E.: The Sea, Vol. 4: New concepts of sea floor evolution, part 1 and 2, New York: Wiley Interscience **1970**.
70 M 2 Menard, H.W., Chase, T.E., in: The Sea, Vol. 4 (Maxwell, A.E., ed.), New York: Wiley Interscience **1970**, p. 421–443.
70 P Pfannenstiel, M.: Geol. Rdsch. **60** (1970) 3.
70 R Reimnitz, E., Gutierrez-Estrada: Mar. Geol. **8** (1970) 245.
70 W Wilson, J.T., in: The Sea, Vol. 4 (Maxwell, A.E., ed.), New York: Wiley Interscience **1970**, p. 664.

71 C 1 Chase, T.E.: Map: Topography of the North Pacific. Scripps Institution of Oceanography. IMR Technical Report, Series TR-17, 1 : 13 Mill., 10 sheets in 1 : 6 500 000. Inst. of Marine Resources, San Diego: Univ. of California **1971**.
71 C 2 CNEXO: Bull. Int. CNEXO **34**, Brest-Paris (1971) 7–10.
71 C 3 Curray, J.R., Moore, D.G.: Geol. Soc. Am. Bull. **82** (1971) 563.
71 D 1 Deacon, M.: Scientists and the sea 1650–1900, London: Academic Press **1971**, p. 445.
71 D 2 Decker, R.W., Einarsson: Trans. Am. Geophys. Union **52** (1971) 352.
71 D 3 Delany, F.M. (ed.): The geology of the East Atlantic continental margin. 3 Vols. Natural Environment Research Council. Inst. of Geol. Sciences, Report No. 70/13, 14, 15, ICSU/SCOR working party, 31st symposium, Her Majesty's Stat. Off., London **1971**, p. 175, 106, 170.
71 H Heezen, B.C., Hollister, Ch.D.: The face of the deep, New York: Oxford Univ. Press **1971**, p. 656.
71 I International Hydrographic Bureau: Nomenclature of ocean bottom features. Int. Hydrogr. Rev. **48** (1971) 203.
71 K Karig, D.E.: J. Geophys. Res. **76**, (1971) 2542.
71 L Lonardi, A.G., Ewing, M., in: Physics and chemistry of the earth, Vol. 8, New York, London: Pergamon Press **1971**, p. 123–251.
71 N National Geographic Society, USA. Physiographic maps of the oceans, Washington **1967**, **1968**, **1969**, **1971**.
71 S Sclater, J.G., Andersen, R.N., Bell, M.L.: J. Geophys. Res. **76** (1971) 7889.
71 V Vine, F.J., Hess, H.H., in: The Sea, Vol. 4 (Maxwell, A.E., ed.), New York, London: Wiley Interscience **1971**, p. 587–622.
71 W Wagner, K.H.: Atlas zur Physischen Geographie. Bibliogr. Inst. Mannheim **1971**, p. 59.
72 B 1 Belderson, R.H., Kenyon, N.H., Stride, A.H., Stubbs, A.R.: Sonographs of the sea floor. A picture atlas, Amsterdam, London, New York: Elsevier Scientific Publ. Comp. **1972**, p. 185.
72 B 2 Bogorov, G.V., Ilyin, A.V.: Microrelief of the ocean-floor on the basis of analysis of photo- and stereo-material. Oceanology International **1972**, p. 363.
72 D Dewey, J.F.: Sci. Am. **226** (1972) 56.
72 E Emery, K.O., Uchupi, E.: Am. Ass. Petr. Geol. Mem. Tulsa **17** (1972) 532.
72 H Heezen, B.C., Tharp, M., Bentley, C.: Antarctic Map Folio Ser. **16**, Am. Geol.Soc., New York (1972).
72 J Jackson, E.D., Silver, E.A., Dalrymple, G.B.: Geol. Soc. Am. Bull. **83** (1972) 601.
72 K Kerr, A.J., Kordick, A.: Oceanographic carthography. Intern. Cartographic Assoc., Ottawa **1972**, p. 132.
72 M Morgan, W.J.: Plate motions and deep mantle convection, in: Geol. Soc. Am. Mem. 132 (Shagam, R., ed.), **1972**, 7–22.
72 S Swift, D.J.P., Duane, D.B., Pilkey, O.H. (ed.): Shelf sediment transport, process and pattern, Stroudsburg: Dowden, Hutchinson, Ross **1972**, p. 656.
73 C Cox, A. (ed.): Plate tectonics and geomagnetic reversals, San Francisco: Freeman & Co. **1973**, p. 702.
73 H 1 Hallam, A.: A revolution in earth sciences; from continental drift to plate tectonics, Oxford: Clarendon Press **1973**, p. 127.
73 H 2 Heezen, B.C., Matthews, J.L., Catalano, R., Rawson, M., in: Initial Reports of the Deep Sea Drilling Project, Vol. 20, Washington: US Government Printing Office **1973**, p. 653–723.
73 L 1 Laughton, A.L., Roberts, D.G., Graves, R.: Int. Hydr. Rev. **L**, 1 (1973).
73 L 2 LePichon, X., Francheteau, J., Bonin, J.: Plate tectonics, Amsterdam, London, New York: Elsevier Scientific Publ. Comp. **1973**, p. 300.
73 N Nairn, A., Stehli, F.G. (eds.): The ocean basin and margins. 7 Vols.: 1 The South Atlantic, 2 The North Atlantic, 3 The Gulf of Mexico and the Caribbean, 4 The Mediterranean, 5 The Arctic Ocean, 6 The Indian Ocean, 7 The Pacific Ocean, New York: Plenum Press **1973–1981**.
73 R 1 Reineck, H.E., Singh, I.B.: Depositional sedimentary environments, Berlin, Heidelberg, New York: Springer **1973**, p. 439.
73 R 2 Rust, U., Wieneke, F.: Münchener Geogr. Abh. **9** (1973) 43.
73 S Sigl, W., Hinz, K., Garde, S.: Meteor Forsch. Erg., R. C, **14** (1973) 54.
73 T Tarling, D.H., Runcorn, S.K.: Implications of the continental drift to earth sciences. 2 Vols., London, New York **1973**, p. 622, 559.
73 W Wilson, J. Tuzo: Tectonophysics **19** (1973) 149
74 A American Geographical Society: Map. The floor of the oceans (based on studies by B.C. Heezen and M. Tharp). 1 : 40 000 000, equatorial scale, New York: American Geographical Society **1974**.
74 B 1 Berger, W.H.: Classification of deep sea sediments, in: The geology of continental margins (Burk, C.A., Drake, C.L., eds.), Berlin, Heidelberg, New York: Springer **1974**.

74 B 2 Burk, C.A., Drake, C.L. (eds.): The geology of continental margins, Berlin, Heidelberg, New York: Springer **1974**, p. 1069.
74 C 1 Cox, D.P., Cox, H.R.: Geology principles and concepts, San Francisco: Freeman **1974**.
74 C 2 Cruickshank, M.J., in: Continental margins (Burk, C.A., Drake, C.L., eds.), Berlin, Heidelberg, New York: Springer **1974**, p. 965–1000.
74 E Emiliani, C., Shackleton, N.J.: Science **183** (1974) 511.
74 H 1 Herman, Y.: Marine geology and oceanography of the Arctic Seas, Berlin, Heidelberg, New York: Springer **1974**, pp. 397, 194 Figs.
74 H 2 Hinz, K.: Meteor Forsch. Erg. R.C, **17**, (1974) 67.
74 I IOC: Metallogenesis, hydrocarbons and tectonic patterns in Eastern Asia. Rep. of the IDOE Work Shop, Bangkok, Thailand, Sept. 1973, Comm. for Coordin. of Joint Prosp. for Miner. Resourc. in Asia Offshore Areas (CCOP/IOC), **1974**, p. 150.
74 P Pitman, W.C., Larson, L., Herron, E.M.: Geol. Soc. Am. Bull. Map and Chart Ser. MC-**6** (1974).
74 S 1 Schlee, S.: Die Erforschung der Weltmeere, Oldenburg-Hamburg: Stalling **1974**, p. 307.
74 S 2 Seibold, E., in: Lehrbuch der Allgemeinen Geologie, Vol. 1 (Brinkmann, R., ed.), Stuttgart: Enke **1974**, p. 290–511.
74 S 3 Smith Walton, F.G. (ed.): Handbook of marine science, Vol. 1, Cleveland, Ohio: CRC-Press **1974**, p. 627.
74 T Tarr, S.C.: World seismicity map, 1 : 39 000 000. US Geol. Survey, from data of the Nat. Oceanic and Atmospheric Admin., Washington D.C.: Dept. of the Interior **1974**.
75 A Arcyana: Science **190** (1975) 108.
75 B Berger, H., Bé, A.W.H., Sliter, W.: Cushman Foundation for Foraminiferal Res., Spec. Pub. **13**, 1 (1975) 9.
75 D Dietrich, G., Kalle, K., Krauss, W., Siedler, G.: Allgemeine Meereskunde. Eine Einführung in die Ozeanographie, Berlin, Stuttgart: Borntraeger **1975**, p. 593.
75 H Hallam, A.: Sci. Am. **232** (2) (1975) 88.
75 I 1 Ingham, A.E.: Sea surveying, London, New York: Wiley Interscience **1975**, p. 306.
75 I 2 International Indian Ocean Experiment (IIOE): Geological-geophysical atlas of the Indian Ocean, Moscow: Academy of Sciences of the USSR **1975**, p. 151.
75 K King, C.A.M.: Introduction to marine geology and geomorphology, London: Arnold **1975**, p. 309.
75 R 1 Reineck, H.E., Singh, S.B.: Depositional sedimentary environments, Berlin, Heidelberg, New York: Springer **1975**, p. 439.
75 R 2 Renard, V., Schrumpf, B., Sibuet, J.C.: Bathymetrie detaillée d'une partie de vallée du rift et de la fault transformate, près de 36°50′ de l'Océan Atlantique. Project FAMOUS, CNEXO, Brest, 4 feuilles **1975**.
75 S Swift, D.J.P.: Mar. Geol. **18**, 2 (1975) 105.
75 V Vinogradov, A.P., Udintsev, G.B.: The rift zones of the world ocean, Cambridge **1975**, p. 540.
76 A Agence de cooperation culturelle et technique: Vocabulaire de l'océanologie. Paris, **1976**, p. 433.
76 B Burke, K.C., Wilson, J.Tuzo: Sci. Am. **235** (1976) 46.
76 C CLIMAP Project Members: Science **191** (1976) 1131.
76 D 1 Dewey, J.F.: Plate tectonics, in: Continents adrift and continents aground (Wilson, J. Tuzo, ed.), Sci. Am., San Francisco: W.H. Freeman & Co. **1976.**
76 D 2 Drake, E.T.: Geology **1** (1976) 41.
76 D 3 Drake, D.E., in: Marine sediment transport and environmental management (Stanky, D.J., Swift, D.J.P., eds.), New York: Wiley Interscience **1976**, p. 127–158.
76 F Freiesleben, H.Ch.: Geschichte der Navigation, Wiesbaden: F. Steiner **1976**, p. 151.
76 G 1 General Bathymetric Chart of the Oceans (GEBCO): IHO, Monaco **1976** onward.
76 G 2 Geological World Atlas: 1 : 10000000; in progress since 1976. Ocean sheets 1 : 36 000 000, 22 sheets, Commission for the Geological Map of the World (CGMW), Int. Union of Geological Sciences, UNESCO, Washington **1976** onward.
76 J Johnson, G.L., Vogt, P.R., Hey, R., Campise, J., Lowrie, A.: Mar. Geol. **21** (1976) 81.
76 L Loncarevic, B.D., in: Mid-Atlantic Ridge (Rona, P.A., ed.), Geol. Soc. Am. Int. Publ. 5, **1976**, 451–470.
76 V Vanney, J.R., Johnson, G.L.: Mar. Geol. **22** (1976) 71.
76 W 1 Whiteaker, J.H.McD: Submarine canyons and deep sea fans. Modern and ancient benchmark papers in geology, 24, Stroudsburg: Dowden, Hutchinson, Ross **1976**.
76 W 2 Wilson, J. Tuzo (ed.): Continents adrift and continents aground. Sci. Am., San Francisco: W.H. Freeman **1976**, p. 230.
76 W 3 Withmarsh, R.B., Laugthon, A.S.: Deep Sea Res., **23** (1976) 23.

77 A Angenheister, G., Bartels, J.: Geophysik, in: Fischer-Lexikon, Frankfurt: S. Fischer **1977**, p. 374.

77 B Ballard, R.D., Moore, J.G.: Photographic atlas of the Mid-Atlantic Ridge rift valley, Berlin, Heidelberg, New York: Springer **1977**, p. 114, 183 Figs.

77 C 1 Coleman, J.M., Garrison, L.E.: Mar. Technol. **2** (1977) 9.

77 C 2 Cronan, D.S., in: Marine manganese deposits (Lusby, G.P., ed.), Amsterdam, London, New York: Elsevier Scientific Publ. Comp. **1977**, p. 11–44.

77 E Embley, R.W., Langseth, M.G.: Mar. Geol. **25** (1977) 279.

77 H 1 Heezen, B.C. (ed.): Mar. Geol., Spec. Issue, **23**, 1, 2 (1977) 1–200.

77 H 2 Heezen, B.C., Tharp, M.: World ocean floor, Mercator-projection, scale 1:23 230 000. US Navy, Office of Naval Research, Arlington, Va. **1977**.

77 H 3 Holcombe, T.L.: Geo-Journal **1**, 6, Wiesbaden (1977) 25–48.

77 I IHO: Precise positioning system for hydrographic surveys. Spec. Publ. 39 (Series 1977), nos. 39-2-1 to 39-2-7, Monaco: IHO **1977**.

77 N NOAA: Icosahedron world map, earthquake data services and publication, Boulder, Colorado: Nat. Geophys. Solar-Terrestrical Data Center **1977**.

77 P 1 Parsons, B., Sclater, J.C.: J. Geophys. Res. **82** (1977) 819.

77 P 2 Piper, D.Z., Williamson, M.E.: Mar. Geol. **23** (1977) 285.

77 S 1 Seely, D.R., Dickinson, W.R.: Geology of continental margins. Continuing Education Course, Ser. 5, Am. Ass. Petr. Geol. **1977**.

77 S 2 Shepard, F.P.: Geological oceanography, New York: Crane-Russack **1977**, p. 214.

77 T Talwani, M., Pitman, W.C.: Island arcs, deep-sea trenches and back arc basins, Maurice Ewing Ser. 1, Washington, D.C.: AGU **1977**.

77 V Vanney, J.R.: Geomorphologie des plates-formes continentales, Paris: Doin-Editeurs **1977**, p. 302.

77 W Weissel, J.K., Hayes, D.E., Herron, E.M.: Mar. Geol. **25** (1977) 231.

78 E Ellis, M.Y. (ed.): Coastal mapping handbook, US Dep. of the Interior, NOAA, Washington **1978**, p. 200.

78 F Flemming, B.W.: Mar. Geol. **26** (1978) 177.

78 H 1 Hayes, D.E., Taylor, B.: A geophysical atlas of the East and Southeast Asian Seas. Geol. Soc. Am. **1978**.

78 H 2 Heath, G.R.: Oceanus **21**, 1 (1978) 60.

78 H 3 Hinz, K., Schlüter, H.U.: Erdöl-Erdgas Z. **94**, 8 (1978) 271.

78 K Kümmerly & Frei: Atlas of seiscimity and volcanism. Zürich: Kümmerly & Frei **1978**, p. 40.

78 L Lucchini, L., Voelckel, M., in: Notes et etudes documentaires, nos. 4451–4452, Paris: La Documentation Français **1978**, p. 463.

78 R Reimnitz, E., Toimil, L., Barnes, P.: Mar. Geol. **28** (1978) 179.

78 S 1 Schweller, W.J., Kulm, L.D.: Mar. Geol. **28** (1978) 271.

78 S 2 Stanley, D.J., Kelling, G. (eds.): Sedimentation in submarine canyons, Stroudsburg: Dowden, Hutchinson, Ross **1978**.

78 S 3 Suhayda, J.N., Prior, D.B.: Explanation of submarine landslide morphology by stability analysis and rheological models, Coastal Studies Institue, Louisiana State University **1978**.

78 S 4 Summerhayes, C.P., Sestini, G., Misdorp, R., Marks, N.: Mar. Geol. **27** (1978) 43.

78 U UNESCO Technical Papers in Marine Science: Eighth report of the joint panel on oceanographic tables and standards, Woods Hole, USA, Sponsored by Unesco, ICES, SCOR, IAPSO **1978**.

79 C Corliss, J.B., Dymond, J., Gordon, L.I., Edmond, J.M., v. Herzen, R.P., Ballard, R.D., Green, K., Williams, D., Bainbridge, A., Crane, K., v. Andel, T.H.: Science **203** (1979) 1073.

79 D Dickinson, W.R., Seely, D.R.: Bull. Am. Ass. Petr. Geol. **63** (1979) 2.

79 E Ellwood, B.B., Ledbetter, M.T.: Science **203**, (1979) 1335.

79 K Keen, C.E. (ed.): Crustal properties across passive margins, New York: Elsevier Scientific Publ. Comp. **1979**, p. 390.

79 L 1 Lowman, P.D., Allenby, R.J., Frey, H.V.: Proposed satellite laser ranging and very long baseline interferometry sites for crustal dynamics investigations, Greenbelt, Maryland: Geophysics Branch, Goddard Space Flight Center **1979**.

79 L 2 Lowman, jr., P.D., Frey, H.V. (eds.): A geophysical atlas for interpretation of satellite derived data, Greenbelt, Maryland: Goddard Space Flight Center, NASA technical memorandum 79722, **1979**

79 M McCaffrey, E.K.: Int. Hydrogr. Rev., IHO, Monaco **58**, 1 (1979) 19.

79 P 1 Pitman, W.C.: Am. Assoc. Pet. Geol., Mem. **29** (1979) 453.

79 P 2 Prior, D.B., Coleman, J.M.: Z. Geomorphol. **23** (1979) 415.

79 S 1 Sclater, J.G., Tapscott, C.: Sci. Am. **240** (1979) 156.
79 S 2 Sharma, G.D.: The Alaskan shelf. Hydrographic, sedimentary and geochemical environment, Berlin, Heidelberg, New York: Springer **1979**, p. 498, 345 Figs., 15 Tabs.
79 T Temple, D.G., Scott, R.B., Rona, P.A.: J. Geophys. Res. **84** (1979) 7453.
79 U 1 UNESCO Technical Papers in Marine Science: Ninth report of the joint panel on oceanographic tables and standards. Unesco, Paris, 11.–13. Sept. 1978, **1979**.
79 U 2 United Nations Ocean Economics and Technologies Office: Manganese nodules: dimensions and perspectives. Dordrecht: Reidel **1979**, p. 194.
80 A Aubouin, J., Azema, J., Renard, V., Stephan, J.F.: La fosse d'Amérique Centrale: données tectoniques nouvelles issues des forages IPOD et d'un levé du Jean-Charcot au sondeur multifaisceaux (sea-beam). Ibid. **1980**, p. 186–190.
80 B 1 Ballard, R.D.: Mapping of the mid-ocean ridge. Publ. of the Technol. Conf. Houston **1980**, 55–64.
80 B 2 Blatt, H., Middleton, G., Murray, R.: Origin of sedimentary rocks, 2nd ed., Englewood Cliffs, N.J.: Prentice-Hall Inc. **1980**, p. 782.
80 F 1 Flood, R.D.: Mar. Geol. **38** (1980) 77.
80 F 2 Fornari, D.J., Lockwood, J.P., Lipman, P.W., Rawson, M., Malahoff, A.: J. Vulcanol. Geot. Res. **7** (1980) 323.
80 G Gierloff-Emden, H.G.: Geographie des Meeres; Ozeane und Küsten, 2. Vols., in: Lehrbuch der Allgemeinen Geographie. Vol. 5, Berlin, New York: W. de Gruyter **1980**, p. 1455.
80 H 1 Howell, D.G., v. Huene, R.: Tectonics and sediment along active continental margins. Soc. of Econom. Paleontol. and Mineral. Short Course, San Francisco: **1980**.
80 H 2 Hayes, D.E.: The tectonic and geologic evolution of the Southeast Asian seas and islands, Monogr. Ser. 23, Washington, D.C.: AGU **1980**.
80 I IHO/IOC: Summary report on the 7th session of the Joint IHO/IOC Guiding Committee for GEBCO, Monaco: IHO, **1980**, p. 148.
80 L Lefort, J.P.: Mar. Geol. **37** (1980) 355.
80 M McDougall, I., Duncan, R.A.: Tectonophysics **63** (1980) 280.
80 R 1 Rona, P.A. (ed.): NOAA atlas 3, the central North Atlantic ocean basin and continental margins: geology, geophysics, geochemistry and resources including Trans-Atlantic geotraverse (TAG), Washington D.C.: US Government Printing Office **1980**, p. 102.
80 R 2 Ross, D.A.: Opportunities and uses of the ocean, Berlin, Heidelberg, New York: Springer **1980**, p. 320.
80 S 1 Sears, U., Merriman, D.: Oceanography: the past, Berlin, Heidelberg, New York: Springer **1980**, p. 812.
80 S 2 Shipley, T.H., McMillen, K.J., Watkins, J.S., Moore, J.C., Sandoval-Ochoa, J.H., Worzel, J.L.: Mar. Geol. **35** (1980) 65.
80 S 3 Sarntheim, M., Seibold, E., Rognon, P. (eds.): Sahara and surrounding seas, palaeo-ecology of Africa, Vol. 12, Rotterdam: Balkema **1980**.
80 U Ulrich, J.: Dtsch. Hydrogr. Z. **33** (1980) 236.
81 B 1 Ben Avrahami, Nur, A.: J. Geophys. Res. **85** (1981) 4325.
81 B 2 Ballard, R.D., van Andel, T.H., Holcomb, R.T.: J. Geophys. Res. **85** (1981).
81 B 3 Bellaiche, G., Droz, L., Aloisi, J.C., Bouye, C., Got, H., Monaco, A., Maldonado, A., Serra-Raventos, J., Mirabile, L.: Mar. Geol. **43** (1981) M75–M85.
81 C Coleman, J.M.: Deltas, processes of deposition and models for explorations. 2nd ed., Minneapolis: Burgess Publ. Comp. **1981**, p. 124.
81 E Emiliani, C., in: The Sea, Vol. 7 (Emiliani, C., ed.), New York, London: Wiley Interscience **1981**, 1687–1728.
81 G Garrett, W.E. (ed.): World ocean floor, physiographic map in 1:42440000, Suppl. of Nat. Geogr. Magazine, Washington, D.C.: Nat. Geogr. Soc. **1981**.
81 H 1 Hay, W.W.: Oceanol. Acta **4** (suppl.) (1981) 135.
81 H 2 Hinz, K.: Geol. Jahrb. E, Geophysik, Hannover **22** (1981) 3.
81 H 3 Hinz, K., Dostman, H., Fritsch, J.: Proceedings 26th International Geological Congress, Geology of Continental Margins Symposium, Paris, 7.–17. July, 1980, Oceanologia Acta, **21** (1981).
81 I IHO: Extracts from International Hydrographic Bulletin, **I**, Monaco **1981**.
81 J Johnson, G.L., Vanney, J.R., Elverhoi, A., La Brecque, J.L.: Dtsch. Hydrogr. Z. **34** (1981) 263.
81 M Martin, A.K.: Mar. Geol. **39** (1981) 259.
81 N National Geographic Society, USA: World ocean floor. Physiographic map, Washington **1981**.
81 S Smoot, N.C.: Mar. Geol. **43** (1981) M87–M94.

81 V Vanney, J.R., Falconer, R.K.H., Johnson, G.L.: Mar. Geol. **41** (1981) 73.

82 B Beiersdorf, H., Gundlach, H., Heye, D., Marchig, V., Meyer, H., Schnier, C., in: The dynamic environment of the ocean floor (Fanning, K.A., Manheim, F.T., eds.), Lexington, Mass.–Toronto: Lexington Books D.C. Heath Comp. **1982**, p. 359–368.

82 F Fanning, K.A., Manheim, F.T. (eds.): The dynamic environment of the ocean floor, Lexington, Mass.-Toronto: Lexington Books, D.C. Heath Comp. **1982**, p. 502.

82 H Hinz, K., Krause, W.: Geol. Jb., Reihe E **23** (1982) 17.

82 K Kennett, J.P.: Marine Geology, Englewood Cliffs, N.J.: Prentice-Hall Inc. **1982**, p. 813.

82 M Mrozowski, C.L., Lewis, S., Hayes, D.E.: Tectonophysics **82** (1982) 1.

82 N Norton, J.H.: Int. Hydrogr. Rev. IHO Monaco **59**, 1 (1982) 23.

82 P 1 Panza, G.F., Calcagnile, G., Scandone, P., Mueller, S.: Spektrum der Wissenschaft **1** (1982) 18.

82 P 2 Parke, Stavert, Hussey: Topographic relief from Seasat altimeter mean sea surface, July-Oct. 1978, Pasadena: Jet Propulsion Laboratory **1982**.

82 R 1 Rad, U., Hinz, K., Sarntheim, M., Seibold, E.: Geology of Northwest African continental margin, Berlin, Heidelberg, New York: Springer **1982**, p. 700.

82 R 2 Roeser, H.A., Rilat, M.: Geol. Jb., Reihe E **23** (1982) 71.

82 S 1 Seibold, E., Berger, W.H.: The sea floor. An introduction to marine geology, Berlin, Heidelberg, New York: Springer **1982**, p. 288.

82 S 2 Smoot, N.C.: Mar. Geol. **47** (1982) 153.

82 S 3 Spindel, R.C.: Oceanus **25**, 2 (1982) 12–27.

82 S 4 Springer, V.G.: Pacific plate biogeography, with special reference to shorefishes. Smithsonian Contr. Zoology, 367, Washington: **1982,** pp. 182, 65 Figs.

82 S 5 Scott, R.B., Temple, D.G., Peron, P., in: The dynamic environment of the ocean floor (Fanning, K.A., Manheim, F.T., eds.), Lexington, Mass., Toronto: Lexington Books D.C. Heath Comp. **1982**, 381–416.

82 U Ulrich, J., Kögler, F.C.: Dtsch. Hydrogr. Z. **35**, 6 (1982) 239.

82 V Vink, G.E.: J. Geophys. Res. **87** (1982) 10677.

82 W Wissmann, G., Roeser, H.A.: Geol. Jb., Reihe E **23** (1982) 43.

83 A 1 Agterberg, F.P.: Nature and Resources **19**, 4 (1983) 20.

83 A 2 Adams, C.E., jr., Prior, D.B., Coleman, J.M.: Soc. Petrol. Eng. AIME, Febr. **1983**, 177.

83 A 3 Allan, T.D. (ed.): Satellite microwave remote sensing, Ellis Horwood Ser. in Mar. Sci., Chichester, UK: Ellis Horwood **1983**, p. 526.

83 B Barnes, N.E., Normark, W.R.: Geo-Marine Lett. **3**, 2–4 (1983) 353.

83 C 1 Carey, S. Warren (ed.): The expanding earth, a symposium: Sydney 1981, Hobart: University of Tasmania, **1983**, p. 423.

83 C 2 Coleman, J.M., Prior, P.B., Lindsay, J.F.: Soc. Econ. Paleontol. Mineralog. Spec. Publ. **33**, 6 (1983) 121.

83 C 3 Courtillot, V., Vink, G.E.: Spektrum der Wissenschaft **9** (1983) 50.

83 E 1 Edmond, J.M., v. Damm, K.: Spektrum der Wissenschaft **6** (1983) 74.

83 E 2 Enas, D.B., Niela, V.F.: Int. Hydr. Rev. IHO, Monaco **57**, 2 (1983) 93.

83 F Francheteau, J.: Spektrum der Wissenschaft **11** (1983) 80.

83 G 1 Gierloff-Emden, H.G.: Geowissenschaften in unserer Zeit, Weinheim **1**, 6 (1983) 191.

83 G 2 Gierloff-Emden, H.G.: Geowissenschaften in unserer Zeit, Weinheim **1**, 2 (1983) 46.

83 H 1 Hay, W.W., in: Proc. of the Joint Oceanogr. Assembly 1982-General Symposia, Ottawa: Canadian National Committee/Scientific Committee on Oceanic Research (CNC/SCOR), **1983**, 9–16.

83 H 2 Hinz, K., in: Proc. of the Third Int. Sympos. held in Hannover, Fed. Rep. of Germany, at the Federal Institute for Geosciences and Mineral Resources, Oct. 27–29, 1982, on New Paths to Mineral Exploration, Stuttgart: E. Schweizerbart'sche Verlagsbuchhandlung **1983**.

83 H 3 Hinz, K., Block, M.: Antarctica, **1982**, 79–91.

83 J 1 Johns, B.: Elsevier Oceanogr. Ser. **35** (1983) 470.

83 J 2 Proceedings of the Joint Oceanographic Assembly (JOA) 1982. General Symposium. Dalhousie University, Halifax, Nova Scotia, Canada 1982. Ottawa **1983**.

83 J 3 Jones, D.L., Cox, A., Coney, P., Beck, M.: Spektrum der Wissenschaft **1** (1983) 66.

83 L 1 Laughton, A.S.: Int. Hydr. Rev. IHO, Monaco **60**, 1 (1983) 13.

83 L 2 Laughton, A.S., Roberts, D.G., Graves, R.: Bathymetric of the Northeast Atlantic, 1:2400000, 5 sheets. Taunton, UK: Inst. of Oceanographic Sciences, **1975–1983**.

83 M 1 Molnia, B.F.: Glacial marine sedimentation. New York, London: Plenum Press, **1983**, p. 650.

83 M 2 Morgan, W.J.: Tectonophysics **94**, 1–4 (1983) 123.

83 N 1 Nelson, C.H., in: Revolution in the earth sciences, advances in the past half-century (Boardman, S.J., ed.), Dubuque, Iowa: Kendall-Hunt **1983**, 148–166.
83 N 2 Normark, W.R., Mutti, E., Bouma, A.H.: Geo-Marine Lett. **3**, 2–4 (1983) 353.
83 O Owen, H.G.: Atlas of the continental displacement 200 million years to the present, Cambridge: Cambridge University Press **1983**, p. 180.
83 R Rona, P.A., Bostrom, K., Laubier, L., Smith, K.L., jr. (eds.): Hydrothermal processes at sea-floor spreading centers, Nato Conf. Ser. IV, Marine Sci., Vol. 12, New York: Plenum Press **1983**, p. 796.
83 S 1 Shanmugan, G., Moiola, R.J.: Geo-Marine Lett. **3**, 2–4 (1983).
83 S 2 Smith, A.G., Briden, J.C.: Mesozoic and Cenozoic palaeocontinental maps, Cambridge: Cambridge University Press **1983**.
83 T Thiede, J. (ed.): Whither the oceanic geosciences. A report by CMG of IUGS, III. Int. Workshop on Marine Geosciences, Heidelberg, July 1982. Oslo: **1983**.
84 B 1 Belderson, R.H., Jones, E.J.W., Gorini, M.A., Kenyon, N.H.: Mar. Geol. **56** (1984) 65.
84 B 2 Bettac, W.: Automated cartography. In: [84 L 1], **1984**, 540–542.
84 B 3 Boccaletti, M., Nicolich, R., Tortorici, L.: Mar. Geol. **55** (1984) 219.
84 B 4 Bonatti, E., Crane, K.: Spektrum der Wissenschaft **7** (1984) 62.
84 B 5 Broadus, J.M., Bowen, R.E.: Oceanus **27** (1984) 26.
84 D 1 Dixon, J.E., Robertson, A.H.F. (eds.): The geological evolution of the Eastern Mediterranean. Spec. Publ. of the Geol. Soc. Oxford, London, Edinburgh: Blackwell Scientific Publ. **1984**, p. 825.
84 D 2 Duin, E.J.Th., Mesdag, C.S., Kok, P.T.J.: Mar. Geol. **56** (1984) 299.
84 E Emery, K.O., Uchupi, E.: The geology of the Atlantic Ocean. Accompanied by: Geology of the Atlantic Ocean map set with 22 charts, New York: Springer **1984**, p. 1050.
84 H 1 Hékinian, R.: Spektrum der Wissenschaft, **9** (1984) 62.
84 H 2 Hollister, D., Nowell, A.R.M., Jumars, P.A.: Spektrum der Wissenschaft **5** (1984) 84.
84 I IOC-Report: Ocean science for the year 2000, Strassbourg: UNESCO **1984**.
84 J Johnson, D.A.: Mar. Geol. **58** (1984) 1.
84 L 1 Langeraar, W.: Surveying and charting of the seas, Amsterdam: Elsevier Scientific Publ. Comp. **1984**, p. 612.
84 L 2 Ledbetter, M.T.: Mar. Geol. **58** (1984) 137.
84 M Masson, D.G., Miles, P.R.: Mar. Geol. **56** (1984) 279.
84 P Powell, R.D.: Mar. Geol. **57** (1984) 1.
84 R 1 Rehault, J.P., Boillot, G., Mauffret, A.: Mar. Geol. **55** (1984) 447.
84 R 2 Ryan, R.: Oceanus **27** (1984) 78.
84 S 1 Sawkins, F.J.: Metal deposits in relation to plate tectonics. Minerals and rocks, Vol. 17, Berlin, Heidelberg, New York: Springer **1984**, pp. 340, 173 Figs., 6 Tables.
84 S 2 Schneider, G., in: Landolt-Börnstein **V**, Vol. 2a (Fuchs, K., Soffel, H., eds.), Berlin, Heidelberg, New York, Tokyo: Springer **1984**, 47–59.
84 S 3 Sleath, J.F.A.: Ocean engineering, New York, London: Wiley Interscience **1984**, p. 335.
84 S 4 Stanley, St., M.: Spektrum der Wissenschaft **8** (1984) 92.
84 U Ulrich, J.: Kartograph. Nachr. **34**, 2 (1984) 41.
84 V 1 Van Loon, H. (ed.): Climates of the oceans, Amsterdam, New York: Elsevier Scientific Publ. Comp. **1984**.
84 V 2 Vink, G.E.: J. Geophys. Res. **89**, B 12 (1984) 9949.
84 W Worsley, T.R., Nance, D., Moody, J.B.: Mar. Geol. **58** (1984) 373.
84 Y Yuru, L. (ed.): Proceedings of the International Symposium on the Continental Shelf, with special reference to the East China Sea. 2 Vols. **1984**, pp. 1000.
85 B 1 Bäcker, H., in: DFG/Geowissenschaften: Mitt. XIV der Kommission für Geowissenschaftliche Gemeinschaftsforschung, Weinheim: VCH Verlag, **1985**, 43–55.
85 B 2 Bosum, W., Pucher, R., Roeser, H., in: Landolt-Börnstein, **V**, Vol. 2b (Fuchs, K., Soffel, H., eds.), Berlin, Heidelberg, New York, Tokyo: Springer **1985**, 74–99.
85 E ESA: Looking down, looking forward. Earth observation – sciences and application – a perspective, Paris: European Space Agency, **1985**, p. 53.
85 G Gierloff-Emden, H.G., in: Landolt-Börnstein, **V**, Vol. 2b, (Fuchs, K., Soffel, H., eds.), Berlin, Heidelberg, New York, Tokyo: Springer **1985**, p. 272.
85 J 1 Jacoby, W.R., in: Landolt-Börnstein, **V**, Vol. 2b (Fuchs, K., Soffel, H., eds.), Berlin, Heidelberg, New York, Tokyo: Springer **1985**, 298–369.
85 J 2 Jones, M.L.: The hydrothermal vents of the eastern Pacific: an overview, Bull. Biol. Soc., Washington **1985** (in press).

85 L Larson, R.L., Pitman, W.C., III, Golovchenko, X., Cande, S.C., Dewey, J.F., Haxby, W.F., La Brecque, J.L.: The bedrock geology of the world, map in 1:23 230 000 (Larson, R.L., Pitman, W.C., III, eds.), New York: Freeman **1985**.

85 N NASA: Crustal dynamics research. Space science and applications. 4, Greenbelt, Maryland: NASA, Goddard Space Flight Center **1985**, p. 15.

85 R Robinson, I.S.: Satellite oceanography, Ellis Horwood Ser. in Mar. Sci., Chichester, UK: Ellis Horwood **1985**, p. 455.

85 S Soffel, H., in: Landolt-Börnstein, **V**, Vol. 2b (Fuchs, K., Soffel, H., eds.), Berlin, Heidelberg, New York, Tokyo: Springer **1985**, 184–243.

85 T Thiel, H., Weikert, H., Karbe, L., in: DFG/Geowissenschaften: Mitt. XIV der Kommission für Geowissenschaftliche Gemeinschaftsforschung, Weinheim: VCH Verlag **1985**, 17–42.

85 V Vink, G.E., Morgan, W.J., Vogt, P.R.: Spektrum der Wissenschaft **6** (1985) 62.

2 In-situ instruments and measuring techniques

Oceanographical measuring techniques have developed significantly during the past twenty years. With the advent of microprocessors for in-situ instrumentation continuously new devices are becoming available commercially. They permit not only faster and more convenient data acquisition, but also extensions in time and space. The last decade has also seen a tremendous increase in utilizing satellites for remote sensing of the ocean surface and for improving position finding of ships and drifting buoys.

This chapter restricts itself to in-situ measurements, and it is centered around methods of measurement rather than presenting a large number of instruments.

2.1 The nature of oceanographic measurements

2.1.1 Variables to be measured, time and space scales

Using the continuum hypothesis, the equation of state of seawater and the concept of constant relative composition of sea-water salts, theoretical description of oceanographic processes is possible on the basis of 7 variables. These are 3 components of velocity (u, v, w), density ϱ, temperature T, salinity S and pressure p. Additionally one has 3 space coordinates ϕ, λ, z, where ϕ is the geographical latitude, λ the longitude; z is the depth coordinate and time is denoted by t.

This list is based on the traditional equation of state

$$\varrho=\varrho(S, T, p) \tag{1}$$

Today other forms of this equation are also known. They may involve conductivity or sound velocity.

The basic task of experimental oceanography is to measure combinations of the mentioned variables for different scales in time and space. If a variable is generally denoted by G, the aim is to obtain the four-dimensional function

$$G=G(\phi, \lambda, z, t) \tag{2}$$

by measurements. The respective strategies, called *methods of observations*, are presented in sect. 2.6. In contrast, in dealing with measuring techniques one has to find ways to measure a variable G quantitatively.

The space scales of oceanic movements range from the order of dimensions of ocean basins (1000 km horizontal, 1000 m vertical) down to the millimeter range. Time scales have to be considered from less than a second to a decade or longer.

This vast range of scales explains the great variety of devices used in oceanography. Occasionally, instruments which measure the same variable in different scales have even different names, e.g., a wave rider buoy and a tide gauge, both measure the elevation of the sea surface.

In earlier times oceanographers have set out to explore water masses and currents in the unknown sea. Now it is possible, based on the acquired regional knowledge for large parts of the oceans, to investigate processes on the basis of theoretical assumptions and models with the intention to verify or reject them by measurements. Corresponding observation schemes are called *experiments* although this is not correct in a strictly physical sense.

2.1.2 Environmental conditions and accuracy of instruments

Oceanographic instrumentation has to cope with a number of unfavourable environmental conditions. Sea water is a very corrosive fluid. The outmost care with the choice of materials for instrument housings and sensors is of vital importance for successful measurements. The same holds for an array of instruments in a mooring.

High hydrostatic pressure is a factor which prescribes to build containers for electronic circuits as small as possible, and it has to be taken into account for the protection of pressure-dependent sensor elements.

Shock and vibration are further characteristics experienced by under-water instrumentation. The mechanical wear and tear on moorings and instruments in the reach of orbital wave motion determines greatly the lengths of time series to be measured in the upper layers of the ocean.

Fowling of sensors by marine growth in the euphotic zone is a great problem, particularly in the warm water sphere and during particular summer months in the moderate latitudes.

By the rapidly advancing use of low-power integrated circuits including solid state memories for data storage, the power supply of self-contained instruments poses less problems than before. Nevertheless, further development of suitable batteries for extensive periods of measurement at low temperatures is still desirable.

A detailed treatment of the mentioned engineering aspects is beyond the scope of this chapter. Literature on this topic is found in [69 B, 69 F, 71 U].

2.1.3 Accuracy and standards

Inspite of adverse environmental conditions high resolution and accuracy of measurements is required, especially for the variables entering the equation of state. Specifications are given in sect. 2.5. The important point to be made here is that the international data exchange and the maintenance of the World Data Centers requires a fixed absolute accuracy of temperature, pressure and salinity data and comparable measurements of currents. The standard for temperature and pressure measurements is provided by the national bureaus of standards. For salinity the oceanographic community has its own standard, the "Copenhagen Normal Water". Measurements of anchored current meters of different design have to be compared in intercomparison experiments from time to time. Problems of calibration are discussed in [85 S 1].

It is important to realize that the total accuracy of an oceanographic measurement includes precision of position finding, depth determination of the instrument and of time measurement. This is readily seen from the total derivative of eq. (2).

$$dG = \frac{\partial G}{\partial \phi} d\phi + \frac{\partial G}{\partial \lambda} d\lambda + \frac{\partial G}{\partial z} dz + \frac{\partial G}{\partial t} dt . \tag{3}$$

2.2 Platforms

Platforms are required to perform measurements, to carry instruments, and occasionally they also represent a measuring device as in the case of drifters or buoys to measure wind waves and swell. The most important platforms are research ships and instrument moorings.

2.2.1 Ships

The types of vessels range from ordinary surface ships to submarines, deep-sea vehicles and some very special constructions. Ship-shaped vessels without propulsive power are considered as instruments.

From a purely scientific view a research ship is a carrier of winches to handle underwater instruments. In order to reduce effects of ship motions on lowered instrumentation the preferable place for the winches is the midship area. Very good manoeuverability of the ship is required to keep the instrument wires or cables vertical during station work. Construction of research ships has pioneered additional manoeuvring aids such as bow and stern thrusters, active rudder, Becker rudder and other devices to move a ship in any direction for this purpose. The midship area is also the preferable location for the laboratories.

Ship's speed is important, but it is not so decisive as for cargo ships because most of the time is spent on station work. This type of operation requires high precision of position finding. Quality and extend of navigational aids and echo sounding devices distinguish research and merchant ships.

Because of high running costs almost all of the larger research ships are built to be used by many disciplines of the marine sciences. There is also an increasing trend to using container laboratories to make operations as flexible as possible.

The more than 500 existing research ships all over the world can roughly be devided into the classes of large, multipurpose ships for world wide operations, special ships (fisheries research, exploration, drilling) and a rather large fleet of small vessels for coastal waters and training purposes.

In the framework of this article only a few examples of practical solutions can be given.

Table 1 gives a survey on older research ships. Most of these ships are closely related to the large national and international ocean surveys during the time when oceanography emerged as a special science [64 W].

Table 1. Older research vessels. t: displacement of water; BRT: gross register ton.

Name of vessel	Years of expeditions	Country	Size	Operating area
Challenger	1873/76	Gt. Britain	2306 t	all oceans
Fram	1893/96	Norway	530 t	North Arctic Ocean
Valdivia	1898/99	Germany	2176 BRT	Atlantic and Indian Ocean
Carnegie	1928/29	USA	568 t	Pacific Ocean
Will. Snellius	1929/30	Netherlands	1055 t	Indian Ocean
Pourqui-pas	1911/36	France	500 BRT	North Atlantic Ocean
Dana II	1921/35	Denmark	360 BRT	all oceans
Sedow	1938/40	USSR	1538 BRT	North Arctic Ocean
Albatross	1947/48	Sweden	1450 t	all oceans
Armauer Hansen	since 1913	Norway	57 BRT	North Atlantic Ocean
Meteor	1925/38	Germany	1178 t	Atlantic Ocean
Discovery II	1930/62	Gt. Britain	2100 t	all oceans
Atlantis	1931/64	USA	460 BRT	North Atlantic Ocean
Ryofu Maru	1937/66	Japan	1206 BRT	Pacific Ocean

Since the 1960s a rather large fleet of research vessels was commissioned. A selection of this generation of larger ships is listed in Table 2. More information is found in [71 N]. These ships were designed for research whereas many of the older ships had been converted from merchant, military or fishing vessels.

Table 2a. Some research vessels of the 1960's.

Name of vessel	Year built	Country	Size	Operating area	Number of scientists
Meteor	1964	Germany	2740 t	all oceans	24
Jean Charcot	1965	France	2200 t	all oceans	29
Hakuho Maru	1967	Japan	4207 t	all oceans	32
Atlantis II	1963	USA	2100 t	all oceans	25
Knorr	1970	USA	2075 t	all oceans	23
Discovery III	1962	Gt. Britain	2800 t	all oceans	21
Michail Lomonossow	1957	USSR	5960 t	all oceans	65

Table 2b. Some recent oceanographic research vessels.

Name of vessel	Year built	Size	Length	Beam	Draught	Number of scientists	Country
Hakurei Maru	1980	2111 t	88.85 m	13.8 m	5.4 m		Japan
Johann Nordenanker	1980	2140 t	73.0 m	14.0 m	4.0 m		Sweden
Dana	1981	2483 t	78.4 m	14.8 m	5.9 m	12	Denmark
Justo Sierra	1982	770 t	50.0 m	10.3 m	4.4 m		Mexico
Akademik Shuleykin	1982	1754 t	71.6 m	12.8 m	4.7 m		USSR
Capitan Oca Balda	1983	1180 t	65.0 m	11.4 m	4.2 m	9	Argentina
Sagar Kanya	1983	4209 t	100,3 m	16.3 m	5.6 m	32	India
Bimasakti	1984	1374 t	59.7 m	13.0 m	3.6 m	10	Indonesia
Ekteshaf	1984	804 t	55.0 m	11,0 m	3.5 m	21	Iran
Charles Darwin	1984	1936 t	69,3 m	14,4 m	4,8 m	18	United Kingdom
Franklin	1985		55,2 m	11,8 m	3,8 m	12	Australia
Meteor	1986	3990 t	97,5 m	16,5 m	4,8 m	29	Germany

A modern medium size research vessel is presented here (Fig. 1) as an example of a ship designed to suit the needs of the German Hydrographic Office (DHI) in marine research. It shows side view and main deck of the "Gauß" who was commissioned 1980. Like for many other research ships the main deck is asymmetric. This allows handling of instruments in the mid-ship area next to the laboratories. Working area: North Sea, Baltic Sea.

Dimensions:

length overall:	68.9 m
beam:	13.0 m
draught:	4.25 m
displacement:	1813 t

Power generation:

3 Diesel engines MAK, 588 kW each at 750 rpm, combined with ac generators AEG, 910 kVA.

Propulsion:

dc motor AEG 2 × 650 kW directly coupled to the propeller shaft.
Gill's bow thruster 533 kW.
Fin stabilisation. The frontal view of the ship in Fig. 2 shows the stabilizers.

Winches and cranes:
1. Reversable A-frame, load 30 kN, auxiliary winch 20 kN
2. Crane 30 kN, outreach 1.8···14.0 m
3. Jib boom 80 kN, outreach 2.5 m over the side, with auxiliary winch and derrick
4. Heavy duty winch 80 kN, 1200 m wire
5. Single-conductor cable winch 25 kN, 1100 m cable
6. Hydrographic winch 27 kN, 1100 m wire
7. Wire rope winch 25 kN, 1000 m
8. Spill
9. Space for additional winches

Laboratories:

1. Filling room	12.4 m^2
2. Wet laboratory	29.7 m^2
3. Dry laboratory	17.0 m^2
4. Multi-purpose laboratory	19.4 m^2
5. Recording room	8.3 m^2
6. Photographic laboratory	4.5 m^2
7. 3 spaces for container laboratories (20 feet)	
8. 2 workshops	

Hold: 90 m^2/250 m^3

Crew: nautical 19, scientific 12.

An overview on the computer-controlled navigation system is shown in Fig. 3, and Fig. 4 displays the central data processing and information distribution system

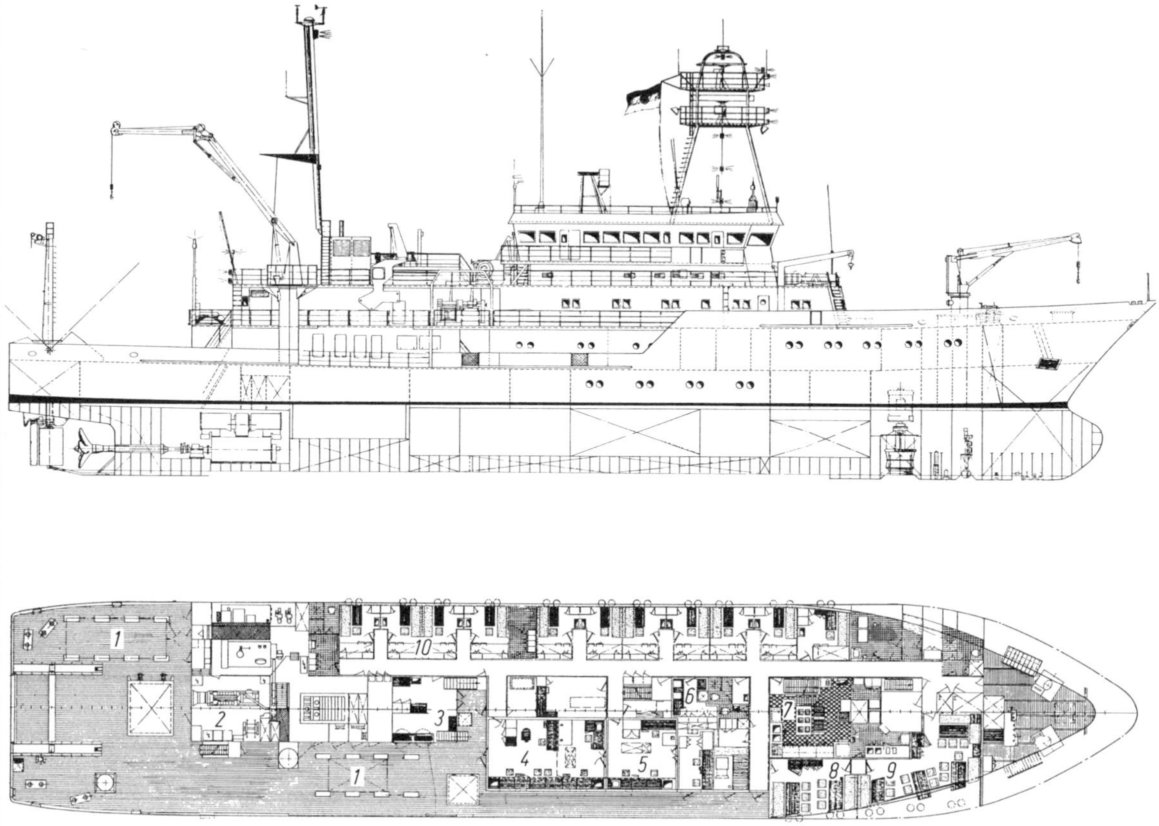

Fig. 1. Sideview and maindeck of the "Gauss" [80 V 2]. *1* Space for container laboratories; *2* Emergency and harbour diesel; *3* Filling room; *4* Wet laboratory; *5* Dry laboratory; *6* Foto laboratory; *7* Galley; *8* Crew's mess; *9* Officer's mess; *10* Crews' quarters.

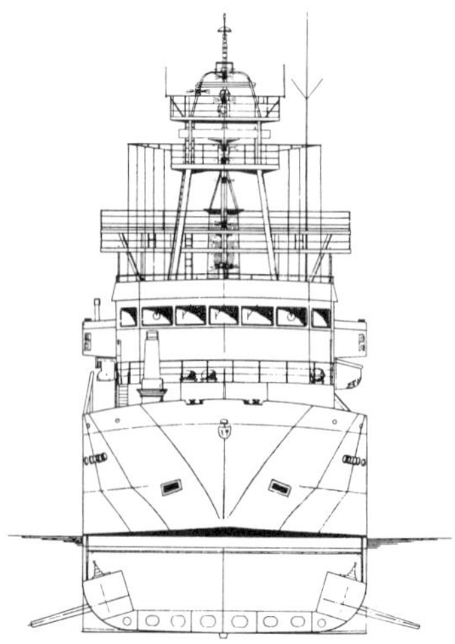

Fig. 2. Frontal view of the "Gauss" showing the stabilizers in working position [80 V 2].

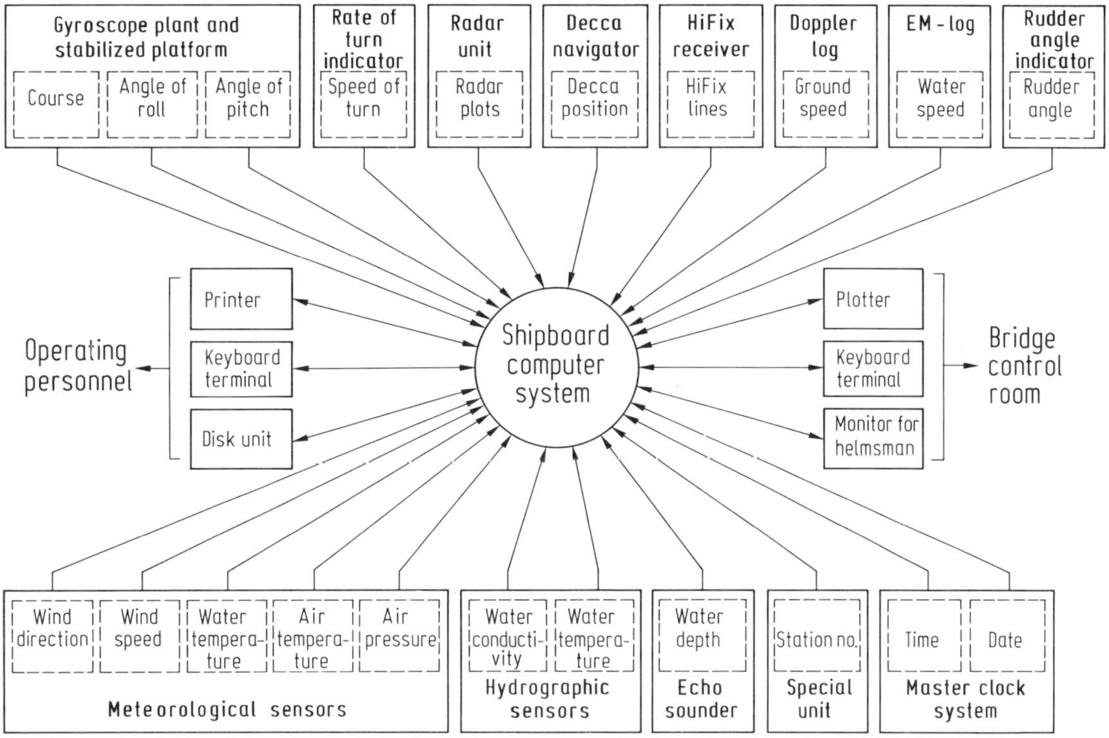

Fig. 3. Navigation system and environmental data acquisition aboard FS "Gauss" [80 V 2].

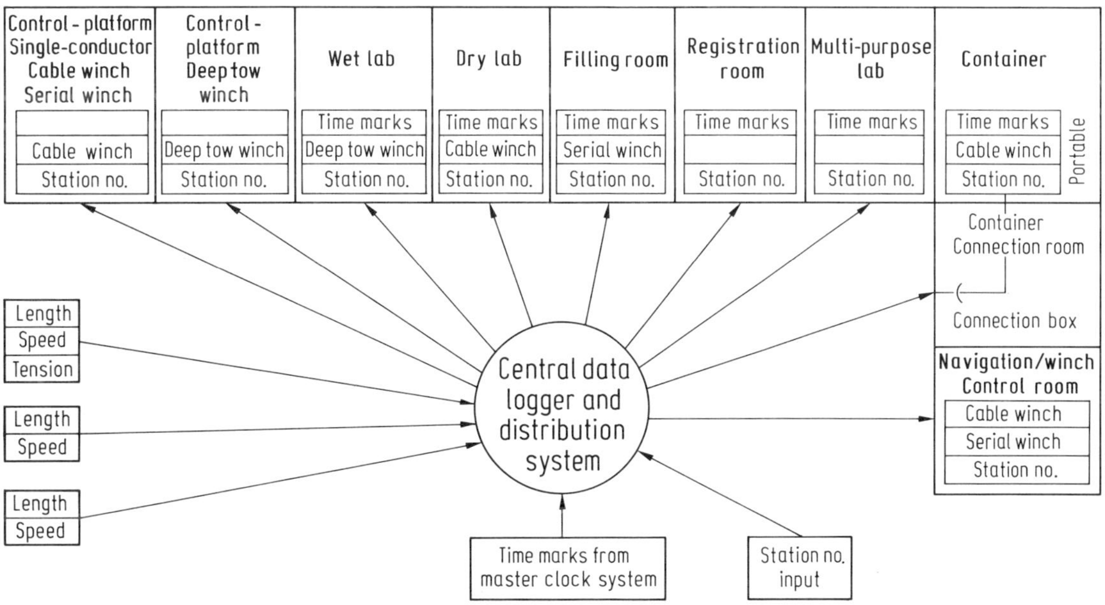

Fig. 4. Data logging and information distribution system aboard FS "Gauss" [80 V 2].

Krause

2.2.2 Instrument moorings

Instrument moorings are used for getting time series and to obtain synoptic data from an area of the sea. The basic mooring configurations are sketched in Fig. 5. They can be subdivided into surface and subsurface moorings. Regardless whether a buoy is seen at the surface or not, the classification is based on the location of the main buoyancy element for the instrumentation. If it is at the surface it is a surface mooring, otherwise it is a subsurface mooring. In the latter case the main buoyancy element should be out of the reach of orbital motion of surface waves.

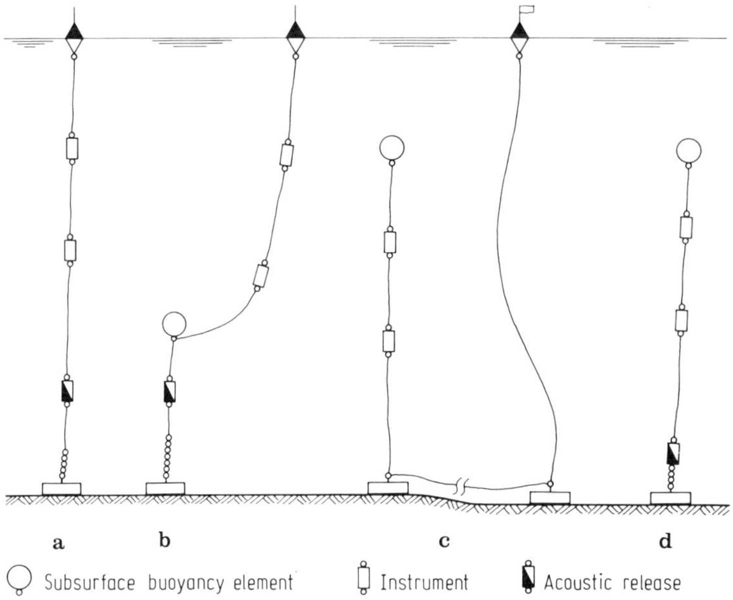

Fig. 5. The basic instrument moorings. Surface moorings: **a** taut wire mooring, **b** slack wire mooring. Subsurface moorings: **c** ground-line mooring, **d** single point subsurface mooring.

2.2.2.1 Surface moorings

Surface moorings are subdivided into taut wire and slack wire moorings depending on whether the length of the mooring line is approximately equal or larger than the water depth.

All surface moorings have the disadvantage of high dynamical loading on all mooring components and instruments. This is particularly true for taut wire moorings, but these have the advantage of small buoy watch circles and small horizontal excursions of instruments under varying current conditions.

The dynamical loading of a surface mooring depends on the type of buoy used. Surface-following buoys (types b to d in Fig. 6) provide large buoyancy to support long deep-sea moorings but they move heavily under the action of waves thus inducing severe stresses to the system [70 C] and "noise" to the time series to be measured. With spar buoys wave action on the mooring can be suppressed, but the buoyancy safety margin can be a problem in areas with strong currents. Large spar buoys are necessary for measurements near the surface, both in air and in water.

Practical and engineering aspects of buoys are treated in [76 B]. A survey article on platforms for air-sea interaction studies can be found in [80 B].

The Figs. 7 and 8 show a spar buoy and a surface-following buoy which have been successfully used.

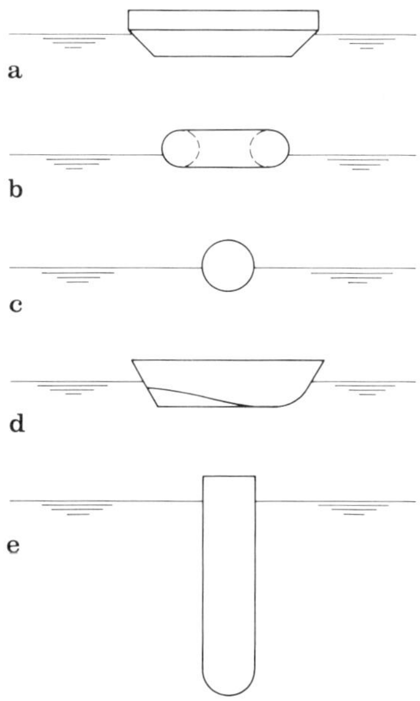

Fig. 6. Generic buoy forms [80 B]. **a** Disc; **b** Torus; **c** Sphere; **d** Boat; **e** Spar.

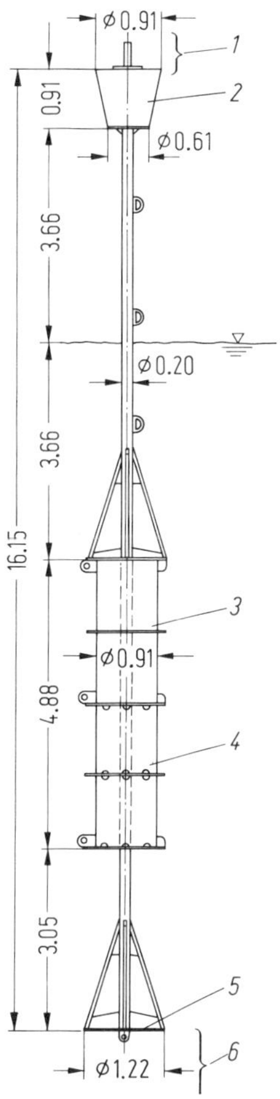

Fig. 7. Woods Hole Oceanographic Institution (WHOI)-spar buoy. All dimensions in [m]. Note the reserve buoyancy. This is necessary because this buoy is nearly neutrally buoyant [80 B]. *1* Antenna, navigation warning light; *2* Reserve buoyancy (polyurethane foam, covered by fibreglass); *3* Foam filled buoyancy tank; *4* Ballast tank; *5* Ballast, damping plate; *6* Instrument line.

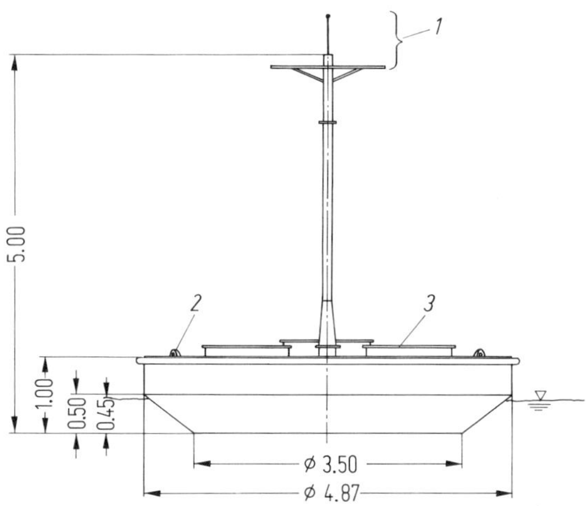

Fig. 8. General Dynamics SODS/16-buoy (Small Ocean Data Station) [80 B]. All dimensions in [m]. *1* Antenna, navigation warning light, meteorological sensors; *2* Lifting lug (3); *3* Equipment hatch (3).

Krause

2.2.2.2 Under-water moorings

Current measurements in the deep sea are normally performed with a number of self-contained current meters in a subsurface mooring line. Together with instruments and acoustic controlled release devices such moorings are expensive, and they require great care in designing and handling.

To describe design criteria and safety considerations we consider a single mooring consisting of only one anchor weight, a steel wire and a subsurface buoy. The tension in the wire would increase towards the buoyancy element as sketched in Fig. 9, which also shows the residual buoyancy of this mooring.

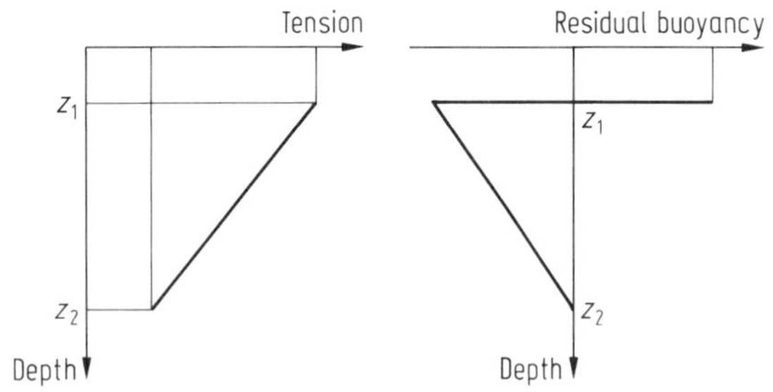

Fig. 9. Tension in the mooring line and residual buoyancy of a primitive mooring consisting of an anchor weight at depth z_2, a wire, and a buoyancy element at depth z_1.

This mooring has two disadvantages: If the subsurface buoy tears off due to damage the mooring is lost. The larger the buoyancy at top of the mooring the greater is the area exposed to the current which normally is stronger in the upper layers. Therefore there is an increasing trend towards greater deflections from the vertical.

If one uses many subsurfaces buoys for a great safety margin such a mooring has other disadvantages: The lifting force at the anchor gets larger, heavier anchor weights would be necessary. The tension in the wire increases. The mooring gets more expensive.

In practice a compromise between the two extremes has to be found depending on the number of instruments, the water depth and the current profile to be expected. [63 R, 67 B].

Solutions are normally found by computations of the static behavior of moorings under the action of assumed current profiles.

Most of the oceanographic institutions have developed computer programs to perform respective calculations [73 B 1, 76 M, 80 G 1, 82 B 1, 82 S 3]. The principle approach is as follows:

Under the action of a current the mooring line will be deflected from the vertical. If one assumes that tension in the wire and the buoyancy is not much changed by the current, equilibrium with drag forces is reached when the mooring line assumes a respective shape. To obtain this shape, the total mooring line is subdivided into a number of segments. Each segment is thought to be a rigid cylinder connected to the neighboring segments by universal joints.

The following definitions (see Fig. 10) are necessary for practical computations. A vertical line through the anchor and the deflected mooring line (direction φ with respect to North) define a plane which is referred to as the reference plane. A current vector pointing in a direction $(\gamma-\varphi)$ with respect to this reference plane causes drag forces $\boldsymbol{F}$ which tend to bend and to rotate the mooring line into the current direction γ with respect to North.

The velocity components relevant to the calculation of the drag forces are

V_F: velocity component in direction of the reference plane, and

V_N: velocity component perpendicular to the reference plane,

$$V_F = V \cdot \cos(\gamma-\varphi) \tag{4}$$

$$V_N = V \cdot \sin(\gamma-\varphi)\,. \tag{5}$$

V is the magnitude of velocity vector.

With respect to a a point P (Fig. 10) on the mooring line we have

V_T: velocity component of V_F along the mooring line, and

V_{TN}: velocity component of V_F perpendicular to the mooring.

If Θ is the angle of deflection of the mooring line from the vertical we have

$$V_T = V \cdot \cos(\gamma-\varphi)\sin\Theta\,, \tag{6}$$

$$V_{TN} = V \cdot \cos(\gamma-\varphi)\cos\Theta\,. \tag{7}$$

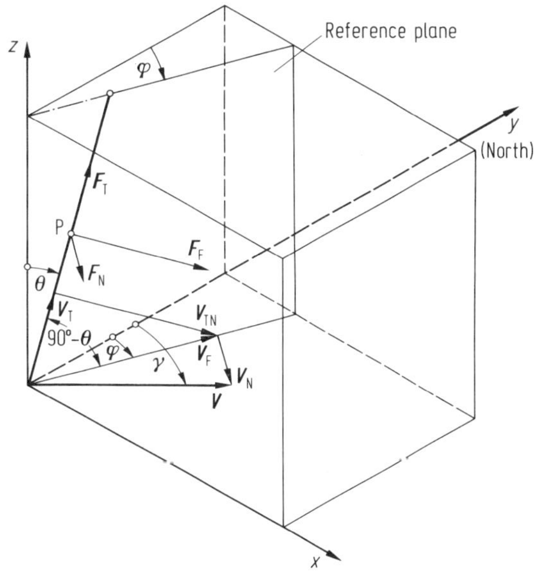

Fig. 10. Definition sketch showing the reference plane and the splitting-up of the velocity vector V into components. Also shown are the components of the drag force. In order to avoid confusion the velocity and force vectors have been separated. The thick line is a segment of the mooring [82 S 3].

The respective drag forces are

F_T: component tangential to the mooring line in the reference plane,

F_F: component perpendicular to the mooring line within the reference plane,

F_N: component perpendicular to the reference plane. This component causes rotation of the mooring line.

Using the quadratic law for the drag force one obtains

$$F_T = \tfrac{1}{2}\varrho \cdot C_{DT} \cdot A_L \cdot l \cdot |V_T| V_T, \tag{8}$$

$$F_F = \tfrac{1}{2}\varrho \cdot C_{DN} \cdot A_L \cdot l \cdot |V_F| V_{TN}, \tag{9}$$

$$F_N = \tfrac{1}{2}\varrho \cdot C_{DN} \cdot A_L \cdot l \cdot |V_N| V_N, \tag{10}$$

where C_{DT} is the drag coefficient along, and C_{DN} the respective coefficient perpendicular to the mooring line [71 C, 84 F, 86 F]. A_L is the area per unit length of a mooring segment and l is the length of a segment. ϱ denotes density.

Finally we have to consider buoyancy and weight.

F_S: sum of buoyancy and weight.

It has the normal and tangential components

$$F_{SN} = F_S \cdot \sin\Theta, \tag{11}$$

$$F_{ST} = F_S \cdot \cos\Theta. \tag{12}$$

In Fig. 11 the drag forces acting on a point P between two adjacent segments n and n−1 are shown in a vertical plane and seen from above. In the equilibrium state the vectorial sum of the forces must be zero. This means in components

$$F_{SN} + F_F = F_{zn-1} \cdot \sin(\Theta_n - \Theta_{n-1}), \tag{13}$$

$$F_{ST} + F_T = F_{zn} - F_{zn-1} \cdot \cos(\Theta_n - \Theta_{n-1}), \tag{14}$$

$$F_N = F_{zn-1} \cdot \sin\Theta_{n-1} \cdot \sin(\varphi_n - \varphi_{n-1}), \tag{15}$$

$$F_{zn} \sin\Theta_n = F_{zn-1} \cdot \sin\Theta_{n-1} \cdot \cos(\varphi_n - \varphi_{n-1}), \tag{16}$$

where F_{zn} is the force due to the tension in the n-th segment of the mooring line. These equations can be used for an iteration calculation of the equilibrium state. The force at the topmost segment is used as the starting value. If Θ' and φ' denote the vertical and azimuth angles, respectively, resulting from the previous iteration step, equilibrium is reached when

$$\Delta\Theta = \Theta_n - \Theta', \tag{17}$$

$$\Delta\varphi = \varphi_n - \varphi', \tag{18}$$

are smaller than given values. Replacing Θ_n, φ_n by Θ', φ' in eqs. (13)···(16) the final equations are

$$F_{zn} = F_{zn-1} \cos(\Theta' - \Theta_{n-1}) + F_{ST} + F_T, \quad (19)$$

$$\Theta_n = \arctan\left[\frac{F_{SN} + F_F - F_{zn-1} \sin(\Theta' - \Theta_{n-1})}{F_{zn}}\right] + \Theta', \quad (20)$$

$$\varphi_n = \arctan\left[\frac{F_N - F_{zn-1} \cdot \sin\Theta_{n-1} \sin(\varphi' - \varphi_{n-1})}{F_{zn} \cdot \sin\Theta_n}\right] + \varphi'. \quad (21)$$

As a practical application a subsurface mooring of the Institut für Meereskunde Kiel is presented here in some detail [82 S 3].

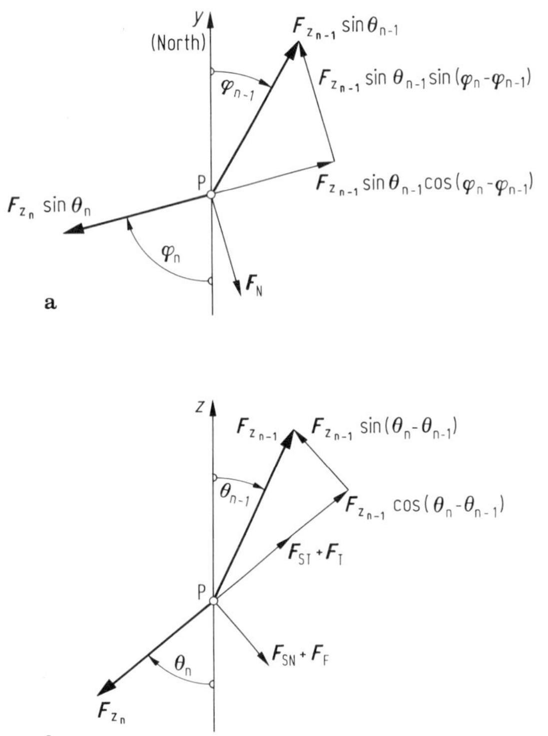

Fig. 11. Forces acting on the connection P between two segments n and n−1 of the mooring line. In the equilibrium state the vectorial sum of all forces must be zero. Using this sketch eqs. (13)···(16) are derived [82 S 3]. **a** Seen from above; **b** Seen from the side.

The mooring configuration is shown in Fig. 12. It supports 8 Aanderaa current meters on a water depth of 1500 m. The mooring is deployed with the subsurface buoy first. Supported by the buoyancy elements the line floats at the surface while instruments and wires are shackled together aboard of the ship. At last the anchor weight is slipped together with a parachute to reduce its sinking speed.

The distribution of the residual buoyancy with depth has been calculated using the scheme presented above, and the results are shown in Fig. 13.

The response of the mooring to 2 simplified velocity profiles (Fig. 14) has also been computed. Results are summarized in Table 3. It presents a survey on the vertical displacements and horizontal excursions of selected points of the mooring line. Some of the model results on the shape which the mooring line assumes under the action of chosen velocity profiles are shown in Fig. 15.

The examples may serve as a guideline for similar mooring configurations. They show that strong surface currents have the greatest effect on the displacements of moored instruments. If the current is variable the vertical up and down excursions can cause instruments to record, e.g., above or below the main thermocline. In order to separate between real temporal changes and mooring-induced variability, the upper instruments should always record pressure for reconstruction of mooring movements.

Only the most fundamental approach to the calculation of a mooring has been presented. Additional attention must be given to the elasticity of mooring wires and ropes and many technical details of mooring components [72 E, 76 S 3]. The large field of mooring dynamics [80 G 1] cannot be presented here.

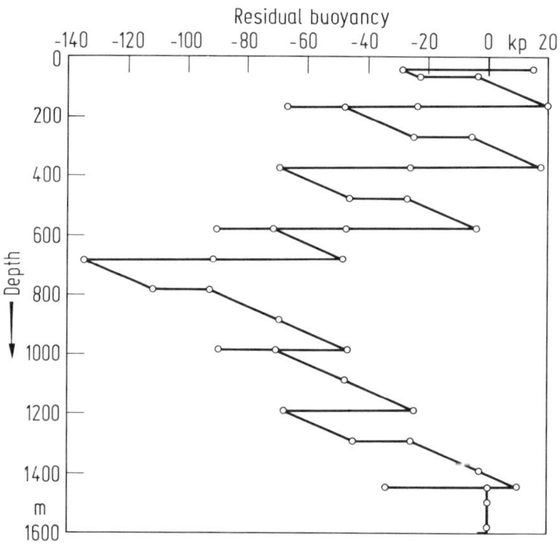

Fig. 13. Distribution of residual buoyancy with depth calculated for the mooring in Fig. 12 [82 S 3]. The large contribution of the subsurface buoy (273 kp) is not indicated. 1 kp ≙ 9.81 N.

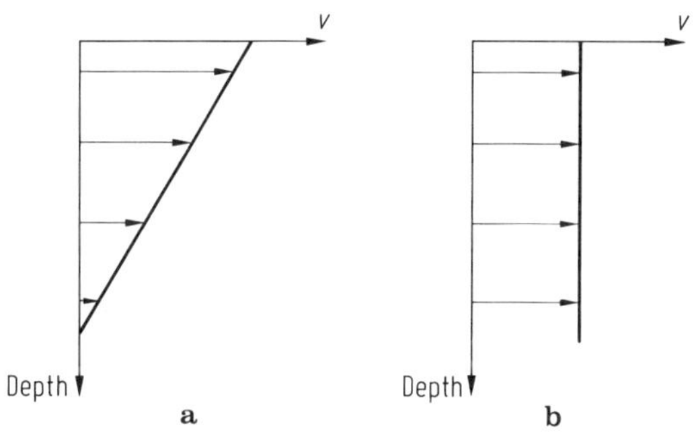

Fig. 14. Two model velocity profiles used for calculation of the vertical and horizontal excursions of instruments in the mooring of Fig. 12. **a** Profile 1 "LIN"; **b** Profile 2 "CON".

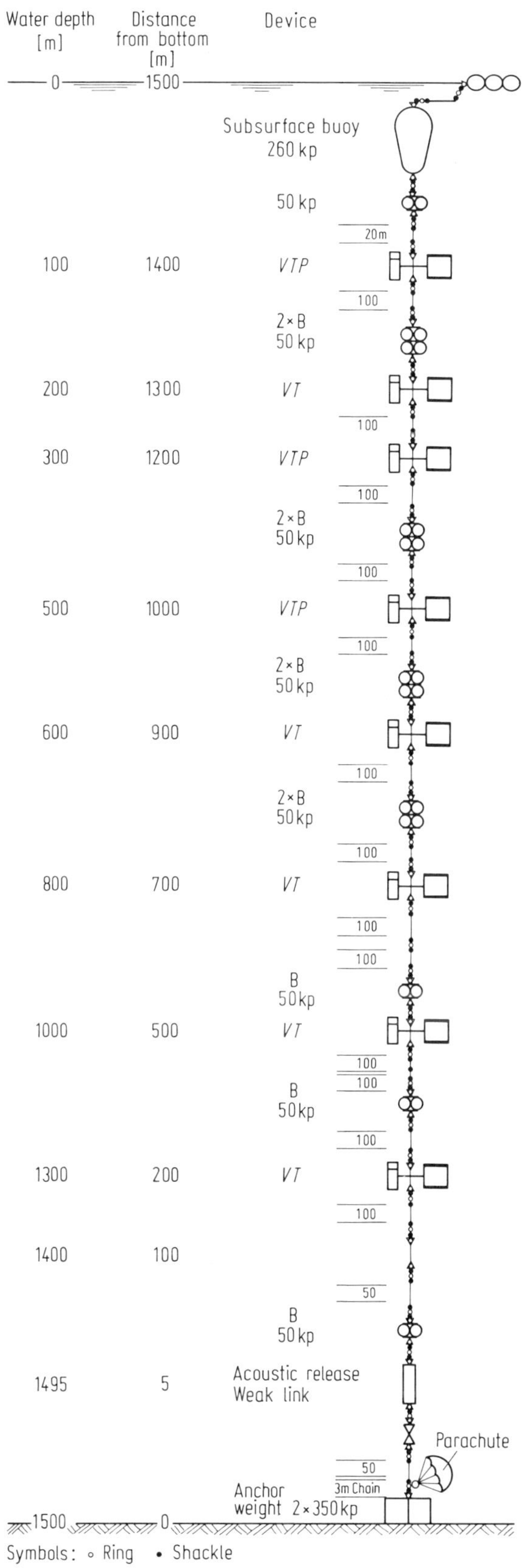

Fig. 12. A single point subsurface mooring as used by the Institut für Meereskunde Kiel [82 S 3]. 1 kp ≙ 9.81 N. B = buoyancy element consisting of two glass spheres; VTP = recording instruments, V = velocity, T = temperature, P = pressure; wire length in m.

Table 3. Horizontal and vertical excursions of selected points of the mooring of Fig. 12 under the action of the velocity profiles sketched in Fig. 14 [82 S 3]. z: depth of a point on the mooring line; EXC: horizontal excursion of a point on the mooring line; LIN: linearly decreasing velocity profile followed by velocity in $cm\,s^{-1}$ at the surface; CON 20: constant velocity of $20\,cm\,s^{-1}$ in all depths.

Zero current	LIN 20		LIN 30		LIN 40		LIN 70		CON 20	
z [m]	z [m]	EXC	z [m]	EXC	z [m]	EXC	z [m]	EXC	z [m]	EXC
45	47	84	57	185	121	455	233	703	56	160
67	70	84	80	185	143	453	255	700	78	160
171	173	81	183	180	246	440	357	679	182	157
273	275	78	285	173	346	423	455	651	283	154
476	477	69	488	153	544	374	643	574	487	143
580	582	64	590	141	643	344	735	527	590	135
784	786	54	793	119	839	289	920	440	793	119
987	989	42	994	93	1032	225	1098	342	996	98
1291	1291	22	1295	50	1317	120	1354	180	1296	57

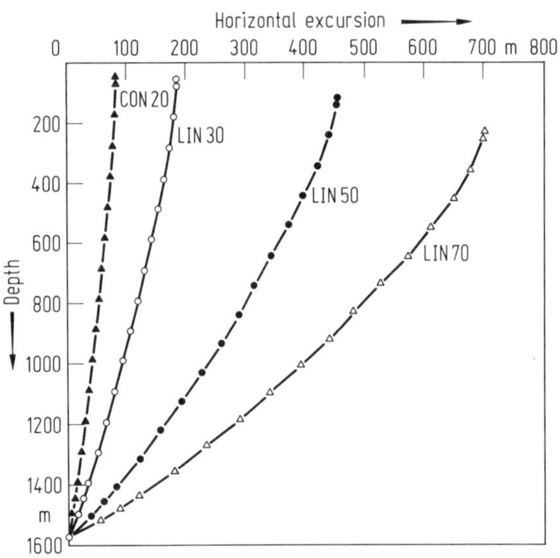

Fig. 15. Shapes of the mooring line from Fig. 12 computed for the velocity profiles as given by Fig. 14. The numbers following LIN and CON are in [$cm\,s^{-1}$] [82 S 3].

2.3 Measuring techniques for oceanic motions

According to their time and space scales, motions of water in the sea are called turbulence, ocean currents, waves, tides, etc. It is useful to distinguish between currents and the displacement of the sea surface. This section deals with horizontal movements.

2.3.1 Current sensors and current meters for Eulerian measurements

Currents can be considered to represent the most sensitive indicator for forces acting in a fluid. The basic hydrodynamic equations are expressed in terms of velocity (see sect. 4.1).

In view of the broad range of oceanic motions in time and space, there exists no ideal current meter for all purposes. Numerous measuring devices have been developed for different applications. The following discussion on current sensors and their principal limitations is based on a physical classification outlined in Fig. 16.

Firstly, sensors are discussed which utilize the kinetic energy of fluid motion and transport properties of currents. The basic flow diagram of respective current meters is sketched in Fig. 17. The basic equations for momentum and transport involve physical properties of the fluid. Consequently, measurements by sensors based on these equations must depend on these properties as well. In principle, readings obtained by sensors will depend on density, i.e. on temperature and salinity, and on transport coefficients in the case of transport sensors.

Most of a second class of current meters, which utilize general effects of motion like induction or Doppler effects, need an exciting device to generate magnetic fields, sound waves or light. Their basic flow diagram is presented in Fig. 18.

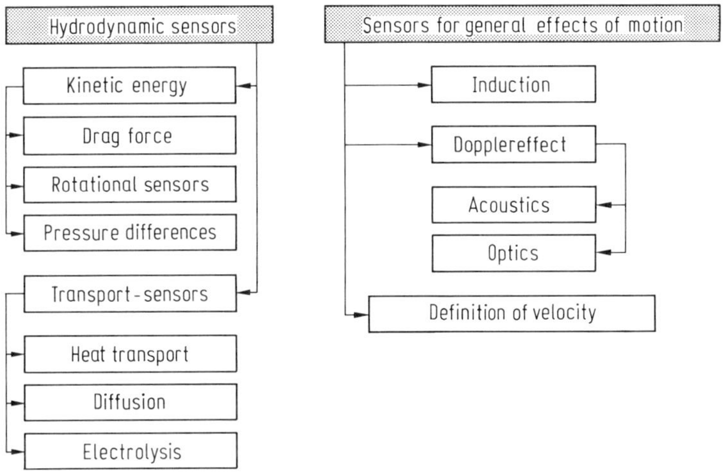

Fig. 16. Principles of Eulerian current measurements.

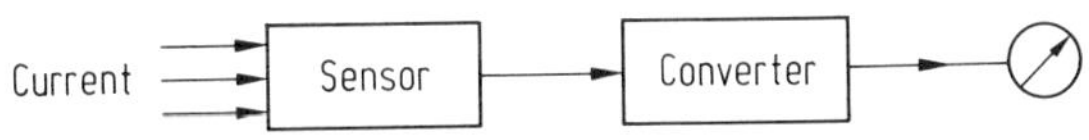

Fig. 17. Flow diagram of current meters utilizing the kinetic energy of fluid flow.

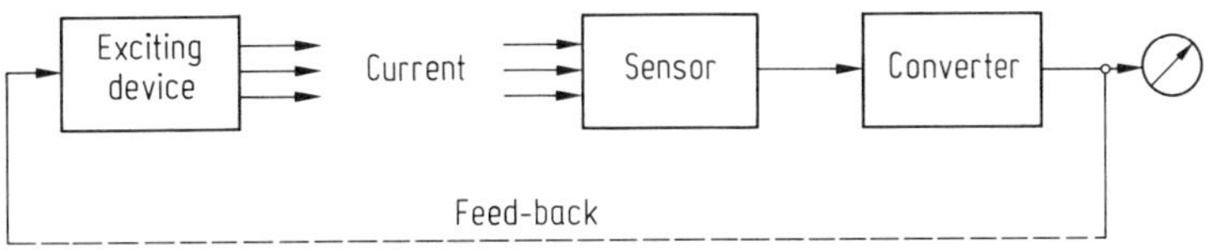

Fig. 18. Flow diagram of current meters utilizing general effects of motion.

2.3.1.1 Current meters with hydrodynamic sensors

2.3.1.1.1 Rotating elements and current vanes

Theory:

Independently of their form rotating elements share the following characteristics which describe satisfactorily their response to currents. They are devices which rotate about an axis; they possess a moment of inertia; they are subject to friction in bearings; their resistance to the current is characterized by an asymmetric distribution. In place of all practical designs, Fig. 19 represents a possible form of a rotating element. Its behavior is described by the scalar differential equation [67 F]

$$\Theta\dot{\Omega}+B\Omega+A=F_u-F_l \tag{22}$$

where Θ is the (scalar) moment of inertia, Ω is the angular frequency, A and B are the coefficients of friction of the bearings and F_u-F_l is the difference between the hydrodynamic resistances in the direction of the current and opposite to it.

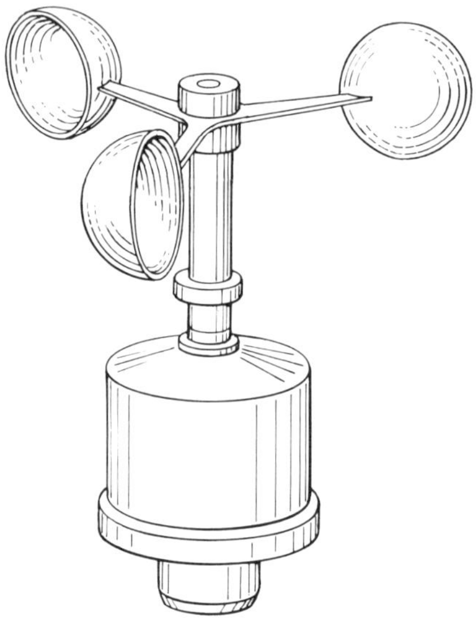

Fig. 19. A rotational sensor.

If V is the velocity of the current and R the effective radius of the element, then, assuming the fluid friction to be proportional to the square of the velocity, one has

$$\Theta\dot{\Omega}+B\Omega+A=L(V-R\Omega)^2-M(V+R\Omega)^2, \tag{23}$$

where L and M are constants which involve the size of the rotating element, the drag coefficient and the density of the fluid. On rewriting one obtains

$$\dot{\Omega}=k_1V^2-2k_2V\Omega-k_3\Omega^2-r_1-r_2\Omega, \tag{24}$$

where the k_i are constants containing R, L, M and Θ, while the r_i depend on the frictional coefficients A and B.

The stationary solution is the linear relationship

$$\Omega=\frac{V}{k}-\Omega_s, \quad \text{with} \quad k=\frac{k_2+\sqrt{k_2^2+k_1k_3}}{k_1}. \tag{25}$$

Ω_s corresponds to a starting velocity which depends on L, M, R, A and B. Its reduction to the smallest possible value is the more difficult the smaller is the radius of the sensor.

Time constant and frequency response are found by linearizing eq. (24). Letting

$$V=\bar{V}+v, \qquad \dot{\Omega}=\dot{\bar{\Omega}}+\dot{\omega},$$

it follows from eq. (24) neglecting friction

$$\dot{\omega}+2C\bar{V}\omega=\frac{2C}{k}\bar{V}v \tag{26}$$

where C is a constant, $\bar{V}$ denotes the mean velocity and v is a small deviation from V. For the initial condition,

$$\omega=0 \quad \text{at} \quad t=0,$$

one obtains as solution of eq. (26)

$$\omega=\frac{v}{k}(1-\mathrm{e}^{-2C\bar{V}t}). \tag{27}$$

It is seen that the time constant,

$$\tau=\frac{1}{2C\bar{V}}, \tag{28}$$

is a function of the mean velocity. This is a most undesirable feature of rotating elements. The physical explanation is found by rewriting eq. (28) in the form

$$\bar{V}\tau=\text{const.} \tag{29}$$

The constant has the dimension of a length and describes the length of a water column which must have passed the rotating element before the element has adjusted to a sudden change in the current velocity.

The directional sensitivity of a current sensor is defined as the change in the values measured which occurs when, for constant current velocity, the angle of attack, φ, relative to a space axis of the sensor, changes. Between the indicated velocity V_i and the true velocity V of the stream the cosine law

$$V_i=V\cdot\cos\varphi \tag{30}$$

is desirable. Some screw and propeller sensors, if suitably designed, satisfy eq. (30). Savonius rotors are insensitive to changes in direction vertical to their axis of rotation, but small tilts of the axis lead to strong variations in the indicated values.

Most of the current meters for application in the open ocean sense the current vector by two sensors: Rotating elements for speed and vanes for direction. Two methods are in use:

(a) The complete instrument together with the speed sensor is aligned into the direction of the "mean" current. Direction is obtained from a compass housed in the instrument.

(b) By means of small vanes, the direction of the current relative to the instrument housing is measured, while the orientation of the housing relative to North is registered inside it.

Although method (a) is very common, it has the disadvantage that speed and direction sensors are coupled. In case of strong lateral turbulence or under the action of orbital currents of surface waves, a combination of a large vane and an omnidirectional Savonius rotor yields grossly overestimated mean currents and spectral intensities derived by time series analysis.

Recent developments aim at two- or three-component measurements of the velocity vector.

Typical instrumentation

From the large variety of current meters three instruments, A, B and C, of different design philosophy have been selected for a more detailed discussion of practical problems associated with velocity measurements in the field. Table 4 summarizes their specifications, and short descriptions follow below.

Table 4. Instrument characteristics of three current meters utilizing magnitude averaging (type A), vector averaging (type B) and component measurement (type C). Dimension of length: [cm], dimension of speed: [$cm\,s^{-1}$].

Instrument type	A	B	C
Velocity sensor	Savonius rotor	Savonius rotor	2 cosine propellers
diameter	10.5	16.5	22
height	7	17	–
starting velocity	2	2.57	2.0
digitizing interval	0.3	0.5	
range	2.5 to 200	2.57 to 308	2.0–500
accuracy	$\pm 1\,cm\,s^{-1}$ or 2% of speed whichever is higher		$\pm 1\,cm\,s^{-1}$ or ± 1% of speed whichever is higher
response length		33.6	37.5
Direction			
vane	trailing fin	freely rotating	–
dimensions	100 × 37	9.3 × 17.2	–
compass	magnet + potentiometer	magnet + encoder disk	flux gate
digitizing interval	0.35°	2.81°	
accuracy	$\pm 7.5°$ for $2.5 < v < 5$, $100 < v < 200$; $\pm 5°$ for $5 < v < 100$		$\pm 5°$
Manufacturer	Aanderaa Instruments Bergen, Norway RCM 4	AMF Electrical Products Development Division Alexandria, Virginia USA	Bay Technical Associates St. Louis, Mississippi USA

Instument A (Fig. 20)

Compass and velocity sensor are mounted in one housing which is aligned into the flow by a large vane. The velocity sensor is a Savonius rotor in connection with an electronic counter. It accumulates the rotor revolutions between two successive recordings. The selectable sampling interval is determined by a quartz clock. At the time of a measurement the compass needle, which is suspended in silicon oil of high viscosity, is clamped on to a potentiometer ring.

The heart of the data acquisition system is an electromagnetic encoder which is used to balance a Wheatstone bridge or to sample data from a shift register. Its electromotor also drives the magnetic tape.

The instrument has six channels. Additional sensors are provided for temperature, conductivity and pressure. [64 A].

Instrument B (Fig. 21)

The velocity sensor is a Savonius rotor, direction is sensed by a small vane relative to the pressure case. Its orientation with respect to magnetic North is measured by a magnetic compass in connection with an encoder disk. [63 R].

There are eight magnets equally spaced around the circumference of the Savonius rotor. Passing of a magnet at a magnetodiode pair triggers a vector computation, i.e. magnitude and direction are converted into North and East components. These are added to the previous components. The summing process depends on the speed of the rotor. The sums are recorded on magnetic tape at a preselected time interval (usually 15 min). Thus they represent a time average. It is also possible to install additional sensors for physical properties of sea water [73 A].

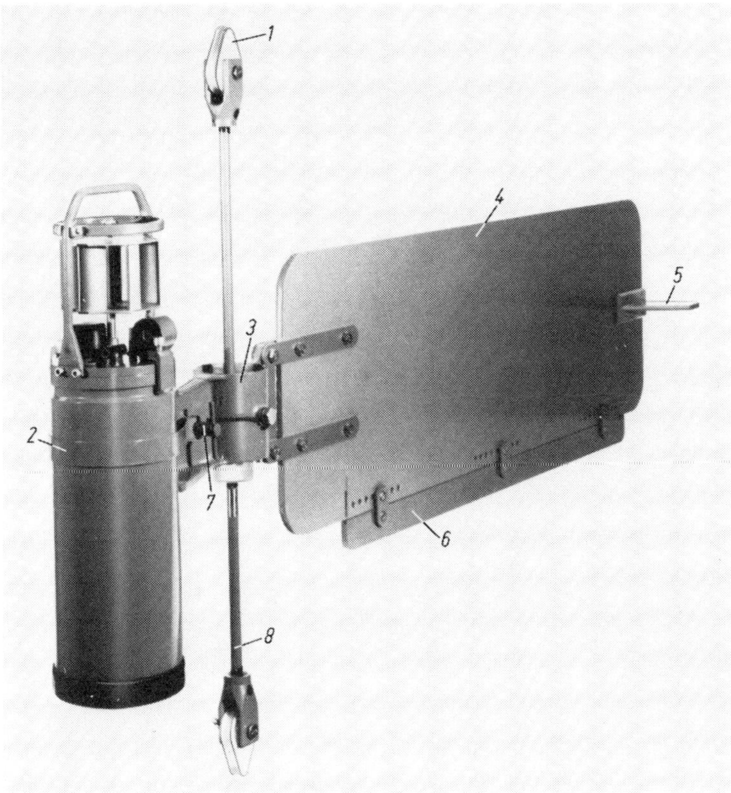

a

b

Fig. 20a, b. Aanderaa current meter [79 A]. *1* Thimble; *2* Fastening band; *3* Gimbal housing; *4* Vane plate; *5* Tail fin; *6* Balance weight; *7* Screw for fastening band; *8* Spindle assembly.

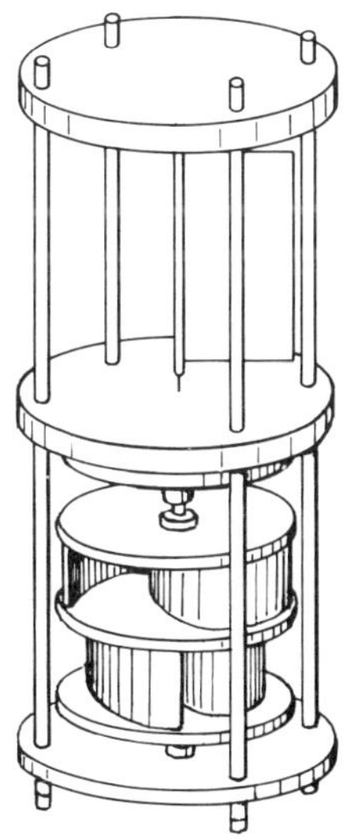

Fig. 21. The sensors of a vector averaging current meter.

Instrument C (Fig. 22)

This instrument has two orthogonal cosine response propellers which measure directly the components of horizontal velocity with respect to the instrument case. Orientation with respect to magnetic North is measured by a flux gate compass: The measured components are rotated electronically into the conventional North and East components, averaged and stored on magnetic tape [80 W].

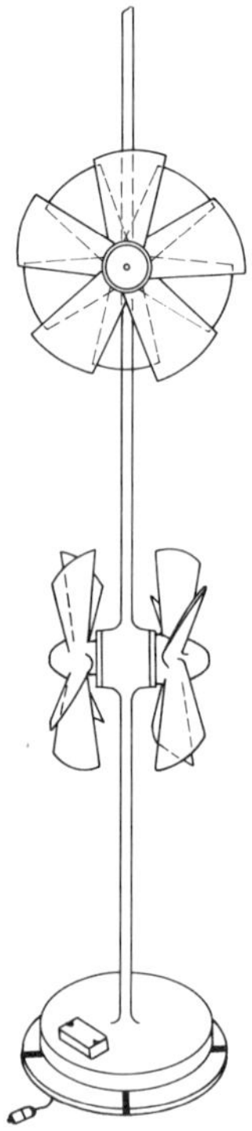

Fig. 22. Sketch of a biaxial propeller sensor. Propeller diameters are 22 cm; the distance between the propellers is 40 cm. Pick-ups to sense propeller revolution are mounted internally, and the wires from the pick-ups lead down inside the tube between the propeller sensors and out the base of the sensor. [80 D 1].

Field performance

The specifications displayed in Table 4 result from measurements in a towing tank. In the field, normally additional errors occur which exceed considerably the values obtained under defined laboratory conditions.

In a weak current regime all of the instruments provide erroneous residual speeds due to the finite starting velocity. However, the main sources of errors result from unsatisfactory mooring behavior in connection with the dynamical response characteristics of the current meters and the averaging procedures employed. In the upper layers of the ocean orbital wave motion can be a serious source of error for some of the sensors.

For many years, the UNESCO Scientific Committee on Oceanic Research (SCOR) [69 S 1, 74 S 2, 75 S 2] maintained a working group on the intercalibration of commonly used current meters. In the meantime additional information is available in the oceanographic literature (Table 5). Characteristic results of thoroughly conducted field tests will be discussed for the typical instruments A, B and C. Many other current meters have been tested, but the presentation of results for these three meters is sufficient to discuss most of the problems arising in the field.

For demonstration a very simple model time series has been constructed according to Fig. 23. There is a very small mean current velocity which gradually changes direction superimposed by a strong oscillatory motion. Such motion could arise from orbital velocities or mooring motions. Additionally, some random noise has been added.

This artificial time series is now resampled according to the sampling methods adopted for real instruments, but perfect sensor response is assumed.

Table 5. Instrument characteristics of current meters used in the UNESCO intercomparison experiment [74 S 2].

	Alexaev BPV-2	Alexaev BPV-6	Bergen RCM-4	Braincon 381	Geodyne 850	LSK 801.1	Plessey MO 21
Speed sensor:							
Type	shielded flat-bladed paddle wheel	unshielded 4-cup rotor	6-cup Savonius-like rotor	Savonius rotor	Savonius rotor	propeller in tube	propeller
Diameter [cm]	16.9		10.5	10.5	16	12	11
Height [cm]	11.5		7	8.6	17		8
Axis	vertical	vertical	vertical	vertical	vertical	horizontal	horizontal
Threshold [$cm\,s^{-1}$]	2.0	2.0	2.6	2.0	1.8	4.0	3.6
Digitizing [$cm\,s^{-1}$] Interval	1.0	1.0	0.3	1.0	0.5	1.3	0.4
Sample Duration	3 min	3 min	15 min	14 min	29 each 2.47 min	9.8 min	15 min
Vane:							
Type	double trailing fin	double trailing fin	trailing fin	trailing fin	freely rotating	crossed trailing fins	crossed trailing fins
Dimensions [cm]	35 × 49		95 × 37	56 × 61	9.1 × 17.2	32 × 30 32 × 30	40 × 60 30 × 17
Digitizing Interval	5°	5°	0.351°	2°	2.81°	5°	0.351°
Recording Medium	paper strip	paper strip	digital magnetic tape	analog 16 mm film	digital magnetic tape	waxed paper	digital magnetic tape

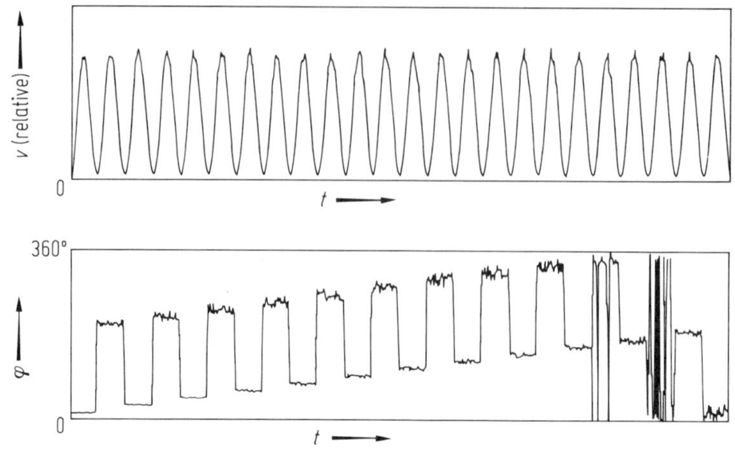

Fig. 23. Model time series for illustration of response characteristics of the three types of current meters. V: magnitude, φ: direction, t: time. Speed and time in arbitrary units.

In Fig. 24 the first vertical column shows scatter plots of one against the other velocity component, and the second column displays respective progressive vector diagrams. The first line is for the "real" current, and the successive lines show the results as they appear after vector averaging (e.g. by instruments of type B and C), when only velocity is averaged (as e.g. by instruments of type A), and when only direction is averaged. The latter case would occur if a fast-response velocity sensor would be mounted in front of a very large vane.

It is seen, that vector averaging provides a simplified but undistorted picture of the true situation. The most striking feature of magnitude averaging is the absence of low speeds in the scatter plot, whereas direction averaging gives serious errors in size and direction of the residual current.

For the current meter intercomparison exercises carried out in the field, the real situation was always unknown. However, by placing different current meters in one mooring and using the same suspension technique for all of them, a clear picture of the different performances could be obtained.

Fig. 25 shows the suspension technique and the current meters used by the SCOR working group 21 during their experiment in 1970. Short characteristics of the instruments are summarized in Table 6.

Although some of these current meters are no longer widely used today, the results of this and many other experiments display convincingly the characteristic limitations of current meters with rotational elements.

As a general result it has been demonstrated that all of the instruments perform more or less badly in the upper layers of the ocean where orbital velocities are present. Similarly, current meters mounted in a surface mooring show unsatisfactory results even at greater depths due to mooring motions.

Some of the differences between paired instruments during intercalibration experiments can be explained in terms of the known characteristic of the current meters. As in the previous model example, separate magnitude and direction averaging current meters tend to overestimate low true speeds under the presence of mooring noise. Due to the omnidirectional response of Savonius rotors the effect is more pronounced for current meters with this sensor than for those with propellers or shielded paddle wheel sensors. The best results are obtained with vector averaging current meters.

→

Fig. 24. Left column: scatter plots of the velocity components of the model time series displayed in Fig. 23. Right column: corresponding progressive vector diagrams. B = begin, E = end. First line: original data, equivalent to the result which a current meter of perfect response would record if all data were sampled. The following lines show respective diagrams as recorded by a vector averaging, a magnitude averaging and a direction averaging current meter. Note that for the progressive vector diagram on the bottom line the scale had to be reduced by a factor of 5.

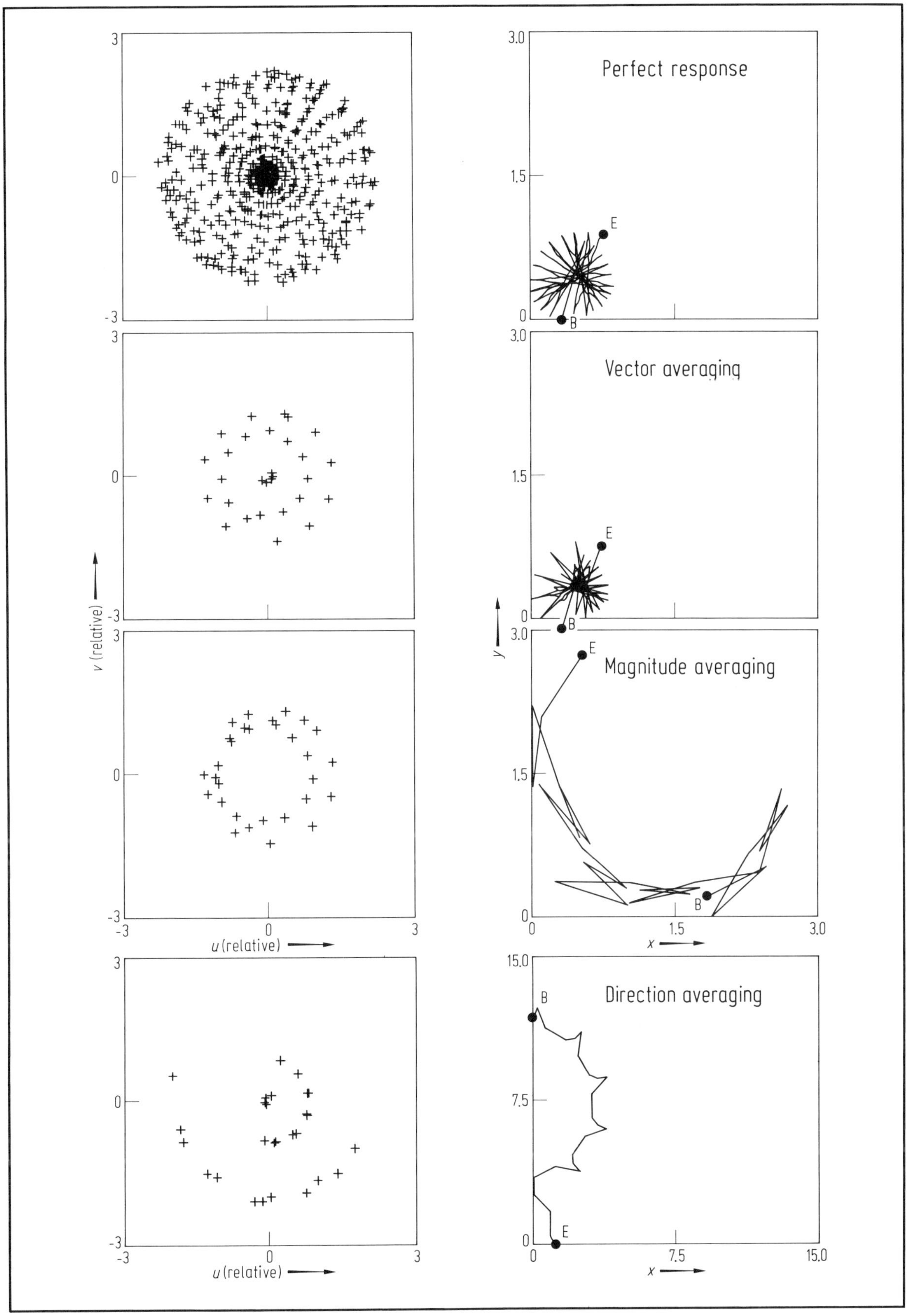
Perfect response
Vector averaging
Magnitude averaging
Direction averaging
v (relative)
u (relative)
x
y
B
E
3
0
-3
3.0
1.5
0
15.0
7.5

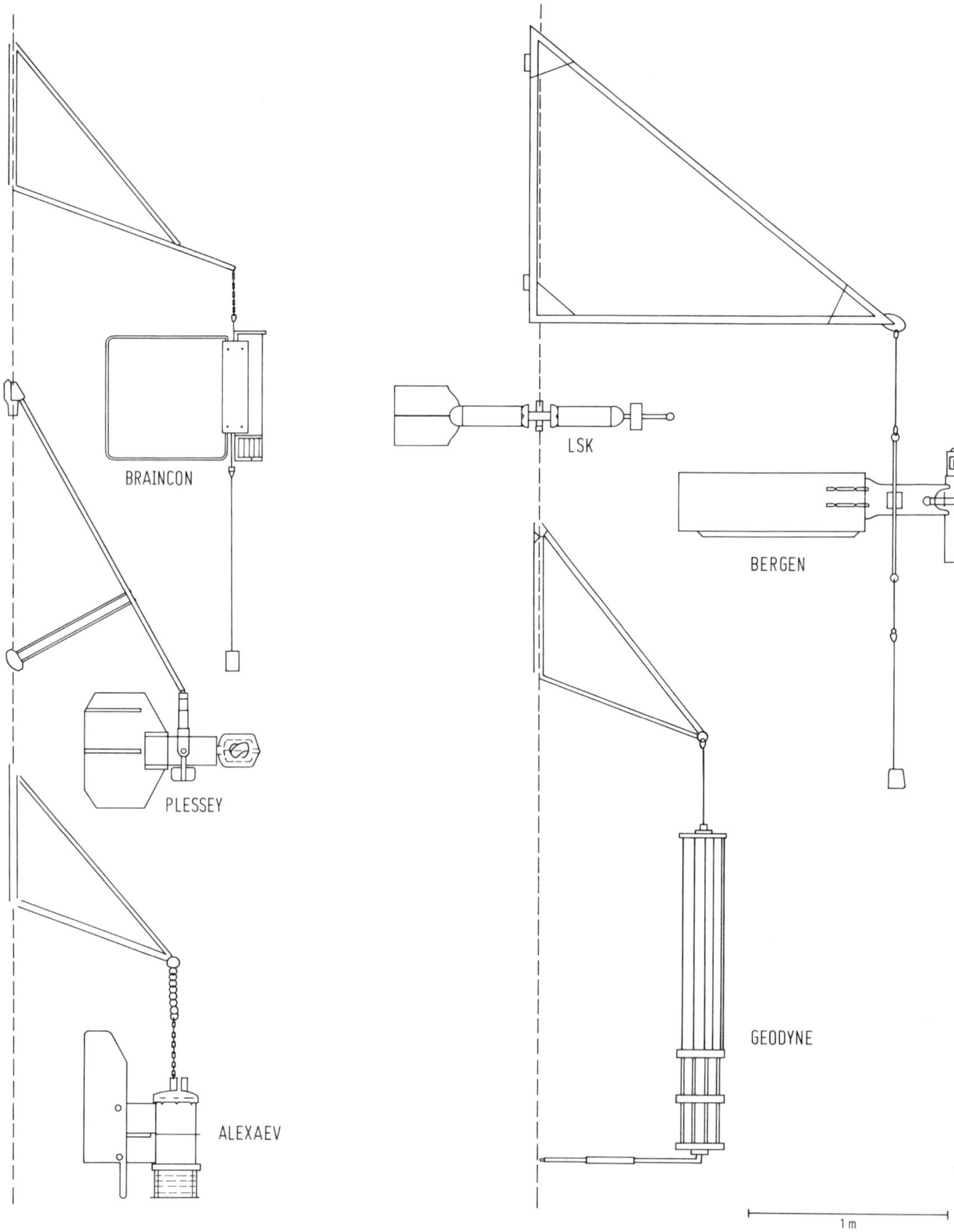

Fig. 25. Sketch of instruments used in the UNESCO intercomparison experiment showing means of mounting and relative instrument configurations [74 S 2].

Table 6. List of intercomparison tests of moored current measurements [80 H 1]. z: current meter depth, H: water depth, SPEC: specialized mooring such as a rigid tower or spar-buoy, SUR: single-point taut-line surface mooring, SUBSUR: subsurface mooring.

Reference	Current meter	z [m]	H [m]	SPEC	SUR	SUBSUR
Beardsley et al. [77 B 2]	Aanderaa RCM4, EG & G CT/3, Endeco 105, Geodyne 850, Marsh McBirney 711, VACM	7···25	28	×	×	×
Boicourt [73 B 2]	Braincon 1381	3···13	15	×		×
Briscoe and Hayes [73 B 3]	Geodyne 850, VACM	50···1000	2600		×	×
Cheng [78 C]	Aanderaa RCM4, Endeco 174, Oceanics 6011, Marsh McBirney 711	3	8	×		
Chiocchio [76 C]	Geodyne 850, Plessey M021	15···68	28···161			×
Gonella and Lamy [74 G 1]	Aanderaa	4	85	×	×	
Gould et al. [74 G 3]	Geodyne 850, VACM	1500	2600		×	×
Gould and Sambuco [75 G]	Geodyne	100···2500	2600		×	×
Halpern et al. [74 H]	Aanderaa RCM4, Geodyne 850, VACM	18···20	100		×	×
Halpern and Pillsbury [76 H 1]	Aanderaa RCM4	43	50			×
Halpern and Pillsbury [76 H 2]	Aanderaa RCM4, VACM	8	50			×
Halpern [78 H]	VACM	9	4200	×		×
Kullenberg and Woods [75 K]	Laser Doppler, acoustic, hot-film, electromagnetic	2···9	20	×		
Loeng [76 L]	Aanderaa RCM4, Ekman, propeller	4···22	23	×	×	
Magnell [77 M 1]	Endeco 105, EG & G CT/3, Geodyne 102, VACM	4···5, 10	13			×
Paramonov and Kushnir [77 P 1]	Alexaev, Geodyne	575···4000	4300		×	×
Petrie [77 P 2]	Braincon 381, Endeco 174	17···20	170			×
Pollard [74 P]	VACM	10, 32	3000	×	×	
Ramster and Howarth [75 R 1]	Aanderaa RCM4, Plessey M021	16···80	38···86			×
Saunders [76 S 1]	Aanderaa RCM4, VACM	10, 12	3000	×	×	
Saunders [76 S 2]	Acoustic VACM	35	4000	×		
SCOR WG 21 [69 S 1]	Bergen 42-66, Geodyne A100 and 850, Plessey M021, Tiefenstrommesser, Hydrowerkstätten	500	2600			×
SCOR WG 21 [74 S 2]	Alexaev BPV, Bergen RCM4, Braincon 381, Geodyne 850, LSK 801, Plessey M021	50···100	5100		×	
SCOR WG 21 [75 S 2]	Alexaev, Geodyne 850, LSK, VACM	50···1000	2600		×	×
Winant et al. [78 W 2]	S10 VMCM, electromagnetic	6	20			×
	S10 VMCM, VACM	10.5	4500	×		

The overestimation of speeds is important for the assessment of kinetic energy density spectra obtained from time series of measured ocean currents, especially if one is interested in the slope of the high-frequency end. Fig. 26 shows how different spectral properties were recorded by different current meters in the same depth during an intercalibration exercise.

On the other hand there is encouraging agreement between paired instruments concerning integral quantities like mean currents and progressive vector diagrammes.

Thus the key to successful use of current meters with rotational sensors is the mooring configuration. Subsurface moorings with the topmost float out of the reach of orbital motion are reasonable platforms for such current meters. If vector averaging current meters are employed, some mooring noise or orbital velocities can be tolerated.

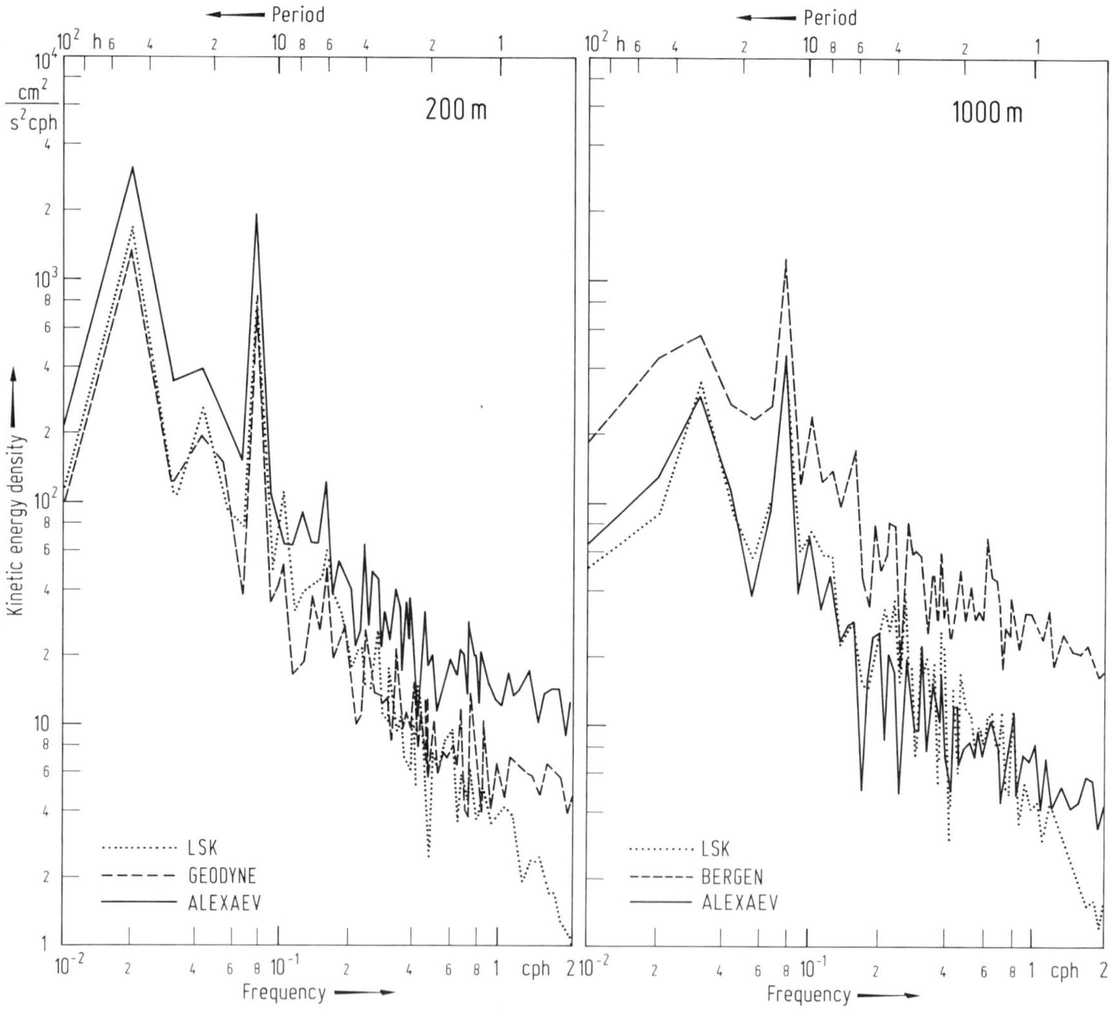

Fig. 26. Spectra of current meter records from an intercalibration experiment. Depth is 200 m and 1000 m, respectively. For the instruments used, see Fig. 25.

2.3.1.1.2 Drag force sensors

The use of drag force to measure current speed allows an easy splitting up into components. The quality of a drag force sensor mainly depends upon its hydrodynamical part.

In a nonstationary current the force on a solid body can be described by

$$\boldsymbol{F}=\tfrac{1}{2}C_{\mathrm{D}}(Re)\varrho A|\boldsymbol{V}|\cdot\boldsymbol{V}+C_{\mathrm{Dy}}(Re,S)\varrho Q\frac{\mathrm{d}\boldsymbol{V}}{\mathrm{d}t} \tag{31}$$

with $\boldsymbol{F}$: force
C_{D}: drag coefficient
C_{Dy}: dynamical drag coefficient
ϱ: density of fluid
Re: Reynolds number
A: cross sectional area (perpendicular to current direction or total area depending on the shape)
$\boldsymbol{V}$: velocity
ϱQ: mass of water displaced by the sensor
S: Strouhal number
t: time.

Eq. (31) displays the most important aims for practical design of a sensor:

(i) the 2nd term on the r.h.s. of eq. (31) should be made negligibly small in order to avoid solving a differential equation

(ii) the drag coefficient should be constant within the measuring range.

The second demand is easily fulfilled in a one-dimensional current. A holed flat plate with its normal pointing into current direction fulfills this need [66 F]. The explanation is that the turbulence in the wake is only determined by sharp edges and because regions of low pressure are avoided by flow of water through the holes. The drag force would be measured by strain gauges on the rod.

In the ocean several types of pendulum current meters have been used [60 K 2]. In this case a two-dimensional tilt meter was surrounded by a holed cone (Fig. 27).

A recent drag force meter is the Niskin winged current meter (Fig. 28). The instrument is tethered with a swivel and hangs vertically in the absence of a current. It is fitted with a wing-like structure which stabilizes and orients the instrument in the direction of a current. A tilt sensor and three orthogonally mounted flux-gate sensors provide the necessary data for the calculation of speed and direction. They are recorded on cassette tape together with time, temperature, conductivity and pressure. In a burst sampling mode of operation data sampled at intervals of a second can be used for vector averaging.

Attempts have also been made to use a vertically anchored rope as a drag sensor by measuring its shape under the influence of a vertical current profile [68 N 1].

For a long time the drag force principle was the only reasonable method to measure oscillatory currents [63 B 1, 69 S 2, 69 K, 73 K 1]. Still today it is a low-cost solution.

In oscillatory currents it is important to select a suitable drag sensor to avoid a contribution by the acceleration term in eq. (31). A rigid net of half-spherical shape (Fig. 27) is one of the possibilities. It has a large surface but deplaces very few water so that ϱQ in the 2nd term on the r.h.s. of eq. (31) becomes very small.

Sensors of this type are mounted on rigid platforms, and the horizontal components of the drag force are recorded. It is important to note that the instruments indicating the force components cannot be calibrated in terms of velocity.

In a rectangular coordinate system the components of the force vector $\boldsymbol{F}$ are related to the velocity vector $\boldsymbol{V}$ by

$$\begin{aligned} F_x &= c\cdot V\cdot u \\ F_y &= c\cdot V\cdot v \\ F_z &= c\cdot V\cdot w \end{aligned} \tag{32}$$

where $V=|\boldsymbol{V}|=\sqrt{u^2+v^2+w^2}$ and c is a constant. It is seen that the force components involve the magnitude of the current which must be computed by

$$V=\frac{1}{\sqrt{c}}\sqrt[4]{F_x^2+F_y^2+F_z^2}\,. \tag{33}$$

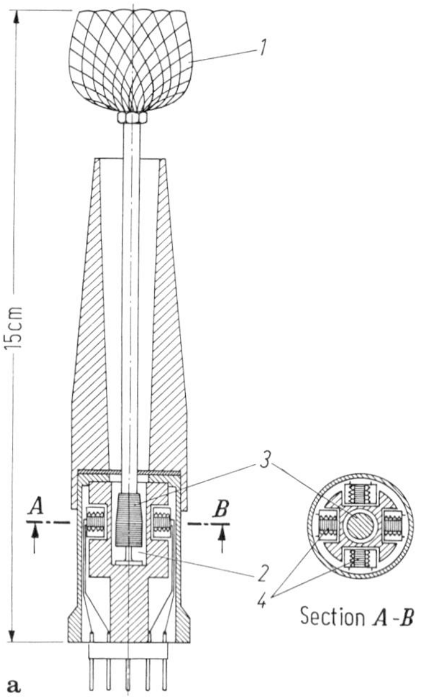

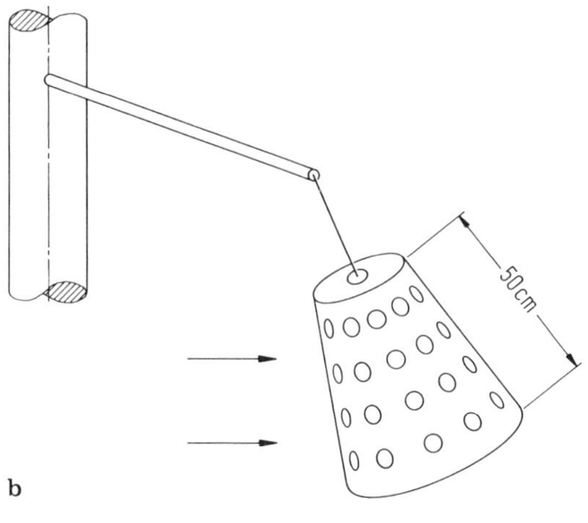

Fig. 27. (a) A two-dimensional drag force current sensor. The half-spherical mesh *1* at the top is exposed to the current. The drag force acts against a stiff spring *2* and changes the position of an iron core *3*. This is sensed by two pairs of coils *4*. (b) This pendulum sensor uses a holed cone to make its drag coefficient independent of velocity within the measuring range.

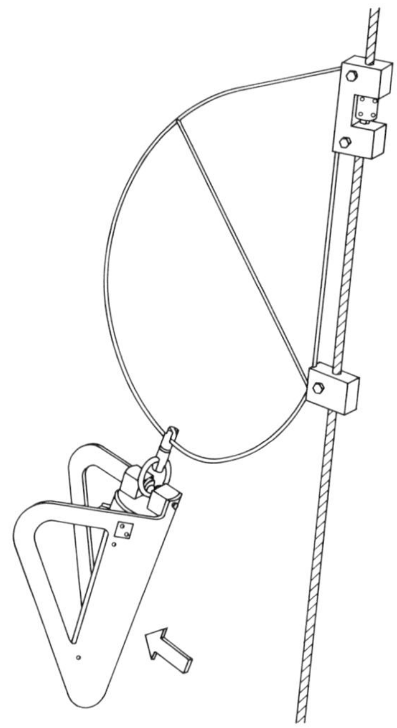

Fig. 28. Niskin winged current meter [84 G 1].

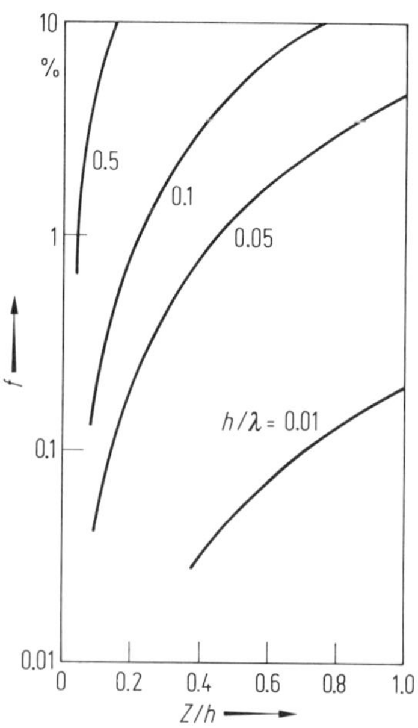

Fig. 29. Relative error f of horizontal velocity indicated by a two-dimensional drag force sensor when used in the 3-dimensional velocity field of shallow water waves (wavelength λ) of small amplitude, h: water depth, z: height above bottom.

If only the horizontal components are measured, the relative error f of the measured velocity magnitude to the true velocity is

$$f = 1 - \sin \arctan \frac{u}{w} \tag{34}$$

where v has been assumed to be zero. Inserting the relationships for shallow water waves of small amplitude yields (see sect. 6.1)

$$f = 1 - \sin \arctan \frac{1}{2\pi \frac{h}{\lambda} \cdot \frac{z}{h}} \tag{35}$$

where h is the water depth, λ denotes the wavelength, and z is the depth of the current meter. Fig. 29 shows this relationship. It is seen that two-dimensional drag force current meters can be used in the vicinity of the bottom but not close to the sea surface.

The winged current meter described before has a range from 0 to 300 cm s^{-1} according to 0°···90° degrees tilt. To achieve this the tilt sensor has a resolution 0.001% of full scale and an accuracy of ±0.5°.

2.3.1.1.3 Hydrodynamic pressure sensors

The hydrodynamic pressure p in a current with velocity V and density ϱ is

$$p = \tfrac{1}{2} \varrho V^2 .$$

It can be measured by a Pitot tube. Within the range of velocities in the ocean the pressures are very small. Only high performance differential pressure sensors can be used. This principle has found only a few special applications up to now [53 M].

2.3.1.1.4 Heat transport sensors

In the following the methods are briefly outlined and for more detailed information a list of references is presented. Survey articles are [58 H 3] and [68 R], while [80 F] and [83 F] are the best up-to-date complete literature reference lists in the field of thermal anemometry. The principle of heat transport sensors appears quite simple: An electrically heated solid is exposed to a current, the heat transport is a measure for its velocity.

In case of a wire, being heated at a temperature T_D everywhere on its length l, the flux ϕ of heat into the surrounding medium of temperature T_m is [58 H 3]

$$\phi = \alpha \cdot \pi \cdot d \cdot l (T_D - T_m) \tag{36}$$

with α: heat transport coefficient
d: diameter of the wire.

For the heat transport coefficient one has established an empirical relationship which holds under the assumption that the wire is only being cooled by direct heat conduction, but not by radiation and convection. Eq. (37) shows such a relationship:

$$\alpha = \frac{K_{mf}}{d} (0.42 Pr_f^{0.2} + 0.57 Pr_f^{0.33} \cdot Re_f^{0.50}) \tag{37}$$

with K_{mf}: heat conductivity of surrounding medium at temperature $T_f = \frac{T_D + T_m}{2}$
$Pr_f = \frac{c_{pm} \mu_{mf}}{K_{mf}}$: Prandtl number
c_{pm}: specific heat at constant pressure
μ_{mf}: viscosity of surrounding medium at temperature T_f
$Re_f = \frac{\varrho_{mf} V d}{\mu_{mf}}$: Reynolds number
ϱ_{mf}: density of surrounding medium at temperature T_f
V: velocity of surrounding medium.

In a steady state the electrically generated heat must be transported away by the current

$$\phi = I^2 R_D, \tag{38}$$

where I is the electric current through the wire which has the resistance R_D.

Eq. (38) suggests that there will be two methods of measuring ϕ. In a suitable circuit one can keep constant the electric current I and measure the resistance. One could also keep constant the wire resistance (i.e. the temperature of the wire) and measure the electric current which is necessary for this purpose.

(i) Constant-current method (Fig. 30a).

If there exists a linear relationship between resistance and temperature of the wire, what will be assumed for simplicity, the temperature difference in eq. (36) can be written in terms of resistances

$$T_D - T_m = \frac{R_D - R_m}{a R_0} \tag{39}$$

where a is the temperature coefficient of the wire resistance and R_0 the resistance at a reference temperature. From eq. (38) one obtains with eqs. (36), (37) and (39)

$$A + B\sqrt{V} = \frac{I^2 R_D}{R_D - R_m} \tag{40}$$

where A and B stand for combinations of quantities mentioned before. This equation can be used to calibrate the hot-wire anemometer in water free of turbulence [67 D].

In case of a turbulent current it is useful to split up V into a mean velocity $\bar{V}$ and the turbulent fluctuations v.

$$V = \bar{V} + v. \tag{41}$$

The same procedure can be done for the wire resistance

$$R_D = \bar{R}_D + r_D. \tag{42}$$

For weak turbulence one has

$$\frac{v}{\bar{V}} \ll 1, \qquad \frac{r_D}{\bar{R}_D - R_m} \ll 1. \tag{43}$$

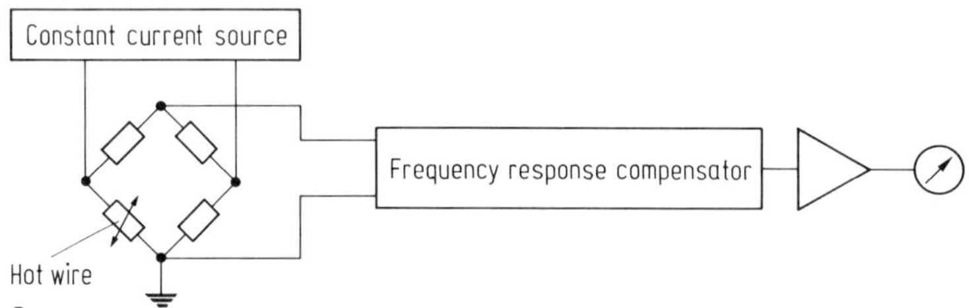

a

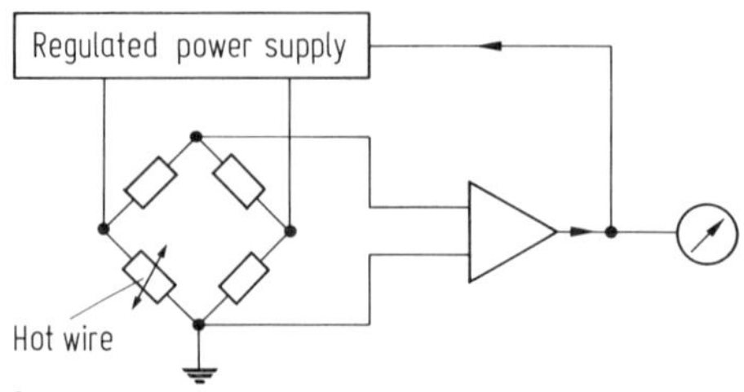

b

Fig. 30. Methods for measuring the electric current. **a** Constant-current method; **b** Constant-temperature method.

With these assumptions one obtains from eq. (40)

$$u = -s_i v \tag{44}$$

where u is the fluctuating part of the voltage drop across the wire. The sensitivity s_i is

$$s_i = \frac{(\bar{R}_D - R_m)^2}{2IR_m} \cdot \frac{B}{\sqrt{\bar{V}}} \tag{45}$$

with

$$B = 0.57 \frac{\pi K_{mf}}{aR_D} Pr_f^{0.33} \left(\frac{\varrho_f d}{\mu_f} \right). \tag{46}$$

Eq. (44) displays the advantage of the method. There holds a linear relationship between the voltage and velocity fluctuations. On the other hand, eqs. (45) and (46) indicate that the sensitivity only is constant if the heat transport occurs under constant conditions, i.e., in particular, if no changes in medium temperature [65 E] and mean velocity take place. These assumptions can be met with in the laboratory. At sea, the mean quantities will have to be measured simultaneously with a turbulence measurement.

(ii) Constant-temperature method (Fig. 30b).

If the power supply to the Wheatstone bridge with the sensor is controlled in such a way that the hot wire resistance remains constant, the temperature of the probe will also be kept at a constant value. Perturbation analysis similar to that outlined before yields for the electric current fluctuations i in the presence of turbulent fluctuations in the water

$$i = -g_L I r_D \tag{47}$$

where g_L is the transconductance of the regulator circuit. The corresponding voltage fluctuations are

$$u = s_t \cdot v \tag{48}$$

with

$$s_t = \frac{\bar{R}_D - R_m}{4I} \cdot \frac{B}{\sqrt{\bar{V}}}. \tag{49}$$

Comparison of eq. (49) with eq. (45) results in

$$s_t = \tfrac{1}{2} \frac{R_m}{\bar{R}_D - R_m} s_i \tag{50}$$

which shows that the constant-temperature method has a lower sensitivity for $\dfrac{\bar{R}_D}{R_m} > \tfrac{3}{2}$.

There are several advantages of this method. First of all the frequency response extends to much higher frequencies. In case of the constant-current method the time constant is practically given by that of the hot wire, because the response of the electronics can be made negligibly small compared with it. Now, in the other method the temperature is kept constant. Any small perturbation is sensed by the electronic regulator. In this case the response will be the faster the higher the sensitivity of the regulator is. The effective time constant therefore involves the transconductance of the electronic circuit besides properties of the wire. In practice the ratio of the time constants of the constant-temperature and constant-current method is in the order of 1 : 200. A second advantage is that the quantities A and B are constants.

Difficulties common to both methods can be seen from the relation for the time constant of a wire in flowing water which is

$$\tau = \frac{c_w}{aR_0(A + B\sqrt{\bar{V}} - I^2)}. \tag{51}$$

c_w is the heat capacity. To achieve a small time constant, it is necessary to use very thin and short hot wires or hot film probes. From the heated parts, however, flux of heat also takes place into the supports of the probe. Its relative proportion will be the greater the shorter the heated element is. Eventually, mechanical stiffness of the probes has to be taken into consideration in designing a sensor with a small time constant.

A further common difficulty is concerned with the directional sensitivity of heat transport sensors. With regard to the axis of a hot wire it is not possible to decide whether heat transport takes place to the "left" or the "right" side. This can be overcome by pairs of probes because directional sensitivity exists with respect to a direction perpendicular to the wire axis.

When used in the marine environment many additional difficulties arise [73 F 2]. Hot wire probes cannot be used due to mechanical problems. Instead quartz coated hot film sensors and thermistors [68 C] have been applied. There are problems with fouling by plankton, dirt and air bubbles [67 R]. Nevertheless, there are available rugged hot film probes of high technological standard for special investigations. An example of such a probe is shown in Fig. 31. For investigation of motions in the deep-sea a modified quartz thermometer has been used [72 R].

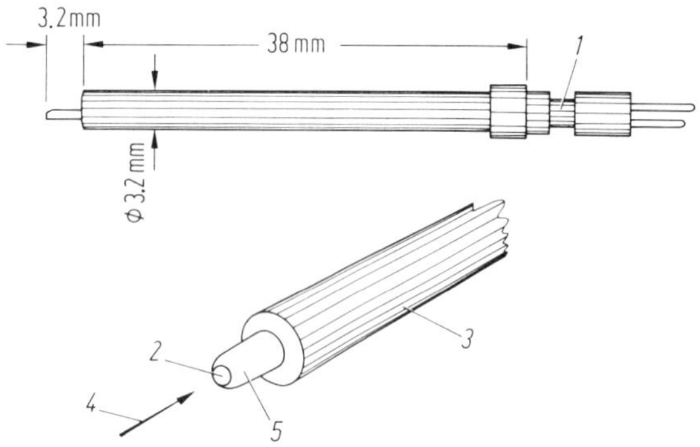

Fig. 31. A commercial hot film sensor for field measurements of turbulence [84 T]. *1* Quartz coated hot film sensor on end of hemisphere; *2* Hemispherical tip; *3* Stainless steel body; *4* Flow; *5* Quartz rod 1.5 mm diameter [84 T 1].

2.3.1.2 Current meters based on general effects of motion

2.3.1.2.1 Electromagnetic induction by the Earth magnetic field

Theory

When an electrolyte with velocity $\boldsymbol{V}$ moves through a magnetic field of flux density B, an electrical potential gradient $\nabla\phi$ is generated, which causes an electrical current density $\boldsymbol{i}$ depending on the conductivity C

$$\nabla\phi = \boldsymbol{V} \times \boldsymbol{B} - \frac{1}{C}\boldsymbol{i}. \tag{52}$$

Because of the presence of the Earth's magnetic field any ocean current generates potential differences which may be measured to derive the speed of moving sea water. A comprehensive treatment of the theoretical background is given by [48 S, 54 L]. Although the interaction of hydrodynamic and magnetic fields is complex, the situation for real ocean currents allows for considerable simplifications. As the vertical velocity is small compared to the horizontal components, the cross product of $\boldsymbol{V}$ and $\boldsymbol{B}$ produces only horizontal fields depending on u, v, and B_z, and the conductivity does not vary much in the open ocean.

Thus the evaluation of eq. (52) yields for the horizontal component

$$\nabla_h\phi = (\boldsymbol{V}_h \times \boldsymbol{e}_z B_z) - \frac{1}{C}\boldsymbol{i} \tag{53}$$

where h stands for horizontal, and $\boldsymbol{e}_z$ is the unit vector in z direction.

The potential gradient can be measured by two electrodes separated by a distance L as in Fig. 32 which shows the situation of a current flowing through a strait [55 B, 52 *M*].

The two electrodes can also be towed behind a ship [50 A, 62 A] as in Fig. 33. This arrangement is known as the GEK, the geomagnetic electro-kinetograph. Assuming the electrodes to be perfectly aligned on the ship's course and at equal depths the circuit does not cut field lines in the direction of the moving ship. This provides an elegant method to measure currents perpendicular to the ship's course if the leeway of the towing vessel can be neglected or measured by other means.

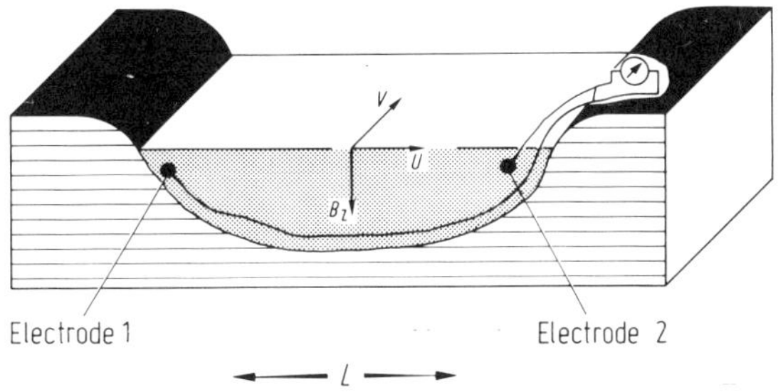

Fig. 32. Principle of geomagnetic current measurement in a strait. V is velocity, U indicates the potential difference, and B_z is the vertical component of the flux density of the Earth's magnetic field.

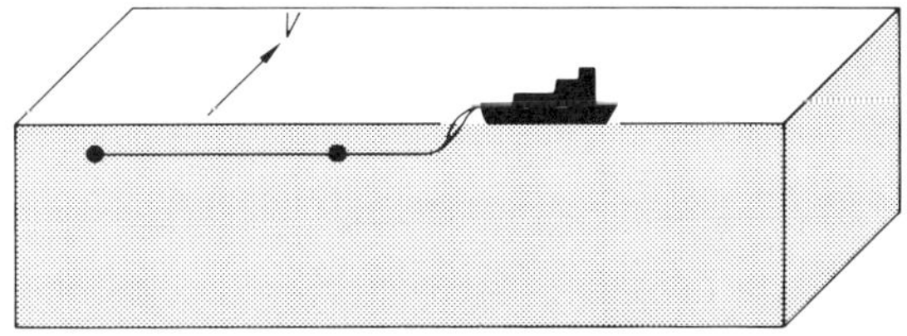

Fig. 33. The method of towed electrodes to determine the velocity component perpendicular to the ship's course. Distance between electrodes is 50···100 m.

The fixed electrodes measure directly the motional electromagnetic force minus the voltage drop due to the resistances of the conducting water and sea floor according to eq. (53).

When the measuring circuit also drifts with the water at equal speed, the signal received per unit length is

$$\nabla_h \phi = \left(\boldsymbol{V}_h \times \boldsymbol{e}_z B_z - \frac{1}{C} \boldsymbol{i}_h \right)_o - \left(\boldsymbol{V}_h \times \boldsymbol{e}_z B_z - \frac{1}{C} \boldsymbol{i}_m \right)_m \tag{54}$$

where index o refers to the ocean and m to the measuring circuit. By a continuous balance potentiometer or other electronical means it can be achieved that $i_m = o$ when measuring the potential difference. As the measuring circuit drifts with the water, the two cross products are equal. What is measured per unit length is then

$$\boldsymbol{E}_m = -\frac{1}{C} \boldsymbol{i}_h. \tag{55}$$

Using eq. (53) it follows that

$$\boldsymbol{E}_m = -(\boldsymbol{V}_h \times \boldsymbol{e}_z B_z - \nabla_h \phi). \tag{56}$$

The value of B_z is known. By towing the electrodes along two different courses one could also determine the vector $\boldsymbol{V}_h$. The electric potential in the sea is not initially known, but depends on the velocity distribution and other factors discussed below.

In the very idealized situation sketched in Fig. 34 we assume a uniform stream of width D moving perpendicular to the section with velocity v. The electrical current density lines occupy the water and the bottom down to a depth of the order of the width of the current.

The equivalent curcuit for such a case is sketched in Fig. 35. The specific resistance of the water path is approximately

$$R_w = \frac{1}{C_w} \cdot \frac{D}{h}$$

while that of the bottom is

$$R_B = \frac{1}{C_B}.$$

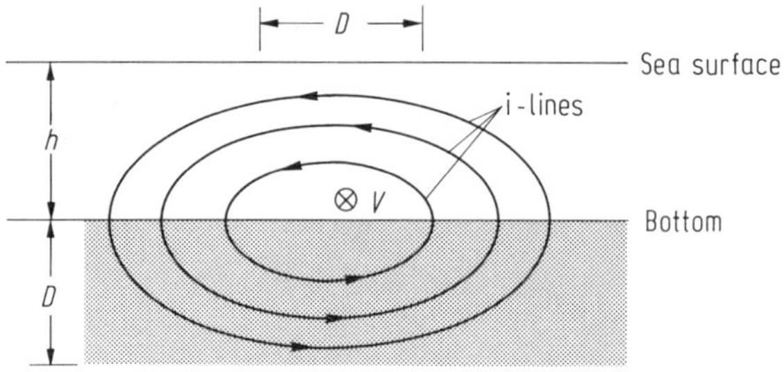

Fig. 34. A very simple model for the electric current lines associated with an ocean current of width D flowing into a direction perpendicular to the section [72 W].

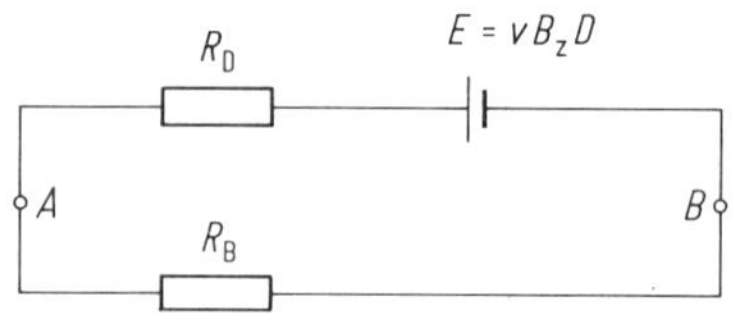

Fig. 35. The equivalent circuit to the situation in Fig. 34.

The potential difference between points A and B in Fig. 35 is

$$\phi_A - \phi_B = iR_B = \frac{E_o}{1 + \frac{R_w}{R_B}}$$

and the gradient

$$\frac{\phi_A - \phi_B}{D} = \frac{vB_z}{1 + \frac{DC_B}{hC_w}}. \tag{57}$$

It is seen that the potential gradient can vary greatly depending on the width of the flow compared to the water depth. The ratio of bottom to water conductivity can vary between unity and $2.5 \cdot 10^{-5}$.

For a stratified ocean it is necessary to extend the simple model of Fig. 35 in such a way that one assumes an array of batteries and series resistors connected in parallel to represent the values of a number of levels in the ocean.

According to Fig. 36 the analysis shows [78 S], that the potential gradient can be expressed as

$$\frac{\phi_A - \phi_B}{D} = B_z \bar{V}^* \tag{58}$$

where

$$\bar{V}^* = \sum_{i=1}^{M} C_i V_i \Big/ \sum_{i=1}^{N} C_i \tag{59}$$

is the conductivity-weighted, vertically averaged velocity. The components of the measured signal can then be written as

$$E_x = -B_z(u - \bar{u}^*)\,, \qquad E_y = B_z(v - \bar{v}^*)\,. \tag{60}$$

If one considers a GEK device mounted on the bottom where $u = v = o$, eq. (60) shows that this would provide $\bar{V}^*$. A GEK towed in the vicinity of this site would allow current velocities to be calculated on the basis of this model.

In practice, however, one often uses an empirical constant k defined by

$$k = \frac{|\boldsymbol{V} \times \boldsymbol{B}|}{|\boldsymbol{V} \times \boldsymbol{B} - \nabla\phi|}.$$

It is the ratio of the desired signal, from which the current velocity could be calculated directly, to the actual signal observed, which contains the unknown potential $\nabla\phi$. As shown above this depends on the velocity distribution and the conductivity ratio of the sea water to that of the sea floor in the first place.

The average open ocean value of k is 1.04. On the continental shelf it averages less than 2, and it may range up to 15 in shallow water. The factor is determined by observations of speeds indicated by non-electromagnetic and electromagnetic devices [61 S].

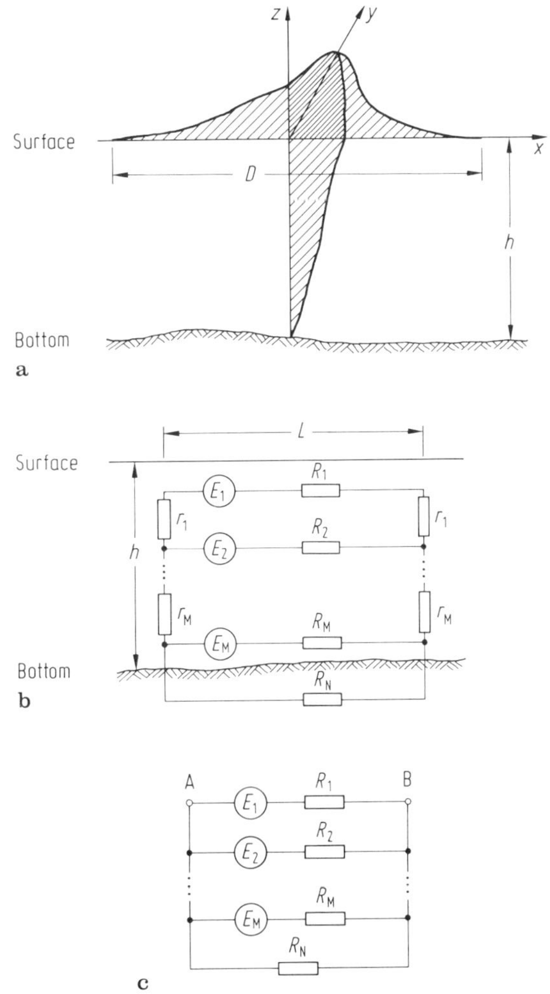

Fig. 36. Electric network equivalent of motional induction in the deep sea: (**a**) an ocean current having horizontal (D) and vertical (h) variations; (**b**) the equivalent network; (**c**) simplified electric circuit for $D \gg h$ [78 S].

Applications

One of the severe problems connected with GEK measurements concerns the electrodes. With an ocean current of 100 cm s^{-1} and electrode distance of 20 m, a typical signal ranges is in the order of 1 mV. The offset potential between electrodes in the same sea water can sometimes reach the same level.

Tests have found Ag–AgCl electrodes to be best suited for use in GEK measurements, but even the most careful fabrication methods cannot produce an electrode pair of identical properties.

The first operational GEK instrument (Fig. 33) [50 A, 62 A] dealt with the offset problem by towing the electrodes back and forth on the same ship's track. Assuming the ocean current to be constant for the short time during this procedure, the zero point could be determined. The velocity was calculated using the k-factor method described above. This GEK instrument has played an important role in Gulf Stream research.

Another approach to cope with the electrode bias problem is the switched electrode system [72 W] sketched in Fig. 37. The pair of electrodes is mounted very closely together and is connected to the sea water via salt bridges. A soleneoid-operated valve separates the two electrodes during measurement while a second valve is normally open. To determine the offset potential the first valve is opened and the second is closed. This potential can then be subtracted from the measuring signal. The method is also suitable for a bottom mounted GEK.

The free-fall GEK velocity profiler [71 S, 74 S 1, 78 S] copes with the electrode problem by converting the dc into an ac signal. This is achieved by letting the free-fall instrument spinning about its vertical axis using constant-pitch fins (Fig. 38). The electrodes are connected to the sea water by salt bridges with openings on opposite sides of the instrument housing.

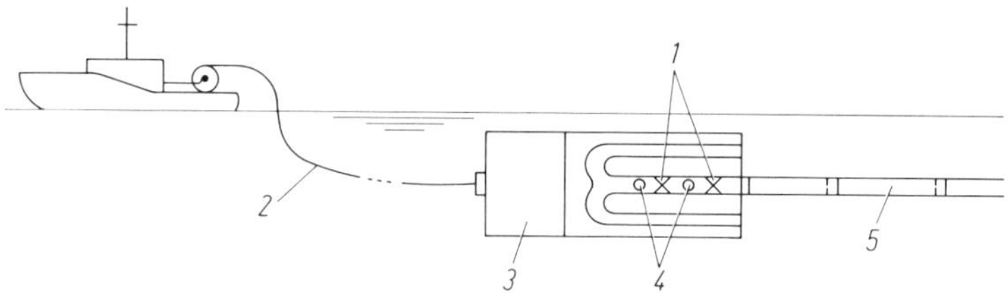

Fig. 37. The towed GEK with switched electrodes [72 W]. *1* Solenoid controlled valves; *2* 800 ft eight conductor towing cable: *3* Electronics; *4* Ag-AgCl electrodes; *5* Tygon tubing, variable length salt bridge.

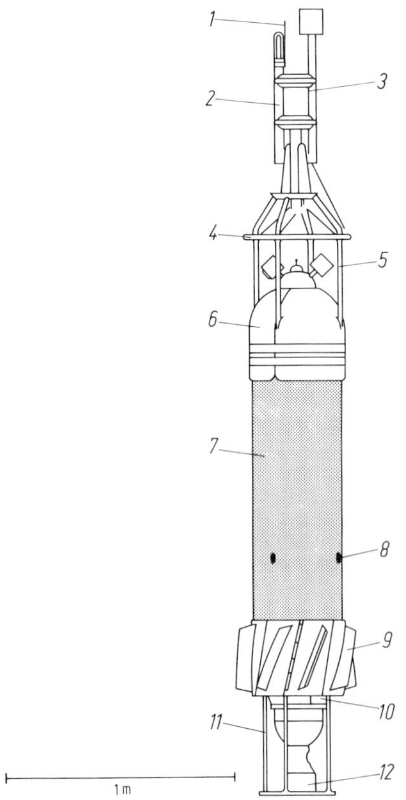

Fig. 38. The free-fall Electro-Magnetic Velocity Profiler (EMVP). The instrument rises from the sea floor as depicted in this figure, but is usually drawn towards the bottom in the reverse orientation with the sensor cage in the lead. A balloon keeps the instrument on the surface until the preset time when it is released by the surface relase mechanism and the instrument begins its down profile [78 S]. *1* Temperature probe; *2* Xenon flasher; *3* Radio; *4* Sensor cage; *5* Pressure transducer; temperature sensor; conductivity sensor; *6* Nose cone; *7* Polyethylene skin; *8* Electrode ports; *9* Rotation fins; *10* Surface release; *11* Projection cage; *12* Acoustic projector.

In recent years GEK measurements have experienced a renaissance for several reasons. Firstly, modern electronics cope much more easily with the processing of small dc signals than it was the case in the 1950's. Secondly, the new methods of position finding at sea together with the Doppler sonar and a GEK device allow to operate an integrated system for the determination of absolute velocities. Thirdly, it has been realized [82 S 1, 82 S 2] that the GEK measures the product of velocity and conductivity, i.e. the conductivity flux. This offers the opportunity to determine heat fluxes from a ship under way whereby large areas of the upper ocean can be covered.

The conductivity-weighting of GEK measurements can also be effectively used in observations on submarine cables to determine water and heat transports across large sections of the ocean. An example of such an application for the Florida Current is described in [82 S 2].

In all electromagnetic methods using the Earth's magnetic field magnetic disturbances and Earth currents will effect the measurement. However, variations of the magnetic field can be observed separately.

2.3.1.2.2 Electromagnetic current sensors

Theory

It is obvious from the small field strength of the natural magnetic field that it cannot be used in connection with small sensors on the centimeter scale. Artificial magnetic fields have to be applied in this case and ac fields will be used to avoid difficulties with the electrodes.

In Fig. 39 a number of possible shapes of electromagnetic velocity sensors is presented. The aim of the designs is to generate a magnetic field in which pairs of electrodes can sense components of the flow. Most of the current meters have a vertical magnetic axis in order to measure the horizontal components of ocean currents.

The general law of induction applies as for the devices described in the previous chapter. Specific details can be found in [70 B] and [62 S].

A second goal in constructing such sensors concerns the interference of the flow with the rigid sensor body. It is necessary to have the electrodes outside the boundary layer of the housing. Only if the pattern of fluid flow around the sensor is similar for all velocities, the output will be a linear function of speed and the electrical voltage induced will change sign with the velocity vector. Therefore the sensor housing should be shaped in a manner that within the measuring range sensitivity and directional response of the current meter are independent of the Reynolds number calculated for the typical diameter of the sensor.

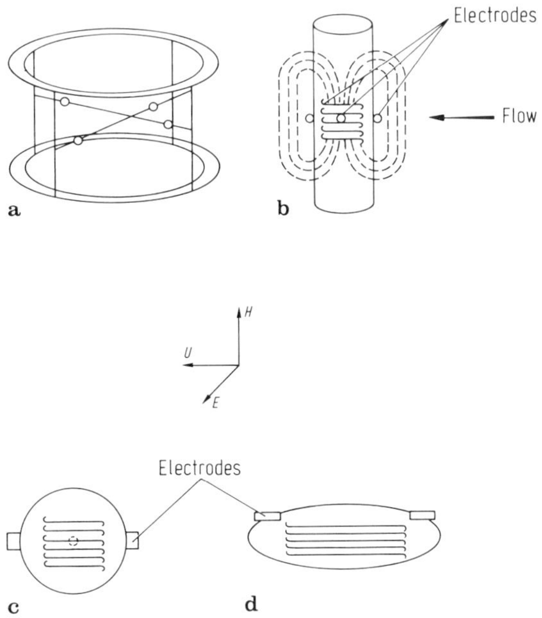

Fig. 39. A number of electromagnetic current meter shapes that have been developed. (U: current, H: magnetic field, E: induced voltage.) **a** Helmholtz coil; **b** Cylinder; **c** Sphere; **d** Ellipsoidal [80 J].

Also the power consumption of the coil is an important design factor, especially for self-contained instruments. At a magnet power consumption of 200 mW, the output signal ranges in the order of $10^{-2} \cdots 10^{-1}$ μV for each $cm\,s^{-1}$ increase of velocity. Thus the signal to noise ratio of the electrode amplifier determines the threshold velocity of the current meter. The low signals explain why sensors of this type may be sensitive to interference with electrical signals of other origins.

Applications

No moving parts, two-component measurement and fast response characteristics are the most attractive features of electromagnetic current meters which are not only used in oceanography [44 K]. In this field they have found first applications in shallow water, in connection with the measurement of orbital velocities [66 B] and with turbulence studies near the sea floor [73 T], and a number of instruments for this purpose have been developed [56 C, 72 O].

Most of the instruments used the ellipsoidal configuration shown in Fig. 40. However, signals from sensors of this shape are easily disturbed by a third velocity component which can cause vortex shedding of one or more electrodes [70 T]. This restricts its application to the vicinity of the sea floor where it should be mounted strictly vertical to the flow direction.

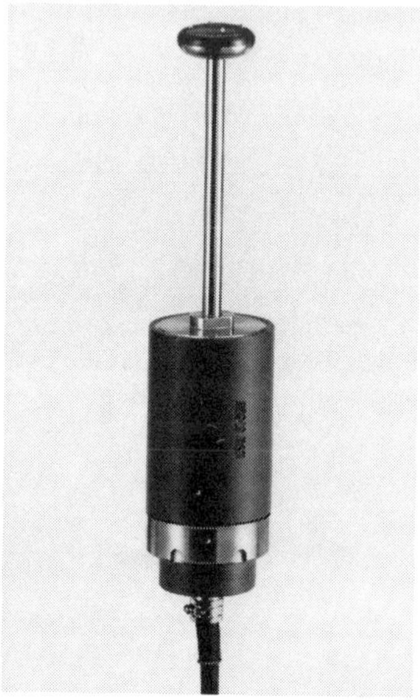

Fig. 40. Two-component electromagnetic current sensor with ellipsoidal sensor body [84 N 2]. Diameter of the disk is 11.5 cm. Measuring range: $\pm 100\,cm\,s^{-1}$, threshold $\pm 1\,cm\,s^{-1}$.

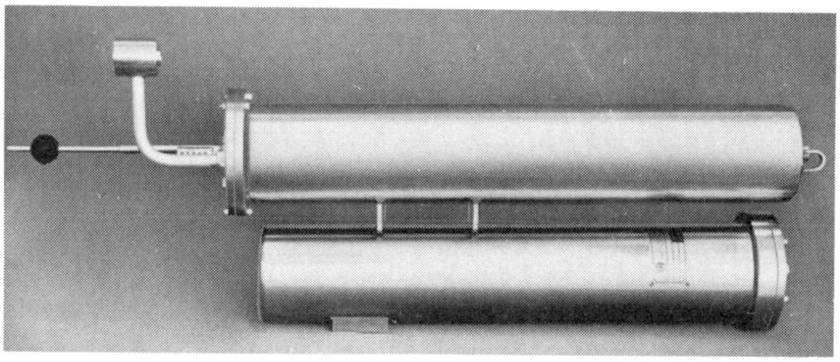

Fig. 41. A self-contained electromagnetic current meter with additional package for conductivity and temperature. Diameter of the spherical sensor is 3.8 cm. Measuring range: $0 \cdots 300\,cm\,s^{-1}$, threshold: $0.6\,cm\,s^{-1}$. Direction with respect to North is measured by a flux gate compass. [84 I].

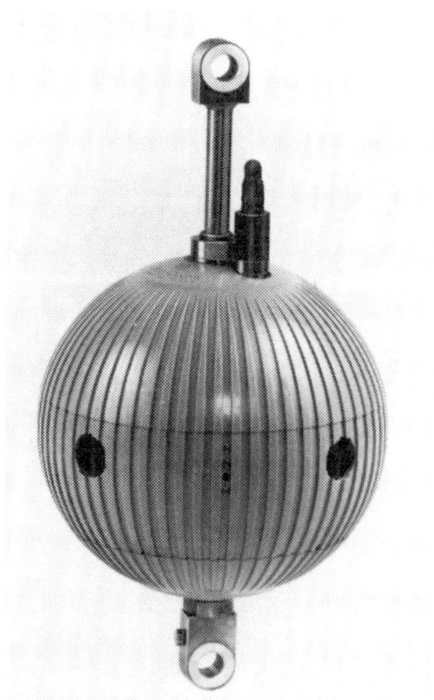

Fig. 42. A self-contained electromagnetic current meter. The spherical housing contains the electronics and a solid state memory for data recording. Simultaneously, the housing carries the electrodes and represents the sensor. Measuring range: $0 \cdots 350\,cm\,s^{-1}$, resolution $0.2\,cm\,s^{-1}$, accuracy 2% of reading $\pm 1\,cm\,s^{-1}$. Diameter of sphere: 25 cm, weight (air): 8 kg [84 I].

With the advent of modern electronics and microprocessors, electromagnetic current meters have become available to be used in moorings like rotor current meters. This is not only due to reduction of the magnetic coil power consumption but also due to the possibilities in fast data processing in situ. Because of the fast response characteristics it is necessary to sample the component signals at intervals in the order of a second, to perform averaging procedures and to calculate averaged velocities with reference to North prior to data storage. Two modern instruments are shown in Figs. 41 and 42.

2.3.1.2.3 Acoustic Doppler effect

Doppler sensors can measure velocity at some distance away from the instrument. Thus the flow to be measured needs not to be affected by the presence of a sensor.

The acoustic version of the application of the Doppler effect is sketched in Fig. 43. Controlled by a crystal oscillator, an acoustic transducer generates sound waves into the flowing water. The directional sensitivity of an acoustic receiver overlaps with that of the transducer within a small volume. If particles with acoustic impedances different from water move within this control volume, the sound waves are reflected. The receiver "sees" reflectors as moving sound sources, provided their sizes being at least greater than half the wavelength of the sound emitted.

When there is a cloud of particles of equal size, all moving with the same speed v towards the receiver, then the Doppler frequency would be ($v \ll c$)

$$f_D = \frac{2v}{c} f_{os}$$

where c is the speed of sound in sea water, and f_{os} is the oscillator frequency. If one assumes the particles to move at the same speed as the flowing water the Doppler frequency is a measure of velocity. Therefore it is desirable to make use of very small particles, and hence to use high-frequency sound waves. At a sound frequency of 5 MHz the smallest reflectors have a characteristic length of 0.15 mm.

In reality, size and concentration of particles varies in wide ranges in ocean water. A Doppler frequency band rather than a single frequency appears at the receiver output and the amplitude fluctuates considerably. There may also occur short time spans during which no particles are present. The absence of any signal causes "drop-out" problems [68 W].

A few small-scale acoustic Doppler current meters have been tried out in oceanography [61 L, 62 C, 65 K, 68 S 3], but the method has found much more fruitful application for remote sensing techniques and in the Doppler sonar ship's log. A survey article on acoustic Doppler techniques is provided by [80 P].

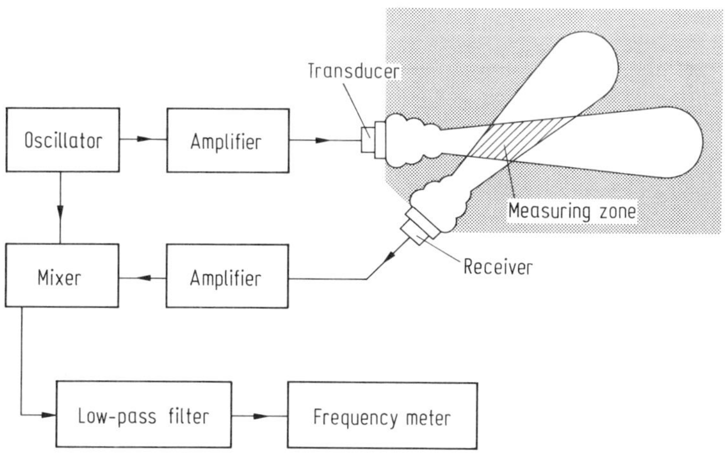

Fig. 43. Principle of an acoustic Doppler current meter.

2.3.1.2.4 Optical Doppler effect

Theory

The principle arrangement of transmitter and receiver to detect Doppler signals in moving water is the same in acoustics and optics. The optical method has become feasible after the emergence of the Laser light source. Besides other advantages the temporal and spatial coherence of its light allows its concentration in very small volumes (1 mm^3) at a distance away from an instrument housing. The optical Doppler effect is a very promising technique to study turbulence and boundary layer flow in the ocean.

In the laboratory the laser Doppler anemometry (LDA) is a very well developed technique [80 R, 84 W, 76 D 2, 76 D 3].

The principle of measurement is outlined in Fig. 44. The light emitted from a laser with frequency f_0 is scattered by a particle moving in the fluid with velocity V. The moving particle receives a frequency

$$f_p = f_0\left(1 - \frac{\boldsymbol{V}\cdot\boldsymbol{k}_0}{C}\right). \tag{61}$$

The receiver "sees" this particle as a moving light source and would detect

$$f_r = f_p\left(1 + \frac{\boldsymbol{V}\cdot\boldsymbol{k}_r}{C}\right) \tag{62}$$

where

$$C = \frac{C_0}{n}$$

is the light velocity in the fluid with refraction index n. $\boldsymbol{k}_0$ and $\boldsymbol{k}_r$ are unit vectors along the laser beam and the receiver axis, respectively.

The frequency f_r in eq. (62) is too high to be detected by a photomultiplier. Therefore, a reference light beam is necessary to form the Doppler frequency (optical heterodyning) $f_D = f_r - f_0$.

For $V \ll C$ it follows from eqs. (61) and (62)

$$f_D = \frac{n}{\lambda_0}\boldsymbol{V}\cdot(\boldsymbol{k}_r - \boldsymbol{k}_0) \tag{63}$$

where

$$\lambda_0 = \frac{C}{f_0}.$$

If Θ is the angle between the direct beam and the reference beam (Fig. 44), then

$$|\boldsymbol{k}_r - \boldsymbol{k}_0| = 2\sin\frac{\Theta}{2}$$

and the direction of this vector points perpendicular to the bisector of the angle Θ. The Doppler frequency is therefore [80 R]

$$f_D = \frac{n}{\lambda_0}\cdot V\cdot 2\cdot\sin\frac{\Theta}{2}\cos\phi \tag{64}$$

where ϕ is the angle between $\boldsymbol{V}$ and $\boldsymbol{k}_r - \boldsymbol{k}_0$.

The Doppler shift is proportional to the particle velocity. As all other quantities in eq. (64) are known, there is no need for calibration in a towing tank as it is the case for most of other velocity sensors.

The basic optical configurations are classified by the location of the detector in the system (Fig. 45). It determines which velocity component is measured and it is important for the practical design of a current meter to be used under field conditions.

The forward scattering fringe mode uses two beams of equal intensity which are crossed by means of a lense. With a diaphragm the direct beams are prevented from entering the detector, only scattered light is received. Typical scattering functions of sea-water suggest the use of forward scattering methods because 50···60% of the total scattered intensity falls into a range of $\pm 5°$ in forward direction, while in the same direction backwards intensities are lower by a factor $10^{-3}\cdots10^{-4}$.

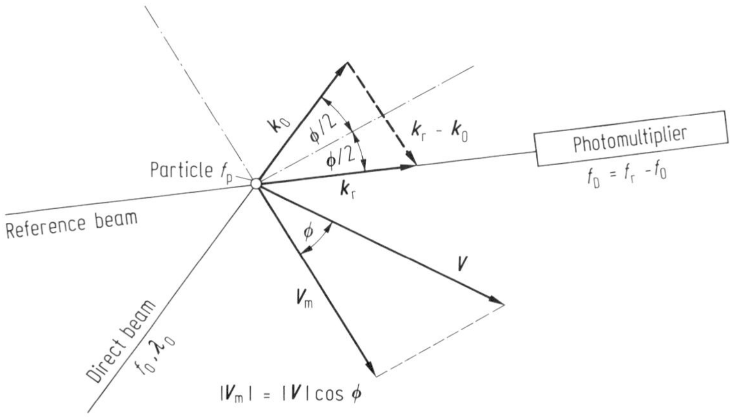

Fig. 44. Explanation of the optical Doppler effect velocity sensor. The geometrical arrangement corresponds to the reference beam mode shown in Fig. 45. The illuminated particle moves with velocity. V. The measured velocity V_m is the projection of V into the direction perpendicular to the bisector of the crossing direct and reference beams. The photomultiplier receives the reference beam light (frequency f_0) together with the Doppler-shifted scattered light (frequency f_r) from the moving particle, and it indicates the difference frequency.

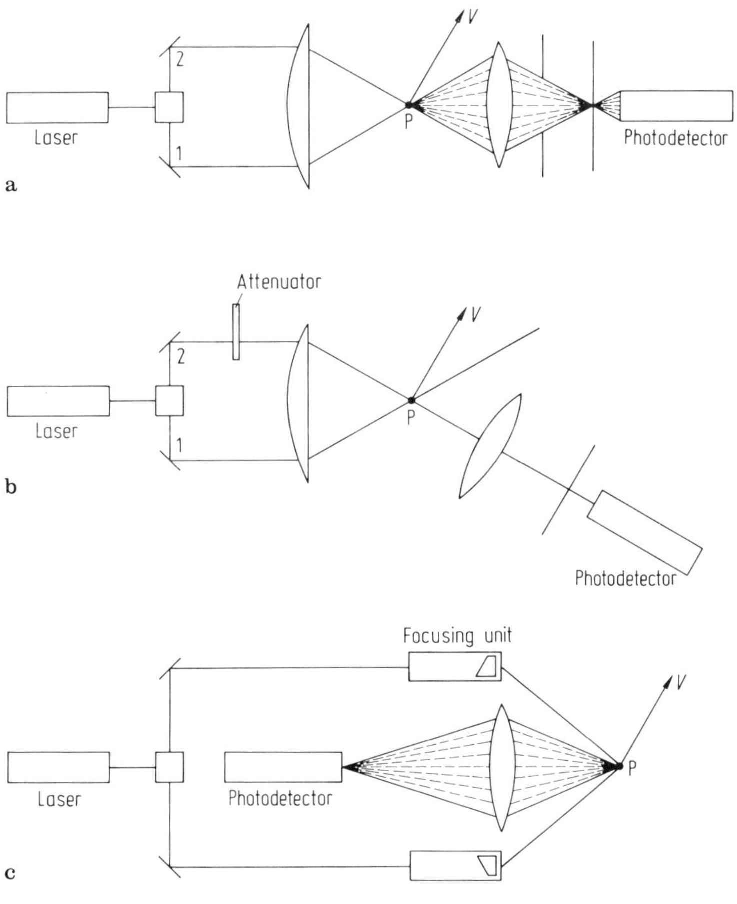

Fig. 45. Three basic optical configurations of LDA systems. P: particle. **a** Forward scattering fringe mode or differential mode; **b** Reference beam mode; **c** Backward scattering fringe mode.

The forward scattering fringe mode is also named differential method. Fig. 46 provides an alternative for the derivation eq. (64). The two crossing beams generate a system of interference stripes with a distance

$$d = \frac{\lambda_0}{2\sin\frac{\Theta}{2}}. \tag{65}$$

When a particle crosses this system of dark and light stripes it is illuminated by a frequency

$$f_D = \frac{V}{d} \tag{66}$$

which is detected by the photomultiplier. The combination of these two equations yields eq. (64).

In the reference beam mode the detector faces a weak reference beam and receives scattered light from a strong beam.

The backscattering fringe mode is a very promising method for field use. Although the backscattered light is of low intensity all components can be mounted rigidly in one watertight housing.

A recent interesting mode of operation uses a fibre glass cable to duct the laser beam to a very small probe (order of 10 cm) to be used under water whereas light source and receiver are mounted separately. The same fibre glass cable is also used to transmit the backscattered light in a different plane of polarisation [84 D, 84 T].

Laser Doppler anemometers do not provide a priori the sign of the velocity component measured. This is evident from Fig. 46. Therefore, the two original laser beams are modulated with an extra frequency by means of Bragg cells. Then the interference stripes move against each other which is equivalent to a known constant velocity. This can be added or substrated from the velocity (frequency) measured.

If the concentration of particles is low the Doppler signal output consists of individual bursts corresponding to the passing of particles. In this case the frequency is determined by a counter technique. For high concentration of particles a frequency tracking method is used [84 D, 84 T].

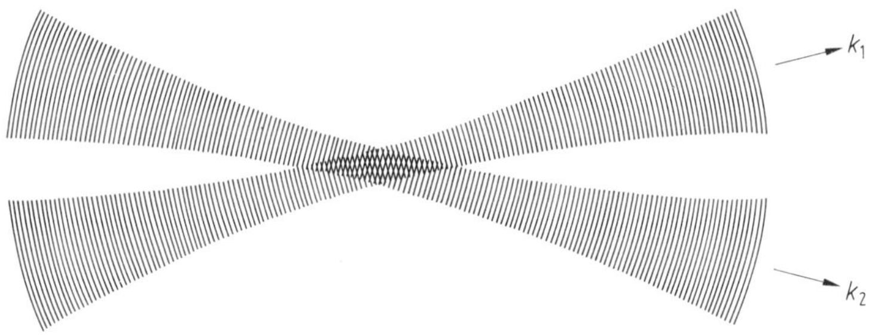

Fig. 46. Interference stripes generated by 2 crossing Laser beams in the velocity measuring zone. [84 D].

Applications

Compared with rotor, electromagnetic and acoustic current meters the application of laser Doppler anemometers in oceanography [74 F 2] is still in an experimental stage. LDA measurements carried out in the sea used an underwater optical unit while the main data processing was done on board of a ship. The underwater unit was either mounted on a frame to be used on the bottom [74 K 3, 78 K 2] or towed behind a ship [80 V 1] for measuring turbulence.

The mounted version [78 K 2] is shown in Fig. 47. It is a two-component velocimeter, which measures the velocity in two orthogonal directions in a plane perpendicular to the axis of the instrument. Two sets of beams are used for this purpose. Each is frequency shifted by a Bragg cell by different amounts, in this case 38 MHz and 42 MHz. The scattered light is received by only one photodetector, which has a band pass covering both channels. The two components are later separated by electrical filters and mixed independly with the outputs of the local oscillators. The velocity is obtained by frequency trackers (Fig. 48).

An instrument of this type has a measuring volume of about 0.5 mm^3, several measuring ranges from ± 2.5 cm s^{-1} to 80 cm s^{-1}, and it can detect fluctuations of about 0.01 cm s^{-1} with frequencies up to some kHz.

The towed turbulence meter [80 V 1] is sketched in Fig. 49. It uses the reference beam mode and is mounted in a two-hull towed body. It has been successfully used to measure small-scale turbulence in the North Sea.

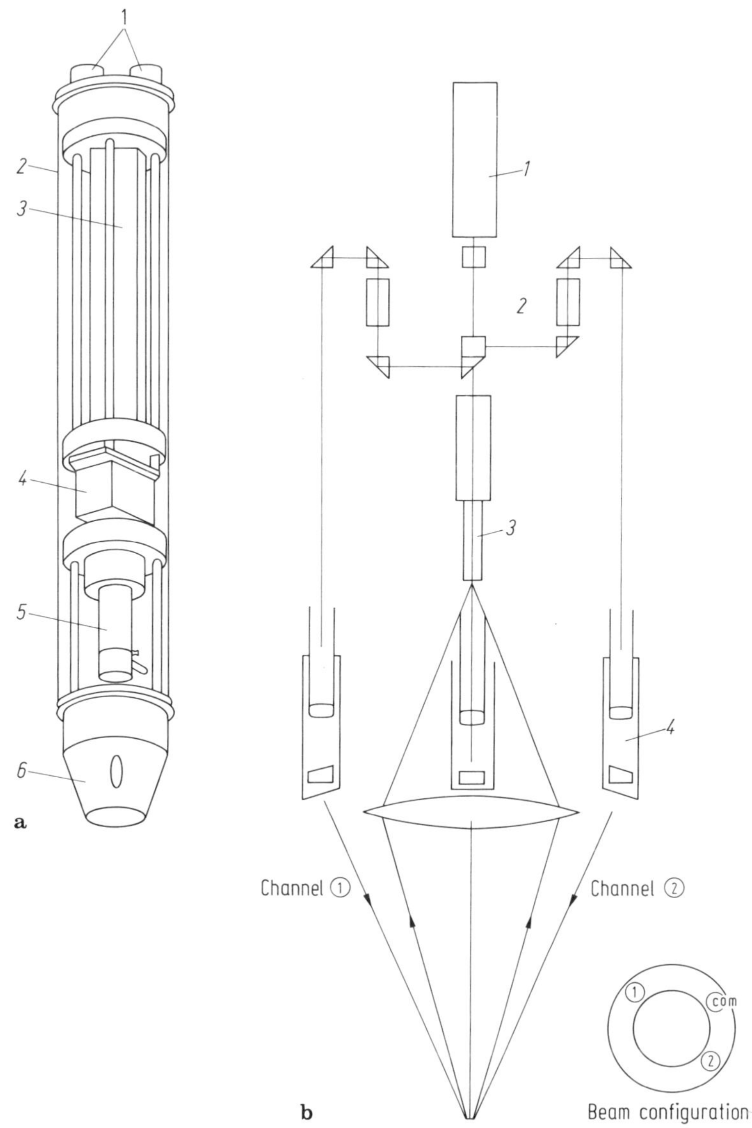

Fig. 47. An oceanographic LDA (a) and its optical configuration (b). [76 D 2]. This instrument can also be used in the forward scattering fringe mode if a mirror is placed in the optical axis some 100 cm away from the lense. a: *1* Cable connections; *2* Tube; *3* Laser; *4* Bragg cells; *5* Photomultiplier; *6* Lenses. b: *1* Laser; *2* Bragg cells; *3* Photomultiplier; *4* Focusing unit; com = common for channel 1 and 2.

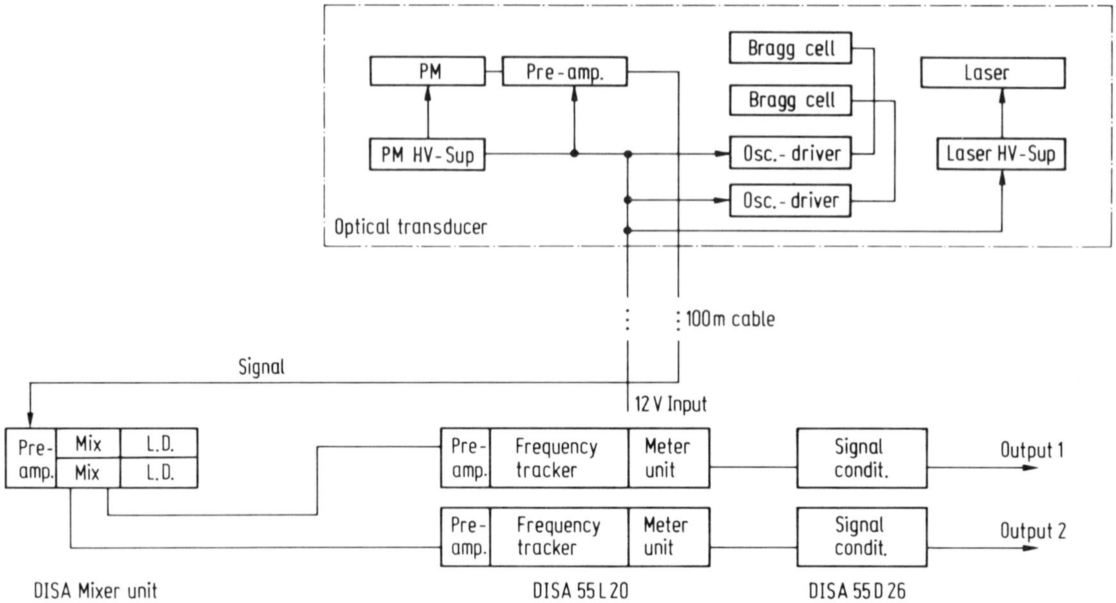

Fig. 48. Block diagram of the oceanographic LDA of Fig. 47 [76 D 2].

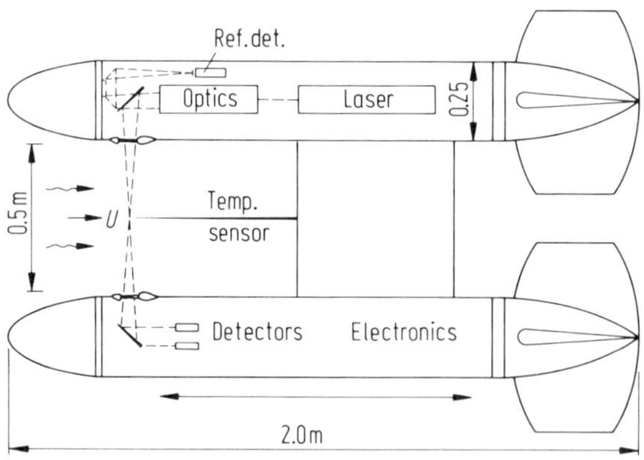

Fig. 49. An oceanographic LDA system which is towed by a ship to measure turbulence [80 V 1].

2.3.1.2.5 Travel time current meters

The definition of velocity suggests measurements of travel times at a constant measuring distance. When the distance is sufficiently small one has an Eulerian sensor.

A very simple device of this kind is sketched in Fig. 50. It could be used in a one-dimensional current of a laboratory tank or could be aligned into the current direction by a vane. It consists of a small heating coil, a thermistor and a simple electronic circuit. If an electric pulse is applied to the heater, a small volume of warm water is generated. It flows downstream, is sensed by the thermistor and the change in resistance could be used to generate a new pulse for the heating coil.

In this "sing-around circuit" the pulse frequency is

$$f=\frac{V}{L}$$

if one neglects time constants of heater, thermistor and electronic circuit. One can also construct two-dimensional devices of this kind [76 K 2] or use photoelectric methods [65 N].

More successful for field applications is the use of ultrasonic sound pulses or waves. The principle is that the propagation velocity of an acoustic wave in a moving fluid is the vectorial sum of fluid velocity and sound velocity in the fluid at rest.

The principle arrangement is sketched in Fig. 51. There are two transducers in A and B in a current flowing with speed $V(y)$ in the xy-plane.

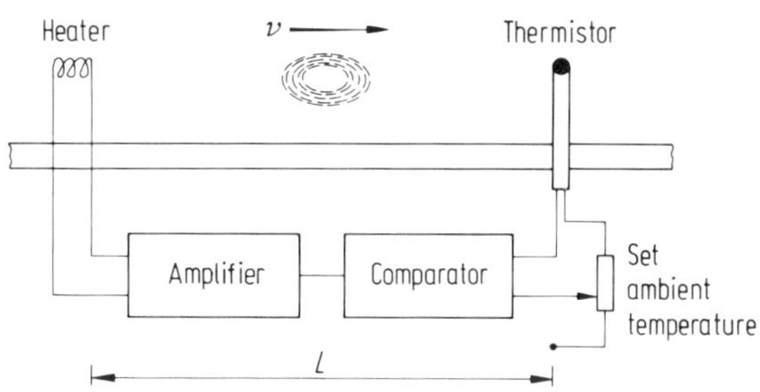

Fig. 50. A simple sing-around circuit to measure velocity in a one-dimensional current.

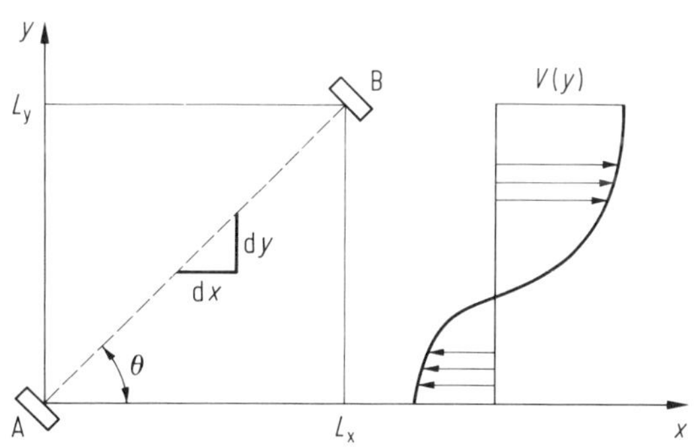

Fig. 51. Principle of operation for acoustic current meters. Two transducers in a one-dimensional flow field $V(y)$ emit acoustic signals to each other at an angle θ with the current direction [80 G 2].

In the travel-time mode A and B are simultaneously excited to produce a burst of exponentially damped acoustic oscillations. The respective wave trains travel to the opposite transducers, now switched to be acoustic receivers, with velocities $c \pm \boldsymbol{V} \cdot \boldsymbol{K}$, where c is the sound velocity of the fluid at rest, and $\boldsymbol{K}$ is a unit vector along the direction AB.

The travel time difference of the two acoustic pulses is

$$T_{BA} - T_{AB} = \frac{2L}{c^2} \bar{V} \cos\Theta$$

where $\bar{V}$ is the mean velocity along AB. In this mode the travel-time difference contains c, which is variable in the ocean. Current meters using this principle have to compensate for changing c. This can be done by measuring the average travel-time simultaneously [80 G 2].

In the sing-around mode the role of transducers is also switched. First, the sing-around frequency is generated with A as transducer and B as receiver to yield a frequency f_{AB}, and after changing roles f_{BA} is observed. The difference between up-stream and down-stream frequencies is

$$f_{AB} - f_{BA} = \bar{V} \cos \Theta$$

where c cancels out.

In phase difference systems one uses continuous signals and measures phase lags by various techniques. Again the velocity of sound appears in the equation of the quantity measured.

The basic problem with the electronic and transducer design of acoustic current meters of this type is the detection of very small travel time differences, high precision frequency measurement or phase detection depending on the mode of operation [74 G 2]. If one aims at a resolution of 0.1 cm s^{-1} using a distance of 10 cm and assumes a sound velocity of 1500 m s^{-1}, then travel times of the order of 10^{-10} s have to be detected. In order to achieve fast transducer response piezo-electric transducers with resonant frequencies in the MHz-range are used.

A typical transducer with a resonant frequency at 3.5 MHz has a diameter of 10 mm and is 0.5⋯1 mm thick. The probe must protect this sensitive element and its physical shape has to combine mechanical strength with a minimum interference with the flow field. It is important to minimize the interference of the wake of an upstream probe with the flow in the measuring path. Fig. 52 shows some probe arrangements. Configuration **a** can only be used for longer distances between transducers. Method **b** is most commonly used. Whereas in Fig. 52 only the wakes due to horizontal flow are sketched, interference of a vertical component, e.g. induced by orbital wave or mooring motion, can affect the measurement.

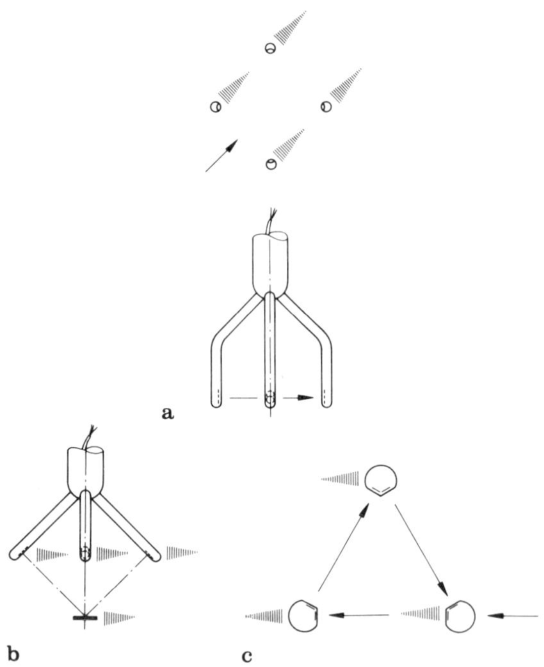

Fig. 52. Three possibilities of transducer arrangements. **a** direct transmission, **b** transmission using a reflector, **c** three direct transmission paths [80 G 2].

Application

Acoustic travel-time current meters are available commercially today. In addition to two-component measurement of flow they normally contain a flux gate compass and a microprocessor to compute conventional North and East components, to perform vector averaging and to organize data storage on a magnetic cassette recorder (Fig. 53). Simultaneous measurements of temperature, conductivity and pressure are optional.

The velocity range covers ±250⋯±300 cm s^{-1} [84 M, 84 N 1]. Resolution is in the order of 0.1 cm s^{-1}. The overall accuracy is determined by the errors in velocity component measurement, including possible drifts in the electronic circuits and the magnetic compass used for direction finding. In practice, field instruments have typical velocity magnitude errors of ±1 cm s^{-1} or 5%, whichever is higher, and a directional error of ±5°. The response time is in the order of 0.2 s. The noise increases with increasing velocity and is about 2 cm s^{-1} at a speed of 250 cm s^{-1}.

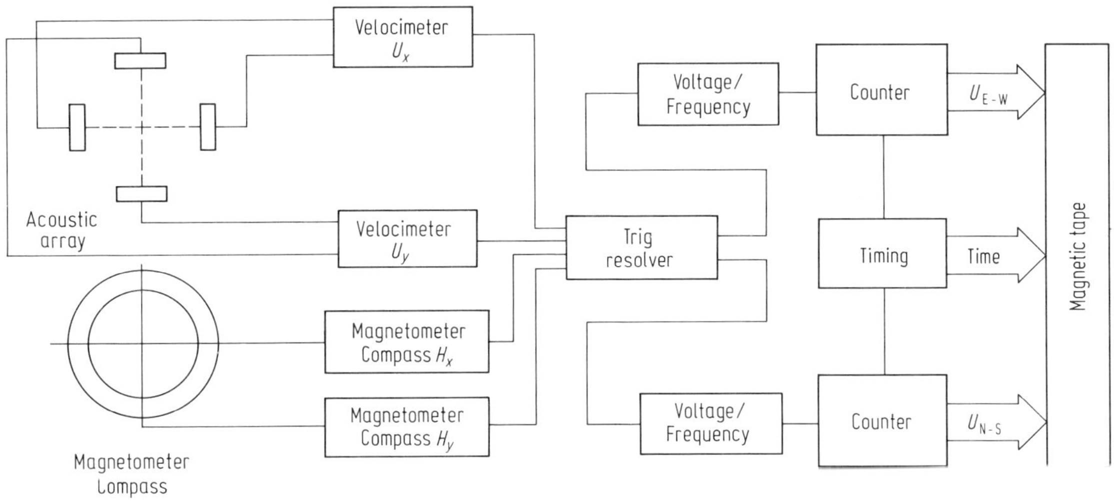

Fig. 53. Block diagram of an acoustic travel time current meter [84 N 1].

a

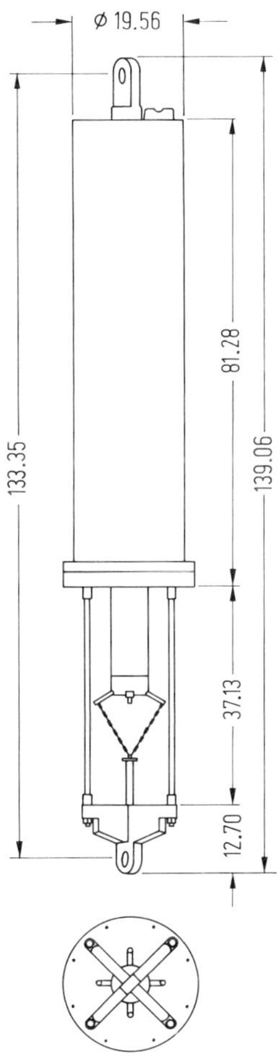

b

Fig. 54. Acoustic travel time current meters. **a**: [80 G 2], **b**: [84 N 1].

2.4 Elevation of the sea surface

Sensing of surface elevations may be carried out by geometrical measurements. One can use further pressure measurements and the hydrostatic equation. Use can be made also of the fundamental relationships between distance, speed and acceleration. The time scale of the process under study is the decisive factor for selection of one or the other of the principles.

2.4.1 Surface elevation for tidal and subtidal periods

In this range of periods water level measurements are required in studying tides, seiching, wind set-up, tsunamis, planetary waves and mean sea levels. Amplitudes of these phenomena range from several meters down to less than one cm. Classical shore based tide gauges cover the whole range if properly designed and installed. Much more sophisticated instruments are required to perform respective measurements on the continental shelf and in the deep-sea. Here the pressure associated with these large-scale phenomena can be measured from the sea floor and converted into surface elevations using the hydrostatic equation.

2.4.1.1 Geometrical measurement of surface elevations

If it is desired only to register tidal heights and to reject other oscillations such as swell and wind waves the measurements are done inside of a well which is connected to the outside by a small opening. If the diameter of this opening is very small compared to that of the well, rapid outside water level change will have little effect inside the well. Inside a float could follow the water level. By means of a pulley, a thin wire and a counter weight, level heights are transformed into turning of an axis. This movement can be used to move a pen over paper, a potentiometer or any other device [70 M].

According to local conditions one has found several solutions to connect the well with the sea. In a simple case it is only necessary to have a little hole in the well which itself stands close to the shore or in a harbour basin. Where heavy surf is present, it might be more practicable to locate well and recorder at the shore and to connect it to the sea via a long pipe. This little difference has a fundamentally different influence on the measurement. The two solutions are shown in Fig. 55. The dashed line indicates the datum for all relevant heights. It is differently defined in different countries, but normally connected to the datum for geodetic surveying.

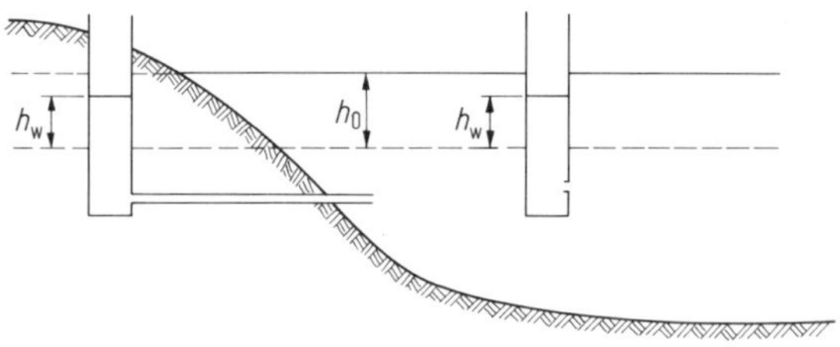

Fig. 55. Left: Shore-based well for the measurement of water level. It is connected to the sea by a long pipe. Right: This well has a small hole for damping of surface waves. Although otherwise identical, the response of the inside water level h_w to changes of h_0 is described by a linear differential equation for the left well whereas it is non-linear for the right.

If the water level in the well is lower than in the sea, water will flow into it with a speed v which is some function of the height difference H between outside and inside.

$$v = v(H) \tag{67}$$

$$H = h_0 - h_w . \tag{68}$$

The speed of the rising water level in the well is $\frac{dh_w}{dt}$. Due to continuity reasons we have

$$A_w \frac{dh_w}{dt} = A'_p v . \tag{69}$$

A_w is the (constant) cross section of the well, and A'_p denotes the effective area of the orifice, which may be different from the geometrical area A_p. We assume

$$A'_p = C_c \cdot A_p \tag{70}$$

where C_c is a constant.

With these equations we obtain the basic relationship between inside and outside water level changes [68 N 2]

$$\frac{dH}{dt} + \frac{A'_p}{A_w} v(H) = \frac{dh_0}{dt}. \tag{71}$$

It is seen, that the solutions of this differential equation will critically depend on the function $v(H)$.

First we consider the gauge with the long pipe. L_p may denote its length. In order to neglect complicated flow conditions at its inlet and outlet it should be very long compared with its diameter D_p, say

$$L_p \geqq 200 D_p. \tag{72}$$

It is desirable that the flow in the pipe has to be laminar. This can be achieved by keeping a Reynolds number Re_p below a critical value.

$$Re_p = \frac{D_p |v|}{\nu} < 2300 \tag{73}$$

where ν is the (laminar) viscosity of water. If this is fulfilled, a linear relationship between the speed in the pipe and the water level difference across it holds

$$v = C_i H \tag{74}$$

with

$$C_i = \frac{\varrho A'_p}{8\pi\nu L_p}. \tag{75}$$

The differential equation (71) is linear in this case

$$\frac{dH}{dt} + CH = \frac{dh_0}{dt}. \tag{76}$$

The range of height differences within which eq. (76) holds is found by combining eqs. (73)···(75)

$$H \leqq \frac{73\,600 \nu^2 L_p}{\varrho D_p^3}. \tag{77}$$

If greater differences would occur, caused by heavy swell for instance, it would be necessary to use a longer pipe or a smaller diameter.

Frequency and phase response of this linear tidal well are readily obtained by assuming a sinusoidal outside water level oscillation

$$h_0 = \hat{h}_0 \sin \Omega t. \tag{78}$$

The solution of the differential eq. (76) is

$$H = \hat{h}_0 a(\Omega) \sin(\Omega t + \varphi) \tag{79}$$

with

$$a(\Omega) = \frac{1}{\sqrt{1 + \left(\frac{c}{\Omega}\right)^2}} \tag{80}$$

and

$$\varphi = \arctan\left(\frac{c}{\Omega}\right). \tag{81}$$

The water level inside the well is

$$h_w = h_0 - H\,, \tag{82}$$

$$h_w = \hat{h}_0 A(\Omega) \sin(\Omega t - \psi)\,, \tag{83}$$

where

$$A(\Omega) = \sqrt{1 + (a(\Omega))^2 + 2a(\Omega)\cos\varphi} \tag{84}$$

and

$$\psi = \arctan\left(\frac{a(\Omega)\sin\varphi}{1 - a(\Omega)\cos\varphi}\right), \tag{85}$$

or

$$A(\Omega) = \frac{1}{\sqrt{1 + \left(\frac{\Omega}{c}\right)^2}} \tag{86}$$

and

$$\psi = \arctan\left(\frac{\Omega}{c}\right). \tag{87}$$

As frequency and phase response, like in any linear system, are independent of amplitude, corrections of amplitudes and phases can be made for any frequency within the operating range of the gauge.

Water level meters without pipes have different characteristics. If there is a small orifice between the well and the outside, the velocity in the opening as a function of H is

$$v = \sqrt{2g|H|}\,\mathrm{sgn}(H) \tag{88}$$

where sgn(H) is the sign of H. The basic differential equation is nonlinear in this case,

$$\frac{\mathrm{d}H}{\mathrm{d}t} + C_2\sqrt{|H|}\,\mathrm{sgn}(H) = \frac{\mathrm{d}h_0}{\mathrm{d}t}, \tag{89}$$

where

$$C_2 = \sqrt{2g}\,\frac{A'_p}{A_p}.$$

If one uses a circular orifice, the ratio $\frac{A'_p}{A_p}$ is constant and independent of speed.

Results of numerical calculations [68 N 2] and of perturbation analysis show that the inside water level response depends on outside amplitudes and that it may occur that frequencies are recorded which are not present outside. However, with suitable dimensions acceptable solutions can be found [67 M, 67 S, 68 M, 68 S 1, 69 M 1, 69 M 2].

It has to be mentioned that the survey on the problems of water level measurement has been demonstrated here only on the basis of a quasi-stationary approach. Perturbations caused by outside sea waves or swell result in oscillatory motions through the openings. Dynamics of oscillatory flow would have to be used to assess quantitatively the effects of fast outside water level oscillations.

2.4.1.2 Deep-sea tide gauges

The main problem with measuring surface elevations from the sea floor in the tidal and subtidal period range is the presence of the high hydrostatic pressure. In order to resolve surface elevations of 0.5 cm at a depth of 5000 m the resolution of the pressure sensor has to be $1 \cdot 10^{-6}$.

Besides high resolution also high stability of the same order of magnitude is required for times in excess of at least one month.

The most obvious approach to this problem is a differential measurement using the pressure at begin of a time series as a reference. This method requires a reference medium where the initial value can be stored either with high stability or in which the initial pressure can be recovered from knowing its time history throughout the measurement.

The first tide gauges used air as a reference medium. Later designs preferred an inert gas to avoid problems with the solution of air in water. Fig. 56 shows the principle design of a deep-sea tide gauge of this type [71 B 1].

By applying the equation of state for the enclosed gas the reference pressure can be calculated if temperature has been recorded parallel to the pressure measurement. As the thermal expansion coefficient for a gas is large high-precision temperature measurements are necessary. In a depth of 1000 m a temperature change of the gas by 0.1 °C would correspond to a water level change of 36.6 cm. This shows that precise temperature measurements of the order of 10^{-3} °C are required to resolve surface elevations in the above mentioned range. Even if this can be achieved there are still principal problems with the correction procedure.

Many other reference media have been investigated. Instead of gas one could use pure water or one can make use of eleastic properties of solids.

The various possibilities have been investigated in an international intercomparison experiment [75 U]. Its aims were to compare the performance of various tidal pressure sensors, mooring and recovery techniques, and various methods of analysis of pressure time series.

Some characteristics of the deep-sea tide gauges are summarized in Table 7. The instruments ranged from elaborate high-cost equipment to less expensive devices. Some of the instruments were lost during the experiment.

Results of the different sensors were found in relatively close agreement. In regard to all-around performance, drift, temperature coefficient and noise level the quartz crystal transducer was superior to other sensors. It is now used in many offshore tide gauges. A handy instrument is described by [78 A].

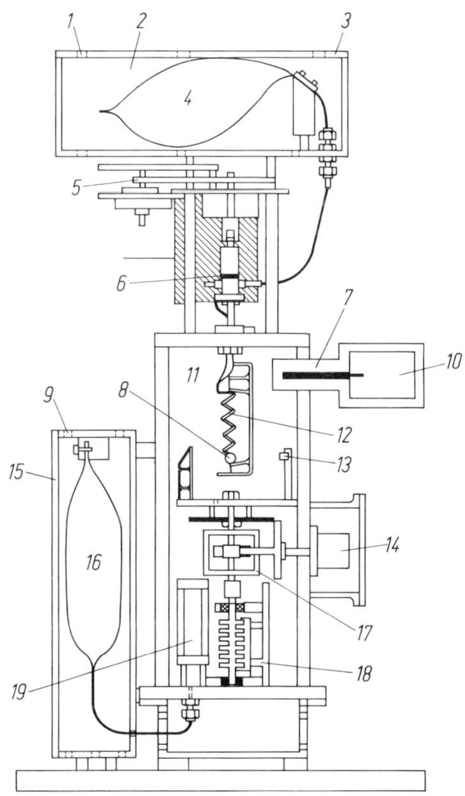

Fig. 56. Operation principle of the Harvard deep-sea tide gauge. On its way down to the sea floor the valve is open. Air is compressed into the inside of a quartz Bourdon tube while oil of equal pressure surrounds the pressure transducer. After deployment the valve is closed by a motor. Changes in sea pressure are transmitted by the oil to the Bourdon tube and cause a deflection of a mirror at its lower end. By means of a light beam and a photocell the deflection is sensed and it causes a servo motor to follow the deflected beam. Coupled to the gear of this motor is a variable capacitor which changes the frequency of an RC oscillator. This frequency is thus dependent on the differential pressure between inside and outside of the Bourdon tube. Temperature is measured by a quartz thermometer. *1* Hole; *2* Sea water; *3* Plastic housing; *4* Air; *5* Valve motor and gear; *6* Valve; *7* Quartz thermometer; *8* Mirror; *9* Hole; *10* Electronics; *11* Oil; *12* Quartz Bourdon tube; *13* Photocell; *14* Servo motor; *15* Plastic housing; *16* Oil; *17* Gear; *18* Variable capacitor; *19* Oscillator.

Table 7. Characteristics of deep-sea tide gauges which have been compared in a UNESCO intercomparison experiment [75 U].

Type	Max. depth	Pressure sensor	Reference elastic element	Order of magnitude of drift	Mooring
IGPP	5500 m	Resonating quartz crystal	Quartz	1 mbar/month	Free fall/acoustic release
NOAA	5000 m	Oil-filled Bourdon-tube	NiSpanC alloy	logarithmic creep 100 mbar/month	Free fall/acoustic release
COB	6000 m	Differential, vibrating wire	Nitrogene		Free fall/acoustic release
IOSW 1	4000 m	Membrane + strain gauges	Alloy	30 mbar/40 days	Free fall/acoustic release
IOSW 2	300 m	1. Membrane + condenser	Alloy	20 mbar/month	Free-fall for the last 10 m, acoustic relase
		2. Membrane + strain gauges		10 mbar/month	
IOSB	200 m	1. Vibrating quartz crystal	Quartz		Ground-line mooring
		2. 2 vibrating wire sensors	Alloy	100 mbar/month [1])	
		3. 2 strain-gauge sensors			
SHOM	200 m	Vibrating wire	Alloy		Subsurface buoy, acoustic release
CANADA	183 m	Differential, potentiometric output	Water		Subsurface buoy, acoustic release

[1]) Probably not typical.

Table 8. Origin of the deep-sea tide gauges listed in Table 7 and literature.

IGPP	Institute of Geophysics and Planetary Physics, University of California, La Jolla, USA [68 S 2, 72 I].
NOAA	Scripps Institute of Oceanography, La Jolla, USA. [71 F].
COB	Centre Oceanologique de Bretagne, Brest, France. [64 E].
IOSW 1, 2	Institute of Oceanographic Sciences, Wormley, UK. [72 C].
IOSB	Institute of Oceanographic Sciences, Bidston Observatory, Birkenhead, UK. [73 S].
SHOM	Service Hydrographique et Oceanographique de la marine, Brest, France. [75 U].
CANADA	Marine Science Directorate, Ottawa, Canada. [75 U].

2.4.2 Surface elevations in the range of wind waves and swell

This chapter considers the measurement of surface waves with periods between 1 and 30 s. In this range one usually assumes the dispersion relationship

$$\omega^2 = gk \tanh(kh) \tag{90}$$

for small amplitude gravity waves. In eq. (90) ω denotes the angular frequency, k is the wavenumber, h is the water depth and g the acceleration of gravity. The complicated structure of the sea surface is then described as a superposition of plane waves having varying wavelengths and directions α, the energy distribution of which is given by the directional spectrum $\phi(\omega, \alpha)$.

There are numerous possibilities to obtain the directional spectrum by measurements [73 H], e.g. by arrays of wave height measuring devices [68 H, 77 D, 77 F], by combined measurements of wave heights and orbital velocities or [64 N, 66 B] slopes of the sea surface [63 L].

This section restricts itself to the measurement of wave heights at one point which is a basic type of measurement in wave research.

Very many instruments have been developed for this purpose. They can be classified as surface piercing instruments, pressure measuring devices and instruments which perform inertial measurements [80 S]. Alternatively, the type of platform which carries the sensor offers itself as useful to distinguish various types of instruments. The latter will be used here.

On *shallow water* rigid platforms can be single posts, oil rigs or other structures mounted on the sea floor. Sensors for such platforms including the sea floor are the following ones.

Resistance wires:

A resistance wire is supsended between supports above and below the water. The total resistance of the wire in air plus the sea-water path is a function of water level. It is measured with audio-frequency current. Usually a very thin (0.2···0.5 mm diameter) resistance wire is wound around a rigid support or a flexible and strong cable [63 F].

Alternatively, a pair of thick wires of high conductance can hang into the water, and the resistance of the sea water path between them is measured. This is a very rugged device, but it relies on constant conductivity of sea water or correction procedures if this changes with temperature and salinity.

Capacitance wires:

Because of the high dielectric constant of water the capacitance of an insulated wire is a function of water level. Although this principle appears very promising it has several shortcomings. With falling water level surface films of water draining downwards are left on the cable. This contributes to the capacitance measurement and distorts the measured wave height. There are also long term changes of the dielectric properties of the insulations material by slow absorption of water [60 K 1].

Pressure measurements:

Pressure measuring devices can be placed at the sea floor. No need for surface piercing supports is very attractive in rough waters.

The amplitude of the pressure signal Δp varies with wave height a as

$$\Delta p = \varrho g a \frac{\cosh k(z+h)}{\cosh kh} \tag{91}$$

where ϱ is density of sea water and z the depth where the pressure is measured.

For shallow water waves ($kh \ll 1$) the pressure fluctuations are directly proportional to wave height. For the intermediate case each Fourier component in the pressure spectrum can be converted to wave heights according to eq. (91) and related to the same frequency component in the wave height spectrum [70 E]. It has to be noted that eq. (91) is valid for waves of small amplitude, i.e. $a/h \ll 1$. Caution is therefore necessary near the surf zone where the wave amplitude comes into the order of the water depth.

For deep-water waves the pressure fluctuations decrease rapidly with depth leaving no chance for pressure measurements.

Other devices:

Any distance measuring device suitable for the ocean environment could also be used for the measurement of wave heights. This includes the inverted echo sounder mounted on the sea floor or laser altimeters facing the sea surface from above. The inverted echo sounder is particularly useful in the near-shore zone [73 F 1].
On *deep water* the most common in-situ techniques are based on inertial measurements [59 T, 76 D 1, 77 S].
On the average, the sea surface has no mean acceleration. If the vertical accelerations associated with surface waves are measured and integrated twice, one obtains the elevation of the sea surface relative to the mean sea level.

It is not easy to construct accelerometers which respond linearly in the whole spectral range of wind waves and swell. Especially difficult is the detection of small swell at presence of large wind waves. Nevertheless, the technical expense even for sophisticated inertial measurements is still much smaller than for a stable platform.

The measurement of vertical accelerations requires a vertical reference. This can be procided by a damped pendulum with a long period. When supported close to its center of mass, it will respond to the average acceleration, i.e. it will remain vertical. Thus the pendulum serves as a stabilized platform on which the accelerometer can be attached.

This principle has been successfully applied as sensor in the "waverider" buoy. This is a spherical surface following buoy which is moored with a rubber cord at the buoy end of the mooring. Its response as a function of wave frequency is shown in Fig. 57. The peak at high frequencies results from the natural frequency of the buoy, and it is slightly influenced by the force of the mooring line.

The decreasing response towards low frequencies is the result of the double integration of the accelerometer output. The natural period of the pendulum stabilized platform is about 40 s. The response curve shows that amplitudes within the frequency range between 0.065 Hz and 0.03 Hz can be measured with an accuracy of 3%. The maximum wave height is 40 m, and the noise level in a 1 Hz band is 0.02 m.

The wave amplitude signal is radio transmitted and recorded at the shore.

Technically, this principle of wave measurement can be varied in many ways. The pendulum stabilized platform e.g. can additionally be supported by a vertical gyro. With modern electronics and sensors it is possible

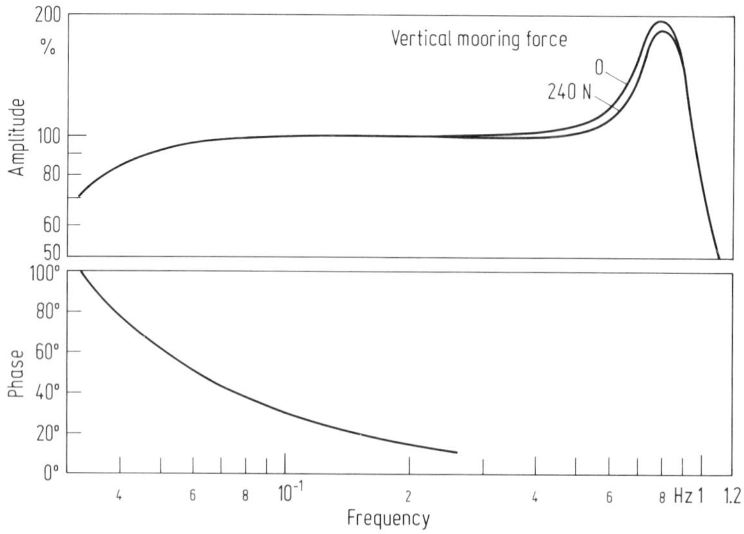

Fig. 57. Amplitude and phase response of the waverider buoy [76 D 1].

to measure additionally horizontal accelerations. Toroidal buoys can be equipped with sensors for heave, pitch, roll and magnetic north to provide the necessary information on the directional spectrum. Modern radio transmission can also make use of satellites to make measurements in the open ocean. There is a great variety of instrumentation available from oceanographie instrument manufactures.

Other devices:

The simple techniques of wave measurements at one point at the sea surface cover just a very small range of instruments used in modern wave research. More sophisticated techniques are available to measure the wavenumber-frequency spectrum with the aid of radio waves, satellite based radar altimeters and photographic techniques. By these techniques millions of observations on the wave climate of the oceans have become available. The treatment of these remote sensing techniques is beyond the scope of this article.

A source of information on ocean wave sensing in general is provided by [79 N].

2.5 Variables of state

Temperature T, salinity S and pressue p determine the fundamental properties of sea water like density, conductivity, sound speed, viscosity, freezing point, etc. (see ch. 3). They also govern the limits of existence of marine organisms. Measurements of the variables of state range under the most important tasks in the marine sciences.

In general, applications in marine biology require less resolution of temperature and salinity measurements than in physical oceanography. Highest precisions are necessary for calculations of geostrophic currents and stabilities from density distributions. In these fields accuracies of $\Delta T = \pm 0.01$ °C, $\Delta S = \pm 0.01$ and $\Delta p = \pm 1$ dbar are necessary, but in large areas of the ocean like in the central water masses and in the polar seas a higher resolution by one order of magnitude is desirable.

2.5.1 Temperature

Temperature measurements in the ocean have to be based on the International Practical Temperature Scale (IPTS-68) [69 C 2, 74 K 1]. Temperatures on the IPTS-68 are defined by the resistance values of precise platinum resistance thermometers. Standard thermometers of this kind are kept by the National Bureaus of Standards. Oceanographic institutes may have substandards or have thermometers calibrated by a Bureau of Standards. In any case the international data exchange requires absolute accuracy of temperature data.

The classical instrument to measure temperature in the sea is the reversing thermometer sketched in Fig. 58. Together with a water sampler it is lowered to the desired depth. The time constant of this mercury thermometer is in the order of 1 min. Therefore the adjustment time to the ambient temperature with a deviation of less than 0.01 °C is approximately 5 min.

Triggered by a messenger weight, the water sampler is closed while the thermometer is turned by 180°. The mercury above the break-off point is separated from the reservoir and flows into a calibrated capillary tube. The length of the mercury column is proportional to the in-situ temperature T, and this value can be recovered if the thermometer is read at a different temperature T_a which is indicated by the auxiliary thermometer.

The correction procedure is based on the equation for thermal expansion

$$V_t = V_1(1 + \alpha \Delta T) \tag{92}$$

where V_t is the observed volume of mercury, V_1 the original volume, α the coefficient of thermal expansion, and ΔT the temperature change.

The total volume of mercury within the thermometer can be thought of as the sum of the volume below the zero mark, V_0, and the volume above. It can be expressed in terms of the graduation of the thermometer scale, i.e. in [°C].

If the main thermometer indicates T' at reading it is necessary to find

$$\Delta T = T - T'. \tag{93}$$

In terms of observable quantities it follows from eq. (92)

$$V_0 + T' = (V_0 + T)\left(1 + \frac{1}{K}(T_a - T)\right) \tag{94}$$

where K is the thermal expansion coefficient for mercury in a quartz glass capillary tube.

Typical values are $V_0 \cong 100\,°C$ and $K = 6100\,°C$.

Depending on the neglections introduced when solving eqs. (93) and (94) for ΔT various correction formulae can be found in the literature. They all are accurate for measurements in the order of $\pm 0.01\,°C$.

A formula which is widely used is

$$\Delta T = \frac{(V_0 + T')(T' - T_a)}{K - (\frac{1}{2}(T' - T_a) + (V_0 + T'))}. \qquad (95)$$

In most cases the term $\frac{1}{2}(T' - T_a)$ in the denominator can be neglected.

Although most of the temperature measurements are carried out with electrical sensors the classical reversing thermometer is still used to check possible drift and calibration of electronical equipment.

Amongst the many principles to measure temperature [66 M, 72 P] the platinum thermometer and thermistors are most commonly used. Also Silicon transistors have been considered [77 O]. After introducing a special angle of cut through a quartz crystal [67 H, 69 C 1] to make its resonant frequency temperature dependent, quartz thermometers are also used in oceanography [68 S 2]. Very high resolution of the order 0.001 °C and excellent linearity can be achieved by this sensor.

Of particular importance to oceanographic applications is the time constant of temperature sensors. The measurement of temperature profiles requires very small time constants whereas adjustable time constants are necessary for time series measurements.

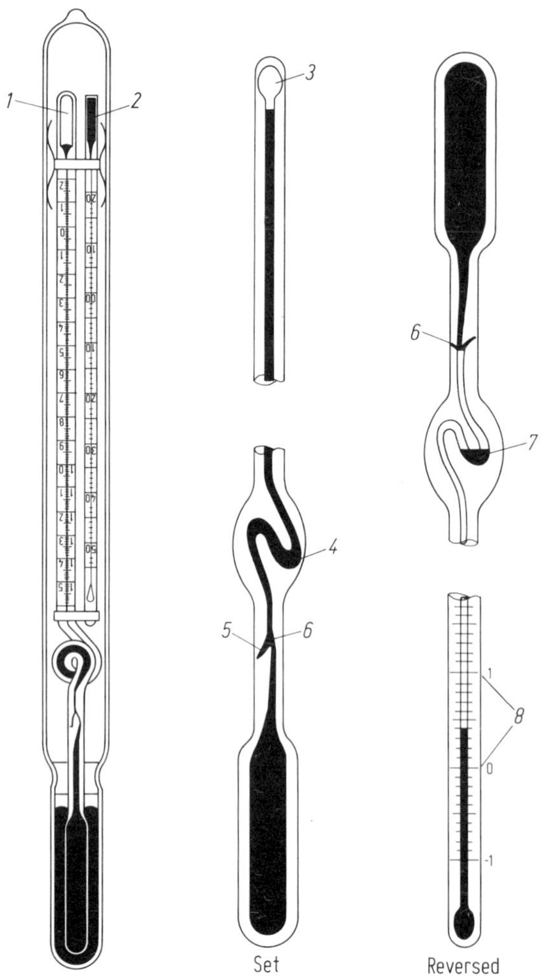

Fig. 58. The reversing thermometer. On the left the complete instrument is shown. It houses the pressure protected main and auxiliary thermometers. The principle of operation is sketched on the right. In the set-position the thermometer is lowered and is given time to adjust to the in-situ temperature. After reversing by 180° the mercury column beyond the break-off point fixes the in-situ temperature. *1* Main thermometer; *2* Auxiliary thermometer; *3* Void; *4* Trap; *5* Appendix; *6* Break-off point; *7* Mercury extruded after reversal; *8* Scale graduations.

2.5.1.1 Temperature profiles

If the time response of an electronic temperature sensor can be adequately described by only one time constant k, the indicated temperature T adjusts itself to the water temperature T_w in a manner that can be described by the differential equation

$$\frac{dT}{dt} = -\frac{1}{k}(T - T_w) \tag{96}$$

where t denotes time. If the sensor is lowered into the ocean (Fig. 59) with a velocity v along the vertical coordinate z, eq. (96) can be rewritten in the form

$$\frac{dT}{dz} = -\frac{1}{kv}(T - T_w). \tag{97}$$

If the probe is adjusted to the air temperature T_L and then lowered continuously through the surface ($z=0$) into water of constant temperature T_w, the integration of eq. (97) yields

$$T = T_w + (T_L - T_w)\, e^{-\frac{z}{kv}}. \tag{98}$$

Instead of the time constant the product of time constant and lowering speed controls the adjustment of the probe to new conditions (Fig. 59).

If the temperature sensor is lowered further (Fig. 59) into a region of constant temperature decrease the asymptotic reading is

$$T = T_w - kv\frac{dT_w}{dz}, \tag{99}$$

i.e. at a given depth it is too high during lowering and too low during rising of the probe. The distortion of thermocline measurements as a result of time constant and lowering speed are often incorrectly interpreted as elastic hysteresis of the pressure sensor used to indicate the depth of the temperature probe.

If temperature profiles are taken during rough seas or high swell the constant lowering speed is superimposed by oscillatory motions of the ship. The up- and downward movements of the underwater unit may result in temperature "structures" which have nothing to do with reality. Such "instrumental structures" are discussed in more detail in [71 P].

From these considerations it follows that only fast response temperature sensors can be used [75 H]. Aiming at an accuracy of 0.01 °C at temperature gradients of the order of 0.1 °C m^{-1} and a lowering speed of 1 m s^{-1} the time constant should not exceed 0.1 s.

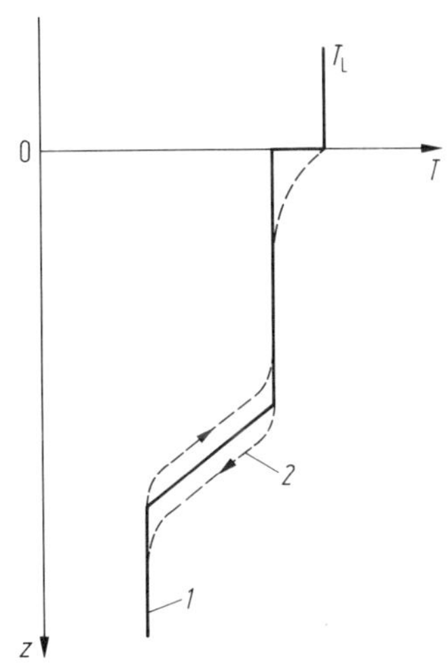

Fig. 59. Schematic diagram to demonstrate distortions of a temperature profile by a thermometer with a time constant k which is moved along the depth axis z with velocity v. At beginning the thermometer is in air of temperature T_L. During lowering it indicates a virtual temperature layer of half-thickness $k \cdot v$ under the surface. In the constant-gradient regime it produces a hysteresis. *1* True profile; *2* measured profile.

A successful temperature sensor with a platinum wire is shown in Fig. 60. The time constant of 60 ms is acceptable for most purposes, and this sensor has a very good stability over long times.

Due to their small geometrical dimension thermistor probes [63 C, 72 B, 77 L] can have shorter time constants, but their long-term stability is not satisfactory for many purposes.

A method to combine the stability of the platinum wire with the fast response of a thermistor [82 B 2] is sketched in Fig. 62. The equivalent circuit displayed in the sketch only serves to illustrate the principle of operation. In an actual circuit the thermistor sensor is part of an electronically balanced bridge. The output of the bridge is precisely zero for steady state conditions. Whenever temperature changes rapidly, an integrator in the control loop of the bridge allows the adjustment of an exponentially decaying signal which is combined with the slow response of the platinum thermometer.

The fastest possible temperature sensor is one which can be directly immersed in sea water. It must have such a small electrical resistance that the shunt by sea water can be neglected or be taken into account as a small correction. This is the case for a platinum wire with a resistance of the order of 10 mΩ. Aiming at a resolution of 0.001 °C such a sensor requires an electronic circuit which can resolve $4 \cdot 10^{-8}\,\Omega$. This problem has been solved [81 K] and a respective sensor with a time constant of 1 ms has been successfully tested at sea. One design of the fast temperature sensor is shown in Fig. 62 together with two fast response thermistor probes.

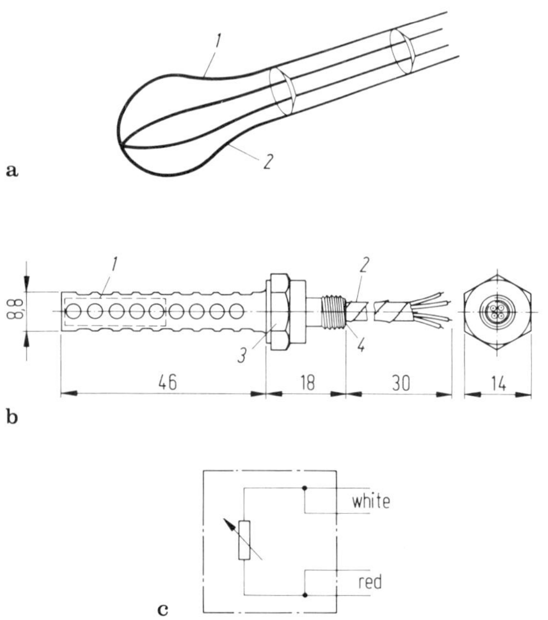

Fig. 60. A fast response platinum wire thermometer [83 R]. Operating range: −5 °C ··· +40 °C. Resistance at 0 °C: 200 Ω. The calibration is provided by a table in increments of 1 °C, and it is accurate to 0.01 °C. The time constant is 60 ms in water flowing at 90 cm s^{-1} transverse to the sensor. At this speed the self-heating is less than 0.005 °C per 1 mA excitation current. In a water depth of 6000 m the pressure error is less than 0.01 °C. Stability is better than 0.01 °C for one year. **a** Sensing element shown without protective shield. *1* Support tube; *2* Element within Inconel tube. **b** Design. *1* Element location; *2* Cable, 4 Conductors, copper, Teflon insulated; *3* Identification data; *4* Epoxy resin seal. **c** Schematic diagram. All dimensions in [mm].

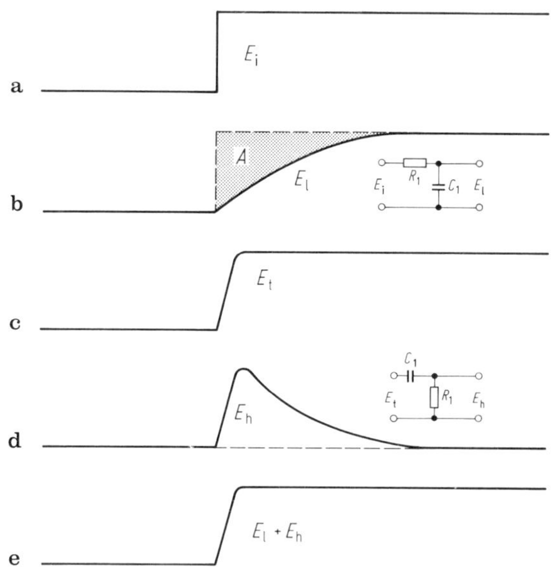

Fig. 61. A method to speed up the response of a platinum thermometer with a fast thermistor [82 B 2]. E_i is the voltage which would appear at the output of a perfect sensor in response to a step in temperature. E_l is the output of the real platinum thermometer, and E_t is the thermistor output. After high-pass filtering the voltages E_h and E_l are added thus combining the fast response of the thermistor with the precision of the platinum sensor. **a** Input; **b** Platinum thermometer response; **c** Thermistor response; **d** Thermistor response after high-pass filtering; **e** Combined output.

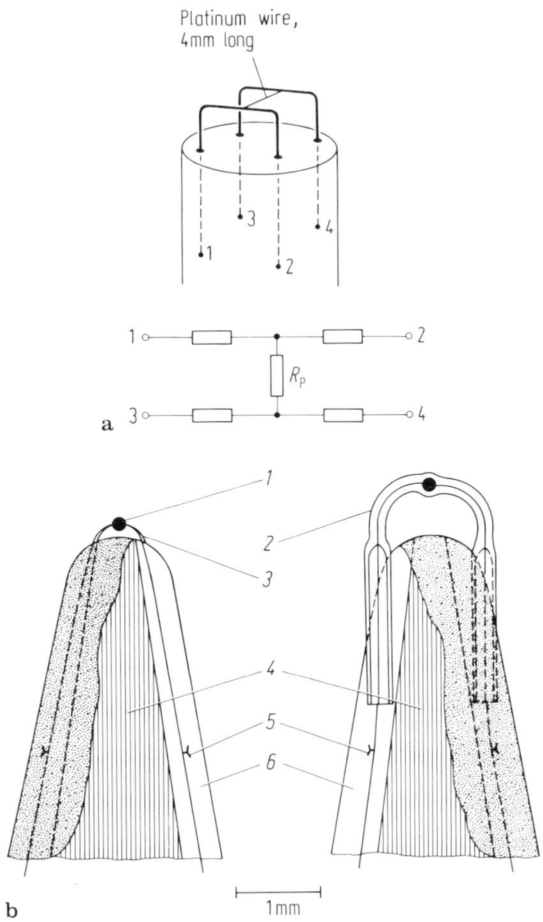

Fig. 62. Fast response platinum sensor with a time constant of about 1 ms together with the electrical equivalent circuit. R_p is the resistance of the platinum wire, the other resistors represent the resistances of the support [81 K]. The small thermistor sensor with epoxy coating has a time constant of 20 ms. Glass coating increases the time constant to 60–80 ms [63 C]. *1* Glazed thermistor bead; *2* Pb glass capillary fused around bead and bent; *3* Thin epoxy coating over all; *4* Phenolic nose peaces; *5* Solder joints to #36 wire; *6* Grooves filled with epoxy resin.

Finally, an improvement of a measured temperature profile with a thermometer of time constant k is also possible on the basis of a numerical integration of eq. (96). It can be written as

$$T_r(t) = T_i(t) + k\frac{dT_i(t)}{dt} \quad (100)$$

where r refers to the real and i to the indicated temperature. If data are measured with a sampling rate Δt, one of the possible approximations is (n integer)

$$T_r(n\Delta t) = T_i(n\Delta t) + k\frac{T_i[(n+1)\Delta t] - T_i[(n-1)\Delta t]}{2\Delta t}. \quad (101)$$

Whether this formula can be applied directly to the data or after some initial smoothing depends on individual circumstances [75 S 1, 72 G].

Time constant and overall calibration stability are of primary concern to the oceanographer. The important role of electronics and data transmission in in-situ temperature measurements is reviewed in [83 D].

Typical Instrumentation

Profiling temperature sensors are rarely used as single instruments in oceanography [63 H]. The most important exception is the Expendable Bathythermograph (XBT) [65 F]. It provides temperature profiles from a ship under way.

The probe (Fig. 63) is a ballistically shaped, spin-stabilized body. It carries a thermistor, and it is designed to fall at a precalculated speed without being affected by the forward speed of the ship. This is achieved by a dual spooling technique of a fine insulated wire. One spool is housed in the falling probe, the other pays out wire in a launcher aboard the ship to cover the distance which the ship travels during the descent of the probe.

By this technique a constant fall rate of the probe is achieved, thus time after launching is a measure for depth. Today most installations are controlled by a computer for on-line data processing and storage [78 M, 83 S, 86 E].

Over the last decade the XBT has been greatly improved. Malfunctions of the system have been analyzed [78 K 1]. The depth accuracy has been estimated [65 H, Table 9], and there were a number of intercomparison experiments [82 S 4, 83 H] between XBT and CTD (conductivity-temperature-depth-recorder) measurements with regard to isotherm depth determination.

The dynamics of the XBT probe has also been modelled [84 G 2]. The theory identifies the critical physical parameters of the falling probe.

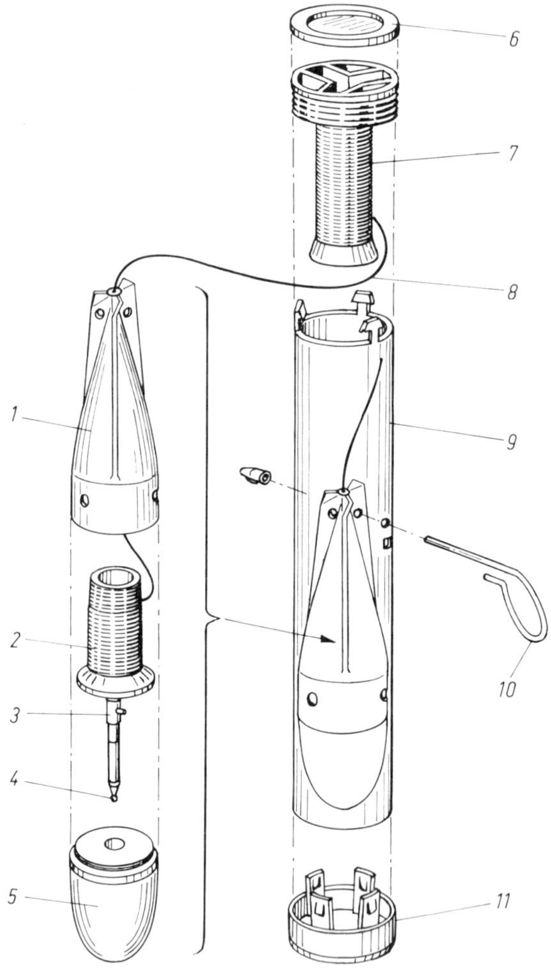

Fig. 63. Essential parts of the expendable bathythermograph (XBT). After release the canister together with spool *7* remains in the launcher aboard the ship. The wire on spool *7* is used mainly to cover the horizontal distance which the ship traverses during descent of probe. The probe spool *2* pays out wire while the probe falls spinning about its vertical axis. *1* Afterbody; *2* Probe spool; *3* Sea electrode; *4* Thermistor; *5* Zinc nose; *6* Label; *7* Shipboard spool; *8* Signal wire; *9* Canister; *10* Retaining pin; *11* Shipping cap.

Table 9. Factors influencing the depth accuracy of an XBT [65 H].

Group	Variation	Range	Resulting error	Remarks
Manufacturing tolerances	Weight	±0.5%	±0.34% depth	
	Dimensional	±0.5%	±0.4% depth	
	Roughness	±0.5%	±0.23% depth	
			±0.97% depth	Worst case
Launching conditions	Launch height	0···10 m	+5···−2 m	Can be reduced
	Impact		Negligible	
	Wave height	0···5 m	±0.5 m	
			+7.5···−4.5 m	Worst case
"In-situ" effects	Wire drag		Negligible	
	Weight loss due to wire payout	8%	±1.2% depth	Predictable
	Density	1.01···1.03	±0.6% depth	
	Vertical currents		Undetermined	
	Viscosity	±50%	±1.2% depth	
			±3.0% depth	Worst case

2.5.1.2 Slow-response thermometers

If temperature time series are measured over extended periods of time the sampling interval of the recording instrument is normally chosen much larger than the time constant of a temperature sensor. In order to avoid aliasing it is necessary to increase the time constant, e.g. by mounting the sensor on the inside of the instrument's housing. Experience has shown that careful studies of the response characteristics are necessary to predict the behavior of such arrangements.

In many cases the pressure-case end plate, the inside of which is a suitable location to mount a temperature sensor, can be modelled as an infinite plate of thickness b. The contact with the water is at $y=0$ and the heat flux at the inside boundary, $y=b$ is neglected. The equation of heat conduction

$$\frac{\partial T}{\partial t}=k\frac{\partial^2 T}{\partial y^2}, \quad 0\leqq y\leqq b \tag{102}$$

for sinusoidal forcing of the form $e^{i\omega t}$ at $y=0$ yields for the squared response [81 L]

$$G(f)=\frac{\sinh^2[(b-y)/L]+\cos^2[(b-y)/L]}{\sinh^2(b/L)+\cos^2(b/L)}, \tag{103}$$

where

$$L=\sqrt{\frac{k}{\pi f}} \tag{104}$$

is the penetration length which depends on the frequency f and the thermal conductivity k.

Depending on the relative magnitudes of y, b and L the exact solution, eq. (103), can be simplified in one of three regimes

(I) $$b\ll L:\ G(f)\approx 1+\tfrac{2}{3}\frac{b^4}{L^4}\left[\left(1-\frac{y}{b}\right)^4-1\right] \tag{105}$$

(II) $$b\gg L,(b-y)\gg L:\ G(f)\approx e^{-2y/L} \tag{106}$$

(III) $$b\gg L,(b-y)\ll L:\ G(f)\approx 4e^{-2y/L}. \tag{107}$$

Fig. 64 displays the exact solutions together with the areas in which the above approximations hold within an accuracy of 10%. It can serve for rough estimates of the behavior of a commonly used technique to slow down the response of temperature sensors. Thermal insulation can not be recommended because of self-heating of the sensor during measurement.

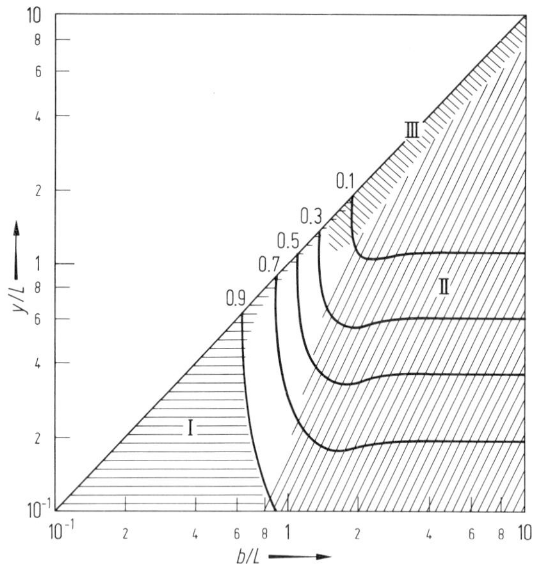

Fig. 64. Exact gain-squared response of plate as a function of thickness b and distance y from the water boundary. Both lengths are scaled by the penetration depth L. The regions where the approximation of regimes I, II, and III are within 10% of the exact solution are shaded [81 L1].

2.5.2 Conductivity

The electrical conductivity of an electrolyte is a measure of ion migration. It is directly related to the concentration of salts which are dissociated in sea water. This is one of the reasons why conductivity has been adopted as the standard for salinity determinations (see sect. 3.1).

The fundamental relationship between conductivity and salinity, which also involves temperature and pressure, shows that very precise measurements of conductivity are required to meet the demands in physical oceanography. A change of salinity by 0.01 in the vicinity of 35, 15 °C and atmospheric pressure results in a change of conductivity by 0.011 $\mathrm{mS\,cm^{-1}}$. The total range of conductivities in the ocean is between 0 and 70 $\mathrm{mS\,cm^{-1}}$, in the open sea it is typically between 20 and 55 $\mathrm{mS\,cm^{-1}}$.

In view of the high resolution, accuracy and long-term stability the first phase of modern in-situ determinations of conductivity [56 D, 57 H, 58 H 1, 58 H 2] aimed at probe designs without electrodes to be exposed to sea water in order to avoid problems with uncontrolled impedances at the electrode-electrolyte interfaces.

The simplest possible device for this purpose is a ring core transformer as sketched in Fig. 65 where the sea water path acts as a single secondary winding. The path for the electrical current induced in sea water occupies a rather large volume which is not exactly defined. The measured sea water resistance R_w has to be converted into the true conductivity C by using the "cell constant" k

$$C=\frac{k}{R_w}. \tag{108}$$

For simple geometrical shapes the cell constant can be calculated, but the high accuracy requires a final experimental determination of this constant.

There are several possibilities to measure the sea water resistance R_w. For an ideal transformer (no dissipative impedance, no capacitive reactance) the complex current through the coil is

$$I=\left(\frac{1}{n_1^2 R_w}+\frac{1}{i\omega L}\right)U \tag{109}$$

where U is the voltage, L is the inducatance, ω denotes the frequency and n_1 is the number of windings of the coil. For

$$\omega L \gg n_1^2 R_w$$

it follows

$$I \approx \frac{1}{n_1^2 R_w}\cdot U. \tag{110}$$

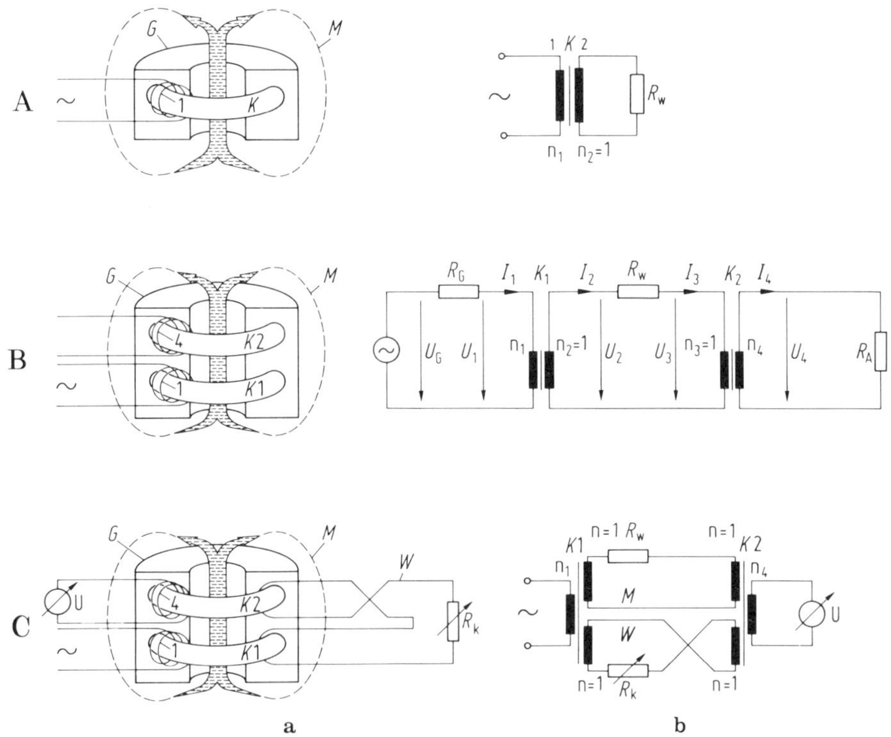

Fig. 65. Basic designs of inductive conductivity sensor [84 S 2]. A: Single transformer, B: Double transformer, C: Double transformer with compensation loop. (a) Mechanical arrangement, (b) circuit diagram. K, K_1, K_2: iron cores of transformers; M: sea water loop; G: housing; W: compensation loop; n_1, n_2: number of windings; R_w: sea water resistance; R_k: compensation resistance.

Fig. 66. An inductive conductivity bridge [63 S]. Both, the sensor (S) and reference transformer (SK) are identically constructed and immersed in the same environment, but the hole of the reference ring core is insulated by a plate to avoid induction of a current in sea water. The bridge is balanced if the compensation resistor R_k equals the resistance of the water path R_w.

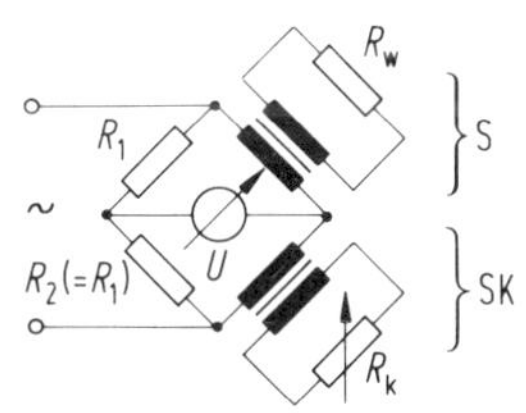

There are other possibilities which are discussed in [84 S 2] in great detail.

A practical solution [63 S] is shown in Fig. 66 where a reference transformer is exposed to the same environment as the probe transformer.

The double transformer (Fig. 65) is most commonly used with a compensation loop. If $R_k = R_w$ the bridge is balanced. For remote measurements it is more practicable to use other modes of operation. If one chooses a fixed resistor R_k it can be shown [57 H, 58 H 2, 84 S 2] that the amplitude of the output short-circuit current is a measure of conductivity. If R_k is a pure reactance the phase of the output voltage can be used.

Inductive conductivity sensors are mainly used in moored instruments. For profiling equipment one prefers electrodes after overcoming the electrode problems. Cells with electrodes are attractive because the test volume for the conductivity measurement can be made much smaller than for inductive cells. For a typical inductive conductivity cell 65% of the reading results from the interior of the cell, the rest is due to the surrounding medium [77 B 3]. This can be a problem during calibration in a laboratory container.

One approach to reduce drastically the measuring volume is the needle-probe sketched in Fig. 67a. It takes advantage of the strong convergence of electric stream lines near a needle-like electrode with respect to a distant ground plate. Test volumes are of the order of 5 times the radius of the electrode tip. With wires between 10 and 50 µm [80 G 3] excellent spatial resolution with a free-flushing device can be achieved. The disadvantages of this construction are the uncontrolled electrode impedances which fully affect the measurement. Its use is restricted for measurements of fluctuations where absolute accuracy is not necessarily required.

Electrode problems are also characteristic of the 3-electrode cell (Fig. 67b) which restrict its application for instruments of moderate accuracy [79 P]. As the outer electrodes are connected, this cell has the advantage that the measuring volume is completely inside the cell.

The 4-electrode cell (Fig. 67c) avoids the electrode problems in the following way. A constant-current source drives an alternating current through the outer electrodes, thus the resistance at the electrode-electrolyte interfaces has no influence on the magnitude of the current. If the voltage U at the voltage electrodes is measured with an amplifier of very high input impedance, also these electrodes do not affect the measurement. The ratio of the current impressed to the voltage sensed is directly proportional to conductivity. The disadvantage of the cell is that some part of the constant current flows through the surroundings of the probe and can be changed in inhomogeneous water.

This disadvantage is avoided by the modification in Fig. 67d. Here the outer electrodes have the same potential, and the electric stream lines are restricted to the interior of the cell. The two voltages U_1 and U_2 can be added to increase the signal.

Conductivity cells described so far need great care during manufacturing and concerning the choice of materials. Thermal expansion and the effects of pressure on the cell could slightly change the position of electrodes which has a rather large effect if they are located in large gradients of the electric potential.

Both effects can be described as a change of the cell constant

$$\frac{\Delta k}{k} = -\alpha(T-T_0)+\varkappa(p-p_0), \tag{111}$$

where α is the coefficient of linear thermal expansion and $\varkappa$ the respective coefficient of compressibility. T_0 and P_0 are temperature and pressure at a reference level. Table 10 gives some numerical values of α and $\varkappa$. Under high pressure (5000 dbar) and large temperature changes (20 °C) conductivity errors of 0.01 mS cm^{-1} may occur [74 F 1, 77 G] as a result of cell deformations.

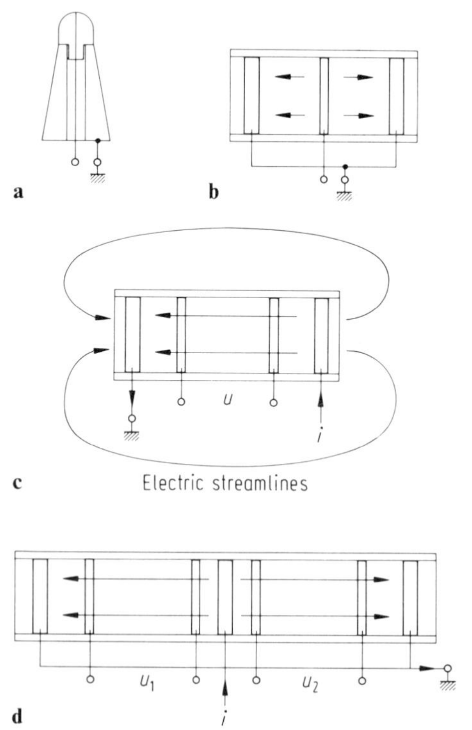

Fig. 67. Basic conductivity cells. **a**: Needle probe, **b**: 3-electrode cell, **c**: 4-electrode cell, **d**: modified 4-electrode cell.

Table 10. Expansion and compressibility coefficients of material suitable for conductivity cells.

	α [°C^{-1}]	$\varkappa$ [dbar^{-1}]
Quartz-glas	$5.1 \cdot 10^{-7}$	$9 \cdot 10^{-8}$
Duran, Pyrex	$3.2 \cdot 10^{-6}$	$1 \cdot 10^{-7}$
Ceramic, Alumina (Al_2O_3)	$6.5 \cdot 10^{-6}$	$1.5 \cdot 10^{-8}$

An electrode configuration which allows easier manufacturing and is less sensitive to mechanical deformations is shown in Fig. 68. In this 4-electrode conductivity cell the sensing electrodes are placed in regions of minimal potential gradients.

The characteristic response time τ of a conductivity cell to a sudden change of conductivity, as it occurs when the cell is lowered through a discontinuity layer, is of the type

$$\tau = \frac{l}{v} \tag{112}$$

where l denotes the length of the measuring volume and v the lowering speed. A more realistic treatment involves the hydrodynamics of flow through the cell including boundary layer effects. Not only the length but also the diameter, the ratio of these quantities and the boundary layer thickness determine the response time, which can be much larger than estimated by eq. (112).

Laboratory experiments on the response of conductivity cells were performed by [78 W 1] and [80 D 2].

In an attempt to reduce the response time drastically [84 K] has constructed a conductivity cell whose principle of operation is presented in Fig. 69.

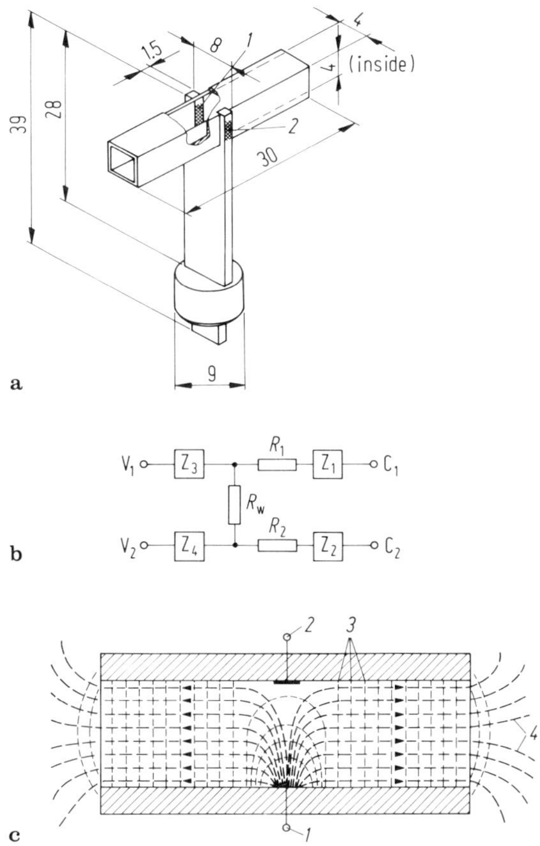

Fig. 68. **a**: General purpose conductivity cell [82 B 2]. Dimensions are in [mm]. *1* Internal electrode (1 each side); *2* External electrode (1 each side). **b**: Electrical equivalent of the cell. C_1, C_2 are the current electrodes, V_1, V_2 are the sensing electrodes. $Z_1 \cdots Z_4$ are polarization impedances at each electrode surface. **c**: Electric field in the interior of the cell. External electrodes are omitted. It is seen that the voltage electrode is at a location of minimal potential gradient. *1* "Current" electrode; *2* "Voltage" electrode; *3* Equipotential lines: *4* Current stream lines.

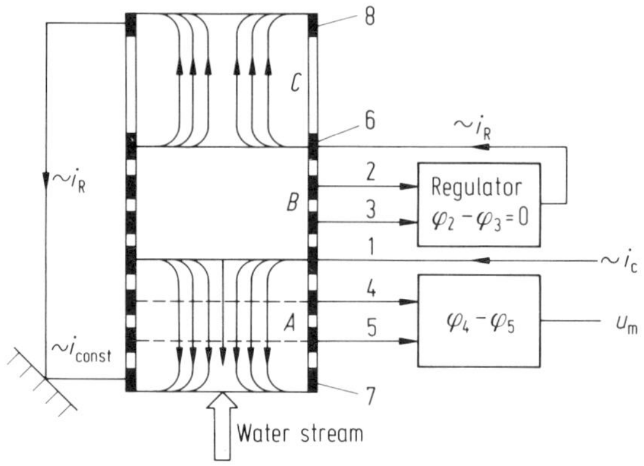

Fig. 69. Fast-response conductivity cell [84 K]. The lower part of the cell is a normal 4-electrode configuration. At electrode 1 a constant current i_c is impressed with respect to electrode 7. Measuring electrodes are 4 and 5. The field-free volume B is generated by a regulator which controls a current i_R into electrode 6 such that the potential difference between electrodes 2 and 3 is kept zero. A sudden change of conductivity is measured in the small and well-defined measuring volume A. As the change advances through volumes B and C it has no influence on the measurement. A Measuring volume; B Field free volume; C Auxiliary volume; U_M Measuring voltage; $-i_R$ Regulation current; φ Potential.

2.5.3 Pressure

Pressure sensors in oceanography are used to determine the depth of instruments and to measure the elevation of the sea surface from bottom-based instruments after applying the hydrostatic equation. Pressure as a variable of state enters directly into the equation of state and the conversion formulae for calculating salinity, sound velocity, specific volume and other derived quantities.

Most pressure sensors transform the pressure force into a deformation of an elastic material which is sensed by various electronic means, e.g. by strain gauges, vibrating wires (Vibroton), inductive and capacitive pick-ups or by a potentiometer if the mechanical excursion is large enough as in the case of a Bourdon tube. They all suffer from the problems associated with the elastic properties of matter: hysteresis, sensitivity to temperature and long-term drift. The quality of a pressure transducer depends mainly on the material which is exposed to the pressure force.

One of the most successful pressure-measuring device in classical oceanography was the combination of a protected and an unprotected reversing thermometer, see e.g. [68 U]. It utilizes the excellent elastic properties of a quartz glass capillary tube. With respect to a protected thermometer the increase of the indicated (apparent) temperature is about 1 °C per 100 m increase of depth. A water depth of 5000 m can be determined with an accuracy of ± 20 m.

The quartz crystal is also the most successful element of modern pressure transducers. One of the best investigated sensors is the Digiquartz pressure sensor [82 W]. The internal configuration is illustrated in Fig. 70. The fluid pressure at the sensor port is transformed into a force on a lever arm at the bellows' end cap. Through the lever it is transmitted and axially loads the quartz crystal. The central beam of the crystal is induced to vibrate in its lowest, flexural mode within an oscillator circuit. A change in the axial load also changes the resonance frequency of the quartz. The oscillator circuit tracks this frequency which is a measure for pressure. The output frequency or the period can be calibrated in terms of pressure. A suitable equation is

$$p = A\left(1-\frac{\tau_0}{\tau}\right) - B\left(1-\frac{\tau_0}{\tau}\right)^2 \tag{113}$$

where p is the pressure applied, τ is the vibrational period at this pressure, τ_0 is the period at zero pressure, and A and B are constants.

As the geometrical dimensions of the vibrating quartz crystal and hence its resonance frequency depend on temperature, the coefficients A, B and τ_0 are temperature-dependent. The set of polynomial series

$$\begin{aligned} \tau_0 &= K_1 + K_2 T + K_3 T^2, \\ A &= K_4 + K_5 T, \\ B &= K_6, \end{aligned} \tag{114}$$

models the temperature dependence to within the accuracy that eq. (113) describes the calibration of the sensor.

Besides the temperature effect a long-term drift of the calibration is characteristic of almost all pressure sensors. Its physical cause is elastic creep of the materials exposed to changing pressure. In case of the Digiquartz sensor elastic creep of the sensor bellows under stress is the dominant factor.

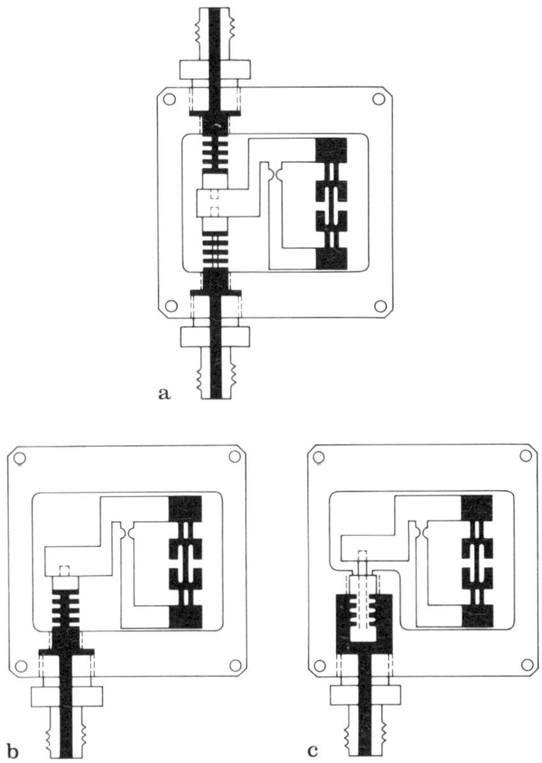

Fig. 70. The digiquartz pressure sensor [82 W]. **a**: Differential pressure sensor with dual internally pressurized bellows. **b**: Single internally pressurized bellows. **c**: Externally pressurized bellows. In all cases the pressure is transformed at the bellows' end cap into a force on the lever arm. From there it is transmitted and axially loads the vibrating quartz crystal and changes its natural frequency.

Three mathematical models can be used to account for this type of drift. They describe the difference between the indicated pressure p_i and the applied pressure p_a as a function of time t after applying the pressure at time t_0 [10 A, 53 M 1]:

$$p_i - p_a = A(t - t_0)^b + C \tag{115}$$

[05 P, 77 M 2, 77 W]:

$$p_i - p_a = A \log(t - t_0) + C \tag{116}$$

[54 M, 53 P, 48 A]:

$$p_i - p_a = C + \sum_i A_i \exp(-(t - t_0)/\tau_i)\,. \tag{117}$$

In these formulae A, A_i, C and τ_i are free constants. They are determined by lengthy laboratory tests. Which of the three equations fits the best depends on the type of sensor. Experience has shown that the power law is preferred because it usually represents the observed drift well with only few free parameters. If the parameters are known the above equations can be used to correct observed pressure time series.

Concerning the effects of elastic hysteresis it is useful to distinguish between the modes of operation. When used at the sea floor to record elevations of the surface the pressure fluctuations are within less than 3 m whereas a depth-indicating pressure sensor in profiling equipment experiences the full range of sea pressures in a short time.

Quantitative data on hysteresis, temperature sensitivity and drift for the Digiquartz pressure sensor are summarized in Tables 11 and 12 while Fig. 71 compares the drift data to other sensors.

Table 11. Summary of sensor properties [82 W]. The range of measured values of several sensitivities of Digiquartz pressure sensors are compiled for sensors of three operating ranges. All values are in [mbar] unless otherwise indicated.

	280 dbar Sensors Serial # < 1629	620 dbar Sensors Serial # ≧ 1629	4100 dbar Sensors Serial # ≧ 1629
Voltage sensitivity, 9.0···14.0 V		−0.09···−0.22 mbar/V	−0.3···+0.3 mbar/V
Hysteresis, 3 m tide range	≦ +0.03 [1])	< +0.03 [1])	< +3.5 [1])
full-scale range	+0.06···+1.9	−2.4···+1.2	−2···−83
Temperature sensitivity, 0···25° C zero to full-scale pressure	−1.8···+1.7 mbar/°C	−1.9···+1.1 mbar/°C	−38···−2.5 mbar/°C
Drift, under sustained pressure	152 dbar	565 dbar	2100 dbar
0.01···0.1 day	+0.3···+1.5	−0.8···−2.5	
0.1···1.0 day	+0.1···+1.1	−1.3···−2.8	
1···10 days	≦ +1.5	−2.1···−3.1	−5···−75
10···100 days	−0.4···+1.1	−2.9···−3.8	−20···−110

[1]) Upper limits to the expected hysteresis.

Table 12. Typical errors of Digiquartz sensors in profiling and moored instruments [82 W]. For the profiling mode a test profile of 1 h to 2 h duration has been assumed. It extends to the pressure limit of the sensor and includes a temperature change of 10 °C. The drift is assumed to be totally reversible. The drift errors in the lower part of the table are for one-year deployment at maximum operating depth. Temperature fluctuations are assumed to be 5 °C at 100 m, 3 °C at 500, and 0.2° at 5000 m.

	Sensor range dbar	Drift mbar	Temperature sensitivity mbar	Hysteresis mbar
Profiling	280 [1])	3	20	2
instrument	620	3	20	2.5
applications	4100	50	400	40···80
Moored	280 [1])	1···10	10	0.0
instrument	620	10···20	10	0.0
applications	4100	50···500	10	< 4

[1]) Only early model 280 dbar sensors were tested.

Fig. 71. Comparison of drift of different types of sensors during one month at sustained pressure. Letters denote the type of instrument and numbers give the data source from Table 13. Shaded areas show the approximate range of drift for different types of sensors [82 W].

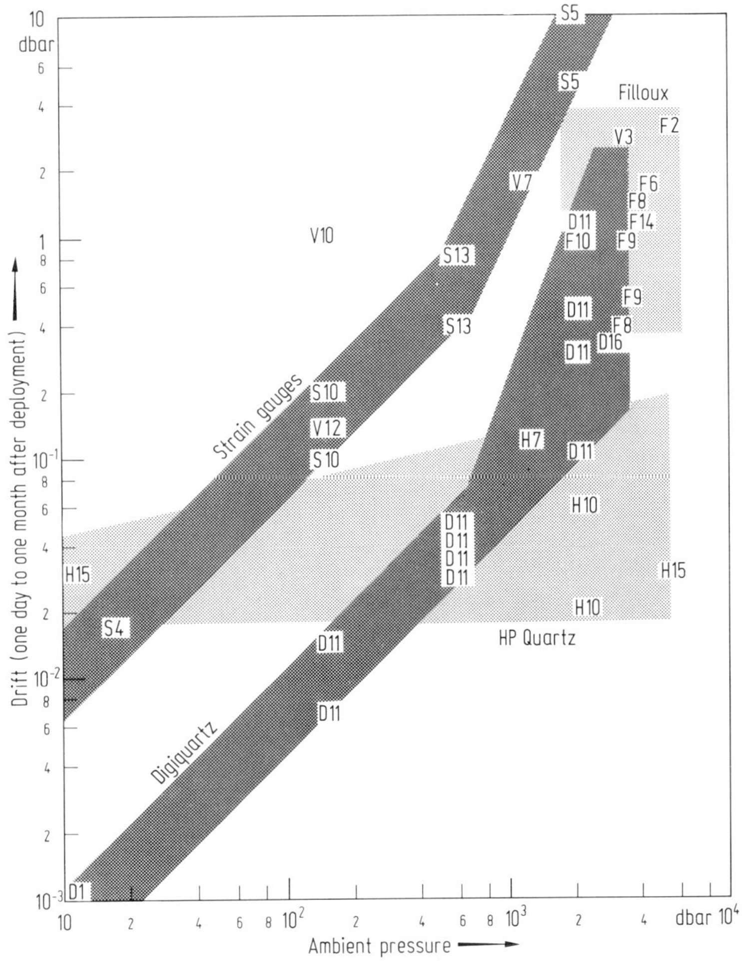

Table 13. Data sources for Fig. 71.

Instrument
- D Digiquartz
- F Filloux metal Bourdon tube
- H Hewlett-Packard sonde
- S Strain gauge
- V Vibrotron

Data sources
1 Barry [77 B 1]
2 Brown, Munk, Snodgrass, Mofjeld and Zetler [75 B]
3 Caldwell, Snodgrass and Wimbush [69 C 1]
4 J. Dahlen, Charles Stark Draper Laboratory (personal communication)
5 J. Dean, Woods Hole Oceanographic Institution (personal communication)
6 Filloux [71 F]
7 Irish and Snodgrass [72 I]
8 H.O. Mofjeld, Pacific Marine Environmental Laboratory, Seattle, Washington (personal communication)
9 Pearson [75 P]
10 SCOR, working group report [75 S 2]
11 Wearn and Larson [80 W 1]
12 Wimbush [77 W]
13 Wunsch and Wimbush [77 W 2]
14 Zetler and Cummings [72 Z]
15 Snodgrass, Brown and Munk [75 S 3]
16 S.P. Hayes, Pacific Marine Environmental Laboratory, Seattle, Washington (personal communication)

2.5.4 STD/CTD-instrumentation

The basic configuration of an instrument for the measurement of salinity and temperature as a function of depth (STD) consists of the sensors for conductivity, temperature and pressure as described in the previous sections. In order to indicate that not salinity but conductivity is directly measured, respective devices are also called CTD instruments.

Since the times of the first commercially available CTD systems [57 H, 58 H 1, 63 B 2, 68 B] considerable progress has been achieved especially with the electronic components and with data handling. A survey article on the principles of electronic circuits for digitizing of sensor outputs in the underwater unit is given in [83 D].

A general flow diagram for modern CTD systems is given in Fig. 72. The design is governed by the fact that the underwater unit has to be lowered on a single conductor steel cable and that the transmission path contains a slip ring at the winch drum. In order to exclude any influence on the quality of the measurement it is necessary to choose transmission methods on the basis of frequency or pulse code modulation. In the underwater unit the sensor outputs are digitized. A variety of analogue-digital converter (A/D) principles are used, e.g. comparator methods or frequency modulation in connection with counters [74 K 2]. The outputs of the converters are connected to a shift register (SR) or multiplexer (MUX) to form a sequential signal. The frequency shift modulator (FSKM) prepares the signal to be transmitted (AMP) through the single conductor steel cable. The separation circuit (S) (e.g. a transformer) is necessary to separate the power supply (PS) for the underwater unit from the signal.

The deck unit shown in Fig. 72 receives and decodes the FSK-signal (FSKD). The serial raw data can be directly stored on tape. The conversion into physical units and salinity computation is done by a computer which also organizes the printout and graphical representations of the data (XY).

Modern CTD systems normally have a command unit (COM) to operate various equipment lowered together with the CTD underwater unit. This can be a rosette-water sampler as shown in Fig. 72, single samplers, plankton nets or other devices.

This description has assumed only the 3 basic sensors. The circuit can be extended to include measurements of transparency, oxygen content, sound velocity and other variables. Fig. 73 displays the sensor package of a larger system.

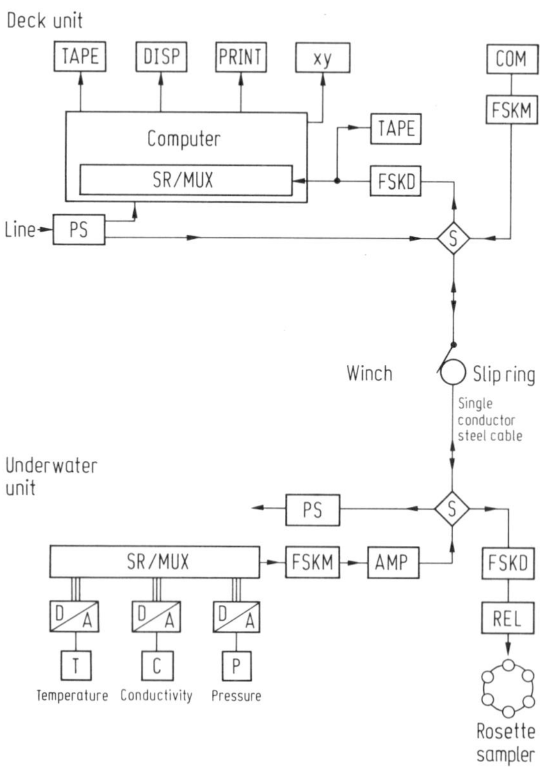

Fig. 72. Basic flow diagram of an STD instrument.

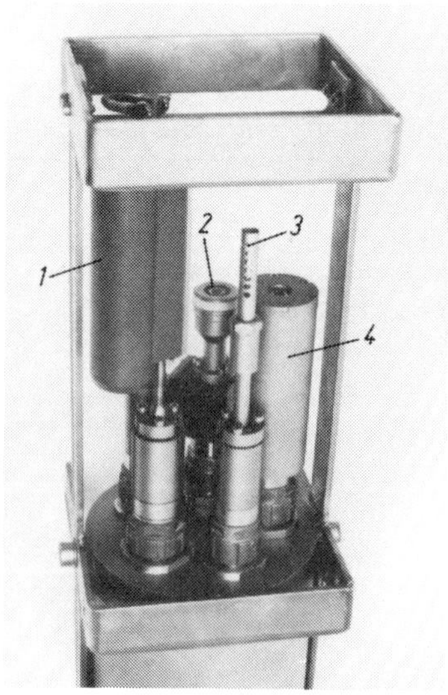

Fig. 73. Sensor package of a CTD system [83 M] *1* Conductivity; *2* Oxygen; *3* Temperature; *4* Pressure.

From the oceanographic point of view hardly any CTD system is yet felt to be so accurate and reliable as to make unnecessary the task for checking calibration during a cruise at sea. Desirable accuracies for work in the deep sea are $\Delta T = \pm 0.001$ °C, $\Delta S = \pm 0.001$ and $\Delta p \leqq 0.5$ dbar. In the harsh ocean environment this is in fact very difficult to achieve. Also the calibration of CTD instruments in the laboratory requires a considerable amount of expensive equipment.

A survey on practical aspects of vertical profiling is given in Fig. 74. It is based on some of the results of a questionaire which had been sent out to 150 users of CTD systems [73 O]. Although it reflects the state of the art

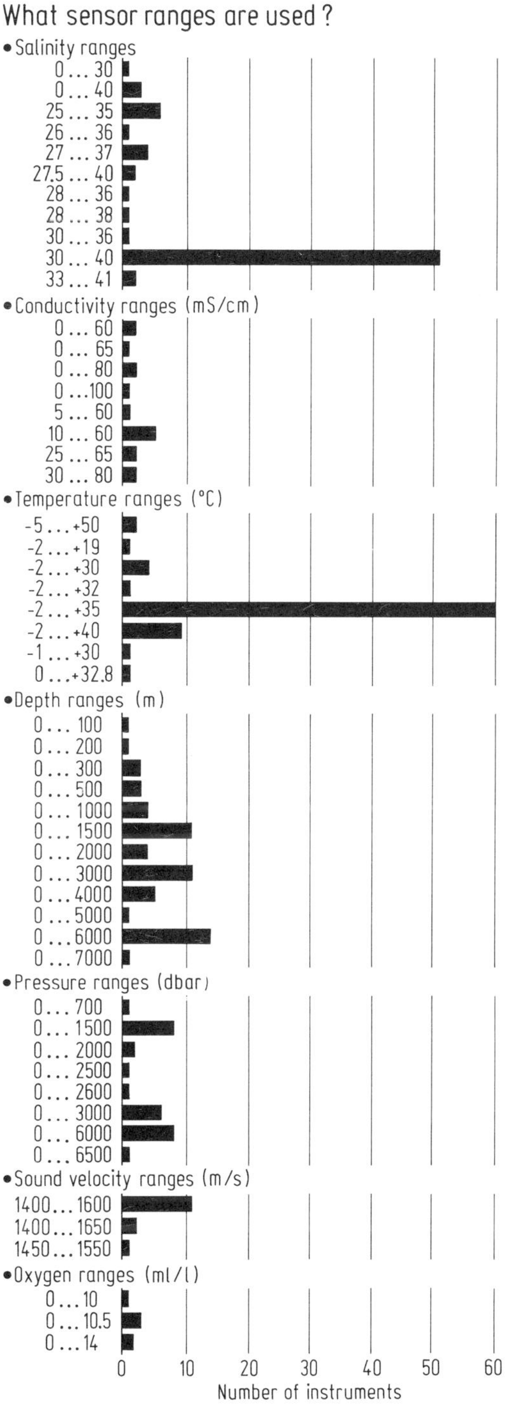

Fig. 74. Sensor ranges, calibration procedures, techniques of deployment and data quality of STD instruments [73 O].

in 1972/1973, it is still informative with respect to ranges, calibration, techniques of deployment and even accuracies achieved under field conditions. If a similar survey were carried out today, the histograms on absolute salinity accuracies would probably show a more distinct peak somewhere between 0.005 and 0.02.

Fig. 74 (continued).

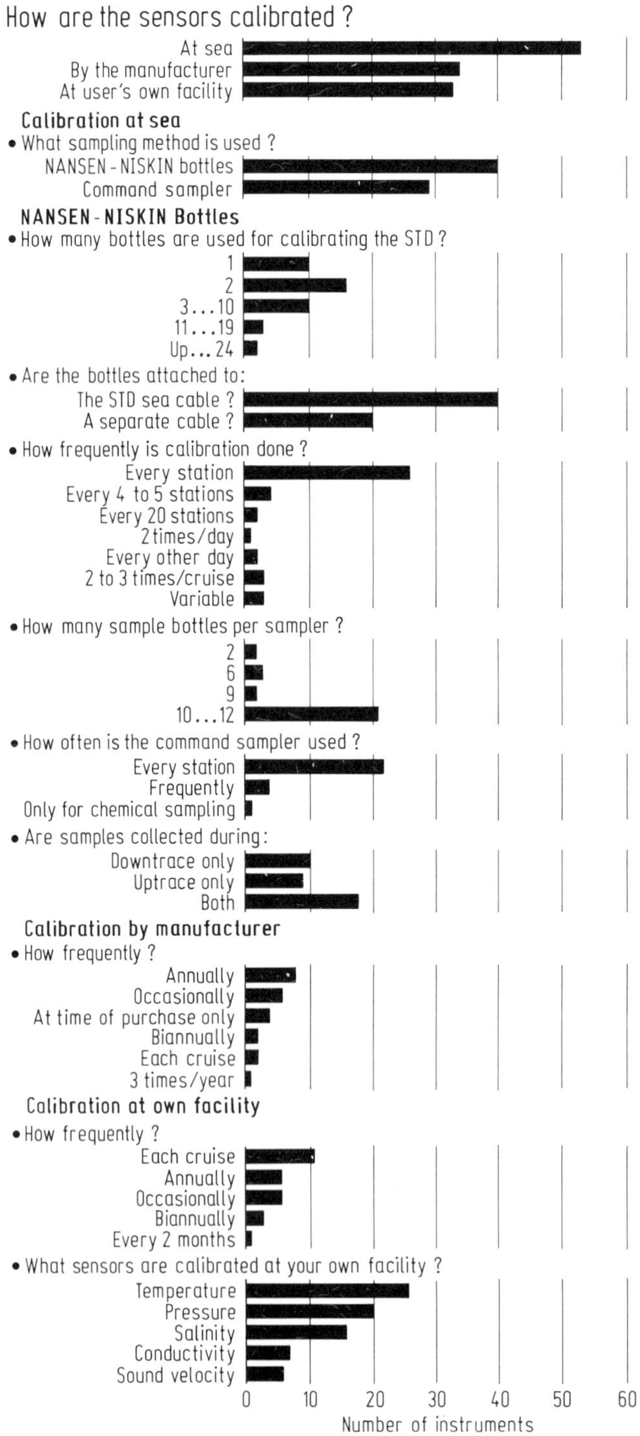

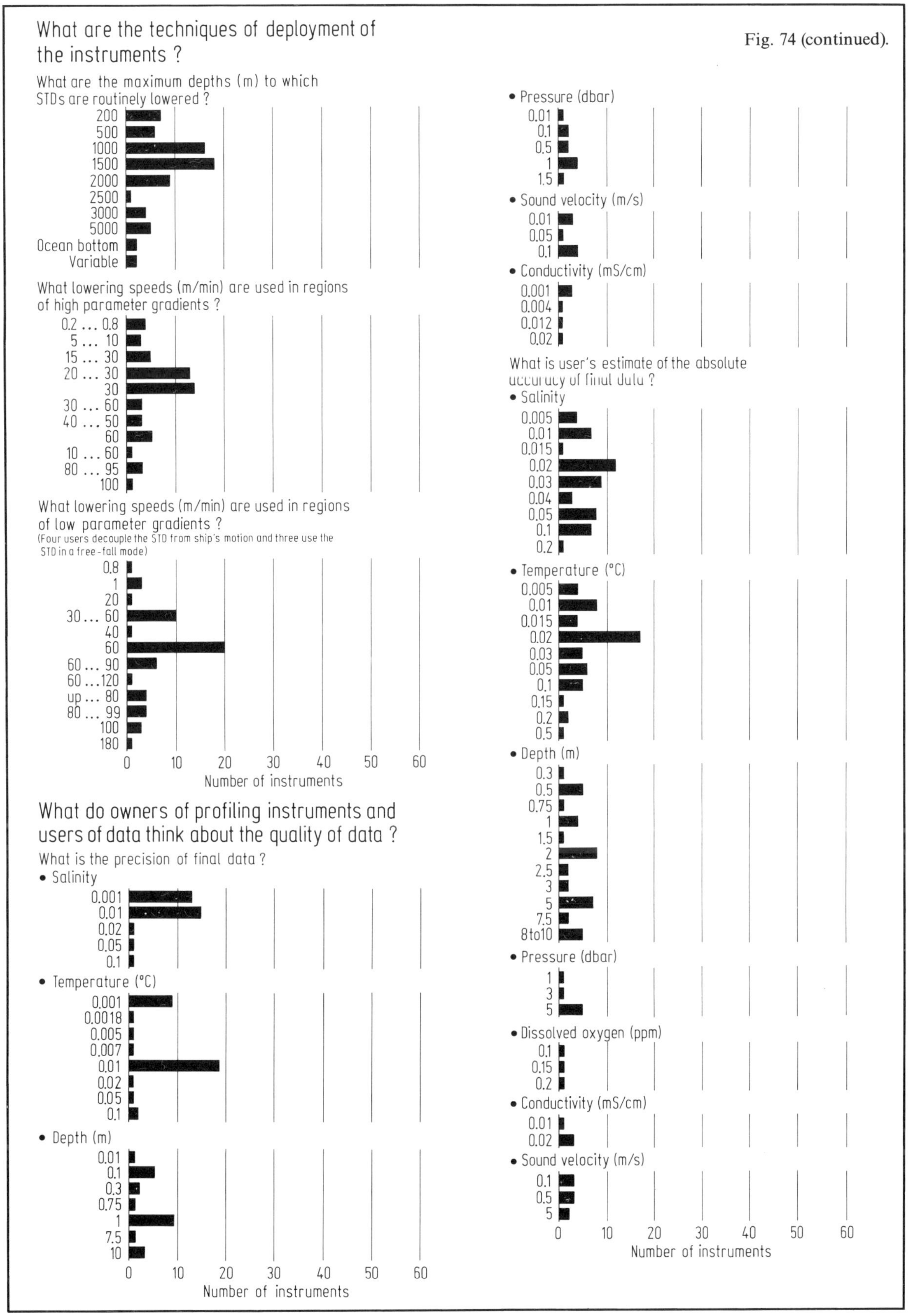

Fig. 74 (continued).

Next to the calibration and accuracy problems range difficulties with sensor response mismatch. Salinity is a result of measurements of 3 sensors. If this sensor package is lowered through a water stratification with steep gradients only equal response characteristics of all would ensure a correct derived quantity.

The major sensor reponse mismatch in CTD instruments is due to the temperature probe, whose signal normally lags behind the output of the conductivity sensor in time. The effect of the different behavior in space is sketched in Fig. 75. A step change of temperature and salinity has been assumed to occur in a vertical profile. If a CTD instrument is lowered through this sharp interface its conductivity sensor responds rapidly across the step while the thermometer records a lower temperature just below it. The salinity calculated in this region is therefore too high. It appears as a spike on the calculated salinity profile. In a TS-diagram the data points lie on a curve towards higher densities rather than on a straight line connecting the TS values of the two water masses. If temperature decreases with depth the salinity spike points towards lower salinities, and density seems to indicate a small region of instability. In actual records such situations may occur many times. Bad records can result if the constant lowering speed of the instrument is superimposed by oscillatory motions of a rolling ship. An overview on resulting virtual step structures and loops in vertical profiles is given in [71 P].

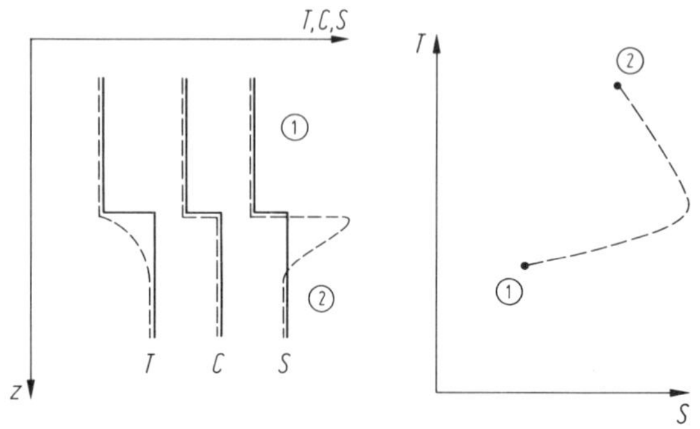

Fig. 75. Assumed vertical profiles of temperature T, conductivity C and salinity S (solid lines) together with the STD output (dashed lines) in case of "salinity spiking". In the corresponding TS-diagram the data from the transition zone do not lie on a straight line between the water masses 1 and 2.

Various attempts have been made to correct for the sensor response mismatch. The most obvious approach is to correct the temperature record by solving the differential equation of heat conduction as described in subsect. 2.5.1. However, this model with only one time constant is a rough approximation to the real response of a thermometer and, additionally, it also contains contributions by the AD-converters of the underwater unit [75 R 2] and possible integration times by the read-out module. More recent techniques rely on observed response functions from which correction procedures are derived, not only for temperature but for all 3 sensors employed in salinity measurements.

One of the possibilities [74 D] is to split up the in-situ salinity S, which is recorded as a function of time,

$$S = f(C, T, t) \tag{118}$$

into

$$S(t) = \overline{S(t)} + S'(t) \tag{119}$$

and similarly conductivity C and temperature T. The overbar indicates centered running averages, and the prime the fluctuations within the averaging period. Salinity fluctuations about the mean can be represented as

$$S'(t) = \left(\frac{\partial S}{\partial C}\right)_{T(t)} C'(t) + \left(\frac{\partial S}{\partial T}\right)_{C(t)} T'(t) \tag{120}$$

if the influence of dynamical pressure fluctuations is neglected.

The output of the conductivity sensor can be written as

$$C^*(t) = \int_{-\infty}^{0} g(\tau) C(t-\tau) \mathrm{d}\tau \tag{121}$$

where $g(\tau)$ is the filter function comprising the flushing characteristics of the conductivity cell and all other properties of the transmission path which influence the measurement. A similar expression holds for the indicated temperature $T^*(t)$.

It can be shown [74 D] that the averaged in-situ salinities $S(t)$ and the averaged indicated salinities $S^*(t)$ are related as

$$\overline{S(t)} = \overline{S^*(t)} - \left(\frac{\partial S}{\partial T}\right)_{\overline{C(t)}} \cdot \frac{\mathrm{d}T}{\mathrm{d}t} (\tau_C - \tau_T) \qquad (122)$$

where τ_C and τ_T are characteristic delays in the conductivity and temperature measurement. The change in salinity with respect to temperature at constant conductivity is known and the mean temperature gradient can be determined from the measurement. By repeated profile measurements at different lowering speeds it is possible to determine $(\tau_C - \tau_T)$, and then eq. (122) can be used to correct actual measurements.

Whereas this technique was successfully employed for slightly averaged temperature and salinity profiles, especially fine-structure measurements require a correction procedure which comprises a much wider spectral range. Instead of the characteristic times in eq. (122) the response as a function of frequency must be known. In [80 H 2] a method is outlined how to construct correction filters to match the inverse of the sensor's frequency response. In time space this corresponds also to a running average procedure but with a set of weights. For good filters the number of weights is between 20 and 40. Therefore it is important to record raw CTD data with a rather high sampling rate.

Users of CTD equipment have developed extensive processing methods to include the above mentioned elements [74 F 1, 75 S 1, 81 M, 82 P].

Typical instrumentation

One can roughly distinguish between three classes of CTD instruments, Table 14. For work in the open ocean and polar seas costly instruments of high precision are used which have to be deployed from large ships with suitable winches and laboratory facilities. Fig. 76a [68 B, 72 D, 73 K 2, 74 B, 76 K 1, 83 M, 85 N].

On the other hand work in estuaries and coastal waters requires robust, preferably battery operated, equipment to be used from small ships and boats with restricted winch capacities and working space. As temperature and salinity variations are much larger in such waters respective instruments with lower resolution and accuracies are built as handy and light weight devices, Fig. 76b [83 M, 85 N].

Finally, the expendable bathythermograph has been extended by a conductivity sensor to form the XCTD [84 S 1]. As a low-cost solution such instruments have modest accuracies. They are adequate for many oceanographic applications where the salinity signal is large in space or time. In this field improvements by manufacturers are still expected.

Table 14. Range, resolution and accuracies for the 3 categories of CTD instruments. Typical values are given. Not all of the commercially available instruments can be exactly classified as type A or type B.

		Type A open ocean	Type B coastal waters	Type C expandable
Conductivity	Range	5···55 mS/cm	1···70 mS/cm	20···75 mS/cm
	Resolution	0.001 mS/cm	0.01 mS/cm	0.03 mS/cm
	Accuracy	±0.002 mS/cm	0.05 mS/cm	0.03···0.1 mS/cm
Temperature	Range	−2···+40 °C	−2···+40 °C	−2.2···30 °C
	Resolution	0.001 °C	0.01 °C	0.01 °C
	Accuracy	±0.005 °C	0.02 °C	0.03···0.1 °C
Pressure (depth)	Range	0···6000 dbar	0···500 dbar	0···1000 m
	Resolution	0.1 dbar	1 dbar	1 m
	Accuracy	0.25% of full range	0.5% of full range	2···3% of depth

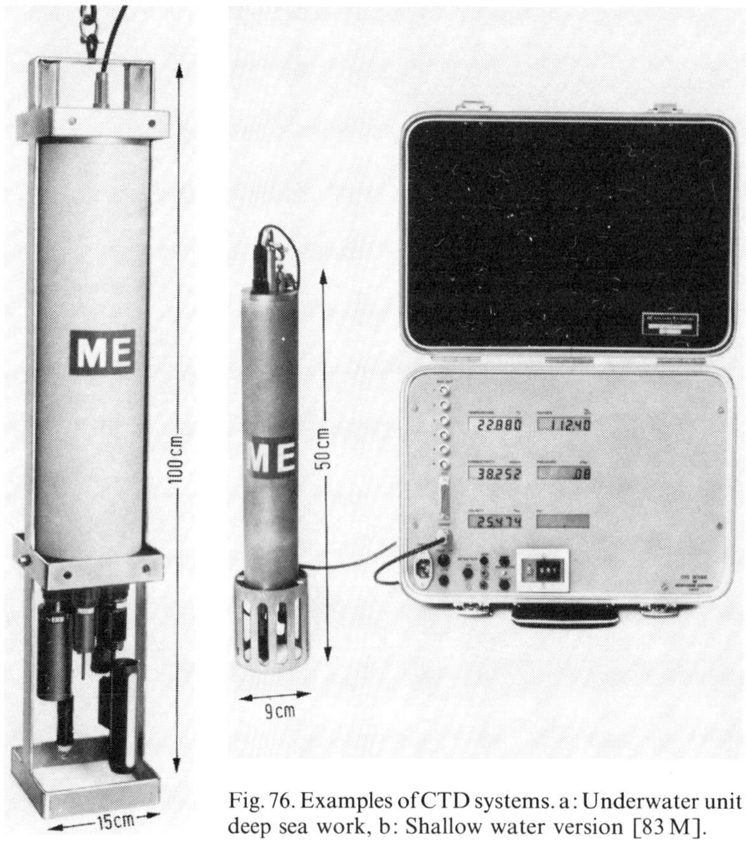

Fig. 76. Examples of CTD systems. a: Underwater unit for deep sea work, b: Shallow water version [83 M].

2.6 Basic methods of observation

This section does not deal with the design of experiments [76 B 1, 77 W 1] and large expeditions in which many ships and some hundreds of instruments may be involved. It gives an overview on elements of such surveys by presenting methods which have been used sucessfully to explore certain features of an oceanographic variable G, which is a function of geographical latitude ϕ, longitude λ, depth z and time t, i.e.

$$G = G(\phi, \lambda, z, t) \tag{123}$$

It depends on the problem whether Eulerian or Lagrangian methods are selected. The first requires measurements at grid points, the second has to rely on drifting platforms which follow a water parcel on its way through space.

2.6.1 Observations at grid points

Again the problem to be studied is the decivise factor for the decision on one of the following methods.

2.6.1.1 Spatial observations along hydrographic sections

When measuring spatial distributions from a ship or any other moving platform it is necessary to have a working hypothesis or knowledge on the temporal variability of G. Usually one assumes a quasistationary state during the time of observation.

If a ship is used one proceeds to a position ϕ_1, λ_1 and measures G as a function of depth. The measuring site is called a hydrographic station, the data series $G(z)$ is a vertical profile.

Usually a vertical profile, e.g., of temperature and salinity, shows several water layers. If Δz is the vertical sampling distance and d the thickness of a water layer, a measure of the spatial resolution is

$$d \geqq 2\Delta z \qquad (124)$$

i.e. that at least two data points are to be obtained in it.

By occupying several stations along a line one gets a hydrographic section, and by working on several sections one gets an insight into the spatial distribution of G. A measure of horizontal resolution can be chosen similar to eq. (124). Only Δz has to be replaced by the distance between stations and, depending on the problem, d could be the width of a current or the diameter of an eddy.

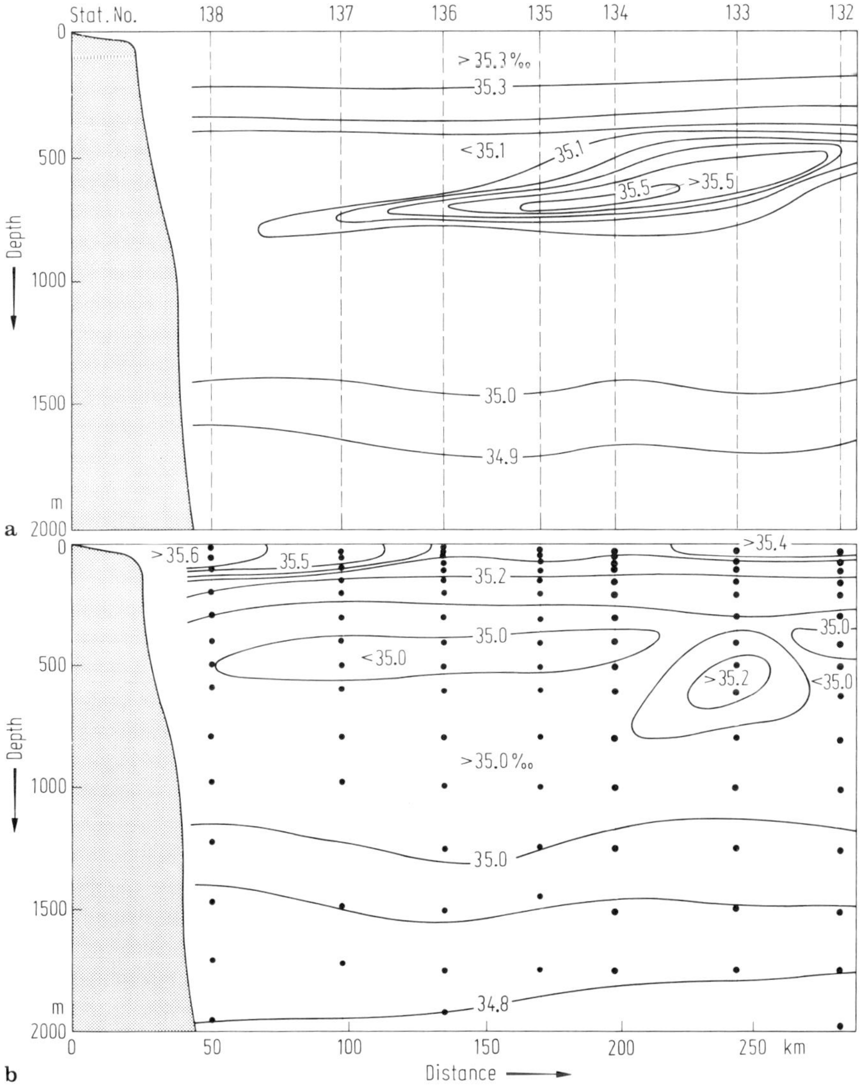

Fig. 77. An example to demonstrate the importance of the vertical resolution of instruments used to measure spatial distributions along hydrographic sections. The two salinity distributions were obtained simultaneously from the same ship on a section perpendicular to the Somali Coast [68 K]. (a) Data measured by an electronic CTD-unit at 9 m depth interval and slightly averaged. The core of the Red Sea outflow is at station 135 in a depth of 700 m and the outflow is depicted as a contiguous area. (b) Data obtained by Nansen bottles at depths indicated by dots. The core appears at station 133, the maximum value is lower, and the pattern does not contain much information on layer thickness.

The considerable progress achieved by the use of electronic CTD-instruments with respect to large-scale mapping of a water mass on a hydrographic section is demonstrated in Fig. 77 by a comparison with data obtained by Nansen bottles from the same ship at the same stations. It is obvious that with respect to the detection of the salty layer, eq. (124) is not fulfilled for the old technique.

Today the spatial resolution can be selected as desired. The real restrictions of the section technique arise from the assumption of a quasi-steady state which is never really fulfilled in the ocean. The predominant variability arises from the tides and internal waves, the periods of which are in the order of the time which a research ship needs to cover only a part of a hydrographic section.

The following simple example shows that the measured spatial structures are dependent on the ship's speed.

Assume that there is an internal wave propagating in positive x-direction such that the salinity in a certain depth is given by

$$S(x,t)=S_0+\hat{S}\cos\left[\frac{2\pi}{T}\left(t-\frac{x}{c}\right)\right] \qquad (125)$$

where S_0 is the mean salinity, $\hat{S}$ is the amplitude of the salinity variations, T is the period, c is the phase velocity, x denotes space, and t time.

If this space/time-dependent salinity distribution is measured by a ship, starting at x_0 in the same direction as the internal wave, the ship's position as a function of time is

$$x=x_0+ut \qquad (126)$$

where u is the average ship's speed including time for observations. The measured salinity in the same depth is now a function of u

$$S'(x,u)=S_0+\hat{S}\cos\left[\frac{2\pi}{uT}\left(x\left(1-\frac{u}{c}\right)-x_0\right)\right]. \qquad (127)$$

Depending on the sign of u (with or against the wave) and the relative magnitudes of u and c, a large variety of salinity distributions could have been measured and the Doppler effect deceives longer or shorter wavelengths.

In reality there is never a single plane internal wave as in this simple example but a complicated time-dependent three-dimensional salinity structure through which a section is obtained by instruments operated from a ship moving from station to station. It is clear that great care is necessary to avoid erroneous conclusions. Only under very favourable circumstances, corrections for the Doppler effect are possible [50 D, 66 K].

Another important source of error arises from breaks during a survey. Bad weather, e.g., can force a ship to stand by for some time until station work can be resumed. One typical example is given below.

Salinity distributions in part of a lagoon, 100 km long, 3 m deep and 2 km wide were measured. The area was a part of an estuary, and there was fresh water inflow from the left hand side in Fig. 78. Because of unfavourable weather conditions the observations had been carried out on two successive days, but in Fig. 78a all of the data obtained were used to plot a salinity section. The brakish water up to 10‰ is indicated by a dot pattern. From this section one could draw the conclusion that time-varying fresh-water input may have caused the brakish-water lense seen around the station CAM. However, when the data is separated and sections are drawn on the basis of salinities obtained on the first (Fig. 78b) and second day (Fig. 78c) of observation, it is readily seen that the fresh-water tongue only extends more or less far into the lagoon.

In order to obtain spatial distributions as fast as possible a variety of instruments have been designed which can be used from a ship under way. They include the towed thermistor chain [67 G] and undulating towed vehicles known as Batfish [76 D 4, 81 D] or Dolphin [62 J]. Also the GEK, expendable instruments and Doppler sonar techniques have been designed to obtain as synoptic data as possible.

Some of the modern research vessels are equipped with extensive systems to monitor the upper ocean layers while steaming. One example is the "Kiel Sea-Rover System" [85 F 1, 85 L] sketched in Fig. 79.

In spite of its shortcommings the strategy to measure spatial distributions of water masses along sections through the sea has provided not only most of our regional knowledge on watermasses in the ocean, but also on a large range of important physical processes which take place in the ocean.

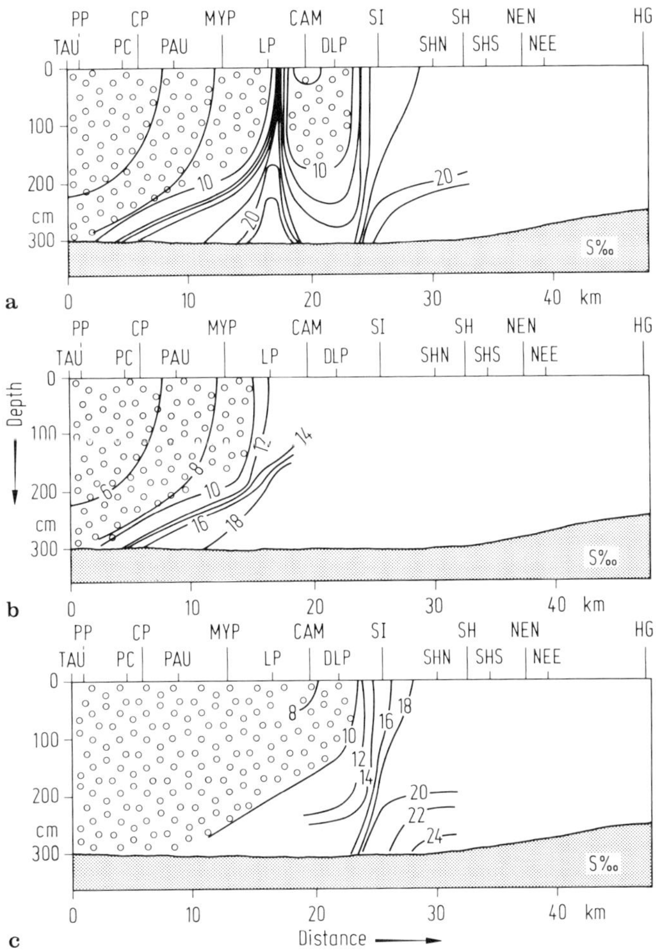

a b c

Fig. 78. An example on the effect of a break during surveying of a salinity section in an estuary. (a) Salinity distribution in the Coorong lagoon in South Australia, drawn on the basis of data gathered on two successive days. The measuring stations are indicated by letters on the top of the section. (b) Salinity distribution after the data set of the first day only. (c) Salinity distribution after the data set of the second day only.

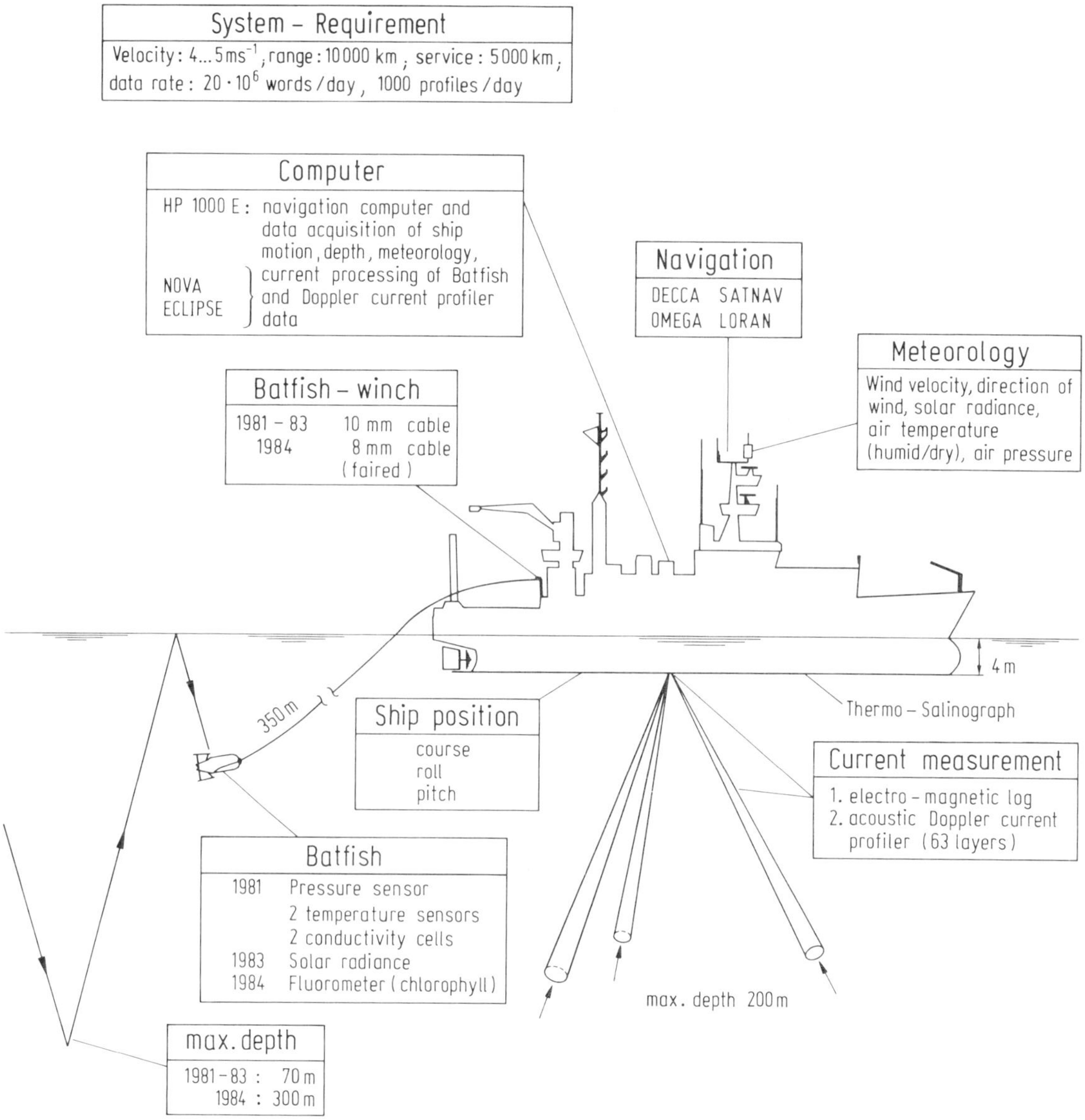

Fig. 79. A modern system for data acquisition along hydrographic sections [85 A, 85 F 1]. Instrument carrier is the "Batfish", an undulating platform towed behind the ship. Simultaneously current velocity profiles are obtained by a Doppler sonar current profiler.

2.6.1.2 Repeated profile measurements at a fixed position

Besides measurements along hydrographic sections observations at a fixed position are needed to get a better insight into the time variability of oceanographic variables. Reducing the first technique to one station means measuring series of vertical profiles as a function of time. Referring to the four-dimensional function, eq. (123), to be investigated, here ϕ and λ are constants, $G(z)$ is measured and time t can be regarded as a parameter. Again, quasi-stationary conditions have to be assumed but only over a time span necessary to measure two successive vertical profiles.

The high information content of respective data sets is demonstrated in Fig. 80 where the individual profiles are drawn in the same sequence as they have been measured. Alternative representations are isopleth diagrams and time series in selected depths.

Series of vertical profiles of temperature and salinity are readily obtained from an anchored or drifting ship, using any CTD-instrument ("yo-yo-stations"), but as velocity profiles [76 R] are needed additionally, a large variety of profilers has been developed as special instruments. They can be classified (Table 15) according to the mode operation: *Winched profilers* are operated from a stable platform (large ship or buoy, oil rig, seafloor) where sufficient power for raising and lowering of the instrument package is available. Near-shore sites could be connected by cable to the shore to have power supply for an underwater winch, which moves vertically an instrument package with buoyancy [64 K]. This arrangement can also be regarded as a *moored automatic profiler* as the "cyclesonde" [74 V] depicted in Fig. 81.

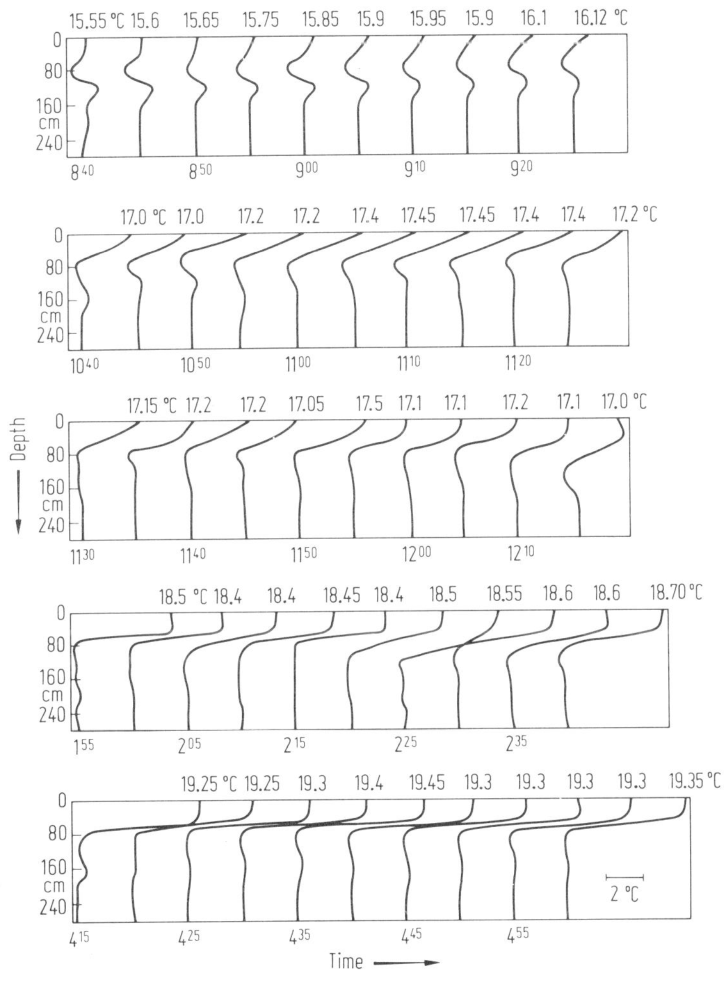

Fig. 80. A series of vertical temperature profiles for the study of daily heating of a shallow lagoon as an example to demonstrate the detailed information provided by the repeated vertical profiling technique. The numbers on top of the profile indicate surface temperature. On the lower end one finds the time of observation. With the additional information that at 11.30 h a sea breeze followed calm conditions, the total process of thermocline and mixed-layer development is represented.

Krause

Table 15. Survey on profiling equipment [80 L].

	Type of profiler	Means of profiling	Method of velocity sensing and accuracy (out of wave noise)	Sensors and resolution	Limitations	Comments on use	Ref.
Winched profilers	FLIP profiler (winched)	Winch-lowered and raised at $4\,m\,s^{-1}$	Doppler acoustic (accuracy unknown)	3 temperatures $\pm 0.01\,°C$ pressure ± 1 dbar	Can be used from FLIP only to depths of 440 m	Extremely fast profiling rate: 30 profiles/h with 3 independent temperature/pressure probes	71 B 3 75 P 1 64 K 78 E
Moored automatic profiler	Cyclesonde (autonomous, moored profiling and telemetry package)	Compressed helium inflated bladder periodically giving 100···1000 profiles/tank full (depending on depth) $0.5 \cdots 0.15\,m\,s^{-1}$	Twin Aanderaa rotors horizontally oriented $\pm 2\,cm\,s^{-1}$ at 1···5 m vertical resolution. Speed estimated by pressure change between samples.	Temperature $\pm 0.005\,°C$ Conductivity $\pm 10^{-5}\,\Omega^{-1}\,cm^{-1}$ Pressure ± 0.1 dbar Direction $\pm 0.5°$	Depth limitation 200···300 m; current speed limitation MK I and MK II $60\,cm\,s^{-1}$ (est.); 0.25···2 profiles/h typical. Profile schedule and rest periods and telemetry transmissions are controlled by a programmed clock.	Used on taut wire moorings to gather time series profiles lasting typically 4 days to 1 month, ≈15000 profiles.	74 V 76 L 1 76 J 78 V
Wire-guided winch-recovered profilers	Profiling current meter (PCM) (free-falling roller coupled and winched recovered)	Descends at $0.1 \cdots 0.3\,m\,s^{-1}$ with 1 kg overweight; returned to surface by ship's winch.	Aanderaa rotor vertical oriented $\pm 5\,cm\,s^{-1}$; ±ship navigational errors during profile at 5 m vertical resolution.	Temperature $\pm 0.025\,°C$ Conductivity $\pm 5 \cdot 10^{-5}\,\Omega^{-1}\,cm^{-1}$ Pressure $\pm 0.5\%$ full scale Direction $\pm 3°$	Must be lowered from ship with good navigation or have radar reference buoys installed	Has been widely used to survey strong currents on the equator and in Gulf Stream; about 1000 profiles to 750 m.	72 D 1

Table 15 (continued)

	Type of profiler	Means of profiling	Method of velocity sensing and accuracy (out of wave noise)	Sensors and resolution	Limitations	Comments on use	Ref.
Acoustic profilers	Rossby/acoustic (acoustic)	Free-fall $\approx 0.3\,m\,s^{-1}$ expendable weight and excess buoyancy	Acoustic range on bottom transponders ± 1 or $2\,cm\,s^{-1}$	Temperature $\pm 0.02\,°C$ Pressure ± 5 dbar	Must be used near hydrophone array and retrieved by ship after each profile	Can be used over entire ocean depth for surveys near hydrophone array as in Bermuda.	72 P 2
	White horse (acoustic)	Free-fall $\approx 0.4\,m\,s^{-1}$ expendable weight and excess buoyancy	Acoustically selfnavigating dropsonde $\pm 2\cdots 3$ m; gives $\pm 2\cdots 3\,cm\,s^{-1}$ accuracy with 100 m resolution	Temperature $\pm 0.001\,°C$ Conductivity $\pm 10^{-6}\,\Omega^{-1}\,cm^{-1}$ Pressure ± 3 dbar	Must be used with bottom transponders and retrieved by ship after each profile Can't be used during rainstorms due to acoustic noise.	High quality combined sensor profiler for survey of time series use. Coarse vertical-resolution.	78 E 76 L 1
Electro-magnetic profiler	Electromagnetic velocity profiler (EMVP) (electromagnetic)	Free-fall $\approx 1\,m\,s^{-1}$ with expendable weight release and excess buoyancy	Senses with electrode pairs sea water potential caused by motion in the earth's magnetic field; velocity $\pm 1.3\,cm\,s^{-1}$ and 10 m resolution.	Temperature $\pm 0.01\,°C$ Conductivity $\pm 2\cdot 10^{-5}\,\Omega^{-1}\,cm^{-1}$ Pressure ± 3 dbar Direction $\pm 3°$.	Must be retrieved by ship after each profile pair; velocity measured includes an unknown constant ($4\,cm\,s^{-1}$ in MODE).	Makes 6000 m profile in ≈ 1.5 h; vertical resolution of velocity 10 dbar, has recorder, profiles in time series or survey mode. Use near magnetic equator or anomalies may be restricted.	70 D 77 S1 78 S

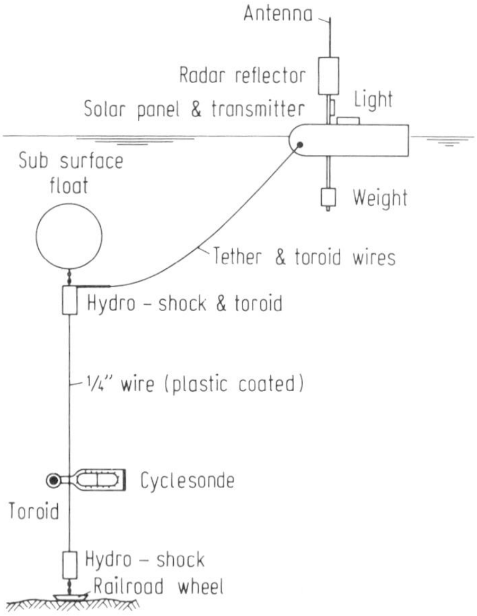

Fig. 81. A cyclesonde mooring for shallow water (300 m) [80 L].

Wire-guided profilers are widely in use to decouple vertical ship motions from the profiler which slides down slowly a weighted wire while measuring temperature, salinity and velocity relative to the ship. After reaching the maximum depth the profiler is recovered by hauling in the wire by a winch. With modern navigation equipment the relative velocity profiles can be converted into absolute profiles.

One of the first profilers of this type [72 D 1, 74 M] was simply an Aanderaa current meter with buoyancy elements to give the package a sinking velocity of some 6···10 m/min.

A better data quality is obtained by *free-falling profilers*. [80 L]. The first instrument of this type [65 R] was used for the direct measurement of transport through the Straits of Florida. The instrument was given free at a location A. It descended down to the sea-floor, dropped ballast and was recovered at a location B. The distance AB is a direct measure of transport. More modern instruments record various parameters during descend and ascend, and this method is particularly suited for fine-structure measurements [74 J, 74 W, 75 C, 76 M 1, 77 O 1].

Acoustically tracked profilers are instruments which sink at rates of 0.2···0.4 m s^{-1} and rely on 3 transponders on the bottom for precise determinations of position. The differences between pairs of positions are used to derive velocity. The chief advantage is that absolute velocity profiles are obtained, but very precise positions are needed [78 E, 76 V, 77 H, 81 H].

Electromagnetic profilers need no additional means for velocity determination. These profilers [78 S] rotate while sinking and they sense velocity by the voltage induced by the earth's magnetic field. (See subsect. 2.3.1.2.1, Fig. 38).

Profilers, especially free-falling types, have demonstrated convincingly the capability of extensive information gathering in the ocean. The more sophisticated instruments, however, require much attention and ship time. Therefore the trend is towards simple profilers and expendable instruments.

2.6.1.3 Measurement of time series

If a self-recording measuring device is anchored then ϕ, λ and z are constant and only the temporal fluctuations $G(t)$ are observed. The measurement of time series at grid points is the only in-situ method for obtaining real synoptic data, and it is a powerful tool for the testing of mathematical models.

Most of this article has been devoted to this type of measurement (sects. 2.2.2, 2.3 and 2.4). Only a few comments are added to reveal the restrictions of this method.

(i) It is important to realize that in the ocean the time change of any quantity indicated by a recording device at a fixed, place is the sum of the advective and the local change. E.g., an anchored temperature recorder indicates the temperature T on the left hand side of the equation

$$\frac{\mathrm{d}T}{\mathrm{d}t}=\frac{\partial T}{\partial t}+u\frac{\partial T}{\partial x}+v\frac{\partial T}{\partial y}+w\frac{\partial T}{\partial z},$$

where u, v and w are the velocity components along the space axes x, y and z, respectively. With a single instrument it is not possible to find out whether water is cooled locally or by advection of water of different temperature.

This is one of the reasons why temperature and salinity are hardly recorded separately, but mostly together with velocity in self-recording current meters. Only arrays of instruments are capable to resolve the causes of observed temporal fluctuations.

(ii) Concerning the space scales covered by such arrays the information content depends mainly on the number of instruments [77 W 1] available. The high expenses of instrument moorings explains why only pooling of resources of several institutes within national and international co-operation is a suitable approach to make reasonable use of time series measurements at grid points.

(iii) Both, sampling interval and desired length of a time series, determine the most important instrumental characteristics: power supply and data storage capacity.

With the advent of CMOS electronics, improved batteries, digital cassette type and rapidly increasing solid state memories, the limiting factors for the length of time series are quality of mooring components, reliability of acoustic releases and marine fouling of sensors in the upper layers of the ocean.

The repeated measurement of data at grid points is very closely linked with time series analysis [68 J, 71 B 2, 75 R 4]. Its requirements provide the guide to selecting the important experimental parameters, although many of the basic assumptions of time series analysis are not always fulfilled in the ocean where many processes are of intermittent or sporadic nature.

2.6.2 Lagrangian observations

With a Lagrangian drifter one aims at marking of a small water volume and at following its movements by periodical determinations of its position.

The primary goal is to obtain a trajectory, but modern Lagrangian platforms can also carry sensor packages for the measurement of additional quantities like temperature, salinity or air pressure.

Since the midfifties [55 S] we can distinguish between surface and subsurface drifters.

2.6.2.1 Surface drifters

Surface drifters are the oldest means to visualize and measure water velocities. It is a low-cost solution to obtain trajectories of horizontal flow patterns with the possibility of using expendable devices.

Although this method appeals for its simplicity, there are some problems associated with this type of obtaining information on ocean currents.

As surface drifters must have buoyancy they cannot follow the originally marked volume when water sinks into the interior of the ocean as it is the case in any convergence the drifter encounters on its way. As it is forced to remain at the surface it might or might not be caught by a different current pattern. Because no drifter can be perfectly locked to the water also the finite velocity of the drifter through water can cause this phenomenon. This makes it difficult to interpret drifter trajectories.

Besides these general limitations which occasionally give rise to astonishing surprises after starting of more than one drifter at the same location and time, the problems of measurement center around the possibilities of locking them to the water and around the methods of position finding.

Before the arrival of satellite- or radio-tracked buoys single surface drifters required much effort in position finding. Relying on a reasonable recovery statistics, drift bottles and drift cards [54 H, 75 M] were released in large batches to derive general information on currents and likely drift of oil patches [66 N].

A modern surface drifter consists of a buoy with a HF-transmitter, a long-life battery, a sensor package and a "sea-anchor" to provide very large drag resistance in water.

Four of the buoy forms used for such drifters are shown in Fig. 82. They are of the surface-following type, and they must very strongly be built to survive for extended periods at the sea surface while being loaded by the large drag of a drogue.

Sufficient reserve buoyancy, antenna as high above sea-level as possible and minimum wind drag are the important design criteria for such buoys.

Suspended from the buoy is a drogue [73 M] plus weight, and in some designs the drogue has a separate line which is fastened to the weight or to another suitable point along the connecting link.

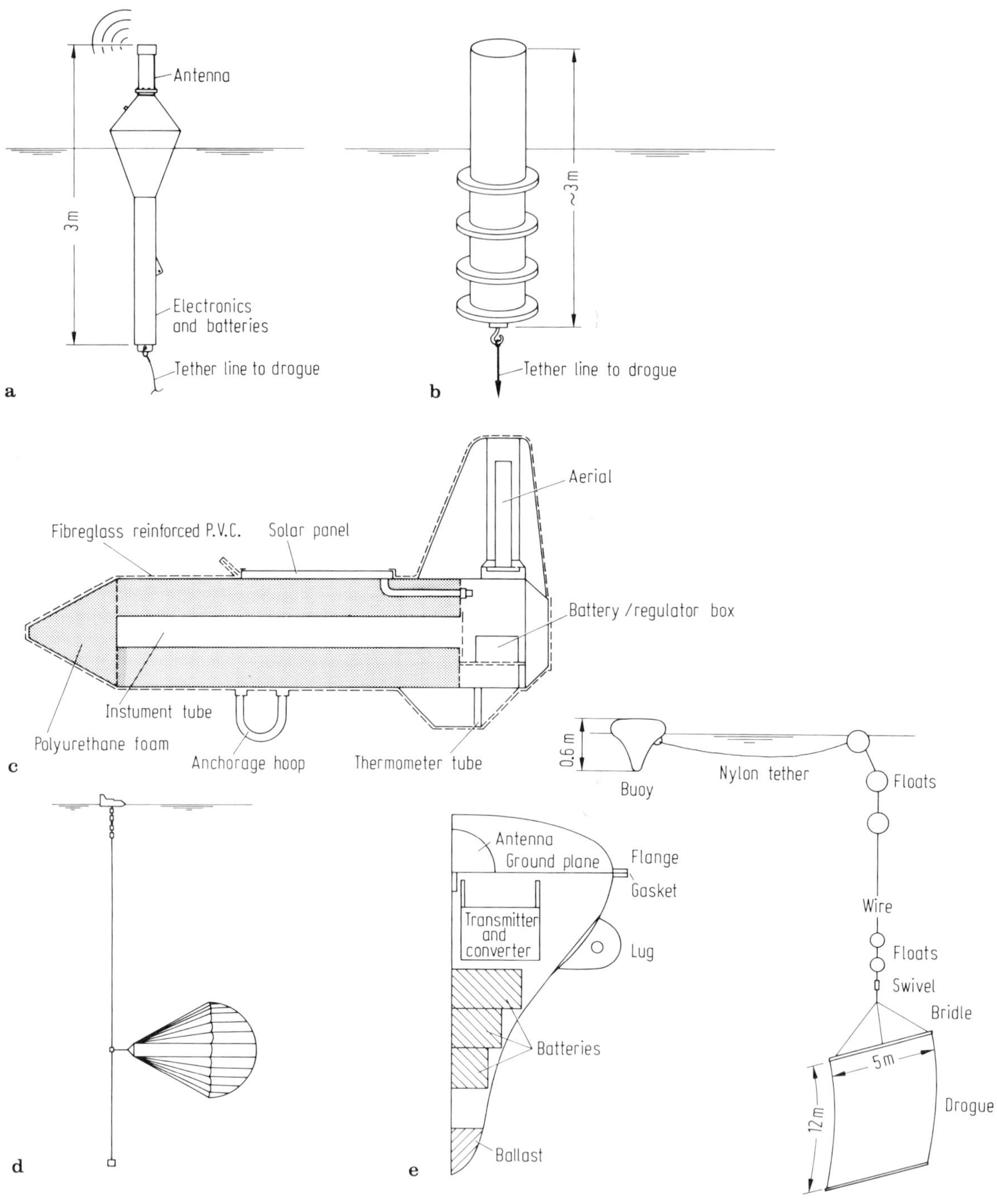

Fig. 82. Four buoy forms used for satellite tracked buoys. (a) A combination of cylinder and double-cone [80 V 3]. (b) Cylinder with damping rings [80 V 3]. (c) Torpedo buoy to reduce wind drag [78 C 1]. (d) Torpedo buoy with parachute drogue [78 C 1]. (e) Very small buoy with a displacement of only 25 kg and a reserve buoyancy of 0.4 kg. Dynamically it behaves almost like the water mass which it displaces [83 B].

The drogue force can be estimated from the vector equation

$$F = \tfrac{1}{2}\varrho_w \cdot C_D \cdot A\,|V_c - V_B|(V_c - V_B)$$

where F is the force vector, ϱ_w is water density, C_D is the drag coefficient and A is the lateral area of the drogue. The difference between the true water velocity V_c and the buoy velocity V_B over ground is often called the buoy slippage.

Table 16 displays the drag coefficients of some drogues. Surprisingly the very simple window shade drogue turns out to have the highest drag coefficient. It has, however, to be noted that these numbers are based on steady uniform flow. A drogue, suspended from a surface-following buoy in a rough sea state performs complex movements which makes it almost impossible to predict the overall drag as well as the actual behavior of a drogue. In the meantime several intercomparison tests and experience from extensive field operations are available [54 C, 75 K 1, 78 K 3, 75 M, 72 P 1, 77 R, 76 S 4, 82 K]. Concerning drogue performance [80 V 3] reaches the following conclusions on the instruments depicted in Figs. (83)···(85):

– Under steady flow conditions a window shade drogue (Fig. 83) tethered in line with the center of pressure, aligns itself perpendicularly to the prevailing flow and has a high drag coefficient C_D. This is not always the case under dynamic conditions where classical designs as crossed vanes (Fig. 84) or porous cylinders are recommended. They provide a more uniform product $C_D \cdot A$ regardless of direction.

– Parachutes and similar shapes (Fig. 85) should be held open by rings or spreaders in order to enhance their effectiveness at low relative velocities. They should be made as neutrally buoyant as possible.

– Simple shapes are not nearly as susceptible to fouling as complicated forms like the parachute.

– All components of a surface drifter must be of very strong design because of mechanical wear and tear. The drogue system should be given elasticity to extend lifetime.

The less expensive buoys are positioned by an attending ship by visual observation or radar measurements.

Positioning of unattended drifters is based on one of the following methods:

The buoy is equipped with a receiver for an existing radio navigation system like LORAN, DECCA or OMEGA and it retransmits the received signals to the shore or to a ship where the position is determined.

The buoy employs a buoy transmit terminal (BTT) with a UHF link to a satellite. Position is automatically computed at least twice a day and printouts or tapes of subsequent positions together with meteorological or oceanographical data which the buoy may measure are available from respective operation centers.

Table 17 summarizes accuracies of position finding and other characteristics of the various location methods.

After the world-wide coverage of platform positioning had been established, the old Lagrangian drifter technique has experienced a tremendous development. The overall performance of drifters and data buoys was evaluated by SCOR Working Group 66. A comprehensive guidance into this field is provided by [82 G].

Table 16. Summary of drogue drag coefficients measured during scale model tests. (Reynolds No. $\cong 10^4 \cdots 10^5$, model scale factor $\cong 1/16$) [80 V 3]. C_D: average drag coefficient.

Drogue description	C_D
Parachute	1.35
Conical muslin sea anchor [1])	1.53
Muslin sea bucket	1.54
Crossed vane, 2 solid walls	1.2 [2])
Crossed vane, 3 solid walls	1.18 [2])
Crossed vane, 3 plastic walls	1.4 [2])
Buoyed fishing net, streaming parallel to flow	0.2 [3])
Cylinder	1.1
Window shade drogue	1.93

[1]) Downstream spill hole area = 5% of inlet area.
[2]) Based on full area of one wall.
[3]) Based on full area of net.

Table 17. Comparison of characteristics of some buoy locating systems [85 S].

	Accuracy	Data renewal time	Availability time	Range	Advantages	Drawbacks	Costs (1985)
Radar	2···3% of the range	continuous availability	immediate	variable depending on radar used	no specific on-board equipment; data available on ground	problems of clutter; availability of radar	
Radiolocation systems (Decca...)	from a few tens to a few hundred meters depending on the system, buoy position and environmental conditions	continuous availability	immediate	variable, depending on the system of data retransmission	accuracy independent of range; other data can be transmitted using the same medium	on-board data availability requiring a data retransmission system; land-based transmitting stations; performances vary according to conditions	from FF 100 000 to FF 250 000; as an example, a Decca Tracer buoy-location system costs FF 240 000

Table 17 (continued)

	Accuracy	Data renewal time	Availability time	Range	Advantages	Drawbacks	Costs (1985)
Transit satellite	from 200···400 m	from 30 min to 7 h, depending on respective satellite positions	immediate	variable, depending on the system of data retransmission	accuracy independent of range; other data can be transmitted using the same medium; no land station apart from the receiver station	data-renewal period; on-board data availability requiring a data retransmission system	from FF 10000 to FF 25000 (excluding the data retransmission system)
Argos satellite	on the order of 350 m	from 30 min to 6 h, depending on buoy latitude and time of day	from 3···6 h	worldwide coverage	accuracy independent of range; other data can be transmitted using the same medium; data available on ground by phone line	data-renewal period; delay in providing data	from FF 30000 to FF 45000 (location + remote control)

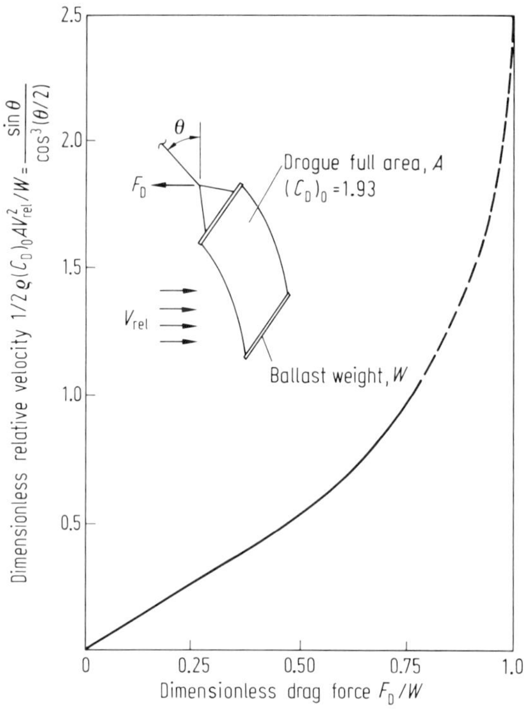

Fig. 83. Window shade drogue estimated performance curve [80 V 3].

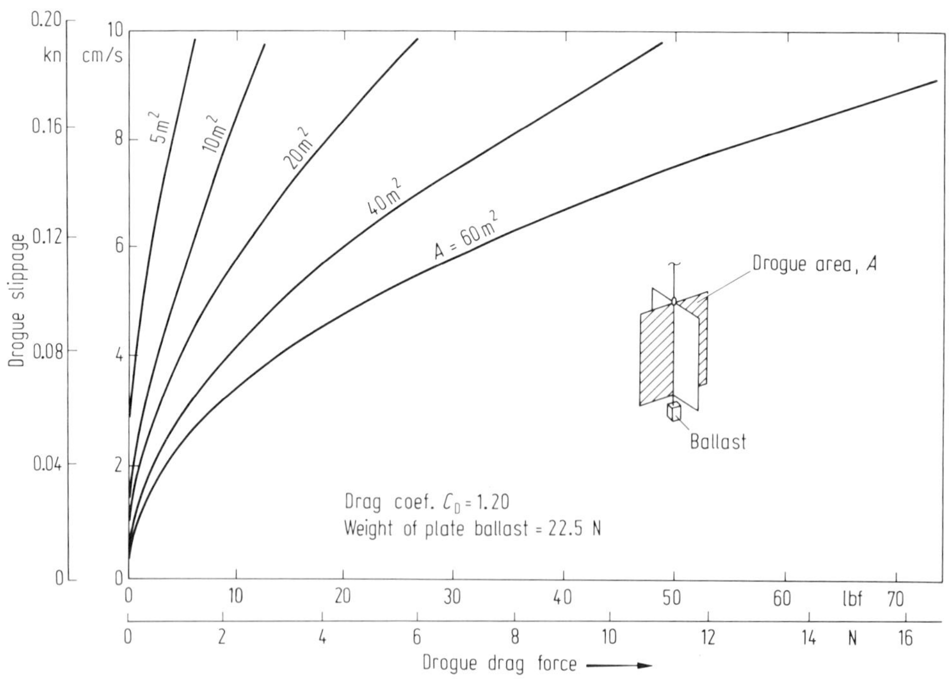

Fig. 84. Bi-planar crossed vane performance curves [80 V 3].

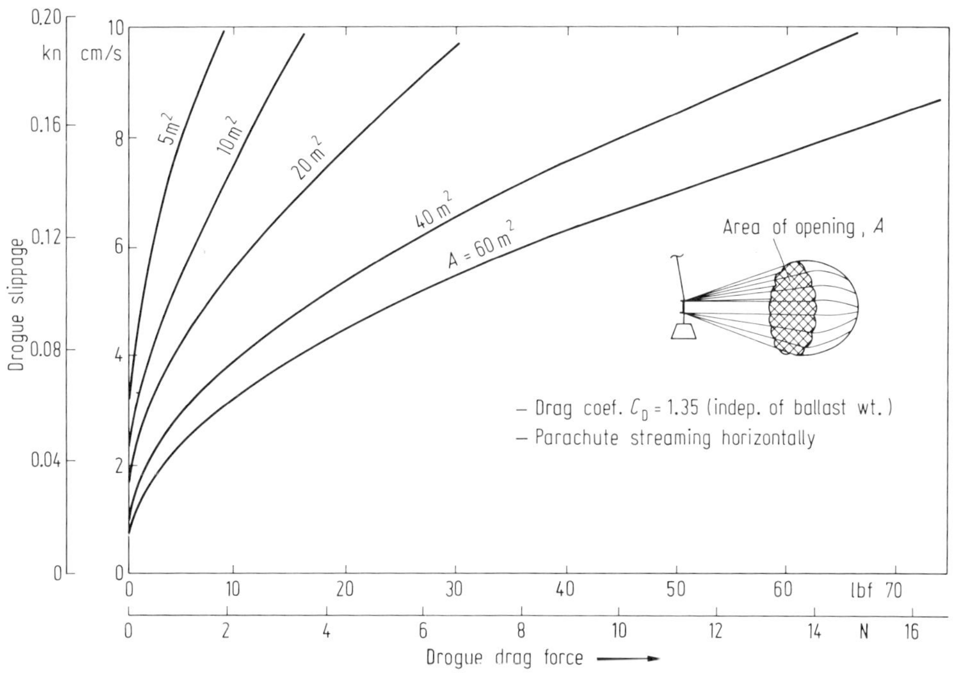

Fig. 85. Parachute drogue performance curves (after Vachon, 1973) [80 V 3].

2.6.2.2 Subsurface drifters

A subsurface drifter is a platform which is neutrally buoyant in a certain depth of the ocean. It can be located by acoustic positioning methods from board of a ship [55 S] or by hydrophones with shore connections as in case of "SOFAR-floats" which utilize the long-range sound transmission channels [61 H, 63 B 3, 70 R, 71 R] of the ocean for acoustic tracking and data transmission. There are also systems which can track a number of floats simultaneously [74 S 3].

Subsurface drifters have been successfully used for direct deep-current measurements [55 S, 72 P 1, 77 R, 83 R 1], small-scale [63 P] and vertical [68 V, 71 V] motion and in batches during large experiments [74 S 3, 75 R 3].

Not all of the subsurface drifters are Lagrangian platforms. It depends on whether they follow isobaric or isopycnal surfaces. The latter is required from a Lagrangian measurement and it can be achieved by a respective design of the float [85 R].

Neutral buoyancy of a body in sea water is a very critical state and it can be changed by only very small variations of thermal expansion and compressibility of the drifting instrument. In Table 18 some material properties are compared with the compressibility and thermal expansion of sea water.

Table 18. Physical constants for various materials compared to sea-water.

Material	Wall thickness		Diameter		Compressibility	Coefficient of thermal expansion
	inch	cm	inch	cm	10^{-7} bar^{-1}	10^{-4} °C^{-1}
Aluminum tubing	5/8	1.60	12	30.5	−2.5	0.71
Glass sphere	3/8	0.95	10	25.4	−2.6	0.096
Glass tube	13/64	0.52	3	7.6	−2.8	0.096
Seawater	–	–	–	–	−4.4	2.26 (@ 15 °C, 36.0‰) 0.99 (salt adjusted)
Total float	–	–	–	–	−3.6	0.18

If a float (subscript F) is displaced a small vertical distance Δz, it experiences a change in volume ΔV_F due to changing pressure p and temperature T

$$\Delta V_F = \frac{\partial V_F}{\partial p}\Delta p + \frac{\partial V_F}{\partial T}\Delta T$$

$$= V_F\left(\frac{1}{V_F}\frac{\partial V_F}{\partial p}\right)_T \frac{\partial p}{\partial z}\Delta z + V_F\left(\frac{1}{V_F}\frac{\partial V_F}{\partial T}\right)_p \frac{\partial T}{\partial z}\Delta z. \qquad (128)$$

The first term in brackets is the compressibility $\varkappa_F$ of the float at constant temperature, and the second characterizes the thermal expansion coefficient α_F.

The corresponding changes of a sea water parcel (subscript W) would be

$$\Delta V_W = V_W \varkappa_W \frac{\partial p}{\partial z}\Delta z + V_W \alpha_W \frac{\partial T}{\partial z}\Delta z + V_W \beta_W \frac{\partial S}{\partial z}\Delta z \qquad (129)$$

where the last term is due to the vertical salinity gradient $\partial S/\partial z$.

Therefore the restoring force per unit volume and unit displacement is

$$F = \varrho g\left[(\varkappa_W - \varkappa_F)\frac{\partial p}{\partial z} + (\alpha_W - \alpha_F)\frac{\partial T}{\partial z} + \beta_W \frac{\partial S}{\partial z}\right]. \qquad (130)$$

In order to approximate an isopycnal float, the pressure term in eq. (130) should vanish in the first place. This can be accomplished by an arrangement as in Fig. 86. Attached to the depicted float is a "compressee" (E) whose compressibility coefficient $\varkappa_C$ is selected so that

$$\varkappa_W - \varkappa_F + \varkappa_C = 0. \qquad (131)$$

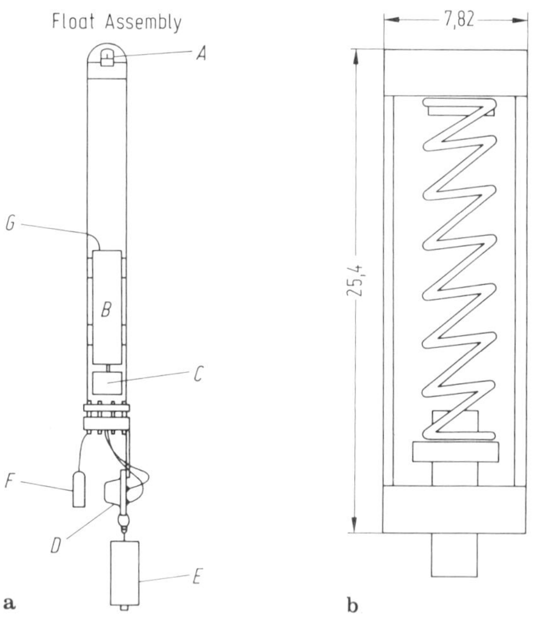

Fig. 86. (a) Schematic diagram of the cylindrical isopycnal follower: (A) flasher; (B) electronics; (C) batteries; (D) transducer; (E) compressee; (F) expendable ballast; (G) thermistor. (b) Schematic diagram of compressee. The diameter of the piston and the spring constant are adjusted so that the compressibility of the float assembly matches that of seawater. [85 R].

Thus, no Archimedian buoyancy is created due to changes in depth.

If further the thermal expansion coefficient of the float is chosen significantly smaller than that of sea water α_w, the restoring force in eq. (130) would be approximately the same as for a parcel of water, i.e., such a float would follow an isopycnal surface as it is required from a Lagrangian platform.

In [85 R] field tests with respective drifters are described which demonstrate that Lagrangian measurements are indeed feasible in the interior of the ocean.

With the existing and rapidly progressing art of acoustic data transmission in the ocean, Lagrangian subsurface drifters equipped with oceanographic sensors have a great potential in providing three-dimensional path information including studies on vertical velocities and relative vorticities with batches of such floats.

2.7 References for 2

05 P Phillips, P.: Philos. Mag. **9**, (1905) 513.
10 A Andrade, E.N., Da, C.: Proc. R. Soc. London Ser. A **84** (1910) 112.
44 K Kolin, A.: J. Appl. Phys. **15** (2) (1944).
48 A Alfrey, T.: Mechanical behavior of high polymers, New York: Interscience Publishers **1948**, 581 pp.
48 S Stommel, H.: J. Mar. Res. **7** (3) (1948) 386.
50 A Arx, v., W.S., in: Papers in Physical Oceanography and Meteorology, Vol. XI, No. 3, Woods Hole, Massachusetts: Woods Hole Oceanographic Institution and the Massachusetts Institute of Technology **1950**.
50 D Defant, A.: J. Mar. Res. **9** (1950) 120.
52 M Malkus, W.V.R., Stern, M.E.: J. Mar. Res. **11** (1952) 97.
53 M Mott, N.F.: Philos. Mag. **44** (1953) 742.
53 M Malkus, W.V.R.: J. Mar. Res. **12** (1953) 51.
53 P Pao, Y.H., Marin, J.: Trans. Am. Soc. Mech. Eng. **75** (1953) 245.
54 C Cromwell, T.R., Montgomery, R.B., Stroup, E.D.: Science **119** (1954) 648.
54 H Herdman, H.F.P.: The Trident **16** (1954) 196.
54 L Longuet-Higgins, M.S., Stern, M.E., Stommel, H., in: Papers in Physical Oceanography and Meteorology, Vol. 13, Woods Hole, Massachusetts: Woods Hole Oceanography Institution and the Massachusetts Institute of Technology **1954**, p. 1–37.
54 M Marin, J., Webber, A.C., Weissmann, G.F.: Proc. Am. Soc. Testing Materials **54** (1954) 1313.
55 B Bowden, F.F.: Philos. Trans. R. Soc. London Ser. A **248** (953) (1955) 517.
55 S Swallow, J.C.: Deep-Sea Res. **3** (1955) 74.
56 C Cushing, V.: Rev. Sci. Instrum. **29** (8) (1956) 692.
56 D Dorrestein, R.: Int. Hydrogr. Rev., Nov. **1956**, p. 1–15.
57 H Hinkelmann, H.: Z. Angew. Phys. **IX** (10) (1957) 505.
58 H 1 Hamon, B.V., Brown, N.L.: J. Sci. Instrum. **35** (1958) 452.
58 H 2 Hinkelmann, H.: Z. Angew. Phys. **X** (11) (1958) 500.
58 H 3 Hinze, J.O.: Turbulence. An introduction to its mechanism and theory, New York, Toronto, London: McGraw-Hill IX **1958**, 586 pp.
59 T Tucker, M.J.: Deep-Sea Res. **5** (1959) 185.
60 K 1 Kinsman, B.: Surface waves at short fetches and low wind speeds – a field study, Vol. 1, John Hopkins University, Chesapeake Bay Institute Technical Report **19** (1960) 581 pp.
60 K 2 Krauss, W.: Kiel. Meeresforsch. **16** (1) (1960) 13.
61 H Hale, F.E.: J. Acoust. Soc. Am. **33** (1961) 456.
61 S Solovien, V.J., Stern, J.M.: Tr. Gos. Okeanogr. Inst. **65** (1961) 159.
62 A Arx v., W.S.: An introduction to physical oceanography, Amsterdam: Addison-Wesley **1962**, 422 pp.
62 C Chalupinik, J.D., Green, P.S.: Mar. Sci. Instrum. **1** (1962) 194.
62 J Joseph, J.: Dtsch. Hydrogr. Z. **15** (1962) 16.
62 L Lester, R.A.: Mar. Sci. Instrum. **1** (1962) 200.
62 S Shercliff, J.A.: The theory of electromagnetic flow-measurements, Cambridge: Cambridge University Press **1962**.
63 B 1 Beardsley, G.F., jr.: Rev. Sci. Instrum. (1963) 516.
63 B 2 Brown, N.L.: Mar. Sci. Instrum. **2** (1963) 19.
63 B 3 Bryan, G.M., Truchan, M., Ewing, J.I.: J. Acoust. Soc. Am. **35** (1963) 273.
63 C Cook, J.C., Kenyon, K.E.: Rev. Sci. Instrum. **34** (5) (1963) 496.
63 F Farmer, H.G., in: Ocean Wave Spectra, Englewood Cliffs, New Jersey: Prentice-Hall Inc. **1963**, p. 227–233.
63 H Hinkelmann, H., Ihme, H.M.: Z. Angew. Phys. **15** (1963) 429.
63 L Longuet-Higgins, M.S., Cartwright, D.E., Smith, N.D., In: Ocean Wave Spectra, Englewood Cliffs, New Jersey: Prentice-Hall Inc. **1963**, p. 111–132.
63 P Pochapsky, T.E.: Tellus **15** (1963) 352.
63 R Richardson, W.S., Stimson, P.B., Wilkins, C.H.: Deep-Sea Res. **10** (1963) 369.
63 S Skinner, D.D.: Mar. Sci. Instrum. **2** (1963) 25.
64 A Aanderaa, I.R.: A recording and telemetering instrument, technical report 16, fixed buoy project, NATO Subcommittee for Oceanographic Research **1964**, 53 pp.
64 E Eyries, M., Dars, M., Erdely, L.: Cah. Oceanogr. **16** (9) (1964) 781.

64 K Krause, G., Siedler, G.: Kiel. Meeresforsch. **20** (1964) 130.
64 N Nagata, Y.: J. Oceanogr. Soc. Japan **19** (1964) 169.
64 W Wüst, G.: Progr. Oceanogr. **2** (1964) 1.
65 E English, J., Fletcher, B., Steckelmacher, W.: J. Sci. Instrum. **42** (1965) 77.
65 F Francis, S.A., Campbell, G.C.: Mar. Sci. Instrum. **3** (1965) 82.
65 H Haynes, A.H., Reid, W.L., Appell, G.F.: Mar. Sci. Instrum. **3** (1965) 149.
65 K Kronengold, M., Vlasak, W.: Mar. Sci. Instrum. **3** (1965) 237.
65 N Nakamura, S.: Bull. Disaster Prevention Res. Inst. **15** (1) (1965) 63.
65 R Richardson, W.S., Schmitz, W.J., jr.: J. Mar. Res. **23** (2) (1965) 172.
66 B Bowden, K.F., White, R.A.: Geophys. J. **12** (1966) 33.
66 F Führböter, A.: Elektrischer Geber für elektronische Meßgeräte im wasserbaulichen Versuchswesen, Mitt. **27** (1966) p. 259–276.
66 K Krauss, W.: Methoden und Ergebnisse der Theoretischen Ozeanographie, II, Interne Wellen, Berlin-Nikolassee: Gebrüder Borntraeger **1966**, pp. 210–215.
66 M Matthäus, W.: NTM, Schriftenr. Gesch. Naturwiss. Tech. Med. **8** (1966) 85.
66 N Neumann, H.: Dtsch. Hydrogr. Z. **19** (1966) 253.
67 B Berteaux, H.O., Walden, R.G.: Analysis and experimental evaluation of single point moored buoy systems, WHOI Technical Report, Ref. No. 69-36, Woods Hole, Massachusetts: Woods Hole Oceanographic Institution **1967**.
67 D Dreyer, G.F.: Calibration of hot-film sensors in a towing tank and application to quantitative turbulence measurements, Annapolis, Maryland: U.S. Naval Academy **1967**.
67 F Fofonoff, N.P., Ercan, Y.: Response characteristics of a Savonius rotor current meter, Technical Report 67-33, Woods Hole, Massachusetts: Woods Hole Oceanographic Institution **1967**, 36 pp.
67 G Grafa, J.B.: The thermistor chain, Undersea Technology, May **1967**.
67 H Hewlett-Packard: Quartz Thermometer, Instrument Data Sheet **1967**, 12 pp.
67 M Matthäus, W.: Monatsber. Dtsch. Akad. Wiss. Berlin **9** (11) (1967) 816.
67 R Rasmussen, C.G.: The air bubble problem in water flow hot-film anemometry, DISA information No. 5, **1967**.
67 S Sager, G., Matthäus, W.: Monatsber. Dtsch. Akad. Wiss. Berlin **9** (11) (1967) 809.
68 B Brown, N.L., in: Proc. 4th Natl. ISA Marine Sciences Instrumentation Symp. **4** (1968) p. 563–577.
68 C Caldwell, D.R.: Rev. Sci. Instrum. **39** (1968) 1865.
68 H Haubrich, R.: Bull. Seismol. Soc. Am. **58** (1968) 977.
68 J Jenkins, G.M., Watts, D.C.: Spectral analysis and its applications, San Francisco: Holden Day **1968**, 525 pp.
68 K Krause, G.: Meteor Forschungsergeb. Reihe A **4** (1968) 77.
68 M Matthäus, W., Sager, G.: Monatsber. Dtsch. Akad. Wiss. Berlin **10** (2) (1968) 81.
68 N 1 Niskin, S.J.: Mar. Sci. Instrum. **4** (1968) 15.
68 N 2 Noye, B.J.: The frequency response of a tide-well. Proceedings of the Third Australasian Conference on Hydraulics and Fluid Mechanics, Sydney, Australia: The Institute of Engineers **1968**, pp. 65–71.
68 R Rasmussen, C.G., Madsen, B.B.: Hitzdraht- und Heißfilm-Anemometrie, Archiv für Technisches Messen, A.T.M. 385 **1968**, R 17–R 28.
68 S 1 Sager, G., Matthäus, W.: Beitr. Meereskd. **22** (1968) 61.
68 S 2 Snodgrass, F.E.: Science **162** (1968) 78.
68 S 3 Squier, E.D.: Mar. Sci. Instrum. **4** (1968) 581.
68 U U.S. Naval Oceanographic Office: Instruction manual for obtaining oceanographic data, Washington, D.C.: U.S. Government Printing Office **1968**, Publ. No. 607.
68 V Voorhis, A.D.: Deep-Sea Res. **15** (1968) 599.
68 W Wisemann, W.J.: Doppler-shift current meter signals and the drop-out problem, Cheasapeake Bay Institute, The John Hopkins University, Technical Report 41 **1968**.
69 B Bretschneider, C.L.: Topics in Ocean Engineering, Houston, Texas: Gulf Publishing Company **1969**.
69 C 1 Caldwell, D.R., Snodgrass, F.E., Wimbush, M.H.: Phys. Today **22** (1969) 34.
69 C 2 Commité international des poids et mesures: Meteorologia **5** (2) (1969) 35.
69 F Fofonoff, N.P.: Buoy-system motions, in: Handbook of Ocean and Underwater Engineering, New York, McGraw-Hill **1969**, 9-109/9-115.
69 K Kestner, A.P.: Oceanology **9** (1969) 585.
69 M 1 Matthäus, W.: Beitr. Meereskd. **26** (1969) 59.

69 M 2 Matthäus, W., Sager, G.: Der Einfluß nichtperiodischer Wasserstandsstörungen auf zweistufige Pegel mit poröser Durchflußöffnung, Warnemünde: Institut für Meereskunde der Dt. Wiss. Forschungsgem. **1969**, p. 307–316.

69 S 1 SCOR Working Group 21: UNESCO Tech. Pap. Mar. Sci. **11** (1969) 70 pp.

69 S 2 Sleath, J.F.A.: J. Sci. Instrum. **2** (1969) 446.

70 B Bevir, M.K.: J. Fluid Mech. **43** (1970) 577.

70 C Casarelly, M.J., Parsons, M.: Mar. Technol. Soc. J. **4** (1970) 27.

70 D Drever, R.G., Sanford, T.B.: Proc. IEEE Conf. Electronic Eng. Ocean Technol., London **1970**, p. 353–370.

70 E Esteva, D., Harris, D.L.: Comparison of pressure and staff wave gauge records, Proceedings of 12th Coastal Engineering Conference, Washington, D.C., 1970, New York: American Society of Civil Engineers **1970**, p. 101–116.

70 M Matthäus, W.: NTM, Schriftenr. Gesch. Naturwiss. Tech. Med. **7** (2) (1970) 71.

70 R Rossby, T.E., Webb, D.: Deep-Sea Res. **17** (1970) 359.

70 T Tucker, M.J., Smith, N.D., Pierce, F.E., Collins, E.P.: J. Inst. Navigation **23** (3) (1970) 302.

71 B 1 Baker, D.J., jr.: The Harvard deep-sea pressure gauge; progress report Nov. 1971. Rep. Meteorol. Oceanogr. No. 4 (1971) Cambridge.

71 B 2 Bendat, J.S., Piersol, A.G.: Random data: analysis and measurement procedures, New York: John Wiley & Sons **1971**.

71 B 3 Brown, M.D., Isaacs, J.D., Sessions, M.H.: Deep-Sea Res. **18** (1971) 845.

71 C Choo, Y.I., Casarella, M.J.: J. Hydronaut. **6** (1971) 126.

71 F Filloux, J.H.: Deep-Sea Res. **18** (1971) 275.

71 N Nelson, S.B.: Oceanographic Ships, Fore and Aft, Office of the Oceanographer of the Navy, Washington, D.C.: U.S. Government Printing Office **1971**.

71 P Pingree, R.D.: Deep-Sea Res. **18** (1971) 841.

71 R Rossby, T., Webb, D.: Deep-Sea Res. **18** (1971) 1035.

71 S Sanford, T.B.: J. Geophys. Res. **76** (1971) 3476.

71 U Uhlig, H.H.: Corrosion and Corrosion Control, New York: John Wiley & Sons **1971**, 381 pp.

71 V Voorhis, A.D.: Response characteristics of the neutrally buoyant float, WHOI Ref. No. 71–73 **1971**, 56 pp.

72 B Bennett, A.S.: Deep-Sea Res. **19** (1972) 157.

72 C Collar, P.G., Cartwright, D.E.: Deep Sea Res. **19** (1972) 673.

72 D Dauphinee, T.M., in: Oceanology, Int. Conf. Proc. **1972**, p. 53–57.

72 D 1 Düing, W., Johnson, D.: Deep-Sea Res. **19** (1972) 259.

72 E Engelmann, H.: Kiel. Meeresforsch. **28** (2) (1972) 119.

72 G Goulet, J.R., Culverhouse, B.J.: J. Geophys. Res. **77** (1972) 4588.

72 I Irish, J.D., Snodgrass, F.E.: Deep-Sea Res. **19** (1972) 165.

72 O Olsen, J.A.: Mar. Technol. Soc. J. **6** (1972) 19.

72 P Plumb, H.H. (ed.): Temperature, its measurement and control in science and industry, Vol. 4, Pittsburgh, Pennsylvania: Instrument Society of America **1972**.

72 P 1 Parker, C.E.: Deep-Sea Res. **19** (1972) 879.

72 P 2 Pochapsky, T.E., Malone, F.D.: J. Mar. Res. **30** (1972) 163.

72 R Resch, F.J., Irish, J.D.: Deep-Sea Res. **19** (1972) 171.

72 W Williams, A.J., Jaffee, R.F., Poranski, P.F., Simoneuti, P.J.: Short arm electric field measurements of ocean currents, Ref. No. 72-13, Woods Hole, Massachusetts: Woods Hole Oceanographic Institution **1972**, 81 pp. (unpublished).

72 Z Zetler, B.D., Cummings, R.A.: Geophys. Surveys **1** (1972) 85.

73 A AMF: Vector averaging current meters (1973), Alexandria, Virginia, U.S.A.: AMF Electrical Products Development Division of AMF Incorporated **1973**.

73 B 1 Berteaux, H.O., Chhabra, N.K.: Computer programs for the static analysis of single point moored surface and subsurface buoy systems, WHOI Blue Cover Report No. 73-22, Woods Hole, Massachusetts: Woods Hole Oceanographic Institution **1973** (unpublished).

73 B 2 Boicourt, W.: Current measurements from a subsurface mooring and a rigid tower. Mode Hot-Line News, No. 44, Woods Hole, Massachusetts: Woods Hole Oceanographic Institution **1973** (unpublished):

73 B 3 Briscoe, M., Hayes, S.: Current measurements from surface and subsurface moorings. Mode Hot-Line News, No. 40, Woods Hole, Massachusetts: Woods Hole Oceanographic Institution **1973** (unpublished).

73 F 1 Fahrentholz, S.R.: Proc. Interocean **1973**, contribution No. 657.

73 F 2 Frey, H.R., McNally, G.J.: J. Geophys. Res. **78** (9) (1973) 1449.

73 H Hasselmann, K., Barnett, T.P., Bouws, E., Carlson, H., Cartwright, D.E., Enke, K., Ewing, J.A., Gienapp, H., Hasselmann, D.E., Richter, K., Sell, W., Walden, H.: Dtsch. Hydrogr. Z. A **8** (1973) 95 pp.

73 K 1 Krause, G.: Die Küste **22** (1973) 39.

73 K 2 Kroebel, W.: „Meteor" Forschungsergeb. Reihe A **12** (1973) 53.

73 M Monahan, E.C., Monahan, E.A.: Limnol. Oceanogr. **18** (1973) 981.

73 O OCEAN AFFAIRS BOARD: Continuously sampled oceanographic data, Ocean Affairs Board, National Academy of Sciences, 2101 Constitution Avenue, N.W., Washington, D.C. 20418, U.S.A. **1973**.

73 S Skinner, L.M., Rae, J.B.: Proc. Interocean Conf. Düsseldorf **2** (1973) 895.

73 T Thorpe, S.A., Collins, E.P., Gaunt, D.I.: Deep-Sea Res. **20** (1973) 933.

74 B Brown, N.L.: A precision CTD microprofiler, in: IEEE Int. Conf. Eng. Ocean Environment (Ocean 74) **2** (1974) p. 270–278.

74 D Dantzler, H.L., jr.: Deep-Sea Res. **21** (1974) 675.

74 F 1 Fofonoff, N.P., Hyes, S.P., Millard, R.C.: WHOI/Brown CTD microprofiler: methods of calibration and data handling, Technical Report WHOI-74-89, Woods Hole, Massachusetts: Woods Hole Oceanographic Institution **1974**, p. 270–278.

74 F 2 Fowlis, W.W., Thompson, J.D., Terry, W.E.: J. Mar. Res. **32** (1) (1974) 93.

74 G 1 Gonella, J., Lamy, A.: Lab. Oceanogr. Phys. Paris **1974**, 11 pp.

74 G 2 Gytre, T.: Operating experiences with an ultrasonic current meter, International Council for the Exploration of the Sea, ICES, C.M. **1974**/C: 29 (mineogr.).

74 G 3 Gould, W.J., Schmitz, W.M., Wunsch, C.: Deep-Sea Res. **21** (1974) 911.

74 H Halpern, D., Pillsbury, R.D., Smith, R.L.: Deep-Sea Res. **21** (1974) 489.

74 J Johnson, B.P.: Proc. 10'ʰ Annual Conf. Mar. Technol. Soc., Washington D.C. **1974**, p. 651–662.

74 K 1 Kraus, K.: Temperaturmessung im Bereich von 13.81 bis 273.15 K nach der IPTS-68, ATM, Blatt V 210-3 **1974**, p. 23–26.

74 K 2 Kroebel, W.: A two phase ac-bridge as a basis for in-situ measurements of different physical parameters, in: IEEE Int. Conf. Eng. Ocean Environment (Ocean 74) **2** (1974) p. 270–278.

74 K 3 Kullenberg, G.E.B., Buchhave, F.: An oceanographic in-situ Laser-Doppler-Anemometer, ICES, C.M. **1974**/C: 25 (mimeogr.).

74 M Müller, T.J., Schott, F.A., Siedler, G., Koltermann, K.P.: Meteor Forschungsergeb. Reihe B **15** (1974) 49.

74 P Pollard, R.T.: Mem. Soc. R. Sci. Liège **6** (VI) (1974) 17.

74 S 1 Sanford, T.B., Drever, R.G., Dunlap, J.H.: The design and performance of a free-fall electromagnetic velocity profiler (EMVP), Ref. No. 74-46, Woods Hole, Massachusetts: Woods Hole Oceanographic Institution **1974**, 114 pp. (unpublished).

74 S 2 SCOR Working Group 21: UNESCO Tech. Pap. Mar. Sci. **17** (1974) 116 pp.

74 S 3 Swallow, J.C., McCartney, B.S., Millard, N.W.: Deep-Sea Res. **21** (1974) 573.

74 V Van Leer, J.C., Düing, W., Erath, R., Kennelly, E., Speidel, A.: Deep-Sea Res. **21** (1974) 385.

74 W Williams, A.J.: Free sinking temperature and salinity profiler for ocean microstructure studies. Ocean 74; IEEE Intern. Conf. Eng. Ocean Environment, Halifax, Nova Scotia 1974, New York: IEEE **1** (1974) p. 279–283.

75 B Brown, W., Munk, W., Snodgrass, F., Mofjeld, H., Zetler, B.: J. Phys. Oceanogr. **5** (1975) 75.

75 C Caldwell, D.R., Wilcox, S.D., Matsler, M.: Limnol. Oceanogr. **20** (1975) 1035.

75 G Gould, W.J., Sambuco, E.: Deep-Sea Res. **22** (1975) 55.

75 H Hurst, W.S.: J. Geophys. Res. **80** (1975) 2663.

75 K Kullenberg, G., Woods, J.D.: Preliminary report on current meter intercomparisons at Stareso, August 1975, Institut de Fysisk Oceanografic, Copenhagen: Københavns Universitet **1975**, 37 pp.

75 K 1 Kirwan, A.D., jr., McNally, G., Chang, M.S., Molinary, R.: J. Phys. Oceanogr. **5** (1975) 361.

75 M Monahan, E.C., Higgins, B.J., Kaye, G.T.: Limnol. Oceanogr. **20** (1975) 141.

75 P Pearson, C.A.: Deep-sea tide observations off the southeastern United States, NOAA Technical Memorandum NOS 17, Oceanographic Division, National Ocean Survey, Rockville, Maryland: National Oceanic and Atmospheric Administration **1975**, 13 pp. (unpublished).

75 P 1 Pinkel, R.: J. Geophys. Res. **80** (1975) 3892.

75 R 1 Ramster, J.W., Howarth, M.J.: Dtsch. Hydrogr. Z. **28** (1975) p. 1–25 (Sonderdruck).

75 R 2 Roden, I.G., Irish, J.D.: J. Phys. Oceanogr. **5** (1975) 195.

75 R 3 Rossby, H.T., Voorhis, A.D., Webb, D.: J. Mar. Res. **33** (1975) 355.

75 R 4 Rabiner, L.R., Gold, B.: Theory and application of digital signal processing, Englewood Cliffs, New Jersey: Englewood Cliffs **1975**, 762 pp.
75 S 1 Scarlet, R.I.: Deep-Sea Res. **22** (1975) 509.
75 S 2 SCOR Working Group 21: UNESCO Tech. Pap. Mar. Sci. **23** (1975) 42 pp.
75 S 3 Snodgrass, F., Brown, W., Munk, W.: J. Phys. Oceanogr. **5** (1975) 63.
75 U UNESCO: An intercomparison of open sea tidal pressure sensors, report of SCOR working group 27: "Tides of the open sea", UNESCO technical papers, in: Mar. Sci. **21** (1975) 67 pp.
76 B Berteaux, H.O.: Buoy Engineering, New York: Wiley & Sons **1976**, 314 pp.
76 B 1 Bretherton, F., Davis, R.E., Fandry, C.B.: Deep-Sea Res. **23** (1976) 559.
76 C Chiocchio, F.: Evaluation of the use of moored current meters in the Great Lakes, Burlington, Ontario: Applied Research Division, Canada Centre for Inland Waters **1976**, 154 pp. (unpublished).
76 D 1 Datawell: Waverider, instrument brochure, Datawell bv, laboratory for instrumentation, Zomerluststraat 4, Haarlem, The Netherlands **1976**.
76 D 2 DISA: Type 55L Laser-Doppler-Anemometer Mark II, instruction manual, DISA Elektronik GmbH, Schinnrainstr. 9, 7500 Karlsruhe, FRG **1976**.
76 D 3 Durst, F., Melling, A., Whitelaw, J.H.: Principles and practice of Laser-Doppler anemometry, New York: Academic Press **1976**.
76 D 4 Dessureault, J.G.: Ocean Eng. **3** (1976) 99.
76 H 1 Halpern, D., Pillsbury, R.D.: Limnol. Oceanogr. **21** (1976) 611.
76 H 2 Halpern, D., Pillsbury, R.D.: Mar. Technol. Soc. J. **10** (1976) 32.
76 J Johnson, W.R.: Cyclesonde measurements in the upwelling region off Oregon, Technical Report, Miami, Florida, USA: University of Miami **1976**.
76 K 1 Kroebel, W., Diehl, P., Ginzkey, L., Mahrt, K.H., Rathlev, J., Siara, R., Schulz, T.: Die Kieler Multisonde der Jahre 1975/76, ihre Sensoren, Parameter mit Ergebnissen von Datenaufnahmen und Perspektiven für ihre Auswertungen, Proc. Interocean **1976** p. 1034–1046.
76 K 2 Kuhn, H.: Strömungsmessung durch thermische Markierung, Proc. Interocean **1976**, p. 1091–1095.
76 L Loeng, H.: A comparison of current meters and mooring methods in the Kattegat. Technical Report 43, Geophysical Institute, Bergen: University of Bergen **1976**, 17 pp.
76 L 1 Luyten, J.R., Swallow, J.C.: Deep-Sea Res. **23** (1976) 999.
76 M Moller, D.A.: A computer program for the design and static analysis of single point subsurface mooring systems, WHOI Technical Report, Ref. No. 76-59, Woods Hole, Massachusetts: Woods Hole Oceanographic Institution **1976**.
76 M 1 Mortensen, A.C., Lange, R.E.: Deep-Sea Res. **23** (1976) 1231.
76 R Rossby, H.T., Sanford, T.B.: J. Phys. Oceanogr. **6** (1976) 766.
76 S 1 Saunders, P.M.: Deep-Sea Res. **23** (1976) 249.
76 S 2 Saunders, P.M.: Intercomparison of a VACM and acoustic current meter, Polymode News, No. 12, Woods Hole, Massachusetts: Woods Hole Oceanographic Institution **1976** (unpublished).
76 S 3 Siedler, G., Gerlach, E.: Verankerte Meßsysteme für die Tiefsee, Interocean **1976**, IO 76-355, NOWEA, Düsseldorf.
76 S 4 Saunders, P.M.: Drifting buoy Lagrangian test. Final Report of Woods Hole (Contract NAS 13-5 with NOAA Data Buoy Office, NSTL Station, Mississippi 39529, **1976**.
76 V Van Leer, J.C.: Instrum. Aerosp. Ind. **23** (7626 B) (1976) 489.
77 B 1 Barry, W.A.: Barometric sensor evaluation program, Applied Physics Laboratory, University of Washington, APL-UW 7714 **1977**, 32 pp. (unpublished).
77 B 2 Beardsley, R.C., Scott, J., Boicourt, W.: CMICE 76, Technical Report 77-62, Woods Hole, Massachusetts: Woods Hole Oceanographic Institution **1977**, 123 pp.
77 B 3 Breitenbach, J.: Der Einfluß der Wannendimensionen auf die Kalibrierung von induktiven Meßfühlern zur Bestimmung der elektrischen Leitfähigkeit des Meerwassers, Diplomarbeit, Kiel University **1977**.
77 D Davis, R.E., Regier, L.A.: J. Mar. Res. **35** (1977) 453.
77 F Fryer, D.K., Thomas, M.W.S.: J. Phys. E **10** (1977) 683.
77 G Ginzkey, L.: Der Leitfähigkeitssensor der Kiel Multisonde für in-situ-Messungen der spezifischen elektrischen Leitfähigkeit, Ph.D.-Thesis, Kiel, University **1977**.
77 H Hendricks, P.J., Rodenbush, G.: Interpretation of velocity profiles measured by freely sinking probes, Woods Hole Oceanographic Contribution No. 4006, Woods Hole, Massachusetts: Woods Hole Oceanographic Institution **1977**, 15 pp.

77 L Lueck, R.G., Hertzman, O., Osborn, T.R.: Deep-Sea Res. **24** (1977) 951.

77 M 1 Magnell, B.: Intercomparison of the EG & G model 102 current meter with EG & G model CT/3, Endeco model 105 and AMF VACM current meters in shallow coastal waters, Environmental Consultants Monograph B-4518, Waltham, Massachusetts: EG & G, Environmental Consultants **1977**.

77 M 2 Mofjeld, H.O., Wimbush, M.: Deep-Sea Res. **24** (1977) 987.

77 O Ohte, A., Yamagata, M.: IEEE Trans. Instrum. Meas. **IM-26(4)** (1977) 335.

77 O 1 Oakey, N.S.: An Instrument to measure oceanic turbulence and microstructure, Dartmouth, Nova Scotia: Bedford Institute of Oceanography Report Series, BI-R-77-3, May **1977**, 52 pp.

77 P 1 Paramonov, A., Kushnir, V.: An intercomparison of ocean current meters, Polymode News, No. 38, Woods Hole, Massachusetts: Woods Hole Oceanographic Institution **1977** (unpublished).

77 P 2 Petrie, B.: An intercomparison of data from Endeco 174 and Braincon 381 current meters, Technical Report BI-R-77-10, Dartmouth, N.S., Canada: Bedford Institute of Oceanography **1977**, 18 pp.

77 R Richardson, P.L., Cheney, R.E., Mantini, L.A.: J. Phys. Oceanogr. **7** (1977) 580.

77 S Stewart, R.H.: Ocean Eng. **4** (1977) 101.

77 S 1 Sanford, T.B.: Bericht Inst. Meereskd. Nr. 30, Kiel: Christian-Albrechts-Universität **1977**, 25 p.

77 W Wimbush, M.: Deep-Sea Res. **24** (1977) 493.

77 W 1 Woods, J.D., in: Modelling and prediction of the upper layers of the ocean (Krauß, E.B., ed.), New York: Pergamon Press **1977**, pp. 263–283.

77 W 2 Wunsch, C., Wimbusch, M.: J. Mar. Res. **35** (1977) 75.

78 A Aanderaa Instruments: Water level recorders, data sheet D150N, Bergen, Norway **1978**.

78 C Cheng, R.T.: Proceedings of a Working Conference on Current Measurement, Technical Report DEL-SG-3-78, College of Marine Studies, Newark, Delaware: University of Delaware **1978**, p. 293–304.

78 C 1 Cresswell, G.R., Richardson, G.T., Wood, J.E., Watts, R.: The CSIRO satellite-tracked "Torpedo" buoy. CSIRO Division of Fisheries and Oceanography. Reprint No. 966, **1978**.

78 E Evans, R.H., Leaman, K.: J. Geophys. Res. **83** (C 11) (1978) 5515.

78 H Halpern, D.: Proceedings of a Working Conference on Current Measurement, Technical Report DEL-SG-3-78, College of Marine Studies, Newark, Delaware: University of Delaware **1978**, p. 69–76.

78 K 1 Kroner, S., Blumenthal, B.: Guide to common shipboard expendable bathythermography (SXBT) recording malfunctions, Bay St. Louis, Mississippi: Naval Oceanographic Office, Reference Publication 21, **1978**.

78 K 2 Kullenberg, G.: Finn. Mar. Res. **244** (1978) 42.

78 K 3 Kirwan, A.D., jr., McNally, G., Pazan, S.: J. Phys. Oceanogr. **8** (1978) 1146.

78 M McDowell, S., Dorson, D.: An XBT digital-cassette recording system, Polymode News, No. 50, Woods Hole, Massachusetts: Woods Hole Oceanographic Institution **1978**.

78 S Sanford, T.B., Drever, R.G., Dunlap, J.H.: Deep-Sea Res. **25** (1978) 183.

78 V Van Leer, J.C., Leaman, K.D.: Proc. Working Current Measurements, Technical Report DEL-SG-3-78, University of Delaware, Newark: College of Marine Studies **1978**, p. 77–93.

78 W 1 Walker, E.R.: Laboratory determination of the response of thermometer and conductivity cell of the Guideline 8101A, Patricia Bay, Sidney, B.C., Canada: Techn. Rep. Institute of Ocean Sciences 1978, IOS Note 7, 33 pp. (unpublished).

78 W 2 Winant, C.D., Davis, R.E., Weller, R.: Shallow current measurements. Proceedings of a Working Conference on Current Measurement, Technical Report DEL-SG-3-78, College of Marine Studies, Newark, Delaware: University of Delaware **1978**, p. 129–136.

79 A Aanderaa instruments: Operating manual RCM 4/5, Bergen, Norway: Aanderaa instruments **1979**.

79 N National Technical Information Service: Ocean wave sensing, a bibliography with abstracts, National Technical Information Service, U.S. Departement of Commerce, Springfield, Vt. 22161, U.S.A. **1979**.

79 P Pederson, A.M., Gregg, M.C.: IEEE J. Ocean Eng. Vol. **OE-4**, No. 3, 1979.

80 B Blendermann, W., in: Air-sea interaction, instruments and methods (Dobson, F., Hasse, L., Davis, R., eds.), New York: Plenum Press **1980**, p. 645–679.

80 D 1 Davis, R.E., Weller, R.A.: Propeller current sensors, in: Air-sea interactions, instruments and methods (Dobson, F., Hasse, L., Davis, R., eds.), New York: Plenum Press **1980**, 801 pp.

80 D 2 Der, J.C.: Flow response tests of a prototype profiling sonde configuration, Exposure Vol. **7**, Nr. 6 (1980) p. 1–7.

80 D 3 Dobson, F., Hasse, L., Davis, R. (eds.): Air-sea interaction, instruments and methods, New York: Plenum Press **1980**, 801 pp.

80 F Freymuth, P.: A bibliography of thermal anemometry, TSI Corporation, St. Paul, Minnesota 55164 – U.S.A. **1980**.
80 G 1 Gerlach, E.: Zur Berechnung dynamischer Vorgänge in Trossensystemen bei Einsätzen auf See, Hamburg: Inst. f. Schiffbau der Univ. Hamburg, **1980**, Bericht No. 400.
80 G 2 Gytre, T., in: Air-sea interaction, instruments and methods (Dobson, R.L., Hasse, L., Davis, R., eds.), New York: Plenum Press **1980**, 480 pp.
80 G 3 Gregg, M.C., Pederson, A.M., in: Air-sea interaction, instruments and methods (Dobson, R.L., Hasse, L., Davis, R., eds.), New York: Plenum Press **1980**, pp. 319–348.
80 J Jones, U.S.F., in: Air-sea interaction, instruments and methods (Dobson, F., Hasse, L., Davis, R., eds.), New York: Plenum Press **1980**, 801 pp.
80 H 1 Halpern, D., in: Air-sea interaction, instruments and methods (Dobson, F., Hasse, L., Davis, R., eds.), New York: Plenum Press **1980**, p. 127–140.
80 H 2 Horne, E.P.E., Toole, J.M.: J. Phys. Oceanogr. **10** (1980) 1122.
80 L Lange, R.E., in: Air-sea interactions, instruments and methods (Dobson, F., Hasse, L., Davis, R., eds.), New York: Plenum Press **1980**, p. 725–738.
80 P Pinkel, R., in: Air-sea interaction, instruments and methods (Dobson, R., Hasse, L., Davis, R., eds.), New York: Plenum Press **1980**, 480 p.
80 R Rückauer, C.: Mess. Prüfen **16** (10) (1980) 672.
80 S Steward, R.H., in: Air-sea interaction, instruments and methods (Dobson, F., Hasse, L., Davis, R., eds.), New York: Plenum Press **1980**, p. 447–470.
80 V 1 Veth, C., in: Marine Turbulence (Nihoul, J.C.J., ed.), Amsterdam: Elsevier **1980**, p. 303–317.
80 V 2 Vollbrecht, K., Enslin, H., Kunstmann, H., Koehler, H., Dobinsky, H., Hoffmann, K.: Hansa **117** (22) (1980) 1731.
80 V 3 Vachon, W.A., in: Air-sea interactions, instruments and methods (Dobson, F., Hasse, L., Davis, R., eds.), New York: Plenum Press **1980**, p. 201–217.
80 V 4 Van Leer, J.C., in: Air-sea interactions, instruments and methods (Dobson, F., Hasse, L., Davis, R., eds.), New York: Plenum Press **1980**, p. 701–724.
80 W Weller, D., Davis, R.E.: Deep-Sea Res. **27** (1980) 565.
80 W 1 Wearn, R.B., jr., Larson, N.G.: The paroscientific pressure transducer, measurement of its sensitivities and drift. APL-UW 8011. Applied Physics Laboratory, University of Washington **1980**, 92 pp. (unpublished document).
81 D Dettmann, E.: Konzeption, Gestaltung und Anwendung eines ozeanographischen Schleppsystems, Ph.D. Thesis, University of Hannover **1981**.
81 H Hendricks, P.J., Rodenbusch, G.: Deep-Sea Res. **28** (1981) 1199.
81 K Kroebel, W.: IEEE J. Ocean. Eng. **OE-6(4)** (1981) 118.
81 L Levine, M.D.: Deep-Sea Res. **28** A (1981) 1401.
81 M Miyake, M., Emery, W.J., Lovett, J.: J. Phys. Oceanogr. **11** (1981) 1159.
82 B 1 Breitenbach, J., Schröder, M.: Ber. Inst. Meereskd., Universität Kiel, No. **109** (1982) 90 pp.
82 B 2 Brown, N.: Mark IIIB Conductivity, Temperature, Depth Profiler Under Water Unit, Operation and Maintenance Manual, Neil Brown Instrument Systems, Inc., P.O. Box 498, Cataumet, Massachusetts 02534, U.S.A. **1982**.
82 G Garret, J.: Oceanographic application of drifting buoys. Report to the intergovernmental oceanographic commission on oceanographic drifting buoys. SCOR working group 66 **1982**. Division of Marine Sciences Unesco, Place de Fontenoy, 75700-Paris, France.
82 K Krauss, W., Meincke, J.: Nature, **296** (1982) 737.
82 P Perkin, R.G., Lewis, E.L.: Can. Tech. Rep. Hydrogr. Ocean Sci. **7** (1982) 47 pp.
82 S 1 Sanford, T.B.: J. Mar. Res. **40** (1982) Supplement, 621.
82 S 2 Sanford, T.B., Spain, P.F.: Meteor Cruise 57: Scientific Results Applied Physics Laboratory, University of Washington APL-UW 8213, **1982**.
82 S 3 Schröder, M.: Ber. Inst. Meereskd., Universität Kiel, No. **108** (1982) 164 pp.
82 S 4 Seaver, A.G., Kuleshov, S.: J. Phys. Oceanogr. **12** (1982) 592.
82 W Wearn, R.B., jr., Larson, N.G.: Deep-Sea Res. **29** (1 A) (1982) 111.
83 B Booth, D.A., Richie, D.: SMBA Satellite tracked buoy and drogue. Marine Physics Group Report No. 22. **1983**. Scottish Marine Biological Association, Dunstaffnage Marine Research Laboratory, P.O. Box 3, Oban, Argyll, Scotland.
83 D Dauphinee, T.M.: IEEE J. Ocean Eng **OE-8(3)** (1983) 184.
83 F Freymuth, P.: A bibliography of thermal anemometry (1983 Addendum), TSI Corporation, St. Paul, Minnesota 55164 – U.S.A. **1983**.

83 H Heinmiller, R.H., Ebbesmeyer, C.C., Taft, B.A., Olson, D.B., Nikitin, O.P.: Deep-Sea Res. **30** (1983) 1185.

83 M Meerestechnik-Elektronik GmbH: Kiel Multisonde, Kompaktsystem, ME Meerestechnik-Elektronik GmbH, Postfach 1129, 2351 Trappenkamp, FRG, Instrument brochures **1983**, photographs 2.5.16, 2.5.19.

83 R ROSENMOUNT Engineering Company: Product data sheet, P.O. Box 35129, Minneapolis, Minnesota 55435, U.S.A. **1983**.

83 R 1 Rossby, H.T., Dorson, D.: Deep-Sea Res. **30** (12A) (1983) 1279.

83 S Sippican: Model SA-180XBT Data Acquisition System Instrument Brochure, Sippican Ocean Systems, Inc., Seven Barnabas Road, Marion, MA 02738, USA **1983**.

84 D DISA: Laser-Doppler anemometry and instrument brochures, DISA Elektronik GmbH, Schinnrainstr. 9, 7500 Karlsruhe, FRG **1984**.

84 F Finke, M.: Ber. Inst. Meereskd., Universität Kiel, No. **108** (1984) 96 pp.

84 G 1 General Oceanics: Niskin winged current meter, Company Brochure, 1295 N.W. 163rd Street, Miami, Florida 33169, U.S.A. **1984**.

84 G 2 Green, A.W.: Deep-Sea Res. **31** (1984) 415.

84 I Inter Ocean Systems: Instrument brochures. 3540 Aero Court, San Diego, California, 92123, U.S.A. **1984**.

84 K Kroebel, W., von Bosse, N.: A new concept of very fast and stable conductivity probes of small size for modern high-speed CTD-systems, Proc. Interocean **1984**, p. 214–218.

84 M Meerestechnik-Elektronik GmbH: Strömungsmesser SM 21 F, vorläufiges Datenblatt, ME Meerestechnik-Elektronik GmbH, Postfach 1129, 2351 Trappenkamp, FRG **1984**.

84 N 1 Neil Brown Instrument System Inc.: Vector averaging acoustic current meter, Neil Brown Instrument Systems Inc., P.O. Box 498, Cataumet, Massachusetts 02534, U.S.A., instrument brochure **1984**.

84 N 2 Norddeutsche Seekabelwerke AG: Technische Informationsmappe „Sondensystem“, Postfach 80, 2890 Nordenham, FRG **1984**.

84 S 1 Sippican ocean systems Inc.: A system for the calibration of expandable conductivity, temperature and depth (XCTD) probes, Horizon Vol. **8** No. 3 (1984); Sippican Ocean Systems, INC., Seven Barnabas Road, Marion MA 02738, USA.

84 S 2 Striggow, K., Dankert, R.: Beitr. Meereskd. **50** (1984) 25.

84 S 3 Saure, G.: Verhalten der Freifallprofilsonde FPS: Beobachtungs- und Meßmethoden 1984, Ber. Institut für Meereskunde, Kiel: Christian-Albrechts-Universität **1984**, 68 p.

84 T TSI: Laser velocimeter systems and instrument brochures, TSI Deutschland Inc., Schmiedstr. 3, 5100 Aachen, FRG **1984**.

84 T 1 TSI (1984): Hot wire/hot film anemometry, probes and accessories, TSI Deutschland Inc., Schmiedstr. 3, 5100 Aachen, FRG, **1984**.

84 W Wiedemann, J.: Laser-Doppler-Anemometrie, Berlin, Heidelberg, New York, Tokyo: Springer **1984**.

85 A Arbeits- und Ergebnisbericht 1983–1985, SFB 133, Kiel: Christian-Albrechts-Universität **1985**, p. 149.

85 F 1 Fischer, J., Meinke, C., Minnett, P.J., Rehberg, V., Strass, V.: A description of the Institut für Meereskunde Schleppfisch-System. Technischer Bericht Nr. 1 der Abt. Regionale Ozeanographie des Instituts für Meereskunde, Kiel **1985**, 2. Auflage.

85 L Leach, H., Minnett, P.J., Woods, J.D.: Deep-Sea Res. **32** (1985) 575.

85 N Neil Brown Instrument System, Inc. Mark III CTD system: The smart instruments. Neil Brown Instrument Systems Inc., 1140 Route 28A, Cataumet, Massachusetts, 02534, U.S.A., Instrument brochures **1985**.

85 R Rossby, H.T., Levine, E.R., Connors, D.N.: Progr. Oceanogr. **14** (1985) 511.

85 S Service Argos: Monitoring Marine Buoys. Newsletter No. 23 **1985**. Service Argos, Centre National D'Études Spatiales, 18 Av. Édouard-Belin, 31055 Toulouse Cédex France.

85 S 1 Society for Underwater Technology: Advances in Underwater Technology and Offshore Engineering, Vol. 4, London: Graham and Trotman **1985**, 267 pp.

86 E Emery, W.J., Lee, W., Zenk, W., Meincke, J.: J. Atmos. Oceanic Technol. **3** (1986) 75.

86 F Finke, M., Siedler, G.: J. Atmos. Oceanic Technol. **3** (2) (1986) 255.

3 Properties of sea water

3.1 Physical properties (general) of sea water

3.1.0 List of symbols and abbreviations

a	surface tension, in [$N m^{-1}$]
C	electrical conductivity, as function of salinity, temperature, and pressure: $C(S, T, p)$, in [$\Omega^{-1} cm^{-1}$]
Cl	chlorinity, in [‰]
Cl_v	volume chlorinity, in [$kg m^{-3}$]
c	sound velocity
c_p	specific heat at constant pressure, in [$J kg^{-1} K^{-1}$]
c_v	specific heat at constant volume, in [$J kg^{-1} K^{-1}$]
D	Fickian diffusion coefficient, in [$m^2 s^{-1}$]
D_S	Soret coefficient, in [K^{-1}]
E_i	Chebyshev polynomial of degree i
g	gravitational acceleration
I	ionic strength
J	flux density of substance, in [$kg m^{-2} s^{-1}$]
K	secant bulk modulus, in [dbar]
K_{15}	electrical conductivity ratio of sea water and standard KCl solution, $K_{15}=C(S, 15, 0)/C(35, 15, 0)$
L	latent heat, heat of evaporation, in [$J kg^{-1}$]
M	air concentration, in [$mol m^{-3}$]
N	Brunt-Väisälä frequency
n	concentration of substance
p	pressure, in [dbar], excess over atmospheric pressure (101325 Pa)
p_r	reference pressure
p_v	vapour pressure
q	heat flux
R	electrical conductivity ratio of sea water and standard KCl solution, $R=C(S, T, p)/C(35, 15, 0)$
R_p	electrical conductivity ratio, $R_p=C(S, T, p)/C(S, T, 0)$
R_T	electrical conductivity ratio, $R_T=C(S, T, 0)/C(35, T, 0)$
R_{15}	electrical conductivity ratio of sea water and Normal Water, $R_{15}=C(S, 15, 0)/C(35, 15, 0)$
r_T	electrical conductivity ratio, $r_T=C(35, T, 0)/C(35, 15, 0)$
Δr	relative isotopic difference, eq. (24)
S	salinity
T	temperature, in [°C], differences also in [K]
T_a	thermodynamic temperature, in [K]
T_b	boiling point temperature
T_f	freezing point temperature
T_m	density maximum temperature
U'	shear, in [s^{-1}]
v	specific volume, in [$m^3 kg^{-1}$]
z	depth, in [m]
α	thermal expansion coefficient, in [K^{-1}]
Γ	adiabatic lapse rate, in [$K dbar^{-1}$]
γ	specific heat ratio, $\gamma=c_p/c_v$
δ	specific volume anomaly, $\delta=v-v(35, 0, p)$
η	entropy
Θ	potential temperature
$\varkappa_\eta$	adiabatic compressibility, in [$dbar^{-1}$]
$\varkappa_T$	isothermal compressibility, in [$dbar^{-1}$]

Siedler, Peters

λ	thermal conductivity, in [W m^{-1} K^{-1}]
μ	viscosity, in [N s m^{-2}]
ν	kinematic viscosity, in [m^2 s^{-1}]
Π	osmotic pressure, in [MPa]
ϱ	density, in [kg m^{-3}]
ϱ_m	maximum density
τ	tangential stress, in [Pa]
ϕ	geographic latitude
IAPSO	International Association for the Physical Sciences of the Ocean
IPTS	International Practical Temperature Scale
EOS	International Equation of State of Sea Water
PSS	Practical Salinity Scale
SMOW	Standard Mean Ocean Water

3.1.1 Thermodynamic variables

Most physical properties of sea water are similar to those of fresh water which can usually be described by functions of temperature and pressure. Since the composition of salt in water from the open ocean is constant to a good approximation (see sect. 3.4), the addition of the variable salinity is generally sufficient to obtain a complete set of 3 variables determining the physical properties of sea water. The effect of non-ideal composition is briefly discussed in subsect. 3.1.3. Most properties will differ if suspended material is contained in the sample. Only the pure sea water case, however, will be considered here. The selection of the above 3 variables is arbitrary and results from practical considerations related to measurement methods. Any of these variables could be replaced by others, like sound velocity or refractive index. Only material constants will be discussed here, thus disregarding properties that depend on the motion of the liquid, such as turbulent diffusion coefficients or turbulent viscosity.

The properties of fresh water are strongly influenced by the fact that the H_2O molecule presents an electric dipole. This results in a high dielectric constant and in the formation of water polymers that change in composition with temperature. Water, therefore, behaves abnormally compared to similar hydrogen compounds, e.g. having higher boiling and freezing temperatures, a greater specific heat, and a density maximum with respect to temperature. Properties of fresh water are summarized in various volumes of Landolt-Börnstein, „Zahlenwerte und Funktionen". A review of the density of fresh water is given by Kell [75 K].

Differences between fresh and sea water properties, although usually being small, may have important effects, such as the density increase, the freezing temperature decrease, and the density maximum temperature decrease with growing salinity. Electrical conductivity increases enormously when adding sea salt to fresh water, and a nonvanishing osmotic pressure only occurs with salinity differences (Dietrich et al. [80 D]). Reviews of the thermodynamics of sea water can be found in Fofonoff [62 F] and Millero [74 M, 82 M, 83 M].

The presentation of physical properties in the following subsections starts with a discussion on salinity. This basic variable had to be redefined several times in the development of oceanography due to progress in measurement techniques. The chemical aspects of this topic are discussed in sect. 3.4. The second variable, the temperature, was defined by the *International Practical Temperature Scale* of 1968 (IPTS-68), on which recent formulas and tables are based (Barber [69 B]). The third variable, the pressure, refers to the excess over atmospheric pressure, being zero at the ocean surface. The presentations of sea water properties in this section only include items that are not discussed in detail later, such as acoustical (see sect. 3.2), optical (see sect. 3.3), and sea ice properties (see ch. 8).

SI units or their derivatives are used in the presentations, following the recommendations given in the SUN Report (IAPSO [79 I]). The units, decibar [dbar] or Pascal [Pa] are taken for pressure values. A pressure change of one dbar closely corresponds to a depth change of one meter in the ocean (see subsect. 3.1.3). The functional dependence on the three variables will always be written in the order salinity, temperature, and pressure, e.g. density $\varrho(S, T, p)$, and symbols T and T_a will be used for Celsius and thermodynamic temperatures, respectively.

To facilitate the use of computers in calculating sea water properties from the basic variables, all functions are presented in formulas that can easily be programmed, and check values are provided to test these programmes.

The check values were computed using 35 bit words for real variables. Other word lengths will cause small deviations.

3.1.2 Salinity

Salinity is basically a measure of the mass of salt per unit mass of sea water. In oceanography salinity differences of order $10^{-5} \cdots 10^{-6}$ have to be resolved for identifying deep-sea water masses or for computing geostrophic currents from the density field. Data have to be referred to an internationally accepted standard with the above resolution to render accurate differences when comparing different sets of observations.

The classical definitions of salinity S and chlorinity Cl are given in sect. 3.4. Based on the assumption of the constancy of composition of sea salt, salinity was defined by Forch et al. [02 F] as a function of chlorinity which can be obtained from a volumetric titration (S and Cl in [‰]):

$$S = 0.03 + 1.805\,Cl\,. \tag{1}$$

Recently measurements showed that salinity may be obtained from electrical conductivity with nearly an order of magnitude better precision than from chlorinity. Salinity was therefore redefined in terms of the ratio R_{15} of sea water conductivity C to Normal Water conductivity at 15 °C, and chlorinity was related to salinity by a revised formula (Wooster et al. [69 W]), henceforth called ISS-69:

$$S = -0.08996 + 28.29720\,R_{15} + 12.80832\,R_{15}^2 - 10.67869\,R_{15}^3 + 5.98624\,R_{15}^4 - 1.32311\,R_{15}^5\,, \tag{2}$$

$$R_{15} = C(S, 15, 0)/C(35, 15, 0)\,, \tag{3}$$

$$S = 1.80655\,Cl \tag{4}$$

with S and Cl in [‰].

In order to eliminate ambiguities under conditions of ionic ratio variations, an international *Practical Salinity Scale* 1978 (PSS-78) was later introduced that relates salinity by the conductivity ratio to a standard KCl solution (Lewis [80 L], UNESCO/ICES/SCOR/IAPSO [81 U 1, 81 U 3]). The practical salinity S of a sample of sea water is defined in terms of the ratio K_{15} of the electrical conductivity of the sea water sample at the temperature of 15 °C and the total pressure of one standard atmosphere (101 325 Pa = 10.1325 dbar, $p = 0$) to that of a KCl solution in which the mass fraction of KCl is $32.4356 \cdot 10^{-3}$, at the same temperature and pressure. The K_{15} value exactly equal to 1 corresponds, by definition, to a practical salinity S exactly equal to 35:

$$S = a_0 + a_1 K_{15}^{1/2} + a_2 K_{15} + a_3 K_{15}^{3/2} + a_4 K_{15}^2 + a_5 K_{15}^{5/2} \tag{5}$$

with

$a_0 = 0.0080$
$a_1 = -0.1692$
$a_2 = 25.3851$
$a_3 = 14.0941$
$a_4 = -7.0261$
$a_5 = 2.7081\,.$

Range of validity:

$$2 \leqq S \leqq 42\,.$$

In situ conductivity/temperature/pressure instruments measure the value of a conductivity ratio R for ambient salinity S, temperature T, and pressure p to that of the standard KCl solution at 15 °C.

$$R = \frac{C(S, T, p)}{C(35, 15, 0)} = R_p \cdot R_T \cdot r_T\,, \tag{6}$$

where

$$R_p = \frac{C(S, T, p)}{C(S, T, 0)}\,, \quad R_T = \frac{C(S, T, 0)}{C(35, T, 0)}\,, \quad r_T = \frac{C(35, T, 0)}{C(35, 15, 0)}\,. \tag{7}$$

R_p, r_T, and R_T obtained in sequence constitute the procedure to reduce in situ instrument data. The terms are given by the following equations

$$R_p = 1 + \frac{p(e_1 + e_2 p + e_3 p^2)}{1 + d_1 T + d_2 T^2 + (d_3 + d_4 T)R}, \tag{8}$$

where

$e_1 = 2.070 \cdot 10^{-5}$ $d_1 = 3.426 \cdot 10^{-2}$

$e_2 = -6.370 \cdot 10^{-10}$ $d_2 = 4.464 \cdot 10^{-4}$

$e_3 = 3.989 \cdot 10^{-15}$ $d_3 = 4.215 \cdot 10^{-1}$

$d_4 = -3.107 \cdot 10^{-3}$

$$r_T = c_0 + c_1 T + c_2 T^2 + c_3 T^3 + c_4 T^4, \tag{9}$$

where

$c_0 = 0.6766097$

$c_1 = 2.00564 \cdot 10^{-2}$

$c_2 = 1.104259 \cdot 10^{-4}$

$c_3 = -6.9698 \cdot 10^{-7}$

$c_4 = 1.0031 \cdot 10^{-9}$

$$R_T = \frac{\mathrm{R}}{\mathrm{R}_p \mathrm{r}_T}, \tag{10}$$

$$S = a_0 + a_1 R_T^{1/2} + a_2 R_T + a_3 R_T^{3/2} + a_4 R_T^2 + a_5 R_T^{5/2} + \Delta S \tag{11}$$

with coefficients $a_0 \cdots a_5$ given by eq. (5), and

$$\Delta S = \frac{(T-15)}{1+K(T-15)}(b_0 + b_1 R_T^{1/2} + b_2 R_T + b_3 R_T^{3/2} + b_4 R_T^2 + b_5 R_T^{5/2}) \tag{12}$$

$b_0 = 0.0005$ $K = 0.0162$

$b_1 = -0.0056$

$b_2 = -0.0066$

$b_3 = -0.0375$

$b_4 = 0.0636$

$b_5 = -0.0144$

Units:

S	T	p
PSS-78	IPTS-68	
	°C	dbar

Range of validity:

$2 \leq S \leq 42$

$-2\,°\mathrm{C} \leq T \leq 35\,°\mathrm{C}$

$0 \leq p \leq 10\,000\,\mathrm{dbar}$

Only combinations of S, T, p in oceanographic range.

Check values:

R	T[°C]	p[dbar]	R_p	r_T	R_T	S
1	15	0	1.000000	1.000000	1.000000	35.00000
1.888091	40	10000	1.040399	1.613508	1.124739	39.99999

Estimates of accuracy:

Standard deviation of mathematical fit from measurements and maximum deviation in oceanographic range of values at low pressure equivalent to 0.0015, accuracy at high pressure estimated as 0.003.

Much of the data in the oceanic range of salinities archived before the introduction of PSS-78 are within ± 0.01 of salinity computed by the new formulas. In certain cases it may be necessary to convert earlier data to the new scale. Recommendations for data conversion procedures are given by Lewis and Perkin [81 L].

The variation of electrical conductivity at atmospheric pressure as a function of salinity and temperature is shown in Fig. 1, and its temperature and pressure dependence is represented for $S=35$ in Fig. 2. Different values of the absolute conductivity of standard sea water at 15 °C have been reported, $C(35, 15, 0) = 42.914 \cdot 10^{-3}\,\Omega^{-1}\,\mathrm{cm}^{-1}$ according to Culkin and Smith [80 C 2]. In these and the following figures the validity is restricted to the liquid phase (see subsect. 3.1.8).

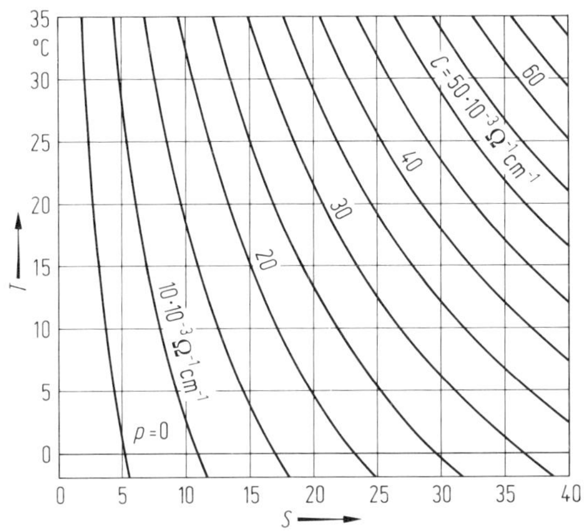

Fig. 1. Electrical conductivity C in $[10^{-3}\,\Omega^{-1}\,\mathrm{cm}^{-1}]$ as function of practical salinity and temperature at atmospheric pressure using $C(35, 15, 0) = 42.914 \cdot 10^{-3}\,\Omega^{-1}\,\mathrm{cm}^{-1}$. Contour interval $5 \cdot 10^{-3}\,\Omega^{-1}\,\mathrm{cm}^{-1}$.

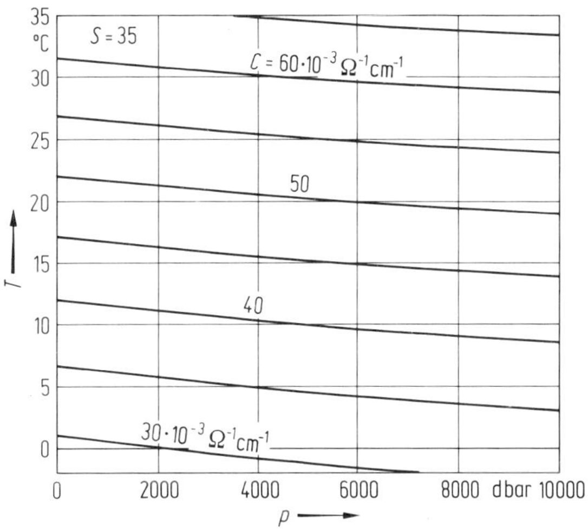

Fig. 2. Electrical conductivity C in $[10^{-3}\,\Omega^{-1}\,\mathrm{cm}^{-1}]$ as function of pressure and temperature for $S=35$ using $C(35, 15, 0) = 42.914 \cdot 10^{-3}\,\Omega^{-1}\,\mathrm{cm}^{-1}$. Contour interval $5 \cdot 10^{-3}\,\Omega^{-1}\,\mathrm{cm}^{-1}$.

3.1.3 Density

Density ϱ is defined as mass per unit volume. In the oceanographic literature the specific gravity or relative density (the dimensionless ratio of the density of a sea water sample to that of a reference fresh water sample) was mostly used and often called "density". This led to the dimensionless quantity $\sigma(S, T, p) = (\varrho/\varrho_m - 1) \cdot 10^3$, see Striggow [83 S] for a discussion. Following the SUN Report (IAPSO [79 I]) the concept of relative density should be abandoned in oceanography, and density should be expressed in $[\mathrm{kg\,m^{-3}}]$. This results in the following expressions:

$v = \frac{1}{\varrho}$ specific volume of a sea water sample, in $[m^3 kg^{-1}]$,

$v_0 = \frac{1}{\varrho_0}$ specific volume of a reference sea water, in $[m^3 kg^{-1}]$, $v_0 \equiv v(35, 0, p)$,

$\delta = v - v_0$ specific volume anomaly, in $[m^3 kg^{-1}]$.

An equivalent to the quantity σ is not defined in connection with the newly defined equation of state. The density of sea water can be related to 3 basic variables, usually salinity, temperature, and pressure. The mathematical expression for this calculation is the equation of state of sea water which has to be determined empirically. The earlier, internationally accepted equation of state has been based on direct measurements of density, chlorinity, and salinity by Forch et al. [02 F], and of sea water compressibility by Ekman [08 E]. From new measurements in different laboratories an improved equation was derived (UNESCO [81 U 2]), the *International Equation of State of Seawater*, 1980 (EOS-80). Density is given as a function of practical salinity S, temperature T, and excess pressure p. For one standard atmosphere (101325 Pa = 1.01325bar, $p=0$) the density $\varrho(S, T, 0)$ is described by the following *One Atmosphere International Equation of State of Seawater*, 1980 (Millero and Poisson [81 M 1]).

$$\begin{aligned}\varrho(S,T,0) = \varrho_w &+ (8.24493 \cdot 10^{-1} - 4.0899 \cdot 10^{-3}\,T \\ &+ 7.6438 \cdot 10^{-5}\,T^2 - 8.2467 \cdot 10^{-7}\,T^3 \\ &+ 5.3875 \cdot 10^{-9}\,T^4)S \\ &+ (-5.72466 \cdot 10^{-3} + 1.0227 \cdot 10^{-4}\,T \\ &- 1.6546 \cdot 10^{-6}\,T^2)S^{3/2} \\ &+ 4.8314 \cdot 10^{-4}\,S^2, \end{aligned} \qquad (13)$$

where ϱ_w, the density of the Standard Mean Ocean Water (SMOW) taken as pure water reference, is given by Bigg [67 B]:

$$\begin{aligned}\varrho_w = 999.842594 &+ 6.793952 \cdot 10^{-2}\,T \\ &- 9.095290 \cdot 10^{-3}\,T^2 + 1.001685 \cdot 10^{-4}\,T^3 \\ &- 1.120083 \cdot 10^{-6}\,T^4 + 6.536332 \cdot 10^{-9}\,T^5. \end{aligned} \qquad (14)$$

Units:

S PSS-78	T IPTS-68	p	ϱ, ϱ_w
	°C	dbar	$kg\,m^{-3}$

Range of validity:

$0 \leqq S \leqq 42$

$-2\,°C \leqq T \leqq 40\,°C$.

Check values:

$S = 40$, $T = 40\,°C$

$\varrho_w = 992.2204\,kg\,m^{-3}$, $\varrho = 1021.6788\,kg\,m^{-3}$.

Estimate of accuracy relative to fresh water:

Standard deviation of mathematical fit: $3.6 \cdot 10^{-3}\,kg\,m^{-3}$; differences between data by Millero et al. [76 M 2] and Poisson et al. [80 P]: typically no more than $\pm 5 \cdot 10^{-3}\,kg\,m^{-3}$, maximum deviation $-7.8 \cdot 10^{-3}\,kg\,m^{-3}$ at 30 °C and salinities between 15 and 20.

The *High Pressure International Equation of State of Seawater*, 1980 (Millero et al. [80 M]) is then given by:

$$\varrho(S,T,p) = \frac{\varrho(S,T,0)}{1 - p/K(S,T,p)}, \qquad (15)$$

where the secant bulk modulus K is obtained from:

$$K(S,T,p)=K(S,T,0)+A(S,T)p+B(S,T)p^2\,, \tag{16}$$

$$\begin{aligned}K(S,T,0)=K_w&+(546.746-6.03459\,T\\&+1.09987\cdot10^{-1}\,T^2\\&-6.1670\cdot10^{-4}\,T^3)S\\&+(7.944\cdot10^{-1}+1.6483\cdot10^{-1}\,T\\&-5.3009\cdot10^{-3}\,T^2)S^{3/2}\,,\end{aligned} \tag{17}$$

$$\begin{aligned}A(S,T)=A_w&+(2.2838\cdot10^{-3}-1.0981\cdot10^{-5}\,T\\&-1.6078\cdot10^{-6}\,T^2)S\\&+1.91075\cdot10^{-4}\,S^{3/2}\,,\end{aligned} \tag{18}$$

$$\begin{aligned}B(S,T)=B_w&+(-9.9348\cdot10^{-8}+2.0816\cdot10^{-9}\,T\\&+9.1697\cdot10^{-11}\,T^2)S\,.\end{aligned} \tag{19}$$

The pure water terms K_w, A_w, and B_w of the secant bulk modulus are given by

$$\begin{aligned}K_w=&\,1.965221\cdot10^5+1484.206\,T\\&-23.27105\,T^2+1.360477\cdot10^{-1}\,T^3\\&-5.155288\cdot10^{-4}\,T^4\,,\end{aligned} \tag{20}$$

$$\begin{aligned}A_w=&\,3.239908+1.43713\cdot10^{-3}\,T\\&+1.16092\cdot10^{-4}\,T^2-5.77905\cdot10^{-7}\,T^3\,,\end{aligned} \tag{21}$$

$$\begin{aligned}B_w=&\,8.50935\cdot10^{-6}-6.12293\cdot10^{-7}\,T\\&+5.2787\cdot10^{-9}\,T^2\,.\end{aligned} \tag{22}$$

Units:

S	T	p	ϱ, ϱ_w	v	K, K_w	A, A_w	B, B_w
PSS-78	IPTS-68						
	°C	dbar	$kg\,m^{-3}$	$m^3\,kg^{-1}$	dbar		$dbar^{-1}$

Check values:

$S=40$, $T=40\,°C$, $p=10000$ dbar

$v=9.4355613\cdot10^{-4}\,m^3\,kg^{-1}$,

$B_w=-7.536450\cdot10^{-6}\,dbar^{-1}$, $A_w=3.446154$,

$B=-2.311202\cdot10^{-6}\,dbar^{-1}$, $A=3.465376$,

$K_w=2.260440\cdot10^5$ dbar, $K(S,T,0)=2.434421\cdot10^5$ dbar,

$K(S,T,p)=2.778648\cdot10^5$ dbar.

Range of validity:

$0\leqq S\leqq 42$

$-2\,°C\leqq T\leqq 40\,°C$

$0\leqq p\leqq 10000$ dbar.

Estimate of accuracy:

Standard deviation of mathematical fit $9\cdot10^{-9}\,m^3\,kg^{-1}$, in oceanic range of values $5\cdot10^{-9}\,m^3\,kg^{-1}$.

International Oceanographic Tables for the equation of state of sea water are found in UNESCO [82 U]. The salinity dependence of the density maximum temperature is discussed in subsect. 3.1.5.

Sea water density at atmospheric pressure ($p=0$) is represented in Figs. 3 and 5, and the density of sea water with $S=35$ as a function of temperature and pressure is shown in Fig. 4.

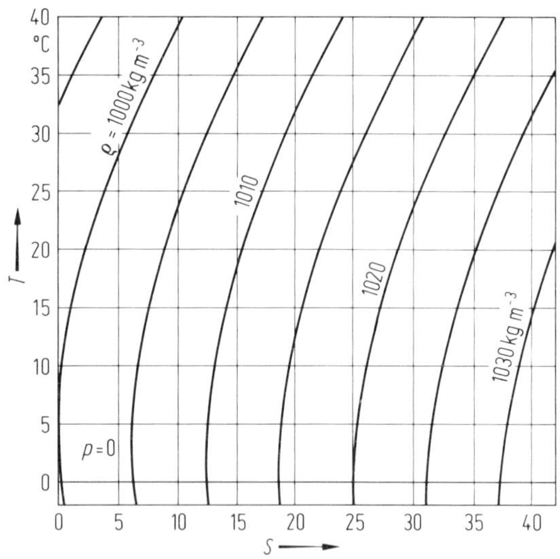

Fig. 3. Density ϱ in [$kg\,m^{-3}$] as function of salinity and temperature at atmospheric pressure. Contour interval $5\,kg\,m^{-3}$.

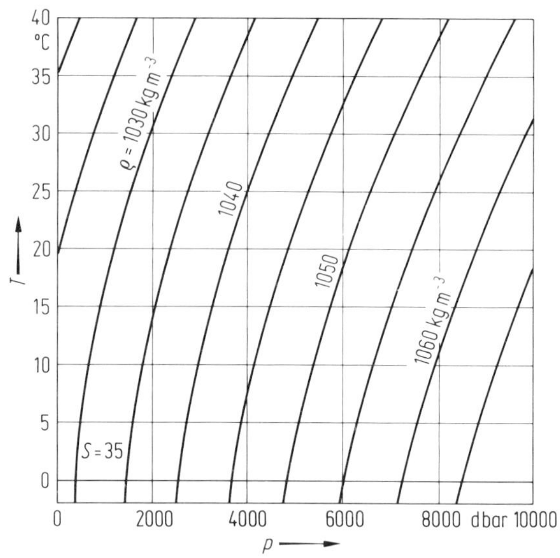

Fig. 4. Density ϱ in [$kg\,m^{-3}$] as function of pressure and temperature at $S=35$. Contour interval $5\,kg\,m^{-3}$.

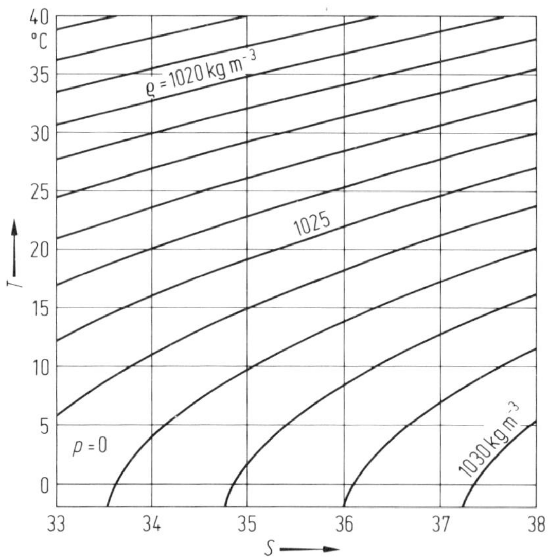

Fig. 5. Density ϱ in [$kg\,m^{-3}$] as function of salinity and temperature at atmospheric pressure, oceanic range. Contour interval $1\,kg\,m^{-3}$.

Even in the open ocean *variations in the composition* of salts cannot always be disregarded. For the North Pacific deep ocean a correction of $0.016\,kg\,m^{-3}$ to the value calculated from the algorithm given above is necessary to obtain the true density (Millero et al. [76 M 2]). In general, such corrections are much smaller. The effect of variations in composition is larger in coastal areas. In the Baltic Sea, e.g., corrections of up to $0.045\,kg\,m^{-3}$ are required (Millero and Kremling [76 M 4]; Lewis and Perkin [78 L]).

The effect of natural variations in isotopic composition on the density of water can be approximated by the following equation (Ménaché [71 M]):

$$\Delta\varrho = 2.1 \cdot 10^{-4}\,\Delta r(^{18}O/^{16}O) + 1.5 \cdot 10^{-5}\,\Delta r(D/H)\,, \qquad (23)$$

where $\Delta\varrho$ is the density difference at 4 °C between a water sample and Standard Mean Ocean Water (SMOW), in [$kg\,m^{-3}$], and Δr the isotopic difference defined by

$$\Delta r = \left(\frac{R_{sample}}{R_{standard}} - 1\right) \cdot 10^3, \tag{24}$$

with R representing, for sample and standard water, the respective isotopic ratios of, respectively, $^{18}O/^{16}O$ and D/H. In oceanic systems the density differences due to changes in the isotopic composition remain below $1 \cdot 10^{-4}\,kg\,m^{-3}$.

Dissolved air in sea water leads to a density decrease which is small at high temperatures and when the sea water is undersaturated, and is high at saturation and at low temperatures. The density decrease $\Delta\varrho$ as a function of air concentration M can be described (Millero and Emmet [76 M 1]) by:

$$\Delta\varrho = -2.6 \cdot 10^{-6} M \tag{25}$$

with $\Delta\varrho$ in [$kg\,m^{-3}$] and M in [$mol\,m^{-3}$].

For saturation, the temperature dependence of the air concentration results in a variation of $\Delta\varrho$ from $3.6 \cdot 10^{-3}$ to $1.8 \cdot 10^{-3}\,kg\,m^{-3}$ for temperatures increasing from 0 to 40 °C.

As pointed out by Fofonoff and Millard [84 F], oceanic regions of very small gradients in temperature and salinity require care in calculating the density or related properties. It is desirable to calculate the specific volume anomaly $\delta(S, T, p)$ directly and not as difference of $v(S, T, p)$ and $v(35, 0, p)$ in order to increase numerical precision. Using the notations

$$v_{35} = v(35, 0, 0) = 9.7266204 \cdot 10^{-4}\,m^3\,kg^{-1},$$
$$\varrho_{35} = \varrho(35, 0, 0) = 1028.106331\,kg\,m^{-3},$$
$$\varrho'(S, T) = \varrho(S, T, 0) - \varrho_{35},$$
$$K'(S, T, p) = K(S, T, p) - K(35, 0, p)$$

the specific volume anomaly is computed using

$$\delta(S, T, 0) = -\frac{\varrho'(S, T)v_{35}}{\varrho'(S, T) + \varrho_{35}}, \tag{26}$$

$$K(35, 0, p) = 2.158227 \cdot 10^5 + 3.359406\,p + 5.03217 \cdot 10^{-6} p^2, \tag{27}$$

$$\delta(S, T, p) = \delta(S, T, 0)\left[1 - \frac{p}{K(35, 0, p)}\right] + \frac{[v_{35} + \delta(S, T, 0)]pK'(S, T, p)}{K(35, 0, p)\,[K(35, 0, p) + K'(S, T, p)]} \tag{28}$$

with $K(35, 0, p)$ and p in [dbar].

$\varrho'(S, T)$ is calculated from eq. (13) by changing in eq. (14) for ϱ_w the coefficient 999.842594 to −28.263737. $K'(S, T, p)$ is similarly computed from eqs. (16)···(22) by changing the following coefficients from

$1.965221 \cdot 10^5$ to $-1.93006 \cdot 10^4$ in eq. (20),

3.239908 to −0.1194975 in eq. (21),

$8.50935 \cdot 10^{-6}$ to $3.47718 \cdot 10^{-6}$ in eq. (22).

Fofonoff and Millard [84 F] have developed an algorithm for the conversion of pressure to depth for the standard ocean ($S = 35$, $T = 0$ °C) which is consistent with EOS-80 [eqs. (13)ff.]. The depth z as function of geographic latitude ϕ and pressure p is given by the following formulas:

$$z = a/b, \tag{29}$$

$$a = a_1 p + a_2 p^2 + a_3 p^3 + a_4 p^4, \tag{30}$$

$$b = b_0(1 + b_1 \sin^2\phi + b_2 \sin^4\phi) + b_3 p, \tag{31}$$

where

$a_1 = 9.72659$ $b_0 = 9.780318$

$a_2 = -2.2512 \cdot 10^{-5}$ $b_1 = 5.2788 \cdot 10^{-3}$

$a_3 = 2.279 \cdot 10^{-10}$ $b_2 = 2.36 \cdot 10^{-5}$

$a_4 = -1.82 \cdot 10^{-15}$ $b_3 = 1.092 \cdot 10^{-6}$.

Units:

p	z
dbar	m

Range of validity:

$$0 \leqq p \leqq 10000\,\text{dbar}.$$

Check value:

$$p = 10000\,\text{dbar}, \quad \phi = 30°: \quad z = 9712.654\,\text{m}.$$

Estimate of accuracy:

The systematic error for the standard ocean is about 0.1 m. The correction for the actual density distribution is generally less than ± 2 m in the open ocean. Fig. 6 displays the difference $|p| - |z|$ (p in [dbar] and z in [m]) calculated from eqs. (29)···(31) of the standard ocean as function of p and ϕ.

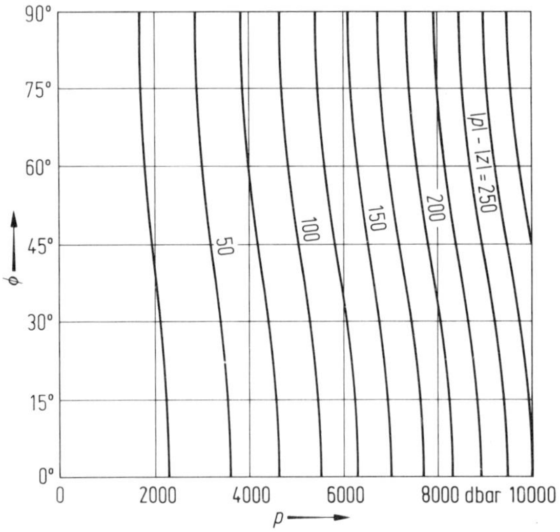

Fig. 6. Numerical difference $|p| - |z|$ of pressure p in [dbar], and depth z in [m], for the standard ocean ($S = 35$, $T = 0\,°\text{C}$) as function of pressure and geographic latitude ϕ. z is calculated from eqs. (29)···(31). Contour interval 25.

3.1.4 Compressibility

The *isothermal* compressibility $\varkappa_T$ is defined by:

$$\varkappa_T = -\frac{1}{v}\left(\frac{\partial v}{\partial p}\right)_{S,T}. \tag{32}$$

With the secant bulk modulus K (the inverse "mean isothermal compressibility") in EOS-80 [see eq. (15)] one obtains:

$$\varkappa_T(S, T, p) = \frac{v(S, T, 0)}{v(S, T, p)K(S, T, p)}\left[1 - \frac{p}{K(S, T, p)}\left(\frac{\partial K(S, T, p)}{\partial p}\right)_{S,T}\right], \tag{33}$$

where

$$\left(\frac{\partial K(S, T, p)}{\partial p}\right)_{S,T} = A(S, T) + 2pB(S, T). \tag{34}$$

The isothermal compressibility can thus be calculated from eqs. (13)···(22).

Units:

S	T	p	$\varkappa_T$
PSS-78	IPTS-68		
	°C	dbar	dbar^{-1}

Range of validity:

$0 \leq S \leq 40$

$-2\,°C \leq T \leq 40\,°C$

$0 \leq p \leq 10000$ dbar.

Check value:

$S=40$, $T=40\,°C$, $p=10000$ dbar

$\varkappa_T = 3.273850 \cdot 10^{-6}\,\mathrm{dbar}^{-1}$.

Estimate of accuracy:

Assuming an error of $\Delta\varrho = 10^{-2}\,\mathrm{kg\,m^{-3}}$ distributed over the full pressure range yields $\Delta\varkappa_T = 10^{-9}\,\mathrm{dbar}^{-1}$. This is a rough estimate. It is possible that the error increases with pressure.

With entropy η, the *adiabatic* compressibility $\varkappa_\eta$ is defined by:

$$\varkappa_\eta = -\frac{1}{v}\left(\frac{\partial v}{\partial p}\right)_{S,\eta}. \tag{35}$$

It can be obtained from the thermodynamic relation (T_a: thermodynamic temperature, α: thermal expansion coefficient at constant salinity, c_p: specific heat at constant pressure and constant salinity):

$$\varkappa_\eta = \varkappa_T - \frac{T_a v \alpha^2}{c_p}. \tag{36}$$

With $\varkappa$ in [dbar^{-1}] and SI units otherwise (i.e. T_a in [K], α in [K^{-1}], and c_p in [$\mathrm{J\,kg^{-1}\,K^{-1}}$]):

$$\varkappa_\eta = \varkappa_T - 10^4\frac{T_a v \alpha^2}{c_p}. \tag{36a}$$

The formulas for $v(S,T,p)$, $\alpha(S,T,p)$, and $c_p(S,T,p)$ are given by eqs. (13)···(22), (38)···(43), and (46), (47), respectively.

Note that $\varkappa_T/\varkappa_\eta = c_p/c_v$; see subsect. 3.1.6.

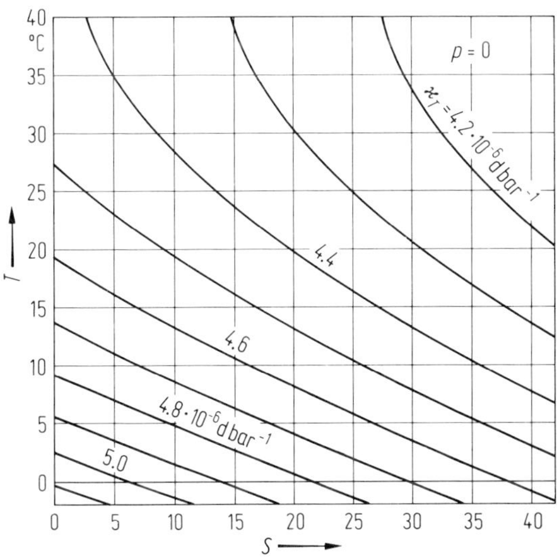

Fig. 7. Isothermal compressibility $\varkappa_T$ in [$10^{-6}\,\mathrm{dbar}^{-1}$] as function of salinity and temperature at atmospheric pressure. Contour interval $10^{-7}\,\mathrm{dbar}^{-1}$.

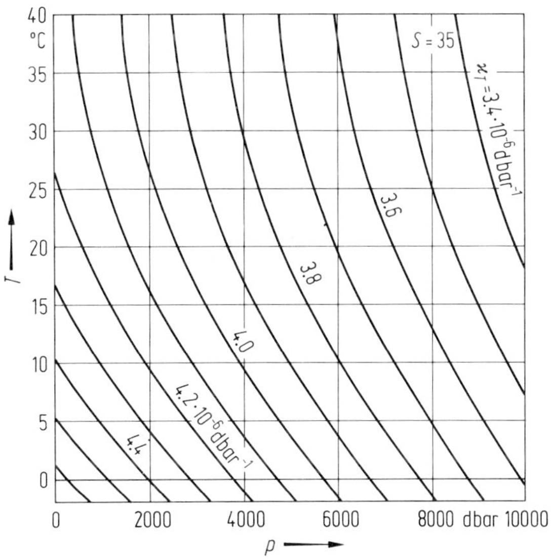

Fig. 8. Isothermal compressibility $\varkappa_T$ in [$10^{-6}\,\mathrm{dbar}^{-1}$] as function of pressure and temperature at $S=35$. Contour interval $10^{-7}\,\mathrm{dbar}^{-1}$.

Check value:

$$S=40,\quad T=40\,^{\circ}\mathrm{C},\quad p=10000\,\mathrm{dbar}$$
$$\varkappa_\eta=3.139704\cdot10^{-6}\,\mathrm{dbar}^{-1}.$$

The isothermal compressibility for atmospheric pressure is displayed in Fig. 7, and the isothermal compressibility of sea water with $S=35$ is shown in Fig. 8 as a function of temperature and pressure.

3.1.5 Thermal expansion, density maximum

The thermal expansion coefficient α is defined by:

$$\alpha=\frac{1}{v}\left(\frac{\partial v}{\partial T}\right)_{S,p}. \tag{37}$$

Introducing the secant bulk modulus K from eq. (15) leads to:

$$\alpha(S,T,p)=pK^{-2}[1-p/K(S,T,p)]^{-1}\left(\frac{\partial K(S,T,p)}{\partial T}\right)_{S,p}$$
$$-\varrho^{-1}(S,T,0)\left(\frac{\partial\varrho(S,T,0)}{\partial T}\right)_S. \tag{38}$$

The terms are obtained from EOS-80, i.e. from eqs. (13) and (14):

$$\left(\frac{\delta\varrho(S,T,0)}{\delta T}\right)_S=\sum_{i=0}^{4}a_iT^i+S\sum_{i=0}^{3}b_iT^i+S^{3/2}\sum_{i=0}^{1}c_iT^i \tag{39}$$

where

$a_0=6.793952\cdot10^{-2}$	$b_0=-4.0899\cdot10^{-3}$
$a_1=-1.819058\cdot10^{-2}$	$b_1=1.52867\cdot10^{-4}$
$a_2=3.005055\cdot10^{-4}$	$b_2=-2.47401\cdot10^{-6}$
$a_3=-4.480332\cdot10^{-6}$	$b_3=2.155\cdot10^{-8}$
$a_4=3.268166\cdot10^{-8}$	$c_0=1.0227\cdot10^{-4}$
	$c_1=-3.3092\cdot10^{-6}$

and, according to eq. (16),

$$\left(\frac{\partial K(S,T,p)}{\partial T}\right)_{S,p}=\left(\frac{\partial K(S,T,0)}{\partial T}\right)_S+\left(\frac{\partial A(S,T)}{\partial T}\right)_S p+\left(\frac{\partial B(S,T)}{\partial T}\right)_S p^2, \tag{40}$$

while eqs. (17)···(22) yield

$$\left(\frac{\partial K(S,T,0)}{\partial T}\right)_S=\sum_{i=0}^{3}d_iT^i+S\sum_{i=0}^{2}e_iT^i+S^{3/2}\sum_{i=0}^{1}f_iT^i, \tag{41}$$

where

$d_0=1.484206\cdot10^{3}$	$e_0=-6.03459$
$d_1=-4.65421\cdot10^{1}$	$e_1=2.19974\cdot10^{-1}$
$d_2=4.081431\cdot10^{-1}$	$e_2=-1.8501\cdot10^{-3}$
$d_3=-2.0621152\cdot10^{-3}$	$f_0=1.6483\cdot10^{-1}$
	$f_1=-1.06018\cdot10^{-2}$

$$\left(\frac{\partial A(S,T)}{\partial T}\right)_S=\sum_{i=0}^{2}g_iT^i+S\sum_{i=0}^{1}h_iT^i, \tag{42}$$

where

$$g_0 = 1.43713 \cdot 10^{-3} \qquad h_0 = -1.0981 \cdot 10^{-5}$$
$$g_1 = 2.32184 \cdot 10^{-4} \qquad h_1 = -3.2156 \cdot 10^{-6}$$
$$g_2 = -1.733715 \cdot 10^{-6}$$

$$\left(\frac{\partial B(S,T)}{\partial T}\right)_S = \sum_{i=0}^{1} k_i T^i + S \sum_{i=0}^{1} l_i T^i \qquad (43)$$

where

$$k_0 = -6.12293 \cdot 10^{-7} \qquad l_0 = 2.0816 \cdot 10^{-9}$$
$$k_1 = 1.05574 \cdot 10^{-8} \qquad l_1 = 1.83394 \cdot 10^{-10}$$

Units:

S	T	p	α
PSS-78	IPTS-68		
	°C	dbar	K^{-1}

Range of validity:

$$0 \leq S \leq 40$$
$$-2\,°C \leq T \leq 40\,°C$$
$$0 \leq p \leq 10000\ \text{dbar}.$$

Check value:

$S = 40$, $T = 40\,°C$, $p = 10000$ dbar

$\alpha = 4.1810597 \cdot 10^{-4}\,K^{-1}$.

Estimate of accuracy:

Assuming an error of $\Delta\varrho = 5 \cdot 10^{-3}\,kg\,m^{-3}$ occurring over an interval of 10 °C (see subsect. 3.1.3) yields $\Delta\alpha = 5 \cdot 10^{-7}\,K^{-1}$.

This is a rough estimate. Also see below.

The thermal expansion coefficient of sea water for atmospheric pressure is represented in Fig. 9, the temperature and pressure dependence of $S = 35$ sea water is given in Fig. 10.

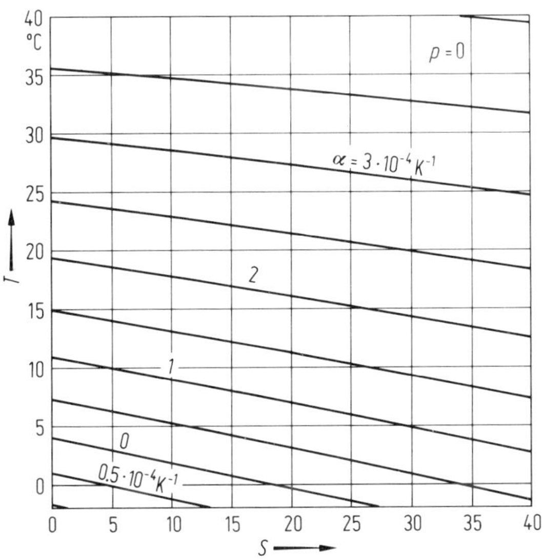

Fig. 9. Thermal expansion coefficient α in $[10^{-4}\,K^{-1}]$ as function of salinity and temperature at atmospheric pressure. Contour interval $5 \cdot 10^{-5}\,K^{-1}$.

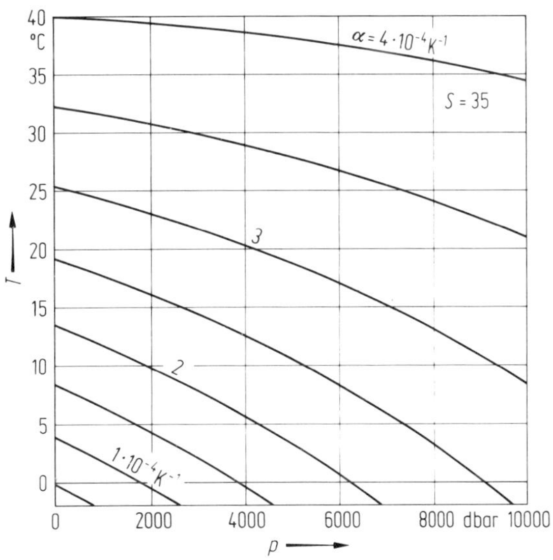

Fig. 10. Thermal expansion coefficient α in $[10^{-4}\,K^{-1}]$ as function of pressure and temperature for $S = 35$. Contour interval $5 \cdot 10^{-5}\,K^{-1}$.

For a number of reasons the above equation of state cannot be expected to yield accurate values of the density maximum temperature. The principal problem is the slow change of density and specific volume with temperature near ϱ_m. It is therefore suggested to follow Caldwell [78 C], who used the fact that, because of the thermal expansion coefficient α becoming zero, the adiabatic lapse rate Γ (see subsect. 3.1.7) also becomes zero at the density maximum temperature,

$$\Gamma = \frac{T_a v}{c_p} \alpha . \qquad (44)$$

A measurement of Γ versus pressure for given S and T leads to the pressure at which Γ passes through zero and thus to the density maximum temperature T_m for a certain S and the respective p. The following formula is derived for the density maximum temperature:

$$T_m = T_0 - aS - bp(1 + cS)(1 + dp) , \qquad (45)$$

where

$T_0 = 3.982$

$a = 0.2229$

$b = 2.004 \cdot 10^{-2}$

$c = 3.76 \cdot 10^{-3}$

$d = 4.02 \cdot 10^{-5}$.

Units:

S ISS-69	p	T_m
	dbar	°C

Range of validity:

$0 \leqq S \leqq 30$

$0 \leqq p \leqq 4000$ dbar.

Check value:

$S = 25$, $p = 1500$ dbar

$T_m = -5.0773644$ °C.

Estimate of accuracy:
Standard deviation of data from formula 0.044 °C.

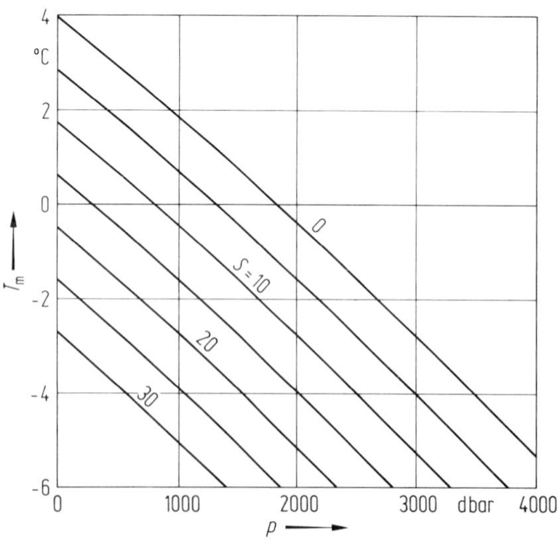

Fig. 11. Density maximum temperature T_m as function of pressure for various salinities between $S = 0$ and $S = 30$.

The values of T_m as derived from Caldwell's formula coincide with those from EOS-80 at $S=0$, $p=0$ and are lower by 0.133 °C at $S=30$, $p=0$. At elevated pressure the differences are even larger. The error introduced by the different salinity scales is only of the order 0.01 °C. The corresponding density maximum temperature T_m is shown in Fig. 11 as a function of pressure p for various salinity values.

3.1.6 Specific heat

The specific heat is the amount of heat required to raise the temperature of a unit mass by one unit of temperature. The values given by Millero et al. [73 M] are based on measurements of the specific heat at *constant pressure* c_p as a function of chlorinity and temperature. Converting with $S=1.80655\,Cl$ (eq. (4)) Millero et al. [81 M 2] arrived at the following formulas for specific heat at atmospheric pressure:

$$c_p(S,T,0)=\sum_{i=0}^{4} a_i T^i + S\sum_{i=0}^{2} b_i T^i + S^{3/2}\sum_{i=0}^{2} c_i T^i, \tag{46}$$

where

$a_0=\ 4.2174\cdot 10^3$	$b_0=-7.643575$	$c_0=\ 0.1770383$
$a_1=-3.720283$	$b_1=\ 0.1072763$	$c_1=-4.07718\cdot 10^{-3}$
$a_2=\ 0.1412855$	$b_2=-1.38385\cdot 10^{-3}$	$c_2=\ 5.148\cdot 10^{-5}$
$a_3=-2.654387\cdot 10^{-3}$		
$a_4=\ 2.093236\cdot 10^{-5}$		

Units:

S ISS-69	T IPTS-68	c_p
	°C	$\mathrm{J\,kg^{-1}\,K^{-1}}$

Range of validity:

$0\leqq S\leqq 40$

$0\leqq T\leqq 35\,°\mathrm{C}$.

Check value:

$S=40$, $T=35\,°\mathrm{C}$

$c_p=3980.0513\ \mathrm{J\,kg^{-1}\,K^{-1}}$.

Estimate of accuracy:

Standard deviation of data from equation $0.5\ \mathrm{J\,kg^{-1}\,K^{-1}}$; typical deviation from former investigations $2\ \mathrm{J\,kg^{-1}\,K^{-1}}$.

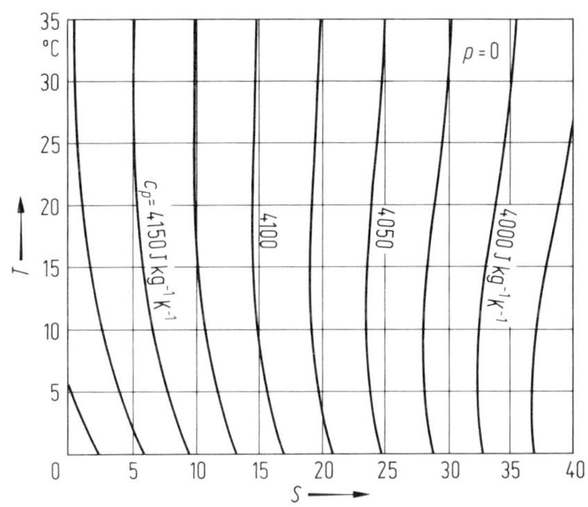

Fig. 12. Specific heat at constant pressure c_p in $[\mathrm{J\,kg^{-1}\,K^{-1}}]$ as function of salinity and temperature at atmospheric pressure. Contour interval $25\ \mathrm{J\,kg^{-1}\,K^{-1}}$.

The dependence of c_p at atmospheric pressure on salinity and temperature is given in Fig. 12.

The pressure dependence of c_p has not been measured directly, but can be obtained from the integration of

$$\left(\frac{\partial c_p}{\partial p}\right)_{S,T} = -T\left(\frac{\partial^2 v}{\partial T^2}\right)_{S,p}.$$

With $v(S, T, p)$ from EOS-80 (eqs. (13)···(22)), Fofonoff and Millard [84 F] obtained:

$$\begin{aligned} c_p(S,T,p) = {} & c_p(S,T,0) + p\sum_{i=0}^{4} a_i T^i \\ & + p^2 \sum_{i=0}^{4} b_i T^i + p^3 \sum_{i=0}^{3} c_i T^i \\ & + Sp \sum_{i=0}^{4} d_i T^i + S^{3/2} p \sum_{i=0}^{2} e_i T^i \\ & + Sp^2 \sum_{i=0}^{3} f_i T^i + g_0 S^{3/2} p^2 \\ & + Sp^3 \sum_{i=0}^{2} h_i T^i + k_1 T S^{3/2} p^3 . \end{aligned} \tag{47}$$

$a_0 = -4.9592 \cdot 10^{-2}$
$a_1 = 1.45747 \cdot 10^{-3}$
$a_2 = -3.13885 \cdot 10^{-5}$
$a_3 = 2.0357 \cdot 10^{-7}$
$a_4 = 1.7168 \cdot 10^{-9}$

$b_0 = 2.4931 \cdot 10^{-6}$
$b_1 = -1.08645 \cdot 10^{-7}$
$b_2 = 2.87533 \cdot 10^{-9}$
$b_3 = -4.0027 \cdot 10^{-11}$
$b_4 = 2.2956 \cdot 10^{-13}$

$c_0 = -5.422 \cdot 10^{-11}$
$c_1 = 2.6380 \cdot 10^{-12}$
$c_2 = -6.5637 \cdot 10^{-14}$
$c_3 = 6.136 \cdot 10^{-16}$

$d_0 = 4.9247 \cdot 10^{-4}$
$d_1 = -1.28315 \cdot 10^{-5}$
$d_2 = 9.802 \cdot 10^{-8}$
$d_3 = 2.5941 \cdot 10^{-9}$
$d_4 = -2.9179 \cdot 10^{-11}$

$e_0 = -1.2331 \cdot 10^{-5}$
$e_1 = -1.517 \cdot 10^{-7}$
$e_2 = 3.122 \cdot 10^{-9}$

$f_0 = -2.9558 \cdot 10^{-8}$
$f_1 = 1.17054 \cdot 10^{-9}$
$f_2 = -2.3905 \cdot 10^{-11}$
$f_3 = 1.8448 \cdot 10^{-13}$

$g_0 = 9.971 \cdot 10^{-10}$

$h_0 = 5.540 \cdot 10^{-13}$
$h_1 = -1.7682 \cdot 10^{-14}$
$h_2 = 3.513 \cdot 10^{-16}$.

$k_1 = -1.4300 \cdot 10^{-15}$

S ISS-69	T IPTS-68	p	c_p
	°C	dbar	$\mathrm{J\,kg^{-1}\,K^{-1}}$

Range of validity:

$0 \leqq S \leqq 40$

$0 \leqq T \leqq 35\,°\mathrm{C}$

$0 \leqq p \leqq 10000$ dbar.

Check value:

$S = 40$, $T = 40\,°\mathrm{C}$, $p = 10000$ dbar

$c_p(S, T, p) = 3940.4995\ \mathrm{J\,kg^{-1}\,K^{-1}}$.

Estimate of accuracy:

The standard deviation of the polynomial from the direct integration is less than 0.1 $J\,kg^{-1}\,K^{-1}$. An estimate of systematic errors is not available in the original literature.

$c_p(35, T, p)$ is represented graphically in Fig. 13.

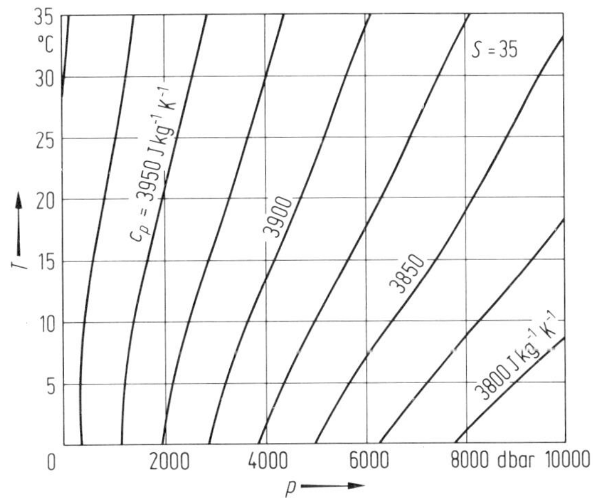

Fig. 13. Specific heat at constant pressure c_p in $[J\,kg^{-1}\,K^{-1}]$ as function of pressure and temperature for $S=35$. Contour interval 25 $J\,kg^{-1}\,K^{-1}$.

The specific heat at *constant volume* c_v can be obtained from the thermodynamic relation

$$c_v = c_p - \frac{T_a v \alpha^2}{\varkappa_T}. \tag{48}$$

For $\varkappa_T$ in $[dbar^{-1}]$ and SI-units otherwise (i.e. T_a in [K], v in $[m^3\,kg^{-1}]$, α in $[K^{-1}]$, and c_p, c_v in $[J\,kg^{-1}\,K^{-1}]$:

$$c_v = c_p - 10^4\,\frac{T_a v \alpha^2}{\varkappa_T}. \tag{49}$$

The corresponding formula results from introducing the specific volume $v(S, T, p)$ corresponding to eqs. (13)···(22), the thermal expansion coefficient $\alpha(S, T, p)$, eqs. (38)···(43), and the isothermal compressibility $\varkappa_T(S, T, p)$, eq. (33).

Check values:

$S=40$, $T=35\,°C$, $p=0$ dbar

$c_v = 3876.9516\,J\,kg^{-1}\,K^{-1}$,

$S=40$, $T=35\,°C$, $p=10000$ dbar

$c_v = 3690.2412\,J\,kg^{-1}\,K^{-1}$.

Estimate of accuracy:

With the difference of c_p and c_v being small, c_v has the same uncertainty as c_p.

The ratio of specific heats $\gamma = c_p/c_v$ is given by

$$\gamma = \left[1 - \frac{T_a v \alpha^2}{\varkappa_T c_p}\right]^{-1} \tag{50}$$

and can be combined from the respective eqs. (13)···(22), (38)···(43), and (33). For the units used in this section a factor of 10^4 occurs with the last term in eq. (50).

Check value:

$S=40$, $T=35\,°C$, $p=10000$ dbar

$\gamma = 1.0341112$.

Estimate of accuracy:

The accuracy may be roughly estimated as $\Delta\gamma = 10^{-4}$ using $\Delta\alpha/\alpha = 5\cdot 10^{-3}$, the largest error occurring in eq. (50).

The ratio γ for atmospheric pressure is represented in Fig. 14, and for $S=35$ at varying temperature and pressure in Fig. 15.

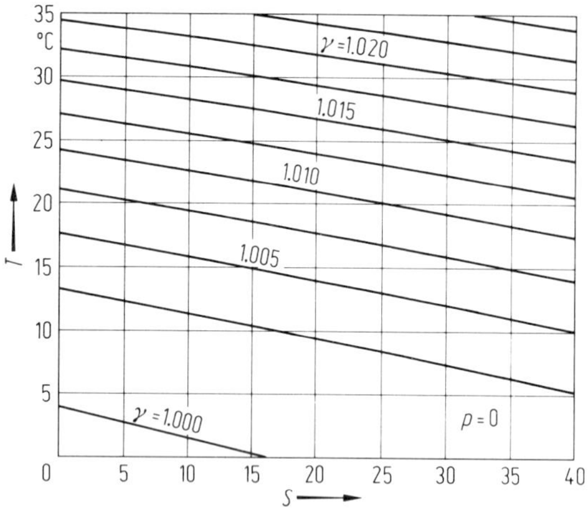

Fig. 14. Ratio of specific heats at constant pressure and constant volume $\gamma=c_p/c_v$ as function of salinity and temperature at atmospheric pressure. Contour interval $2.5 \cdot 10^{-3}$.

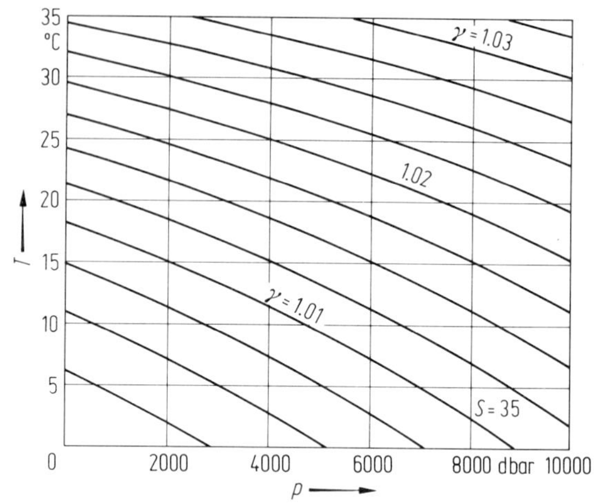

Fig. 15. Ratio of specific heats at constant pressure and constant volume $\gamma=c_p/c_v$ as function of pressure and temperature at $S=35$. Contour interval $2.5 \cdot 10^{-3}$.

3.1.7 Adiabatic lapse rate, potential temperature

The adiabatic lapse rate Γ is the change of temperature due to a unit adiabatic pressure change. It can be obtained from direct measurements or from:

$$\Gamma = \frac{T_a}{c_p}\left(\frac{\partial v}{\partial T}\right)_{S,p}. \tag{51}$$

The formula of Bryden [73 B] resulting from the above equation is given here because it is consistent with EOS-80 (see subsect. 3.1.3).

$$\Gamma(S,T,p) = \sum_{i=0}^{3} a_i T^i + (b_0 + b_1 T)(S-35) + p\left[\sum_{i=0}^{3} c_i T^i + (d_0 + d_1 T)(S-35)\right] + p^2 \sum_{i=0}^{2} e_i T^i \tag{52}$$

$a_0 = 3.5803 \cdot 10^{-5}$ $\quad c_0 = 1.8741 \cdot 10^{-8}$ $\quad e_0 = -4.6206 \cdot 10^{-13}$

$a_1 = 8.52587 \cdot 10^{-6}$ $\quad c_1 = -6.7795 \cdot 10^{-10}$ $\quad e_1 = 1.8676 \cdot 10^{-14}$

$a_2 = -6.83605 \cdot 10^{-8}$ $\quad c_2 = 8.7330 \cdot 10^{-12}$ $\quad e_2 = -2.1687 \cdot 10^{-16}$

$a_3 = 6.6228 \cdot 10^{-10}$ $\quad c_3 = -5.4481 \cdot 10^{-14}$

$b_0 = 1.8932 \cdot 10^{-6}$ $\quad d_0 = -1.1351 \cdot 10^{-10}$

$b_1 = -4.23935 \cdot 10^{-8}$ $\quad d_1 = 2.7759 \cdot 10^{-12}$

Units:

S	T	p	Γ
ISS-69	°C	dbar	K dbar^{-1}

Range of validity:

$$30 \leqq S \leqq 40,$$
$$2\,^\circ\mathrm{C} \leqq T \leqq 30\,^\circ\mathrm{C},$$
$$0 \leqq p \leqq 10000\ \mathrm{dbar}.$$

Check value:

$$S=40, \quad T=30\,^\circ\mathrm{C}, \quad p=10000\ \mathrm{dbar}$$
$$\Gamma=2.8774579 \cdot 10^{-4}\ \mathrm{K\ dbar^{-1}}.$$

Estimate of accuracy:

Systematic errors range from $\pm 2.7 \cdot 10^{-6}$ K dbar^{-1} to $\pm 5.3 \cdot 10^{-6}$ K dbar^{-1}.

Caldwell and Eide [80 C 1] have given an algorithm for Γ based on direct measurements which covers a larger range of validity.

The potential temperature Θ is the temperature resulting from an adiabatic displacement of an element of sea water at pressure p and temperature T to a reference pressure p_r. It can be obtained from

$$\Theta(S, T, p, p_r) = T + \int_p^{p_r} \Gamma(S, T', p')dp' \qquad (53)$$

with integration along an adiabat, using Γ as given above. T' and p' are temperature and pressure, respectively, along the adiabatic line. It is common to select $p_r=0$. Following Fofonoff [77 F], a Runge-Kutta integration algorithm is given which is superior to the available polynomial relation (Bryden [73 B]).

Potential temperature is given by:

$$\Delta\theta_1 = \Delta p \Gamma(S, T, p) \qquad \theta_1 = T + \Delta\theta_1/2$$
$$\Delta\theta_2 = \Delta p \Gamma(S, \theta_1, p+\Delta p/2) \qquad \theta_2 = \theta_1 + (1-\sqrt{2}/2)(\Delta\theta_2 - q_1)$$
$$\Delta\theta_3 = \Delta p \Gamma(S, \theta_2, p+\Delta p/2) \qquad \theta_3 = \theta_2 + (1+\sqrt{2}/2)(\Delta\theta_3 - q_2)$$
$$\Delta\theta_4 = \Delta p \Gamma(S, \theta_3, p+\Delta p/2) \qquad \theta_4 = \theta_3 + (\Delta\theta_4 - 2q_3)/6$$

$$q_1 = \Delta\theta_1 \qquad (54)$$
$$q_2 = (2-\sqrt{2})\Delta\theta_2 + \left(-2+\frac{3}{\sqrt{2}}\right)q_1$$
$$q_3 = (2+\sqrt{2})\Delta\theta_3 + \left(-2-\frac{3}{\sqrt{2}}-q_2\right)$$
$$\Delta p = p_r - p.$$

Then the potential temperature Θ is obtained from:

$$\Theta(S, T, p, p_r) = \theta_4.$$

Units:

Use consistent units, e.g. p in [dbar], Γ in [K dbar^{-1}].

Range of validity:

Δp may exceed 10000 dbar.

Check value [using Γ from eq. (52)]:

$$S=40, \quad T=40\,^\circ\mathrm{C}, \quad p=10000\ \mathrm{dbar}, \quad p_r=0\ \mathrm{dbar}$$
$$\Theta=36.89072\,^\circ\mathrm{C}.$$

Estimate of accuracy:

Integration error is less than 10^{-4} K for $\Delta p=10000$ dbar. Systematic error in conjunction with eq. (53) is less than 0.024 K for $\Delta p=10000$ dbar.

The potential temperature difference $\Delta\Theta = \Theta - T$ for $S = 35$ and $p_r = 0$ dbar is shown in Fig. 16.

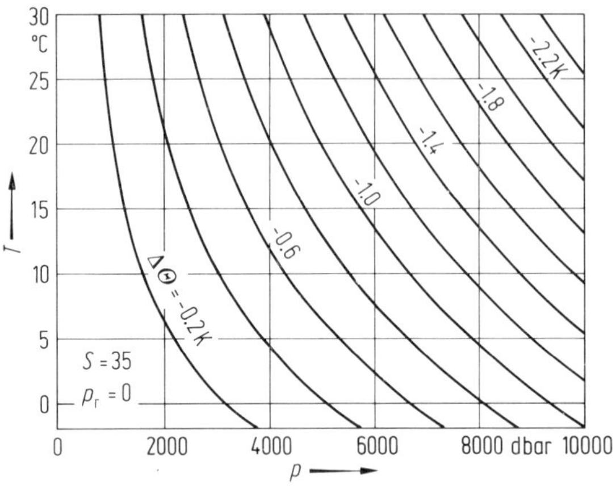

Fig. 16. Difference between potential and in-situ temperature in [K] as function of pressure and in-situ temperature; $S = 35$, $p_r = 0$. Contour interval 0.2 K.

3.1.8 Vapour pressure, boiling point, freezing point

Increasing salinity causes a lowering of vapour pressure at the water surface. This leads to a decrease in freezing temperature and an increase in boiling temperature.

An equation for the vapour pressure p_v of fresh water was derived from several investigators' data, taking account of IPTS-68, by Ambrose and Lawrenson [72 A]:

$$\log p_v = \left[\frac{a_0}{2} + \sum_{i=1}^{11} a_i E_i(x)\right] T_a^{-1}, \tag{55}$$

$$x = [2T_a - (T_{a2} + T_{a1})]/(T_{a2} - T_{a1}), \tag{56}$$

where p_v is the vapour pressure, T_a the thermodynamic temperature, and E_i the Chebyshev polynomial of degree i

$a_0 = 2794.0144$ $a_3 = 7.6875$ $a_6 = 0.1371$ $a_9 = 0.0200$

$a_1 = 1430.6181$ $a_4 = -0.0328$ $a_7 = 0.0629$ $a_{10} = 0.0117$

$a_2 = -18.2465$ $a_5 = 0.2728$ $a_8 = 0.0261$ $a_{11} = 0.0067$

$T_{a2} = 648$ K, $T_{a1} = 273$ K.

Units:

T_a IPTS-68	p_v
K	kPa or 10^{-1} dbar

Range of validity:

$273.15\,\text{K} \leqq T_a \leqq 647.31\,\text{K}$.

Check value:

$T_a = 647.31$ K

$p_v = 2.2106469 \cdot 10^4$ kPa.

Estimate of accuracy:
Not available in original literature.

The change Δp_v in vapour pressure due to sea salt is described by Millero [83 M]:

$$\Delta p_v = AS + BS^{3/2} + CS^2 , \quad (57)$$

$$A = -3.7433 \cdot 10^{-3} - 1.6537 \cdot 10^{-4}\, T - 1.9667 \cdot 10^{-6}\, T^2 - 2.4350 \cdot 10^{-7}\, T^3 , \quad (58)$$

$$B = 5.2556 \cdot 10^{-4} - 7.7266 \cdot 10^{-6}\, T^3 , \quad (59)$$

$$C = -4.9535 \cdot 10^{-5} . \quad (60)$$

Units:

S ISS-69	T IPTS-68	Δp_v
	°C	Pa or 10^{-4} dbar

Range of validity:

$0 \leqq T \leqq 40$ °C

$0 \leqq S \leqq 40$.

Estimate of accuracy:

Standard deviation of fit: 0.13 Pa; values of Δp_v are derived indirectly, thus a number of the accuracy cannot easily be given.

Check value:

$S = 40, \quad T = 40$ °C

$\Delta p_v = -1.5836562 \cdot 10^2$ Pa .

The graph in Fig. 17 represents the change in vapour pressure as a function of salinity and temperature.

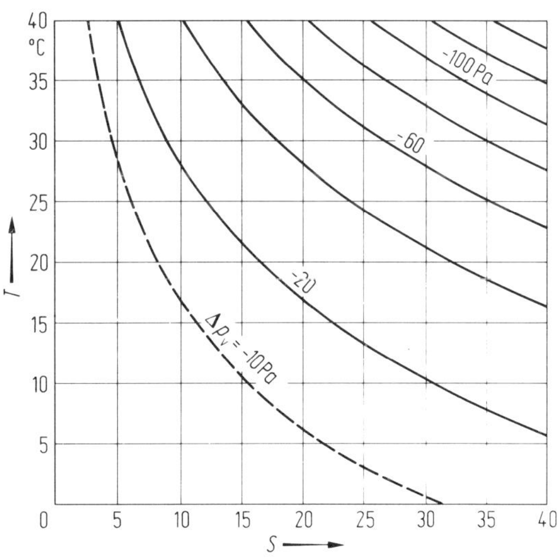

Fig. 17. Lowering of vapour pressure due to sea salt in [Pa] as function of salinity and temperature. Full lines: Contour interval 20 Pa. Dashed line: $\Delta p_v = -10$ Pa.

The boiling point temperature T_b is obtained by inverting

$$p_v = p_v(S, T) \quad \text{to} \quad T_b = T_b(S, p_v)$$

which can be done numerically. For high temperatures and high salinities the boiling point elevation ΔT_b due to sea salt was given by Stoughton and Lietzke [67 S]

$$\Delta T_b = I(A + BI + CI^2) D^{-1} , \quad (61)$$

Siedler, Peters

where

$$A = 565.757\, T_a^{-1} - 9.81559 + 1.54739 \ln T_a$$
$$B = -337.178\, T_a^{-1} + 6.41981 - 0.922753 \ln T_a$$
$$C = 32.681\, T_a^{-1} - 0.55368 + 0.079022 \ln T_a$$
$$D = 266919.6\, T_a^{-2} - 379.669\, T_a^{-1} + 0.334169$$

and where the ionic strength

$$I = \frac{1.9915 \cdot 10^{-2} S}{1 - 1.00487 \cdot 10^{-3} S}. \tag{62}$$

Units:

S	T_a IPTS-68	ΔT_b
	K	K

Range of validity:

$20 \leqq S \leqq 280$,

$298\,\text{K} \leqq T_a \leqq 530\,\text{K}$.

Check value:

$S = 280$, $T_a = 530\,\text{K}$: $\Delta T_b = 15.61977\,\text{K}$.

Estimate of accuracy:

$\pm 2\%$ relative error at low temperature and low concentration, $\pm 1\%$ at $300\,\text{K} \leqq T_a \leqq 520\,\text{K}$ and $34.5 \leqq S \leqq 200$.

In determining the ionic strength, the equation given by Millero and Leung [76 M 3] for the relation between the portion of dissolved solids and salinity was introduced here.

The boiling point temperature as a function of salinity and pressure is shown in Fig. 18.

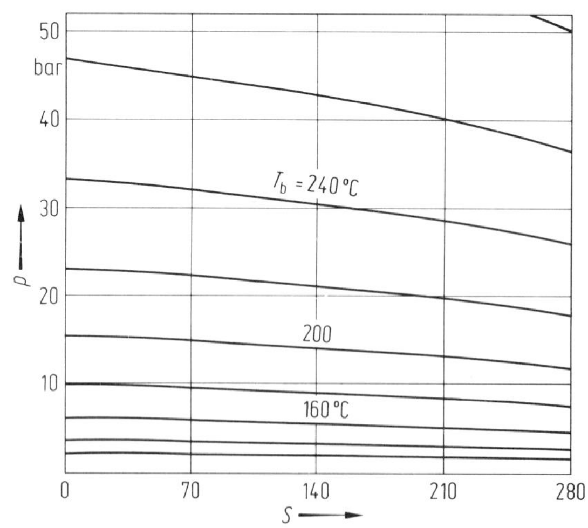

Fig. 18. Boiling point temperature T_b of concentrated sea salt solutions in [°C] as function of salinity and pressure. Contour interval 20 °C.

From several independent measurements of the freezing point dependence on salinity and pressure, Millero [78 M] selected the data of Doherty and Kester [74 D] to establish the following equation for the freezing point temperature T_f. In this case salinities correspond to ISS-69 and not to PSS-78, and the pressure dependence is sufficiently accurate only for oceanic salinities.

$$T_f = -0.0575\,S + 1.710523 \cdot 10^{-3}\,S^{3/2} - 2.154996 \cdot 10^{-4}\,S^2 - 7.53 \cdot 10^{-4}\,p\,. \qquad (63)$$

Units:

S ISS-69	p	T_f IPTS-68
	dbar	°C

Range of validity:

$0 \leqq S \leqq 40$

$0 \leqq p \leqq 500$ dbar.

Check value:

$S = 40, \quad p = 500$ dbar

$T_f = -2.588566\,°C$.

Accuracy:
$\pm 0.004\,°C$.

Fig. 19 represents the freezing point temperature as a function of salinity at atmospheric pressure and as a function of pressure for $S = 35$. The cross-over point for the freezing point temperature and the density maximum temperature T_m (see subsect. 3.1.5) at atmospheric pressure is found at $S = 23.6$ with an uncertainty of ± 0.2. When T_m is derived from EOS-80 a value of $S = 24.6$ results.

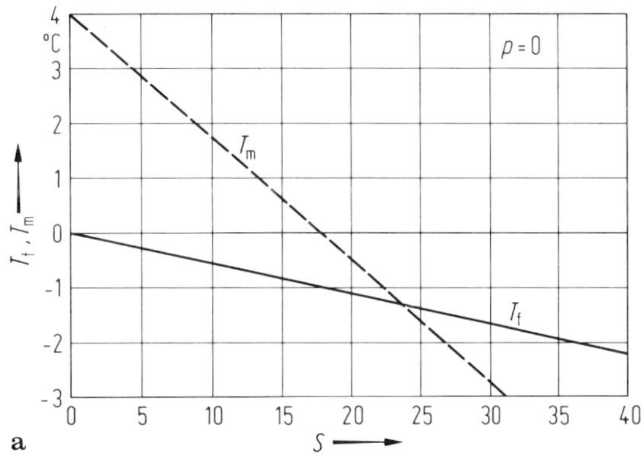

Fig. 19a. Freezing point temperature T_f as function of salinity at atmospheric pressure and maximum density temperature T_m.

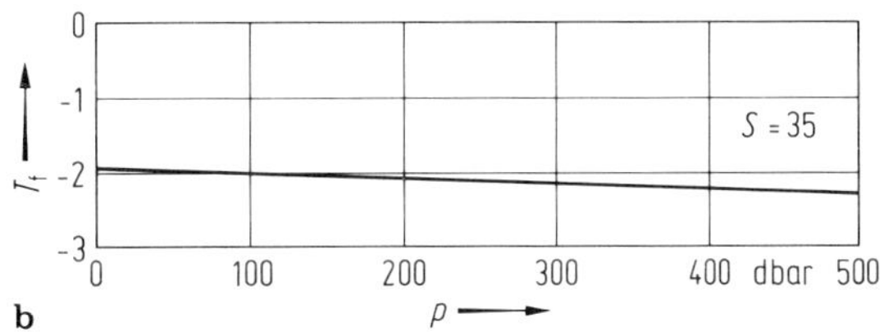

Fig. 19b. Freezing point temperature as function of pressure for $S = 35$.

3.1.9 Heat of fusion, heat of evaporation

The heat of fusion is the amount of heat required to transform a unit mass from the solid to the liquid state. The heat of fusion of fresh water at atmospheric pressure is $3.334 \cdot 10^5\,\mathrm{J\,kg^{-1}}$ according to Landolt-Börnstein, 6th ed., vol. II, part 4, p. 180 [61 S]. It is not possible to give a simple function of salinity for sea water because fusion and crystallization of sea ice do not occur at a fixed temperature, but in a range of temperatures below the freezing point (Doronin and Kheisin [77 D]). The heat required for the complete fusion of sea ice is given in ch. 8.

The heat of evaporation or latent heat L is the amount of heat required to transform a unit mass from the liquid to the gaseous state. For fresh water it depends on temperature, and a least-squares fit to data from Osborne et al. [39 O] yields the following formula for the heat of evaporation for fresh water:

$$L(T)=2.50039 \cdot 10^6-2.3683 \cdot 10^3\,T+4.31 \cdot 10^{-1}\,T^2-1.131 \cdot 10^{-2}\,T^3 \tag{64}$$

Units:

T	L
°C	$\mathrm{J\,kg^{-1}}$

Range of validity:

$$0 \leqq T \leqq 100\,°\mathrm{C}.$$

Check value:

$$T=100\,°\mathrm{C}: \quad L=2.2566 \cdot 10^6\,\mathrm{J\,kg^{-1}}.$$

Estimate of accuracy:

Osborne et al. [39 O] do not give any estimate of accuracy. The standard deviation of eq. (64) from their data is $140\,\mathrm{J\,kg^{-1}}$. According to [61 S] $L(100\,°\mathrm{C})=2.257 \cdot 10^6\,\mathrm{J\,kg^{-1}}$. The above formula is not based on IPTS-68, however, deviations caused by the choice of temperature scales are well within accuracy limits.

No formulas are available for the change of the heat of evaporation as function of salinity. The deviation $\Delta L(S, T)$ from $L(T)$ for selected values can be derived from Stoughton and Lietzke [67 S]:

$$S=34.33, \quad T=25\,°\mathrm{C}: \quad \Delta L=2.57 \cdot 10^3\,\mathrm{J\,kg^{-1}}$$
$$S=34.33, \quad T=100\,°\mathrm{C}: \quad \Delta L=3.18 \cdot 10^3\,\mathrm{J\,kg^{-1}}.$$

Their paper contains data on the minimum energy necessary for the recovery of water from sea water.

3.1.10 Thermal conductivity

The thermal conductivity λ is a measure of heat flux $\boldsymbol{q}$ per unit cross section for a given temperature gradient:

$$\boldsymbol{q}=-\lambda\,\mathrm{grad}\,T. \tag{65}$$

The following formulas for the thermal conductivity are recommended:

$$\lambda(S, T, p)=\lambda(S, T, 0)\,[1+p\,f(S, T, p)], \tag{66}$$

$$\lambda(S, T, 0)=0.565+(1.8 \cdot 10^{-3}-3.2 \cdot 10^{-6}\,S)T-9.5 \cdot 10^{-5}\,S-5.9 \cdot 10^{-6}\,T^2, \tag{67}$$

$$f(S, T, p)=6.9 \cdot 10^{-6}-8 \cdot 10^{-9}\,T-2 \cdot 10^{-15}\,p-10^{-8}\,S. \tag{68}$$

Units:

S ISS-69	T IPTS-68	p	f	λ
	°C	dbar	$\mathrm{dbar^{-1}}$	$\mathrm{W\,m^{-1}\,K^{-1}}$

Siedler, Peters

Range of validity:

$$0 \leqq S \leqq 45, \quad 0 \leqq T \leqq 60\,^\circ\mathrm{C},$$
$$0 \leqq p \leqq 14000\,\mathrm{dbar}.$$

Check value:

$$S = 45, \quad T = 60\,^\circ\mathrm{C}, \quad p = 14000\,\mathrm{dbar}$$
$$\lambda = 0.6897343\,\mathrm{W\,m^{-1}\,K^{-1}}.$$

Estimate of accuracy:

Overall accuracy $\pm 1.5\%$. Internal consistency better than $\pm 0.005\,\mathrm{W\,m^{-1}\,K^{-1}}$. The accuracy of the coefficients in eqs. (67) and (68) is low.

Eqs. (66) and (68) are given by Caldwell [74 C 1], while eq. (67) reduces to Riedel's formula (Powell [58 P]) for fresh water. For $S = 31.5$ it differs from Caldwell's formula [74 C 1] by a constant offset of $-0.004\,\mathrm{W\,m^{-1}\,K^{-1}}$. Within the accuracy stated and for the range of validity given, the above algorithm is consistent with Castelli et al. [74 C 3] and Jamieson and Tudhope [70 J]. The last paper gives a formula for $\lambda(S, T, 0)$ which covers wider ranges in S and T.

Fig. 20 displays λ as function of temperature for $p = 0$ and $S = 35$. The pressure and temperature dependence is represented in Fig. 21 for $S = 35$.

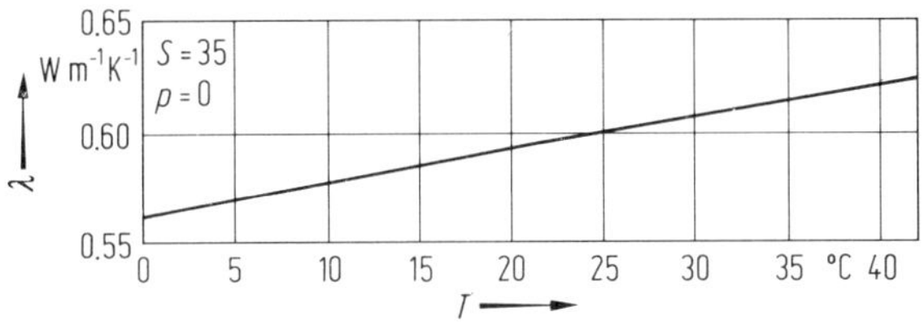

Fig. 20. Thermal conductivity λ in [$\mathrm{W\,m^{-1}\,K^{-1}}$] for $S = 35$ as function of temperature at atmospheric pressure.

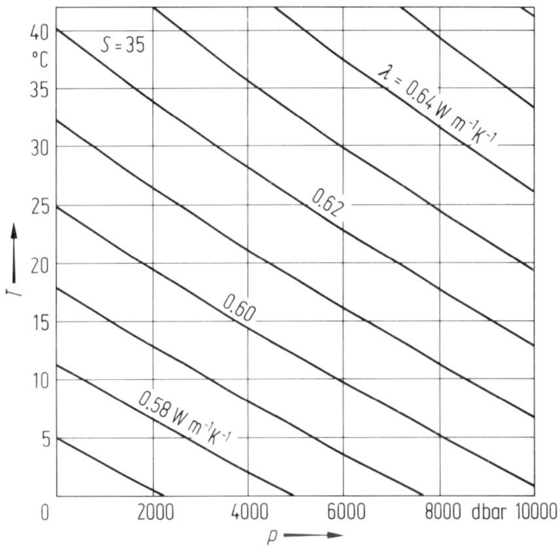

Fig. 21. Thermal conductivity λ in [$\mathrm{W\,m^{-1}\,K^{-1}}$] as function of pressure and temperature at $S = 35$. Contour interval $10^{-2}\,\mathrm{W\,m^{-1}\,K^{-1}}$.

3.1.11 Diffusion

The diffusion coefficient is a measure of the flux of substance per unit cross section. The flux depends on the concentration gradient of the substance (Fickian diffusion), but also to a lesser extent on the temperature gradient (Soret effect, thermal diffusion). The flux J of the substance for a concentration n (mass of solute per unit mass of solution) is given by:

$$\boldsymbol{J} = -\varrho D \nabla n - \varrho D D_s n (1 - n) \nabla T, \tag{69}$$

where D denotes the Fickian diffusion coefficient, and D_s the Soret coefficient.

It is not possible to determine D and D_s as functions of salinity since individual salts have different coefficients. Results are given here for the diffusion coefficient of a NaCl 0.5 normal aqueous solution (Caldwell [73 C, 74 C 2]) corresponding to a concentration $n = 28.5 \cdot 10^{-3}$, which would be found in sea water with $S = 32.9$.

$$D = 10^{-11}(62.5 + 3.63\,T). \tag{70}$$

Siedler, Peters

Units:

T IPTS-68	D
°C	$m^2 s^{-1}$

Range of validity:

$$2\,°C \leqq T \leqq 52\,°C.$$

The pressure dependence of D is small ($0 \leqq p \leqq 7000$ dbar).

Check value:

$$T = 52\,°C: \quad D = 2.513 \cdot 10^{-9}\,m^2 s^{-1}.$$

Estimate of accuracy:

The standard deviation of the measurements from the formula is $9 \cdot 10^{-11}\,m^2 s^{-1}$.

Caldwell [74 C 2] has given the following formula for the Soret coefficient of a 0.5 normal solution of NaCl:

$$D_S = 10^{-5}[-119 + 11.05\,T - 0.064\,T^2 + 10^{-4}\,p(320 - 8.48\,T + 0.0532\,T^2)]. \tag{71}$$

Units:

T IPTS-68	p	D_S
°C	dbar	K^{-1}

Range of validity:

$$2\,°C \leqq T \leqq 52\,°C$$
$$0 \leqq p \leqq 7000\,\text{dbar}.$$

Check value:

$$T = 50\,°C, \quad p = 7000\,\text{dbar}: \quad D_S = 2.938 \cdot 10^{-3}\,K^{-1}.$$

Estimate of accuracy:

The standard deviation of the measurements from the formula is $5 \cdot 10^{-5}\,K^{-1}$. Caldwell [75 C] showed that the pressure dependence of D_S for a 0.1 normal solution of NaCl at 17.65 °C is significantly nonlinear.

3.1.12 Viscosity

In a fluid with molecular friction, the tangential stress τ is proportional to the shear U'. The proportionality factor is called (molecular) viscosity μ

$$\tau = \mu U'. \tag{72}$$

Often the factor divided by density is used, the kinematic viscosity $\nu = \mu/\varrho$. For distilled water, $\mu(20\,°C) = (1.002 \pm 0.0003) \cdot 10^{-3}\,N\,s\,m^{-2}$ (Swindells et al. [52 S]).

According to Korson et al. [69 K] the viscosity of fresh water at atmospheric pressure can be calculated from

$$\log \frac{\mu(T)}{\mu(20\,°C)} = \frac{a(20 - T) - b(T - 20)^2}{T + c}, \tag{73}$$

where

$$a = 1.1709, \quad b = 1.827 \cdot 10^{-3}, \quad c = 89.93$$

and T is in [°C].

Check value:

$$T=70\,^\circ\mathrm{C}: \quad \mu=4.038703\cdot10^{-4}\,\mathrm{N\,s\,m^{-2}},$$

with $\mu(20\,^\circ\mathrm{C})$ as given above, $\mu(20\,^\circ\mathrm{C})=1.002\cdot10^{-3}\,\mathrm{N\,s\,m^{-2}}$.

Range of validity:

$$10\,^\circ\mathrm{C}\leqq T\leqq 70\,^\circ\mathrm{C}.$$

Estimate of accuracy:

The maximum deviation from previously reported values is $5.1\cdot10^{-6}\,\mathrm{N\,s\,m^{-2}}$ at low temperatures.

According to Millero [74 M, p. 65f.] and Riley and Skirrow [75 R, p. 576] the relative viscosity of sea water at atmospheric pressure can be obtained from

$$\mu_s=\frac{\mu(S,T)}{\mu(0,T)}=1+A\,Cl_v^{1/2}+B\,Cl_v \tag{74}$$

$$A=5.185\cdot10^{-5}\,T+1.0675\cdot10^{-4}$$

$$B=3.300\cdot10^{-5}\,T+2.591\cdot10^{-3}.$$

Cl_v is the *volume chlorinity*, which is related to salinity by

$$Cl_v=\frac{\varrho S}{1806.55}, \tag{75}$$

while $\mu(0,T)=\mu(T)$, eq. (73).

Units:

S ISS-69	T IPTS-68	ϱ
	°C	$\mathrm{kg\,m^{-3}}$

Range of validity:

$$0\leqq S\leqq 40$$
$$5\,^\circ\mathrm{C}\leqq T\leqq 25\,^\circ\mathrm{C}.$$

Check value:

$$S=40, \quad T=25\,^\circ\mathrm{C}$$
$$\mu=9.654093\cdot10^{-4}\,\mathrm{N\,s\,m^{-2}},$$

with $\mu(0,T)$ as given above, eq. (73).

Estimate of accuracy:

Not available in original literature.

For low temperatures an algorithm for the pressure dependence of viscosity is available only for standard sea water ($S=35$). According to Matthäus [72 M]:

$$\mu(35,T,p)=\mu(35,T,0)+\Delta\mu, \tag{76}$$

$$\Delta\mu=p_a[a_1+a_2T+a_3T^2+p_a(a_4+a_5T+a_6T^2)], \tag{77}$$

where

$$p_a=p+10.13$$

$a_1=-1.8266\cdot10^{-8}$	$a_4=9.8965\cdot10^{-13}$
$a_2=1.3817\cdot10^{-9}$	$a_5=-6.325\cdot10^{-14}$
$a_3=-2.6362\cdot10^{-11}$	$a_6=1.2115\cdot10^{-15}.$

Units:

T IPTS-68	p	$\Delta\mu$
°C	dbar	$\mathrm{N\,s\,m^{-2}}$

Range of validity:

$0 \leq T \leq 20\,°\mathrm{C}$,

$0 \leq p \leq 10000\,\mathrm{dbar}$

$20\,°\mathrm{C} \leq T \leq 30\,°\mathrm{C}$ for low pressure only.

The algorithm eqs. (73)···(77) can be used for salinities differing from 35 without introducing much error. At 10 °C and 10000 dbar a value is obtained for fresh water which is about $4 \cdot 10^{-5}\,\mathrm{N\,s\,m^{-2}}$ too high.

Check value:

$S = 35, \quad T = 10\,°\mathrm{C}, \quad p = 10000\,\mathrm{dbar}$

$\mu = 1.363401 \cdot 10^{-3}\,\mathrm{N\,s\,m^{-2}}$,

with $\mu(35, T, 0)$ as given above, eqs. (73)···(75).

Estimate of accuracy:

The formula deviates from the measurements by less than $\pm 4 \cdot 10^{-6}\,\mathrm{N\,s\,m^{-2}}$. Stanley and Batten [69 S] estimate the accuracy of their data as $\pm 1.5 \cdot 10^{-5}\,\mathrm{N\,s\,m^{-2}}$.

The viscosity for atmospheric pressure is shown in Fig. 22, and Fig. 23 represents the viscosity for $S = 35$ and varying temperature and pressure.

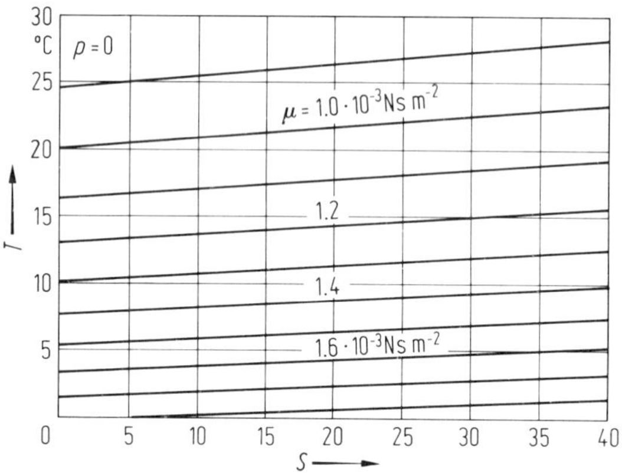

Fig. 22. Viscosity μ in $[10^{-3}\,\mathrm{N\,s\,m^{-2}}]$ as function of salinity and temperature at atmospheric pressure. Contour interval $10^{-4}\,\mathrm{N\,s\,m^{-2}}$.

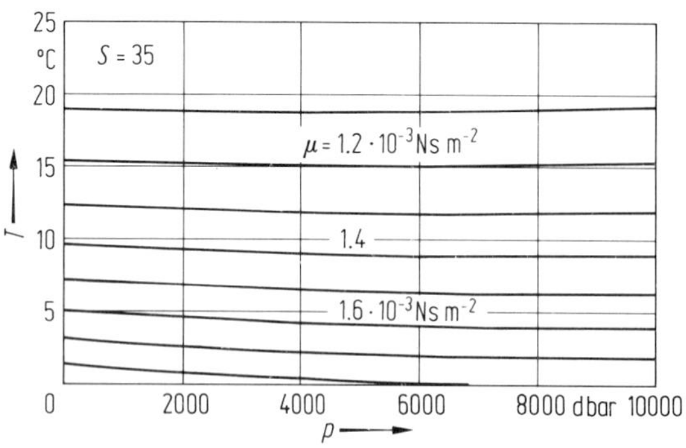

Fig. 23. Viscosity μ in $[10^{-3}\,\mathrm{N\,s\,m^{-2}}]$ as function of pressure and temperature at $S = 35$. Contour interval $10^{-4}\,\mathrm{N\,s\,m^{-2}}$.

3.1.13 Surface tension

If a free surface element is extended by δF, a work δA has to be performed to balance the surface tension:

$$\delta A = a\delta F\,. \tag{78}$$

Here a is the constant of surface tension. Data given in Landolt-Börnstein, 6th ed., vol. II, part 3, p. 421 [56 S], for distilled water can be represented by

$$a(T) = 7.562 \cdot 10^{-2} - 1.3928 \cdot 10^{-4}\,T - 3.063 \cdot 10^{-7}\,T^2. \tag{79}$$

Units:

T IPTS-48	a
°C	$\mathrm{N\,m^{-1}}$

Range of validity:

$0 \leqq T \leqq 60\,°\mathrm{C}$.

Check value:

$T = 60\,°\mathrm{C}$: $a = 6.616052 \cdot 10^{-2}\,\mathrm{N\,m^{-1}}$.

Estimate of accuracy:

Not available in literature cited.

No recent measurements are available for sea water. According to Krümmel [01 K] a is about 1% lower for $S = 40$ than for fresh water. Due to surface films, mostly of organic origin, on the sea surface, large variations occur in the actually observed values of a.

3.1.14 Osmotic pressure

If water volumes with different salinities are separated by a porous wall that is permeable for water and not for salt, diffusion of water molecules will change salinities and cause a pressure difference. The pressure difference is the osmotic pressure Π. Following Millero and Leung [76 M 3] it can be calculated for sea water with salinity S relative to fresh water by the following equation:

$$\Pi = AS + BS^{3/2} + CS^2\,, \tag{80}$$

$$A = 0.070249 + 2.3938 \cdot 10^{-4}\,T - 3.7170 \cdot 10^{-7}\,T^2\,, \tag{81}$$

$$B = -2.1601 \cdot 10^{-3} + 4.8460 \cdot 10^{-7}\,T - 1.0492 \cdot 10^{-7}\,T^2\,, \tag{82}$$

$$C = 2.7984 \cdot 10^{-4} + 1.552 \cdot 10^{-6}\,T - 2.7048 \cdot 10^{-9}\,T^2\,. \tag{83}$$

Units:

S ISS-69	T IPTS-68	Π
	°C	MPa

Range of validity:

$0 \leqq S \leqq 40$

$0 \leqq T \leqq 40\,°\mathrm{C}$.

Check value:

$S = 40$, $T = 40\,°\mathrm{C}$

$\Pi = 3.1252952\,\mathrm{MPa}$.

Estimate of accuracy:

The standard deviation of the formula from the original data is $3 \cdot 10^{-4}$ MPa. Considerable deviations exist from earlier reported values.

For $p=0$ the osmotic pressure is represented in Fig. 24.

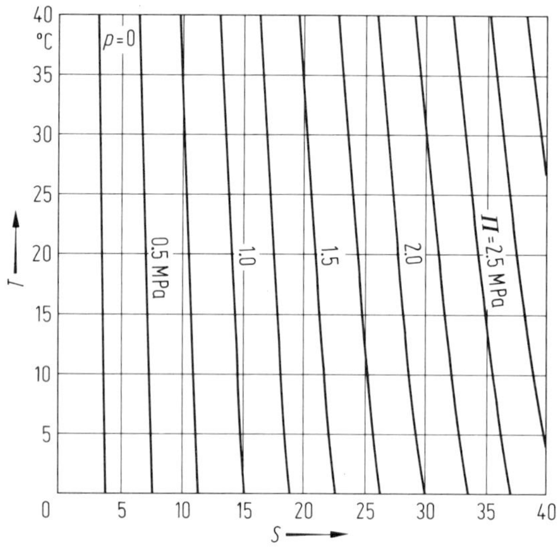

Fig. 24. Osmotic pressure Π in [MPa] as function of salinity and temperature at atmospheric pressure. Contour interval 0.25 MPa.

3.1.15 Brunt-Väisälä frequency

The Brunt-Väisälä frequency N (see sect. 6.1) is often calculated from the expression

$$N^2(p)=g^2\left[\frac{\mathrm{d}\varrho}{\mathrm{d}p}-c^{-2}\right], \tag{84}$$

where g denotes the gravitational acceleration, ϱ the in-situ density. The term containing the sound velocity c accounts for the adiabatic effect. The algorithms used to compute $\varrho(S,T,p)$ and $c(S,T,p)$ have to be thermodynamically consistent. This is largely true for Chen and Millero's formula for sound speed [77 C] and EOS-80 [eqs. (13)···(22)]. However, systematic errors due to the independent fitting procedures for ϱ and c, and errors due to limited numerical precision lead to large relative errors in N for the deep ocean, where

$$\frac{\mathrm{d}\varrho}{\mathrm{d}p}\approx c^{-2}.$$

Owens and Fofonoff (personal communication) proposed a method called adiabatic levelling. Finite differences are used instead of differentiation.

$$N^2(\bar{p})=\bar{\varrho}^2 g^2\left[\frac{\delta(S_2,\theta(S_2,T_2,p_2,\bar{p}),\bar{p})-\delta(S_1,\theta(S_1,T_1,p_1),\bar{p})}{p_2-p_1}\right], \tag{85}$$

where δ is the specific volume anomaly [eqs. (26)···(28)], θ the potential temperature at reference pressure $\bar{p}=(p_2+p_1)/2$, (e.g. eq. (54)), and $\bar{\varrho}$ is to be calculated at pressure level $\bar{p}$.

Acknowledgements. We wish to acknowledge the most helpful advice and numerous suggestions for improving the manuscript from N. P. Fofonoff, R. C. Millard, jr., and F. J. Millero.

3.1.16 References for 3.1

01 K Krümmel, O.: Wiss. Meeresunters. **5**, II (1901) 7.

02 F Forch, C., Knudsen, M., Sørensen, S.P.: K. Dan. Vidensk. Selsk. Skr. 6, Raekke, Naturvidensk. Math. Afd. **12** (1902) 1.

08 E Ekman, V.W.: Publ. circ. cons. perm. int. pour expl. mer **43** (1908) 2.

39 O Osborne, N.S., Stimson, H.F., Ginnings, D.C.: J. Res. Nat. Bur. Stand. **23** (1939) 197.

52 S Swindells, J.F., Coe, J.R., jr., Godfrey, T.B.: J. Res. Nat. Bur. Stand. **48** (1952) 1.

56 S Schäfer, K., Lax, E. (eds.): Landolt-Börnstein, 6th ed., vol. II, part 3: Schmelzgleichgewichte und Grenzflächenerscheinungen. Berlin, Göttingen, Heidelberg: Springer **1956**, 535 pp.

58 P Powell, R.W.: Adv. Phys. **7** (1958) 276.

61 S Schäfer, K., Lax, E. (eds.): Landolt-Börnstein, 6th ed., vol. II, part 4: Kalorische Zustandsgrößen. Berlin, Göttingen, Heidelberg: Springer **1961**, 863 pp.

62 F Fofonoff, N.P.: Physical properties of sea water, in: The Sea, Vol. **1** (Hill, M.N., ed.), New York: John Wiley **1962**, p. 323.

67 B Bigg, P.H.: Br. J. Appl. Phys. **18** (1967) 521.

67 S Stoughton, R.W., Lietzke, M.H.: J. Chem. Eng. Data **12** (1967) 101.

69 B Barber, C.R.: Metrologia **5** (1969) 35.

69 K Korson, L., Drost-Hansen, W., Millero, F.J.: J. Phys. Chem. **73**, No. 1 (1969) 34.

69 S Stanley, E.M., Batten, R.C.: J. Geophys. Res. **74**, No. 13 (1969) 3415.

69 W Wooster, W.S., Lee, A.J., Dietrich, G.: Deep-Sea Res. **16** (1969) 321.

70 J Jamieson, D.T., Tudhope, J.S.: Desalination **8** (1970) 393.

71 M Ménaché, M.: Deep-Sea Res. **18** (1971) 449.

72 A Ambrose, D., Lawrenson, I.J.: J. Chem. Thermodyn. **4** (1972) 755.

72 M Matthäus, W.: Beitr. Meeresk. **29** (1972) 93.

73 B Bryden, H.: Deep-Sea Res. **20** (1973) 401.

73 C Caldwell, D.R.: Deep-Sea Res. **20** (1973) 1029.

73 M Millero, F.J., Perron, G., Desnoyers, J.E.: J. Geophys. Res. **78** (1973) 4499.

74 C 1 Caldwell, D.R.: Deep-Sea Res. **21** (1974) 131.

74 C 2 Caldwell, D.R.: Deep-Sea Res. **21** (1974) 369.

74 C 3 Castelli, V.J., Stanley, E.M., Fischer, E.C.: Deep-Sea Res. **21** (1974) 311.

74 D Doherty, B.T., Kester, D.R.: J. Mar. Res. **32** (1974) 285.

74 M Millero, F.J., in: The Sea, Vol. **5** (Hill, M.N., ed.), New York: John Wiley **1974**, p. 3.

75 C Caldwell, D.R.: J. Phys. Chem. **79**, No. 17 (1975) 1885.

75 K Kell, G.S.: J. Chem. Eng. Data **20** (1975) 97.

75 R Riley, J.P., Skirrow, G.: Chemical Oceanography, vol. I., 2nd edn., London, New York, San Francisco: Academic Press **1975**, 589 pp.

76 M 1 Millero, F.J., Emmet, R.T.: J. Mar. Res. **34** (1976) 15.

76 M 2 Millero, F.J., Gonzalez, A., Ward, G.K.: J. Mar. Res. **34** (1976) 61.

76 M 3 Millero, F.J., Leung, W.H.: Am. J. Sci. 276 (1976) 1035.

76 M 4 Millero, F.J., Kremling, K.: Deep-Sea Res. **23** (1976) 1129.

77 C Chen, C.T., Millero, F.J.: J. Acoust. Soc. Am. **62** (1977) 1129.

77 D Doronin, Y.P., Kheisin, D.E.: Sea Ice (Translated from Russian edition 1975), New Delhi, Bombay, Calcutta, New York: Amerind **1977**, 323 pp.

77 F Fofonoff, N.P.: Deep-Sea Res. **24** (1977) 489.

78 C Caldwell, D.R.: Deep-Sea Res. **25** (1978) 175.

78 L Lewis, E.L., Perkin, R.G.: J. Geophys. Res. **83** (1978) 466.

78 M Millero, F.J.: Unesco Techn. Pap. Mar. Sci. (Paris) **28** (1978) 29.

79 I IAPSO: SUN Report. International Association for the Physical Sciences of the Ocean, IUGG Publications Office, 39 ter rue Gay-Lussac, Paris **1979**.

80 C 1 Caldwell, D.R., Eide, S.A.: Deep-Sea Res. **27** (1980) 71.

80 C 2 Culkin, F., Smith, N.D.: IEEE J. Oc. Eng. **5** (1980) 22.

80 D Dietrich, G., Kalle, K., Krauss, W., Siedler, G.: General oceanography, 2nd ed., p. 54, New York, Chichester, Brisbane, Toronto: John Wiley **1980**.

80 L Lewis, E.L.: IEEE J. Oc. Eng. **5** (1980) 3.

80 M Millero, F.J., Chen, C.T., Bradshaw, A., Schleicher, K.: Deep-Sea Res. **27** A (1980) 255.

80 P Poisson, A., Brunet, C., Brun-Cottan, J.C.: Deep-Sea Res. **27** (1980) 1013.
81 L Lewis, E.L., Perkin, R.G.: Deep-Sea Res. **28** A (1981) 307.
81 M 1 Millero, F.J., Poisson, A.: Deep-Sea Res. **28** A (1981) 625.
81 M 2 Millero, F.J., Chen, C.T., Bradshaw, A., Schleicher, K.: Unesco Techn. Pap. Mar. Sci. (Paris) **38** (1981) 99.
81 U 1 UNESCO/ICES/SCOR/IAPSO: Background papers and supporting data on the practical salinity scale 1978. Unesco Techn. Pap. Mar. Sci. (Paris) **37** (1981).
81 U 2 UNESCO/ICES/SCOR/IAPSO: Background papers and supporting data on the international equation of state of sea water 1980. Unesco Techn. Pap. Mar. Sci. (Paris) **38** (1981).
81 U 3 UNESCO: International Oceanographic Tables, Vol. 3, Unesco Techn. Pap. Mar. Sci. (Paris) **39** (1981).
82 M Millero, F.J.: Ocean Sc. Eng. **7** (1982) 403.
82 U UNESCO: International Oceanographic Tables, Vol. 4, Unesco Techn. Pap. Mar. Sci. (Paris) **40** (1982), in press.
83 F Fofonoff, N.P., Millard, R.C., jr.: Algorithms for computation of fundamental properties of sea water. Unesco Techn. Pap. Mar. Sci. (Paris) **44** (1983).
83 M Millero, F.J.: Ocean Sc. Eng. **9** (1983) 1.
83 S Striggow, K.: Beitr. Meereskd. **49** (1983) 25.

3.2 Acoustical properties of the Ocean

3.2.0 List of symbols and abbreviations

Symbols

Symbol	Quantity	Unit
a	bubble radius	m
a_n	mode coefficients	dimensionless
A	illuminated area	m^2
A_1	temperature dependence coefficient of viscous absorption	$dB\,km^{-1}\,kHz^{-1}$
A_2, A_3	temperature dependence coefficients of relaxation absorption	$dB\,km^{-1}\,kHz^{-2}$
A_1, A_2, B_1-B_3	coefficients of shallow water transmission loss polynomial	dimensionless
c	sound speed	ms^{-1}
D	depth below sea surface	m
DI	directivity index	dB
DT	detection threshold	dB
f, f'	frequency	s^{-1}
f_r	relaxation frequency	s^{-1}
f_R	resonance frequency	s^{-1}
g	gravity	ms^{-2}
H	water depth	m
$H_{1/3}$	characteristic wave height	m
I	sound intensity	$kg\,s^{-3}$
k	wavenumber	m^{-1}
K	coefficients of sound speed polynomials	dimensionless
L	length (fish)	m
M	transducer receiving response	dB
M'	normalized beam pattern function	dimensionless
m	density ratio	dimensionless
n	refraction index	dimensionless
n	porosity	%
NL	noise level	dB
p	sound pressure	Pa
P	atmospheric pressure	atm
pH	acidity	dimensionless
P_1-P_3	coefficients of pressure dependence of α	dimensionless
q	$\lvert \mathrm{grad}\, c(z) \rvert$	s^{-1}
r	distance, range	m
R	radius of ray curvature	m
RL	reverberation level	dB
RL_s	surface reverberation level	dB
RL_v	volume reverberation level	dB
S	salinity	‰
S	transducer transmitting response	dB
SL	source level	dB
S_s	surface scattering strength	dB
S_v	volume scattering strength	dB
S_{vc}	column scattering strength	dB
t	time	s
T	temperature	K, °C
TL	transmission loss	dB
TS	target strength	dB
u	ship, transducer speed	
U	wind speed	ms^{-1}

Symbol	Quantity	Unit
v	particle velocity	ms^{-1}
V	fluctuation coefficient	%
W	wave front	m
x	horizontal coordinate	m
z	depth coordinate	m
z_0	specific acoustic impedance	$kg\,m^{-2}\,s^{-1}$
α	absorption coefficient	$dB\,km^{-1}$
$\alpha, \beta, \varphi, \psi, \vartheta$	angles of sound propagation	degree, rad
δ	damping constant	dimensionless
∇^2	div grad, Laplacian operator	
∂	partial-derivative symbol	
η	efficiency	%
Θ	grazing angle	degree
λ	wave length	m
μ	variance of refraction index	dimensionless
ϱ	density	$kg\,m^{-3}$
τ	propagation time, time shift	s
τ_r	relaxation time	s
Φ	equivalent angle of surface illumination	radian
Φ	grain size	dimensionless
Φ	latitude	degree
Φ_n	mode functions	Pa
Ψ	equivalent solid angle of volume illumination	steradian
ω	angular frequency	s^{-1}

Abbreviations

cw	continuous wave	RS	remote sensing
DWC	deep-water channel	RSM	remote sensing methods
FE	finite element	SOFAR	sound fixing and ranging
FFP	fast field program method	SONAR	sound navigation and ranging
FFT	fast Fourier transform	TNT	trinitrotoluol
PE	parabolic equation	WKB	Wentzel-Kramers-Brillouin
PZT	lead-zirconate-titanate		

3.2.1 Introduction

Sound differs from other conservative or wavelike motion processes in the ocean (sect. 6.1) by the fact that the compressibility of the sea water is the only restoring force in the general equation of motion. As with all nonrigid media only compressional waves can propagate as sound. Their propagation speed is called sound speed c. The sound speed is isotropic and almost without molecular dispersion (subsect. 3.2.5).

The exceptional position of ocean acoustics for wireless signal transmission and remote sensing under the sea surface is due to the low transmission loss of sound waves in relation to electromagnetic waves in sea water. However, the range benefit of underwater sound as against electromagnetic waves applies only to low frequencies (Fig. 1). Due to the attenuation increase by the second power of frequency in several frequency intervals (subsect. 3.2.6) the acoustic penetration depth decreases to about 100 m already at 1 MHz. This is considered the upper frequency limit for acoustic remote sensing and signal transmission in the sea. The lower frequency limit is of the order of 10 Hz and comes out as a compromise between the necessary penetration depth into the sea bottom for its remote sensing and the increasing difficulties to achieve sufficient acoustic power and spatial resolution at low frequencies (subsect. 3.2.4). The altogether 5 frequency decades correspond to a wavelength interval of about 1.5 mm to 150 m which can be used by ocean acoustics.

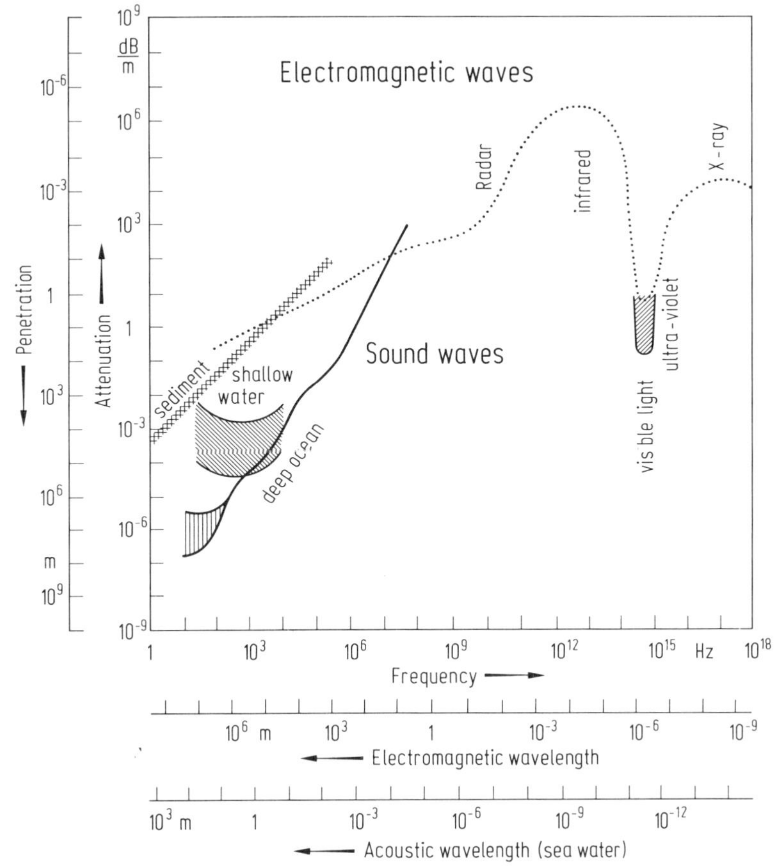

Fig. 1. Comparison of attenuation coefficients of electromagnetic and acoustic waves in sea water.

Sound waves alter the local values of the static pressure p_m, the density ϱ_m, the temperature T_m, and the vector of the particle velocity $\boldsymbol{v}_m$ according to

$$\begin{aligned} p &= p_m + p_\sim \\ \varrho &= \varrho_m + \varrho_\sim \\ T &= T_m + T_\sim \\ \boldsymbol{v} &= \boldsymbol{v}_m + \boldsymbol{v}_\sim . \end{aligned} \qquad (1)$$

Among the 4 sound field quantities on the r.h.s. of eq. (1) the sound pressure $p_\sim$ which can be measured simplest (subsect. 3.2.4) is used almost exclusively (for convenience the index $_\sim$ is omitted). The sound pressure p is defined as the rms-value of a sinusoidal pressure.

The logarithmic ratio $20 \log \frac{p}{p_0}$ related to the reference sound pressure p_0 of 1 μPa is called sound level and is given in decibels [dB]. The rms-value of the particle velocity $\boldsymbol{v}$ of the sound wave is called effective particle velocity, to be distinguished from the sound speed. For plane waves the ratio of sound pressure and effective particle velocity $\frac{p}{|\boldsymbol{v}|} = \varrho c = z_0$ is denoted as specific acoustic impedance. Together with the sound pressure it defines the sound intensity I as time average ratio of sound power per unit area

$$I = \frac{p^2}{\varrho c}.$$

Measurable sound pressure amplitudes in the ocean cover the order of 10^{-5} Pa as the lowest observable noise amplitude (subsect. 3.2.7) up to 10^5 Pa in the near-field of high-power sound transmitters. Still higher sound pressure values occur only in the neighborhood of detonations (subsect. 3.2.4). Apart from these extreme cases the sound propagation in the sea is described by the linearized wave equation [58 O]

$$\nabla^2 p = \frac{1}{c^2}\frac{\partial^2 p}{\partial t^2}, \tag{2}$$

where the space and time changes of state are connected only by the sound speed c, which is therefore the governing specific quantity for all sound propagation processes in the sea. The sound speed contains the compressibility $\varkappa$ [Pa^{-1}] in the sound wave according to

$$c = \frac{1}{(\varrho\varkappa)^{1/2}}. \tag{3}$$

The sound speed in sea water increases with temperature, salinity, and static pressure. This relationship is described by approximation polynomials derived on the basis of precise sound speed measurements (subsect. 3.2.5). The anomalous temperature dependence (sect. 3.1) is caused by a near-field structure of the water molecules at low temperatures. This configuration is destroyed with increased heating, which yields a closer package of the molecules, thus lowering the compressibility of the water. The normal decrease of sound speed with temperature due to heat expansion predominates only at higher temperatures (above 340 K) which do not occur in the open ocean.

Salinity and pressure alter the compressibility of sea water more strongly than its density. The sound speed therefore in principle depends on the amplitude of the sound pressure itself. In the neighborhood of explosive sound sources the corresponding high pressure amplitudes of the order of 10^8 Pa cause deformations of the pressure-time function during propagation and lead to the development of shock fronts (subsect. 3.2.4.1).

However, for sound propagation in the ocean, outside of the near-field of high-power sound sources the sound pressure amplitudes and the resulting density changes are so small that there is no mutual interaction with other mechanical wave types by joint field quantities but only unilateral linear reflection, scattering, and refraction of sound waves at structures of the medium and boundaries where the sound speed or the acoustic impedance change. Further, sound can be originated by special processes of motion in the sea (subsect. 3.2.7). In so far as influence on, and production of, sound are reversibly connected with environmental properties, sound signals are the basis for their nonreactive remote sensing.

With respect to acoustic remote sensing in the ocean by environmentally influenced or originated sound signals (active and passive remote sensing) there are fewer properties for diagnostic purposes both on the object and the signal side than for the electromagnetic remote sensing in the atmosphere. In the ocean resonant objects and structures within the usable acoustic frequency range are rare. Consequently the signal spectra are mostly broadband and thus uncolored both for absorption and scattering as well as emission. For the same reason the sound speed is – apart from wave guide propagation (subsect. 3.2.2.1) – nondispersive. On the signal side the longitudinal sound wave lacks the polarizability of the electromagnetic wave.

Section 3.2 deals with acoustic properties of the ocean volume and the boundaries. The seismic remote sensing of the space below the sea bottom is beyond the scope of this section.

3.2.2 Models of sound propagation

The goal of ocean acoustic models is to describe quantitatively the phenomena of sound propagation whereby the transmission loss (subsect. 3.2.6) between source and receiver is the most important quantity. Propagation models are also incorporated in models of noise (subsect. 3.2.7) and reverberation (subsect. 3.2.8) against which a signal at the receiver has to be discriminated (subsect. 3.2.2.3).

Propagation models represent approximate solutions of the linearized wave equation (2) from which the time-independent Helmholtz wave equation

$$\nabla^2 p + k^2 p = 0 \tag{4}$$

is obtained by applying the harmonic solution of the sound pressure $p = p(x, y, z)\mathrm{e}^{\mathrm{i}(\omega t - kr)}$. Nearly all models treat the problem in the range-depth plane only [81 K].

The separation of the Helmholtz equation into ordinary differential equations requires the unrealistic assumption of horizontally invariant $c(z)$ and ocean boundaries as surfaces of the cartesian coordinates. Depending on the type of simplification the models may be divided into two categories: wave models and ray models (Fig. 2). The former simplify the geometry of the ocean environment and give a formally complete and frequency-dependent solution, including diffraction and scattering without c-gradient limitations and account for any arbitrary source function. Wave models are used to represent the essential propagation features at low frequencies.

Ray models as high-frequency (geometrical acoustic) asymptotic approximations handle range-dependent boundaries and c-distributions easily, yield a simple visualisation of the field distribution by ray diagrams but are restricted to moderate c-gradient conditions $\Delta c/c \ll 1$ in a wavelength distance.

In addition to numerical solutions of the wave equation which are often handicapped by insufficient input data, parametrizations of empirical transmission loss (subsect. 3.2.6), noise (subsect. 3.2.7) and scattering (subsect. 3.2.8) are utilized, per se or in combination with numerical models [81 S 1].

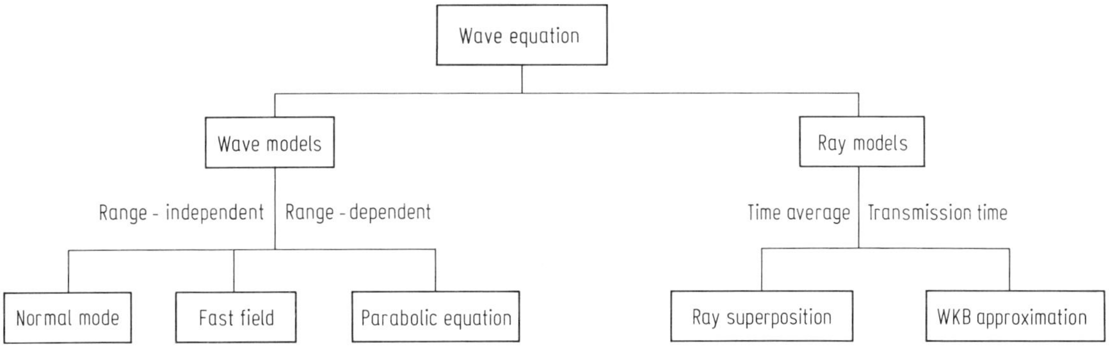

Fig. 2. Types of sound propagation models.

3.2.2.1 Wave models

A solution of the wave equation (2) has to account for the boundary and initial conditions that apply in a particular sound propagation problem. Boundary conditions are given in terms of sound pressure p and particle velocity $\mathbf{v}$, such as $p=0$ at the sea surface. Initial conditions are those of the sound source.

In an idealized horizontally invariant ocean the Helmholtz equation (4) can be separated and the sound pressure then may read [77 K, 60 B]:

$$p(r,z)=\sum_{\mathrm{n}} a_{\mathrm{n}}(r)\Phi_{\mathrm{n}}(z)\,. \tag{5}$$

The mode functions Φ_{n} depend on the boundary conditions and the sound speed stratification and form an orthonormal basis. The coefficients a_{n} are determined by the initial source distribution. The range dependence of $a_{\mathrm{n}}(r)$ is due to the individual mode attenuation. For absent stratification the $\Phi_{\mathrm{n}}(z)$ are trigonometric functions. The modes of order n represent pairs of opposite oblique waves which together fulfill the boundary conditions. The vertical components are standing waves whereas the horizontal components propagate. According to their k-dependent grazing angle the phase and group velocities of the modes are dispersive. The discrete mode spectrum of eq. (5) is a "far-field" solution which neglects the continuous contributions of steep waves near the source which correspond to grazing angles of the wave normals greater than the critical angle (subsect. 3.2.8.4.1) and suffer severe loss [77 K].

Under realistic conditions the wave description of the sound propagation deviates from the ray description in respect to the vertical pressure distribution up to a water depth or effective channel thickness (subsect. 3.2.5.5) of $\approx 20\,\lambda$.

Sound speed profile types $c(z)$ such as tanh(z) (Epstein profile) for which the Helmholtz equation has analytic solutions in terms of special functions [66 D] are less important for practical problems than numerical profile sampling [79 D 2].

Instead by normal modes the acoustic field may be Fourier-decomposed into an infinite set of horizontal waves which can be effectively evaluated using the fast Fourier transform. This direct numerical solution is called fast field program method (FFP), [71 D].

Extensions to range dependence where the wave equation cannot be separated are accomplished either by mode coupling [79 F 2, 81 M 5, 82 M 3], which describes the mutual scattering of modes, or by the "adiabatic" assumption [78 I], which assumes that all energy in a given mode transfers to the corresponding mode in the new gradually changed environment. The mode coupling due to rough surfaces treated by perturbation method yields stationary mode distributions observed at long-range propagation in the sea (subsect. 3.2.6).

A range-dependent modification of normal mode theory which is not limited to a "slowly varying" stratification, water depth and bottom composition divides the entire range interval into segments wherein the sound speed profile and bottom composition are arbitrary functions of depth but do not change with the range over the subinterval. The "trapped modes" within any interval are calculated as if the problem were range-independent. These modes are then summed to give the total field at the junction of the next segment where it is replaced by a vertical distribution of appropriately weighted point sources. The number of required range segments is found empirically [75 K, 76 K 1].

The parabolic equation method (PE) has achieved practical importance for economic range-dependent propagation modelling [77 T 2, 80 J, 83 L]. Under the assumption that the environment is changing slowly with range, the second spatial derivative of the field quantity p in r-direction can be neglected against the product of its first derivative and the wavenumber k. Hence, the parabolic equation differs from the Helmholtz equation by only the first derivative in r-direction. The main advantage is numerical since the Helmholtz equation is an elliptic partial differential equation which must be numerically solved in the entire two-dimensional region simultaneously, whereas the parabolic partial differential equation can be marched in the range dimension. The PE-method treats horizontally changing sound speed conditions $c(z, r)$ by introducing at every new range step Δr another sound speed profile, which may be applied to deterministic horizontal variations as well as to simulate stochastically changing sound speed conditions [81 S 1]. Models to describe the wave propagation in a random medium have been reviewed by Frisch [68 F].

The parabolic equation method which is a far-field and narrow-angle approximation to the wave equation is most accurate when the angle of propagation is confined within a cone of about $\pm 20°$ against the horizontal [81 K].

Despite the simplification, the PE-method as a wave-theoretical approach will properly describe diffraction effects. However, since the sound speed variability is caused by various different processes with a wide range of correlation lengths a simultaneous modelling of small- and large-scale variations may become problematic due to the required step size.

The finite element approach (FE) as a certain competitor to the PE-method has been examined for application to acoustic propagation in a range-dependent environment [79 K 1]. Since the entire fluid and bottom domains are modelled with individual finite elements of own physical constants, the representation of solid media with multi-material properties or the representation of the fluid with variable sound speed profile in three directions is achievable. Irregular boundaries such as sea mounts can be modelled. However, the number of parameters in that both the volume and surface domains have to be discretized may become extremely large (two up to eight elements per wavelength) which is a strong disadvantage for the real ocean.

Models which predict the sound signal wave form to yield the information about the arrival structure or frequency dispersive characteristics are either classical normal-mode theories, in conjunction with stationary-phase techniques [75 P] or ray theory programs (subsect. 3.2.2.2), [71 W 2]. The direct numerical FFT-method has also been implemented for this purpose [77 D] as well as the PE-method [85 N] to model effects of ocean currents on sound transmission.

A careful intercomparison of various model approaches has been made by DiNapoli and Deavenport [79 D 2]. An overview is given in [82 U]. However, since the comparisons against experimental data are limited and often not satisfying due to the incompleteness of the associated environmental data a definite appraisal of models is left for the future.

3.2.2.2 Ray models

Rays are defined as successive normals to wave fronts

$$W(x, y, z) = c(t - \varphi) \tag{6}$$

which represent surfaces of constant phase φ with increasing time t. The time- and frequency-independent form of the phase front is given by the eikonal equation [79 D 1]

$$|\nabla W|^2 = \frac{c_0^2}{c^2(x, y, z)} = n^2(x, y, z)\,. \tag{7}$$

From eq. (7) a set of ordinary differential equations for the ray path can be derived which lead to Snell's law for a stratified ocean $c(z)$:

$$\frac{\cos\Theta}{c(z)} = \frac{\cos\Theta_0}{c_0}, \tag{8}$$

being a constant along any individual ray starting at a depth with c_0 and at grazing angle Θ_0. For a linear sound speed $c = c_0 + qz$ the sound ray is a circle of radius

$$R = \frac{c_0}{q\cos\Theta_0}. \tag{9}$$

For ray tracing, sound speed profiles are represented by constant-gradient polygons corresponding to circle-segment rays [76 S] or, if necessary, are smoothed by spline interpolation [70 C]. The sensitivity of the sound field against profile simplifications depends on the profile type and is critical for profile sections which produce genuine or pseudo sound channels [74 S 3].

The average deep-sea sound speed profile (Fig. 9) yields particular types of sound refraction and channel focussing (subsects. 3.2.5.5, 3.2.6.2), labeled in Fig. 3.

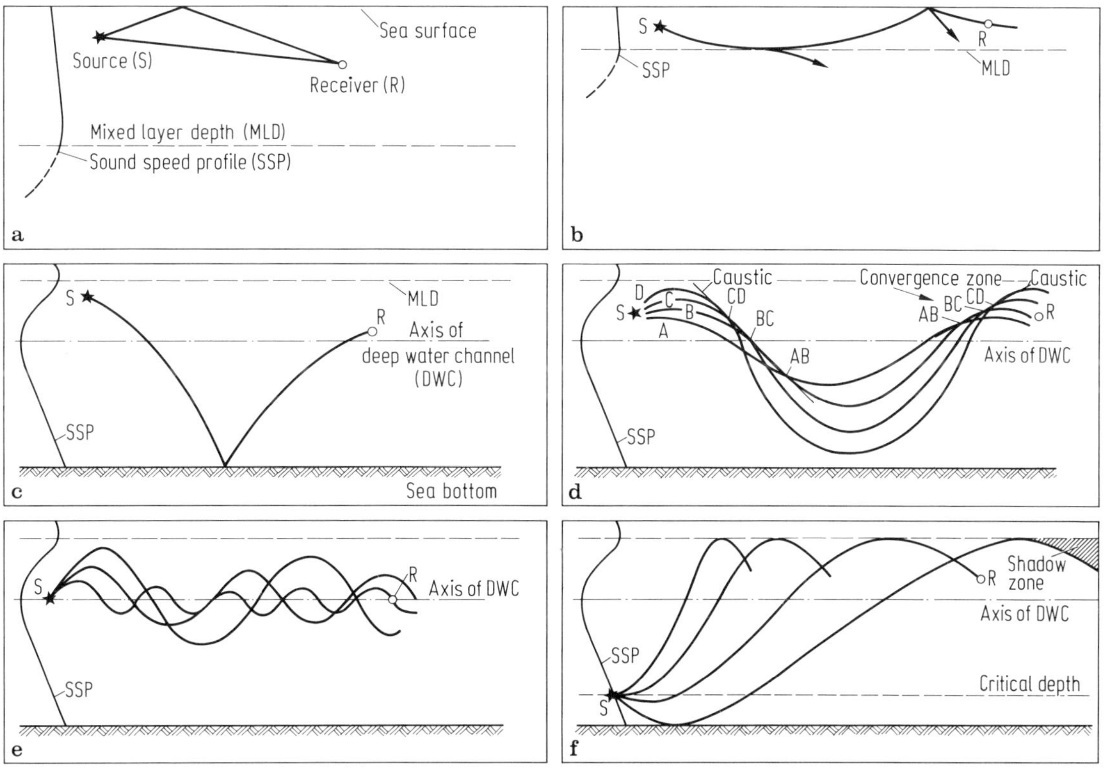

Fig. 3. Types of sound propagation in deep water. (a) Source, receiver (R, S) within mixed layer, short range. Direct and surface-reflected path. (b) R, S within mixed layer, long range. Repeated surface reflections (surface channel by pressure-induced sound speed profile (SSP)-gradient. Channel leakage by high-frequency scattering and low-frequency diffraction. (c) S, R below mixed layer (thermocline), moderate range. Propagation by lossy bottom bounce. (d) Like (c), longer range. Reduced loss by convergence zone. (e) S, R at deep water channel (DWC)-depth, long range, small elevation angles. Lowest average loss by refractive channel propagation. (f) Source or receiver at critical depth. "Reliable acoustic path"-propagation below mixed layer at any depth up to first shadow zone.

According to ray definition, the acoustic energy flux within a bundle of adjacent rays is a constant – except absorption in the sea volume and loss at the boundaries. Since ray-tracing cannot directly account for reflection loss at the ocean bottom, the corresponding coefficients are calculated outside the ray model with wave-theoretical formulae [81 S 1]. By integration of the loss history of a ray, the transmission loss is obtained by counting the rays, weighted by the individual loss, which penetrate through reference windows along the propagation path [76 S]. This method is more simple than ray-divergence calculating, connected with singularity problems at ray crossings [79 D 1].

Deterministic horizontal variations $c(z, x)$ may be treated by horizontal $c(x)$-steps [79 D 1] or by dividing the range-depth plane into triangular regions of constant c-gradients [71 B 3].

The irregular or quasi-stochastic behavior of the sound speed distribution in the upper ocean and in shallow waters (subsect. 3.2.5.6) often predominates acoustically over the mean seasonal thermocline and has led to stochastic propagation models [81 S 1] which produce realistic redistributions of the sound field in shadow zones and avoid pseudo caustics. The effect of the stochastic component of the sound speed can be modelled by applying a joint probability distribution of angular change $\Delta\Theta$, phase change $\Delta\varphi$ and spatial displacement Δr of the sound wave travelling a path length Δs through the stochastic environment δc, in the ray-tracing routine with deterministic profiles after a ray has been traced from r_0 to $r_0 + \Delta s$ to change the ray parameters by a Monte Carlo method according to the particular probability distribution [81 S 1]. The same approach is used to model the scattering of sound from a random surface where the influence function is given by the angular probability density of the reradiated energy caused by the incident plane wave [76 S].

Appropriate probability densities for the ray parameters in a stochastic environment are obtained simplest by diffusion theory [60 C] which also allows to account for the coupling of surface duct and deep water channel as well as channel trapping and leakage due to stochastic refraction. Further, the radiation transport equation [79 W] is applied which describes the deterministic part of the sound speed field by ray approximation and deals with the stochastic component by parabolic equation method.

To account for multipath propagation phenomena where the propagation time of the individual ray path is essential, the time-integral may be obtained by the WKB-method [78 L 1].

Models which are taylored to various applications are reviewed in [79 D 2, 82 U].

3.2.2.3 SONAR equations

To optimize a remote sensing (RS) device or to predict the performance of a given echo sounder set-up, it is convenient to group together the parameters of the medium, the target and the equipment which determine the RS or SONAR range [75 U 1].

The SONAR equations express, in [dB], the ratio of the intensities of the desired and undesired portions of the received signals just sufficient to yield a given probability of detection and which is called the detection threshold DT. The parameters involved are defined in Table 1 and usually denoted by a two-letter symbol.

The medium- and object-related parameters transmission loss TL, noise level NL, target strength TS and reverberation level RL deal with stochastic quantities and are specified in subsects. 3.2.6, 3.2.7, and 3.2.8. The equipment-related parameters source level SL and directivity index DI are specified in subsect. 3.2.4.

For an active RS- or echo-sounding device which is limited by a background of noise as occurring mainly in deep water or at vertical echo sounding, the SONAR equation reads:

$$SL - 2TL + TS - (NL - DI) = DT, \tag{10}$$

where $2\,TL$ is the to and fro loss when source and receiver coincide. The term $(NL - DI)$ implies an isotropic noise field which contributes only through a solid angle according to DI of the receiver. Since the noise spectrum level given in subsect. 3.2.7 refers to a bandwidth of 1 Hz, one has to account for the actual bandwidth Δf, in [Hz], by adding $10 \log \Delta f$ to NL in eq. (10).

The effect of time stretching of a short signal pulse of duration τ_0 to τ_e due to multipath propagation or duration of an unresolved echo is taken into account by adding $10 \log \frac{\tau_0}{\tau_e}$ to SL [75 U 1].

A typical value of DT is 10 dB for 90% detection and 0.1% "false-alarm" probabilities. Signal detection and false-alarm probabilities of known and unknown signals by power-related and sophisticated receivers as a function of DT are given in [54 P, 80 H 3, 60 M 2].

Table 1. Parameters of the SONAR equations, definitions and reference locations [75 U 1].

Parameter	Symbol	Definition	Reference
Source level	SL	$10\log\left[\frac{\text{intensity of source}}{\text{reference intensity}^{1)}}\right]$	1 m from source on its acoustic axis
Transmission loss	TL	$10\log\left[\frac{\text{signal intensity at 1 m}}{\text{signal intensity at target or receiver}}\right]$	1 m from source and at target or receiver
Target strength	TS	$10\log\left[\frac{\text{echo intensity at 1 m from target}}{\text{incident intensity}}\right]$	1 m from acoustic center of target
Noise level	NL	$10\log\left[\frac{\text{noise intensity}}{\text{reference intensity}^{1)}}\right]$	at hydrophone location
Receiving directivity index	DI	$10\log\left[\frac{\text{noise power generated by an equivalent nondirectional hydrophone}}{\text{noise power generated by actual hydrophone}}\right]$	at hydrophone terminals
Reverberation level	RL	$10\log\left[\frac{\text{reverberation power at hydrophone terminals}}{\text{power generated by signal of reference intensity}^{1)}}\right]$	at hydrophone terminals
Detection threshold	DT	$10\log\left[\frac{\text{signal power to just perform a certain function}}{\text{noise power at hydrophone terminals}}\right]$	at hydrophone terminals

[1]) The reference intensity is that of a plane wave of rms pressure 1 μPa.

In some modern SONARs where DI and DT can not be distinguished due to complex signal processing ($DI-DT$) is referred to the increase of signal to background ratio, produced by the entire receiving system.

When the predominant background of the active RS is reverberation instead of noise, as occurring mainly in shallow water and horizontal RS, the parameter DI, defined for isotropic background, is inappropriate; the term ($NL-DI$) is therefore replaced by an equivalent plane wave reverberation level RL (subsect. 3.2.8.1), measured at the receiver terminals. The reverberation-limited active SONAR equation then becomes:

$$SL-2TL+TS-RL=DT_R, \tag{11}$$

where the parameter DT_R is in general different from the one in the noise-limited case [83 U].

For passive RS and one-way propagation, the parameter source level SL_N refers to the radiated noise of the object at the unit distance of 1 m. The passive SONAR equation then reads:

$$SL_N-TL-(NL-DI)=DT_N. \tag{12}$$

Due to the stochastic character of the medium- and target-related parameters of which neither the mean values nor statistical moments are reliably available in the RS practice, the SONAR eqs. (10···12) deal with a "best guess" of the quantities of interest (maximum detection range, maximum admissable noise level or reverberation level, minimum admissable target strength or scattering strength) for which the equations will hold. Further, the equations refer to RS-systems which depend on the signal intensity. Sophisticated SONARs with correlation processing must account for correlation loss when scattering is essentially involved in the signal propagation [85 U].

3.2.3 Applications

3.2.3.1 Remote sensing methods (RSM)

Table 2. Remote sensing methods, related to the object of measurement.

Object type	Object	Active RS monostatic	Active RS bistatic, multistatic	Passive RS, radiated noise
Sea bottom, fixed objects	water depth bathymetry	depth contour, stripchart		
		echo sounder [59 S], fan-shaped sounder [83 K]		
	bottom morphology, man-made objects	relief features, distribution of stones and gravel. Offshore structures, cables, pipelines		
		side scan SONAR, large scale [77 A 2], small scale [82 N], high-frequency SONAR [75 L 2]		
	sediment properties	distribution of hard and soft sediments, sediment thickness, layer structure		
		sediment echo sounder [71 W 1] echo strength sounder [76 D]		
Sea volume	stratification	thermo-halocline depth	three-dimensional sound speed/temperature field	
		vertical echosounder, plankton tracers [77 C]	multistatic tomography [82 B, 79 M 2, 83 S]	
	suspended matter	bubble size spectrometry, grain density	occurence, estimation of bubble size/density, sediment load	occurence of bubble layer, sediment suspension
		echosounder [82 Y], side scan SONAR [82 T]	bistatic range, attenuation due to air [86 W 2] and sediment suspension [83 H 1]	bottom hydrophone, noise screening [85 W 1] sand noise [74 U]

Table 2 (continued).

Object type	Object	Active RS monostatic	Active RS bistatic, multistatic	Passive RS, radiated noise
Moving objects	vehicles	range, bearing, speed	range, bearing, speed	bearing (range), propulsion characteristics
		searchlight SONAR [57 H], panorama SONAR [63 W]	bistatic, multistatic arrays [83 G]	passive SONAR [62 A]
	fish, marine mammals	detection, depth, shoal size, biomass, motion features	occurence, estim. population density, estim. fish size	classification of species, behavior, communication signals, detection signals
		vertical echo sounder [70 A], horizontal echo sounder [66 T]	bistatic range, frequency-dependent attenuation [67 W 2]	single hydrophone, arrays [64 T]
	crustacea			classification (snapping noise) bottom hydrophone [64 T]
	plankton	detection, depth, estim. biomass, motion features vertical echo sounder [70 F]		
Motion processes	currents	vertical current profile oblique echo sounding of volume scattering [84 S]	mesoscale three-dimensional current field bistatic range, multistatic tomography [77 W]	boundary layer current bottom hydrophone, self-noise [62 W]
	tidal level, sea level	instantaneous surface level upward looking bottom echo sounder [73 F]	spectrum of infragravity motion bistatic range [79 F 2]	mean surface level pressure gauge [62 W]
	internal waves	vertical motion of density horizon bottom/buoy mounted vertical echo sounder [77 C]	spectrum features bistatic range [83 D]	
	surface gravity waves	scalar spectrum, directional spectrum upward looking echo sounder [73 F], sounder array [79 R]	spectrum features bistatic range [70 D 1]	
	capillary waves	spectrum features oblique echo sounder [80 L 1]		
	seaquakes			localisation coastal hydrophones [56 S]
	wind			surface wind speed (friction velocity) bottom hydrophone [81 E]
	precipitation			occurence of local rain shower, cloud burst bottom hydrophone [69 B 1]

3.2.3.1.1 Measuring quantities and parameters of remote sensing (RS)

The active acoustic RSM have in common that the object scene is illuminated by (impulse) modulated sound waves, their grazing angle ϑ and carrier frequency f representing the essential measuring parameters (Fig. 4). The available information from the reradiation at a chosen receiving position consists of the distribution of the received signal (amplitude and phase) in the three domains: direction φ, propagation time τ, and frequency f'. Passive RS basically lacks the transmission time information; the distance must be concluded from the direction information by triangulation.

The information on the object scene can be extended only by expansion of the parameter domains f and ϑ and by the receiver direction ψ as additional third parameter up to the entire amplitude-phase distribution along a closed surface surrounding the object (the entire acoustic hologramm).

The active RSM differ by which of the three measuring quantities φ, τ, f' are used, which corresponding resolution is achieved by the measuring setup and the transmitted signal (subsect. 3.2.4) and which of the three parameters ϑ, f, ψ are utilized. However, the effort for realizing variable ϑ and ψ, especially for simultaneous measurements (multistatic arrays) is often prohibitive; thus for most applications the same sound transducer (subsect. 3.2.4) is used both as transmitter and receiver. Then follows: $\vartheta=\psi=\varphi$ and $\alpha=\beta$, $r_1=r_2$.

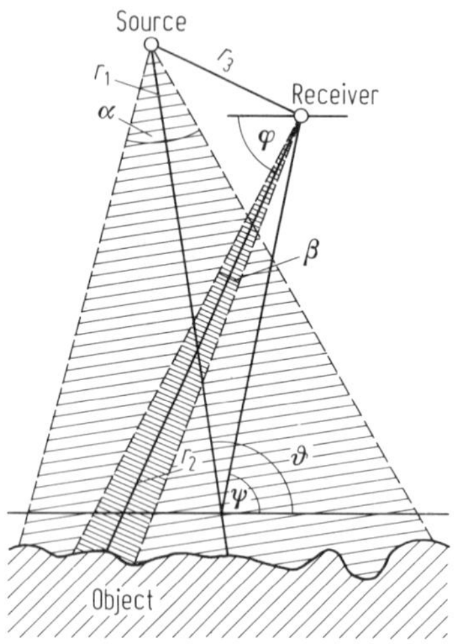

Fig. 4. General configuration of active remote sensing setup. α: source beamwidth, β: receiver beamwidth, ϑ: source angle, ψ: receiver angle, φ: bearing angle. r_1: source-object distance, r_2: receiver-object distance, r_3: source-receiver distance.

3.2.3.1.2 Connection with environmental properties

The acoustically effective properties of the object scene must be concluded from the received signal via inversion of sound propagation models by which the wave equation (2) is approximately solved (subsect. 3.2.2). The propagation model is also a tool to determine the propagation loss (subsect. 3.2.6) which together with the ambient noise (subsect. 3.2.7) and the reverberation (subsect. 3.2.8) is a limiting quantity for the range of RSM (subsect. 3.2.2.3).

The connection between the space-time distribution of sound speed and acoustic impedance as the only environmental properties directly affecting the sound signal and properties of material and state of the subject scenario such as density, salinity, temperature (subsect. 3.2.5) or geological and sedimentological quantities can mostly be determined only empirically and the inverted problem seldom has a unique solution. This ambiguity may at best be diminished by a priori information and by indirect conclusions from the RS of object shape and motion, which itself comprizes the most important application of active RSM. For attribution of the acoustic radiation of objects to processes originating the radiation, one is similarly dependent on empirical parametrization.

The variety of applications listed in Table 2 may be combined into 5 groups which differ with respect to the acoustic object properties predominantly utilized by the RSM.

3.2.3.1.3 Sound reflection/geometrical contour

- bathygraphy
- sediment stratification
- morphology
- thermoclines
- internal waves
- gravity waves
- sea level
- fish shoals
- plankton
- vehicles

Measuring principle

The most utilized configuration of echo sounding applies the same transducer as source and receiver (Fig. 4); thus $r_1 = r_2 = \frac{c}{2}\Delta\tau$ giving the object range within the illuminated surface under the beamwidth α, $\Delta\tau$ being the to and fro travel time. The echo level depends on the impedance jump and on the contour geometry in respect to specular reflection or scattering (subsect. 3.2.8).

The bottom surface and the sediment stratification as well as fish shoals and thermoclines are monitored along sounding lines of the vessel as carrier of a downward looking echo sounding device. Fan-shaped multibeam echo sounders yield a stripwise bathygraphic mapping of the sea bottom by simultaneous sounding at stepped declination angles $\vartheta = \varphi$, transverse to the ship's course. Their strip-width is limited by refraction errors and low bottom backscattering to about twice the water depth. For the strip-scanning of the microrelief, side-scan SONAR systems are utilized which are towed close to the bottom and achieve the highest structure resolution by the use of millimeter waves and a narrow horizontal beam α but without beam-forming in elevation angle. The side-scan method therefore yields only slant range information of the bottom morphology as intensity pattern without direct information on the third dimension.

The vertical motion of the sea surface and of thermo-haloclines as well as plankton scattering layers may be measured from the sea bottom (or a spar buoy) by upward looking echo sounders. By combined echo sounders the surface slope and hence the direction of wave propagation can be determined.

Horizontal sounding is essentially used for the detection of fish shoals, fixed objects, and underwater vehicles. Panorama and sector panorama systems are horizontally fan-shaped. For active SONAR detection purposes transmitter and (several) receivers may be dislocated.

Performance characteristics

The contour resolution is given by the pixel size and the bandwidth of the transmitted signal (subsect. 3.2.4). The highest angular and range resolution of 0.2° and 1.5 cm, respectively, is achieved by side-scan SONAR technique at frequencies of about 500 kHz. For bathygraphy of the continental shelf, signal frequencies of 30 kHz and beamwidths α, β of 1.5°···20° together with range resolution up to 0.1 m are typical. Cargo and fishery ship echo sounders utilize frequencies of > 100 kHz but moderate angular resolution. A deep-sea sounding requires lower frequencies of 10···15 kHz to reduce the propagation loss (subsect. 3.2.6). Corresponding beamwidths are 2°···20°; depth resolutions 0.3···1 m or worse for oblique sounding, depending on the beamwidth. High angular and depth resolution requires vertical stabilization of the sounding device and correction of the vertical ship motion.

The achievable range is of the order of 10 km and 30 km for deep-sea echo sounders and long-range side-scan SONAR, respectively. For shelf echo sounders the ranges are about one order of magnitude less and for high-frequency side-scan SONAR and sea state echo sounders they are 0.1 km and less. The space-time variability of the stratification of the ocean affects the achievable range most severely for the horizontal ranging, less for oblique directions (fan systems), and is almost negligible for vertical sounding. At high sea states, screening by bubble layers limit sounder operation (subsect. 3.2.6).

3.2.3.1.4 Sound refraction/sound speed field

- Ocean sound speed field
- Ocean current field
- Sea bottom sound speed profile
- Tides, internal waves

Measuring principle

The three-dimensional variability of the sound speed in the deep ocean which essentially corresponds to the space-time temperature distribution is remotely measured via the signal travel time along horizontal ray paths. A number of sources and receivers at the edges of the sea area of interest yield a system of one-way transmission ranges which cross inside the area. The travel times provide a set of equations of the average sound speed within adjacent subareas along the ray path (ocean acoustic tomography). Thus the area is sampled by a mesh-width which corresponds to the number and distribution of the propagation path crossings. The vertical stratification is sampled, utilizing the properties of the refractive deep-ocean channel (subsect. 3.2.6.2), where sound rays emanating at different elevation angles pass through different depth intervals when being refracted towards the channel axis. Usually some ten eigenrays through both source and receiver can be discriminated. The inversion of the propagation model assumes small perturbations by the horizontal variability; i.e. the signal arrival pattern allows identification of the rays.

By inverting the direction of propagation (exchanging transmitter and receiver operations) the current field can be determined from the upstream and downstream travel time difference.

The vertical profile of the speed of compressional and shear waves in the sea bottom can be determined by travel time measurements of the refracted (or reflected) signal at several receiving positions, when the sound speed increases with depth. These multiple oblique seismic sounding methods can not be applied to measuring the vertical sound speed profile of the water column, because the sound speed variability in the sea water is too small.

Large-scale motions of sea level and internal stratification can be detected by interference methods even in case of small amplitudes. This is done by measuring the time fluctuation of a low-frequency monochromatic signal propagation along a fixed source-receiver range. The fluctuations are caused by the shift of eigenmodes within the acoustic wave guide due to changes of its geometry. Periodic motions produce a corresponding line spectrum of the signal fluctuation.

Performance characteristics

The first fullscale tomography experiment was realized as $300 \times 300\,\mathrm{km}^2$ area; 5.5 km depth and at 26 °N, 70° W. The array was formed by 4 omnidirectional transmitters at 220 Hz, pulse-length 50 ms and 5 receivers at 2 km depth. The temperature resolution is ≈ 0.2 K; range-depth resolution is $\approx 10\%$.

3.2.3.1.5 Sound scattering/object features

- fish, fish shoals
- plankton
- air bubble layers
- surface waves
- sediments

Measuring principle

Remote sensing by scattered sound is used for object scales smaller than or comparable with the signal wavelength. Contrary to the echo sounding of the contour geometry (subsect. 3.2.3.1.3) where the scattered sound is used merely to indicate the object surface at off-specular aspect angles, the various scatterometry methods additionally utilize the intensity of the scattered sound and its dependence on one or more of the quantities: frequency f and angles ϑ, ψ, φ (Fig. 4). For remote measurement of the motion of the scatterers and scattering structures the frequency spectrum f' of the scattered sound, corresponding to the Doppler shift (subsect. 3.2.3.1.6) and the Doppler spread, is used. The majority of the measuring devices is restricted to backscattering ($\varphi = \vartheta$). The scattering processes in the sea are divided into volume scattering and boundary (surface) scattering, which are treated by different model concepts.

Volume scattering (subsect. 3.2.8.2) predominates inside the water column, originated by biological objects (fish, marine mammals, plankton), air bubbles as product of organic matter, and by sound speed inhomogeneities. However, it is also present at or near the ocean boundaries: air-bubble layers from breaking waves and ship wakes near the sea surface as well as gaseous inclosures and stochastic inhomogeneities from coarse sand, gravel and stones inside the bottom sediment.

For remote sensing of the object features, the volume scattering is conveniently treated as incoherent superposition of the intensity contributions from individual scatterers assumed as independent. The frequency response of the scattered sound is different whether or not the objects contain or consist of gaseous inclosures.

The bubble-type objects act as resonators. Their scattering cross section can exceed the geometrical size by orders of magnitude, thus producing a strong contrast of gaseous scatterers against other objects of comparable size.

The second object type is characterized by higher impedance than the surrounding water which yields acoustic cross sections lower or, at most, equal to the geometrical size. The scattering cross section of such "hard" objects decreases by the second power of frequency ("blue" sound scattering). If the object impedance is moreover close to water (organic tissue) they can be recognized against the ever present background (subsect. 3.2.2.4) only when their size is at least comparable with the wavelength.

Surface scattering occurs at the free and ice-covered sea surface (subsect. 3.2.8.3), the sea bottom and sea-bottom interfaces (subsect. 3.2.8.4). Steep and thin thermoclines or thermo-haloclines in the ocean volume may behave as acoustic interfaces as well.

Surface scattering may be treated as "resonance" (Bragg) scattering (subsect. 3.2.8), the stochastic corrugated surface acting as diffraction grating, superimposed with facet reflection at the swell, sloping bottom and ice cover. The sea surface backscattering which is nearly independent of frequency in a wide wavelength interval rather limits the detection range (subsect. 3.2.2.4) of active RS against other targets than being subject of remote sensing.

The steep decrease of the boundary scattering beyond the Bragg limit allows to discriminate the grain size of objects on the sea bed (boulders, gravel). Particularly useful for remote recognition and classification of sediment types and ice coverage is the combination of structure-resolving methods and scatterometry for the low-ka fractions. Both components are incorporated in the side-scan SONAR method, yielding grain-size information ("gray scale") together with the shape of resolved large objects. By this tool various types of sediment transport and erosion can be recognized from remote which may also allow conclusions on the originating currents.

Performance characteristics

For estimation of population densities of fish shoals which are beaconed by their swim bladders, the echoes are counted or integrated under the assumption of independent scattering. At resonance of the swim bladders population densities down to 1 fish/10^5 m^3 can be determined. The scatterometry of bottom features essentially utilizes empirical data for classification (subsect. 3.2.8.4). Sound speed inhomogeneities of the volume produce extremely low backscattering (subsect. 3.2.8.2), usually masked by scattering from bubble-carrying biological matter. Thus the space-time structure of thermo- and halo-clines is visible for acoustic RS only via biological tracers.

3.2.3.1.6 Doppler shift/object velocity

- vertical current profile
- vehicle speed
- fish motion

The Doppler shift/spread of reflected and scattered signals is used for the RS of the object velocity or velocity distribution (Eulerian description) as supplement or instead of information from the time function of the object position (Lagrangean description). Since the speed of sound in the sea is some 10^5 times lower than the speed of light in the atmosphere, the acoustic measurement of the Doppler frequency is a comparatively sensitive tool for velocity determinations. It is preferably used to measure the eigenspeed of ships (subsect. 3.2.3.2.1) and the radial component of the speed of other vehicles or fish and fish shoals.

The RS of current profiles requires sufficient volume scattering in the ocean and is further identical with the SONAR-Doppler method (subsect. 3.2.3.2.1) by selecting time windows of vertical adjacent scattering horizons instead of the sea bottom as reference. The components of the current vector are derived from the backscattering of three oblique looking echo sounders (subsect. 3.2.3.2.1) placed on the sea bottom or at a ship's hull. An accuracy/resolution of ≈ 0.1 m/s and $5 \cdots 10$ m depth interval requires frequencies > 100 kHz, limiting the profile depth to a few 100 m.

3.2.3.1.7 Sound radiation/motion processes

- wind speed
- precipitation
- seaquakes
- motor vessels
- fish/crustacea

Ambient noise which is produced by a large number of uniform stochastic processes can be described by its broad band spectrum (subsect. 3.2.7). This covers the sound production by the wind-induced breaking sea (subsect. 3.2.7.6) and by precipitation (subsect. 3.2.7.8). Wind speeds beyond the breaking-wave threshold can be determined from the high-frequency noise spectrum level within about 20%, using empirical parametrization. The discrimination against rain noise needs a-priori knowledge.

The description of transient noise events in the time-frequency domain yields clues which may be used for the classification and identification of the noise sources. This concerns the aquisition of seaquakes, localized by triangulation and estimated for magnitude by the received noise level (subsect. 3.2.7.4). Certain transients like the motion of boulders, sea-wave-induced sand transport or breaking surf may be identified simply by listening.

3.2.3.2 Signal transmission methods

Table 3. Methods of acoustic signal transmission.

Application	Acoustical setup	Signal frequency
Navigation SONAR-Doppler, dead reckoning	oblique echo sounder, forward + lateral components, double-sided for pitch-roll compensation integration of ground truth vessel speed, utilizing Doppler shift [75 B 2]	150 kHz
Position indicator	horizontal echo sounder for manoeuvring large ships in harbours [72 L]	100 kHz
Communication underwater telephone	wide-angle transmitter/receiver [72 D 3]	8···10 kHz
Data link	wide-angle transmitter connected to underwater sensor; receiver connected to vessel or buoy [84 B]	1···10 kHz
Beacon identification marker	transponder or omnidirectional pinger connected to measuring equipment or underwater construction [77 H] passive reflector/resonator [66 M 2]	1···10 kHz
Remote control equipment control/release telecommand of un-manned sensor/working vehicles	wide-angle transmitter/receiver on both sides for command and confirmation [79 N, 84 J]	1···10 kHz

3.2.3.2.1 Navigation

Most underwater acoustic navigation systems use the Doppler shift of bottom echoes which do not provide absolute positions but dead reckoning to be combined with reference positions. The method yields both horizontal ship-path components by integrating the ship's speed which is derived from the corresponding Doppler shift of forward and sideward looking oblique echo sounders. The Doppler-shifted frequency f' for each component is connected with the corresponding speed component u and the sound speed c at the transducer by

$$f' = \cos\varphi \cdot f_0 \left(1 \pm \frac{2u}{c} \pm \text{terms negl.} \right), \tag{15}$$

where f_0 is the frequency of the transmitted signal and φ is the grazing angle of the beam axis (Fig. 4). The second-order term in eq. (15) is the result of the different Doppler shift for transmitter and receiver motion, respectively.

The method is not disturbed by sound speed stratification (subsect. 3.2.5) because a change of the cosine of the grazing angle φ according to Snell's law is compensated by the same change of the Doppler shift. Errors from pitch and roll of the carrier ship are corrected by bow-stern and abeam pairs of echo sounders (Janus-setup).

The necessary Doppler resolution requires high frequencies (>100 kHz) which limit the dead-reckoning navigation mainly to continental shelf regions. For the deep ocean only scattering layers (subsect. 3.2.8.2.3) can be used. In combination with satellite navigation a position accuracy of the order of 10 m may be achieved by bottom track mode.

Another acoustic navigation principle is related to fixed reference transmitters using phase difference methods (hyperbola navigation) or propagation time methods (transponders). Both versions are mainly temporally used during the time of construction of large underwater structures and are therefore preferably short range-high accuracy systems (±0.1 m).

3.2.3.2.2 Communication

The underwater sound telephone is the most frequently used communication between divers, diving facilities, submarines and their auxiliary surface vessels. Omnidirectional transducers on both sides are operated, alternating in the transmitting and receiving mode. As a compromise of necessary speech bandwidth, transmission range and power radiation, frequency-modulated signals of 8···10 kHz carrier frequency are customary. The range is of the order of a few km.

3.2.3.2.3 Data telemetry

Acoustic data transmission is a suitable cableless link between distributed deep-sea bottom-mounted or neutrally buoyant drifting oceanographic measuring setups and central recording stations when simultaneous real-time field measurements are required. Combinations with buoy-radio links for far-distance or air- and space-borne reception are possible. The achievable transmission range or duty time, respectively, is mainly restricted by the battery capacity of the self-sufficient stations in the sea.

3.2.3.2.4 Beacons

The retrieval of anchored or neutrally buoyant drifting equipment requires acoustic beacons. There are versions where the code is transmitted at fixed time intervals and others where the cryptographic coding of a request signal releases the beacon signal (transponder) of the equipment to be detected.

Further, passive markers are in use, such as triplane targets which produce strong echoes approximately independent of aspect angle. Gas-filled resonators which strongly increase their scattering cross section at resonance can be used as passive beacons as well by producing amplitude-modulated echoes from frequency-modulated signals.

3.2.3.2.5 Remote control

The field of application covers the acoustic tele-command, control and positioning of equipments, tools and vehicles mainly in connection with off-shore drilling and production rigs. The operational modes of bottom-mounted measuring equipments or their release for return to the sea surface can be activated from remote, including measures against erroneous or intentional abuse.

3.2.3.3 In-situ methods

Table 4. Acoustic in-situ sensors.

Quantities measured	Methods
Sound speed	transmission time (sing around [57 G]) interferometer [73 K]
Current	upstream – downstream transmission time difference 1···3 components [76 W]
Air suspension bubble size spectrum	frequency-selective extinction [80 M 1]

3.2.3.3.1 Sound speed

Two basic measuring principles are in use which differ in robustness and accuracy:

The so-called sing-around method determines the propagation time of a pulsed signal within a probe range, where the received signal triggers the following transmitted pulse. The resulting pulse repetition frequency is proportional to the local speed of sound whereas the acoustic length of the probe range (usually 0.1 m) including internal trigger delay is a constant of the setup. The resulting accuracy is of the order of 0.1 m/s, corresponding to a temperature of about 0.03 °C (subsect. 3.2.5.4).

The complementary measuring method uses interferometry. Operating the transmitter-receiver range in a cw-mode, the signals from both transducers are superimposed electrically. Constructive or destructive interference is adjusted by tuning the frequency which is the measuring quantity. The absolute accuracy is at least one order of magnitude better than the propagation time method.

3.2.3.3.2 Current

Frequently applied acoustic current meters are based on the difference of the up- and downstream signal travel time. The current probe consists of a pair of transmitters and receivers for all three Cartesian coordinates, thus yielding the flow vector at the probe. By probe dimensions of 0.1 m and frequencies of some 100 kHz, a threshold sensitivity of 1 mm/s is achieved which is one order of magnitude better than mechanical current meters.

3.2.3.3.3 Air suspension

Air and gas bubbles generated by breaking waves (subsect. 9.5.3) and by metabolism and decomposition of organic matter (subsect. 3.4.3.4) change the acoustic properties of the sea water more seriously than any other suspension. Since the scattering and absorption cross sections of bubbles at resonance exceeds their geometrical cross section by far (subsect. 3.2.8.2) acoustic methods are a sensitive tool for bubble spectrometry.

By a setup of two opposite plane parallel mirrors, one of them being a transmitter/receiver, the attenuation excess due to the bubbles is determined as a function of frequency from the decrease of the pulse amplitudes within a sequence of multiple echoes. This in-situ method is more accurate than semiquantitative remote sensing methods (subsect. 3.2.3.1.5).

3.2.4 Sound transducer characteristics

3.2.4.1 Sound sources/receivers

Devices that convert electric into sound energy and vice versa are called transducers. Most RS-systems for ocean acoustics apply forced length oscillations of solid state piezoceramic PZT elements (lead-zirconate-titanate [63 S 1]) via periodic electric fields. The necessary volume displacement is achieved by mounting the elements in closed baffles (ship hull) or various diaphragm constructions [75 S].

Receiving response $M = E/p_i$ (ratio of output electric voltage E and incident sound pressure p_i) and transmitting response $S = p_r/I$ (ratio of sound pressure p_r at distance r and input current I) are connected by the reciprocity parameter [47 E]

$$J_{\text{sph}} \equiv \frac{M}{S} = \frac{2r}{\varrho f} \tag{16}$$

which holds for quasi-spherical waves in the far field $r > d^2/2\lambda$ of the transmitter of diameter d [61 H].

The source level SL, in [dB], defined as $SL = 10\log(I_0/I_{\text{ref}})$, where the reference intensity I_{ref} is that of a plane wave of rms pressure 1 µPa and the intensity I_0 refers to 1 m standard distance from the source (Table 1), has to be measured in the far-field and corrected for spherical spreading (subsect. 3.2.6.2.2). The relation between SL, the acoustic power P_a, the electric power P_e (both relative to 1 W) and the efficiency $\eta = P_a/P_e$ is [75 U 1]:

$$SL = 171 + 10\log P_e + 10\log\eta, \tag{17}$$

which holds for a nondirectional source. The efficiency ranges from 20 to 70% for PZT and is highest for sources at mechanical resonance, limiting their half-power bandwidth to 10···40%.

The response M of a receiver element with electric field coupling such as PZT is independent of frequency f below its mechanical resonance whereas S increases $\sim f$ according to eq. (16). Enlarging the range of constant frequency response of M requires the reduction of element size and therefore S [66 T].

The achievable SL is limited by cavitation, corresponding to a sound pressure of ≈ 1 atm or 0.3 W/cm^2 close to the sea surface and near 1 kHz, increasing to >10 atm at 100 kHz (30 W/cm^2) [52 E]. The cavitation threshold increases also by the amount of the static pressure [64 F] and decreases slightly by pulse length [53 L].

Transducers of ship-borne echo sounding devices must be shielded by acoustically transparent "domes" from the turbulent boundary layer, generating self-noise [85 S 3], the spectrum level of which is usually higher than that of ambient noise.

The low transmitting response at low frequencies, eq. (16), for source diameters $d \ll \lambda$ would require diaphragm amplitudes which exceed the elastic range of solid state material. Therefore the transient expansion of gaseous volumes, such as chemical explosives [48 C] or high-pressure "air guns" [84 H] are applied at frequencies below some 10^2 Hz which are necessary for deep penetration of the sea bed (seismic prospecting). The empirical source level value, in [dB], of 1 kg TNT charge is $SL \approx 205 + 10 \log \Delta f$ at 100 Hz, Δf in [Hz], and increases by less than 10 log (charge weight), charge weight in [kg]. The spectrum slope is close to 10 dB per frequency decade between 10^2 and 10^3 Hz for fuse depths close to the surface (40 m) [60 W 1].

3.2.4.2 Directional resolution

The receiving sensitivity M of small transducers (diameter $d \ll \lambda$) is independent of the incident angle of the sound wave. Devices for angular resolution of incident sound are transducer arrays of diameter $d \gg \lambda$ where the elements are positioned within a plane oriented parallel to the incident quasi-plane front of interest, when the sum of the element contributions is maximum. The elements may be distributed in a volume or on a curved or oblique surface. Their contributions are added after phase correction to one or more simultaneous "preformed beams" [77 T 1].

The function $20 \log M(\vartheta, \varphi)$, M in [$V \mu Pa^{-1}$], is called the receiver beam pattern. The direction of M_{max} is called the beam axis. According to reciprocity, eq. (16), the angular dependence of the receiver and source beam pattern of a particular transducer array are identical.

The beam pattern in the domain of the projected wavenumber $k' = (2\pi/\lambda) \sin \Theta$, and the amplitude distribution at the transducer array aperture are Fourier pairs. Thus any undesired side lobes of the beam pattern which produce directional ambiguity may be suppressed by appropriate shading of the transducer aperture [48 W]. The half- and tenth-power beamwidth of circular plane and line transducers as a function of frequency is depicted in Fig. 5. It is based [66 T] on the normalized beam pattern function M'_{cp} of a circular piston of diameter d in an infinite baffle:

$$M'_{cp}(k') = 2 \frac{J_1(k'd/2)}{k'd/2} \tag{18}$$

and of a straight line transducer of length l:

$$M'_l = \frac{\sin(k'l/2)}{k'l/2}. \tag{19}$$

Equations (18) and (19) hold for the unshaded arrays. (The unshaded) beam pattern M'_{lp} of a linear point array of n elements of spacing d, such as towed by a ship for low-frequency sea bottom RS reads [66 T]:

$$M'_{lp} = \frac{\sin(nk'd/2)}{\sin(k'd/2)}. \tag{20}$$

Due to the response benefit of receivers against transmitters at low frequencies, (eq. 16), directional devices for the long-wavelength regime usually consist of omnidirectional sources, combined with high-resolution receiving arrays.

Transducers which utilize the weak nonlinear interaction of two superimposed high-power sound waves of different frequency are called parametric transducers [67 B] and produce a sound signal at the difference frequency. This secondary beam is free of sidelobes whereby the beamwidth and bandwidth correspond to the primary high-frequency signals.

The directivity index DI, defined for a receiving device in Table 1 and derived by integration of eqs. (18) and (19), is depicted for the circular plane and line in Fig. 6 for isotropic noise.

The same DI is valid for the transmitting mode, according to reciprocity, and has to be added on the r.h.s. of eq. (17) for directional sources and coherent propagation.

The equivalent (two-way) beamwidths Ψ and Φ of the circular plane and line array required to describe the echo-sounding illumination of a volume V within the solid angle Ψ, in [sterad], eq. (63), and of a surface A within the angle Φ, in [rad], eq. (64), in the A-plane, are given by the following set of equations [75 U 1]

$$10\log\Psi_{cp}=20\log(\lambda/\pi d)+7.7\,, \qquad (21)$$

$$10\log\Phi_{cp}=10\log(\lambda/\pi d)+6.9\,, \qquad (22)$$

$$\left.\begin{array}{l}10\log\Psi_{l}=10\log(\lambda/2\pi l)+9.2\,,\\ 10\log\Phi_{l}=10\log(\lambda/2\pi l)+9.2\,.\end{array}\right. \qquad (23)$$

Replacing the conventional beam processing of phase correction and sum by advanced filter algorithms for the angle domain [85 S 2] may enhance the achievable angular resolution by a factor of 2···4.

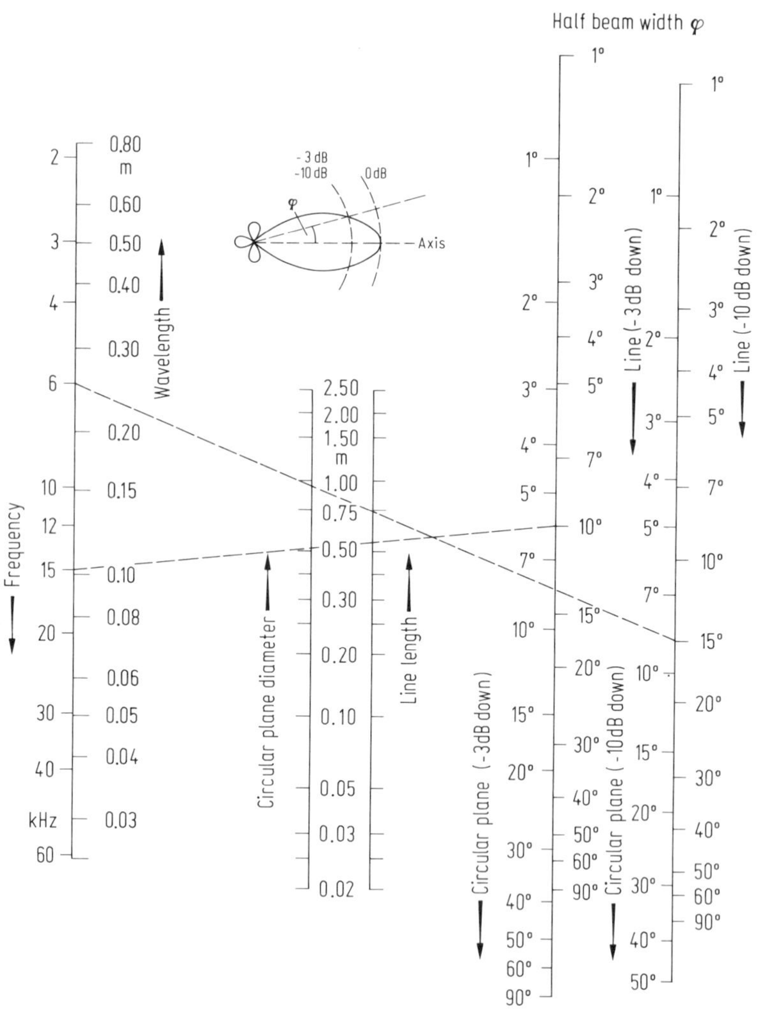

Fig. 5. Half beamwidth nomogram of circular-plane and line transducers [83 U].

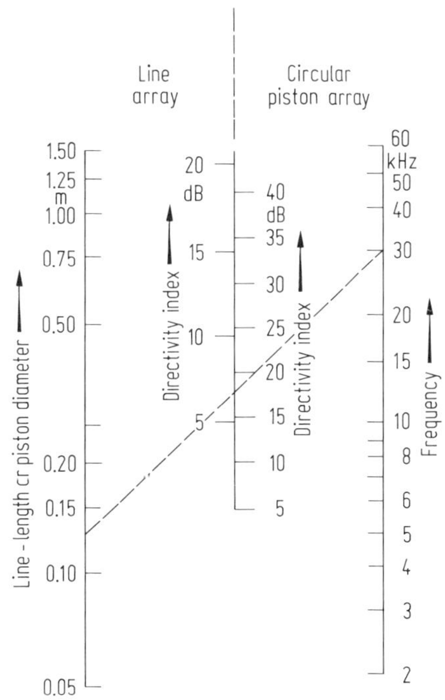

Fig. 6. Directivity-index nomogram for line and circular plane transducers in an infinite baffle [83 U].

3.2.4.3 Time/frequency resolution

The resolution Δf in the frequency domain of a conventional signal pulse (sinusoidal carrier, rectangular or shaded envelope) and the resolution Δt in the time domain are interconnected by the "classical uncertainty relation":

$$\Delta f \, \Delta t \approx 1 \, , \tag{24}$$

resulting from the Fourier transform relation between the descriptions of the signal in both domains. Enhancing the resolution for one domain thus means degrading for the other. The relation (24) also holds for sophisticated signals with time-bandwidth products $\gg 1$, which are utilized to overcome the cavitation limit of the source level (e.g., frequency-modulated, pseudo-random noise-signals). The corresponding value of Δt for eq. (24) is then the duration of the autocorrelation function of the signal at the output of a matched filter [60 L].

3.2.5 Sound speed

3.2.5.1 Definition, data base

The speed of sound in the inhomogeneous ocean is defined and measured as the phase speed parallel to the wavenumber vector in a small volume assumed as homogeneous. Since sound speed in ocean water is nondispersive [62 E 1] within the limits of measuring accuracy and isotropic as to all nonrigid media, both the signal frequency and the direction of propagation may be chosen from viewpoints of accuracy and low disturbance for in-situ sound speed determinations. The required accuracy of the sound speed measurement depends on the type of application. As far as signal travel time and Doppler shift are concerned, absolute values of the sound speed are required. In contrast, for those propagation problems related to sound field qualities, the distribution of sound speed gradients or sound speed differences at different positions are more important than their absolute values.

Sound speed in the sea varies by less than 8% from its nominal value of 1500 m/s. Though this variability is some 8 times smaller than for marine sediments [79 H], the sensitivity of long-range underwater sound propagation with respect to small changes of sound speed may require accuracies up to about 0.1 m/s or 1‰ of the variability. This limit has not yet been achieved for all combinations of temperature, salinity and pressure which occur in the open ocean [71 M 1].

Since the discovery in 1952 [52 D] that the earliest sound speed tables calculated 1939 by Kuwahara [39 K] from the compressibility and density of sea water were erroneous by 3···4 m/s, subsequent investigations have been based exclusively on direct measurements of sound speed. They have in common that they use standard "Copenhagen sea water" probes [01 K] for which sound speed is determined by "sing-around" techniques developed 1957 by Greenspan and Tschiegg [57 G] or interferometer methods [70 D 2] which claim higher accuracy (subsect. 3.2.3.3.1).

Up to now, only two fundamental determinations of the dependence of sound speed on all three quantities, temperature, salinity and pressure, within the main variability range in the sea have been published: Wilson's work of 1960 [60 W 2] which replaced the Kuwahara tables and the publication by Del Grosso and Mader whose data of 1972 [72 D 1] are now considered most reliable. The Wilson data certainly contain systematic deviations from the true sound speed values up to some 0.6 m/s at atmospheric pressure, as comprehensive intercomparisons [69 L 3] have demonstrated, but at high pressure of several km equivalent water depth their validity has not been refuted.

These two fundamental investigations are supported by other primary precision measurements of sound speed in pure water and sea water probes at atmospheric pressure [76 K 3] as well as by secondary measurements for pressure dependence, the latter referring to standard pure-water sound speed values as calibration data [76 C 2]. In-situ measurements of comparable accuracy, especially at great depths, are very difficult and have been realized only by few examples [73 M].

Wilson as well as Del Grosso and Mader have developed approximation formulae (subsect. 3.2.5.3) fitted to their data, the considerable complexity of which has caused other investigators to develop simplified formulae (subsect. 3.2.5.4.3) for practical use which are restricted to realistic combinations of the temperature, salinity and pressure range [81 M 2]. Moreover, merged formulae have been suggested [78 L 2], combining results of other authors to achieve a best estimation of overall accuracy.

3.2.5.2 Specifications of the parameters determining sound speed

3.2.5.2.1 Variability of temperature, salinity and depth combinations; "Neptunian waters"

The variability range of the three single parameters, pressure, temperature and salinity, which determine sound speed bound a variability volume which is only filled out to a small fraction by the combination of all three quantities occurring in the open ocean and marginal seas. Leroy [69 L 1] has therefore recommended to restrict sound speed approximation formulae to that part of the variability volume which corresponds to existing waters. For this purpose he introduced the concept of "Neptunian waters" which identifies all the waters of the world that are interconnected by natural waterways deep enough to allow mediumsize ships to transit freely.

The possible combinations of temperature, salinity and pressure found in the Neptunian waters are represented in Fig. 7, forming a three-dimensional volume as seen looking upwards to the surface. The complex structure of this variability volume is caused by the semi-closed basins of the marginal seas and their incomplete mixing with the open oceans. Fig. 8 presents three sections across this volume at 0 m, 500 m, and 3000 m depth, respectively.

Most of the surface waters (Fig. 8) come from the open oceans and have salinities between 33 and 37‰. Higher salinities are found in the Mediterranean (area A), the Red Sea (area B) and the Persian Gulf (area C). Low salinities occur in a number of different regions. The two most important are the Black Sea (area D) and the Baltic Sea (area E). Area F corresponds to the cold low-saline waters mostly found along the coasts of the Arctic and in the Bering and Okhotsk Seas. The warmer low-saline water of area G corresponds mainly to tropical zones of heavy precipitation and river inflow, primarily the Gulf of Guinea and the Bay of Bengal. Extremely anomalous waters in "hot spots" (e.g., bed of the Red Sea), however, are not considered here.

Below the first 400 m of complicated variability structure only a small number of different water types are left over, each with a particular relationship between salinity and temperature. A list of the semi-closed basins related to the peaks in the variability volume is given in Table 5. The table also gives the isothermal temperature T_i and the constant salinity S_i reached below the sill depth. Special mention must be made of the areas indicated by α and β in Figs. 7 and 8. These correspond to the Mediterranean and Red Sea waters flowing at depth into the Atlantic and Indian Oceans, respectively. Owing to their higher salinities they sink along the bottom slopes, slowly losing their salinity and temperature. All water masses below some 7000 m are located in small basins, troughs or trenches for which only the outer limits of the very narrow variability volume have been indicated in Fig. 7.

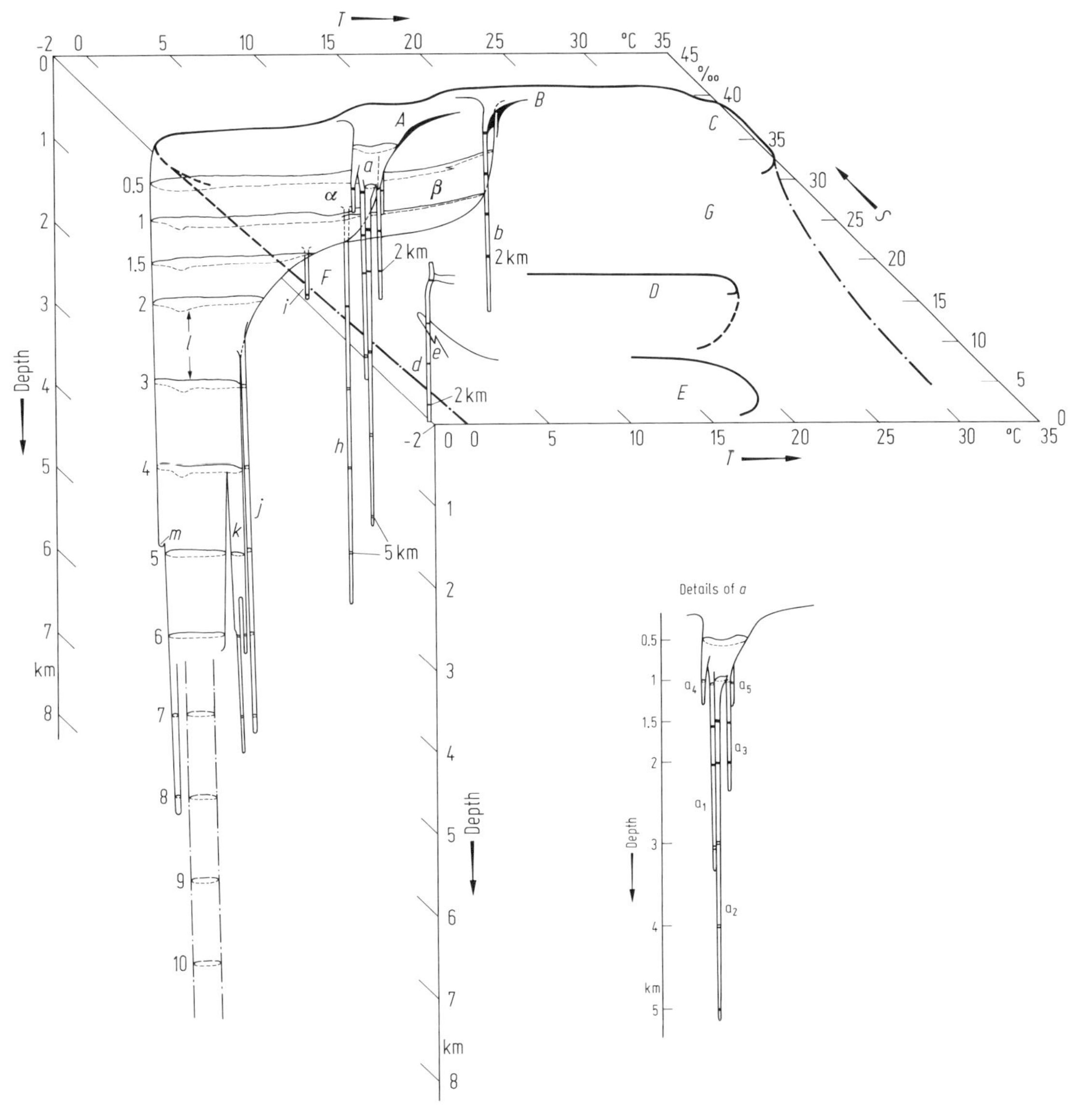

Fig. 7. Three-dimensional representation of the temperature-salinity-depth volume inclosed by real ("Neptunian") waters [69 L 1].

Sea area	Surface water	Deep water	Water flowing into adjacent oceans
Mediterranean	A	a[a])	α
Red Sea	B	b	β
Persian Gulf	C	c	
Black Sea	D	d	
Baltic	E	e	
Arctic coastal waters	F		
Tropical coastal waters	G		
Sulu Sea		h	
Halmahera Basin		i	
"American Mediterranean"		j	
E. Indian Basins		k	
Sea of Japan		l	
Arctic Ocean		m	

[a]) a_1–a_5: see Table 5.

Wille

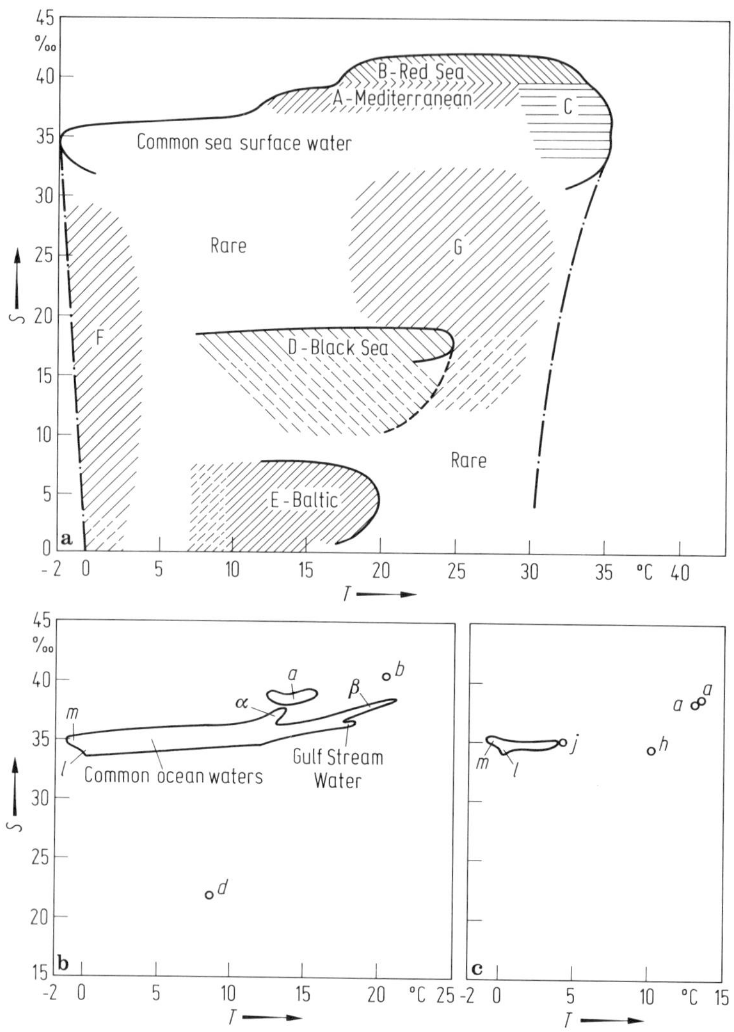

Fig. 8. Cross sections of the Neptunian waters shown in Fig. 7 (a) at surface, (b) at 500 m depth, (c) at 3000 m depth.

Table 5. World's main closed basins (marginal seas) with sill depths much shallower than their maximum depths [69 L 1].

Basin	Ref. on Figs. 7, 8	Sill depth m	Max. depth m	T °C	S ‰
Western Mediterranean	a_1	300	3600	13.0	38.4
Eastern Mediterranean	a_2		5100	13.6	38.7
Sea of Crete	a_3		2500	14.4	38.9
Sea of Thrace	a_4		1500	12.7	38.8
Sea of Marmara	a_5	55	1300	14.3	38.5
Red Sea	b	100	2800	21.5	40.5
Black Sea	d	35	2200	8.8	22.0
Baltic Sea	e	20	400	4···6	12···14
Sulu Sea	h	400	5600	10.3	34.5
Halmahera Basin	i	700	2000	7.8	34.6
"American Med." (Caribbean, etc.)	j	1600	7200	3.8	35.0
Miscellaneous East Indian Basins (Celebes etc.)	k	1400···3000	3700···7400	3.2···4.2	34.6
Sea of Japan	l	160	≈4000	0.2	34.1
Arctic Basin (Polar Basin and Norwegian Sea)	m	2×500	≈4800	≈−1.0	≈35.0

3.2.5.2.2 Temperature scale

The most frequently applied sound speed data of Wilson [60 W 2] and Del Grosso and Mader [72 D 1] have been based on the different international temperature scales T_{48} of 1948 and T_{68} of 1968, respectively [72 D 2]. Both scales which utilize different fixed points do not differ more than 0.01° within the variability range of ocean temperature. This corresponds to ≈0.03 m/s sound speed difference which is negligible with respect to the claimed measuring accuracy.

3.2.5.2.3 Depth-equivalent of pressure

For the calculation of the absolute pressure in the oceans from the knowledge of only the depth z and the geographical latitude Φ, an equation was presented by Leroy 1969 [69 L 1] which yields an error limit of less than 0.1 m/s sound speed, eq. (25). This accuracy corresponds to 8000 m depth and is still better at shallower depths. The equation is of universal use for underwater acoustics within "Neptunian waters" with the exception of the Black Sea and the Baltic Sea for which separate equations are given, eqs. (26) and (27), respectively.

$$P(z,\Phi)=1.04+0{,}102506(1+0.00528\sin^2\Phi)z+2.524\cdot10^{-7}z^2, \tag{25}$$

where P is in kg/cm² gauge (1 kg/cm² ≙ 0.980665 · 10⁵ Pa), z is in [m] and Φ is the latitude in degrees,

$$P(z)_{\text{Black Sea}}=1.03+0.10168z+2.6\cdot10^{-7}z^2 \tag{26}$$

$$P(z)_{\text{Baltic}}=1.03+0.1008z+1.4\cdot10^{-6}z^2. \tag{27}$$

The formulae are based on the simplified "international gravity formula" neglecting gravity anomalies and using a straight-line approximation of the equation for density, thus assuming a standard salinity. Only for the two low-salinity areas of the Black Sea and the Baltic Sea the standard density approximation has been corrected.

3.2.5.2.4 Salinity concept

Employing "salinity" as an input quantity for sound speed equations the assumption is made that the ratios of the component solutes of the sea water are constant [73 D 1]. Lord et al. [65 L] quote the UNESCO reports as indicating that ionic concentration ratios with respect to chlorinity vary by 0.15% for SO_4^{--}, 0.25% for Ca^{++}, and 0.13% for Mg^{++} with no geographical or depth trend for sulfate but with Ca^{++} increasing with depth,

Mg^{++} low in the deep Mediterranean and high in the surface Atlantic. Lord et al., using the variation above and extrapolating from their Mg^{++} and So_4^{--} measurements reinforced by the parallel conclusions obtained from Weissler and Del Grosso [51 W], estimate that at most there would be a 0.08 m/s uncertainty for a real salinity determination. As a proper measure for salinity the conductivity was recommended rather than chlorinity. In-situ sound speed measurements at oceanic depths have been made only as selected examples [73 M].

3.2.5.2.5 Influence of suspensions

The sound speed of gaseous or solid suspensions in water is governed by the effective compressibility and density, eq. (3). Air bubbles as produced by breaking waves change the compressibility of the water more strongly than the density, thus reducing the sound speed. At 4···5 m/s wind speed, reductions up to 1% have been observed 5 m below the sea surface. The effect will be stronger at higher wind speeds. At bubble resonance (subsect. 3.2.8.2) the sound speed becomes dispersive [77 C].

Suspensions of sediment change the density of the water more than the compressibility – the sound speed is also reduced but much weaker than by air suspensions. A 1% reduction requires a sediment concentration of ≈3% volume, several orders of magnitude more than the bubble suspension [83 H 1].

3.2.5.3 Tables of sound speed

Reliable sound speed tables covering the entire range of the Neptunian waters are not available at present. Since pure-water sound speeds have been determined at atmospheric pressure with highest accuracy [57 G, 68 C, 72 D 2, 76 K 3] they are presented separately from the sea water tables which are derived from the original measured data by approximation formulae of the authors.

3.2.5.3.1 Sound speed table for pure water under atmospheric pressure

The high precision data of Del Grosso and Mader [72 D 2] as well as those of Kroebel and Mahrt [76 K 3] appear to be appropriate as standard for distilled water calibration of sound speed meters. The data sets are in closer agreement than others and differ at most by 7 cm/s.

Because the accuracy of the Kroebel-Mahrt pure-water table is possibly slightly superior to that of Del Grosso-Mader it is chosen for Table 6. The table values are calculated by a 6-term approximation formula (subsect. 3.2.5.4.1) fitted with 0.02 m/s standard deviation to the original measurements which cover the temperature interval of 3.3 °C to 33.8 °C (T_{68}-scale) without gaps of more than 2.5 °C.

3.2.5.3.2 Sound speed tables for sea water; temperature, salinity and pressure dependence

According to recent exhaustive intercomparisons by Mackenzie [81 M 1] the primary measurement data of sound speed in sea water published by Del Grosso and Mader 1972 [72 D 1] and their approximation formula (subsect. 3.2.5.4.2) fitted to these data within 0.05 m/s standard deviation are suggested for general use within the range of

30···41‰	salinity
0···40 °C	temperature
0···1000 kg/cm²	pressure

(1 kg/cm² ≙ 0.980665 · 10^5 Pa) which includes the Mediterranean Sea and the Red Sea but not the Black Sea and the Baltic Sea.

For practical use the tables of 1973 [73 D 1] are reproduced by the separate Tables 7···10 corresponding to the sound speed increments by increase of temperature T, salinity S and pressure P and a mixed TSP dependence. The final sound speed value is derived from the tables by the equation

$$c = C_{000} + \Delta C_T + \Delta C_S + \Delta C_P + \Delta C_{TSP}. \qquad (28)$$

C_{000} representing the reference sound speed of 1401.869 m/s at zero salinity, temperature (°C) and pressure.

Table 10 of the interaction term ΔC_{TSP} is to allow linear interpolation and is restricted to the parameter range of Neptunian waters. Thus, the temperature range displayed decreases for the successive tables of increasing pressure.

3.2.5.4 Approximation formulae of sound speed

3.2.5.4.1 Approximation formula of sound speed in pure water under atmospheric pressure

The suggested formula has been developed by Kroebel and Mahrt [76 K 3] and is the basis for Table 6. The range of validity is mentioned in subsect. 3.2.5.3.1.

$$c = \sum_i K_i T^i \tag{29}$$

$K_0 = 0.140238754 \cdot 10^4$ $\quad K_3 = 0.334198834 \cdot 10^{-3}$

$K_1 = 0.503711129 \cdot 10^1$ $\quad K_4 = -0.147800417 \cdot 10^{-5}$

$K_2 = -0.580852166 \cdot 10^{-1}$ $\quad K_5 = 0.314643091 \cdot 10^{-8}$.

c is in [m/s] and T is in [°C].

The formula differs from Del Grosso's of 1972 [72 D 2] only in K_0 and K_1. The claimed accuracy is better than 0.04 m/s.

Table 6. Sound speed [m/s] in pure water at atmospheric pressure [76 K 3].

T_{68} °C		0.0	0.1	0.2	0.3	0.4	0.5	0.6	0.7	0.8	0.9
0	1400+	2.382	2.885	3.387	3.888	4.388	4.887	5.384	5.881	6.376	6.870
1		7.363	7.855	8.346	8.835	9.324	9.811	10.297	10.782	11.266	11.749
2		12.231	12.711	13.191	13.669	14.146	14.622	15.097	15.571	16.044	16.515
3		16.986	17.455	17.924	18.391	18.857	19.322	19.786	20.249	20.711	21.171
4		21.631	22.089	22.547	23.003	23.458	23.913	24.366	24.818	25.269	25.719
5		26.167	26.615	27.062	27.508	27.952	28.396	28.838	29.280	29.720	30.159
6		30.597	31.035	31.471	31.906	32.340	32.773	33.205	33.636	34.066	34.495
7		34.923	35.349	35.775	36.200	36.624	37.046	37.468	37.889	38.308	38.727
8		39.145	39.561	39.977	40.392	40.805	41.218	41.629	42.040	42.450	42.858
9		43.266	43.672	44.078	44.483	44.886	45.289	45.690	46.091	46.491	46.889
10		47.287	47.684	48.080	48.474	48.868	49.261	49.653	50.044	50.434	50.823
11		51.211	51.598	51.984	52.369	52.753	53.136	53.519	53.900	54.280	54.660
12		55.038	55.416	55.792	56.168	56.543	56.916	57.289	57.661	58.032	58.402
13		58.771	59.139	59.506	59.873	60.238	60.602	60.966	61.329	61.690	62.051
14		62.411	62.770	63.128	63.485	63.841	64.197	64.551	64.904	65.257	65.609
15		65.959	66.309	66.658	67.006	67.354	67.700	68.045	68.390	68.733	69.076
16		69.418	69.759	70.099	70.438	70.777	71.114	71.451	71.786	72.121	72.455
17		72.788	73.121	73.452	73.782	74.112	74.441	74.769	75.096	75.422	75.747
18		76.072	76.395	76.718	77.040	77.361	77.681	78.001	78.319	78.637	78.954
19		79.270	79.585	79.899	80.212	80.525	80.837	81.148	81.458	81.767	82.076
20		82.383	82.690	82.996	83.301	83.606	83.909	84.212	84.514	84.815	85.115
21		85.415	85.713	86.011	86.308	86.604	86.900	87.194	87.488	87.781	88.073
22		88.364	88.655	88.945	89.234	89.522	89.809	90.096	90.382	90.667	90.951
23		91.234	91.517	91.799	92.080	92.360	92.640	92.918	93.196	93.474	93.750
24		94.026	94.300	94.575	94.848	95.120	95.392	95.663	95.933	96.203	96.472
25		96.739	97.007	97.273	97.539	97.804	98.068	98.331	98.594	98.856	99.117
26		99.377	99.637	99.896	100.154	100.411	100.668	100.924	101.179	101.433	101.687
27		101.940	102.192	102.444	102.694	102.944	103.194	103.442	103.690	103.937	104.183
28		104.429	104.674	104.918	105.162	105.404	105.646	105.888	106.128	106.368	106.607
29		106.845	107.083	107.320	107.556	107.792	108.027	108.261	108.494	108.727	108.959
30		109.191	109.421	109.651	109.880	110.109	110.337	110.564	110.790	111.016	111.241
31		111.465	111.689	111.912	112.134	112.356	112.577	112.797	113.017	113.235	113.454
32		113.671	113.888	114.104	114.319	114.534	114.748	114.962	115.174	115.386	115.598
33		115.809	116.019	116.228	116.437	116.645	116.852	117.059	117.265	117.470	117.675

3.2.5.4.2 Approximation formula of sound speed in sea water

At present there is no obligatory recommendation of international oceanographic committees (e.g., ICES, IOC) concerning a general-purpose formula of sound speed in sea water. The formula suggested here has been developed by Del Grosso and Mader [72 D 1] and is the basis for Tables 7···10, following the intercomparison of Mackenzie [81 M 1]

$$\begin{aligned} c = &+0.140186937948 \cdot 10^{4} \\ &+0.497381864699 \cdot 10^{1}\, T \\ &-0.555729359526 \cdot 10^{-1}\, T^{2} \\ &+0.317853157875 \cdot 10^{-3}\, T^{3} \\ &-0.204347838124 \cdot 10^{-5}\, T^{4} \\ &+0.134835141594 \cdot 10^{1}\, S \\ &+0.156423586203 \cdot 10^{0}\, P \\ &+0.329512349091 \cdot 10^{-4}\, P^{2} \\ &-0.157792308592 \cdot 10^{-7}\, P^{3} \\ &-0.101845833872 \cdot 10^{-1}\, TS \\ &-0.553689472711 \cdot 10^{-6}\, TS^{3} \\ &+0.184467788305 \cdot 10^{-5}\, T^{3}S \\ &-0.483822628514 \cdot 10^{-6}\, TP^{2} \\ &+0.278897942694 \cdot 10^{-9}\, TP^{3} \\ &-0.189621054524 \cdot 10^{-5}\, T^{3}P \\ &-0.323670573546 \cdot 10^{-6}\, SP^{2} \\ &+0.214187852200 \cdot 10^{-9}\, SP^{3} \\ &+0.130983950829 \cdot 10^{-5}\, T^{2}SP \\ &-0.157719455877 \cdot 10^{-7}\, TSP^{2} \end{aligned} \qquad (30)$$

c is in [m/s], T is in [°C], S is in [‰] and P is in kg/cm² gauge, $1\ \mathrm{kg/cm^2} \mathrel{\hat{=}} 0.980665 \cdot 10^{5}$ Pa. The range of validity is stated in subsect. 3.2.5.3.2.

Reliable error limits cannot be given; however, the agreement with the Kroebel Mahrt data at $S = 35‰$ and 4°···19 °C is better than 0.03 m/s. For high pressure (oceanic depths), deviations of several dm/s cannot be excluded [81 M 1]. When employing the formula for salinities below $S = 30‰$, especially under higher pressures, discrepancies of several dm/s must be taken into account [82 M 1].

Though in 1974 DelGrosso suggested another 19-term formula (NRL II) [74 D 1] claiming to be valid for salinities down to pure water, this NRL II formula should be used with caution, since Del Grosso's extensive comparison with other measurements leaves a gap between 0 and 29‰ salinity.

It is mentioned that several authors assume the high-pressure data of Wilson [60 W 2] to be more realistic than the Del Grosso data [72 D 1], whereas Del Grosso is considered superior for the low-pressure range. Lovett therefore developed a merged sea water sound speed equation 1978 [78 L 2]. His third (13-term) formula may be a reasonable alternative to eq. (30) [81 C] as long as a final decision is not possible.

Table 7. Sound speed in sea water. ΔC_T component for sea water sound speed, in [m/s] [73 D 1].

T °C	0.0	0.1	0.2	0.3	0.4	0.5	0.6	0.7	0.8	0.9
−2	−10.173	−10.693	−11.215	−11.738	−12.262	−12.787	−13.313	−13.841	−14.369	−14.899
−1	− 5.030	− 5.539	− 6.049	− 6.561	− 7.073	− 7.587	− 8.102	− 8.618	− 9.135	− 9.653
0	0.000	− 0.498	− 0.997	− 1.497	− 1.998	− 2.501	− 3.004	− 3.509	− 4.015	− 4.522
0	0.000	0.497	0.993	1.487	1.981	2.473	2.964	3.455	3.944	4.432
1	4.919	5.404	5.889	6.373	6.855	7.337	7.817	8.296	8.775	9.252
2	9.728	10.203	10.677	11.150	11.621	12.092	12.562	13.030	13.498	13.964
3	14.430	14.894	15.357	15.820	16.281	16.741	17.200	17.658	18.115	18.571
4	19.026	19.480	19.933	20.384	20.835	21.285	21.734	22.181	22.628	23.074
5	23.518	23.962	24.404	24.846	25.286	25.726	26.164	26.602	27.038	27.474
6	27.908	28.342	28.774	29.206	29.636	30.066	30.494	30.921	31.348	31.773
7	32.198	32.621	33.044	33.465	33.886	34.305	34.724	35.141	35.558	35.974
8	36.388	36.802	37.215	37.626	38.037	38.447	38.856	39.264	39.670	40.076
9	40.481	40.885	41.288	41.690	42.092	42.492	42.891	43.289	43.687	44.083
10	44.478	44.873	45.266	45.659	46.051	46.441	46.831	47.220	47.608	47.995
11	48.381	48.766	49.150	59.533	49.916	50.297	50.678	51.057	51.436	51.813
12	52.190	52.566	52.941	53.315	53.688	54.060	54.432	54.802	55.172	55.540
13	55.908	56.275	56.640	57.005	57.369	57.733	58.095	58.456	58.817	59.176
14	59.535	59.893	60.250	60.606	60.961	61.315	61.668	62.021	62.372	62.723
15	63.073	63.422	63.770	64.117	64.463	64.808	65.153	65.497	65.839	66.181
16	66.522	66.863	67.202	67.540	67.878	68.215	68.550	68.885	69.220	69.553
17	69.885	70.217	70.548	70.877	71.206	71.534	71.862	72.188	72.514	72.838
18	73.162	73.485	73.808	74.129	74.449	74.769	75.088	75.406	75.723	76.039
19	76.355	76.669	76.983	77.296	77.608	77.919	78.230	78.539	78.848	79.156
20	79.463	79.769	80.075	80.379	80.683	80.986	81.288	81.590	81.890	82.190
21	82.489	82.787	83.084	83.380	83.676	83.971	84.265	84.558	84.850	85.142
22	85.433	85.722	86.012	86.300	86.587	86.874	87.160	87.445	87.729	88.013
23	88.295	88.577	88.858	89.138	89.418	89.696	89.974	90.251	90.528	90.803
24	91.078	91.352	91.625	91.897	92.168	92.439	92.709	92.978	93.246	93.514
25	93.781	94.047	94.312	94.576	94.840	95.102	95.365	95.626	95.886	96.146
26	96.405	96.663	96.920	97.177	97.432	97.687	97.942	98.195	98.448	98.700
27	98.951	99.201	99.451	99.699	99.947	100.195	100.441	100.687	100.932	101.176
28	101.419	101.662	101.904	102.145	102.385	102.625	102.863	103.101	103.339	103.575
29	103.811	104.046	104.280	104.513	104.746	104.978	105.209	105.439	105.669	105.898
30	106.126	106.353	106.580	106.805	107.030	107.255	107.478	107.701	107.923	108.144

Table 8. Sound speed in sea water. ΔC_S component for sea water sound speed, in [m/s] [73 D 1].

S ‰	0.0	0.1	0.2	0.3	0.4	0.5	0.6	0.7	0.8	0.9
30	40.451	40.585	40.720	40.855	40.990	41.125	41.260	41.394	41.529	41.664
31	41.799	41.934	42.069	42.203	42.338	42.473	42.608	42.743	42.878	43.012
32	43.147	43.282	43.417	43.552	43.687	43.821	43.956	44.091	44.226	44.361
33	44.496	44.630	44.765	44.900	45.035	45.170	45.305	45.439	45.574	45.709
34	45.844	45.979	46.114	46.248	46.383	46.518	46.653	46.788	46.923	47.057
35	47.192	47.327	47.462	47.597	47.732	47.866	48.001	48.136	48.271	48.406
36	48.541	48.675	48.810	48.945	49.080	49.215	49.350	49.484	49.619	49.754
37	49.889	50.024	50.159	50.294	50.428	50.563	50.698	50.833	50.968	51.103
38	51.237	51.372	51.507	51.642	51.777	51.912	52.046	52.181	52.316	52.451
39	52.586	52.721	52.855	52.990	53.125	53.260	53.395	53.530	53.664	53.799
40	53.934	54.069	54.204	54.339	54.473	54.608	54.743	54.878	55.013	55.148
41	55.282	55.417	55.552	55.687	55.822	55.957	56.091	56.226	56.361	56.496

Table 9. Sound speed in sea water. ΔC_P component for sea water sound speed, in [m/s] [73 D 1]. $1\ \mathrm{kg/cm^2} \triangleq 0.980665 \cdot 10^5$ Pa.

P kg/cm² gauge	0	10	20	30	40	50	60	70	80	90
0	0.000	1.568	3.142	4.722	6.309	7.902	9.501	11.106	12.717	14.334
100	15.956	17.584	19.218	20.857	22.502	24.152	25.807	27.467	29.132	30.802
200	32.477	34.156	35.840	37.529	39.222	40.919	42.620	44.326	46.036	47.749
300	49.467	51.188	52.913	54.641	56.373	58.108	59.847	61.588	63.333	65.081
400	66.832	68.585	70.341	72.100	73.862	75.625	77.391	79.160	80.930	82.703
500	84.477	86.254	88.032	89.811	91.593	93.375	95.160	96.945	98.732	100.520
600	102.308	104.098	105.888	107.680	109.471	111.264	113.057	114.850	116.643	118.437
700	120.230	122.024	123.817	125.611	127.403	129.196	130.988	132.779	134.570	136.360
800	138.149	139.937	141.724	143.509	145.294	147.077	148.859	150.639	152.417	154.194
900	155.969	157.742	159.513	161.281	163.048	164.812	166.574	168.333	170.090	171.844
1000	173.596	175.344	177.089	178.832	180.571	182.307	184.040	185.769	187.495	189.217
1100	190.935	192.649	194.360	196.066	197.769	199.467	201.161	202.850	204.535	206.216

Table 10. Sound speed in sea water. ΔC_{TSP} component for sea water sound speed, in [m/s] [73 D 1]. 1 kg/cm² ≙ 0.980665 · 10⁵ Pa. Solid line frame: open ocean, dashed line frame: Mediterranean, dotted line frame: Red Sea [69 L 1].

T [°C] \ S [‰]	30	31	32	33	34	35	36	37	38	39	40	41	42
						$P=0$ kg/cm² gauge							
-2	0.641	0.664	0.688	0.711	0.736	0.760	0.784	0.809	0.834	0.860	0.885	0.911	0.937
-1	0.320	0.332	0.344	0.356	0.368	0.380	0.392	0.405	0.417	0.430	0.443	0.456	0.469
0	0.000	0.000	0.000	0.000	0.000	0.000	0.000	0.000	0.000	0.000	0.000	0.000	0.000
1	-0.320	-0.332	-0.344	-0.356	-0.368	-0.380	-0.392	-0.405	-0.417	-0.430	-0.443	-0.456	-0.469
2	-0.641	-0.664	-0.688	-0.711	-0.736	-0.760	-0.784	-0.809	-0.834	-0.860	-0.885	-0.911	-0.937
3	-0.960	-0.995	-1.031	-1.066	-1.102	-1.139	-1.176	-1.213	-1.250	-1.288	-1.326	-1.365	-1.404
4	-1.278	-1.325	-1.372	-1.420	-1.468	-1.517	-1.566	-1.615	-1.665	-1.716	-1.767	-1.818	-1.870
5	-1.596	-1.654	-1.713	-1.772	-1.832	-1.893	-1.954	-2.016	-2.078	-2.141	-2.205	-2.269	-2.334
6	-1.911	-1.981	-2.052	-2.123	-2.195	-2.267	-2.341	-2.415	-2.489	-2.565	-2.641	-2.718	-2.796
7	-2.224	-2.306	-2.388	-2.471	-2.555	-2.639	-2.725	-2.811	-2.898	-2.986	-3.074	-3.164	-3.255
8	-2.536	-2.628	-2.722	-2.817	-2.912	-3.009	-3.106	-3.204	-3.303	-3.404	-3.505	-3.607	-3.711
9	-2.844	-2.948	-3.053	-3.160	-3.267	-3.375	-3.484	-3.594	-3.705	-3.818	-3.932	-4.046	-4.162
10	-3.150	-3.265	-3.381	-3.499	-3.618	-3.737	-3.858	-3.981	-4.104	-4.228	-4.354	-4.482	-4.610
11	-3.452	-3.578	-3.706	-3.835	-3.965	-4.096	-4.229	-4.363	-4.498	-4.635	-4.773	-4.912	-5.053
12	-3.750	-3.888	-4.027	-4.167	-4.308	-4.451	-4.595	-4.741	-4.888	-5.036	-5.186	-5.338	-5.491
13	-4.045	-4.193	-4.343	-4.494	-4.647	-4.801	-4.956	-5.113	-5.272	-5.433	-5.595	-5.758	-5.924
14	-4.335	-4.494	-4.655	-4.817	-4.980	-5.146	-5.312	-5.481	-5.651	-5.823	-5.997	-6.173	-6.350
15	-4.621	-4.790	-4.962	-5.134	-5.309	-5.485	-5.663	-5.843	-6.024	-6.208	-6.393	-6.581	-6.770
16	-4.901	-5.081	-5.263	-5.446	-5.632	-5.819	-6.008	-6.198	-6.391	-6.586	-6.783	-6.982	-7.183
17	-5.176	-5.367	-5.559	-5.753	-5.949	-6.146	-6.346	-6.548	-6.751	-6.957	-7.165	-7.376	-7.589
18	-5.446	-5.646	-5.849	-6.053	-6.259	-6.467	-6.677	-6.890	-7.104	-7.321	-7.540	-7.762	-7.986
19	-5.710	-5.920	-6.132	-6.346	-6.563	-6.781	-7.002	-7.224	-7.450	-7.677	-7.907	-8.140	-8.375
20	-5.967	-6.187	-6.409	-6.633	-6.859	-7.087	-7.318	-7.551	-7.787	-8.025	-8.266	-8.510	-8.756
21	-6.218	-6.447	-6.678	-6.912	-7.148	-7.386	-7.627	-7.870	-8.116	-8.365	-8.616	-8.870	-9.127
22	-6.461	-6.700	-6.941	-7.184	-7.429	-7.677	-7.927	-8.181	-8.436	-8.695	-8.956	-9.221	-9.488
23	-6.698	-6.945	-7.195	-7.447	-7.702	-7.959	-8.219	-8.482	-8.747	-9.016	-9.287	-9.562	-9.839
24	-6.927	-7.183	-7.441	-7.702	-7.966	-8.232	-8.501	-8.773	-9.048	-9.327	-9.608	-9.892	-10.180
25	-7.147	-7.412	-7.679	-7.949	-8.221	-8.496	-8.774	-9.055	-9.340	-9.627	-9.918	-10.211	-10.509
26	-7.360	-7.633	-7.908	-8.186	-8.467	-8.750	-9.037	-9.327	-9.620	-9.917	-10.216	-10.520	-10.826
27	-7.564	-7.844	-8.127	-8.414	-8.703	-8.995	-9.290	-9.588	-9.890	-10.195	-10.504	-10.816	-11.132
28	-7.759	-8.047	-8.338	-8.631	-8.928	-9.228	-9.532	-9.838	-10.148	-10.462	-10.779	-11.100	-11.425
29	-7.944	-8.240	-8.538	-8.839	-9.143	-9.451	-9.762	-10.077	-10.395	-10.717	-11.042	-11.372	-11.705
30	-8.120	-8.423	-8.728	-9.036	-9.348	-9.663	-9.981	-10.303	-10.629	-10.959	-11.292	-11.630	-11.971
31	-8.286	-8.595	-8.907	-9.222	-9.541	-9.863	-10.188	-10.518	-10.851	-11.188	-11.529	-11.874	-12.224
32	-8.442	-8.757	-9.075	-9.397	-9.722	-10.051	-10.383	-10.720	-11.060	-11.404	-11.752	-12.105	-12.462

Table 10 (continued).

T[°C] \ S[‰]	30	31	32	33	34	35	36	37	38	39	40	41	42
						$P=1\ kg/cm^2$ gauge							
−2	0.641	0.664	0.688	0.712	0.736	0.760	0.785	0.809	0.834	0.860	0.885	0.911	0.937
−1	0.320	0.332	0.344	0.356	0.368	0.380	0.392	0.405	0.417	0.430	0.443	0.456	0.409
0	−0.000	−0.000	−0.000	−0.000	−0.000	−0.000	−0.000	−0.000	−0.000	−0.000	−0.000	−0.000	−0.000
1	−0.320	−0.332	−0.344	−0.356	−0.368	−0.380	−0.392	−0.405	−0.417	−0.430	−0.443	−0.456	−0.469
2	−0.640	−0.664	−0.687	−0.711	−0.735	−0.760	−0.784	−0.809	−0.834	−0.859	−0.885	−0.911	−0.937
3	−0.960	−0.995	−1.030	−1.066	−1.102	−1.139	−1.175	−1.212	−1.250	−1.288	−1.326	−1.365	−1.404
4	−1.278	−1.325	−1.372	−1.420	−1.468	−1.516	−1.565	−1.614	−1.664	−1.715	−1.766	−1.817	−1.869
5	−1.595	−1.653	−1.712	−1.772	−1.831	−1.892	−1.953	−2.015	−2.077	−2.140	−2.204	−2.268	−2.333
6	−1.910	−1.980	−2.050	−2.122	−2.194	−2.266	−2.339	−2.413	−2.488	−2.563	−2.640	−2.717	−2.794
7	−2.223	−2.305	−2.387	−2.470	−2.553	−2.638	−2.723	−2.809	−2.896	−2.984	−3.073	−3.162	−3.253
8	−2.534	−2.627	−2.720	−2.815	−2.910	−3.007	−3.104	−3.202	−3.301	−3.401	−3.502	−3.605	−3.708
9	−2.842	−2.946	−3.051	−3.157	−3.264	−3.372	−3.481	−3.592	−3.703	−3.815	−3.929	−4.043	−4.159
10	−3.148	−3.263	−3.379	−3.497	−3.615	−3.735	−3.856	−3.978	−4.101	−4.225	−4.351	−4.478	−4.607
11	−3.449	−3.576	−3.703	−3.832	−3.962	−4.093	−4.226	−4.359	−4.495	−4.631	−4.769	−4.908	−5.049
12	−3.748	−3.885	−4.024	−4.164	−4.305	−4.448	−4.591	−4.737	−4.884	−5.032	−5.182	−5.334	−5.487
13	−4.022	−4.191	−4.340	−4.491	−4.643	−4.797	−4.953	−5.109	−5.268	−5.428	−5.590	−5.753	−5.919
14	−4.332	−4.491	−4.652	−4.814	−4.977	−5.142	−5.308	−5.477	−5.647	−5.818	−5.992	−6.167	−6.345
15	−4.618	−4.788	−4.959	−5.131	−5.305	−5.481	−5.659	−5.838	−6.020	−6.203	−6.388	−6.575	−6.764
16	−4.899	−5.079	−5.260	−5.443	−5.628	−5.815	−6.003	−6.194	−6.386	−6.581	−6.777	−6.976	−7.177
17	−5.714	−5.364	−5.556	−5.750	−5.945	−6.142	−6.342	−6.543	−6.746	−6.952	−7.160	−7.370	−7.582
18	−5.444	−5.644	−5.846	−6.050	−6.256	−6.403	−6.673	−6.885	−7.099	−7.316	−7.535	−7.756	−7.979
19	−5.709	−5.918	−6.130	−6.344	−6.559	−6.777	−6.998	−7.220	−7.445	−7.672	−7.902	−8.134	−8.368
20	−5.967	−6.186	−6.407	−6.631	−6.856	−7.084	−7.315	−7.547	−7.782	−8.020	−8.260	−8.503	−8.749
21	−6.218	−6.447	−6.677	−6.911	−7.146	−7.384	−7.624	−7.867	−8.112	−8.360	−8.610	−8.864	−9.120
22	−6.463	−6.700	−6.940	−7.183	−7.428	−7.675	−7.925	−8.177	−8.432	−8.690	−8.951	−9.215	−9.482
23	−6.700	−6.947	−7.196	−7.447	−7.701	−7.958	−8.217	−8.479	−8.744	−9.012	−9.282	−9.556	−9.833
24	−6.930	−7.186	−7.443	−7.704	−7.966	−8.232	−8.501	−8.772	−9.046	−9.323	−9.604	−9.887	−10.174
25	−7.153	−7.416	−7.682	−7.951	−8.223	−8.497	−8.775	−9.055	−9.338	−9.625	−9.914	−10.208	−10.504
26	−7.367	−7.638	−7.913	−8.190	−8.470	−8.753	−9.039	−9.328	−9.620	−9.915	−10.214	−10.517	−10.823
27	−7.573	−7.852	−8.134	−8.419	−8.707	−8.999	−9.293	−9.590	−9.891	−10.195	−10.503	−10.814	−11.129
28	−7.770	−8.057	−8.346	−8.639	−8.935	−9.234	−9.536	−9.842	−10.151	−10.464	−10.780	−11.100	−11.423
29	−7.958	−8.252	−8.549	−8.849	−9.152	−9.459	−9.769	−10.082	−10.399	−10.720	−11.044	−11.373	−11.705
30	−8.136	−8.437	−8.741	−9.048	−9.359	−9.673	−9.990	−10.311	−10.636	−10.964	−11.296	−11.633	−11.973
31	−8.305	−8.613	−8.923	−9.237	−9.554	−9.875	−10.200	−10.528	−10.860	−11.196	−11.535	−11.879	−12.228
32	−8.464	−8.778	−9.095	−9.415	−9.739	−10.066	−10.397	−10.732	−11.071	−11.414	−11.761	−12.112	−12.468

P = 10 kg/cm² gauge

T [°C] \ S [‰]	30	31	32	33	34	35	36	37	38	39	40	41	42
−2	0.641	0.665	0.689	0.713	0.737	0.761	0.786	0.810	0.835	0.861	0.886	0.912	0.938
−1	0.320	0.332	0.343	0.355	0.367	0.380	0.392	0.404	0.417	0.429	0.442	0.455	0.468
0	−0.001	−0.001	−0.001	−0.001	−0.001	−0.001	−0.001	−0.001	−0.001	−0.001	−0.001	−0.001	−0.001
1	−0.321	−0.333	−0.345	−0.357	−0.369	−0.381	−0.393	−0.406	−0.418	−0.431	−0.444	−0.457	−0.470
2	−0.640	−0.664	−0.687	−0.711	−0.735	−0.760	−0.784	−0.809	−0.834	−0.859	−0.885	−0.910	−0.936
3	−0.958	−0.993	−1.029	−1.064	−1.100	−1.137	−1.173	−1.210	−1.248	−1.286	−1.324	−1.362	−1.401
4	−1.275	−1.321	−1.368	−1.416	−1.464	−1.512	−1.561	−1.610	−1.660	−1.710	−1.761	−1.812	−1.864
5	−1.590	−1.648	−1.706	−1.765	−1.825	−1.885	−1.946	−2.008	−2.070	−2.133	−2.196	−2.260	−2.825
6	−1.902	−1.972	−2.042	−2.113	−2.184	−2.257	−2.329	−2.403	−2.477	−2.552	−2.628	−2.705	−2.782
7	−2.213	−2.294	−2.376	−2.458	−2.541	−2.625	−2.710	−2.795	−2.882	−2.969	−3.057	−3.146	−3.237
8	−2.522	−2.614	−2.707	−2.801	−2.895	−2.991	−3.087	−3.185	−3.283	−3.383	−3.483	−3.585	−3.687
9	−2.828	−2.931	−3.035	−3.140	−3.246	−3.353	−3.462	−3.571	−3.681	−3.793	−3.905	−4.019	−4.134
10	−3.131	−3.245	−3.361	−3.477	−3.594	−3.713	−3.832	−3.953	−4.075	−4.199	−4.323	−4.449	−4.577
11	−3.431	−3.556	−3.683	−3.810	−3.939	−4.068	−4.199	−4.332	−4.465	−4.601	−4.737	−4.875	−5.015
12	−3.729	−3.864	−4.001	−4.139	−4.279	−4.420	−4.562	−4.706	−4.851	−4.998	−5.146	−5.296	−5.448
13	−4.022	−4.168	−4.316	−4.465	−4.616	−4.767	−4.921	−5.076	−5.232	−5.391	−5.550	−5.712	−5.875
14	−4.312	−4.469	−4.627	−4.787	−4.948	−5.110	−5.275	−5.441	−5.608	−5.778	−5.949	−6.122	−6.297
15	−4.599	−4.765	−4.934	−5.104	−5.275	−5.449	−5.624	−5.801	−5.979	−6.160	−6.342	−6.527	−6.713
16	−4.881	−5.058	−5.236	−5.416	−5.598	−5.782	−5.967	−6.155	−6.344	−6.536	−6.729	−6.925	−7.123
17	−5.159	−5.345	−5.534	−5.724	−5.916	−6.110	−6.306	−6.504	−6.704	−6.906	−7.110	−7.317	−7.526
18	−5.432	−5.628	−5.826	−6.026	−6.228	−6.432	−6.638	−6.846	−7.057	−7.270	−7.485	−7.702	−7.922
19	−5.701	−5.906	−6.114	−6.323	−6.535	−6.749	−6.965	−7.183	−7.403	−7.626	−7.852	−8.080	−8.310
20	−5.964	−6.179	−6.396	−6.615	−6.836	−7.059	−7.285	−7.513	−7.743	−7.976	−8.212	−8.450	−8.691
21	−6.223	−6.447	−6.672	−6.900	−7.130	−7.363	−7.598	−7.836	−8.076	−8.319	−8.564	−8.812	−9.064
22	−6.476	−6.708	−6.943	−7.180	−7.419	−7.660	−7.905	−8.151	−8.401	−8.653	−8.908	−9.166	−9.428
23	−6.724	−6.964	−7.207	−7.453	−7.700	−7.951	−8.204	−8.460	−8.718	−8.980	−9.244	−9.512	−9.783
24	−6.966	−7.214	−7.465	−7.719	−7.975	−8.234	−8.496	−8.760	−9.028	−9.298	−9.572	−9.849	−10.129
25	−7.202	−7.458	−7.717	−7.978	−8.243	−8.510	−8.780	−9.053	−9.329	−9.608	−9.890	−10.176	−10.465
26	−7.431	−7.695	−7.961	−8.231	−8.503	−8.778	−9.056	−9.337	−9.621	−9.909	−10.200	−10.494	−10.792
27	−7.654	−7.925	−8.199	−8.475	−8.755	−9.038	−9.323	−9.612	−9.904	−10.200	−10.499	−10.802	−11.109
28	−7.821	−8.148	−8.429	−8.713	−8.999	−9.289	−9.582	−9.879	−10.179	−10.482	−10.789	−11.100	−11.414
29	−8.080	−8.364	−8.652	−8.942	−9.235	−9.532	−9.832	−10.136	−10.443	−10.754	−11.069	−11.387	−11.709
30	−8.283	−8.573	−8.866	−9.163	−9.463	−9.766	−10.073	−10.384	−10.698	−11.016	−11.337	−11.663	−11.993
31	−8.478	−8.774	−9.073	−9.376	−9.682	−9.991	−10.305	−10.621	−10.942	−11.267	−11.595	−11.928	−12.265
32	−8.665	−8.967	−9.272	−9.580	−9.892	−10.207	−10.526	−10.849	−11.176	−11.507	−11.842	−12.181	−12.525

Table 10 (continued).

T [°C] \ S [‰]	30	31	32	33	34	35	36	37	38	39	40	41	42
						P = 20 kg/cm² gauge							
−2	0.641	0.664	0.688	0.712	0.736	0.760	0.785	0.810	0.835	0.860	0.885	0.911	0.937
−1	0.318	0.329	0.341	0.353	0.365	0.377	0.389	0.402	0.414	0.426	0.439	0.452	0.465
0	−0.004	−0.004	−0.004	−0.004	−0.004	−0.004	−0.005	−0.005	−0.005	−0.005	−0.005	−0.005	−0.005
1	−0.324	−0.336	−0.348	−0.360	−0.372	−0.384	−0.397	−0.409	−0.422	−0.434	−0.447	−0.460	−0.473
2	−0.642	−0.666	−0.689	−0.713	−0.737	−0.762	−0.786	−0.811	−0.836	−0.862	−0.887	−0.913	−0.939
3	−0.959	−0.994	−1.029	−1.065	−1.101	−1.137	−1.174	−1.211	−1.249	−1.286	−1.325	−1.363	−1.402
4	−1.274	−1.320	−1.367	−1.414	−1.462	−1.511	−1.559	−1.608	−1.658	−1.708	−1.759	−1.810	−1.862
5	−1.586	−1.644	−1.703	−1.762	−1.821	−1.881	−1.942	−2.003	−2.065	−2.128	−2.191	−2.255	−2.319
6	−1.897	−1.966	−2.036	−2.106	−2.178	−2.249	−2.322	−2.395	−2.469	−2.544	−2.619	−2.695	−2.773
7	−2.205	−2.286	−2.367	−2.449	−2.531	−2.615	−2.699	−2.784	−2.870	−2.957	−3.044	−3.133	−3.222
8	−2.512	−2.603	−2.695	−2.788	−2.882	−2.977	−3.073	−3.170	−3.267	−3.366	−3.466	−3.567	−3.669
9	−2.815	−2.918	−3.021	−3.125	−3.230	−3.336	−3.444	−3.552	−3.661	−3.772	−3.883	−3.996	−4.110
10	−3.117	−3.230	−3.344	−3.459	−3.575	−3.692	−3.811	−3.930	−4.051	−4.174	−4.297	−4.422	−4.548
11	−3.415	−3.539	−3.663	−3.789	−3.916	−4.045	−4.174	−4.305	−4.438	−4.571	−4.706	−4.843	−4.981
12	−3.711	−3.845	−3.980	−4.117	−4.255	−4.394	−4.534	−4.676	−4.820	−4.965	−5.111	−5.260	−5.409
13	−4.004	−4.148	−4.294	−4.441	−4.589	−4.739	−4.890	−5.043	−5.198	−5.354	−5.512	−5.671	−5.833
14	−4.294	−4.448	−4.604	−4.761	−4.920	−5.080	−5.242	−5.406	−5.571	−5.738	−5.907	−6.078	−6.250
15	−4.581	−4.745	−4.911	−5.078	−5.247	−5.417	−5.590	−5.764	−5.940	−6.118	−6.297	−6.479	−6.663
16	−4.865	−5.039	−5.214	−5.391	−5.570	−5.750	−5.933	−6.117	−6.303	−6.492	−6.682	−6.875	−7.069
17	−5.146	−5.329	−5.514	−5.700	−5.889	−6.079	−6.271	−6.466	−6.662	−6.861	−7.062	−7.265	−7.470
18	−5.423	−5.615	−5.809	−6.005	−6.203	−6.403	−6.605	−6.809	−7.016	−7.224	−7.435	−7.649	−7.864
19	−5.697	−5.898	−6.101	−6.306	−6.513	−6.722	−6.934	−7.147	−7.364	−7.582	−7.803	−8.026	−8.252
20	−5.967	−6.177	−6.389	−6.603	−6.819	−7.037	−7.257	−7.480	−7.706	−7.934	−8.164	−8.398	−8.633
21	−6.234	−6.452	−6.672	−6.895	−7.119	−7.346	−7.576	−7.808	−8.042	−8.279	−8.519	−8.762	−9.008
22	−6.497	−6.723	−6.951	−7.182	−7.415	−7.651	−7.889	−8.129	−8.373	−8.619	−8.868	−9.120	−9.375
23	−6.756	−6.990	−7.226	−7.465	−7.706	−7.949	−8.196	−8.445	−8.697	−8.952	−9.209	−9.470	−9.734
24	−7.011	−7.252	−7.496	−7.742	−7.991	−8.243	−8.497	−8.754	−9.015	−9.278	−9.544	−9.814	−10.086
25	−7.262	−7.511	−7.761	−8.015	−8.271	−8.530	−8.793	−9.058	−9.326	−9.597	−9.871	−10.149	−10.430
26	−7.509	−7.764	−8.022	−8.283	−8.546	−8.812	−9.082	−9.354	−9.630	−9.909	−10.191	−10.477	−10.766
27	−7.752	−8.013	−8.278	−8.545	−8.815	−9.088	−9.365	−9.644	−9.927	−10.214	−10.503	−10.797	−11.094
28	−7.990	−8.257	−8.528	−8.802	−9.078	−9.358	−9.641	−9.927	−10.217	−10.511	−10.808	−11.108	−11.413
29	−8.223	−8.497	−8.773	−9.053	−9.335	−9.621	−9.911	−10.204	−10.500	−10.800	−11.104	−11.411	−11.723
30	−8.452	−8.731	−9.013	−9.298	−9.587	−9.878	−10.174	−10.473	−10.775	−11.081	−11.392	−11.706	−12.024
31	−8.677	−8.960	−9.247	−9.538	−9.831	−10.129	−10.430	−10.734	−11.042	−11.355	−11.671	−11.991	−12.316
32	−8.896	−9.185	−9.476	−9.771	−10.070	−10.372	−10.678	−10.988	−11.302	−11.619	−11.941	−12.268	−12.598

T [°C] \ S [‰]	30	31	32	33	34	35	36	37	38	39	40	41	42
						$P = 50\ kg/cm^2$ gauge							
−2	0.630	0.653	0.677	0.700	0.724	0.748	0.772	0.796	0.821	0.845	0.870	0.896	0.921
−1	0.301	0.312	0.324	0.335	0.346	0.358	0.369	0.381	0.393	0.405	0.417	0.429	0.442
0	−0.023	−0.024	−0.025	−0.026	−0.027	−0.027	−0.028	−0.029	−0.030	−0.031	−0.031	−0.032	−0.033
1	−0.344	−0.357	−0.369	−0.382	−0.395	−0.408	−0.421	−0.434	−0.447	−0.461	−0.474	−0.488	−0.502
2	−0.662	−0.686	−0.710	−0.734	−0.759	−0.784	−0.809	−0.834	−0.860	−0.886	−0.912	−0.939	−0.965
3	−0.975	−1.011	−1.047	−1.083	−1.119	−1.156	−1.193	−1.230	−1.268	−1.306	−1.345	−1.384	−1.423
4	−1.286	−1.333	−1.380	−1.427	−1.475	−1.524	−1.573	−1.622	−1.672	−1.722	−1.773	−1.824	−1.876
5	−1.594	−1.651	−1.710	−1.768	−1.828	−1.888	−1.948	−2.009	−2.071	−2.133	−2.196	−2.260	−2.324
6	−1.898	−1.967	−2.036	−2.106	−2.177	−2.248	−2.320	−2.393	−2.466	−2.540	−2.615	−2.691	−2.767
7	−2.201	−2.280	−2.360	−2.441	−2.522	−2.605	−2.688	−2.772	−2.857	−2.942	−3.029	−3.117	−3.205
8	−2.501	−2.590	−2.681	−2.773	−2.865	−2.958	−3.052	−3.148	−3.244	−3.341	−3.439	−3.538	−3.639
9	−2.799	−2.899	−3.000	−3.102	−3.205	−3.309	−3.414	−3.520	−3.627	−3.735	−3.845	−3.955	−4.067
10	−3.095	−3.205	−3.316	−3.428	−3.542	−3.656	−3.772	−3.888	−4.006	−4.126	−4.246	−4.368	−4.491
11	−3.390	−3.509	−3.630	−3.753	−3.876	−4.001	−4.126	−4.254	−4.382	−4.512	−4.644	−4.776	−4.911
12	−3.683	−3.812	−3.943	−4.075	−4.208	−4.343	−4.479	−4.616	−4.755	−4.895	−5.037	−5.181	−5.326
13	−3.975	−4.114	−4.254	−4.395	−4.538	−4.682	−4.828	−4.975	−5.124	−5.275	−5.427	−5.581	−5.737
14	−4.267	−4.414	−4.563	−4.714	−4.866	−5.020	−5.175	−5.332	−5.491	−5.651	−5.814	−5.978	−6.144
15	−4.557	−4.714	−4.872	−5.031	−5.192	−5.355	−5.520	−5.686	−5.854	−6.024	−6.196	−6.370	−6.517
16	−4.848	−5.012	−5.179	−5.347	−5.517	−5.689	−5.862	−6.038	−6.215	−6.394	−6.576	−6.760	−6.915
17	−5.138	−5.311	−5.485	−5.662	−5.840	−6.020	−6.203	−6.387	−6.573	−6.762	−6.952	−7.145	−7.340
18	−5.428	−5.609	−5.791	−5.976	−6.162	−6.351	−6.541	−6.734	−6.929	−7.126	−7.325	−7.527	−7.732
19	−5.719	−5.907	−6.097	−6.289	−6.483	−6.680	−6.878	−7.079	−7.282	−7.488	−7.696	−7.906	−8.119
20	−6.010	−6.205	−6.403	−6.602	−6.804	−7.008	−7.214	−7.422	−7.633	−7.847	−8.063	−8.282	−8.503
21	−6.302	−6.504	−6.708	−6.915	−7.123	−7.334	−7.548	−7.764	−7.983	−8.204	−8.428	−8.654	−8.881
22	−6.595	−6.804	−7.014	−7.227	−7.443	−7.661	−7.881	−8.104	−8.330	−8.558	−8.790	−9.024	−9.261
23	−6.890	−7.104	−7.321	−7.540	−7.762	−7.986	−8.213	−8.443	−8.675	−8.911	−9.149	−9.391	−9.636
24	−7.186	−7.406	−7.628	−7.853	−8.081	−8.311	−8.544	−8.781	−9.020	−9.262	−9.507	−9.755	−10.007
25	−7.483	−7.709	−7.936	−8.167	−8.400	−8.636	−8.875	−9.117	−9.362	−9.610	−9.862	−10.117	−10.375
26	−7.783	−8.013	−8.246	−8.481	−8.720	−8.961	−9.205	−9.453	−9.704	−9.957	−10.215	−10.476	−10.740
27	−8.085	−8.319	−8.557	−8.797	−9.040	−9.286	−9.535	−9.788	−10.044	−10.303	−10.566	−10.832	−11.102
28	−8.389	−8.628	−8.869	−9.113	−9.361	−9.611	−9.865	−10.122	−10.383	−10.647	−10.915	−11.186	−11.462
29	−8.696	−8.938	−9.183	−9.431	−9.683	−9.937	−10.195	−10.457	−10.721	−10.990	−11.262	−11.539	−11.819
30	−9.006	−9.251	−9.500	−9.751	−10.006	−10.264	−10.525	−10.790	−11.059	−11.332	−11.608	−11.889	−12.173
31	−9.319	−9.567	−9.818	−10.072	−10.330	−10.591	−10.856	−11.124	−11.396	−11.673	−11.953	−12.237	−12.526
32	−9.636	−9.886	−10.139	−10.396	−10.656	−10.919	−11.187	−11.458	−11.733	−12.013	−12.296	−12.584	−12.876

Table 10 (continued).

T [°C] \ S [‰]	30	31	32	33	34	35	36	37	38	39	40	41	42
						$P = 100\ kg/cm^2$ gauge							
−2	0.586	0.607	0.628	0.650	0.672	0.694	0.716	0.739	0.762	0.785	0.808	0.832	0.856
−1	0.243	0.252	0.261	0.270	0.280	0.289	0.299	0.308	0.318	0.328	0.338	0.348	0.359
0	−0.091	−0.094	−0.097	−0.100	−0.103	−0.106	−0.109	−0.112	−0.115	−0.118	−0.121	−0.124	−0.127
1	−0.417	−0.431	−0.446	−0.461	−0.476	−0.492	−0.507	−0.522	−0.538	−0.554	−0.569	−0.585	−0.602
2	−0.736	−0.762	−0.788	−0.815	−0.842	−0.869	−0.896	−0.924	−0.952	−0.980	−1.008	−1.037	−1.066
3	−1.048	−1.086	−1.123	−1.162	−1.200	−1.239	−1.278	−1.317	−1.357	−1.397	−1.438	−1.479	−1.520
4	−1.356	−1.404	−1.453	−1.502	−1.551	−1.602	−1.652	−1.703	−1.755	−1.807	−1.859	−1.912	−1.966
5	−1.658	−1.717	−1.777	−1.837	−1.897	−1.958	−2.020	−2.082	−2.145	−2.209	−2.273	−2.338	−2.403
6	−1.957	−2.026	−2.096	−2.166	−2.238	−2.309	−2.382	−2.455	−2.529	−2.604	−2.679	−2.756	−2.833
7	−2.253	−2.332	−2.412	−2.492	−2.574	−2.656	−2.739	−2.823	−2.908	−2.993	−3.080	−3.167	−3.256
8	−2.546	−2.635	−2.725	−2.815	−2.906	−2.999	−3.092	−3.186	−3.281	−3.377	−3.474	−3.573	−3.672
9	−2.838	−2.936	−3.035	−3.135	−3.236	−3.338	−3.441	−3.545	−3.650	−3.757	−3.864	−3.973	−4.083
10	−3.130	−3.237	−3.345	−3.454	−3.564	−3.675	−3.788	−3.901	−4.016	−4.132	−4.250	−4.368	−4.489
11	−3.421	−3.537	−3.654	−3.771	−3.890	−4.011	−4.132	−4.255	−4.379	−4.505	−4.632	−4.760	−4.890
12	−3.714	−3.838	−3.963	−4.089	−4.216	−4.345	−4.475	−4.607	−4.740	−4.875	−5.011	−5.149	−5.288
13	−4.009	−4.140	−4.273	−4.407	−4.542	−4.679	−4.818	−4.958	−5.100	−5.243	−5.388	−5.535	−5.683
14	−4.306	−4.445	−4.585	−4.726	−4.870	−5.014	−5.161	−5.309	−5.459	−5.610	−5.763	−5.919	−6.076
15	−4.606	−4.752	−4.899	−5.048	−5.198	−5.351	−5.504	−5.660	−5.818	−5.977	−6.138	−6.302	−6.467
16	−4.911	−5.063	−5.217	−5.373	−5.530	−5.689	−5.850	−6.013	−6.177	−6.344	−6.513	−6.684	−6.857
17	−5.221	−5.379	−5.539	−5.701	−5.865	−6.030	−6.198	−6.367	−6.539	−6.713	−6.889	−7.067	−7.247
18	−5.537	−5.700	−5.866	−6.034	−6.203	−6.375	−6.548	−6.724	−6.902	−7.083	−7.265	−7.450	−7.638
19	−5.859	−6.028	−6.199	−6.372	−6.547	−6.724	−6.903	−7.085	−7.269	−7.455	−7.644	−7.835	−8.029
20	−6.189	−6.362	−6.538	−6.716	−6.896	−7.078	−7.263	−7.450	−7.639	−7.831	−8.026	−8.223	−8.423
21	−6.527	−6.704	−6.884	−7.067	−7.251	−7.438	−7.627	−7.819	−8.014	−8.211	−8.411	−8.613	−8.819
22	−6.874	−7.055	−7.239	−7.425	−7.614	−7.805	−7.998	−8.194	−8.393	−8.595	−8.800	−9.007	−9.217
23	−7.231	−7.415	−7.602	−7.792	−7.984	−8.179	−8.376	−8.576	−8.779	−8.985	−9.193	−9.405	−9.620
24	−7.598	−7.786	−7.975	−8.168	−8.363	−8.561	−8.761	−8.965	−9.171	−9.380	−9.593	−9.809	−10.027
25	−2.977	−8.167	−8.359	−8.554	−8.751	−8.952	−9.155	−9.361	−9.570	−9.783	−9.998	−10.217	−10.440
26	−8.369	−8.560	−8.754	−8.950	−9.150	−9.352	−9.557	−9.766	−9.978	−10.193	−10.411	−10.633	−10.858
27	−8.773	−8.965	−9.160	−9.358	−9.559	−9.763	−9.970	−10.180	−10.394	−10.610	−10.831	−11.055	−11.283
28	−9.191	−9.381	−9.580	−9.778	−9.980	−10.185	−10.393	−10.604	−10.819	−11.037	−11.259	−11.485	−11.715
29	−9.624	−9.817	−10.013	−10.211	−10.413	−10.618	−10.827	−11.039	−11.254	−11.474	−11.697	−11.923	−12.154
30	−10.073	−10.265	−10.460	−10.658	−10.860	−11.065	−11.273	−11.485	−11.701	−11.920	−12.144	−12.371	−12.602
31	−10.538	−10.729	−10.922	−11.120	−11.320	−11.524	−11.732	−11.943	−12.159	−12.378	−12.601	−12.828	−13.060
32	−11.020	−11.209	−11.401	−11.596	−11.795	−11.998	−12.205	−12.415	−12.629	−12.847	−13.069	−13.296	−13.527

$P = 200\,kg/cm^2$ gauge

S [‰] / T [°C]	30	31	32	33	34	35	36	37	38	39	40	41	42
−2	0.410	0.425	0.439	0.454	0.469	0.485	0.500	0.516	0.532	0.549	0.565	0.582	0.599
−1	0.028	0.029	0.031	0.032	0.034	0.036	0.038	0.040	0.042	0.044	0.047	0.049	0.052
0	−0.337	−0.348	−0.359	−0.371	−0.382	−0.393	−0.404	−0.416	−0.427	−0.438	−0.449	−0.461	−0.472
1	−0.686	−0.709	−0.733	−0.756	−0.780	−0.804	−0.828	−0.852	−0.876	−0.900	−0.924	−0.949	−0.973
2	−1.021	−1.056	−1.091	−1.127	−1.162	−1.198	−1.234	−1.270	−1.307	−1.343	−1.380	−1.417	−1.455
3	−1.345	−1.391	−1.437	−1.483	−1.530	−1.577	−1.625	−1.673	−1.721	−1.770	−1.819	−1.868	−1.918
4	−1.658	−1.714	−1.771	−1.828	−1.886	−1.944	−2.003	−2.062	−2.121	−2.181	−2.242	−2.303	−2.365
5	−1.964	−2.030	−2.097	−2.164	−2.232	−2.300	−2.369	−2.439	−2.509	−2.580	−2.651	−2.724	−2.796
6	−2.263	−2.339	−2.415	−2.492	−2.569	−2.647	−2.726	−2.806	−2.886	−2.967	−3.049	−3.132	−3.215
7	−2.559	−2.643	−2.728	−2.814	−2.900	−2.988	−3.076	−3.165	−3.255	−3.345	−3.437	−3.529	−3.623
8	−2.852	−2.945	−3.038	−3.132	−3.227	−3.323	−3.419	−3.517	−3.616	−3.716	−3.816	−3.918	−4.021
9	−3.145	−3.245	−3.346	−3.448	−3.551	−3.655	−3.759	−3.865	−3.972	−4.080	−4.190	−4.300	−4.412
10	−3.440	−3.547	−3.655	−3.764	−3.874	−3.985	−4.097	−4.211	−4.325	−4.441	−4.559	−4.677	−4.797
11	−3.739	−3.852	−3.966	−4.082	−4.198	−4.316	−4.435	−4.555	−4.677	−4.800	−4.925	−5.051	−5.178
12	−4.043	−4.162	−4.282	−4.403	−4.526	−4.649	−4.775	−4.901	−5.029	−5.159	−5.290	−5.423	−5.558
13	−4.355	−4.479	−4.604	−4.730	−4.858	−4.987	−5.118	−5.250	−5.348	−5.520	−5.657	−5.796	−5.936
14	−4.677	−4.805	−4.934	−5.065	−5.197	−5.331	−5.467	−5.604	−5.743	−5.884	−6.026	−6.171	−6.317
15	−5.010	−5.141	−5.274	−5.409	−5.545	−5.683	−5.823	−5.964	−6.108	−6.253	−6.400	−6.549	−6.701
16	−5.356	−5.491	−5.627	−5.765	−5.904	−6.045	−6.188	−6.334	−6.481	−6.630	−6.781	−6.934	−7.089
17	−5.718	−5.855	−5.993	−6.133	−6.275	−6.419	−6.565	−6.713	−6.863	−7.015	−7.170	−7.326	−7.485
18	−6.097	−6.235	−6.375	−6.517	−6.661	−6.807	−6.955	−7.105	−7.257	−7.412	−7.569	−7.728	−7.890
19	−6.496	−6.635	−6.775	−6.918	−7.063	−7.210	−7.360	−7.511	−7.665	−7.821	−7.980	−8.141	−8.305
20	−6.915	−7.054	−7.195	−7.338	−7.484	−7.631	−7.781	−7.933	−8.088	−8.245	−8.405	−8.568	−8.733
21	−7.358	−7.496	−7.637	−7.779	−7.924	−8.071	−8.221	−8.373	−8.528	−8.686	−8.846	−9.009	−9.175
22	−7.826	−7.963	−8.102	−8.243	−8.387	−8.533	−8.682	−8.833	−8.987	−9.144	−9.304	−9.467	−9.632
23	−8.321	−8.455	−8.592	−8.731	−8.873	−9.018	−9.165	−9.315	−9.467	−9.623	−9.782	−9.943	−10.108
24	−8.845	−8.976	−9.110	−9.247	−9.386	−9.528	−9.672	−9.820	−9.970	−10.124	−10.280	−10.440	−10.603
25	−9.399	−9.527	−9.657	−9.790	−9.926	−10.065	−10.206	−10.350	−10.498	−10.648	−10.802	−10.959	−11.120

Table 10 (continued).

T [°C] \ S [‰]	30	31	32	33	34	35	36	37	38	39	40	41	42
						$P = 300$ kg/cm^2 gauge							
−2	0.149	0.154	0.158	0.163	0.168	0.174	0.179	0.185	0.191	0.198	0.204	0.211	0.218
−1	−0.289	−0.299	−0.309	−0.318	−0.328	−0.337	−0.346	−0.355	−0.364	−0.373	−0.382	−0.391	−0.399
0	−0.700	−0.724	−0.747	−0.770	−0.794	−0.817	−0.841	−0.864	−0.887	−0.911	−0.934	−0.957	−0.981
1	−1.088	−1.124	−1.161	−1.197	−1.233	−1.270	−1.306	−1.343	−1.380	−1.417	−1.454	−1.492	−1.529
2	−1.456	−1.504	−1.552	−1.600	−1.649	−1.698	−1.747	−1.797	−1.846	−1.896	−1.946	−1.997	−2.047
3	−1.805	−1.865	−1.924	−1.984	−2.044	−2.105	−2.166	−2.227	−2.288	−2.350	−2.413	−2.475	−2.539
4	−2.141	−2.211	−2.280	−2.351	−2.422	−2.493	−2.565	−2.637	−2.710	−2.783	−2.857	−2.931	−3.006
5	−2.465	−2.544	−2.624	−2.704	−2.785	−2.866	−2.948	−3.030	−3.113	−3.197	−3.281	−3.366	−3.451
6	−2.781	−2.869	−2.957	−3.046	−3.136	−3.226	−3.317	−3.409	−3.501	−3.595	−3.689	−3.783	−3.879
7	−3.093	−3.188	−3.284	−3.381	−3.479	−3.577	−3.677	−3.777	−3.878	−3.980	−4.083	−4.187	−4.291
8	−3.402	−3.504	−3.607	−3.711	−3.816	−3.922	−4.029	−4.137	−4.246	−4.355	−4.466	−4.578	−4.691
9	−3.712	−3.820	−3.930	−4.040	−4.151	−4.264	−4.377	−4.492	−4.607	−4.724	−4.842	−4.961	−5.082
10	−4.026	−4.140	−4.254	−4.370	−4.487	−4.605	−4.724	−4.845	−4.966	−5.089	−5.213	−5.339	−5.466
11	−4.347	−4.465	−4.585	−4.705	−4.826	−4.949	−5.073	−5.198	−5.325	−5.453	−5.583	−5.714	−5.846
12	−4.679	−4.801	−4.923	−5.047	−5.172	−5.299	−5.427	−5.556	−5.687	−5.819	−5.953	−6.089	−6.226
13	−5.024	−5.148	−5.273	−5.400	−5.528	−5.657	−5.788	−5.921	−6.055	−6.191	−6.328	−6.467	−6.608
14	−5.386	−5.512	−5.638	−5.767	−5.896	−6.028	−6.161	−6.296	−6.432	−6.570	−6.710	−6.852	−6.996
15	−5.767	−5.893	−6.021	−6.150	−6.281	−6.413	−6.547	−6.683	−6.821	−6.961	−7.102	−7.246	−7.392
16	−6.171	−6.297	−6.424	−6.553	−6.684	−6.816	−6.951	−7.087	−7.225	−7.365	−7.508	−7.652	−7.799
17	−6.601	−6.725	−6.851	−6.979	−7.109	−7.240	−7.374	−7.509	−7.647	−7.787	−7.929	−8.073	−8.220
18	−7.059	−7.181	−7.305	−7.431	−7.559	−7.688	−7.820	−7.954	−8.090	−8.229	−8.370	−8.513	−8.658
19	−7.550	−7.668	−7.789	−7.912	−8.036	−8.163	−8.292	−8.424	−8.557	−8.693	−8.832	−8.973	−9.117
20	−8.075	−8.189	−8.306	−8.424	−8.545	−8.668	−8.793	−8.921	−9.051	−9.184	−9.320	−9.457	−9.598
21	−8.638	−8.747	−8.859	−8.972	−9.088	−9.206	−9.327	−9.450	−9.576	−9.704	−9.835	−9.969	−10.106
22	−9.243	−9.345	−9.450	−9.558	−9.668	−9.780	−9.895	−10.012	−10.133	−10.256	−10.381	−10.510	−10.642

T [°C] \ S [‰]	30	31	32	33	34	35	36	37	38	39	40	41	42
						$P = 400\ kg/cm^2$ gauge							
-2	-0.162	-0.170	-0.177	-0.184	-0.191	-0.198	-0.204	-0.210	-0.216	-0.222	-0.277	-0.232	-0.237
-1	-0.670	-0.694	-0.717	-0.740	-0.763	-0.786	-0.808	-0.831	-0.854	-0.876	-0.898	-0.920	-0.942
0	-1.142	-1.180	-1.219	-1.257	-1.295	-1.333	-1.371	-1.409	-1.447	-1.485	-1.523	-1.561	-1.599
1	-1.583	-1.635	-1.687	-1.739	-1.791	-1.843	-1.896	-1.948	-2.001	-2.053	-2.106	-2.159	-2.212
2	-1.997	-2.061	-2.126	-2.191	-2.256	-2.321	-2.387	-2.453	-2.519	-2.585	-2.651	-2.718	-2.785
3	-2.387	-2.463	-2.540	-2.616	-2.693	-2.771	-2.848	-2.927	-3.005	-3.084	-3.163	-3.243	-3.323
4	-2.759	-2.846	-2.933	-3.020	-3.108	-3.196	-3.285	-3.374	-3.464	-3.554	-3.645	-3.736	-3.828
5	-3.116	-3.212	-3.309	-3.406	-3.503	-3.601	-3.700	-3.800	-3.900	-4.000	-4.101	-4.203	-4.306
6	-3.463	-3.567	-3.672	-3.778	-3.884	-3.991	-4.099	-4.207	-4.316	-4.426	-4.537	-4.648	-4.760
7	-3.804	-3.915	-4.027	-4.140	-4.254	-4.369	-4.484	-4.600	-4.718	-4.835	-4.954	-5.074	-5.195
8	-4.142	-4.260	-4.379	-4.498	-4.618	-4.739	-4.861	-4.984	-5.108	-5.233	-5.359	-5.486	-5.614
9	-4.484	-4.606	-4.730	-4.854	-4.980	-5.106	-5.234	-5.362	-5.492	-5.623	-5.755	-5.888	-6.022
10	-4.831	-4.958	-5.085	-5.214	-5.343	-5.474	-5.606	-5.739	-5.873	-6.009	-6.145	-6.284	-6.423
11	-5.190	-5.319	-5.449	-5.580	-5.713	-5.846	-5.981	-6.118	-6.256	-6.395	-6.535	-6.677	-6.821
12	-5.563	-5.694	-5.825	-5.958	-6.093	-6.228	-6.365	-6.504	-6.644	-6.785	-6.928	-7.073	-7.219
13	-5.956	-6.086	-6.219	-6.352	-6.487	-6.623	-6.761	-6.901	-7.042	-7.184	-7.329	-7.475	-7.623
14	-6.372	-6.502	-6.633	-6.766	-6.900	-7.036	-7.173	-7.313	-7.454	-7.596	-7.741	-7.887	-8.036
15	-6.815	-6.943	-7.072	-7.203	-7.336	-7.470	-7.606	-7.744	-7.883	-8.025	-8.168	-8.314	-8.461
						$P = 500\ kg/cm^2$ gauge							
-2	-0.489	-0.509	-0.529	-0.549	-0.568	-0.588	-0.607	-0.626	-0.644	-0.663	-0.681	-0.699	-0.716
-1	-1.079	-1.117	-1.154	-1.192	-1.230	-1.267	-1.304	-1.341	-1.378	-1.415	-1.452	-1.489	-1.525
0	-1.624	-1.678	-1.733	-1.787	-1.841	-1.895	-1.949	-2.003	-2.057	-2.112	-2.166	-2.220	-2.274
1	-2.130	-2.200	-2.269	-2.338	-2.408	-2.477	-2.547	-2.617	-2.687	-2.757	-2.827	-2.897	-2.968
2	-2.603	-2.685	-2.769	-2.852	-2.935	-3.019	-3.103	-3.187	-3.272	-3.356	-3.441	-3.526	-3.612
3	-3.046	-3.141	-3.237	-3.333	-3.429	-3.525	-3.622	-3.720	-3.817	-3.915	-4.013	-4.112	-4.211
4	-3.467	-3.573	-3.679	-3.787	-3.894	-4.002	-4.110	-4.319	-4.329	-4.439	-4.549	-4.660	-4.772
5	-3.869	-3.985	-4.101	-4.218	-4.336	-4.454	-4.573	-4.692	-4.812	-4.932	-5.053	-5.175	-5.298
6	-4.259	-4.383	-4.508	-4.634	-4.760	-4.886	-5.014	-5.142	-5.271	-5.401	-5.531	-5.663	-5.795
7	-4.642	-4.773	-4.905	-5.037	-5.171	-5.305	-5.440	-5.576	-5.712	-5.850	-5.988	-6.128	-6.268
8	-5.023	-5.160	-5.297	-5.435	-5.575	-5.715	-5.856	-5.998	-6.141	-6.285	-6.430	-6.576	-6.723
9	-5.408	-5.548	-5.690	-5.833	-5.976	-6.121	-6.267	-6.414	-6.562	-6.711	-6.861	-7.012	-7.165
10	-5.801	-5.945	-6.089	-6.235	-6.381	-6.529	-6.678	-6.829	-6.980	-7.133	-7.287	-7.442	-7.599
11	-6.209	-6.354	-6.500	-6.647	-6.795	-6.945	-7.096	-7.248	-7.401	-7.556	-7.713	-7.870	-8.030
12	-6.636	-6.781	-6.927	-7.074	-7.223	-7.373	-7.524	-7.677	-7.831	-7.987	-8.144	-8.303	-8.463
13	-7.089	-7.232	-7.376	-7.522	-7.669	-7.818	-7.968	-8.120	-8.274	-8.429	-8.586	-8.744	-8.904
14	-7.571	-7.711	-7.853	-7.996	-8.141	-8.287	-8.435	-8.584	-8.735	-8.888	-9.043	-9.200	-9.358
15	-8.090	-8.225	-8.363	-8.501	-8.642	-8.784	-8.928	-9.074	-9.221	-9.370	-9.522	-9.675	-9.831

Table 10 (continued).

T [°C] \ S [‰]	30	31	32	33	34	35	36	37	38	39	40	41	42
						$P = 600$ kg/cm² gauge							
−2	−0.795	−0.828	−0.860	−0.892	−0.923	−0.955	−0.986	−1.017	−1.048	−1.078	−1.108	−1.138	−1.168
−1	−1.478	−1.530	−1.582	−1.634	−1.686	−1.738	−1.789	−1.840	−1.892	−1.943	−1.994	−2.045	−2.096
0	−2.108	−2.178	−2.248	−2.318	−2.389	−2.459	−2.529	−2.600	−2.670	−2.740	−2.810	−2.881	−2.951
1	−2.690	−2.777	−2.864	−2.951	−3.038	−3.125	−3.213	−3.300	−3.388	−3.476	−3.564	−3.652	−3.740
2	−3.232	−3.333	−3.436	−3.538	−3.640	−3.743	−3.846	−3.950	−4.053	−4.157	−4.261	−4.365	−4.470
3	−3.739	−3.854	−3.970	−4.086	−4.202	−4.319	−4.436	−4.553	−4.671	−4.789	−4.908	−5.027	−5.116
4	−4.219	−4.346	−4.474	−4.602	−4.730	−4.859	−4.988	−5.118	−5.249	−5.379	−5.511	−5.643	−5.775
5	−4.677	−4.815	−4.953	−5.091	−5.230	−5.370	−5.510	−5.651	−5.792	−5.934	−6.077	−6.220	−6.364
						$P = 700$ kg/cm² gauge							
−2	−1.046	−1.089	−1.131	−1.173	−1.215	−1.257	−1.298	−1.340	−1.381	−1.421	−1.462	−1.502	−1.542
−1	−1.831	−1.896	−1.961	−2.025	−2.090	−2.154	−2.218	−2.282	−2.346	−2.410	−2.474	−2.538	−2.601
0	−2.554	−2.639	−2.724	−2.809	−2.894	−2.980	−3.065	−3.150	−3.235	−3.320	−3.405	−3.490	−3.576
1	−3.221	−3.325	−3.429	−3.533	−3.637	−3.741	−3.845	−3.949	−4.054	−4.159	−4.263	−4.368	−4.473
2	−3.842	−3.962	−4.083	−4.203	−4.324	−4.446	−4.567	−4.689	−4.811	−4.933	−5.055	−5.178	−5.301
3	−4.422	−4.557	−4.693	−4.829	−4.965	−5.101	−5.238	−5.375	−5.513	−5.651	−5.789	−5.928	−6.067
4	−4.970	−5.118	−5.267	−5.416	−5.565	−5.715	−5.866	−6.017	−6.168	−6.320	−6.472	−6.625	−6.778
5	−5.494	−5.653	−5.813	−5.973	−6.134	−6.296	−6.458	−6.620	−6.784	−6.947	−7.112	−7.277	−7.443
						$P = 800$ kg/cm² gauge							
−2	−1.207	−1.256	−1.306	−1.355	−1.404	−1.453	−1.502	−1.550	−1.598	−1.646	−1.693	−1.741	−1.788
−1	−2.101	−2.176	−2.251	−2.325	−2.399	−2.474	−2.548	−2.622	−2.695	−2.769	−2.843	−2.916	−2.989
0	−2.925	−3.022	−3.120	−3.217	−3.314	−3.412	−3.509	−3.607	−3.704	−3.802	−3.899	−3.997	−4.094
1	−3.685	−3.803	−3.921	−4.040	−4.158	−4.277	−4.396	−4.515	−4.634	−4.753	−4.872	−4.992	−5.111
2	−4.391	−4.528	−4.665	−4.802	−4.940	−5.078	−5.216	−5.354	−5.492	−5.631	−5.770	−5.909	−6.049
3	−5.052	−5.205	−5.359	−5.513	−5.667	−5.822	−5.977	−6.133	−6.289	−6.445	−6.601	−6.758	−6.916
4	−5.676	−5.844	−6.012	−6.181	−6.350	−6.519	−6.690	−6.860	−7.031	−7.203	−7.375	−7.547	−7.721
5	−6.272	−6.452	−6.633	−6.814	−6.996	−7.178	−7.361	−7.545	−7.729	−7.914	−8.099	−8.285	−8.472

T [°C] \ S [‰]	30	31	32	33	34	35	36	37	38	39	40	41	42
							$P = 900\ kg/cm^2$ gauge						
−2	−1.242	−1.294	−1.346	−1.398	−1.450	−1.501	−1.552	−1.603	−1.654	−1.704	−1.755	−1.805	−1.854
−1	−2.252	−2.332	−2.412	−2.492	−2.572	−2.652	−2.732	−2.812	−2.891	−2.971	−3.050	−3.219	−3.208
0	−3.181	−3.287	−3.393	−3.499	−3.605	−3.711	−3.817	−3.923	−4.029	−4.135	−4.241	−4.347	−4.453
1	−4.040	−4.169	−4.298	−4.428	−4.558	−4.687	−4.817	−4.947	−5.077	−5.208	−5.338	−5.469	−5.599
2	−4.837	−4.988	−5.138	−5.289	−5.440	−5.591	−5.742	−5.894	−6.046	−6.198	−6.350	−6.503	−6.656
3	−5.584	−5.753	−5.922	−6.092	−6.262	−6.432	−6.602	−6.773	−6.944	−7.116	−7.288	−7.461	−7.633
4	−6.290	−6.475	−6.661	−6.846	−7.033	−7.220	−7.407	−7.595	−7.783	−7.972	−8.161	−8.351	−8.541
5	−6.965	−7.164	−7.363	−7.563	−7.763	−7.964	−8.166	−8.368	−8.571	−8.774	−8.979	−9.183	−9.389
							$P = 1000\ kg/cm^2$ gauge						
−2	−1.115	−1.165	−1.241	−1.263	−1.311	−1.360	−1.408	−1.456	−1.503	−1.551	−1.598	−1.645	−1.691
−1	−2.245	−2.325	−2.406	−2.486	−2.567	−2.647	−2.727	−2.807	−2.887	−2.967	−3.046	−3.126	−3.205
0	−3.284	−3.394	−3.503	−3.613	−3.772	−3.832	−3.941	−4.501	−4.160	−4.270	−4.379	−4.489	−4.598
1	−4.246	−4.381	−4.517	−4.653	−4.789	−4.925	−5.061	−5.198	−5.334	−5.471	−5.607	−5.744	−5.881
2	−5.139	−5.298	−5.458	−5.617	−5.777	−5.937	−6.098	−6.258	−6.419	−6.580	−6.742	−6.903	−7.065
3	−5.976	−6.156	−6.337	−6.518	−6.699	−6.880	−7.062	−7.244	−7.427	−7.610	−7.793	−7.977	−8.161
4	−6.768	−6.966	−7.165	−7.364	−7.564	−7.764	−7.965	−8.166	−8.367	−8.570	−8.772	−8.975	−9.179
5	−7.525	−7.739	−7.954	−8.169	−8.384	−8.600	−8.817	−9.035	−9.253	−9.471	−9.690	−9.910	−10.131
							$P = 1100\ kg/cm^2$ gauge						
−2	−0.793	−0.832	−0.871	−0.910	−0.949	−0.987	−1.025	−1.063	−1.100	−1.138	−1.175	−1.212	−1.248
−1	−2.044	−2.119	−2.193	−2.267	−2.341	−2.415	−2.488	−2.562	−2.636	−2.709	−2.782	−2.855	−2.928
0	−3.197	−3.303	−3.410	−3.516	−3.623	−3.730	−3.836	−3.943	−4.049	−4.156	−4.262	−4.369	−4.475
1	−4.263	−4.399	−4.535	−4.671	−4.807	−4.943	−5.080	−5.217	−5.353	−5.490	−5.627	−5.764	−5.901
2	−5.255	−5.417	−5.580	−5.742	−5.905	−6.069	−6.232	−6.396	−6.500	−6.724	−6.889	−7.053	−7.218
3	−6.184	−6.370	−6.556	−6.743	−6.930	−7.117	−7.305	−7.493	−7.681	−7.870	−8.059	−8.249	−8.439
4	−7.064	−7.271	−7.478	−7.685	−7.893	−8.101	−8.310	−8.520	−8.729	−8.940	−9.151	−9.362	−9.574
5	−7.906	−8.130	−8.355	−8.581	−8.807	−9.033	−9.260	−9.488	−9.716	−9.945	−10.175	−10.405	−10.636

3.2.5.4.3 Simplified sound speed formula

Based on Del Grosso's 19-term equation [72 D 1] Mackenzie [81 M 2] suggested 1981 a formula of only 9 terms suitable for programmable pocket calculators. Ranges of validity encompass: temperature $-2^\circ\cdots30\,^\circ$C, salinity 30···40‰ and depth 0···8000 m. The standard deviation against the Del Grosso and Mader formula is 0.07 m/s.

$$c=1448.96+4.591\,T-5.304\cdot10^{-2}\,T^2+2.374\cdot10^{-4}\,T^3+1.340(S-35)+1.630\cdot10^{-2}\,D+1.675\cdot10^{-7}\,D^2$$
$$-1.035\cdot10^{-2}\,T(S-35)-7.139\cdot10^{-13}\,TD^3\,. \qquad (31)$$

The equation yields the sound speed c in [m/s]. The temperature T is in [°C], the salinity S is in [‰] and the depth D is in [m]. For his 9-term equation Mackenzie selected 15 representative world-wide stations for examination which cover the latitude range between 70° N and 10° S.

A very simple formula of only 6 terms based on Del Grosso's NRL II formula [74 D 1] was proposed by Medwin 1975 [75 M 1] which covers realistic combinations of T, S, P in the ranges $0\leqq T\leqq35\,^\circ$C, $0\leqq S\leqq45$‰, $0\leqq D\leqq1000$ m.

$$c=1449.2+4.6\,T-0.055\,T^2+0.00029\,T^3+(1.34-0.010\,T)\cdot(S-35)+0.016\,D\,. \qquad (32)$$

The standard deviation against NRL II is less than 0.2 m/s for temperatures between 0 and 30 °C, 0···1000 m depth and 20···40‰ salinity. Absolute accuracies cannot be given.

3.2.5.5 Deep-sea sound speed profiles

Sound speed profile means the sound speed-depth function. It is measured either directly by sound speed meters (subsect. 3.2.3.3.1) or by determinations of temperature, salinity and depth (subsect. 3.2.5.3).

A typical deep-sea profile is shown in Fig. 9. For the acoustic description it is usually divided into several layers of different characteristics and occurrence. The topmost "surface layer" is susceptible to weather influence corresponding to time scales of hours up to days, the thickness being of the order of several 10 m (see sect. 5.4). The resulting space-time structure of the sound field is similar to shallow water. Homogeneous conditions (mixed layer) due to sufficient wind activities allow horizontal sound propagation within the surface layer (subsect. 3.2.2.1).

Below the surface layer lies the "seasonal thermocline" where the vertical gradient of the sound speed and the resulting refraction of the sound waves are strongest during summer and fall. During winter and spring and in the Arctic it tends to merge with the surface layer.

The adjacent layer below the seasonal thermocline is the "main thermocline", where the temperature decreases to the deep cold depths of the ocean.

Below the main thermocline and extending to the sea bottom is the "deep isothermal layer" having a nearly constant temperature in which sound speed increases with depth due to static pressure. Between the negative sound speed gradient of the main thermocline and the positive gradient of the deep layer there is a sound speed minimum, prerequisite of the so-called deep water channel or SOFAR-channel (sound fixing and ranging). The channel has sound focussing properties (subsects. 3.2.2.1, 3.2.6.2), thus allowing long-range sound propagation without bottom-surface contacts (Fig. 22).

The average seasonal effect on the deep water channel for the North Atlantic is shown in Fig. 10. In general the seasonal effect on the channel depth is weaker than the influence of the geographical latitude which is depicted in Fig. 11 [74 N 2] for a north-south section. The geographical distribution of the channel depth for the Atlantic Ocean is found in Fig. 12 [74 N 2]. Characteristic sound speed profiles for the deep ocean areas of the world are compared in Fig. 13 [70 U].

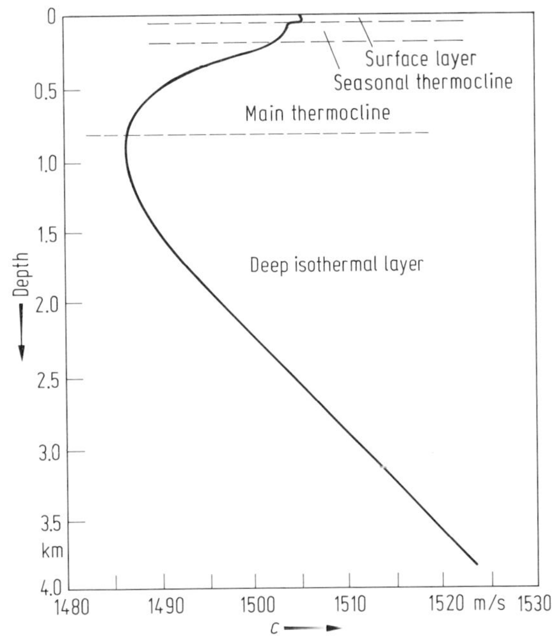

Fig. 9. Typical deep-sea sound speed profile divided into layers of acoustical relevance.

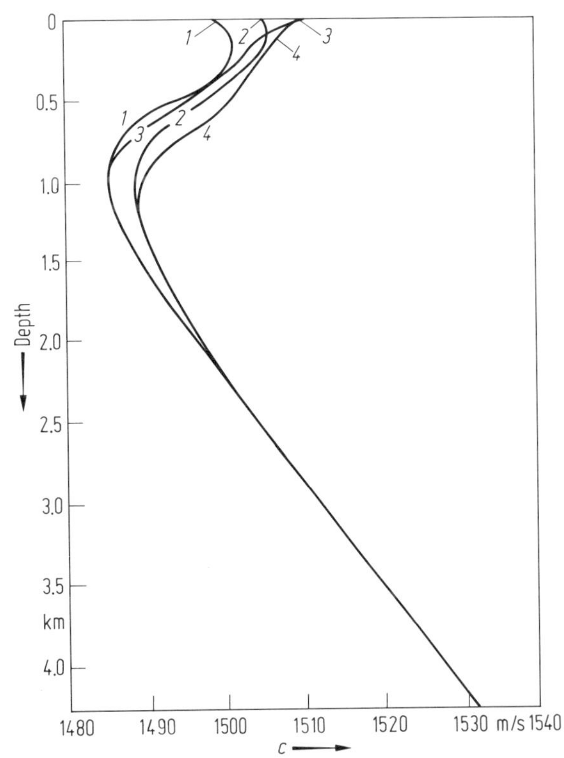

Fig. 10. Average profiles of sound speed in different seasons in North Atlantic areas (latitude 43° to 55° N, longitudes 20° to 40° W), (*1*) winter, (*2*) spring, (*3*) summer, (*4*) fall, [65 U].

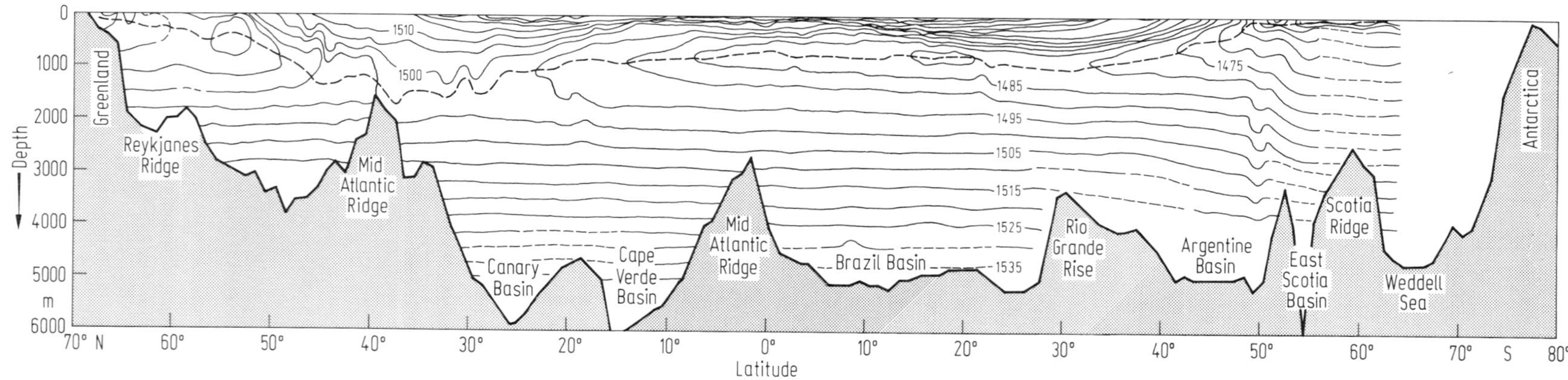

Fig. 11. North-south section of sound channel structure in the North and South Atlantic along the 30.50° W meridian. Sound speeds are in [m/s] and the approximate sound channel axis depth (sound speed minimum) is indicated by a heavy dashed line. Data positions are represented by dots (at surface) and interpolated contours are represented by broken lines [74 N 2].

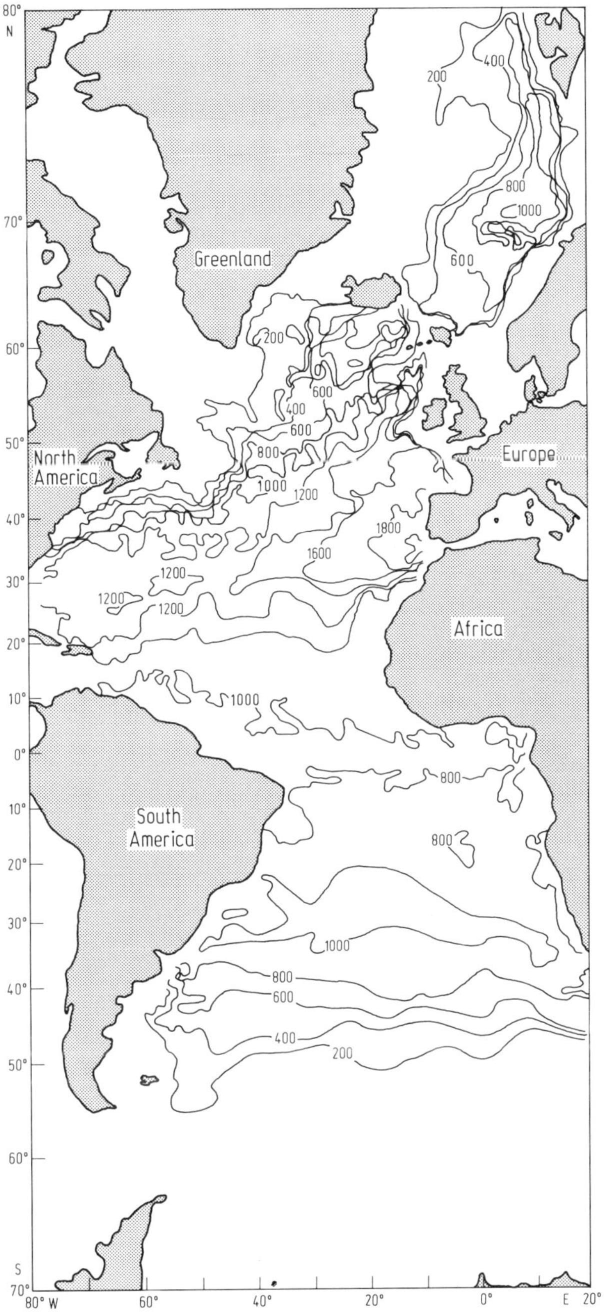

Fig. 12. Deep sound channel axis depth contours for the Atlantic Ocean. Depth in [m]. Channel depths at contour intervals of 200 m range from near the surface in the polar regions to 1800 m in the North Atlantic [74 N 2].

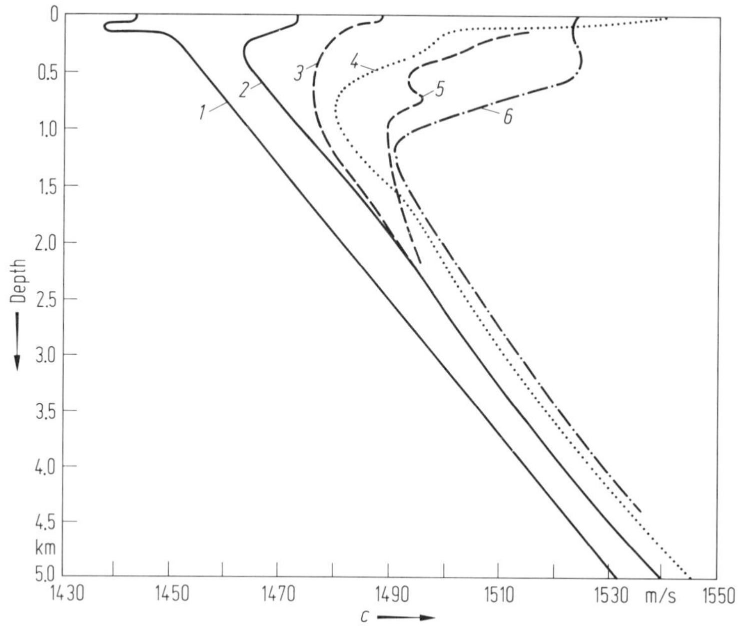

Fig. 13. Characteristic sound speed profiles for deep ocean areas of the world [83 U]. (*1*) Antarctic Ocean (60° S), (*2*) North Pacific (45° to 55° N), (*3*) Southern Oceans (45° to 55° S), (*4*) Pacific and South Atlantic (40° S), (*5*) Indian Ocean under influence of Red Sea outflow, (*6*) North Atlantic under influence of Mediterranean Sea outflow.

3.2.5.6 Shallow-water sound speed profiles

The horizontal variation of the sound speed profile along typical sound transmission ranges often exceeds monthly or even seasonal variations at a given location. A stable sound channel does not exist due to the insufficient sound speed increase with pressure/depth (subsect. 3.2.6.3). Hence, the long-range transmission loss in shallow water is governed by boundary rather than volume losses.

Instead of problematic mean sound speed profiles, typical spatial variations of quasi-synoptic sound speed profiles are represented in Figs. 14···21. For the North Sea examples (Figs. 14/15) the temperature field is the governing quantity; salinity differences have only marginal effects. For the Baltic the near-bottom increase of sound speed due to the higher salinity appears in Fig. 21. The other Baltic Sea examples (Figs. 16···20) represent the temperature structure alone.

Fig. 14⋯20. Two-dimensional temperature fields in shallow water. Towed thermistor cable, 60 sensors, 50° inclination rel. surface. Upper part: temperature profile series. Lower part: isothermes, 1/2° interval (Figs. 14, 15) and 1° interval (Figs. 16⋯20).

Fig.	Sea area	Month	Position (tow start)	Towing speed [m/s]	Depth range [m]	Thermistor time constant [s]	Scanning period [s]	Profile depicted	Features
14	North Sea	May	58°47′ N 01°09′ E	3	7⋯54	15	12	each 5th	weak thermocline, stochastic components dominating [82 S 1]
15	North Sea	May	58°25′ N 00°51′ E	0.7	7⋯65	15	12	each 5th	resolution of fine structure, extremely low towing speed, weak front, stochastic components [82 S 1]
16	Baltic	June	55°29′ N 15°58′ E	3.5	7⋯54	40	30	each 2nd	wavelike declination of thermocline, mixed layer with horizontal structure [81 S 2]
17	Baltic	June	55°20′ N 15°19′ E	3.5	7⋯54	40	30	each 2nd	strong irregularities up to the surface, homogeneous inclosure, horizontal increase of near-bottom temperature [81 S 2]
18	Baltic	August	57°20′ N 20°15′ E	3.5	7⋯54	3	6	each	regular wavelike structure, same periodicity of thermoclines and near-surface temperature [82 S 2]
19	Baltic	August	55°17′ N 16°00′ E	4	7⋯54	3	6	each	weakest horizontal variation observed [82 S 2]
20	Baltic	September	55°17′ N 16°20′ E	3.5	7⋯54	40	30	each	single front, steep inclination of thermocline [80 S]

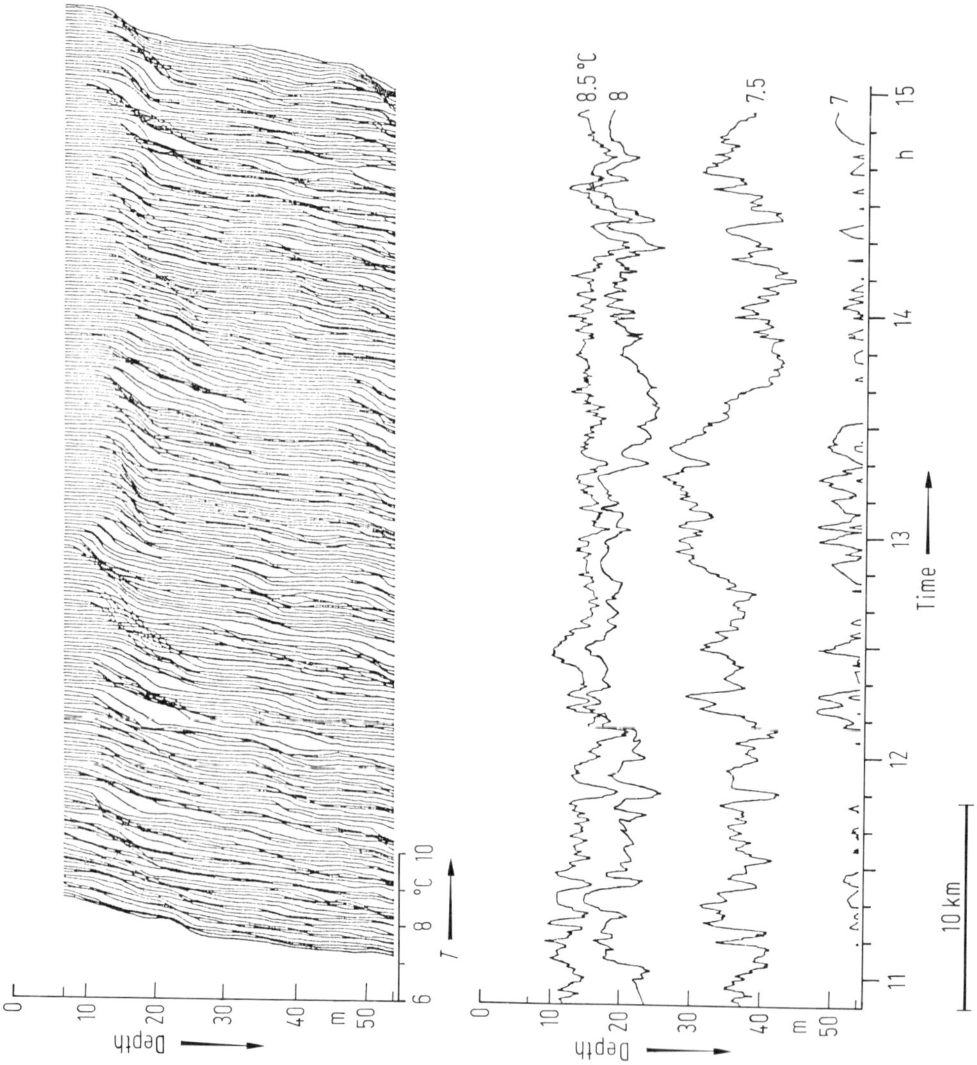

Fig. 14; caption see p. 311.

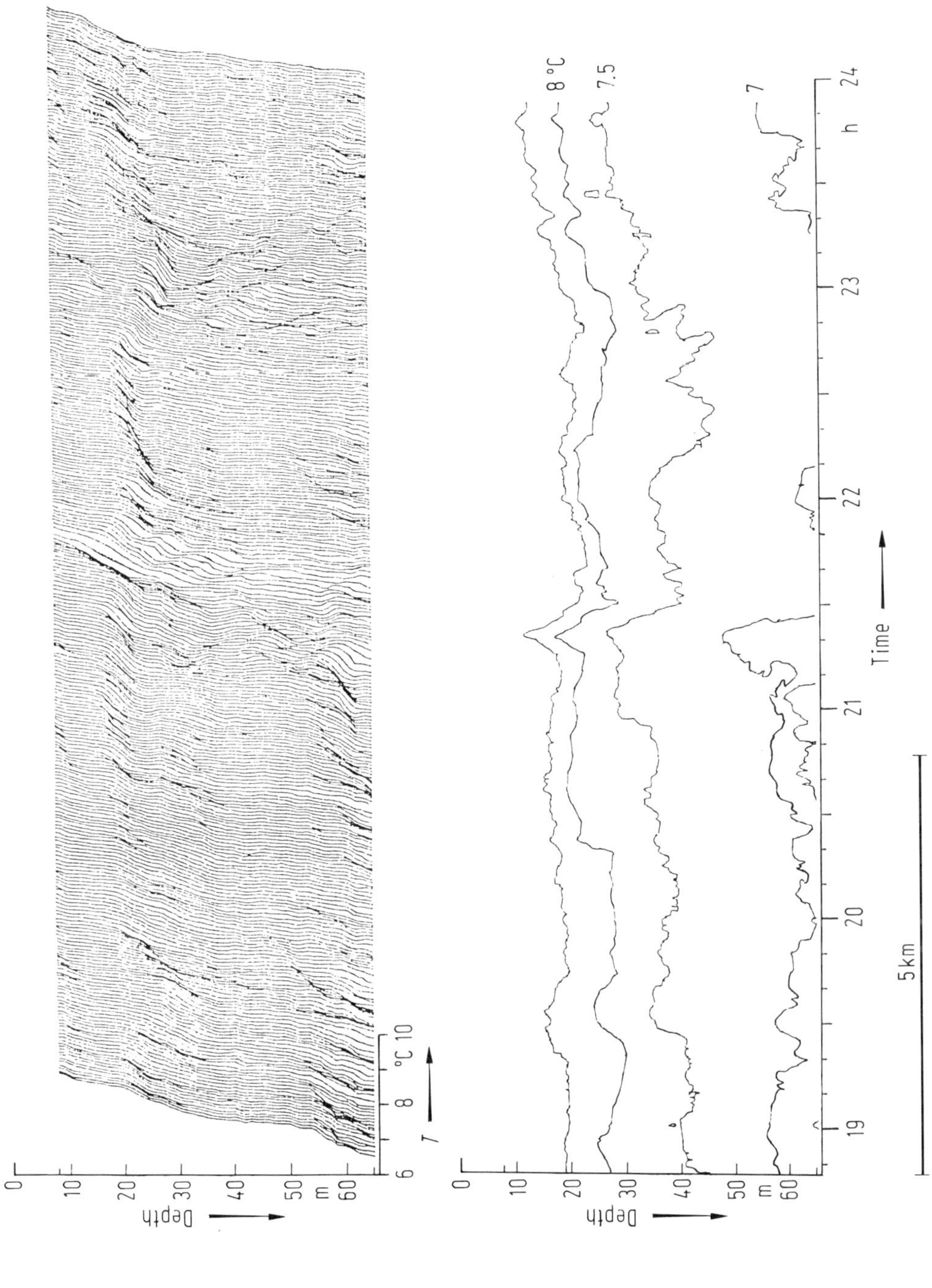

Fig. 15; caption see p. 311.

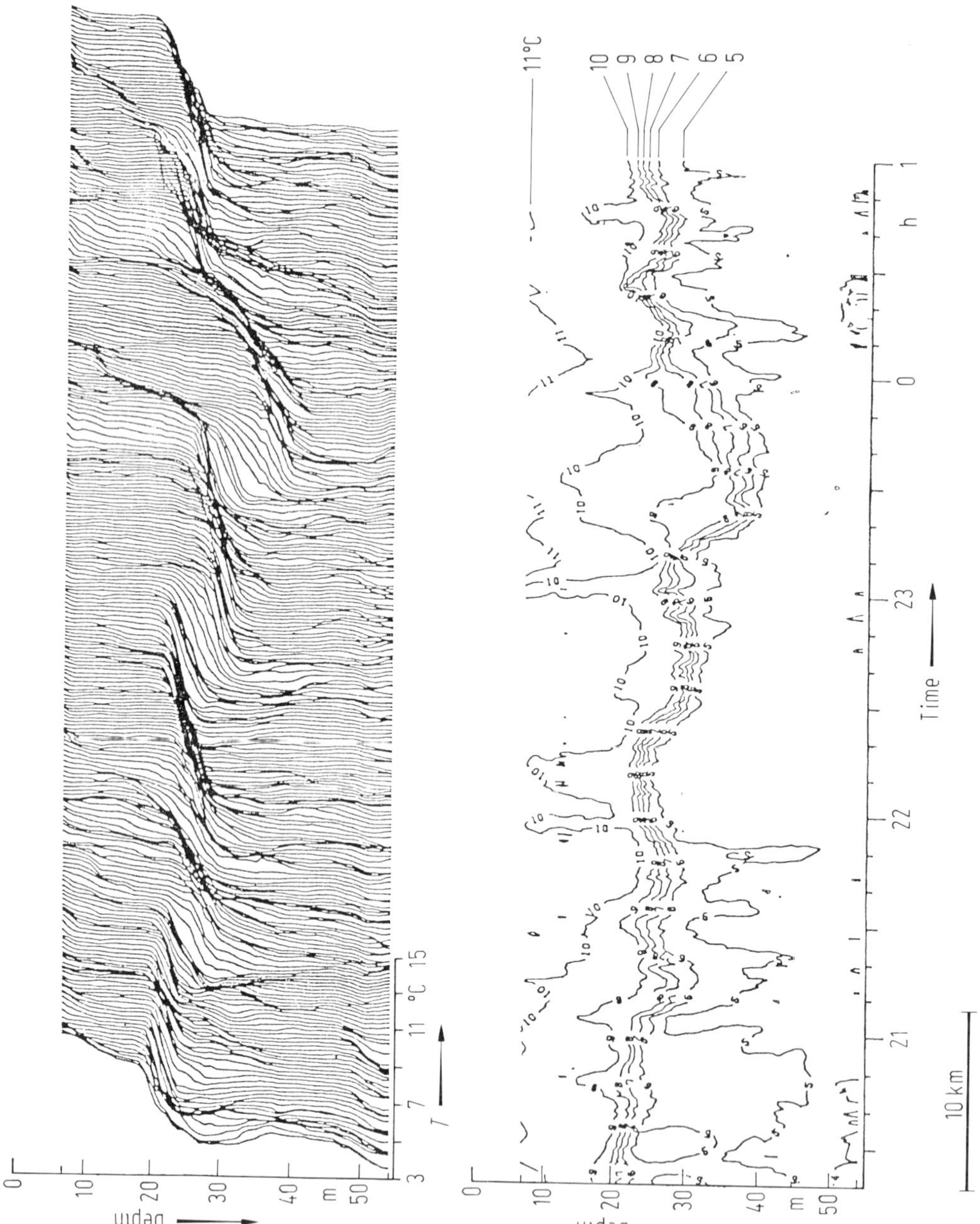

Fig. 16; caption see p. 311.

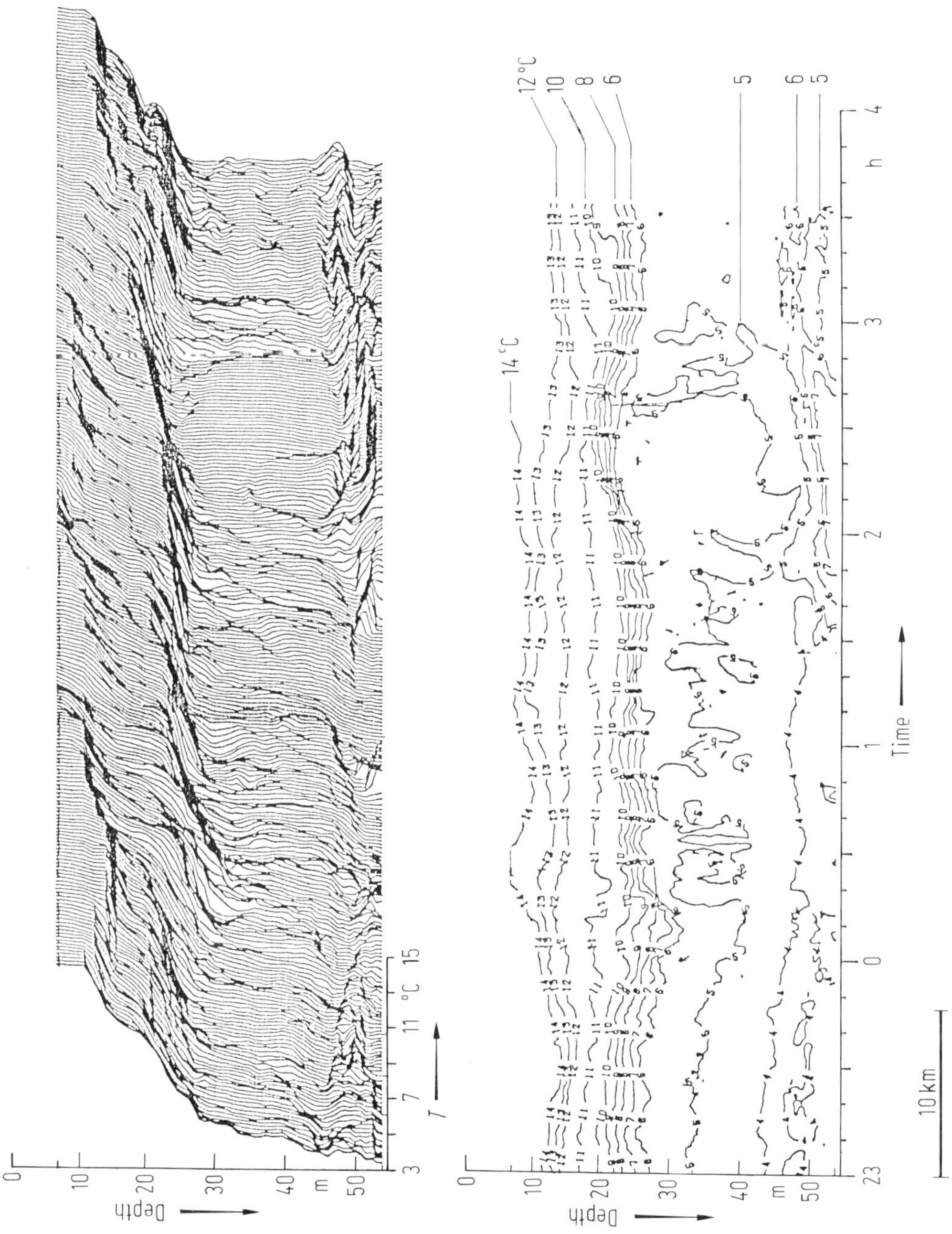

Fig. 17; caption see p. 311.

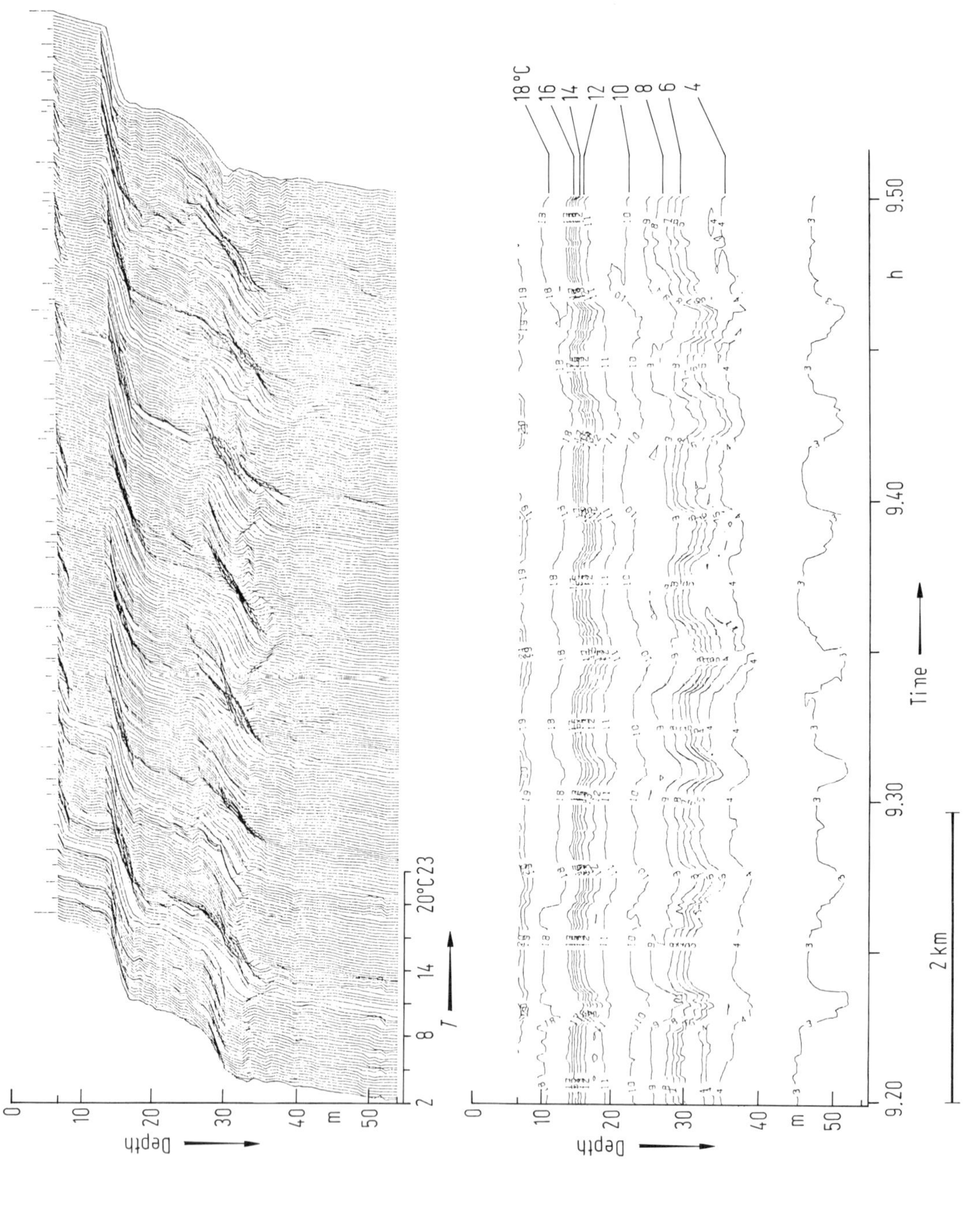

Fig. 18; caption see p. 311.

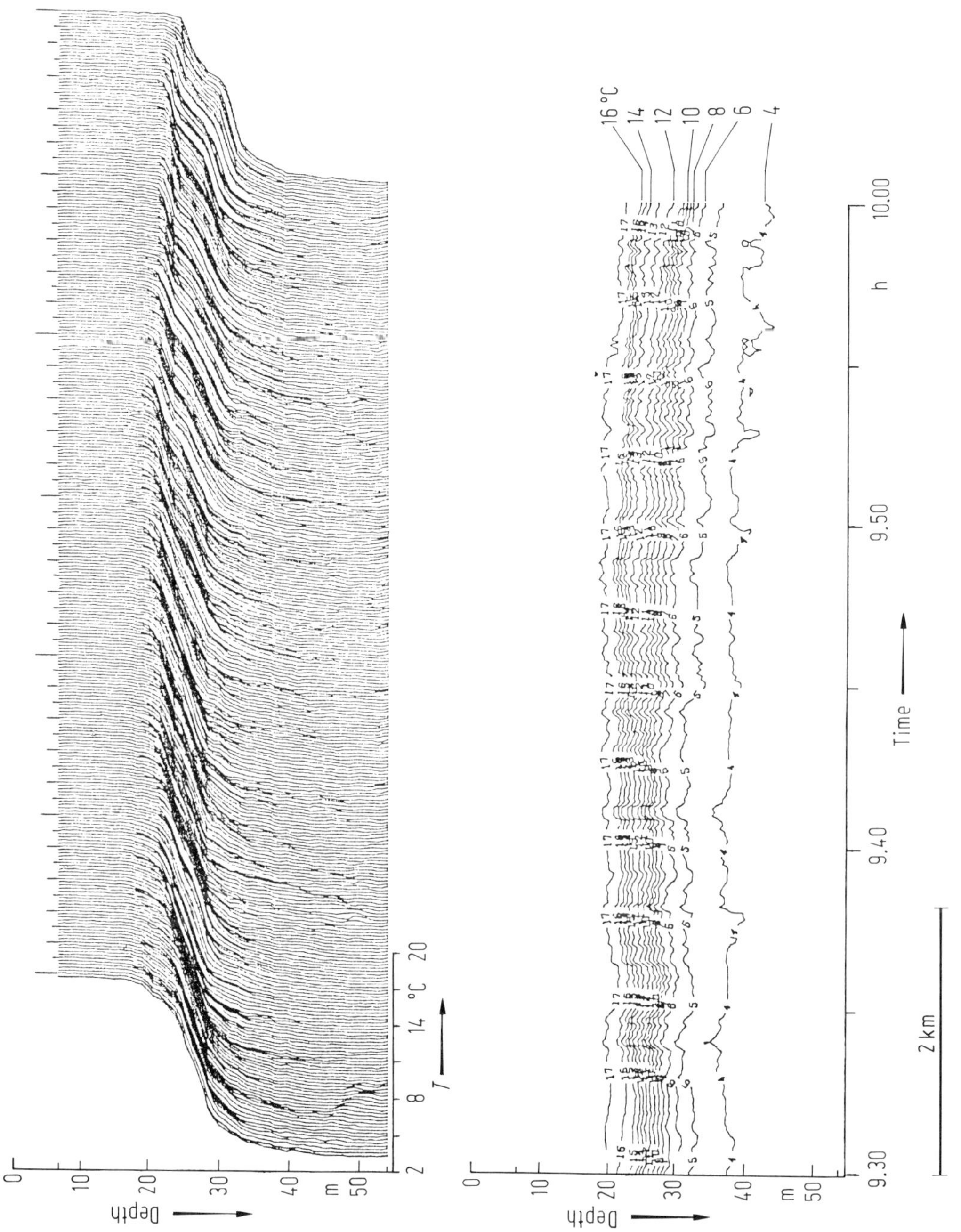

Fig. 19; caption see p. 311.

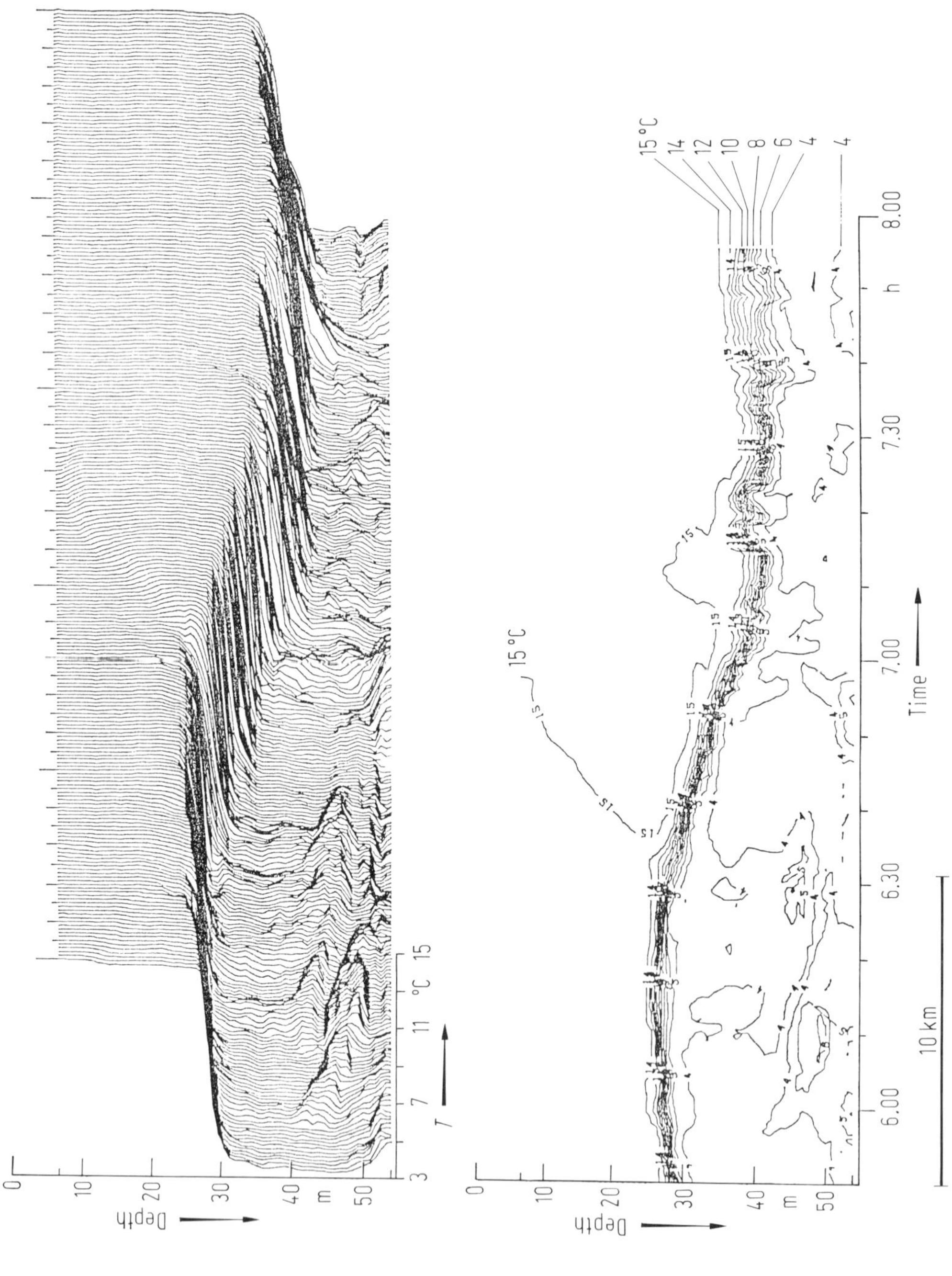

Fig. 20; caption see p. 311.

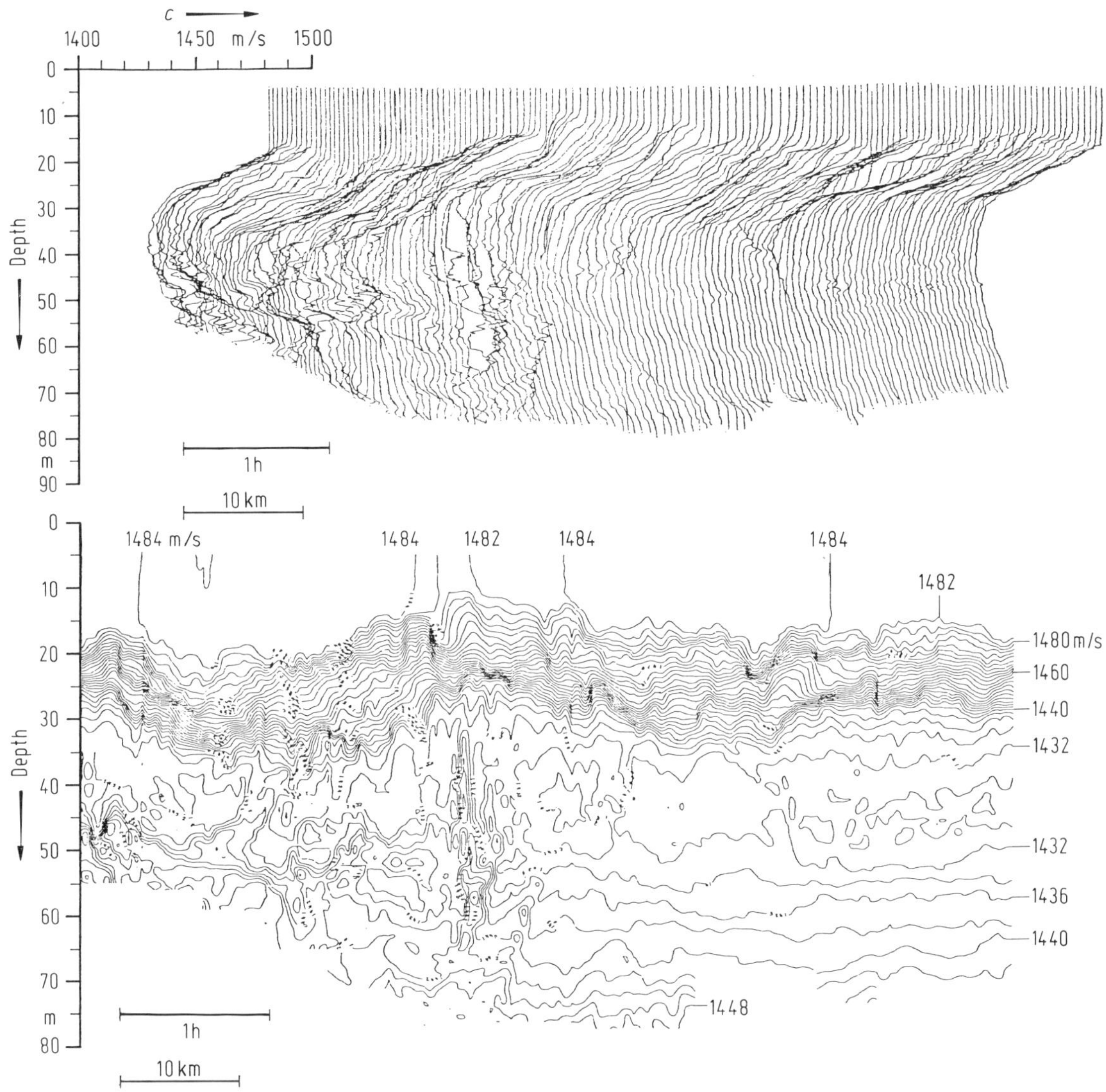

Fig. 21. Two-dimensional sound speed field in shallow water [84 D] refractive channel type (Baltic Sea, August; Batfish towed at 7 kts, sound speed meter), upper part: sound speed profile series, lower part: lines of constant sound speed, 2 m/s interval.

3.2.6 Sound absorption/transmission loss

3.2.6.1 Definitions

The transmission loss (TL) governs the range of signal transmission and any kind of acoustic remote sensing in the ocean. The quantity TL [75 U 1] comprehends both absorption and geometrical spreading/redistribution and describes, in [dB], overall decay of the sound intensity I between the value I_0 at a reference point located 1 m from the "acoustic center" (subsect. 3.2.4.1) of the sound source and the value I_1 at a given distance in the sea by

$$TL = 10\log\frac{I_0}{I_1}. \tag{33}$$

Under the assumption of stationary dissipation along the propagation path $r_2 - r_1$ of a quasi-plane wave, the sound absorption, in [dB km^{-1}], is described by the range-independent quantity

$$\alpha = (10\log I_1 - 10\log I_2)(r_2 - r_1)^{-1} \tag{34}$$

called the logarithmic absorption coefficient (to the base 10) of a quasi-plane wave. The product αr, in [dB], is called the absorption loss or attenuation loss.

By definition the transmission loss is restricted to the spatial distribution of the sound intensity and does not describe the additional spread in the domains of signal frequency and travel time. Hence, the quantity TL specifies neither fast fluctuations due to interference of narrow-band signals nor the time spread of multipath propagation. In general the space-time scale of averaging TL-components is chosen with respect to the scale of processes involved and may reach up to propagation path of some 10^3 km and several days, depending on the measuring procedure [74 B 2].

Both dissipative and conservative processes of sound propagation are associated by reflection, refraction, scattering and diffraction. Deep and shallow waters behave differently in this respect and are treated separately.

3.2.6.2 Deep-ocean transmission loss

The characteristic feature of the deep ocean for long-range sound propagation is the combination of the thermocline and the barocline (subsect. 3.2.5.5), forming the refractive deep water channel (DWC) or SOFAR-channel. Its axis is the horizon of the sound speed minimum. Due to the trapping of sound energy by combined upward and downward refraction, centered at the channel axis (Fig. 22), the highly effective scattering and absorbing boundaries (subsect. 3.2.8) of the sea surface and the bottom do not participate in this type of propagation. Thus, the essential absorption is caused by the water volume itself which is highly transparent for sound waves at low frequencies and enables signal transmission across entire oceans.

The long-range transmission loss of the DWC propagation may be treated simply as additive superposition of the absorption loss αr and the geometrical spreading loss $10\log r$, r in [m], (subsect. 3.2.6.2.2) [74 B 1]. However, for off-axis sound sources the refractive stratification causes alternating zones of convergence and divergence, enhancing and degrading the mean sound intensity by some 10 dB up to more than 20 dB [77 C]. Further complications occur when the vertical profile of the sound speed deviates from the DWC-type and the boundaries partially contribute to the transmission loss. The deviations between the actually measured TL and the value taken from prediction by capable models (subsect. 3.2.2) or empirical parametrisation will often exceed 10 dB and will be at least of the order of 3 dB, due to incomplete knowledge of the space-time distribution of the governing environmental quantities.

The sound absorption by sea water is essentially attributed to three additive terms which are predominant in adjacent frequency ranges (Fig. 23): The classical viscous absorption of pure water (subsect. 3.2.6.2.3) prevails above 500 kHz, followed by the relaxation absorption of $MgSO_4$ (subsect. 3.2.6.2.4), dominating between 200 kHz and 5 kHz. Below 2 kHz the $B(OH)_3$ relaxation (subsect. 3.2.6.2.5) governs the entire sound absorption. The two relaxation processes increase the pure water absorption by a factor of 18 and 16, respectively. Though both constituents represent only a small fraction of the salinity (subsect. 3.4.3), they are decisive for the chemical sound absorption in the ocean.

Two additional mechanisms increase the transmission loss by energy leakage out of the sound channel: scattering by inhomogeneities affect the deep-water propagation below 0.1 kHz (subsect. 3.2.6.2.7) and finally the diffraction loss of the SOFAR-channel in the neighborhood of its cutoff frequency near 2···3 Hz (subsect. 3.2.6.2.8).

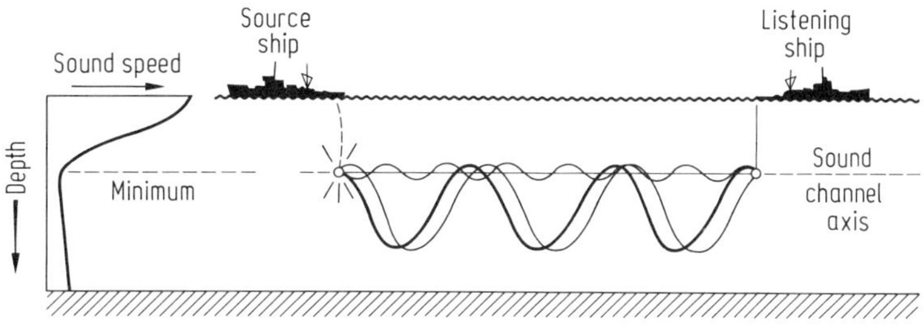

Fig. 22. Configuration of deep water sound channel propagation experiment.

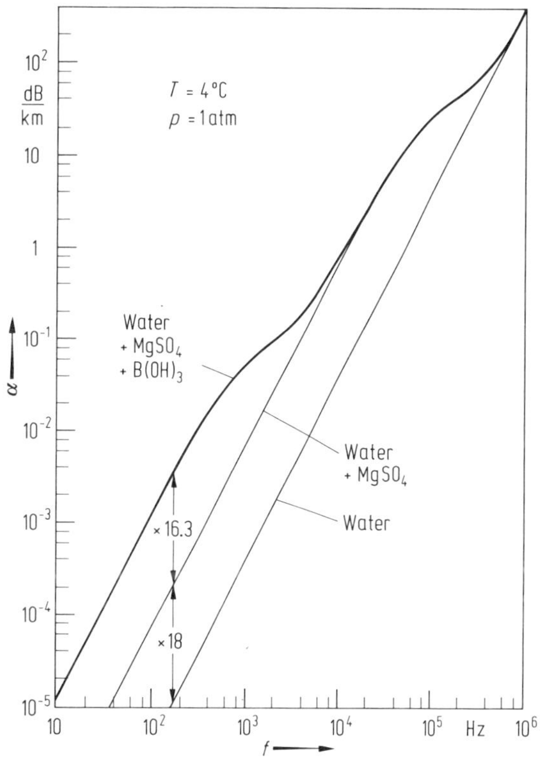

Fig. 23. Sound absorption in sea water of 35‰ salinity and pH = 8.0. Calculated from laboratory absorption measurements from 6···350 kHz in a 200 l spherical resonator [77 F].

3.2.6.2.1 Measuring methods

The free-field transmission loss measurements in the deep water channel are simply realized by a receiving hydrophone suspended to the channel axis from a ship at a fixed position (Fig. 22) and series of explosive sound sources, pressure-fused at the same depth and dropped equidistantly by an aircraft or a source ship along the chosen propagation range [75 U 1, 74 B 2]. The broadband sources simultaneously cover a frequency interval of 10 Hz to some 10^3 Hz, depending on the range. For evaluation, extracts of constant relative frequency bandwidths (third octave) are made from each signal, integrated over the arrival time, to obtain the absorption coefficient α by subtracting the geometrical spreading loss from the total transmission loss and fitting the reduced data of the entire shot series by linear range regression [74 B 1]. An example of the TL spread is shown in Fig. 24.

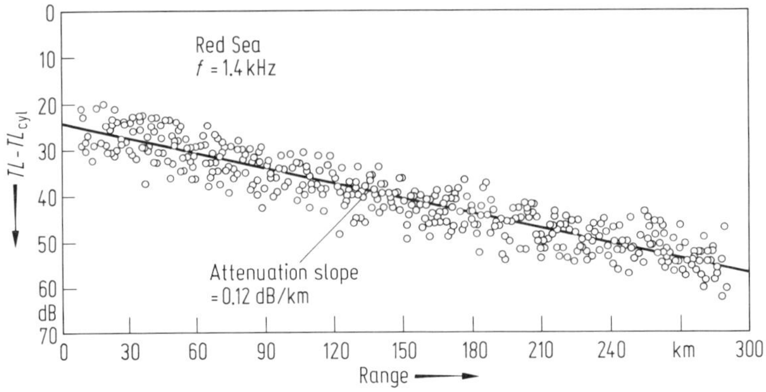

Fig. 24. Regression analysis of transmission loss data, corrected for cylindrical spreading. Red Sea, 1.4 kHz [74 B 1].

The simple regression method becomes problematic at frequencies between 10 Hz and several 10^2 Hz. The extremely small α values near 10^{-3} dB/km require propagation distances of the order of 10^3 km where the assumption of approximately range-independent oceanographic properties will often be violated. An example of sound propagation through adjacent water masses of different properties is given in Fig. 25. The DWC measurements may be modified by several hydrophones in different water depths and other source depths [75 U 1], especially for convergence zone investigation (subsect. 3.2.2.1). Fig. 26 demonstrates the typical convergence periodicity and the corresponding TL variability. For higher frequencies short propagation ranges of a fraction of the water depth are possible which allow to investigate the pressure dependence of the sound absorption [73 B 2].

The free-field measurements are supplemented by laboratory measurements which mainly utilize resonator methods. The absorption coefficient α is usually derived from the decay time of a given sound intensity within a spherical metal or glass resonator [54 W, 61 K] filled with sea water and suspended within a vacuum chamber to achieve extremely low equipment loss. The method is nevertheless limited to higher frequencies.

The data and approximation formulae presented in the following subsections are empirically derived from both free field and laboratory measurements.

Fig. 25: next page.

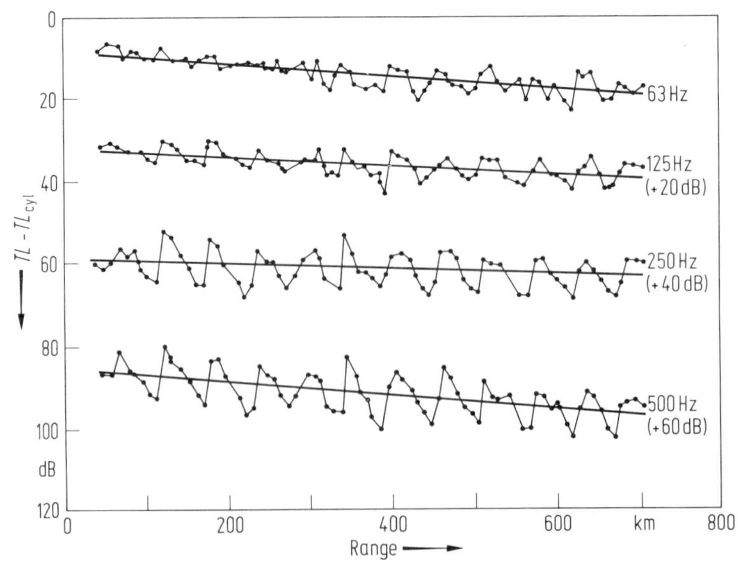

Fig. 26. Measured propagation loss, corrected for cylindrical spreading, for convergence zone experiment [77 B 1].

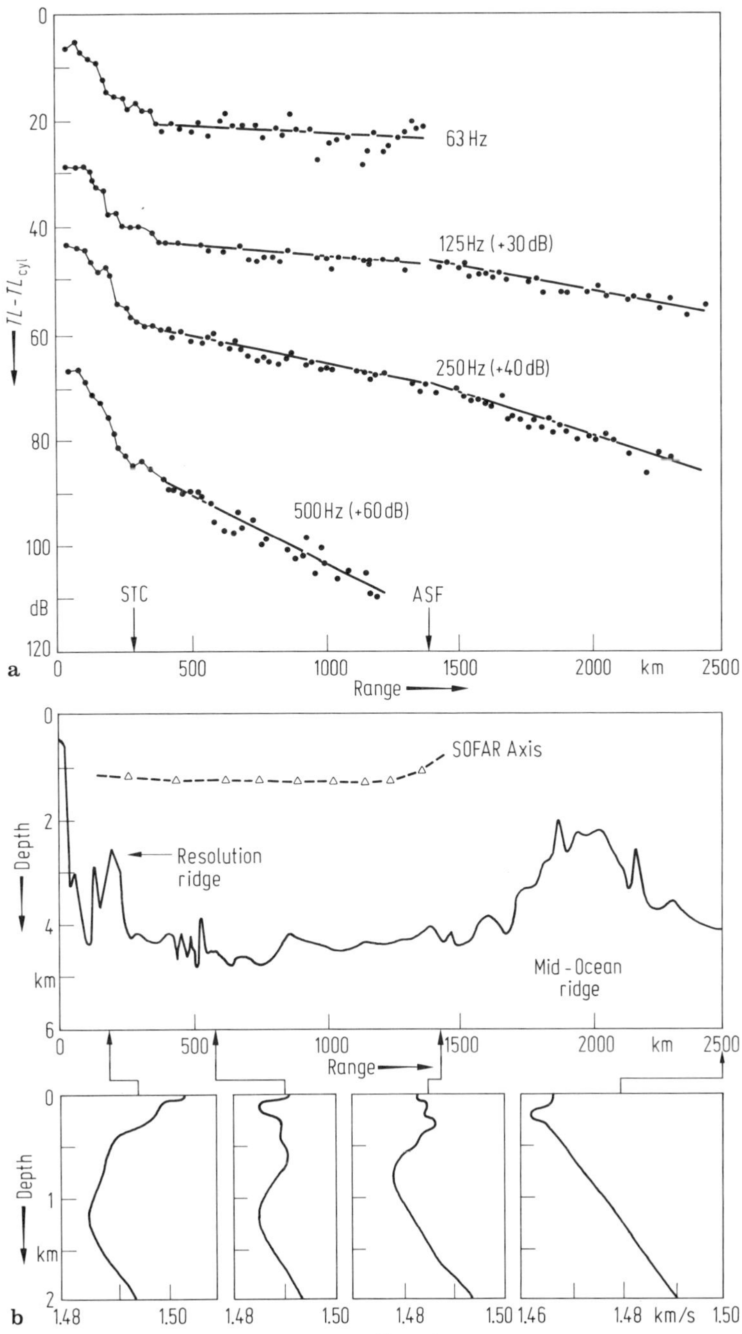

Fig. 25. (a) Measured transmission loss, corrected for cylindrical spreading, across sea areas of different sound speed stratification. STC and ASF indicate positions of the subtropical convergence and Australasia subantarctic front. (b) Bathymetry and typical sound speed profiles along the propagation path [77 B 1].

3.2.6.2.2 Geometrical spreading loss

The spherical-spreading loss, in [dB],

$$TL_{\text{sph}} = 10\log\frac{I_0}{I} = 10\log\frac{r^2}{r_0^2} = 20\log r \tag{35}$$

(r in [m] and r_0 taken as 1 m) valid in the far field of a sound source at distance r imbedded in an infinite homogeneous space is considered an adequate approximation when neither boundary nor refraction processes are essential for the propagation. This is the case for short-range transmission within mixed layers and for echo sounding (subsect. 3.2.3.1.3) at small angles against the stratification gradient where the influence of refraction can be neglected. The spherical spreading loss often appears also a good fit to the measured data of deep-water propagation up to more than 100 km.

The cylindrical spreading loss, in [dB],

$$TL_{\text{cyl}} = 10\log r \tag{36}$$

(r in [m]) is typical of long-range horizontal propagation where the multiple refraction or reflection contribute more than the direct path. The entire geometrical spreading loss TL_{geom}, in [dB], for long-range transmission is the sum of the cylindrical contribution, eq. (36), for large r and a spreading excess H_s for small r which accounts for the initial spherical spreading [74 B 1]

$$TL_{\text{geom}} = 20\log r_s + 10\log\frac{r}{r_s} = 10\log r_s + 10\log r = H_s + 10\log r\,. \tag{37}$$

(r and r_s in [m]). The equivalent range r_s which is of the order of 10 km to 100 km for deep water also accounts for any short and intermediate-range anomalies.

At extremely low frequencies near 2···3 Hz where the deep-water channel loses its trapping properties the cylindrical spreading law is no longer valid (subsect. 3.2.6.2.8).

3.2.6.2.3 Viscous absorption/pure-water absorption

At frequencies above ≈500 kHz the absorption in sea water consists of the classical viscous absorption which is present in pure water also [61 K] and is given, in [dB km^{-1}], by

$$\alpha_1 = A_1 P_1 f^2, \tag{38}$$

where f is the frequency in [kHz], P_1 denotes the dimensionless pressure dependence, and A_1, in [dB km^{-1} kHz^{-2}], stands for the dependence on temperature. Due to the measurements by Litovitz and Carnvale [55 L] the pressure term is obtained as

$$P_1 = 1 - 3.83\cdot 10^{-4} P + 4.9\cdot 10^{-8} P^2, \tag{39}$$

where P is the pressure in [atm], 1 atm ≙ 101325 Pa. The second-degree equation for P_1 has been fitted by Francois and Garrison [82 F 1] and approaches the measured data for 0 °C within 1%. Within these limits P may be replaced by the equivalent water depth of 10 m per atm.

The temperature dependence A_1, in [dB km^{-1} kHz^{-2}], is approximated by the third-degree equation

$$A_1 = 4.937\cdot 10^{-4} - 2.59\cdot 10^{-5} T + 9.11\cdot 10^{-7} T^2 - 1.5\cdot 10^{-8} T^3\,. \tag{40}$$

T is the temperature, in [°C]. The polynomial has been fitted [82 F 1] using Pinkerton's laboratory measurements for $T \leqq 20$ °C. Fig. 27 compares eq. (40) with two earlier publications.

Since the viscous absorption predominates only at the high-frequency regime it is important mainly for narrow-range applications where the distinction between deep and shallow water may be dropped.

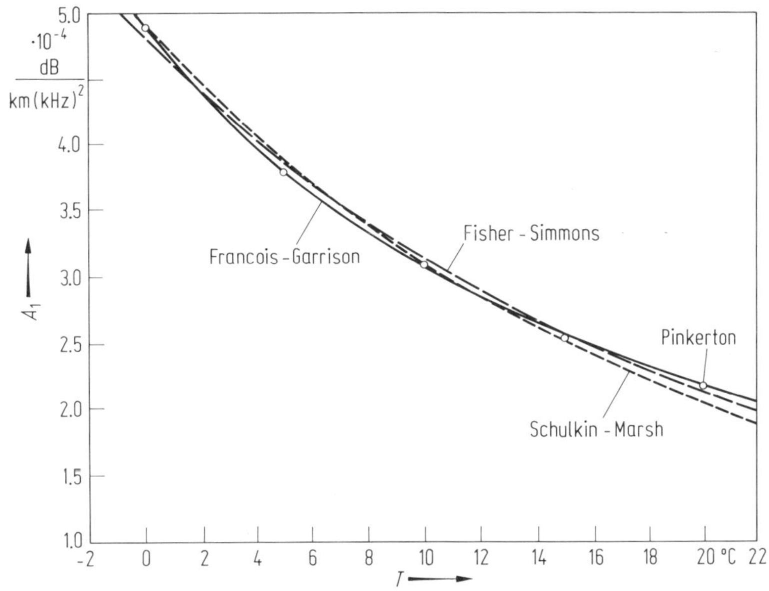

Fig. 27. Temperature dependence of pure-water absorption coefficient A_1. Comparison of approximation equations [82 F 1].

3.2.6.2.4 Absorption by $MgSO_4$

The relaxation mechanism involves pressure-sensitive chemical equilibria. The disturbance of a pressure-dependent equilibrium between two states of the molecules of a liquid by compression yields a re-establishment of the equilibrium, with the relaxation time τ_r. The relaxation frequency f_r is defined as

$$f_r = \frac{1}{2\pi\tau_r}. \tag{41}$$

Because of the time lag of the equilibrium behind the alternating pressure of a sound wave the relaxation causes absorption of the sound energy. Relaxation theory [61 K] leads to a uniform frequency dependence of the resulting attenuation

$$\alpha_i = A_i P_i \frac{f_{ri} f^2}{f_{ri}^2 + f^2}, \tag{42}$$

where f denotes the frequency of the sound wave and A_i, P_i describe the dependence on oceanographic quantities, i denoting the relaxation process. For $f \ll f_{ri}$, α_i increases by f^2 whereas for $f \gg f_{ri}$, α becomes constant. The attenuation per wavelength, $\alpha_i \lambda$, is maximum for $f = f_{ri}$ and goes down to zero for large f because the energy transition becomes more and more incomplete and decreases also for small f because of the decreasing phase lag.

The relaxation responsible for the dominating sound absorption in sea water below 200 kHz has been attributed to a dissociation-reassociation process of $MgSO_4$ interacting with other ion pairs by Kurtze and Tamm [53 K] and Eigen and Tamm [62 E 1] after Leonard [49 L] had identified $MgSO_4$ as the origin of the excess absorption of sea water. References on laboratory and free-field measurements are compiled in [58 F, 63 S 2].

The first approximation formula for practical application in ocean waters which covered the classical viscous absorption and the $MgSO_4$ absorption down to $f \approx 5$ kHz was developed by Schulkin and Marsh 1962 [62 S]. Modifications of the Marsh-Schulkin formula have been suggested in the course of further measurements by Fisher-Simmons [77 F] and Francois-Garrison [82 F 1]. The state of knowledge of the characteristic quantities of the $MgSO_4$-absorption [eq. (42): subscript i = 2] is as follows:

The relaxation frequency f_{r2} does neither depend on the concentration of $MgSO_4$ [53 K], and therefore the salinity, nor on the static pressure [77 F], but f_{r2} increases exponentially with the temperature T, due to laboratory measurements of Kurtze-Tamm [53 K]. The slope of $f_{r2}(T)$ is a measure of the activation energy of the chemical process. However, the existing discrepancies between the data of several investigators (Fig. 28) impede the recommendation of a valid temperature dependence of f_{r2}. The most recent complete approximation formula for sound absorption in sea water has been derived by Francois-Garrison [82 F 1] by comparison of the available data and provides

$$f_{r2}=\frac{8.17\cdot 10^{(8-1990/T+273)}}{1+0.0018(S-35)}, \qquad (43)$$

where T is the temperature in [°C], S is the salinity in [‰] and the frequency f_{r2} is in [kHz].

The coefficient A_2 is proportional to the $MgSO_4$-concentration up to 10^{-1} mol/l as shown by Kurtze and Tamm [53 K] and is therefore coupled to the salinity. How far A_2 depends on temperature appears controversial. Francois-Garrison suggest:

$$A_2=21.44\frac{S}{c}(1+0.025\,T). \qquad (44)$$

The coefficient A_2 is in [dB km^{-1} kHz^{-1}], T is the temperature in [°C], S is the salinity in [‰], c is the sound speed in [ms^{-1}], given approximately by $c=1412+3.21\,T+1.19\,S+0.167\,P$.

The pressure dependence P_2 appears sufficiently described by a first-order approximation

$$P_2=1-b_2(P-1), \qquad (45)$$

where P is the pressure in [atm], 1 atm ≙ 101 325 Pa. The in-situ measurement of Bezdek [73 B 2] up to $P=340$ atm yields $b_2=12.3(11)\cdot 10^{-4}$ (atm)$^{-1}$, Hsu, Fisher [83 H 2] get $b_2=8.06(108)\cdot 10^{-4}$ (atm)$^{-1}$ by laboratory measurements with Lyman-Fleming sea water up to $P=307$ atm. Second-order approximations [82 F 1] by combination of different data sources [73 B 2, 58 F] may not be a better approach to reality.

The claimed validity range for f_{r2}, P_2, and A_2 is $-1.8\cdots 20$ °C and $0\cdots 300$ atm. Though the measurements were made at $S=35$‰, the extrapolation down to very low salinity waters (Baltic Sea) appears permissible [53 K]. It is mentioned that Hsu, Fisher [83 H 2] recently found f_{r2} by 22% lower and A_2 by 15% lower at the single temperature 25 °C for Lyman-Fleming sea water at $P=1$ atm.

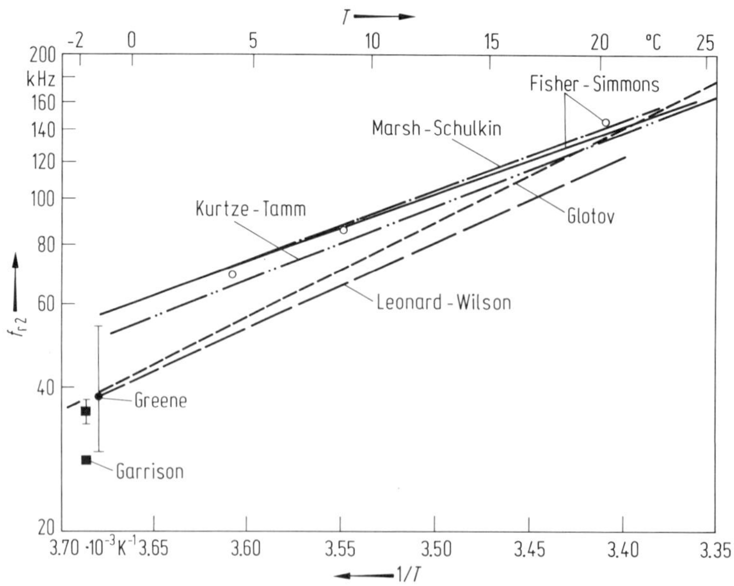

Fig. 28. Temperature dependence of $MgSO_4$ relaxation frequency f_{r2}. Comparison of different measurements and approximation equations [54 W, 82 F 1, 62 S, 77 F, 49 L].

3.2.6.2.5 Absorption by $B(OH)_3$

Numerous free-field attenuation measurements below 1 kHz, compiled by Thorp 1965 [65 T] substantiated the attenuation values to be about 15 times greater (Fig. 23) than those of the $MgSO_4$ attenuation curve. Thorp was also the first to suggest an approximation formula, anticipating 1967 a further relaxation to fit the enhanced attenuation [67 T]. The mechanism of the anomaly was first identified 1972 by Yeager, Fisher [73 Y] as a chemical relaxation of $B(OH)_3$, utilizing temperature jump laboratory measurements. The relaxation time of $\tau_r \approx 1.5 \cdot 10^{-4}$ s being congruent with the relaxation frequency at ≈ 1 kHz, already supposed by Thorp. Fisher, Simmons [75 F] suggested a two-step equilibrium as the essential process

$$B(OH)_3 + OH^- \rightleftharpoons B(OH)_3OH^- \rightleftharpoons B(OH)_4^-,$$

where the formation of the $B(OH)_4^-$ ion involves an intermediate state and the slow second step causes the sound absorption. The absorption is therefore expected to follow the $B(OH)_4^-$ ion concentration, increasing exponentially with the acidity in the sea water (nominally 7.7⋯8.3). This pH-dependence already expected by Fisher corresponds to the available sea data (Fig. 29) analyzed by Mellen and Browning [77 M]. The pH dependence yielded the explanation for the low-frequency attenuation in the Pacific Ocean to be only half of the Atlantic value. Due to the extremely complex chemical reactions the exact relaxation mechanism remains illusive [81 M 4].

The characteristic quantities for the $B(OH)_3$ relaxation f_{r3}, P_3 and A_3 stem from Francois-Garrison [82 F 2] who examined 5 data sets (Table 11) of different ocean areas and took further into account another recent low-frequency approximation of Marsh and Schulkin [78 S]. Laboratory measurements by resonator methods which bracket the relaxation frequency near 1 kHz are not available. Eqs. (46)⋯(48) are based on measurements between 34⋯41 S, $-2\,°C \cdots 22\,°C$ and $P \leqq 150$ atm.

The relaxation frequency f_{r3} is determined as the frequency of the maximum of the attenuation per wavelength, $\alpha_3\lambda$. Similar to f_{r2} the relaxation frequency f_{r3} is essentially dependent on the temperature and increases exponentially with T (Fig. 30). The suggested approximation is

$$f_{r3} = 2.8(S/35)^{0.5} \cdot 10^{(4-1245/T+273)}, \tag{46}$$

where T stands for the temperature in [°C], S denotes salinity in [‰] and the frequency f_{r3} is in [kHz]. Different from the $MgSO_4$ behavior the relaxation frequency f_{r3} appears to depend on the salinity [78 S]. A stronger dependence than the 0.5 power of S, up to power 3/2 as assumed by Mellen, depending on the concentration of cooperating ions, in particular carbonate, is still controversial [85 S 1].

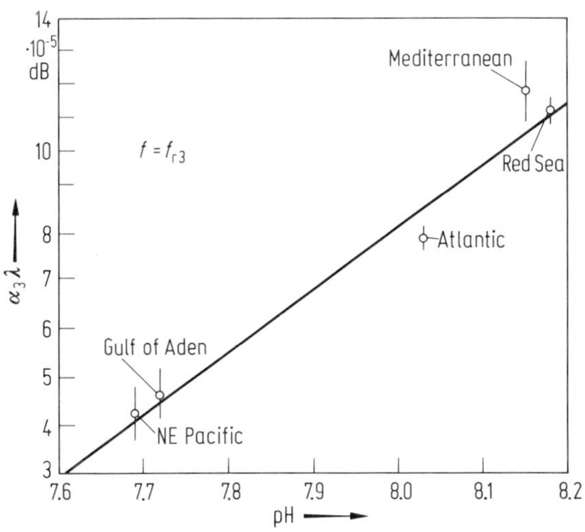

Fig. 29. pH-dependence of sound absorption per wavelength at $B(OH)_3$-relaxation frequency f_{r3}. The best-fit line corresponds to eq. (48) [82 F 2].

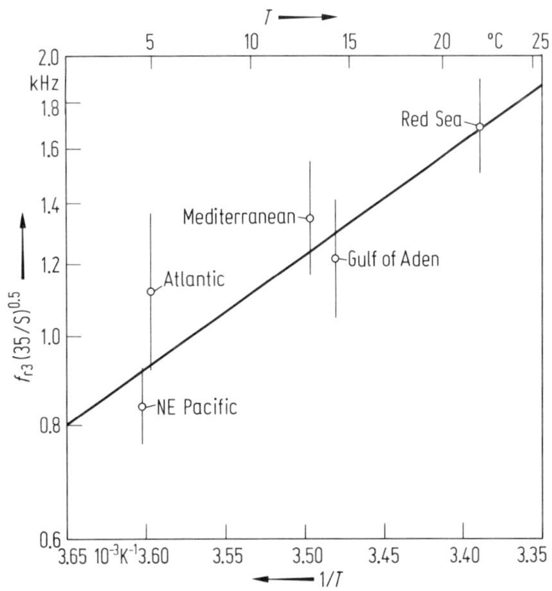

Fig. 30. Temperature dependence of the $B(OH)_3$-relaxation frequency f_{r3} determined from measurements in 5 ocean areas. Salinity S in [‰]. The best-fit line corresponds to eq. (46) [82 F 2].

Table 11. Measurement data sets of low-frequency sound absorption induced by $B(OH)_3$, corrected for cylindrical spreading in comparison to the approximation equations for f_{r3}, A_3 [eqs. (46) and (48)] [82 F 2].

Location and investigator	Year	Sound channel axis depth	Max. range	T	S	pH	Measured		Calculated from eqs. (42) and (46)···(48)	
							f	Total α	Total α	Measured α – calculated α
		m	km	°C			kHz	dB/km	dB/km	dB/km
NE Pacific	1975	753	3000	4.25	34.1	7.67	0.120	0.0011	0.0010	0.0001
Thorp							0.150	0.0018	0.0016	0.0002
$c = 1479$ m/s							0.200	0.0026	0.0028	−0.0002
							0.250	0.0035	0.0043	−0.0008
							0.300	0.0062	0.0061	0.0001
							0.400	0.0094	0.0101	−0.0007
Pacific	1969	700	245	5.0	34.4	7.67	0.750	0.0221	0.0263	−0.0042
Lovett (ISAR)							1.500	0.0547	0.0594	−0.0047
$c = 1481$ m/s							3.000	0.1470	0.1321	0.0149
Gulf of Alaska	1971	75	270	4.0	33.1	7.72	1.500	0.0744	0.0644	0.0100
Lovett							2.500	0.1290	0.1132	0.0158
$c = 1465$ m/s										
NE Pacific	1975	505	2900	4.60	34.1	7.69	0.050	0.0002	0.0002	0.0000
Morris							0.080	0.0004	0.0005	−0.0001
$c = 1476$ m/s							0.100	0.0007	0.0007	−0.0000
							0.125	0.0012	0.0012	0.0000
							0.160	0.0018	0.0019	−0.0001
							0.200	0.0028	0.0029	−0.0001
							0.250	0.0042	0.0044	−0.0002
							0.310	0.0068	0.0066	0.0002
							0.400	0.0092	0.0104	−0.0012
S. Pacific	1971	1250	1150	4.31	35.0	7.96	0.106	0.0007	0.0013	−0.0006
Kibblewhite and							0.212	0.0030	0.0049	−0.0019
Denham (Line PA)							0.424	0.0083	0.0173	−0.0090
$c = 1488$ m/s										
(Line PB1)	1971	1250	1700	4.07	35.0	7.90	0.106	0.0002	0.0012	−0.0010
$c = 1487$ m/s							0.212	0.0032	0.0045	−0.0013
							0.424	0.0084	0.0158	−0.0074

Table 11 (continued).

Location and investigator	Year	Sound channel axis depth	Max. range	T	S	pH	Measured f	Measured Total α	Calculated from eqs. (42) and (46)···(48) Total α	Calculated from eqs. (42) and (46)···(48) Measured α – calculated α
		m	km	°C			kHz	dB/km	dB/km	dB/km
(KIWI WEST)	1974	1250	3000	4.31	35.0	7.96	0.029	0.0015	0.0001	0.0014
$c = 1488$ m/s							0.060	0.0012	0.0004	0.0008
							0.120	0.0021	0.0016	0.0005
							0.250	0.0034	0.0067	−0.0033
							0.424	0.0079	0.0173	−0.0094
(TAS 1)	1963	1300	900	4.35	35.0	7.87	0.450	0.0203	0.0165	0.0038
$c = 1489$ m/s							0.900	0.0246	0.0440	−0.0194
S. Pacific	1975	1350	1800	3.92	35.0	7.87	0.125	0.0016	0.0015	0.0001
Bannister et al.							0.250	0.0063	0.0059	0.0004
(TAS 2, Northwest)							0.500	0.0167	0.0197	−0.0030
$c = 1488$ m/s										
(TAS 2, West)	1975	1100	2800	3.94	35.0	7.87	0.125	0.0012	0.0016	−0.0004
$c = 1484$ m/s							0.250	0.0055	0.0059	−0.0004
			1600				0.500	0.0189	0.0198	−0.0009
Baffin Bay [1])	1972	100	400	− 1.50	33.7	8.01	0.320	0.0170	0.0143	0.0027
Mellen et al.							0.410	0.0198	0.0214	−0.0016
$c = 1442$ m/s							0.500	0.0246	0.0288	−0.0042
							0.640	0.0393	0.0399	−0.0006
							1.000	0.0683	0.0643	0.0040
Bismarck Sea	1974	45	?	30	36	8.20	0.560	0.0700	0.0203	0.0497
Mellen and							1.200	0.1000	0.0776	0.0224
Browning							2.300	0.1900	0.1854	0.0046
$c = 1546$ m/s							4.500	0.3600	0.3367	0.0233

[1]) Measured values have been adjusted by −0.02 dB/km.

A pressure dependence P_3 appears negligible due to measurements of Fisher-Simmons [77 F], therefore

$$P_3 = 1 . \tag{47}$$

The coefficient A_3, once more derived from the maximum values of $\alpha_3\lambda$ (Fig. 31) is adopted from Francois-Garrison [82 F 2]

$$A_3 = \frac{8.9}{c} \cdot 10^{(0.78\,\mathrm{pH} - 5)} , \tag{48}$$

where c stands for the sound speed in [m/s] and the coefficient A_3 is in [dB km^{-1} kHz^{-1}]. The important feature of A_3 is the dependence on the pH-value. The geographical variation of pH enables a world-wide mapping [80 L 3] of the coefficient A_3, based on the world ocean atlas of Gorshkov [78 G 2] (Fig. 32).

The dependence of A_3 on c is the formal consequence of its determination from $\alpha\lambda$. The weak dependence of c on T, S, and P, however, does not correspond to the measured behavior of A_3. According to Fisher, Simmons [77 F] A_3 should rather increase than decrease by T (30% between 4° and 20 °C) but the poor accuracy of the parametrization let any dependence of A_3 other than the pH-effect appear doubtful.

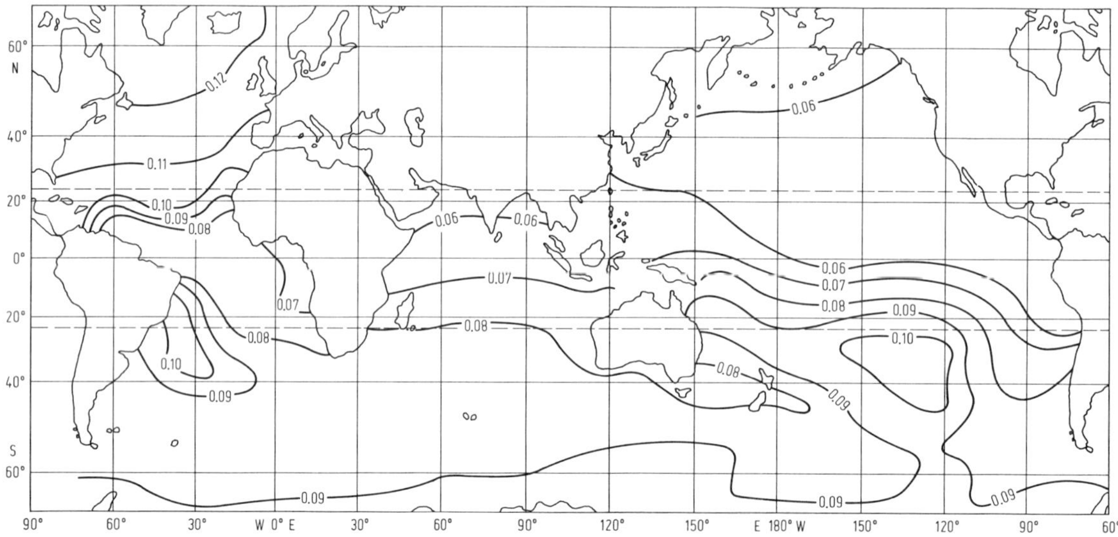

Fig. 32. Global chart of the pH-dependent factor A_3 [dB km^{-1} kHz^{-1}] of the $B(OH)_3$-induced sound absorption [80 L 3, based on 78 G 2].

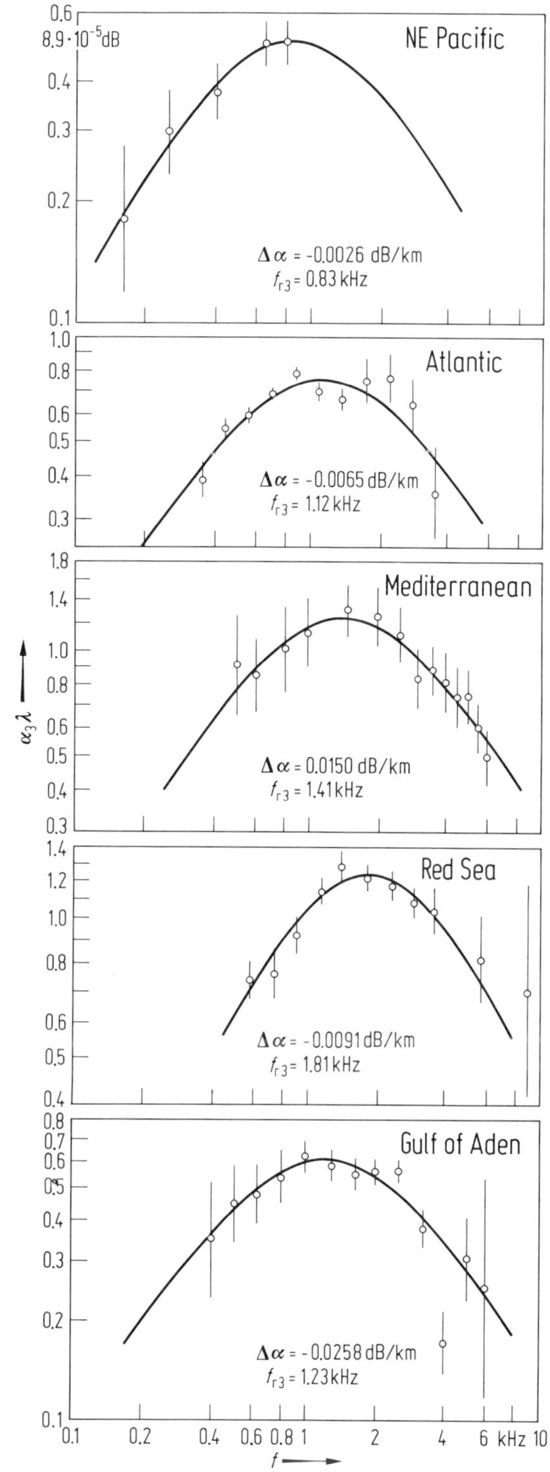

Fig. 31. Determination of the $B(OH)_3$-relaxation frequency f_{r3} [kHz] for 5 water types. Each set of absorption measurements has been adjusted by $\Delta\alpha$ [dB/km] to give the best fit to relaxation theory [82 F 2].

3.2.6.2.6 Absorption by further relaxation processes

Of all relaxations observed by laboratory investigations [80 M 2] the only significant contributors to the sound absorption in sea water are $MgSO_4$, $B(OH)_3$, and $MgCO_3$ as a third constituent. The relaxation involving $MgCO_3$ was discovered latest [79 M 1] and produces a certain excess in the transition region between the two main relaxations.

The logarithmic absorption coefficient α of the $MgCO_3$ process, which increases exponentially with pH like the $B(OH)_4^-$ relaxation, is 2.5 times lower than α_2 of $MgSO_4$ and even smaller above 10 kHz (Fig. 33). This third relaxation was considered partly responsible for the discrepancy [79 F 1] between the $MgSO_4$ absorption given by Schulkin-Marsh [62 S] and Fisher-Simmons [77 F]. Within the low-frequency region governed by the $B(OH)_3$ relaxation the $MgCO_3$-induced absorption is negligible.

Though other chemical equilibria do not directly contribute, the chemical sound absorption cannot simply be considered the sum of independent relaxation processes because many equilibria have common ions [81 M 3]. Especially the $Mg/Ca/CO_2$-system [80 M 3] appears to augment the sound absorption by the borate relaxation via buffer reactions.

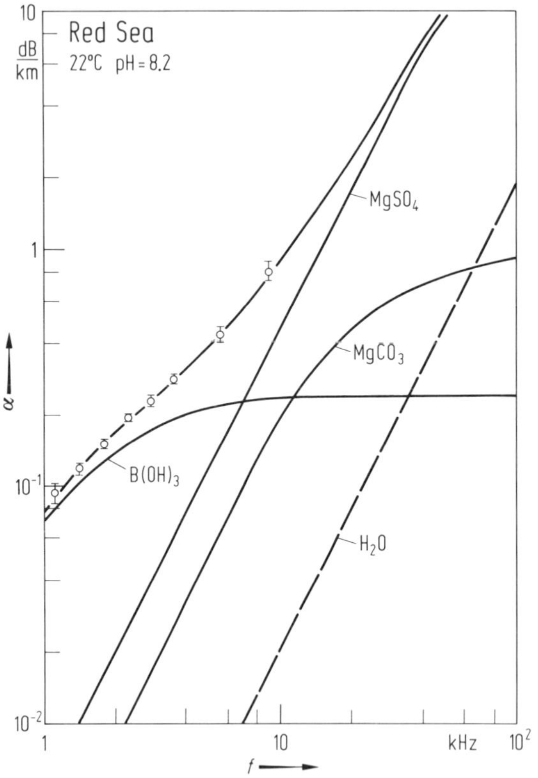

Fig. 33. Sound absorption in sea water by the three dominant chemical relaxations. The upper curve is the total absorption derived from Red Sea experimental data. The pure-water absorption is negligible within the frequency range shown [80 M 2].

3.2.6.2.7 Propagation loss due to scattering

Below 0.1 kHz many independent experiments as compiled in [80 K 1] exhibit a pronounced excess of the attenuation given by the $B(OH)_3$ relaxation. Observed α values cover more than one order of magnitude around 10^{-3} dB/km.

Since the thorough re-examination by Kibblewhite and Hampton [80 K 1] a dependence of the attenuation below 100 Hz on the geographical latitude now appears to be as well established as the geographical relation of the $B(OH)_4^-$-induced absorption. In particular, two southern-hemisphere experiments: KIWI I [74 B 2] and TASMAN II [77 B 1] with 10000 km propagation path have given strong support to the suggestion of an additional attenuation coefficient α_4 which appears essentially frequency-independent but is related to the type of the sound speed profile.

A plausible explanation has been proposed by Mellen et al. [74 M 2, 76 M 2] who attributed the excess attenuation to a leakage of signal energy out of the sound channel by scattering at refractive inhomogeneities, applying Chernov's ray diffusion mechanism [60 C]. The inhomogeneities are simply modelled as isotropic globes of the radius a_0, in [m], corresponding to the correlation radius of the sound speed structure. For large scatterers, $2\pi\lambda^{-1}a_0 > 1$, the variance μ^2 of the refraction index, averaged over the sound channel, yields frequency-independent forward scattering of the sound rays, increasing the variance of their angular distribution linearly with distance. Under the assumption that the leakage portion is totally absorbed by the sea bottom, the resulting attenuation is given by [74 M 2]

$$\alpha_4 = \frac{8 \cdot 10^8 \mu^2}{a_0 \cdot \Delta z}, \qquad (49)$$

where Δz, in [m], is the distance of the channel axis to the sea floor, a_0 is in [m] and α_4 is in [dB km^{-1}]. Along Δz the sound gradient is assumed to be the barocline alone.

According to Mellen, typical values, estimated for the DWC-propagation, are: $\mu^2 = 10^{-7}$, $a_0 = 15$ m, $\Delta z = 3000$ m: $\alpha_4 = 2 \cdot 10^{-3}$ dB/km. A modified model applying normal mode propagation [74 K] yields a similar attenuation. For shallow-duct propagation [78 M 1] and all the more pronounced for shallow water (subsect. 3.2.6.2.7) the attenuation by scattering exhibits values of 1···2 orders of magnitude above those for deep water. In case of high scattering loss conditions the more seldom frequency dependence of the Rayleigh-type for $2\pi\lambda^{-1}a_0 < 1$ [75 M 2] may be also accessible.

Using sound speed profiles characteristic of polar, subpolar, mid-latitude, and tropical waters as suggested by Kibblewhite and Hampton [80 K 1] has produced self-consistent subsets of the attenuation coefficient for these regions (Fig. 34). Typical values for frequencies below 0.1 kHz are (for A, B, and C, see Fig. 34):

Attenuation coefficient	Water classification					
	Polar	Subpolar	Mid-latitude			Tropical
			A	B	C	
α_4 [10^{-3} dB/km]	4.2	1.8	1.3	2.0	0.4	0.2···0.4

This recently observed loss excess by internal scattering represents only another aspect of the equalizing agency of scattering in connection with sound propagation, usually known from the transmission loss reduction by shadow zone illumination (subsect. 3.2.2.1).

The actual oceanic agent responsible for the excess attenuation is still not established. Though the estimation of scattering loss induced by internal waves [76 M 2] using Garrett-Munk spectra [75 G] appears consistent with the lowest measured attenuation coefficient of $5 \cdot 10^{-4}$ dB/km, structures of other origin (turbulence, strong current gradients) may be also important to acoustic propagation at very low frequencies.

Fig. 34

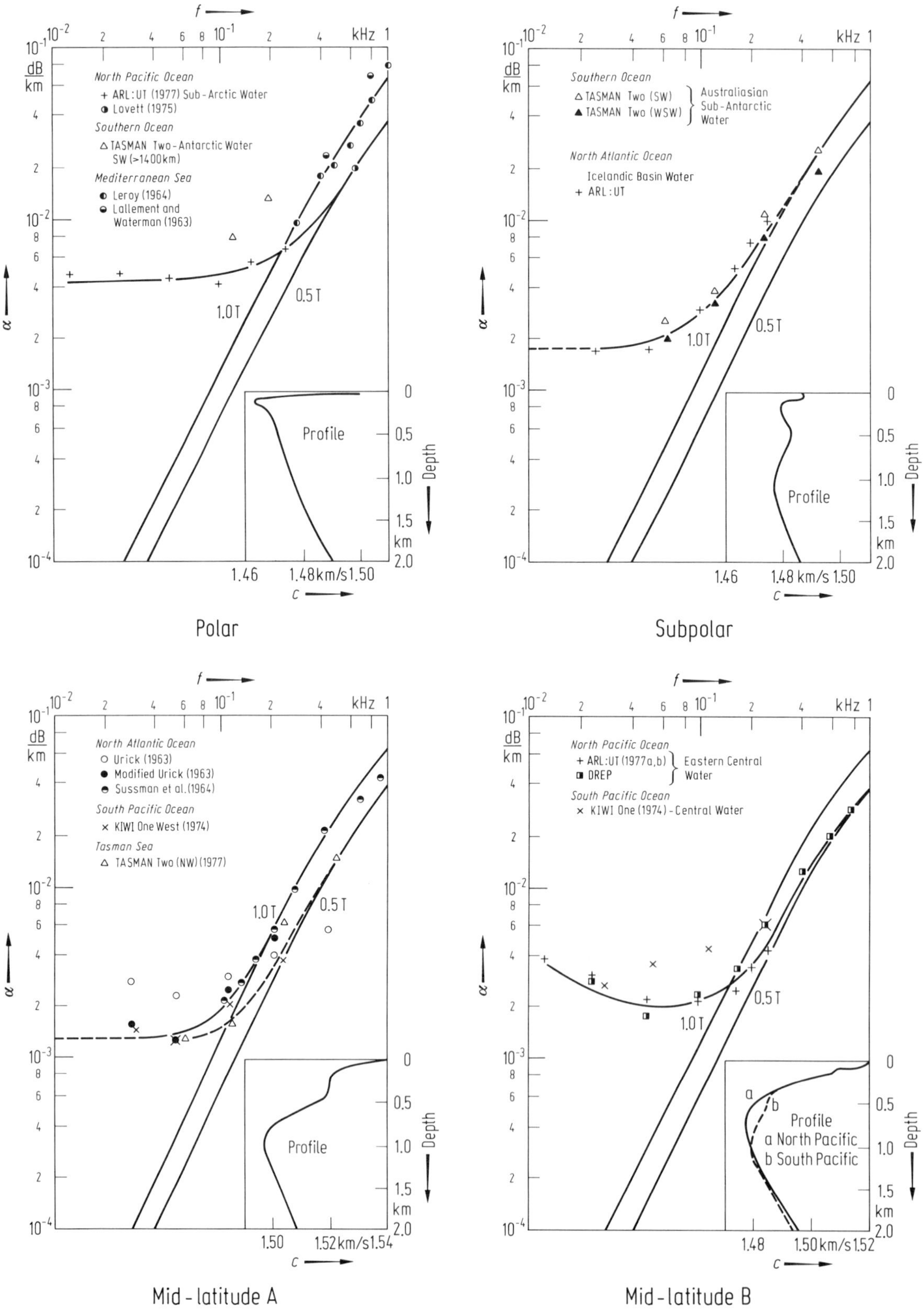
North Pacific Ocean
+ ARL:UT (1977) Sub-Arctic Water
Lovett (1975)
Southern Ocean
△ TASMAN Two-Antarctic Water SW (>1400km)
Mediterranean Sea
Leroy (1964)
Lallement and Waterman (1963)
1.0T
0.5T
Profile
Polar
Southern Ocean
△ TASMAN Two (SW)
▲ TASMAN Two (WSW)
Australiasian Sub-Antarctic Water
North Atlantic Ocean
Icelandic Basin Water
+ ARL:UT
Subpolar
North Atlantic Ocean
○ Urick (1963)
● Modified Urick (1963)
Sussman et al. (1964)
South Pacific Ocean
× KIWI One West (1974)
Tasman Sea
△ TASMAN Two (NW) (1977)
Mid-latitude A
North Pacific Ocean
+ ARL:UT (1977a,b)
DREP
Eastern Central Water
South Pacific Ocean
× KIWI One (1974) - Central Water
a North Pacific
b South Pacific
Mid-latitude B
f
α
dB/km
kHz
Depth
km
c
km/s

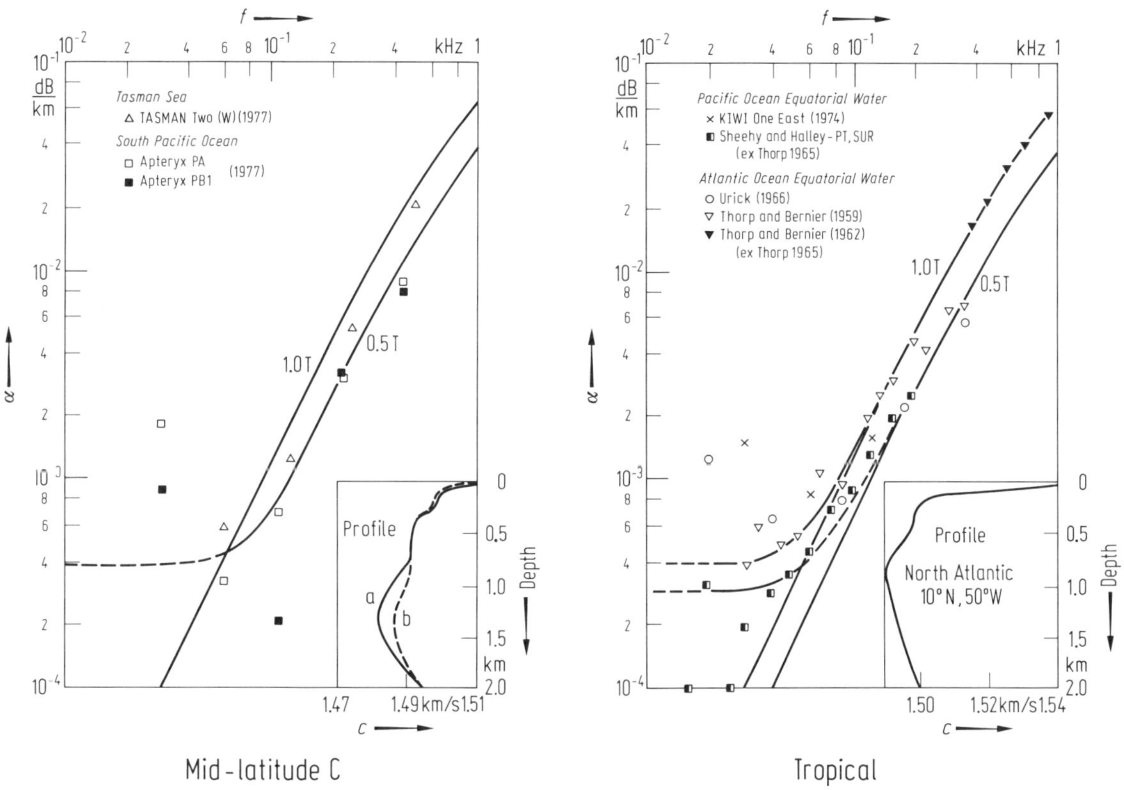

Fig. 34. Very-low-frequency attenuation excess by scattering loss. The data are grouped into 6 subsets corresponding to common sound speed profile types [80 K 1]. 1.0 T and 0.5 T refer to the empirical formula of Thorp [67 T].

3.2.6.2.8 Propagation loss due to diffraction

The deep water channel loses its trapping properties due to sound diffraction into the sea bottom when the signal frequency approaches the low-frequency cutoff of the wave guide (subsect. 3.2.2.1) [77 C]. This diffraction regime becomes effective almost entirely below DWC-propagation measurements which reach down to ≈10 Hz.

The critical frequency for the lowest normal mode of DWC-propagation even for the extremely unfavourable assumption of a totally absorbing bottom is the order of 2⋯3 Hz. The simplified estimation by Urick [63 U] using a bilinear gradient sound speed profile yields $\lambda_{max} \approx 2.5 \left(\frac{\Delta c}{c}\right)^{1/2} d$, where c and Δc are the sound speed in [ms/s] at channel depth d in [m] and its difference from the sea surface value, respectively; λ_{max} is in [m].

3.2.6.2.9 Special phenomena

The deep-water channel propagation may be severely deteriorated when macroscale phenomena violate the presumed range independence of the environmental conditions. Above all, the bottom topography originates transmission loss discontinuities along a DWC propagation behind the intersection of the channel axis by an ocean ridge (Fig. 35). The illumination of the shadow zone behind the ridge [77 B 1] by diffraction is in qualitative agreement with the prediction by range-dependent modelling (subsect. 3.2.2.1).

Other macroscale phenomena influencing sound propagation are oceanographic fronts and large-scale eddies [78 K]. A strong temporary enhancement of transmission loss occurs when the sea water contains air bubble layers [77 C].

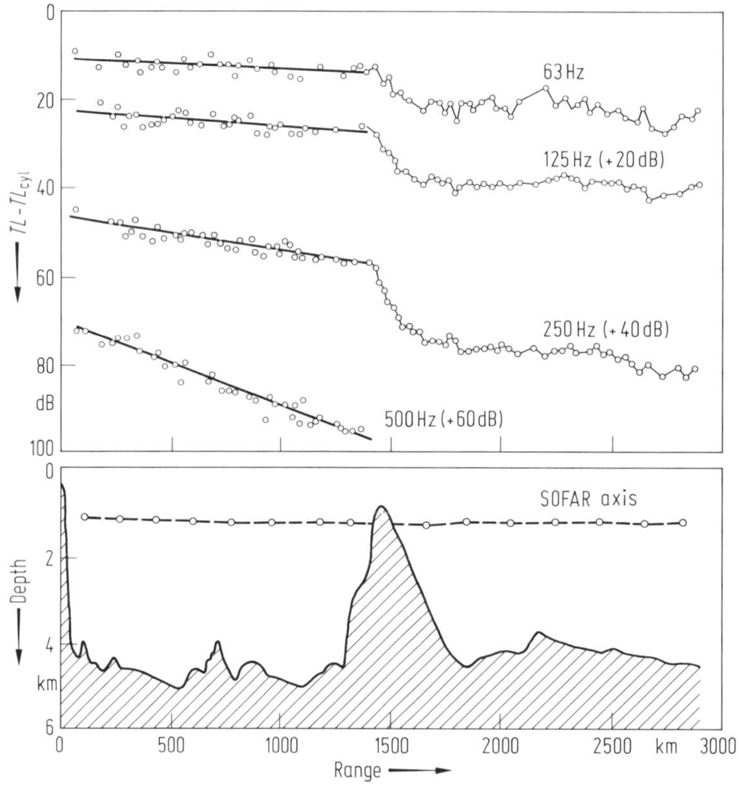

Fig. 35. Transmission loss enhancement by a sea mount bottom profile, crossing the deep water channel axis. The shadowing effect is frequency-dependent [77 B 1].

3.2.6.2.10 Complete approximation formula for sound attenuation in sea water

The most comprehensive review to date of both laboratory and ocean measurements of sound absorption in sea water by Francois, Garrison [82 F 2] is the basis for the following approximation formula for α, in [dB km^{-1}], which is the sum of eqs. (39, 40, 43⋯48) (subsects. 3.2.6.2.3⋯3.2.6.2.5).

$$\alpha = A_1 P_1 f^2 + \frac{A_2 P_2 f_{r2} f^2}{f^2 + f_{r2}^2} + \frac{A_3 f_{r3} f^2}{f^2 + f_{r3}^2}. \qquad (49a)$$

The formula is depicted in Fig. 36 for typical Atlantic and Pacific pH values and for the high-salinity water of the Red Sea. The estimated accuracy of $\alpha = 5\%$ within the parameter range.

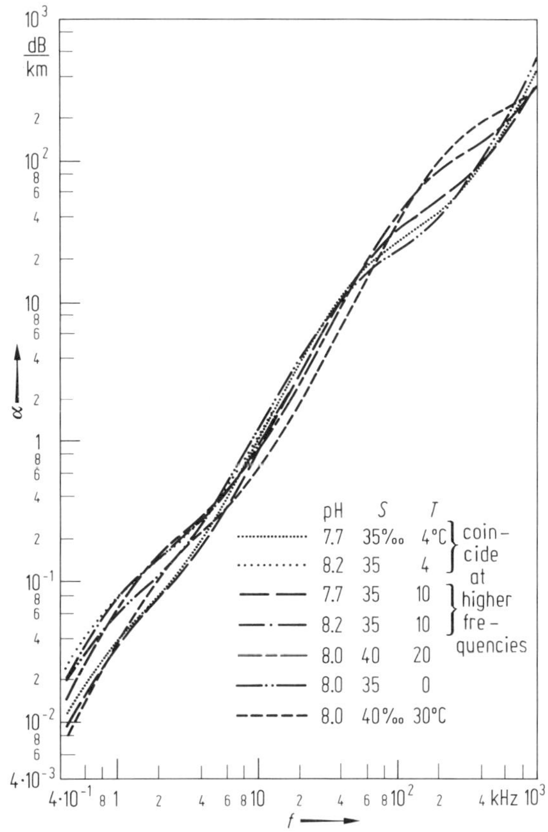

Fig. 36. Evaluation of the sea water sound absorption formula of Francois and Garrison [82 F 2], eq. (49a), for Atlantic waters (pH > 8), Pacific waters (pH < 8) and warm high-salinity waters (Red Sea), at zero depth.

3.2.6.2.11 Simplified approximation formulae for the deep water channel attenuation

The most important frequency interval for moderate to long-range deep-water propagation is between 0.1 kHz and 10 kHz. According to the average temperature, salinity, and pH values of the channel axis the following formulae for selected sea areas may be applied for a rough approximation of the absorption coefficient [77 F] (frequency f is in [kHz].):

Ocean	T [K]	S [‰]	α [dB km^{-1}]
Atlantic	277	35	$0.11 \frac{f^2}{1+f^2} + 0.011 f^2$
Pacific	277	34	$0.055 \frac{f^2}{1+f^2} + 0.011 f^2$
Mediterranean	286	38	$0.15 \frac{1.7 f^2}{1.7^2+f^2} + 0.0066 f^2$
Red Sea	295	40	$0.15 \frac{1.5 f^2}{1.5^2+f^2} + 0.0077 f^2$
Gulf of Aden	287	36	$0.078 \frac{1.2 f^2}{1.2^2+f^2} + 0.0066 f^2$

3.2.6.3 Shallow-water transmission loss

Sea areas which are considered acoustically shallow [75 U 1] are essentially identical with the continental shelf waters and marginal seas where the water depth of less than ≈ 200 m is insufficient to form a stable barocline-thermocline refractive channel (subsect. 3.2.5.5). Without precise definition shallow in this context means a water depth in which sound is propagated to a distance predominantly by repeated reflections from sea surface and bottom. Instead of the refractive trapping of sound energy by the deep-water channel (subsect. 3.2.6.2), where the attenuation is governed by the sea volume, the reflective trapping of the shallow water channel is characterized primarily by boundary losses. The sea volume contributes only a small fraction of the entire shallow-water sound absorption below 10^3 Hz, but controls an essential portion of the loss variability by varying refraction.

The long-range transmission loss may be described by the sum of absorption loss αr and a cylindrical spreading loss $10 \log r$, r in [m], [73 W]. A nearly range-independent α is present as a result of an equilibrium process between higher boundary losses for steeper wave fronts (subsect. 3.2.8.4) and their regeneration by scattering at the boundaries and by volume inhomogeneities. The sound field mixing blurs convergence zones, which are usually not observed at long ranges [77 S].

Characteristic features of the shallow-water transmission loss in comparison to the DWC-propagation are:

(i) higher attenuation at low to moderate frequencies (< 10 kHz), weaker frequency dependence. Typical signal transmission ranges of the order of 10···100 km (Fig. 1);
(ii) higher spatial and temporal variability due to properties of bottom, surface, and volume (Fig. 37);
(iii) minimum transmission loss between 10^2 and 10^3 (optimum frequency) (Figs. 38···43);
(iv) nonadditive superposition of boundary losses.

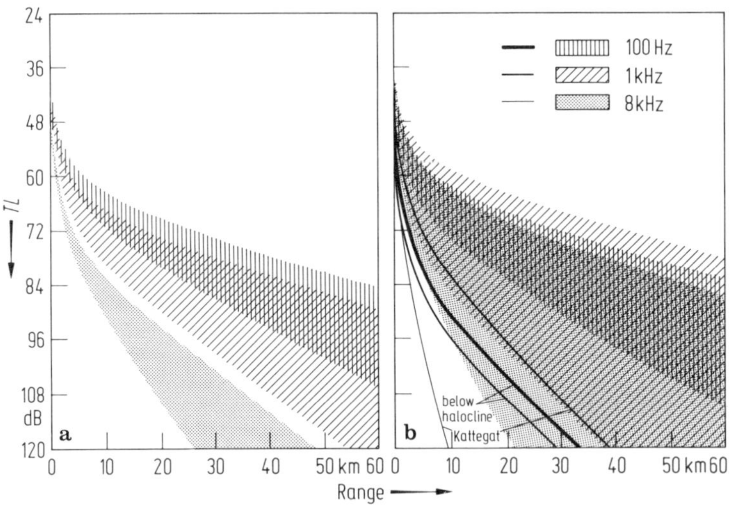

Fig. 37a. Transmission loss variability in the North Sea for selected signal frequencies [80 T].

Fig. 37b. Transmission loss variability in the Baltic Sea for selected signal frequencies [80 T].

3.2.6.3.1 Measuring methods

The transmission loss is usually measured and evaluated in the same way as for DWC propagation (subsect. 3.2.6.1) [73 W]. For studies of the small-scale spatial sound field distribution continuous sources are also in use [74 I]. Single-mode propagation is studied by mode enhancement techniques [74 I] by which the relative stability of the lowest modes has been made evident. Laboratory measurements [74 M 3] are less important for shallow water than for deep water.

3.2.6.3.2 Geometrical spreading loss

The spherical and cylindrical spreading losses of eqs. (35) and (37), respectively, apply under corresponding conditions for shallow water also. The range for the transition between initial spherical spreading and long-range cylindrical propagation above the cutoff frequency, represented by r_s in eq. (37), is of the order of 100···1000 m or 20···30 dB of H_s, respectively. r_s tends to increase by increasing bottom loss and water depth. Substitute spreading laws of, e.g., 15 log r, r in [m], to account for range-dependent attenuation coefficients at intermediate ranges (mode stripping) [60 B] may be sometimes preferable to reduce the data spread [76 M 3].

3.2.6.3.3 Absorption by the sea volume

It is assumed that the deep water viscous and chemical absorption, eqs. (38···48), hold for shallow water also. Due to the f^2 increase of the volume absorption against the f^1 slope of the frictional bottom losses (subsect. 3.2.8.4) the sea volume governs the entire shallow-water absorption for $f > 10···30$ kHz (Fig. 1). Possible effects by chemical pollution appear to be at most marginal [74 S 2].

3.2.6.3.4 Influence of sound refraction by the sea volume

The conservative process of sound refraction by the seasonal thermocline increases the bounce losses at the bottom both by increasing the inclination of the wave front and by decreasing the bounce distance. For similar propagation conditions the sound speed difference between surface and bottom rather than the gradient of the thermocline is considered decisive. The totally mixed winter situation yields the lowest transmission loss which is further reduced by the petty barocline in cases of calm sea [80 A 1].

Typical examples of the transmission loss as a function of range and frequency for winter and summer conditions are shown in Fig. 38. A discrimination into further profile features needs model computation. The optimum frequency of lowest loss is the result of the higher attenuation of the lowest modes near cutoff.

The spatial variability of the volume stratification (subsect. 3.2.5.6) originates forward scattering which contributes to an equilibrium exchange of sound energy between depth intervals separated by a steep thermocline and a sound source below it [77 S].

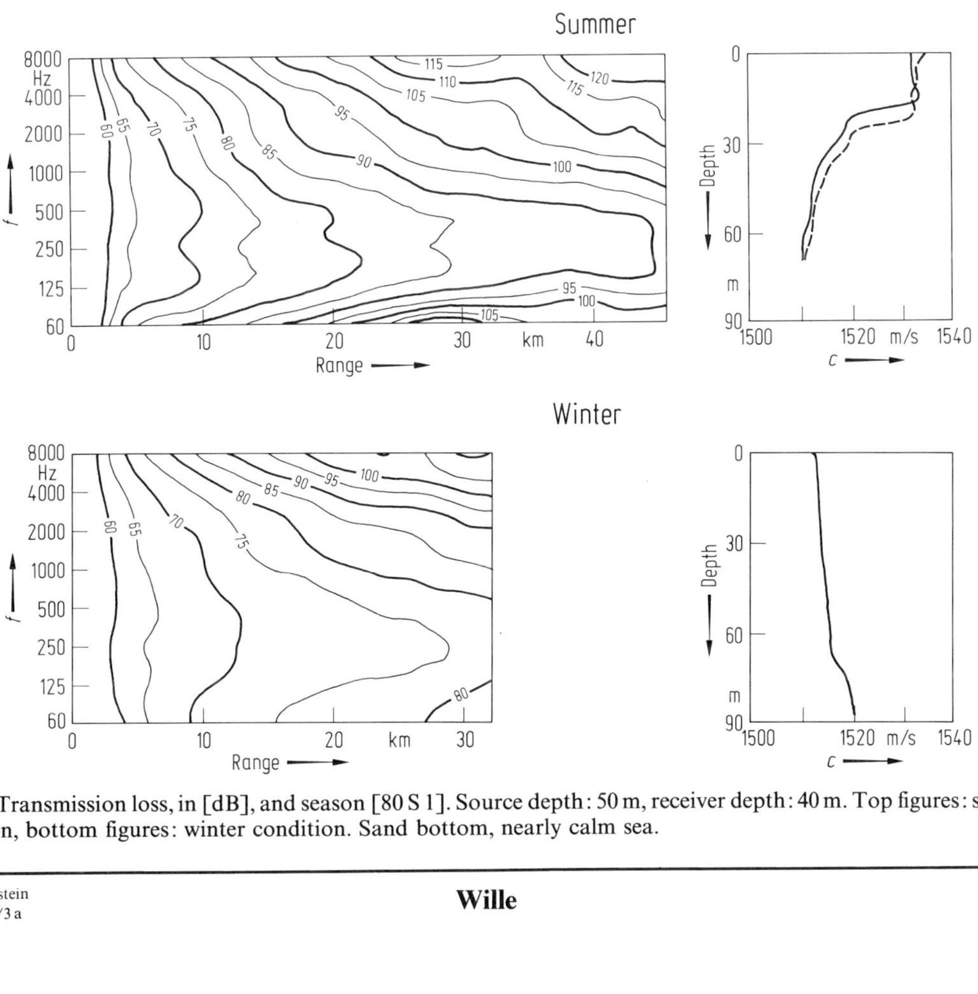

Fig. 38. Transmission loss, in [dB], and season [80 S 1]. Source depth: 50 m, receiver depth: 40 m. Top figures: summer condition, bottom figures: winter condition. Sand bottom, nearly calm sea.

3.2.6.3.5 Absorption by bottom loss

The bottom originates the strongest spatial loss variability of all environmental influences. The contribution of the bottom to the entire attenuation loss is primarily given by its reflection coefficient (subsect. 3.2.8.4). Two essentially different bottom types are distinguished:

(i) high-speed bottom (sand type)
characterized by low reflection loss below the critical grazing angle of $\lesssim 20°$;

$$c_{\text{sand}} > c_{\text{water}} (\Delta c \geqq 10\%), \quad \varrho_{\text{sand}} \approx 2\varrho_{\text{water}}, \quad \text{porosity} \lesssim 40\%;$$

high internal compressional wave attenuation of $\approx 0.35\,\text{dB}/\lambda$

(ii) low-speed bottom (mud type)
characterized by higher reflection loss near the intromission angle of $\approx 15°$;

$$c_{\text{mud}} \lesssim c_{\text{water}} (\Delta c \leqq 5\%), \quad \varrho_{\text{mud}} \approx 1.3 \cdots 1.5\,\varrho_{\text{water}}, \quad \text{porosity} \gtrsim 70\%;$$

low internal compressional wave attenuation of $\approx 0.05\,\text{dB}/\lambda$.

Typical $\alpha(f)$ relations for sand and mud bottom are given in Fig. 39. The optimum frequency seems to increase by increasing reflection coefficient of the bottom (Fig. 40) [83 T].

Transition bottom types [83 M 1] and those with pronounced internal stratification must be treated by numerical models. The vertical profile of the speed of compressional waves within the sediment is critically decisive for the reflection coefficient and therefore the transmission loss within the water volume [73 W].

Mud may contain gaseous inclusions of various origin by which the reflection loss may become extremely small due to impedance mismatch [83 M 4].

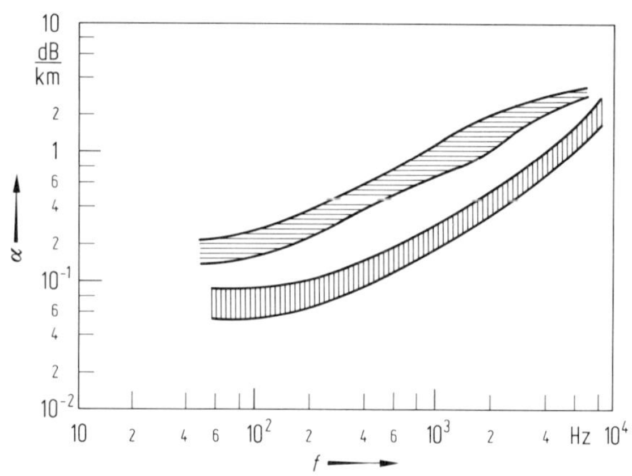

Fig. 39. Absorption coefficient α and North Sea bottom type, 120 m water depth [83 T]. Upper shaded area: mud bottom; Lower shaded area: sand bottom.

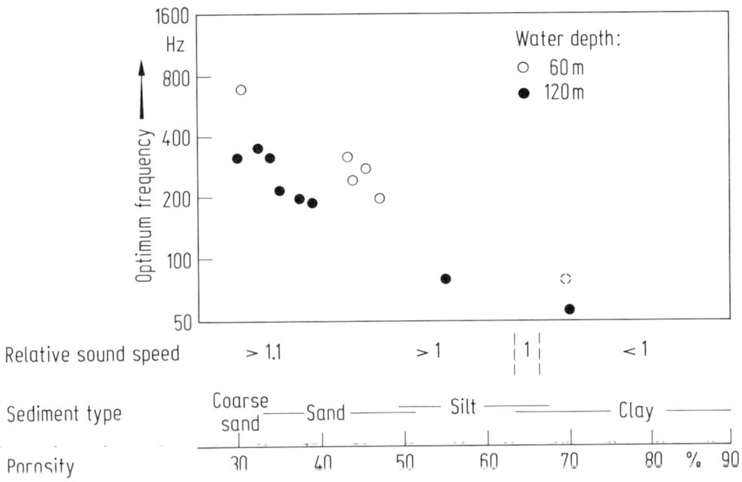

Fig. 40. Empirical relation of optimum frequency and sedimentological parameters [80 A 1].

3.2.6.3.6 Influence of the sea surface

At low sea states the surface enhances the transmission loss indirectly by scattering, producing higher loss by steeper wave incidence at the bottom. Examples of loss excess parametrized by the characteristic wave height $H_{1/3}$ are shown in Fig. 41. Below several 10^2 Hz the roughness influence is hardly significant.

Beyond the onset of pronounced breaking waves (wind speeds > 10 m/s) the attenuation due to bubble layers may increase by the third to the fifth power of the wind speed and the first to the second power of the frequency. Typical values of α are 3···5 dB/km at 8 kHz and 20 m/s [86 W 2].

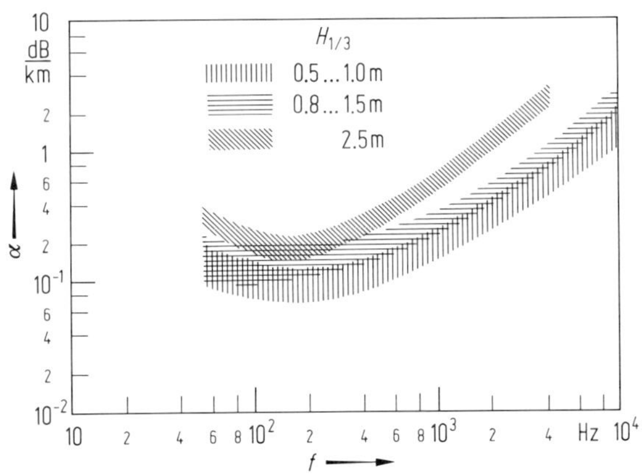

Fig. 41. Absorption coefficient α and wave height $H_{1/3}$. North Sea, sand bottom, 60 m water depth [74 S 2].

3.2.6.3.7 Depth dependence of transmission loss

Due to the decreasing bounce distance by shallower water depth the attenuation tends to increase (Fig. 42). The dependence of α on source and receiver depth is small except for the source below the thermocline [77 S]. At low frequencies (H < 20λ) and winter conditions the wave guide properties of the shallower-water channel become evident by a significant depth dependence of α (Fig. 43a) which may be less effective under summer conditions (Fig. 43b). When selected modes are initiated by the source setup node-like minima may exist between the boundaries (Fig. 44) [74 I].

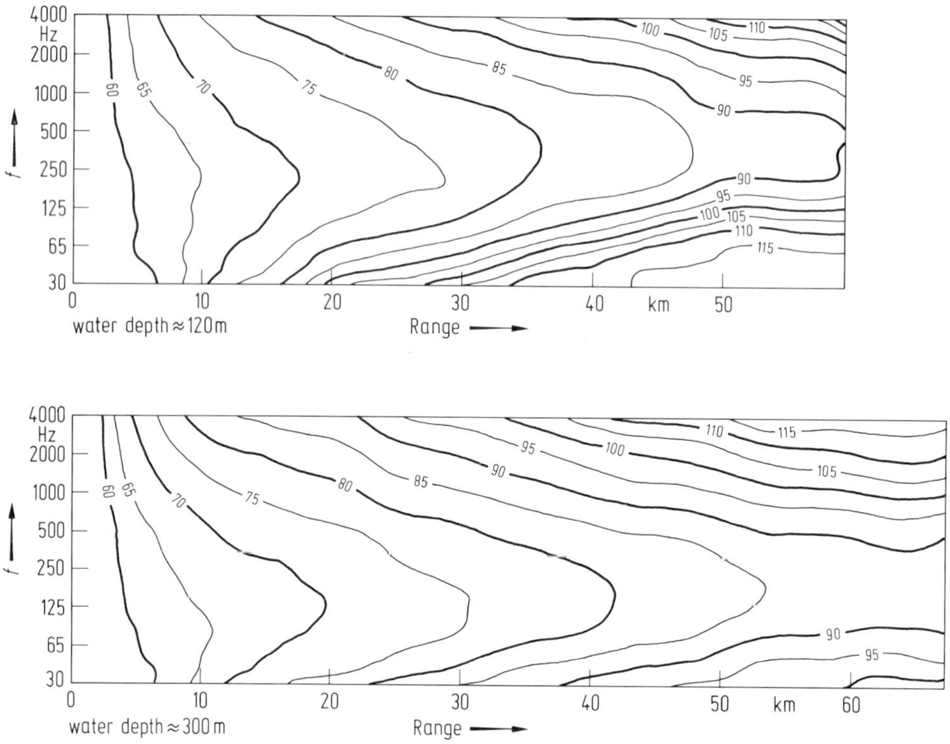

Fig. 42. Transmission loss, in [dB], and water depth [80 A 1]. Source and receiver depth: 50 m. Winter condition, nearly calm sea. Top: silt and clay bottom. Bottom: clay bottom.

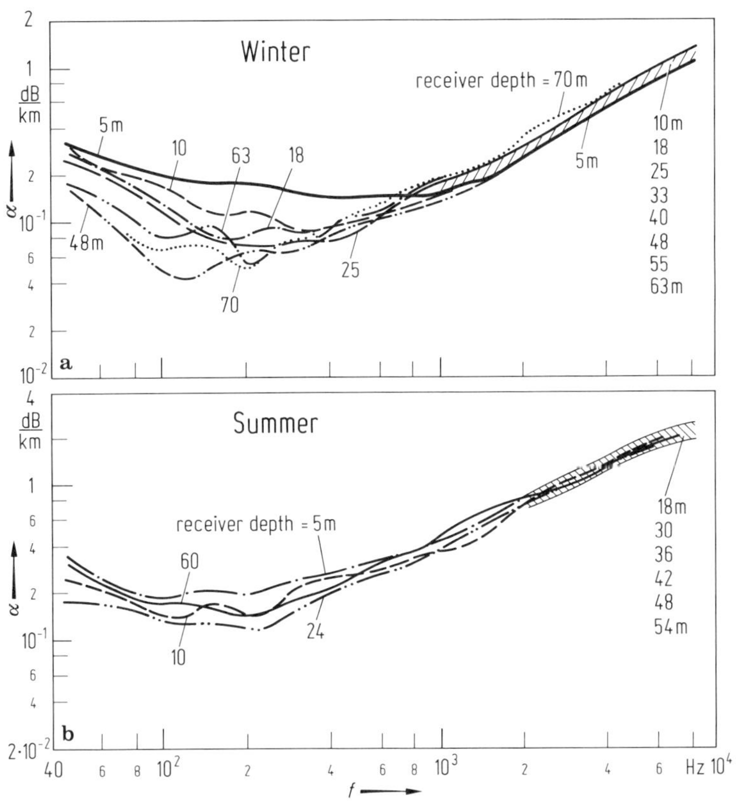

Fig. 43a. Absorption (attenuation) coefficient α and receiver depth – winter condition, North Sea, sand bottom, 70 m water depth [74 S 2].

Fig. 43b. Absorption (attenuation) coefficient α and receiver depth – summer condition, North Sea, sand bottom, 70 m water depth [74 S 2].

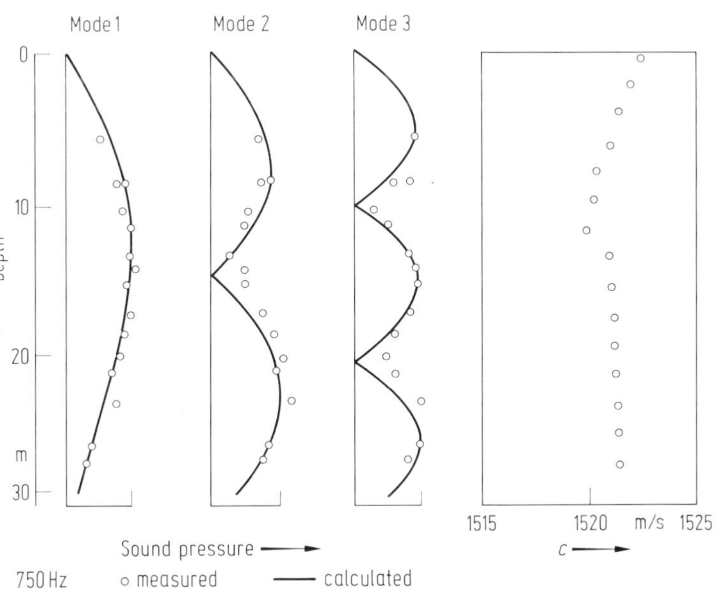

Fig. 44. Vertical distribution of sound pressure amplitude due to mode structure [74 I].

3.2.6.3.8 Special phenomena

Topographic peculiarities [80 A 1] like shallow banks, sloping bottom, and sea mounts (Fig. 45) exhibit a frequency-dependent loss increase due to near or below mode cutoff depths and diffractive shadowing. Frontal structures frequently occurring in shallow water may change summerly into winterly reflective trapping (subsect. 3.2.6.3.4).

Large-size fish shoals produce a pronounced frequency-dependent attenuation by swim-bladder resonance absorption and scattering as shown in Fig. 46 [86 T].

The rare case of shallow water refractive trapping occurs when a pronounced salinity stratification is present. The halocline acts acoustically similar as the deep-sea barocline but within a smaller depth scale. The most prominent example is the Baltic Sea (subsect. 3.2.5.6) where the summerly refractive thermocline-halocline channel is about 20···40 m thick with optimum long-range trapping near 10^3 Hz (Fig. 47a). Depending on the stability of the channel the absorption coefficient may be even below the DWC values (subsect. 3.2.6.2) due to the lower salinity of the Baltic Sea (Fig. 1). The channel leakage at low frequencies (some 10^2 Hz) is due to diffraction, at high frequencies (near 10^4 Hz) it is due to scattering at volume inhomogeneities [74 S 2].

Contrary to the reflective trapping the transmission loss is strongly depth-dependent. The bulk of the sound energy propagates within the duct even for off-axis sources. The depth region below the halocline is weakly illuminated only by diffraction and scattering from the duct (Fig. 37b, 48). The winterly propagation where the thermocline is absent differs also from the reflective trapping type. Together with the sea surface the stable halocline forms a refractive-reflective channel. The near-bottom sound field is similar to the summerly situation in the Baltic Sea whereas the sound intensity within the mixed layer is nearly depth-independent except near the surface at low frequencies where the dominating mode structure reduces the sound intensity (Fig. 49). The winterly transmission loss is comparable to the summerly situation (Fig. 47b) but is more sensitive to sea surface scattering than the regular reflective type (Fig. 50).

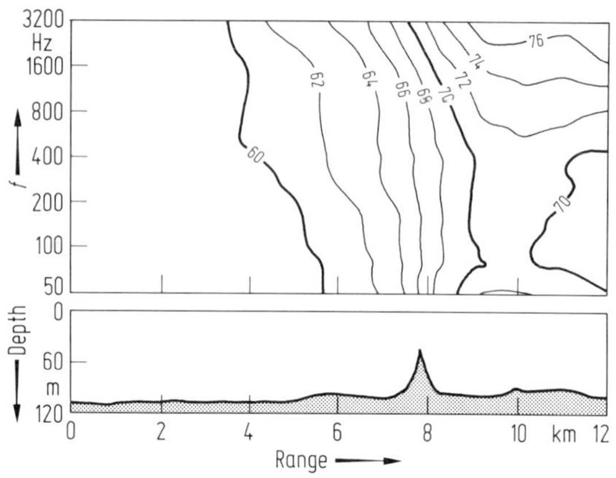

Fig. 45. Effect of a sea-mount on transmission loss [80 A 1]. *TL* in [dB]. Winter condition, nearly calm sea. Bottom silty sand.

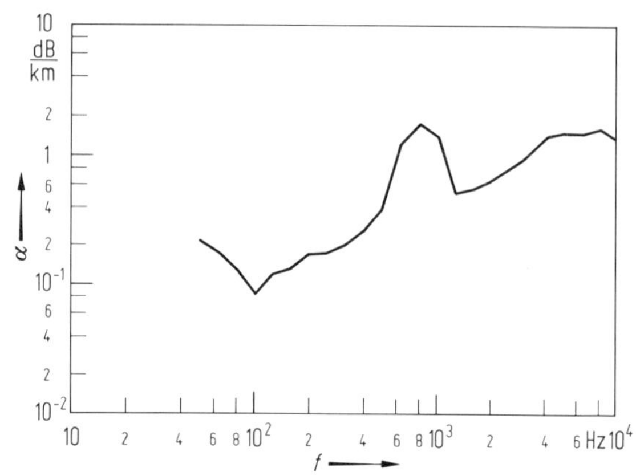

Fig. 46. Frequency-dependent increase of absorption coefficient α due to fish shoal (near 800 Hz). North Sea, summer, 60 m water depth, transmission range 30 nm [86 T].

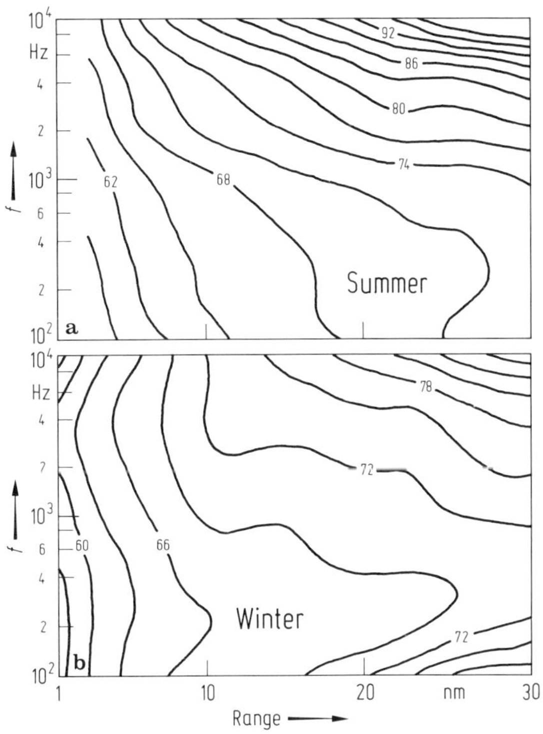

Fig. 47a. Summerly refractive channel; transmission loss dependence on range and frequency, Baltic Sea [80 T]. *TL* in [dB].

Fig. 47b. Winterly reflective-refractive channel; transmission loss dependence on range and frequency, Baltic Sea [80 T]. *TL* in [dB].

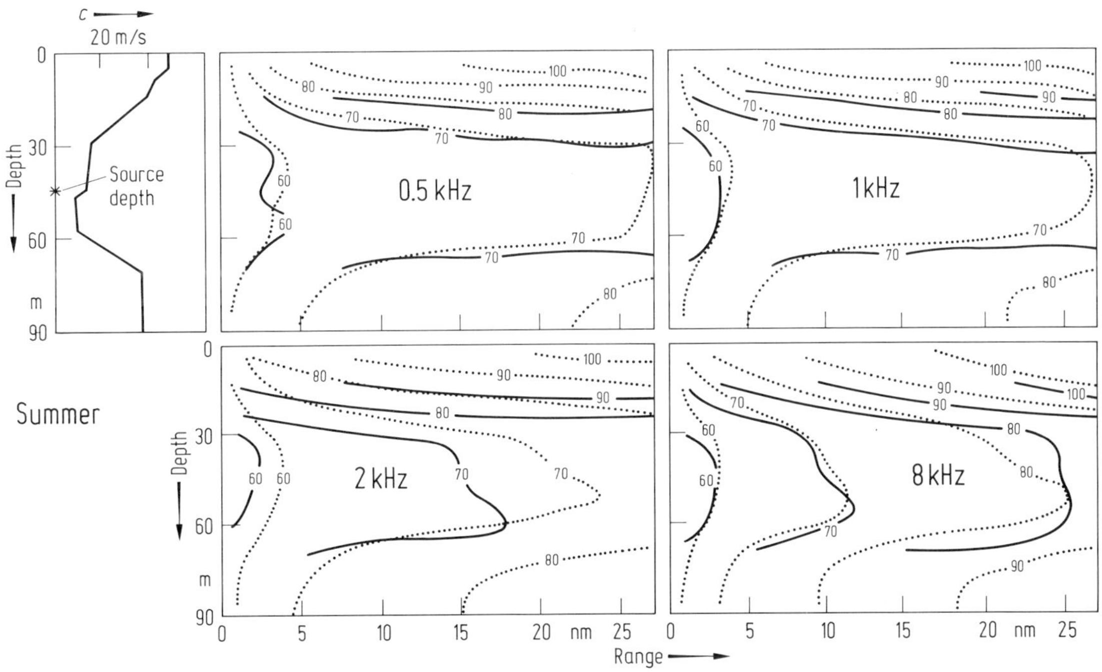

Fig. 48. Summerly refractive channel; transmission loss dependence on range and receiver depth, Baltic Sea [80 T]. *TL* in [dB]. Top: 0.5 kHz (left) and 1 kHz (right), bottom: 2 kHz (left) and 8 kHz (right). Solid lines: measurement (10 × 24 depth-range samples), dotted lines: computed by stochastic model [77 S].

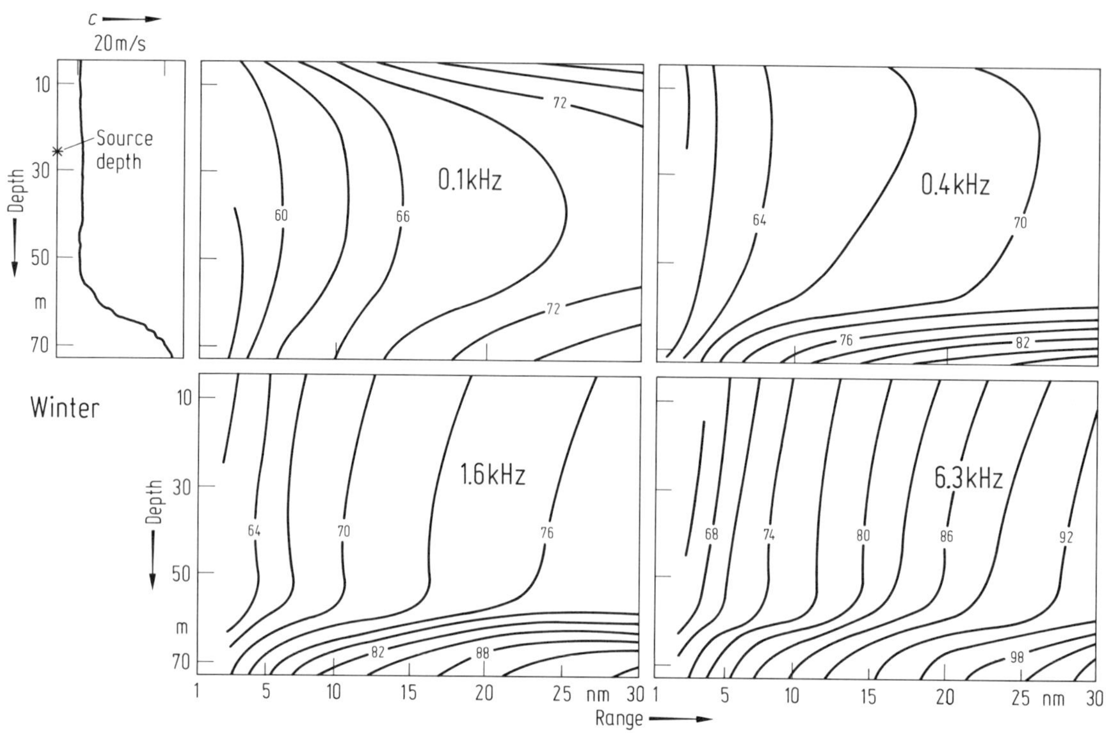

Fig. 49. Winterly reflective-refractive channel; transmission loss dependence on range and receiver depth, Baltic Sea [80 T]. *TL* in [dB]. Top: 0.1 kHz (left) and 0.4 kHz (right), bottom: 1.6 kHz (left) and 6.3 kHz (right). Smoothed curves: measurement (10 × 24 depth-range samples).

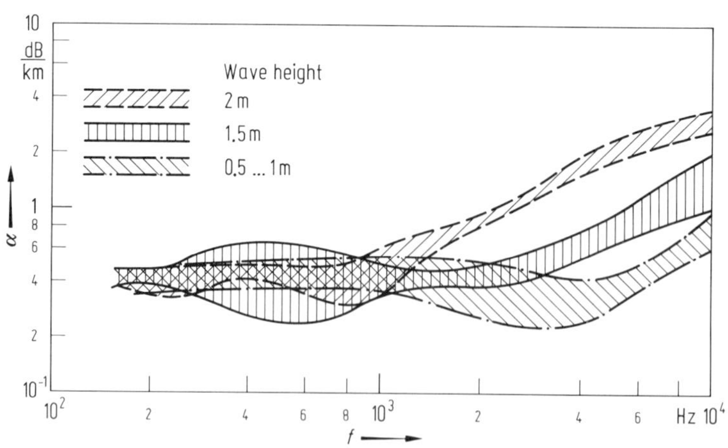

Fig. 50. Absorption coefficient α and wave height; Baltic Sea, late summer condition [74 S 2].

3.2.6.3.9 Approximation polynomials

The transmission loss-range relation for a given shallow-water environment and omnidirectional sound source is sufficiently parametrized by only two coefficients which are determined for each frequency band by least-mean-squares-fit of the measured data. Most frequently used are α, H_s, eqs. (34, 37), on which all absorption coefficients α of subsect. 3.2.6.3 are based. An equivalent data fit is achieved by α', C, where $10 \leq C \leq 20$ stands for a substitute power law between cylindrical and spherical spreading [76 M 3].

A comprehensive approximation for shallow water TL measurements depending on range and frequency was first suggested 1962 by Marsh and Schulkin [62 M 2]. The TL-range tables [69 P] of their Colossus II-formula, based on extensive data mainly from the Scotian Shelf, are subdivided into classes of bottom (sand, mud), frequency 0.1···10 kHz), water/layer depth, and sea state. The volume absorption indicated by a standard salinity and temperature can be adjusted too. However, later measurements in other areas have made evident that the table cannot claim general validity. This is demonstrated by Fig. 51 comparing North Sea-data with Colossus values which may be more than 30 dB away from reality and fail to reproduce the optimum frequency behavior as well. Fig. 51 illustrates further that shallower depths do not necessarily belong to higher losses. A grain size TL-relation for sand bottom as claimed by [83 M 1] has not been verified by other measurements either.

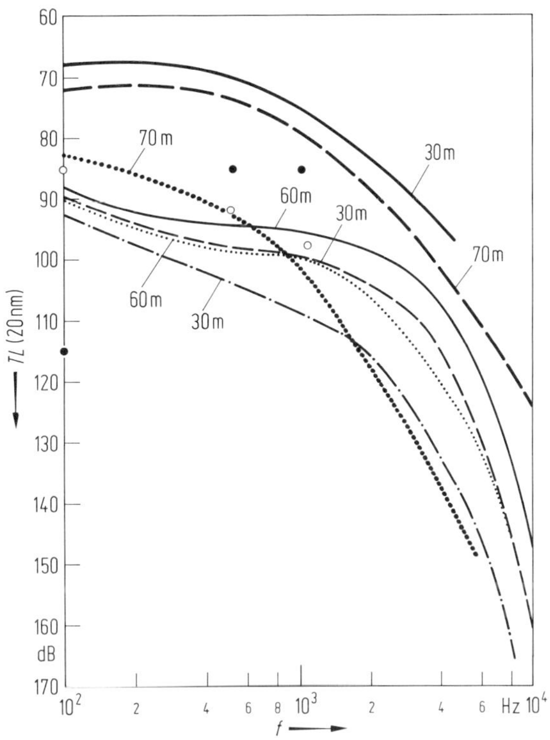

Fig. 51. Transmission loss for 20 nm-reference range; comparison of different "standard" bottom types. Strong dashed line: North Sea, sand bottom, 50···120 m water depth verified for 70 m, thermocline weak or absent [80 T]; strong solid line: North Sea, sand bottom average for 30 m water depth [80 T], coarse sand: 1···3 dB higher loss than average, fine sand: 1···3 dB lower loss than average; strong dotted line: North Sea, mud bottom, reference depth 70 m [80 T]; thin solid line: Scotian Shelf (Colossus II-equation) water depth plus layer depth 60 m [69 P]; thin broken line: as before, but 30 m; thin dotted line: as before, but 60 m, mud bottom; thin dash-dotted line: as before, but 30 m, mud bottom; solid circles: 30 m depth, sand (grain size 0.8···1.5 mm) [83 M 1]; open circles: as before, but grain size 2.5 mm [83 M 1].

A parametrization of a measured $TL(r,f)$ by polynomial fit has been suggested by [80 H 2] and may be useful for data reduction. Based on the cylindrical spreading law used in subsect. 3.2.6, H_s and α can be represented by a first- and second-order polynomial of $\log f$, respectively [80 T]:

$$TL(f,r) = A_1 + A_2 \cdot 10 \log f + [B_1 + B_2 \cdot 10 \log f + B_3(10 \log f)^2] r + 10 \log(1000\,r)\,, \qquad (50)$$

where f, r stand for the frequency in [kHz] and the range in [km], respectively, and TL is in [dB km^{-1}].

For sand and mud bottoms such as occurring in the North Sea, for water depths H of $\approx$50···150 m, weak temperature stratification, and low sea states the following numbers may be realistic within 0.1···10 kHz and $r \gtrsim 30$ H.

	A_1	A_2	B_1	B_2	B_3
Sand	23	0.7	0.3	0.05	0.005
Mud	27	0.4	1	0.1	0.005

An estimate of the wave height contribution may be taken from Fig. 41. A reliable parametrization of the contribution of the seasonal stratification is not available at present.

3.2.6.4 Fluctuation of transmitted sound

3.2.6.4.1 Temporal fluctuation

The time scale of the amplitude and phase variability of sound signals received by a distant hydrophone comprizes long-term quasi-deterministic variations according to seasonal, weather-induced and diurnal changes of the sound-speed stratification up to short-term stochastic fluctuations within seconds due to motions of source/receiver as well as of volume inhomogeneities and the sea surface.

The coefficient V of stochastic fluctuations is defined as the ratio of the standard deviation devided by the mean of a signal sequence. The temporal behavior is described by the autocorrelation function of the fluctuation: $\varrho(\tau)=\overline{p(t)\cdot p(t+\tau)}/\overline{p^2(t)}$ where τ is a delay time and the denominator is the mean square pressure amplitude. The fluctuation spectrum as the Fourier transform of $\varrho(\tau)$ is useful to describe the relation between signal fluctuation and motion processes such as tides, swell, internal waves, wind waves and currents [72 S].

The strongest fluctuations occur under multipath conditions due to the large phase differences [67 W 2]. Depending on the signal bandwidth, the variation of the received signal level may be >40 dB. The signal is blurred in the time and frequency domain, which can be described by a time-varying weighting function [72 T].

The fluctuation under direct-path conditions due to volume patches of size a at distance $r > ka^2/4$ is given by [60 C].

$$V^2 = \frac{\sqrt{\pi}}{2} k^2 a r \mu^2 , \qquad (50a)$$

where μ and a are defined in subsect. 3.2.6.2.7. At $r=1$ km, typical values of V are 20% at about 10^4 Hz [50 S]. A survey on the literature of wave propagation in random media is given by Ishimaru [77 I].

The fluctuation of the surface-reflected path is produced by the interference of signals reflected at wave crests and troughs, where the phase difference is described by the Rayleigh parameter $R=2\,kh\sin\Theta$ which decreases with RMS-roughness h, grazing angle Θ and wavenumber k. Typical values of V for moderate roughness are 40% [57 W], decreasing to values of the direct-path fluctuation at long ranges.

The probability density of V is described by a Rician distribution when a finite unscattered portion of the signal is present. When this component vanishes the distribution is of a Rayleigh type and becomes Gaussian when the constant component is large [75 U 2].

3.2.6.4.2 Spatial coherence

Sound signals at separate hydrophones of a receiving array differ due to nonidentical propagation paths which limit the angular resolution of the array (subsect. 3.2.4.2). However, the resulting angular uncertainty, which is the Fourier transform of the spatial coherence length defined by a correlation decrease between signal pairs [71 W 3], is typically 1° and less for horizontal arrays, transverse to the direction of sound propagation [82 U]. This is below the resolution of most arrays used for long-range acoustic RS in the sea.

3.2.7 Ambient noise

3.2.7.1 Definitions, general remarks

Ambient noise is defined as the naturally occurring noise background in the sea combined with commercial and industrial noise but excluding identifiable localized sources, reverberation and self-noise due to the hydrophone. The ambient noise level is the sound intensity, in [dB], measured with a nondirectional hydrophone and referred to the intensity of a plane wave having an rms pressure of 1 μPa, usually reduced to a 1 Hz frequency band and then called ambient noise spectrum level [75 U 1, 76 R 2].

Numerous noise sources which are mostly of intermittent occurrence and which cover different frequency ranges have been investigated from less than 0.1 Hz to about 10^5 Hz. The most important noise sources which are always present and often predominating in the frequency range from 10 to 10^5 Hz are the ship traffic and the sea surface agitation [62 W]. Most noise sources are empirically related to parametrized environmental quantities whereby the influence of the transmission path may be treated by propagation models (subsect. 3.2.2). The variability of the noise in terms of spectrum level at a given frequency is of the order of 30 dB. This is less than the variability of the other quantities which together limit the signal detection range in the sea (subsect. 3.2.2.3): transmission loss (subsect. 3.2.6), scattering strength and reverberation strength (subsect. 3.2.8).

Table 12 itemizes the different noise sources divided into the three categories of origin: geophysical, technical and biological in the sequence of their main frequency intervals.

Table 12. Noise sources in the sea (in capital letters: predominant processes).

	Noise source	Main frequency range
GEOPHYSICAL	HYDROSTATIC (tides, swell)	< 1 Hz
	HYDRODYNAMIC (wave interaction, turbulence)	0.1···200 Hz
	seismic activity	5··· 30 Hz
	WIND/SEA WAVES	0.1··· 30 kHz
	precipitation	0.1··· 30 kHz
	ice cover	0.01··· 10 kHz
	THERMAL	> 100 kHz
TECHNICAL	oil/gas-rigs	10···100 Hz
	seismic off-shore exploration	10···100 Hz
	DISTANT SHIP TRAFFIC	0.01··· 1 kHz
BIOLOGICAL	marine mammals	10···300 Hz
	fish	100···300 Hz
	crustacea	0.1··· 10 kHz

3.2.7.2 Hydrostatic effects

Hydrostatic pressure changes caused by tides produce the highest infrasonic noise levels of 200 dB, relative to 1 μPa/m [75 U 1], but are far below the frequency range of sound propagation in the sea. The pressure field p_s generated by surface waves of wave height H decreases exponentially with depth z and wavenumber k [65 K].

$$p_s = \varrho g \frac{H}{2} e^{-kz} \cos(kx - \omega t) \tag{51}$$

ϱ, g, x, ω, t are: water density, gravity, horizontal coordinate, angular frequency and time, respectively.

For frequencies > 1 Hz and hydrophone depths $z > 3$ m this contribution is therefore negligible against the noise of other origin.

3.2.7.3 Hydrodynamic effects

The infrasonic spectra of Figs. 52 and 53 [81 N] decrease with the fourth to the fifth power of the frequency between 0.5 and 5 Hz. The data of Fig. 53 belong to the lowest spectrum levels of this frequency decade measured in the open sea. Comparison with theories does not allow conclusive identification of the generating processes. Nonlinear interaction of surface waves [50 L, 63 M 1, 66 B, 68 K, 69 B 2] also presumed as origin of microseism appears to be a probable mechanism for the wind/sea-state-dependent spectrum of Fig. 75. The spectral maximum (0.25 Hz) is expected to occur at twice the frequency of the corresponding peak of the surface wave spectrum.

A similar noise spectrum independent of sea state when far below the surface may be achieved at the hydrophone inside a turbulent boundary flow (Fig. 53) [62 W]. The small-scale turbulent pressure fluctuations do not radiate sound and are therefore usually classified as self-noise.

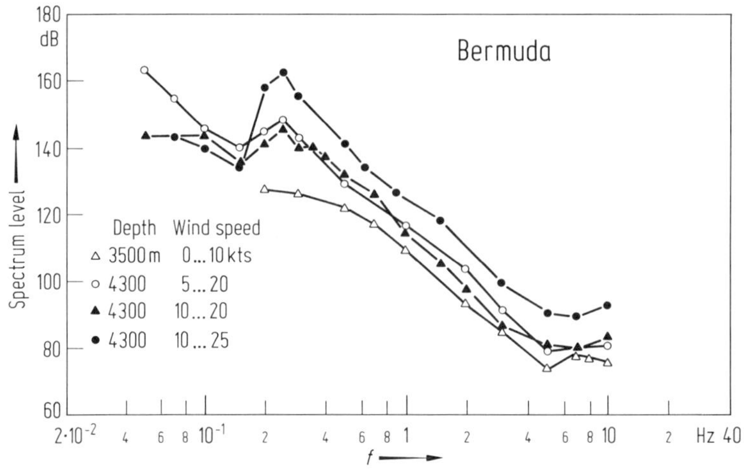

Fig. 52. Spectra of infrasonic noise (relative to 1 μPa) in deep water from bottomed hydrophones off Bermuda [81 N].

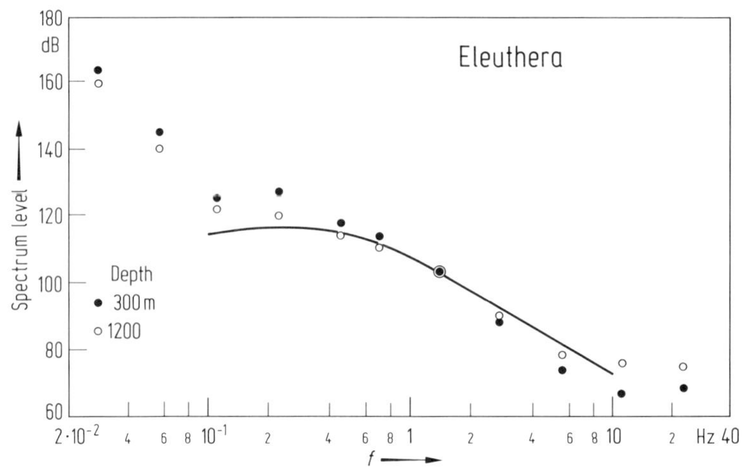

Fig. 53. Comparison of infrasonic noise spectra (relative to 1 μPa) measured at Eleuthera [81 N] with computed ocean turbulence noise spectrum for 0.04 kts turbulence current [62 W], solid line.

3.2.7.4 Seismic activity

Combined measurements of infrasonic noise and sea floor motion in a seismically active area of the Mid-Atlantic Ridge [76 M 1] show shallow seaquakes of magnitude 0···2.5 on the Richter scale and epicenter distances of 3···10 nm to raise the broadband acoustic levels in excess of 20 dB between 5 and 30 Hz (Fig. 54). Due to other acoustical observations [63 N, 65 N, 74 N 1, 74 S 4] the seismicity yields maximum spectrum levels at frequencies between 2 and 25 Hz.

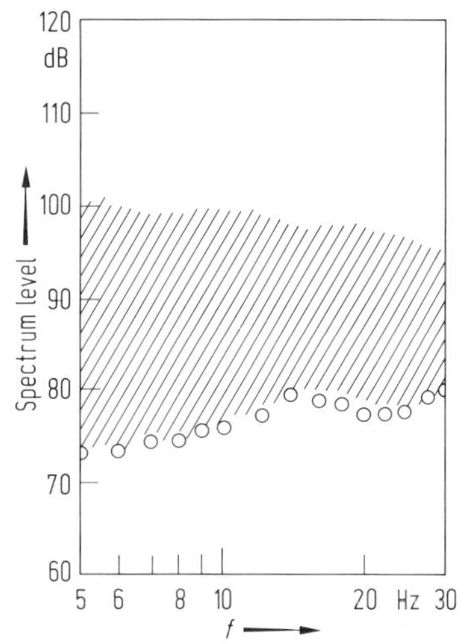

Fig. 54. Infrasonic noise (relative to 1 μPa) in the frequency regime of seismic activity [76 M 1]. Circles: noise from far ship traffic, weather related sources absent. Shaded area: variability of seaquake noise (Mid-Atlantic Ridge).

3.2.7.5 Ship traffic

The noise immission from distant ship traffic is considered predominant in the frequency range from 10 to a few hundred Hz [76 R 2, 62 W, 73 D 2]. Propeller cavitation [80 L 2] being responsible for the high-frequency broadband noise is superimposed by a line spectrum from blade modulations and machinery noise preferably at the lower frequencies [76 R 2]. Together with the sea surface agitation (subsect. 3.2.7.6) the sea traffic is the most important noise source in the sea and more dependent on sound propagation conditions than most other sources.

The radius of traffic noise immission increases with decreasing transmission loss and ship density. The radius varies from some 10 nm in shallow water and traffic densities of the order of 10 per 1000 nm^2 [85 W] up to about 1000 nm in the Atlantic Ocean where the average traffic density is 3 orders of magnitude smaller [73 D2]. The average traffic noise spectra of Fig. 55 [75 U 1] may be used as working graphs though both spectrum level and slope are dependent on the propagation conditions and the receiver depth (subsect. 3.2.7.9.1). Since the cavitation noise rises by the 6th power of the vessel speed [76 R 2], the high-speed vessels usually govern the traffic noise background in a sea area.

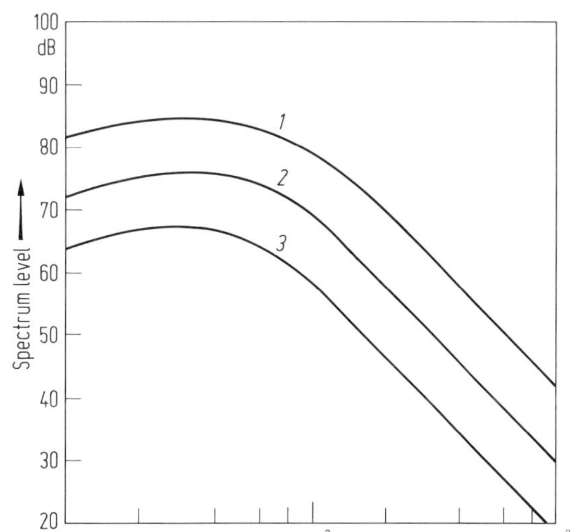

Fig. 55. Noise from ship traffic; average spectra (relative to 1 μPa) for deep water [83 U]. (*1*) heavy shipping (near shipping lanes, North Atlantic), (*2*) moderate shipping, (*3*) light shipping (remote from ship traffic).

3.2.7.6 Sea surface agitation

In the frequency range from 10 Hz to some 30 kHz there is a relation between the local wind speed and the noise spectrum level as far as the spectra are uncontaminated by traffic noise. Empirical parametrizations are given in Fig. 56 [62 W, 64 P, 76 R 2, 84 G].

Splash and spray from the sea surface [62 W] in connection with the radiating cavitation bubbles [66 F] are regarded as dominant mechanism though a quantitative theory does not exist. There seems to be a threshold wind speed of 2···3 m/s [84 G, 76 C 1, 78 M 2] at the onset of microbreaking below which a wind dependence of the noise level is either nonexistent or nonsystematic. The origin of the surface noise below the breaking wave regime is unknown [62 W].

The wind-generated noise increases approximately by the second power of the wind speed (6 dB/octave) and decreases nearly by the second power of the frequency (−5··· −6 dB/octave). Because of the small immission range of the surface noise at frequencies $>10^3$ Hz [85 K] the spectrum level is almost independent of propagation conditions [84 G], water depth and hydrophone depth [78 M 2]. The remaining level variability for constant wind speed [62 W], which may exceed 6 dB, appears mainly due to changes of the wind profile below the reference height, which are also responsible for a spurious noise level dependence on the wave height [84 G].

At very high wind speeds >20 m/s the high-frequency noise spectrum level can be reduced by air bubble screening [85 W].

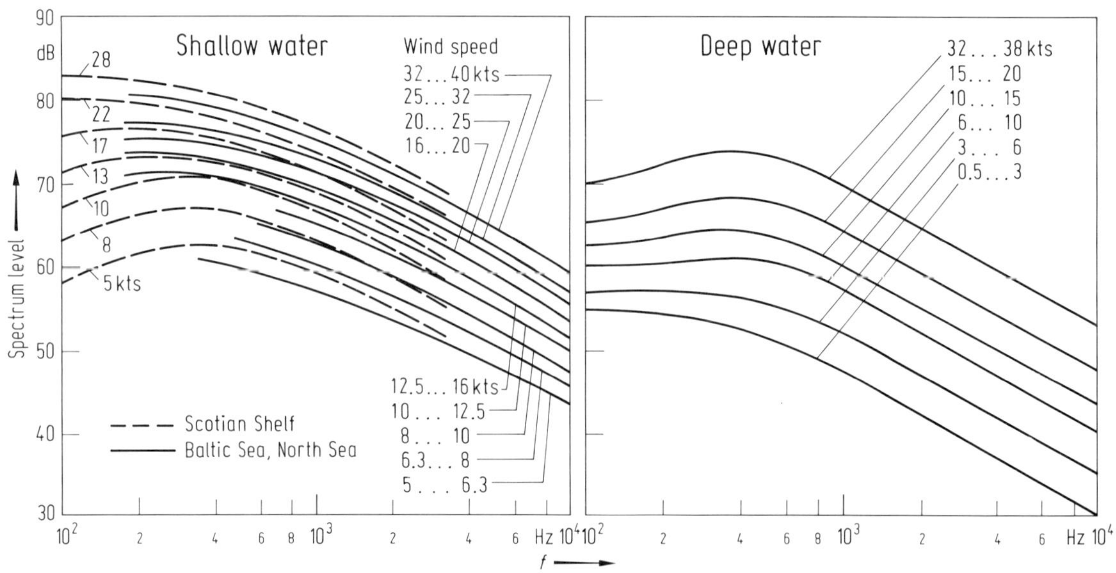

Fig. 56. Wind-dependent ambient noise; average spectra (relative to 1 μPa) for shallow and deep water. Parameter: wind speed. Left: shallow water. Broken lines: Scotian Shelf [64 P]. Solid lines: Baltic Sea, North Sea (extremely low traffic) [84 G]. Right: deep water [76 R 2].

3.2.7.7 Thermal noise

The noise pressure p_{th} from molecular movement [52 M] corresponds to the Nyquist noise developed in the radiation resistance of an omnidirectional hydrophone of the efficiency η [62 E 2] and rises with frequency f at the rate of 6 dB/octave:

$$20 \log p_{th} = -15 + 20 \log f - 10 \log \eta , \qquad (52)$$

p_{th} is in [μPa] and f is in [kHz]. The thermal noise spectrum level which is of the order of 5 dB at 10^4 Hz places a threshold on the minimum observable pressure levels in the sea.

3.2.7.8 Special/intermittent noise sources

3.2.7.8.1 Rain

Rain is the most effective high-frequency noise source in the sea. Moderate to strong rain showers raise higher noise levels in the kHz-region than storms [59 F, 69 B 1] (Fig. 57). Similar to the wind-induced spray (subsect. 3.2.7.6) the rain drops produce noise both by their impact on the surface and the vibration of the generated air bubbles. The ratio of the radiated sound energy to the kinetic energy of the droplet is proportional to the third power of the Mach number relative to the sound speed in water. Due to the high-speed impact, the spectrum slope is flatter (nearly "white noise") than that for noise from breaking waves. Since the horizontal acoustic range of the high-frequency rain noise is small, its predictability is limited in the same way as the occurrence of an individual rain shower.

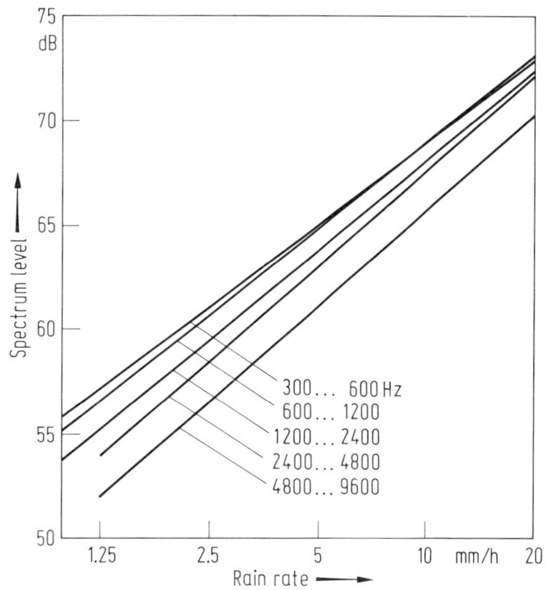

Fig. 57. Spectrum level (relative to 1 μPa) of rain noise vs. rain rate (Lake of Sarzana) [69 B 1]. Parameter: frequency (octave bands).

3.2.7.8.2 Ice

In ice-covered waters the spectrum level and slope is highly variable (up to 40 dB) depending on ice conditions, snow cover, wind speed and air temperature changes. The lowest noise levels (down to 20 dB below calm sea noise [76 K 2]) occur below a continuous ice cover under rising temperature [64 M 4] and become independent of wind speed [62 M 1]. High levels are observed at decreasing temperatures from cracking noise and from movements of noncontinuous ice covers [64 M 4] (Fig. 58). Extremely flat spectra are found in antarctic regions [76 K 2] and near melting icebergs from entrapped air bubbles under pressure [71 U]. For compact ice the noise level has a peak below the ice edge, decreasing stronger towards the ice cover (Fig. 59) [74 D 2] than towards the open sea. Due to the complicated combination of different processes the predictability of the spectra of noise under an ice cover [64 M 4] is rather poor.

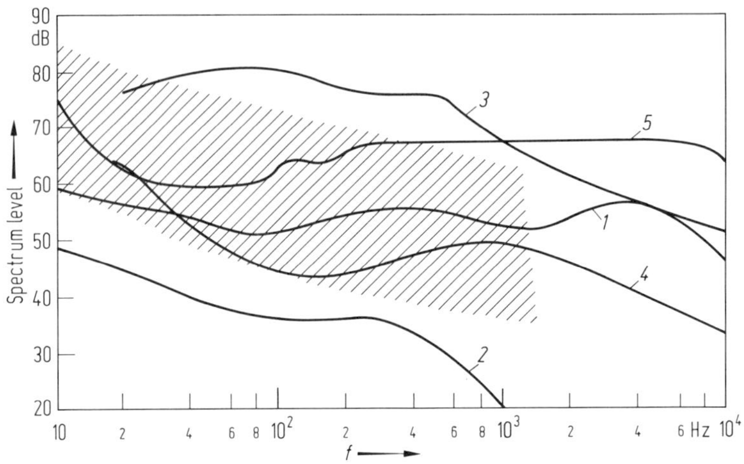

Fig. 58. Spectra of ambient noise (relative to 1 µPa) observed under ice. (1) Old polar ice (frozen leads and pressure ridges), noisy periods [64 M 4], (2) like (1), quiet periods. (3) Shore fast polar ice, cracking noise prevalent [64 M 4], (4) like (3), cracking noise absent. (5) Antarctic ice, thermal cracking [76 K 2]. Shaded area: estimated percentage of the time that a given level will be exceeded, based on all available Arctic data [83 U]. Upper limit: 10%, lower limit: 90%.

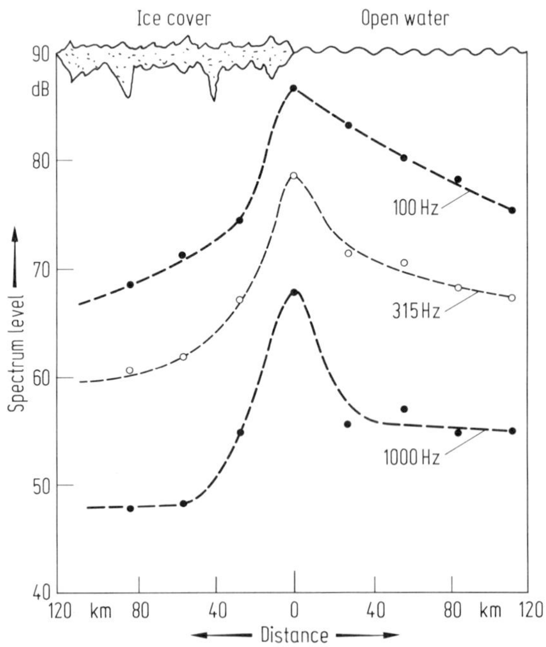

Fig. 59. Variation of average ambient noise spectrum level (relative to 1 µPa) with distance from a compact ice edge at sea state 2 [74 D 2]. Left side: ice cover, right side: open water.

3.2.7.8.3 Biological noise

The marine mammals (whales, dolphins, porpoises) and some kinds of swimbladder fish generate primarily low-frequency noise, containing spectral lines, whereas kinds of crustacea make broadband noise by snapping movements [48 K]. Biological noise mainly occurs in tropical and semi-tropical waters; fish species of commercial importance are not acoustically active. The spectra like the examples of Fig. 60 depend strongly on population densities within the biotopes [64 T] and are hardly predictable.

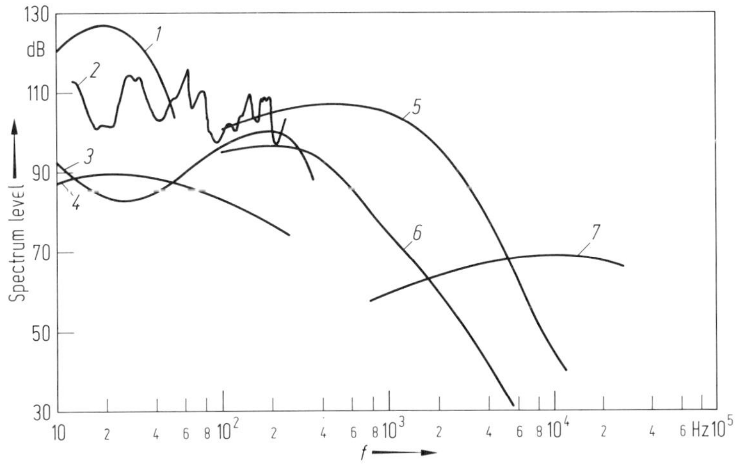

Fig. 60. Examples of biological noise spectral (relative to 1 μPa) [48 K]. 1: toad fish (close to receiver), 2: porpoises, 3: high-pitched drum fish, 4: sea robins, 5: croakers (May/June), 6: croakers (July), 7: shrimps.

3.2.7.8.4 Technical noise of other origin

The main source of this category is noise from seismic exploration by explosive sources [68 S], contributing mainly to the spectrum below 10^2 Hz. Oil rigs and platforms raise the noise background much less than the ship traffic.

Aircrafts contribute to the noise level only when directly passing on top of the hydrophone position [72 U] and can generally be neglected as noise source.

3.2.7.9 Further properties of ambient noise

3.2.7.9.1 Depth dependence

As far as the noise from the sea surface agitation is dominating (subsect. 3.2.7.6) the spectrum level is approximately independent of water depth [76 C 1] and receiver depth [78 M 2] when the volume attenuation (subsect. 3.2.6.2) is taken into account. For the frequency regime below 100 Hz, governed by traffic noise immission, the vertical noise level distribution in deep water depends on the sound propagation conditions and therefore on the sound speed profile (subsect. 3.2.5). Due to the standard graphs of Fig. 61 [78 M 2] the noise level profile below the sound channel axis follows the superposition of traffic noise and surface noise, yielding the strongest level decrease underneath the critical depth of ≈ 10 dB/km.

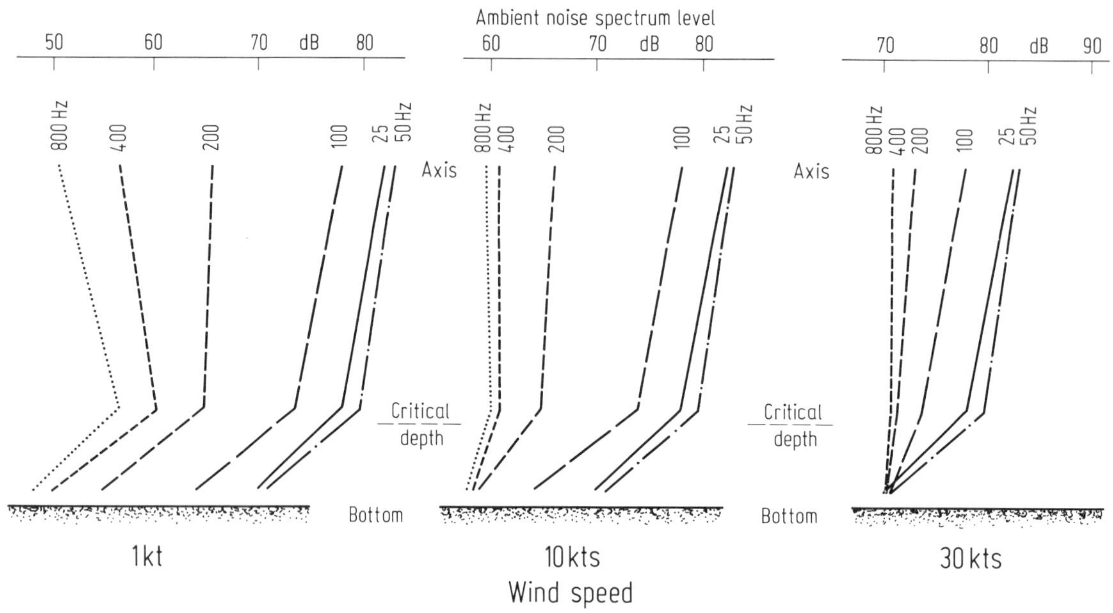

Fig. 61. Standardized depth dependence of ambient noise spectrum levels (relative to 1 μPa) for deep-ocean conditions below the sound channel axis (arbitrary depth scale) [78 M 2].

3.2.7.9.2 Directional characteristics

Within the wind-dependent portion of the noise spectrum the contribution from vertical incidence increases with the wind speed (Fig. 62) [65 A], whereas the horizontal maximum from the traffic noise at low frequencies remains nearly unchanged. The directional characteristics of the wind-induced noise intensity is approximately proportional to $\cos^2 \vartheta$ [63 V] corresponding to dipole sources as a combination of the actual source and its image in the sea surface (ϑ: angle of incidence).

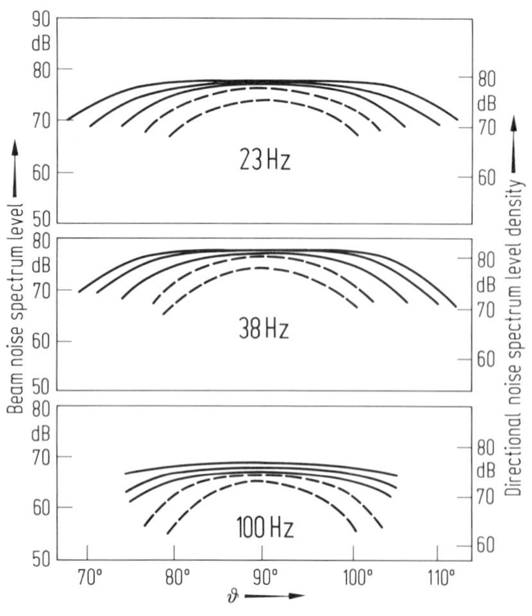

Fig. 63. Depth dependence of vertical directionality of low-frequency ambient noise (deep ocean) [79 A 1]. Spectrum level relative to 1 μPa. ϑ: angle of incidence ($\vartheta = 90°$ horizontal incidence). Parameter: receiver depth, from top to bottom: 700 m, 1250 m, 2500 m, 3800 m, 4800 m.

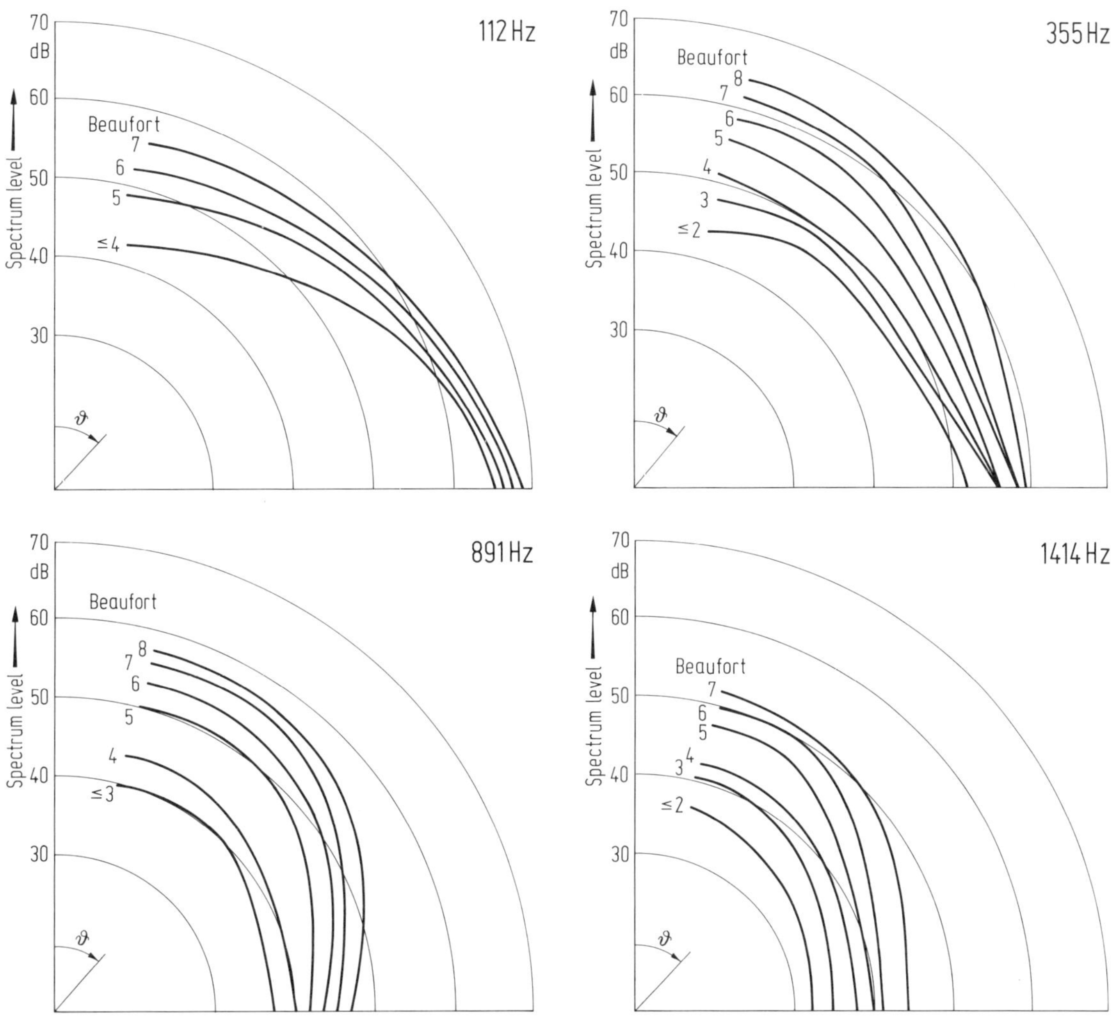

Fig. 62. Vertical directionality of ambient noise (relative to 1 μPa/sr) in the deep ocean (Bermuda) at a bottomed hydrophone [65 A]. ϑ: angle of incidence ($\vartheta = 0$: vertical incidence). Parameter: wind force, frequency.

For the traffic-governed regime below the transition frequency of 100 Hz the directional characteristics are represented as empirical standard curves in Fig. 63 with the receiver depth as parameter. The noise becomes increasingly concentrated towards the horizontal incidence (90°) when the sensor moves downwards [79 A 1].

3.2.7.9.3 Level distribution

As long as the participating processes are stationary, the distribution of the noise spectrum level is approximately Gaussian [72 W]. An important exception is the spiky noise under ice at decreasing temperatures [64 G]. Similar deviations occur directly below the sea surface [75 U 1] or at very high wind speeds where single breakers predominate [85 W].

3.2.8 Scattering and reverberation

3.2.8.1 Definitions

The reradiation of incident sound by discontinuities in the sea is called scattering; the sum of these scattering contributions is called reverberation. Together with the noise (subsect. 3.2.7) the reverberation represents the unwanted background for the remote sensing (RS) of objects or it may itself be subject to RS of features of the reverberation origin.

Volume and surface scattering behave differently. The predominant scatterers in the sea volume are biological organisms. Surface scattering is produced both by the rough boundaries of the ocean surface and bottom – or sometimes by thin layers of biological matter in the volume. The reverberation depends on the scattering strength $S_{s,v}$, in [dB], of the contributing volume or area [75 U 1]

$$S_{s,v} = 10 \log \frac{I_s}{I_i}, \tag{60}$$

where the intensity I_s of the scattered sound is reduced to the distance of 1 m from the center of a reference area of 1 m^2 or volume of 1 m^3, respectively, and divided by the incident plane wave intensity. When source and receiver of the RS setup coincide the backscattering strength is the relevant parameter which is used in the following sections.

The backscattering strength of an individual obstacle or target of arbitrary size in the volume is called target strength, in [dB], and defined as

$$TS = 10 \log \frac{I_r}{I_i}, \tag{61}$$

where I_r is the intensity of return at 1 m distance. In particular a large ($a \gg \lambda$) sphere as an isotropic reflector of radius a will produce the backscattered intensity I_r at distance r (a, r in [m]):

$$I_r = \frac{a^2 \pi I_i}{4\pi r^2} \rightarrow TS_{\text{sphere}} - 10 \log \frac{a^2}{4} \quad \text{at} \quad r = 1\,\text{m}. \tag{62}$$

The ideal sphere of radius 2 m therefore has a TS of 0 dB.

Backscattering strength and target strength depend on the acoustical impedance and shape of the scatterer as well as on frequency and incident angle. The backscattered intensity is measured at a distance $r \gg 1$ m to achieve a quasi-plane wave illumination. The reduction to 1 m reference distance is made by accounting for the transmission loss TL (subsect. 3.2.6).

When the volume contains a mixture of obstacles (e.g. fish) of different TS, the volume backscattering strength S_v is the sum of all contributions under the assumption of random phase and negligible multiple scattering. To describe the backscattering of a water column when illuminated by a vertical echo sounder the column backscattering strengths S_{vc} defined as the integration of S_v from the sea surface to the depth of the deepest contributing scatterer.

The reverberation level of a volume RL_v is the backscattered intensity, in [dB], at the source-receiver unit, defined as (r in [m]):

$$RL_v = SL - 40 \log r + S_v + 10 \log V \tag{63}$$

where $V = \frac{c\tau}{2} \Psi r^2$, in [m^3], which combines the source level SL with the to and fro transmission loss (absorption neglected) and the entire backscattering of the volume V. The quantities c, τ, Ψ being the sound speed, the pulse length and the equivalent solid angle of the beamwidth of the source-receiver combination.

The reverberation level of a surface RL_s is defined analogically (r in [m]):

$$RL_s = SL - 40 \log r + S_s + 10 \log A\,, \tag{64}$$

where $A = \frac{c\tau}{2} \Phi r$, in [m^2]. A is the illuminated area and Φ the equivalent angle of the beamwidth in the A-plane. Equivalent two-way beamwidths Ψ, Φ are given in subsect. 3.2.4.2.

3.2.8.2 Scattering by the sea volume

The dependence of the backscattering quantities of biological objects on frequency and aspect angle, eq. (60), is discussed separately for the short- and the long-wavelength regime, divided at $L/\lambda \approx 1$, where L is taken as the fish length [63 C]. This simplification, which neglects the complicated Mie scattering at the transition between geometric scattering ($L/\lambda \gg 1$) and Rayleigh scattering ($L/\lambda \ll 1$), is reasonable because of the large data spread, characteristic of scattering at biological organisms [69 L 2].

3.2.8.2.1 Backscattering by individual organisms, geometrical region

The regime of high-frequency scattering is preferred for fish echo sounders to resolve shoal structures or individual fish and to achieve biomass information [77 C].

Laboratory measurements of backscattering at fish [77 L] of many different orders and families have shown that the problem of geometric scattering can be simplified by the assumption of a "standard fish", parametrized only by its length L, in [m]. The empirical equation [79 P]

$$TS = 10 \log(3.34 \cdot 10^{-3} L^2), \tag{70}$$

(TS in [dB]) depicted in Fig. 64 for dorsal aspect, has been verified within $1 \leqq L/\lambda \leqq 10^3$, including invertebrates of a few mm length [79 P] up to whales of $L > 10$ m [73 L]. The TS spread is mostly < 10 dB [69 L 2]. The reason for this uniform behavior is an approximately constant impedance ratio of tissue and water $(\varrho c)_{tis}/(\varrho c)_{wa} = 1.10(4)$ [77 C] and similar shape (length/width ratio) for the majority of marine organisms [72 G, 77 B 2]. Thus there is no classification clue to discriminate between different species within the same size class.

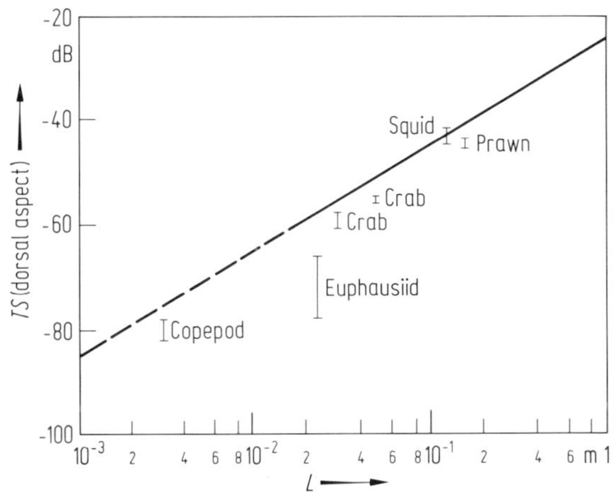

Fig. 64. Target strength TS of single marine organisms at dorsal aspect vs. animal length L. Regime of geometrical scattering: $1 \leqq L/\lambda \leqq 100$. Strong line: representative empirical curve for fish, 14 families in 7 orders, $0.015 \leqq L \leqq 1$ m, compiled by Love [77 L]. Bars: invertebrats/zooplankton, compiled by Penrose and Kaye [79 P].

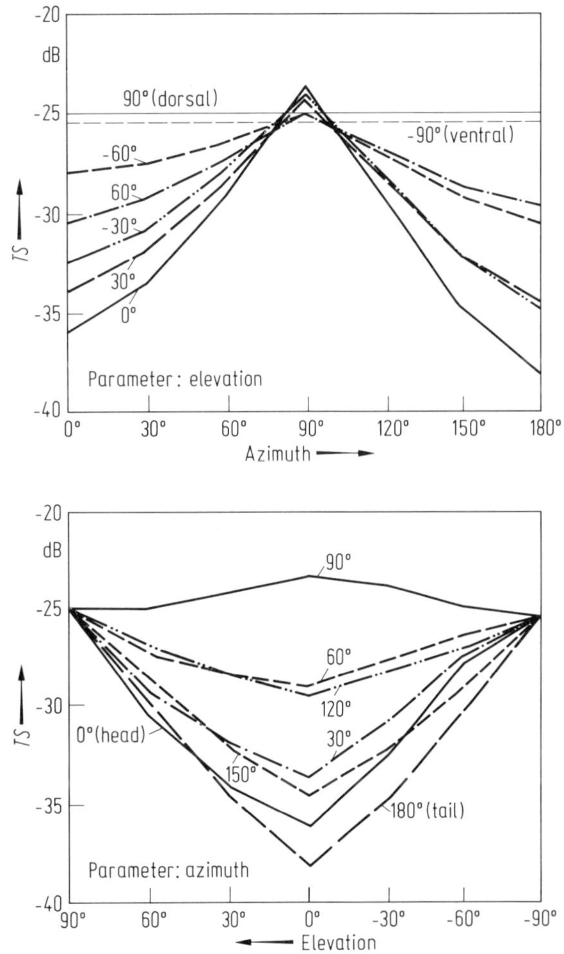

Fig. 65. Target strength TS of an individual fish of reference length $L = 100$ cm at any aspect, after Love [77 L]. Regime of geometrical scattering: $2 \leqq L/\lambda \leqq 100$. Top: dependence on azimuth angle, 30° average. 0°: head aspect, 180°: tail aspect. Bottom: dependence on elevation angle, 30° average. +90°: dorsal aspect, −90°: ventral aspect.

Wille

Due to eq. (70), TS is proportional to the cross section of the projected area A, eq. (60). Eq. (70) ignores the weak frequency dependence measured [77 L] which introduces only small prediction errors of typically 1···4 dB. The target strength of organisms without swimbladder is ≈3 dB lower than the value given by eq. (70) [58 J], which is within the data spread as well.

The dependence of TS on aspect angle due to laboratory measurements is depicted in Fig. 65 [77 L] for two axes of rotation. Due to the dorsal scattering strength maximum in comparison to other aspects, the most consistent backscattering behavior is to be expected from vertical echo-sounding. The aspect dependence of Fig. 65 presents average values; the small-angle fluctuation may be more than 10 dB [74 S 1]. The target strength within a shoal due to aspect angle variability is nearly Rayleigh-distributed [80 K 2]. Biomass estimates are possible by echo integration and by echo counting [77 C, 78 G 1].

3.2.8.2.2 Backscattering by individual organisms; long-wavelength region

For $L/\lambda \ll 1$ the scattering of sound by fish is mainly produced by the air-filled bladder though it amounts only 5% of the entire fish volume [71 B 2]. At the bubble resonance frequency f_R the scatter from the body without bubble is virtually undetectible. The scatter from the excised bladder shows a resonance behavior with a very low damping constant $\delta_B = \Delta f / f_R \approx 0.05$, given by thermal loss and reradiation. Δf is the half-power bandwidth of the scattered sound. The scatter from the life fish with bladder has a high damping constant $\delta_F \approx 0.2$, due to the fish tissue [71 B 2].

The resonance frequency of the bladder is approximately given by [64 A]

$$f_R \approx \frac{3.25}{a}(1+0.1\,z)^{1/2}. \qquad (71)$$

f_R in [Hz], bubble radius a and depth z in [m]. Since $k_R a = 2\pi f_R a/c \approx 0.014$ at sea level, the bubble resonance occurs at very long wavelengths. The resonance for nonspherical prolate swimbladders is approximately given by the radius of a sphere of the same volume. The actual resonance frequency is at most 10% raised when the ratio of the ellipsoid axes increases to 5 [67 W 1].

At resonance the target strength TS which is nearly independent of aspect angle is given by [77 C] (TS in [dB] and a in [m])

$$TS = 10\log\frac{a^2}{(f_R^2/f^2-1)^2+\delta^2}. \qquad (72)$$

The corresponding equation for the volume backscattering strength of 1 m³ water containing n bubbles for negligible multiple scattering is (S_v in [dB] and a in [m])

$$S_v = 10\log\frac{na^2}{(f_R^2/f^2-1)^2+\delta^2}. \qquad (73)$$

The TS enhancement by resonance selects the bubble-carrying organisms when an estimation of biomass is made. Besides of fish any gas bubble carrying plankton can contribute to a resonance scattering. Due to investigations on pneumatophores [63 B 2] typical gas bubble volumes range from 0.3 to 13 mm³ for which the simplified resonance eq. (71) will hold also.

The bubble resonance which represents a rare case of frequency-selective or "colored" acoustic backscattering in the sea yields the most effective clue for size discrimination. The approximate L/a ratio of fish length and equivalent swimbladder radius may be derived from the volume ratio of ≈5% of swimbladder and entire fish tissue within certain shape limits [77 B 2].

Far below resonance the acoustical cross section drops by $\approx(ka)^2$, k being the wavenumber $2\pi/\lambda$, below the geometrical cross section, whereby the coefficient depends on the impedance of tissue and bladder. The governing contribution is given by the air-filled bladder, the tissue appearing nearly transparent. The often used $(ka)^4$-law (Rayleigh scattering) is a limit for infinite rigid spheres only [66 M 1]. Due to experimental problems, the published ka-dependence of the low-frequency scattering of marine organisms is essentially based on theory [77 C].

3.2.8.2.3 Backscattering by distributed organisms (scattering layer)

The scatterers are usually found in the top several hundred meters of the water column where they are concentrated in so-called deep scattering layers. At sonic frequencies deep scattering layers are dominated by mesopelagic fish and other marine organisms which use swimbladders for buoyancy control [54 H, 63 B 1, 69 C]. The comparison of computed volume scattering strength from quasisynoptic biological draught data with measured scattering agrees within 2···6 dB [75 L 1].

Many of the organisms undergo a diurnal migration rising towards the sea surface at sunset and descending to depth at sunrise. Since the composition and density of the populations vary with the environmental conditions [71 B 1, 70 V] the scattering characteristics depend on time of the day, sea area, season, and frequency. Examples of the diurnal and seasonal variability are shown in Fig. 66. The following features of the layer structure have been observed:

– At nighttime the maximum of the scattering is in general near the surface for all frequencies. The diurnal variation of S_v is strong (15···20 dB) for depths $\lesssim$350 m and weak for depths $\approx$400···1000 m. The nighttime values of the column backscattering strength S_{vc} are only 4···5 dB higher than the daytime values [72 V]. The diurnal variability depends on the geographic area and is much smaller for the North Pacific than for North and South Atlantic [74 C]. Cold waters exhibit little day-night difference because fewer resonant scatterers making an appreciable diurnal migration.

– Day to day and night to night differences at the same position are small [72 V]; seasonal variations (Fig. 66) are strongest at 4···5 kHz [72 V] where a pronounced frequency dependence of the average S_{vc} is often found (Fig. 69). The variation between adjacent months is not smaller than for other months.

– The structure of the main scattering layer at 100···350 m is frequency-dependent [72 V]. Diurnal shift of pronounced resonance frequencies (5···8 kHz) correspond to vertical migration of neutrally buoyant constant swimbladder volume ($f_R \sim z^{1/2}$) [69 S]. There is a tendency of lower-frequency scattering layers (5 kHz) to be deeper than higher-frequency scattering layers (12 kHz) [72 V].

– Predominant bubble radii of the cold waters of the northern hemisphere range from 1···2.5 cm and are only 0.4 cm for the Antarctic [74 C]. At higher latitudes the values are in general smaller than those at lower latitudes [71 C].

– Population densities of the marine organisms computed from the column strength at resonance peaks are typically $10^{-5}\,m^{-3}$ up to $10^{-4}\,m^{-3}$, where multiscattering can be neglected [71 C].

Information on the geographical distribution of scattering layer properties is restricted to column backscattering strength spectra and the frequency range of 0.8···25 kHz, due to the predominantly applied explosive source technique. At higher frequencies, where most of the organisms produce geometrical scattering, the variability of S_v and S_{vc} with frequency is slighter or absent.

To achieve the geographical distribution of daytime S_{vc} spectra (Fig. 70) the data of six sources [78 G 1, 74 C, 73 H 1, 73 S, 74 M 1, 75 B 1] have been classified into eight groups which differ by $\approx$10 dB in one or more of the adjacent frequency band 0.8···2.5 kHz, 2.5···8 kHz, 8···25 kHz (Fig. 69). The broad class-width chosen for S_{vc} corresponds to the seasonal and year to year variability of the order of 10 dB. The following features of the geographical distribution of S_{vc} are mentioned:

– The large-scale variability is more dependent on latitude than longitude [74 C]. The strongest variation of S_{vc} coincides with oceanic boundaries (province hypothesis [71 C]) such as polar and subpolar fronts, antarctic and subtropical convergence (Fig. 67). The size of the scattering "provinces" is of the order of 10° latitude/longitude. The provinces differ in their spectra of S_{vc} (Fig. 69). The individual spectra compiled to the shadowed areas often show several pronounced resonance maxima, especially below 8 kHz, which are averaged out by the shaded regimes of Fig. 69.

– The maximum depth of the scattering layers and the separation into sublayers show a trend with latitude without symmetry of both hemispheres (Fig. 68 [73 A]).

– The groups A–H of the S_{vc} spectra of Fig. 69 show a similar behavior in one or two of three frequency intervals: the upper 8···25 kHz regime may be represented by the standard value of −50(6) dB for the classes A, B, C, E, and G or 80% of all published spectra. 17% of the data, i.e., classes D and H, are represented by −60(5) dB within the same frequency band. A similar consistency exists for the lowest frequency band of 0.8···2.5 kHz: the classes A, D, F, and G or 75% of all data are represented by −65(5) dB. The transition frequency near 4 kHz is found for the classes A, C, F, G, and H or 70% of all data.

Since the entire variability within the covered frequency range is about 30 dB and group A already represents more than 50% of the published data, part of the often mentioned feature of scattering provinces is smoothed out.

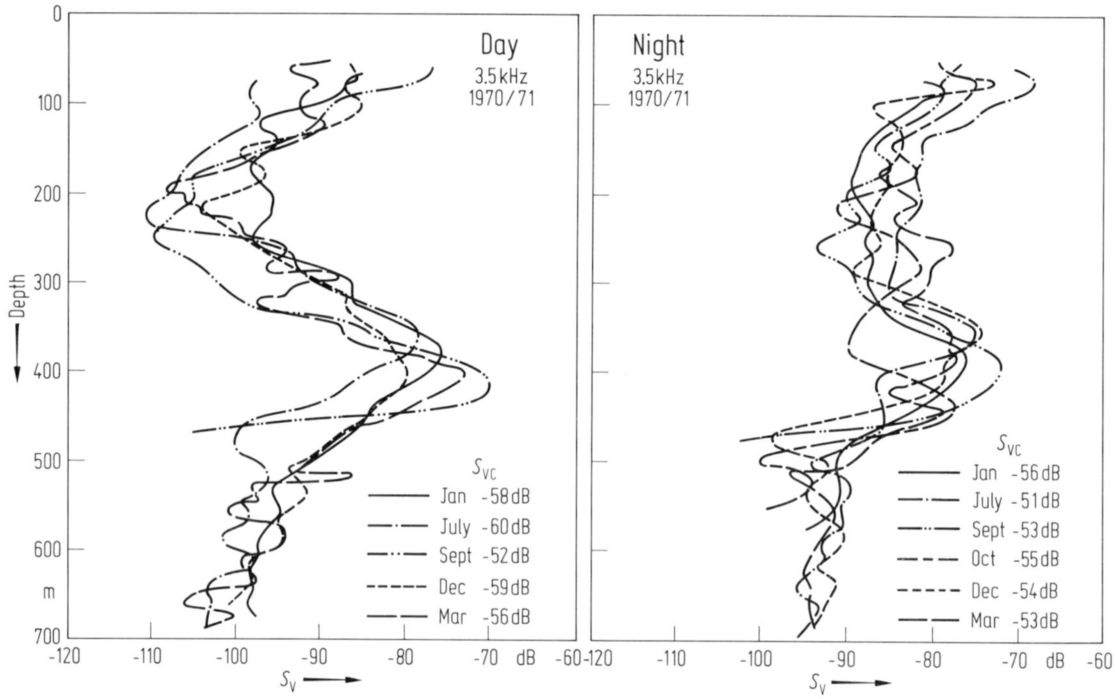

Fig. 66. Seasonal variation of volume-backscattering strength S_v vs. depth. Eastern Pacific Ocean, 3.5 kHz, daytime (left) and nighttime (right). Column backscattering strength data, S_{vc}, refer to entire profiles [72 V].

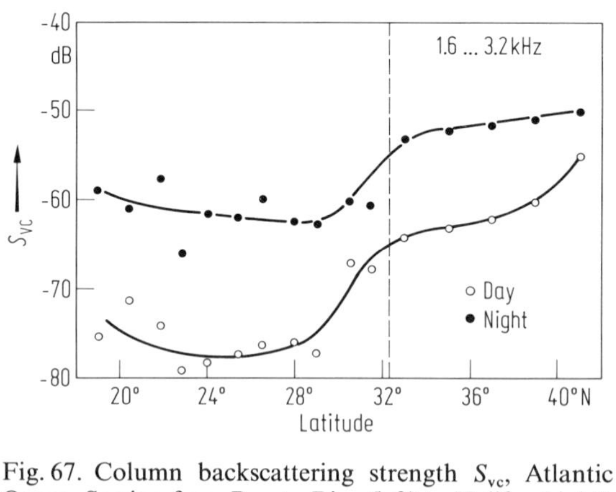

Fig. 67. Column backscattering strength S_{vc}, Atlantic Ocean. Section from Puerto Rico (left) to Halifax (right). Change in scattering strength (1.6···3.2 kHz band) near the location of subtropical convergence (dashed line: latitude of Bermuda) [71 C].

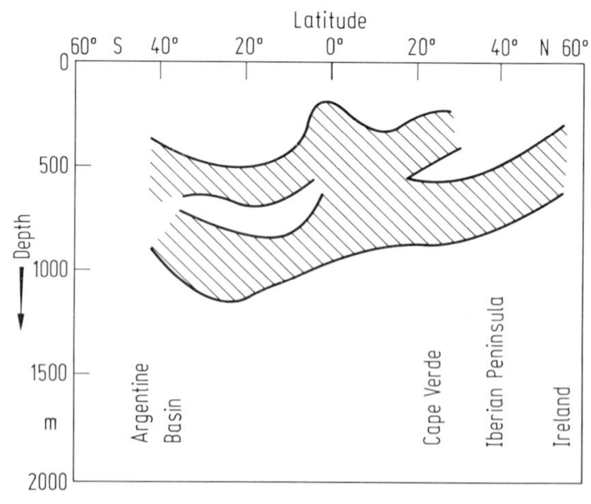

Fig. 68. Average depth of daytime scattering layers. Atlantic Ocean, section from 40° S, 40° W to 55° N, 20° W, compiled by Andreeva [73 A].

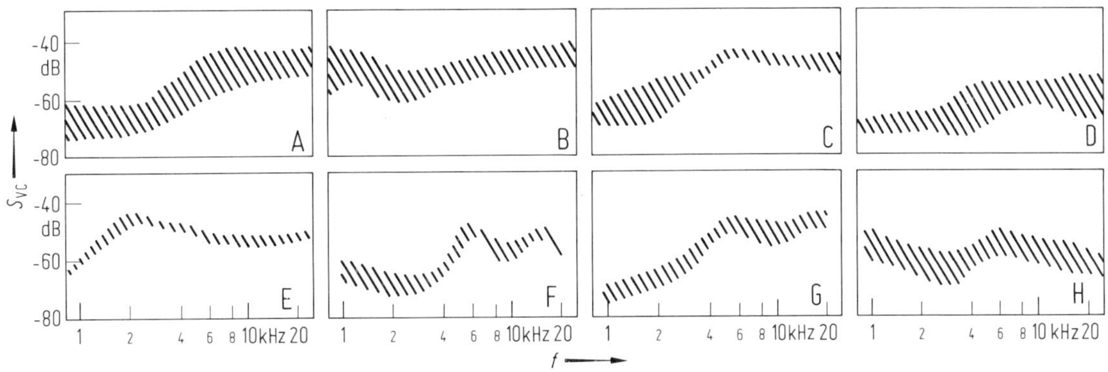

Fig. 69. Types of daytime column backscattering strength S_{vc} vs. frequency f for deep ocean areas. Compilation of explosive source measurements. The geographical occurrence is depicted in Fig. 70. Type A: [78 G 1, 74 C, 73 H 1, 75 B 1], Type B: [74 C, 73 H 1, 75 B 1], Type C: [74 C], Type D: [74 C, 73 S], Type E: [74 C], Type F: [73 S], Type G: [75 B 1], Type H: [74 M 1].

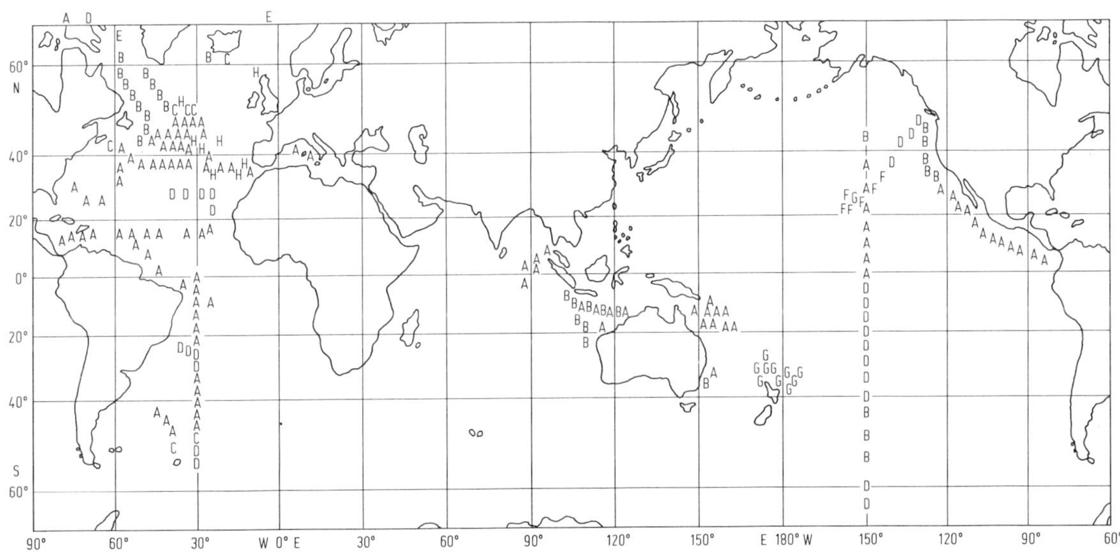

Fig. 70. Geographical distribution of daytime column backscattering strength S_{vc}. The positions indicated by A–H correspond with the types of spectra of Fig. 69.

3.2.8.2.4 Scattering by sound speed structures

The sound speed structure is mainly due to temperature fluctuations and turbulent motion of the medium itself. Kaufman [71 K] has shown that temperature fluctuations cause about 10 times more scattering than the velocity fluctuations. However, even the backscattering at the sound speed microstructure is very weak and can be discriminated against the much stronger biological scatterers only by different wave front curvatures [79 K 2]. The highly directional quasi-reflections at vertical incidence of 8 kHz signals drop by 15 dB at 2° off normal. Corresponding patch sizes of the structures were >6 m in diameter. At higher frequencies near 90 kHz microstructure reflections are totally masked by biological scattering, the scattering strength of which is ranging from –90 to –100 dB at 160 kHz [76 R 1] and may serve as tracer for the RS of turbulence.

The essential influence of sound speed structures concerns forward scattering (subsect. 3.2.6.2.7), producing interference by phase [60 C], the mean-square amplitude of the forward scattered sound being proportional to the transmission path length. The effective patch size responsible for the predominating forward scattering appears to decrease by $f^{-5/2}$, approaching 1 m at 20 kHz [79 A 1]. Spectra of the fluctuation amplitude by forward scattering at internal waves have been investigated by Desaubies [78 D].

3.2.8.2.5 Scattering by air bubble layers

Air bubble layers as generated by breaking waves scatter predominantly by bubbles at resonance, where the scattering cross section exceeds the geometrical value by some 10^2, however, accompanied with an absorption cross section about three times as large [77 C]. Backscattering at bubble layers in the open ocean up to high wind speeds has been measured by Løvik [80 L 4], utilizing vertical echo-sounding at several ultrasonic frequencies. The evaluation of the bubble population densities is essentially treated by eq. (73).

3.2.8.2.6 Scattering by sediment suspensions

Suspended sediment particles scatter within the long-wavelength regime without resonance enhancement and therefore with very low *TS*. Besides of qualitative observations [78 O] quantitative [83 H 1] attempts have been made near 200 kHz at suspensions of mine tailing. Backscattering intensity exhibits a linear relationship within concentrations of $10^{-3} \cdots 10^{-5}$ and $2 \cdots 140\,\mu m$ particle size. Measurements within the geometrical-scattering regime have been made at short range with a bottom-mounted system at 3 MHz [82 Y] to provide profiling of suspension.

3.2.8.3 Scattering by the sea surface

The roughness of the sea surface together with entrapped air bubbles beneath it acts as effective scatterer due to the largest impedance jump possible in the sea. Similar scattering strengths are obtained from ice covers. Due to the motion of the free sea surfaces the backscattered signal frequency is smeared.

3.2.8.3.1 Scattering by the free sea surface

The rough surface acts as diffraction grid where the small-scale ripples of half the effective signal wavelength produce resonant Bragg backscatter. The small-scale surface may be treated by perturbation theory [79 B] when the rms waveheight is much less than the wavelength of the incident radiation and the surface slope is small. This limitation has been overcome by composite roughness models which divide the sea surface into a large-scale and a small-scale wavenumber regime, in relation to the wavelength of the incident sound [83 M 2].

When the wavenumber cutoff of the large-scale surface is chosen at the Bragg wavenumbers, the Kirchhoff physical optics approximation holds for the large-scale portion. The composite roughness model of Bachmann [73 B 1] obtains a frequency-independent backscattering due to the particular slope of the assumed surface wave spectrum (S_s in [dB]):

$$S_s = 10 \log[2.2 \cdot \pi^3 \cdot 10^{-7} U^{7/2} \tan^4(\Theta + \Delta\Theta)]\,. \qquad (74)$$

U is the windspeed in [m/s], Θ is the grazing angle in degrees. $\Delta\Theta$ is the effective increase of Θ due to swell. At low grazing angles the large-scale tilt enhances the Bragg backscattering rather than reduce it by shadowing. The approximate frequency independence beyond several kHz and moderate to steep grazing angles has been observed elsewhere [80 A 3].

However, at lower frequencies and grazing angles $\Theta < 40°$, measurements often exhibit a decrease of S_s by decreasing frequency. The classical data of Chapman and Harris [62 C] obtained by explosive sources still serve as a standard for backscattering at the sea surface for $10° < \Theta < 45°$ and frequencies of several kHz and less (S_s in [dB]):

$$S_s = 3.3\beta \log\frac{\Theta}{30} - 42.4 \log\beta + 2.6\,, \qquad \beta = 158(Uf^{1/3})^{-0.58}\,. \tag{75}$$

The empirical equation contains the frequency f in [Hz], the wind speed U in [m/s] and Θ as grazing angle in degrees. Eq. (75) is depicted in Fig. 71.

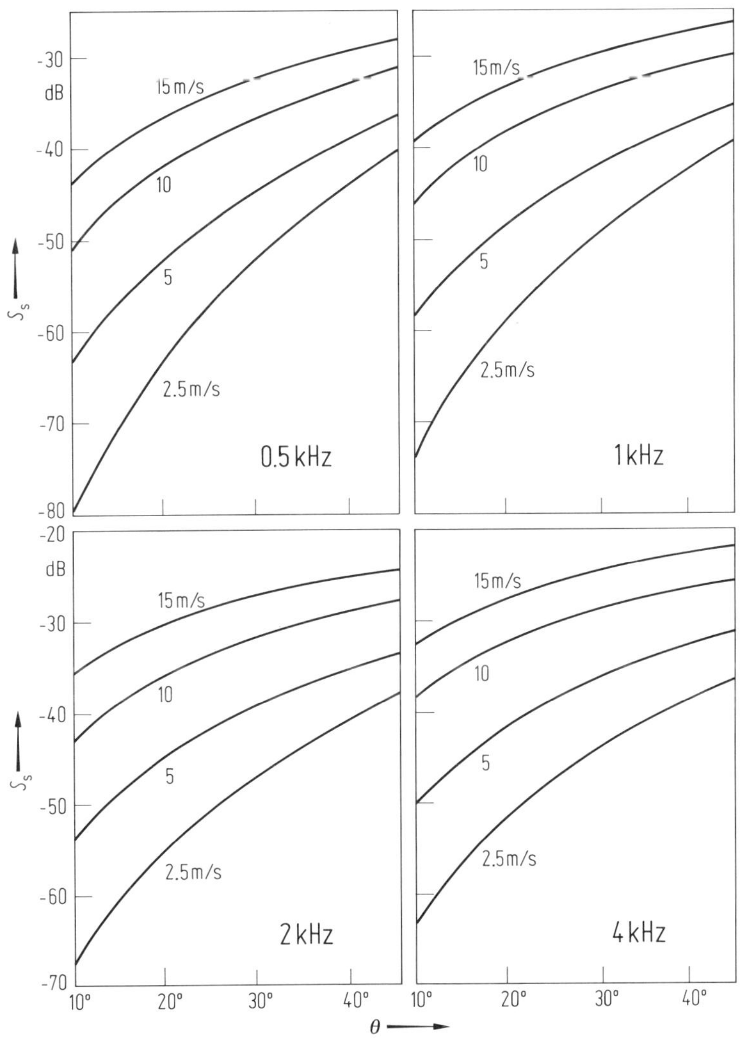

Fig. 71. Backscattering strength S_s of the sea surface at low frequencies, computed by eq. (75). Grazing angle Θ. Parameter: wind speed [62 C].

At higher frequencies S_s becomes nearly independent of grazing angle for $\Theta < 30°$ and moderate to high windspeeds which has been measured at 60 kHz by Urick already in 1956 [56 U]. This feature has been qualitatively attributed to bubble layers beneath the surface. The simplified curves of Fig. 72 are still used as a guide-line for high-frequency sea surface backscattering. The data cover steep angles up to vertical incidence and show the typical crossover behavior at $\Theta \approx 75°$.

The bubble layer assumption has been supported by McDaniel and Gorman [82 M 2] who suggested a modified composite roughness model covered by a sublayer of resonance bubbles which accounts for both the enhanced scattering strength at low grazing angles and the observed saturation of S_s at about 15 m/s wind speed (Fig. 73). Due to the adopted wavenumber spectrum of the sea waves, their model also yields an increase of S_s by frequency at the unscreened surface.

The model applies for $\Theta \lesssim 60°$ where Bragg diffraction predominates. The parametrization of the developed sea wave spectrum by the wind speed requires standard wind profiles and the reduction to a reference height. The prediction accuracy of S_s nevertheless appears hardly better than 5 dB.

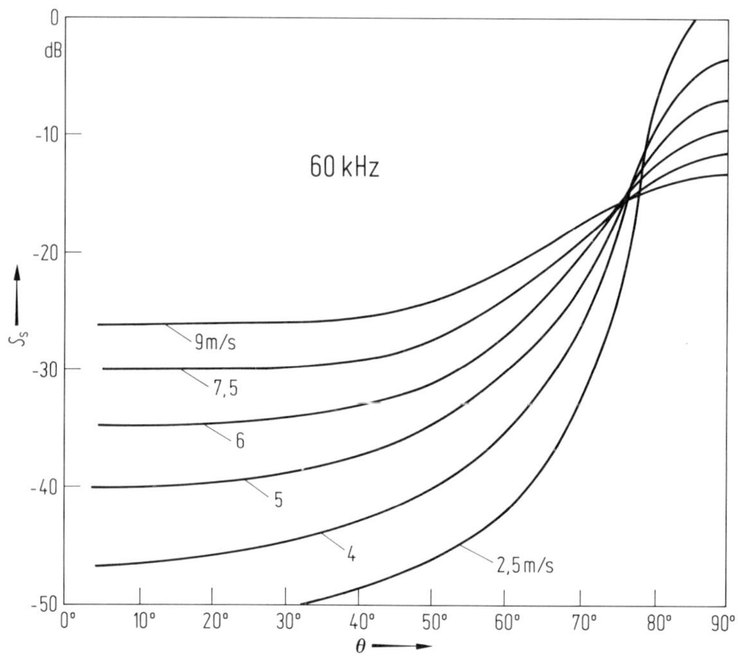

Fig. 72. Backscattering strength S_s of the sea surface at high frequencies (60 kHz) [56 U]. Grazing angle Θ. Parameter: wind speed.

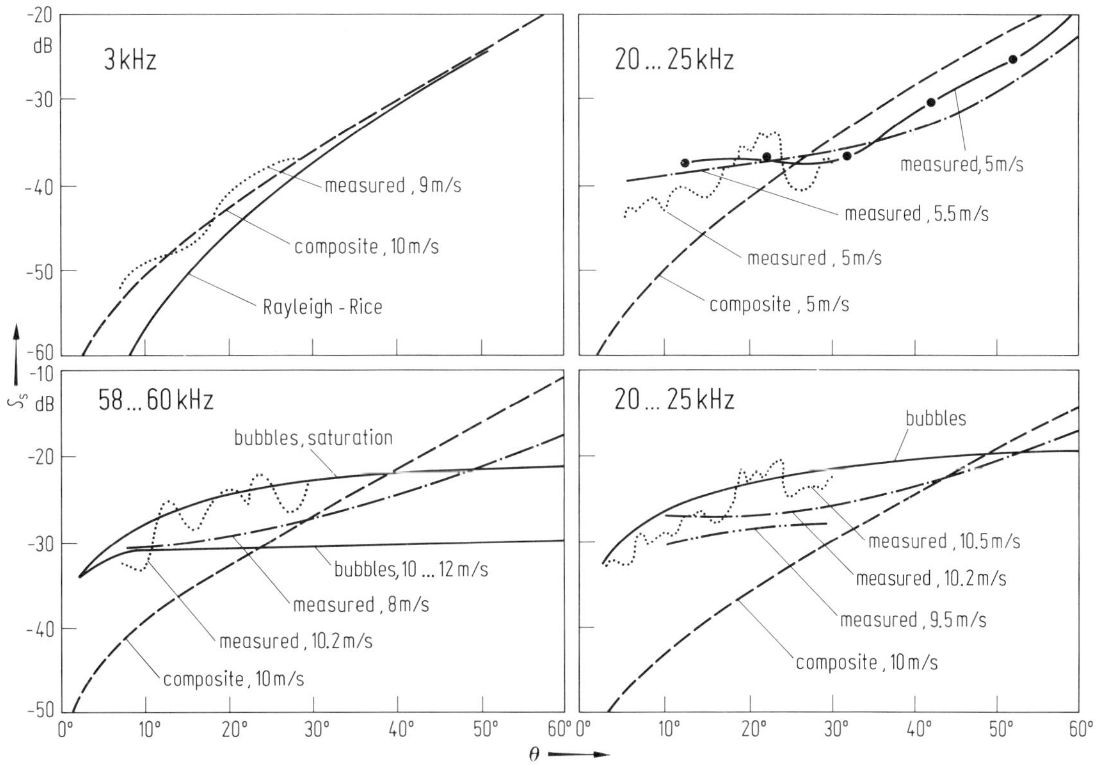

Fig. 73. Backscattering strength S_s of the sea surface; combination of Bragg scattering and bubble-layer scattering [82 M 2]. Grazing angle Θ. Parameter: wind speed. Top left: pure Bragg scattering due to transparent bubble layer. The measurements agree with the composite roughness model which enhances the backscattering strength at low grazing angles against the untilted rough surface (strong line). Bottom right: The measurements of different origin agree with the bubble scattering model (strong line) rather than with the composite model, due to higher frequencies. Bottom left: At high frequencies (60 kHz) the measured data already approach the saturation of bubble layer scattering. Top right: The bubble scattering effect is already apparent at 5 m/s wind speed and sufficiently high frequencies. The strong dot-line is measured by narrow angle setup [84 K].

3.2.8.3.2 Frequency distribution of scattered sound

The surface motion shifts and broadenes the spectrum of the scattered sound. Both frequency shift and spread limit the suppression of unwanted reverberation in the frequency domain.

Frequency shifts in surface reverberation measured by Igarashi and Stern [71 I] correspond to the phase velocity of the surface wave components fullfilling the Bragg condition of first-order resonance theory. The frequency shift is dependent on angle between sound and seawave propagation but is nearly independent of wind speed. The data of Fig. 74 are valid for a fixed echo sounding setup and have to be corrected when the carrier ship is moving.

The spectral broadening of the backscatter is explained by orbital motions of the ultragravity ripples riding on the larger-scale ocean waves and is nearly independent of wave and wind direction. The half-power spectral width normalized by the carrier frequency of the signal scattered at a bubble-covered sea surface has been found nearly proportional to the wind speed (0.002 at 10 m/s) and only weakly dependent on frequency beyond 20 kHz and $10° < \Theta < 40°$ [83 M 3].

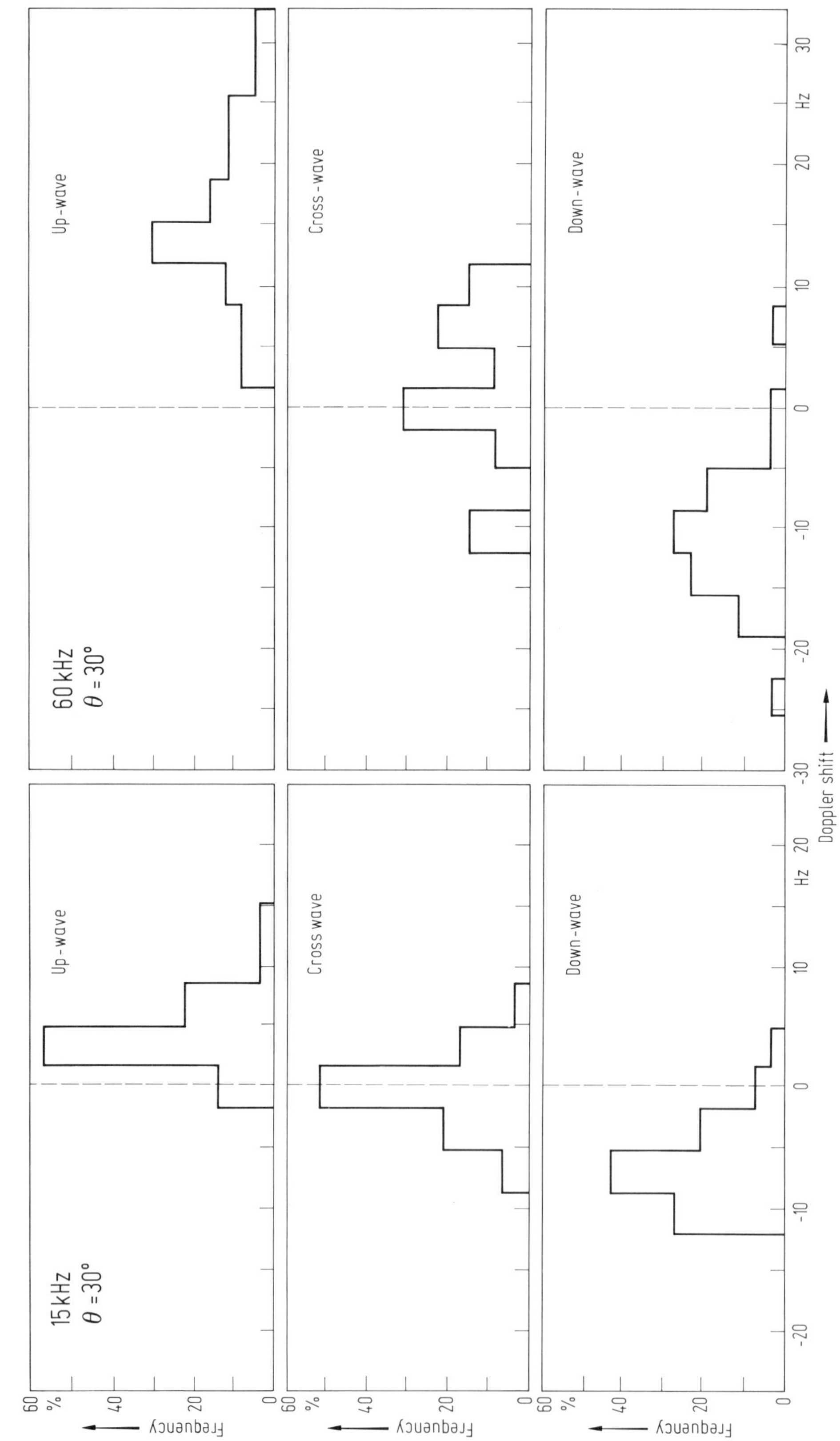

Fig. 74. Histograms of surface-backscattering Doppler shifts for 30° grazing angle. Left: 15 kHz, right: 60 kHz [71 I].

3.2.8.3.3 Scattering by ice covers

The reverberation strongly depends on the under-surface features of the ice cover and appears to be stronger during spring when pressure-packed one-year ice and broken frozen leads dominate [64 M 3]. The summer polar ice produces weaker scattering probably due to the smoother melting undersurface [64 B 1]. Fig. 75 exhibits a pronounced increase of the scattering strength by grazing angle and frequency, though nearly frequency-independent cases may occur as well [63 M 2].

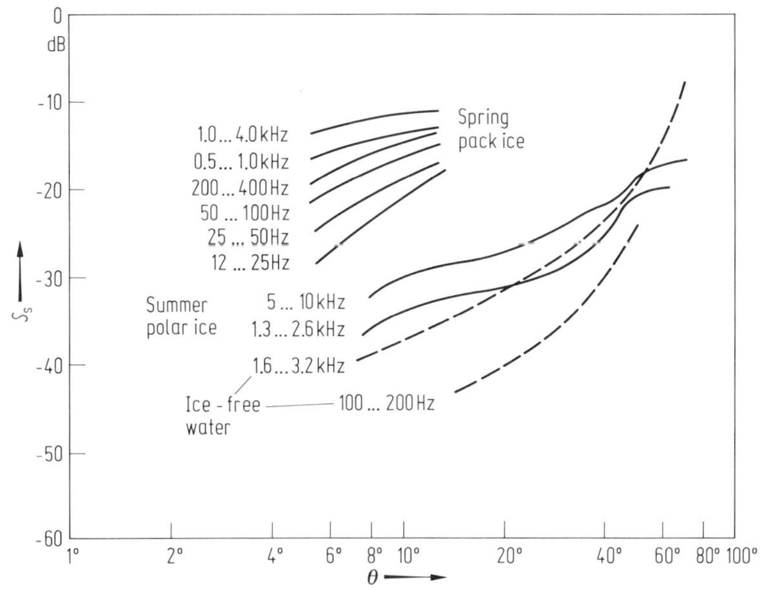

Fig. 75. Backscattering strength of an ice cover. Spring data [64 M 3]; summer data [64 B 1], compiled by [83 U].

3.2.8.4 Reflection and scattering by the sea bottom

The sea bottom interface is treated as a lossy reflector for vertical echosounding when the rms-roughness within the illuminated area is small against the signal wavelength λ. Then the entire geometrical spreading to and fro the bottom is $20\log 2r$, where r is the one-way distance, in [m]. At very rough bottoms or at off-specular angles the bottom has to be treated as a lossy backscatterer whereby the two-way geometrical spreading is $40\log r$, r in [m].

3.2.8.4.1 Reflection at the sea bottom

The reflection loss $10\log\frac{I_R}{I_i}$ of a plane wave incident at grazing angle Θ on a plane interface of a homogeneous unconsolidated sediment of refraction index $n=c_w/c_s$ and density ratio $m=\varrho_s/\varrho_w$ (s and w indicate sediment and water, respectively) is given by the Rayleigh formula [60 B]

$$\frac{I_R}{I_i}=\left[\frac{m\sin\Theta-(n^2-\cos^2\Theta)^{1/2}}{m\sin\Theta+(n^2+\cos^2\Theta)^{1/2}}\right]^2 \tag{76}$$

which appears a realistic approximation close to and at vertical incidence at echo-sounder frequencies and small penetration depths. High-speed bottoms of sand-type, $n<1$, $n<m$ (subsect. 3.2.6.3.5), are indicated by a critical angle at $\cos\Theta_0=n$; low-speed bottoms of mud-type, $1<n<m$, exhibit an intromission angle when the numerator of eq. (76) vanishes. Both types are smeared or complicated at lossy [60 M 1] and stratified bottoms, respectively.

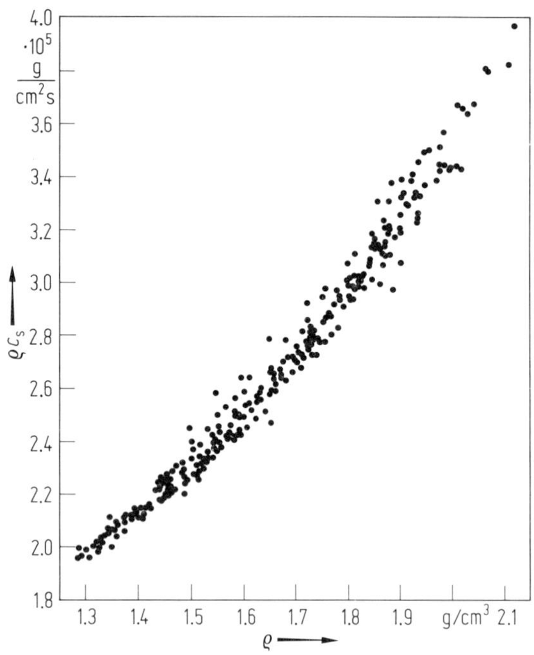

Fig. 76. Impedance ϱc_s vs. saturated bulk density ϱ of silicate sediments, continental terrace (shelf and slope) [82 H 1]. c_s: sediment sound speed.

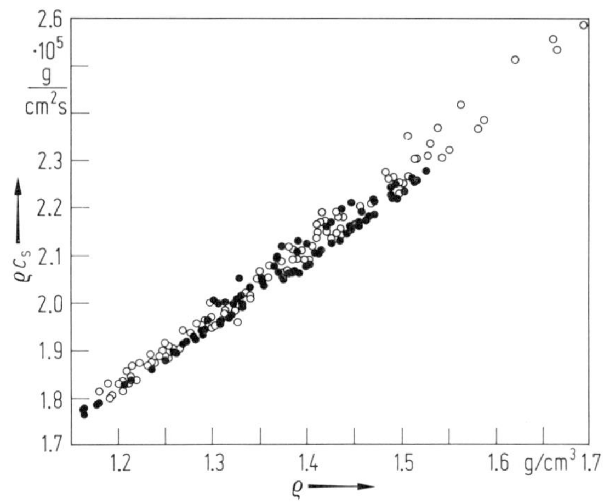

Fig. 77. Impedance ϱc_s vs. saturated bulk density ϱ of silicate sediments, abyssal hill (solid circles) and abyssal plane (open circles) environments [82 H 1]. c_s: sediment sound speed.

Bottom models [80 H 1] usually need more input quantities than available in practice. Estimates of regional reflection loss values at or near vertical incidence may be derived by eq. (76) using Figs. 76 and 77 and the following table [82 H 1], presenting empirical regression equations for sediments of the three general environments: continental terrace (shelf and slope) indicated by (T), abyssal hills (pelagic) indicated by (H) and abyssal plain (turbidite) indicated by (P).

The equations combine the sediment porosity n [%], saturated bulk density ϱ [g/cm³], grain size $\Phi = -\log_2$ (grain diameter, in [mm]), clay size C [%] and the compressional wave sound speed c_s [m/s], corrected for 296 K and 10^5 Pa, using the tables of sea water sound speed (subsect. 3.2.5.3). The standard errors σ apply near the mean values of the validity range.

n vs. Φ:

(T)	$n = 22.01 + 9.24\Phi - 0.365\Phi^2$	$\sigma = 6.5$
(H)	$n = 82.42 - 0.28\Phi$	$\sigma = 4.7$
(P)	$n = 50.0 + 3.34\Phi$	$\sigma = 6.2$

ϱ vs. Φ:

(T)	$\varrho = 2.374 - 0.175\Phi + 0.008\Phi^2$	$\sigma = 0.11$
(H)	$\varrho = 1.327 + 0.005\Phi$	$\sigma = 0.09$
(P)	$\varrho = 1.869 - 0.057\Phi$	$\sigma = 0.11$

c_s vs. Φ:

(T)	$c_s = 1952.5 - 86.26\Phi + 4.14\Phi^2$	$\sigma = 29$
(H)	$c_s = 1594.3 - 10.2\ \Phi$	$\sigma = 12$
(P)	$c_s = 1609.7 - 10.8\ \Phi$	$\sigma = 11$

c_s vs. n:

(T)	$c_s = 2502.0 - 23.45n + 0.14n^2$	$\sigma = 31$
(H)	$c_s = 1410.6 + 1.177n$	$\sigma = 13$
(P)	$c_s = 1564.6 - 0.597n$	$\sigma = 13$

c_s vs. ϱ:

(T)	$c_s = 2330.4 - 1257.0\varrho + 487.7\varrho^2$	$\sigma = 33$
(H)	$c_s = 1591.5 - 63.4\varrho$	$\sigma = 13$
(P)	$c_s = 1476.7 + 29.7\varrho$	$\sigma = 13$

c_s vs. C:

(H)	$c_s = 1549.4 - 0.66C$	$\sigma = 10$
(P)	$c_s = 1554.7 - 0.65C$	$\sigma = 11$

n vs. ϱ:

(T)	$n = 157.6 - 57.8\varrho$	$\sigma = 2.1$
(H)	$n = 150.1 - 51.2\varrho$	$\sigma = 1.1$
(P)	$n = 155.9 - 56.0\varrho$	$\sigma = 0.8$

The limiting values in the regression equations are

Mean grain size Φ:

(T) $1 \cdots 9\Phi$
(H) and (P) $7 \cdots 10\Phi$

Porosity n:

(T) $35 \cdots 85\%$
(H) and (P) $70 \cdots 90\%$

Density ϱ:

(T) $1.25 \cdots 2.10\ \mathrm{g/cm^3}$
(H) $1.15 \cdots 1.50\ \mathrm{g/cm^3}$
(P) $1.15 \cdots 1.70\ \mathrm{g/cm^3}$

Clay size grains C:

(H) and (P) $20 \cdots 85\%$

The equations cover calcareous contents up to 30%. The acoustic properties of higher $CaCO_3$-concentrations are given in [82 H 2]. Gas-bearing sediments may produce a high reflectivity [83 M 4]. Further recent literature on the subject is given in [74 H, 80 K 3].

3.2.8.4.2 Scattering at the sea bottom

The sea bottom acts as scatterer rather by the interface relief features than by sediment particle size or inner inhomogeneities [64 M 2]. The data to follow are restricted to the bottom backscattering strength S_{sb}, defined by eq. (60), to characterize the bottom for oblique echosounding and the bottom reverberation as unwanted background for *RS* of interesting targets.

Depending on the slope of the roughness spectrum [64 M 1], S_{sb} may be often nearly independent of frequency [68 W] at very rough bottoms or may increase by more than a power 3/2 of frequency in case of smooth bottoms [85 B]. The dependence of S_{sb} on the grazing angle Θ of many rough bottom types is close to Lambert's relationship, where the scattered intensity $I_s \sim \sin\Theta$. Since the intensity of the acoustical illumination is $\sim I_i \sin\Theta$, the Lambert relation for a unit area reads $\frac{I_s}{I_i} = \mu \sin^2\Theta$, μ being a proportionality constant. Therefore [61 M]:

$$S_{sb} = 10 \log \frac{I_s}{I_i} = 10 \log \mu + 10 \log \sin^2\Theta . \qquad (77)$$

Fig. 78 represents average values of the quantity $10 \log \mu$, measured at various bottoms, grazing angles and frequencies. The standard deviation is typically ± 5 dB but may be larger when narrow-beam sounding devices are utilized [84 K]. The Lambert correction has been made up to grazing angles of 70° [61 M, 71 S, 68 B]: Fig. 78 exhibits the approximate frequency independence of the rougher bottoms [71 S] as well as frequency, dependent backscattering at smoother sediment interfaces [85 B, 73 H 2]. At heavily dissected bottoms (mid-atlantic ridge) Lambert's rule applies to nearly vertical incidence, giving a value of −18 dB for $10 \log \mu$ [65 J].

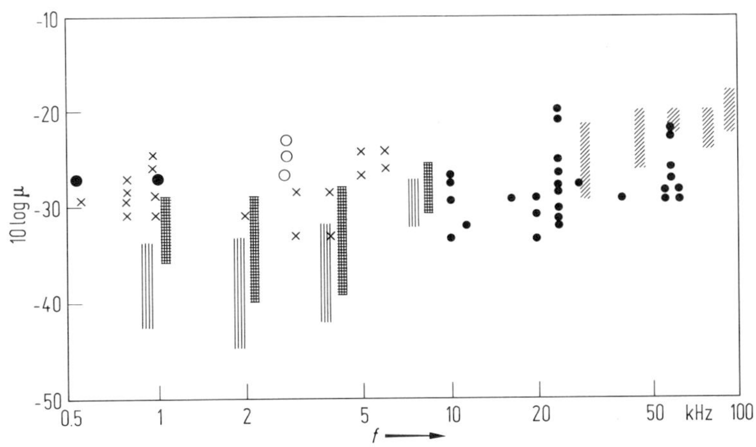

Fig. 78. The quantity $10 \log \mu$ of Lambert's relationship $S_{sb} = 10 \log \mu + 10 \log \sin^2\Theta$. Dots: smooth bottom [61 M], circles: deep ocean [68 B], crosses: North Atlantic and North Pacific Ocean, $\Theta < 50°$ [71 S]. Hatched bars: fine sand, California coast, $2° < \Theta < 10°$ [85 B], light shaded bars: Indian Ocean, $30° < \Theta < 70°$ [73 H 2], dark shaded bars: Tasman sea, $30° < \Theta < 70°$ [73 H 2].

Typical examples of the weak angular dependence of S_{sb} at high frequencies are given in Fig. 79; at low frequencies S_{sb} exhibits a steep increase towards normal incidence (Fig. 80), thus behaving as specular reflection rather than scattering.

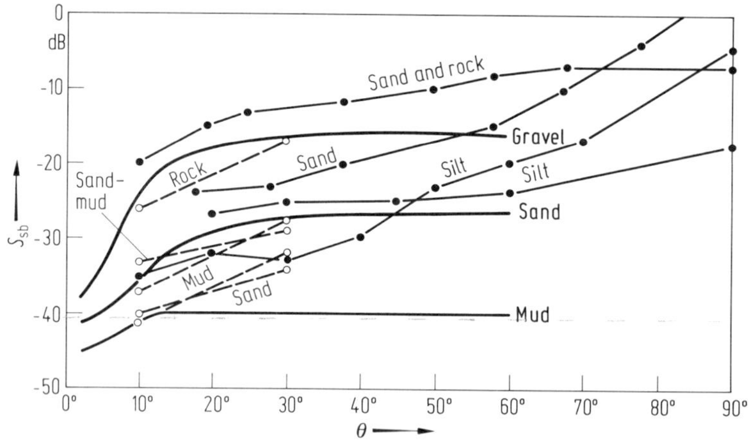

Fig. 79. Backscattering strength S_{sb} of the sea bed of coastal locations at high frequencies. Grazing angle Θ. Open circles: 24 kHz, solid circles: 55 and 60 kHz, solid line: 100 kHz. Compiled by [83 U].

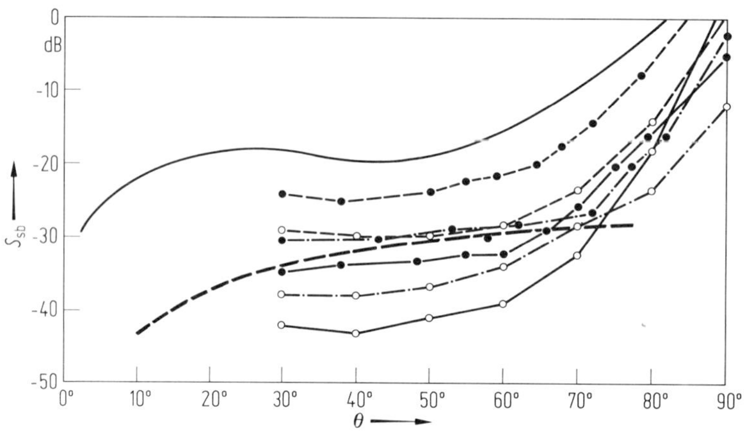

Fig. 80. Backscattering strength S_{sb} of deep water bottoms at low frequencies. Solid line: deep ocean bottom, 2.5 kHz [63 P]. Black dots: continental slope near Bermuda, 0.5···1 kHz, (top) 2.3 km depth, (mid) 2.1 km depth, and (bottom) 2.8 km depth [62 U]. Light dots: deep water Atlantic, 1···3 kHz, (top) 2.8 km mud, (mid) 0.6 km coral, and (bottom) 3.8 km, mud [64 B 2]. Dashed line: abyssal red clay (representing 50% Pacific, 25% Atlantic) 0.5···1 kHz, 4 km depth [61 M], compiled by [83 U].

3.2.8.4.3 Scattering at sediments with gaseous enclosures

Entrapped gas may strongly enhance the scattering strength of high-porosity sediments such as mud [83 T] by 10···20 dB, probably acting as a scattering volume rather than an interface.

3.2.8.4.4 Scattering by morphological features (bedforms)

Features of the sediment relief which are considered the main reason for the variability of the low- and mid-frequency scattering strength rather than particle size may be resolved by very high frequency RS (side scan SONAR, subsect. 3.2.3.1.3) and classified into bedform types. Since they indicate depositional and erosional processes by unidirectional and sea-wave-induced oscillating currents, the acoustic image of the bedforms may serve to label the oceanographic origin. The sonographic bedform records of Fig. 81 are restricted to current-induced structures of shelf areas [86 W 1].

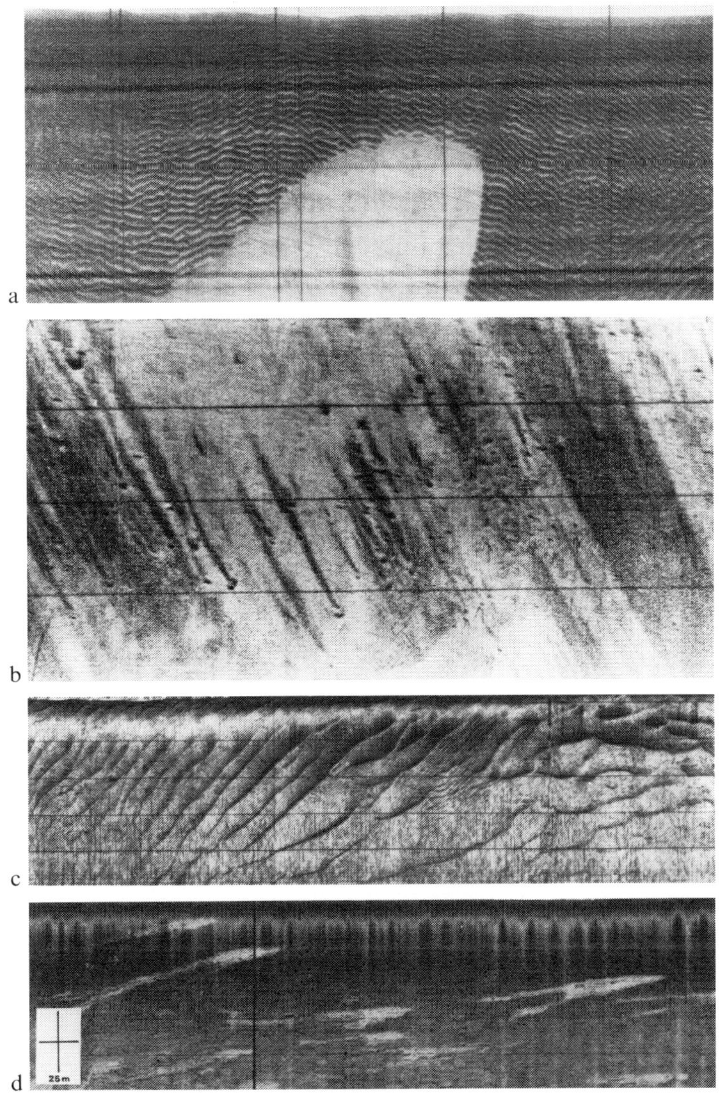

Fig. 81. For caption, see next page.

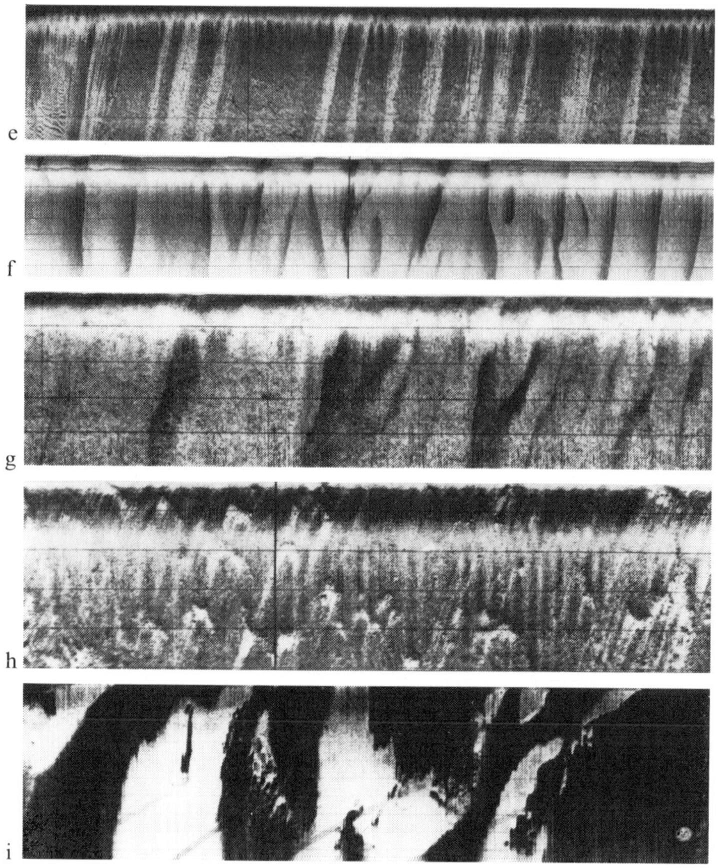

Fig. 81. Side-scan SONAR records of sediment bedforms [86 W 1]. Signal frequency: 100 kHz, line scales: 15 m. (a) Oscillation ripples, due to seawaves (primary bedform), (b) comet marks, storm-induced (secondary bedform), (c) dunes, current-induced (primary bedform), (d) Barkhane (primary bedform), (e) sand ribbons (primary bedform), (f) erosional furrows, asymmetric, storm-induced (primary bedform), (g) erosional furrows, storm-induced (primary bedform), (h) obstacle marks, seawave-current-induced (secondary bedform), (i) patchy sediment, current-induced (secondary bedform).

3.2.9 References for 3.2

01 K Knudsen, M.: Hydrographical Tables (G.E.C. Gad, Copenhagen **1901** and 1948).
39 K Kuwahara, S.: Hydrographic Rev. **16** (1939) 123.
47 E Ebaugh, P.E., Meuser, R.: J. Acoust. Soc. Am. **19** (1947) 695.
48 C Cole, R.H.: Underwater explosions. Princeton, N.J.: Princeton University Press **1948**.
48 K Knudsen, V.O., Alford, R.S., Emling, J.W.: J. Mar. Res. **7** (1948) 410.
48 W Woodward, P.M., Lawson, J.: J. Inst. Elec. Engrs., London **95** (1948) 363.
49 L Leonard, R.W., Combs, P.C., Skidmore, L.R., jr.: J. Acoust. Soc. Am. **21** (1949) 63.
50 L Longuet-Higgins, M.S.: Philos. Trans. R. Soc. London, Ser. A **243** (1950) 1.
50 S Sheehy, M.J.: J. Acoust. Soc. Am. **22** (1950) 24.
51 W Weissler, A., Del Grosso, V.A.: J. Acoust. Soc. Am. **23** (1951) 219.
52 D Del Grosso, V.A.: Naval Res. Lab. Rep. No. 4002, **1952**.
52 E Esche, R.: Akust. Beih., Acustica **4** (1952) AB 208.
52 M Mellen, R.H.: J. Acoust. Soc. Am. **24** (1952) 478.
53 K Kurtzc, G., Tamm, K.: Acustica **3** (1953) 33.
53 L Liddiard, G.E.: Short pulse cavitation thresholds for projectors in open water, U.S. Navy Electron. Lab. Rep. 376, **1953**.
54 H Hersey, J.B., Backus, R.H.: Deep-Sea Res. **1** (1954) 190.
54 P Peterson, W.W., Birdsall, T.G., Fox, W.C.: Trans. IRE (Inst. Radio Eng.) PGIT-**4** (1954) 171.
54 W Wilson, O.B., Leonard, R.W.: J. Acoust. Soc. Am. **26** (1954) 223.
55 L Litovitz, T.A., Carnvale, E.H.: J. Appl. Phys. **26** (1955) 816.
56 S Snodgrass, J.M., Richards, A.F.: Trans. Am. Geophys. Union **37** (1956) 97.
56 U Urick, R.J., Hoover, R.M.: J. Acoust. Soc. Am. **28** (1956) 1038.
57 G Greenspan, M., Tschiegg, C.T.: J. Res. Natl. Bur. Stand. **59** (1957) 249.
57 G Greenspan, M., Tschiegg, C.T.: Rev. Sci. Instrum. **28** (1957) 897.
57 H Horton, J.W.: Fundamentals of SONAR, U.S. Naval Inst. Annapolis **1957**.
57 W Whitmarsh, D.C., Skudrzyk, E., Urick, R.J.: J. Acoust. Soc. Am. **29** (1957) 1124.
58 F Fisher, F.H.: J. Acoust. Soc. Am. **30** (1958) 442.
58 J Jones, F.R.H., Pearce, G.: J. Exp. Biol. **35** (1958) 437.
58 O Officer, C.B.: Introduction to the theory of sound transmission, New York: McGraw Hill **1958**.
59 F Franz, G.J.: J. Acoust. Soc. Am. **31** (1959) 1080.
59 S Shepard, F.P.: The earth beneath the sea, Baltimore: John Hopkins Press **1959**.
60 B Brekhovskikh, L.M.: Waves in layered media, New York: Academic Press **1960**.
60 C Chernov, L.A.: Wave propagation in a random medium, New York: McGraw Hill **1960**.
60 L Lee, Y.W.: Statistical theory of communication, New York: Wiley & Sons 1960.
60 M 1 Mackenzie, K.V.: J. Acoust. Soc. Am. **32** (1960) 221.
60 M 2 Middleton, D.: Introduction to statistical communication theory, chaps. 18–23, New York: McGraw Hill **1960**.
60 W 1 Weston, D.E.: Proc. Phys. Soc. (London) **76** (1960) 233.
60 W 2 Wilson, W.D.: J. Acoust. Soc. Am. **32** (1960) 641.
61 H Horton, C.W., Innis, G.S.: J. Acoust. Soc. Am. **33** (1961) 877.
61 M Mackenzie, K.V.: J. Acoust. Soc. Am. **33** (1961) 1498.
61 K Tamm, K.: Schallabsorption und -dispersion in wäßrigen Elektrolytlösungen, in: Handbuch der Physik, XI/1, Berlin, Göttingen, Heidelberg: Springer **1961**, p. 202–274.
62 A Aleksandrov, I.A.: Sov. Phys. Acoust. **8** (1962) 23.
62 C Chapman, R.P., Harris, J.H.: J. Acoust. Soc. Am. **34** (1962) 1592.
62 E 1 Eigen, M., Tamm, K.: Z. Elektrochem. **66** (1962) 93.
62 E 2 Ezrow, D.H.: J. Acoust. Soc. Am. **34** (1962) 550.
62 M 1 Macpherson, J.D.: J. Acoust. Soc. Am. **34** (1962) 1149.
62 M 2 Marsh, H.W., Schulkin, M.: J. Acoust. Soc. Am. **34** (1962) 863.
62 S Schulkin, M., Marsh, H.W.: J. Acoust. Soc. Am. **34** (1962) 864.
62 U Urick, R.J., Saling, D.S.: J. Acoust. Soc. Am. **34** (1962) 1721.
62 W Wenz, G.M.: J. Acoust. Soc. Am. **34** (1962) 1936.
63 B 1 Barham, E.G.: Science **140** (1963) 826.
63 B 2 Batzler, W.E., Barham, E.G.: J. Acoust. Soc. Am. **35** (1963) 792.
63 C Cushing, D.H., et al.: J. Br. Inst. Radio Eng. **25** (1963) 299.

63 M 1 Marsh, H.W.: J. Acoust. Soc. Am. **35** (1963) 409.
63 M 2 Mellen, R.H., Marsh, H.W.: J. Acoust. Soc. Am. **35** (1963) 1645.
63 N Northrop, J., Raitt, R.W.: Bull. Seismol. Soc. Am. **53** (1963) 577.
63 P Patterson, R.B.: J. Acoust. Soc. Am. **35** (1963) 2010.
63 S 1 Schofield, D.: Transducers, underwater acoustics (Albers, V.M., ed.), New York: Plenum Press **1963**.
63 S 2 Schulkin, M., Marsh, H.W.: J. Br. Inst. Radio Eng. **25** (1963) 493.
63 U Urick, R.J.: J. Acoust. Soc. Am. **35** (1963) 1413.
63 V VonWinkle, W.A.: New London, Connecticut: U.S. Navy Underwater Sound Lab. Rep. 600, **1963**.
63 W Welsby, V.G., Dunn, J.R.: J. Br. Inst. Radio Eng. **26** (1963) 205.
64 A Andreeva, I.B.: Sov. Phys. Acoust. **10** (1964) 17.
64 B 1 Brown, J.R.: J. Acoust. Soc. Am. **36** (1964) 601.
64 B 2 Burstein, A.W., Keane, J.J.: J. Acoust. Soc. Am. **36** (1964) 1596.
64 F Flynn, H.G., in: Physical Acoustics, Vol. 1B, chap. 9 (Mason, W.P., ed.), New York: Academic Press **1964**.
64 G Greene, C.R., Buck, B.M.: J. Acoust. Soc. Am. **36** (1964) 1218.
64 M 1 Marsh, H.W.: Reflection and scattering of sound by the sea bottom, paper presented at the 68th meeting of the Acoustical Society of America, New London, Connecticut: Avco Marine Electronics Office **1964**.
64 M 2 McKinney, C.M., Anderson, C.D.: J. Acoust. Soc. Am. **36** (1964) 158.
64 M 3 Milne, A.R.: J. Acoust. Soc. Am. **36** (1964) 1551.
64 M 4 Milne, A.R., Ganton, J.H.: J. Acoust. Soc. Am. **36** (1964) 855.
64 P Piggott, C.L.: J. Acoust. Soc. Am. **36** (1964) 2152.
64 T Tavolga, W.N.: Marine Bio-Acoustics, Oxford, New York: Pergamon Press **1964**.
65 A Axelrod, E.H., Schoomer, B.A., VonWinkle, W.A.: J. Acoust. Soc. Am. **37** (1965) 77.
65 J Jitkovskiy, Y.Y., Volovova, L.A.: Sound scattering from the ocean bottom, paper E 67, Proc. Fifth Int. Acoust. Congr. Liége, Belg. **1965**.
65 K Kinsman, B.: Wind waves, Englewood Cliffs, N.J.: Prentice-Hall **1965**.
65 L Lord, N.W., Ransone, M.A., Brey, M.E.: J. Acoust. Soc. Am. **38** (1965) 313.
65 N Northrop, J., Johnson, R.J.: J. Geophys. Res. **70** (1965) 311.
65 T Thorp, W.H.: J. Acoust. Soc. Am. **38** (1965) 648.
65 U U.S. Navy: An interim report on the sound velocity distribution in the North Atlantic Ocean, U.S. Navy Oceanogr. Office Technical Report 171, **1965**.
66 B Brekhovskikh, L.M.: Izv. Akad. Nauk SSSR Fiz. Atmos. Okeana **2** (1966) 582.
66 C Chapman, R.P., Marshall, J.R.: J. Acoust. Soc. Am. **40** (1966) 405.
66 D Deavenport, R.L.: Radio Sci. **1** (1966) 405.
66 F Furduyev, A.V.: Izv. Akad. Nauk SSSR Fiz. Atmos. Okeana **2** (1966) 523.
66 M 1 Mechel, F.: Die Streuung ebener Wellen an Zylindern und Kugeln komplexer Impedanz, Göttingen: Vandenhoeck & Ruprecht **1966**.
66 M 2 Mechel, F., Wille, P.: Acustica **16** (1965/66) 305.
66 T Tucker, D.G., Gazey, B.K.: Applied underwater acoustics, Oxford, New York: Pergamon Press **1966**.
67 B Berktay, H.O.: Underwater Acoustics, Vol. 2, chap. 12 (Albers, V.M., ed.), New York: Plenum Press **1967**.
67 T Thorp, W.H.: J. Acoust. Soc. Am. **42** (1967) 270.
67 W 1 Weston, D.E., in: Underwater Acoustics, Vol. 2 (Albers, V.M., ed.), New York: Plenum Press **1967**, pp. 55–88.
67 W 2 Weston, D.E.: On shallow water sound fluctuations, with speculations upon an April evening, NATO Advanced Study Institute on Problems in Sound Propagation, Sept. **1967**.
68 B Buckley, J.P., Urick, R.J.: J. Acoust. Soc. Am. **44** (1968) 648.
68 C Carnvale, A., Bowen, P., Basileo, M., Sprenke, J.: J. Acoust. Soc. Am. **44** (1968) 1098.
68 F Frisch, U., in: Probabilistic methods in applied mathematics, Vol. 1 (Bharucha-Reid, A.T., ed.), New York: Academic Press **1968**, p. 75–198.
68 K Kuo, E.Y.T.: J. Acoust. Soc. Am. **43** (1968) 1017.
68 S Spiess, F.N., Northrop, J., Werner, E.W.: J. Acoust. Soc. Am. **43** (1968) 640.
68 W Wong, H.K., Chesterman, W.D.: J. Acoust. Soc. Am. **44** (1968) 1713.
69 B 1 Bom, N.: J. Acoust. Soc. Am. **45** (1969) 150.
69 B 2 Brekhovskikh, L.M., Goncharov, V.V.: Izv. Akad. Nauk SSSR Fiz. Atmos. Okeana **5** (1969) 374.
69 C Chindonova, Yu.G., Kashkin, N.I.: Oceanology **9** (1969) 430.
69 L 1 Leroy, C.C.: J. Acoust. Soc. Am. **46** (1969) 216.

69 L 2 Love, R.H.: J. Acoust. Soc. Am. **46** (1969) 746.
69 L 3 Lovett, J.R.: J. Acoust. Soc. Am. **45** (1969) 1051.
69 P Podeszwa, E.M.: Tables of shallow water transmission loss based on Colossus II equations, NUSL Report No. 1071, New London, Connecticut: Navy Underwater Sound Laboratory, Dec. **1969**, CT 06320.
69 S Scrimger, J.A., Turner, R.G.: J. Acoust. Soc. Am. **46** (1969) 771.
70 A Ahrens, E.: Geräte für Ultraschall-Fischortung, in: Interocean, Bd. 2, Düsseldorf: VDI-Verlag **1970**, 70.
70 C Cohen, J.S., Einstein, L.J.: Continuous gradient ray tracing system (CONGRATS) II, U.S. Naval Underwater System Center (NUSC), Report No. 1069, **1970**.
70 D 1 DeFerrari, H.A.: J. Acoust. Soc. Am. **48** (1970) 128.
70 D 2 Del Grosso, V.A.: J. Acoust. Soc. Am. **47** (1970) 947.
70 F Farquhar, G.B. (ed.): Proceedings of an international symposium on biological sound scattering in the ocean, Maury Center for Ocean Science, Department of the Navy, Washington, D.C.: Government Printing Office **1970**, 0851–0053.
70 U U.S. Navy: Ocean science program of the U.S. Navy, Alexandria, Virginia: Office of the Oceanographer of the Navy **1970**.
70 V Von Winkle, W.A., Hasse, R.W.: Ocean acre, U.S. Naval Underwater System Center (NUSC), Publication No. 1107, **1970**.
71 B 1 Backus, R.H., et al., in: Proceedings of an international symposium on biological sound scattering in the ocean (Farquhar, G.B., ed.), Maury Center Report 005, Warrenton, Virginia, 1970; Washington, D.C.: U.S. Government Printing Office **1971**.
71 B 2 Batzler, W.E., Pickwell, G.V., in: Proceedings of an international symposium on biological sound scattering in the ocean, pp. 168–179 (Farquhar, G.B., ed.), Maury Center Report 005, Warrenton, Virginia, 1970; Washington, D.C.: U.S. Government Printing Office **1971**.
71 B 3 Bucker, H.P.: Some comments on ray theory with examples from current NUC ray trace models, SACLANTCEN Conf. Proc. No. 5, Part I, **1971**, pp. 32–36.
71 C Chapman, R.P., Bluy, O.Z., Adlington, R.H., in: Proceedings of an international symposium on biological sound scattering in the ocean, pp. 306–317 (Farquhar, G.B., ed.), Maury Center Report 005, Warrenton, Virginia, 1970; Washington, D.C.: U.S. Government Printing Office **1971**.
71 D DiNapoli, F.R.: A fast field program for multy-layered media, U.S. Naval Underwater System Center (NUSC) Report **1971**.
71 F Farquhar, G.B. (ed.): Proceedings of an international symposium on biological sound scattering in the ocean, Maury Center Report 005, Warrenton, Virginia, 1970; Washington, D.C.: U.S. Government Printing Office **1971**.
71 I Igarashi, Y., Stern, R.: J. Acoust. Soc. Am. **49** (1971) 802.
71 K Kaufman, C.: J. Acoust. Soc. Am. **49** (1971) 930.
71 M 1 Mackenzie, K.V.: J. Acoust. Soc. Am. **50** (1971) 1321.
71 S Schmidt, P.B.: J. Acoust. Soc. Am. **50** (1971) 326.
71 U Urick, R.J.: J. Acoust. Soc. Am. **50** (1971) 337.
71 W 1 Walsh, G.M.: Electron. Progr. **13** (1971) 17.
71 W 2 Weinberg, H.: J. Acoust. Soc. Am. **50** (1971) 975.
71 W 3 Wille, P., Thiele, R.: J. Acoust. Soc. Am. **50** (1971) 348.
72 D 1 Del Grosso, V.A., Mader, C.W.: J. Acoust. Soc. Am. **52** (1972) 961.
72 D 2 Del Grosso, V.A., Mader, C.W.: J. Acoust. Soc. Am. **52** (1972) 1442.
72 D 3 Drenkelfort, H., in: Meerestechnologie, München: Thiemig Verlag **1972**, p. 296.
72 G Gibbs, R.H., et al.: Mediterranean biological studies, final report, Washington, D.C.: Smithsonian Inst., July **1972**.
72 L Ludwig, R.: Akustisches Landesystem für Großschiffe, in: Deutsche Gesellschaft für Ortung und Navigation, Beitrag moderner Schallsysteme zur Schiffssicherheit, Vortrag Nr. 10, Hamburg: 11 October **1972**.
72 S Steinberg, J.C., Clark, J.G., DeFerrari, H.A., Kronengold, M., Yacoub, K.: J. Acoust. Soc. Am. **52** (1972) 1521.
72 T Thiele, R.: Anwendung der Theorie zeitveränderlicher Filter zur Beschreibung des Flachwasser-Schallkanals, Dissertation der Technischen Universität Hannover **1972**.
72 U Urick, R.J.: J. Acoust. Soc. Am. **52** (1972) 993.
72 V Vent, R.J.: J. Acoust. Soc. Am. **52** (1972) 373.
72 W Wenz, G.M.: J. Acoust. Soc. Am. **51** (1972) 1010.

73 A Andreeva, I.B.: Oceanology (USSR) **13** (1973) 645; Phys. Abstr. 55466 (15 Aug. 1975).
73 B 1 Bachmann, W.: J. Acoust. Soc. Am. **54** (1973) 712.
73 B 2 Bezdek, H.F.: J. Acoust. Soc. Am. **53** (1973) 782.
73 D 1 Del Grosso, V.A.: J. Acoust. Soc. Am. **53** (1973) 1384.
73 D 2 Dyer, I.: J. Acoust. Soc. Am. **53** (1973) 564.
73 F Fahrentholz, S.R.: Interocean, 73–657, **1973**, pp. 885–894.
73 H 1 Hall, M.: J. Acoust. Soc. Am. **54** (1973) 473.
73 H 2 Hall, M.: J. Acoust. Soc. Am. **54** (1973) 730.
73 K Kroebel, W.: Meteor Forschungsergeb. Reihe A **12** (1973) 53.
73 L Love, R.H.: J. Acoust. Soc. Am. **54** (1973) 1312.
73 M Mackenzie, K.V.: J. Acoust. Soc. Am. **54** (1973) 302.
73 S Scrimger, J.A., Turner, R.G.: J. Acoust. Soc. Am. **54** (1973) 483.
73 W Wille, P., Thiele, R., Schunk, E.: J. Acoust. Soc. Am. **54** (1973) 1708.
73 Y Yeager, E., Fisher, F.H., Miceli, J., Bressel, R.: J. Acoust. Soc. Am. **53** (1973) 1705.
74 B 1 Browning, D.G.: Proc. Intern. Workshop on Low-Frequency Propagation and Noise, Vol. 2, Woods Hole **1974**, 769–799.
74 B 2 Browning, D.G., et al.: Project KIWI ONE: Low-frequency sound attenuation measurements in the South Pacific Ocean, U.S. Naval Underwater System Center (NUSC), Technical Report No. 4949, June **1974**.
74 C Chapman, R.P., Bluy, O.Z., Adlington, R.H., Robison, A.E.: J. Acoust. Soc. Am. **56** (1974) 1722.
74 D 1 Del Grosso, V.A.: J. Acoust. Soc. Am. **56** (1974) 1084.
74 D 2 Diachok, O.I., Winokur, R.S.: J. Acoust. Soc. Am. **55** (1974) 750.
74 H Hampton, L. (ed.): Physics of Sound in Marine Sediments, New York: Plenum Press **1974**.
74 I Ingenito, F., in: Sound propagation in shallow water, SACLANTCEN Conf. Proc. No. 14, La Spezia, Italy: SACLANT ASW Research Center **1974**, pp. 69–78.
74 K Kibblewhite, A.C., in: Physics of Sound in Marine Sediments (Hampton, L., ed.), New York: Plenum Press **1974**, p. 131–160.
74 M 1 McElroy, P.T.: J. Acoust. Soc. Am. **56** (1974) 394.
74 M 2 Mellen, R.H., Browning, D.G., Ross, J.M.: J. Acoust. Soc. Am. **56** (1974) 80.
74 M 3 Muir, T.G., White, R.L., Clynch, J.R., in: Sound propagation in shallow water, SACLANTCEN Conf. Proc. No. 14, La Spezia, Italy: SACLANT ASW Research Center **1974**, pp. 185–200.
74 N 1 Northrop, J.: J. Acoust. Soc. Am. **56** (1974) 837.
74 N 2 Northrop, J., Colburn, J.G.: J. Geophys. Res. **79** (1974) 5633.
74 S 1 Samovolkin, V.G.: Oceanology (USSR) **14** (1974) 655; Phys. Abstr. 33309 (15 Apr. 1976).
74 S 2 Schellstede, G., Wille, P., in: Sound propagation in shallow water, SACLANTCEN Conf. Proc. No. 14, La Spezia, Italy: SACLANT ASW Research Center **1974**, pp. 52–68.
74 S 3 Sluyterman van Langeweyde, W.: Klassifizierung von Schallgeschwindigkeitsprofilen nach akustischen Kriterien, Dissertation der Technischen Universität Hannover **1974**.
74 S 4 Spindel, R.C., Davis, S.B., Macdonald, K.C., Porter, R.P., Phillips, J.D.: Nature, **248** (1974) 577.
74 U Urick, R.J.: J. Acoust. Soc. Am. **56** (1974) 1010.
75 B 1 Batzler, W.E.: J. Acoust. Soc. Am. **58** (1975) 51.
75 B 2 Busch, C.W., Harre, J., Weiß, A.: HANSA **11** (1975) 927.
75 F Fisher, F.H., Simmons, V.P.: IEEE Ocean **75** (1975) 21.
75 G Garrett, C., Munk, W.H.: J. Geophys. Res. **80** (1975) 291.
75 K Kanabis, W.G.: A shallow water acoustic model for an ocean stratified in range and depth, Vol. I, U.S. Naval Underwater System Center (NUSC), Technical Report No. 4887-I **1975**.
75 L 1 Love, R.H.: J. Acoust. Soc. Am. **57** (1975) 300.
75 L 2 Luetkemeyer, M., Diehl, R.: Holographic processing of near-field SONAR data, in: Acoustical Holography, Vol. 6, New York and London: Plenum Press **1975**, pp. 401–418.
75 M 1 Medwin, H.: J. Acoust. Soc. Am. **58** (1975) 1318.
75 M 2 Mellen, R.H., Browning, D.G., Ross, J.M., Merklinger, H.M.: J. Acoust. Soc. Am. **57** (1975) 1201.
75 P Porter, R.P., Leslie, H.D.: J. Acoust. Soc. Am. **58** (1975) 812.
75 S Sherman, C.H.: IEEE Trans. Sonics Ultrason., Sept. **1975**.
75 U 1 Urick, R.J.: Principles of underwater sound, 2nd edition, New York: McGraw-Hill **1975**.
75 U 2 Urick, R.J.: A statistical model for the fluctuation of sound transmission in the sea, White Oak: Naval Surface Weapons Center, Technical Report 75–18 **1975**.
76 C 1 Cato, D.H.: J. Acoust. Soc. Am. **60** (1976) 320.
76 C 2 Chen-Tung Chen, Millero, F.J.: J. Acoust. Soc. Am. **60** (1976) 1270.

76 D Dietrich, R., Kögler, F.C.: Die Anwendung elektroakustischer Verfahren bei geologischen Untersuchungen des oberflächennahen Untergrunds in Schelfgebieten, Kiel: Geologisch-Paläontologisches Institut, Bericht 1976-2 **1976**.
76 K 1 Kanabis, W.G.: A shallow water acoustic model for an ocean stratified in range and depth, Vol. II, U.S. Naval Underwater System Center (NUSC), Technical Report No. 4887-II **1976**.
76 K 2 Kibblewhite, A.C., Jones, D.A.: J. Acoust. Soc. Am. **59** (1976) 790.
76 K 3 Kroebel, W., Mahrt, K.-H.: Acustica **35** (1976) 154.
76 M 1 McGrath, J.R.: J. Acoust. Soc. Am. **60** (1976) 1290.
76 M 2 Mellen, R.H., Browning, D.G., Goodman, L.: J. Acoust. Soc. Am. **60** (1976) 1053.
76 M 3 Murphy, E., Olesen, O.V.: J. Acoust. Soc. Am. **59** (1976) 305.
76 R 1 Rasmussen, R.A., Head, N.E.: J. Acoust. Soc. Am. **59** (1976) 55.
76 R 2 Ross, D.: Mechanics of underwater noise, New York: Pergamon Press **1976**.
76 S Schneider, H.G.: Acustica **35** (1976) 18.
76 W Williams, A.J., Tochko, J.S.: An acoustic sensor of velocity for benthic boundary layer studies. Woods Hole Oceanographic Institution Contribution No. 3843 from bottom turbulence, New York: Elsevier Press **1976**.
77 A 1 Andersen, N.R., Zahuranec, B.J. (eds.): Oceanic sound scattering prediction, New York: Plenum Press **1977**.
77 A 2 Andrews, J.E., Craig, J.D., Hardy, W.A.: Deep-Sea Res. **24** (1977) 975.
77 B 1 Bannister, R.W., Denham, R.N., Guthrie, K.M., Browning, D.G.: J. Acoust. Soc. Am. **62** (1977) 847.
77 B 2 Brooks, A.L., in: Oceanic sound scattering prediction, chap. 3 (Andersen, N.R., Zahuranec, B.J., eds.), New York: Plenum Press **1977**, pp. 565–590.
77 C Clay, C.S., Medwin, H.: Acoustical Oceanography, New York: John Wiley & Sons **1977**.
77 D DiNapoli, F.R.: The inverse fast field program (IFFP): An application to the determination of the acoustic parameters of the ocean bottom, New London, Connecticut: U.S. Naval Underwater Systems Center (NUSC), Technical Memo 771160 **1977**.
77 F Fisher, F.H., Simmons, V.P.: J. Acoust. Soc. Am. **62** (1977) 558.
77 H Heffler, D.E., Locke, D.R.: Mar. Geophys. Res. **3** (1977) 229.
77 I Ishimaru, A.: Proc. IEEE **65** (1977) 1030.
77 K Keller, J.B., Papadakis, J.S.: Wave propagation and underwater acoustics, Berlin, Heidelberg, New York: Springer **1977**.
77 L Love, R.H.: J. Acoust. Soc. Am. **62** (1977) 1397.
77 M Mellen, R.H., Browning, D.G.: J. Acoust. Soc. Am. **61** (1977) 704.
77 S Schneider, H.G.: J. Acoust. Soc. Am. **62** (1977) 871.
77 T 1 Tacconi, G. (ed.): Aspects of signal processing, NATO Advanced Study Institute, Ser. C, Vol. 33 **1977**.
77 T 2 Tappert, F.D., in: Wave propagation and underwater acoustics (Keller, J.B., Papadakis, J.S., eds.), Lecture Notes in Physics, Vol. 70, Berlin, Heidelberg, New York: Springer **1977**.
77 W Worcester, P.F.: J. Acoust. Soc. Am. **62** (1977) 895.
78 D Desaubies, Y.J.F.: J. Acoust. Soc. Am. **64** (1978) 1460.
78 G 1 Gold, B.A., Renshaw, W.E.: J. Acoust. Soc. Am. **63** (1978) 1809.
78 G 2 Gorshkov, S.G.: World Ocean Atlas, Atlantic and Indian Oceans, Vol. 2, New York: Pergamon Press **1978**, pp. 234–235.
78 I Ingenito, F., Ferris, R.H., Kuperman, W.A., Wolf, S.: Shallow water acoustics, summary report (first phase), Naval Research Laboratories, Report 8179 **1978**.
78 K Kibblewhite, A.C., Browning, D.G.: Deep-Sea Res. **25** (1978) 1107.
78 L 1 Leibiger, G.A.: The acoustic propagation model RAYMODE: Theory and numerical treatment, New London, Connecticut: U.S. Naval Underwater Systems Center (NUSC), Technical Report **1978**.
78 L 2 Lovett, J.R.: J. Acoust. Soc. Am. **63** (1978) 1713.
78 M 1 Mellen, R.H., Browning, D.G.: J. Acoust. Soc. Am. **63** (1978) 1624.
78 M 2 Morris, G.B.: J. Acoust. Soc. Am. **64** (1978) 581.
78 O Orr, M.H., Hess, F.R.: J. Geophys. Res. **83** C (1978) 4062.
78 S Schulkin, M., Marsh, H.W.: J. Acoust. Soc. Am. **63** (1978) 43.
79 A 1 Anderson, V.C.: J. Acoust. Soc. Am. **66** (1979) 1446.
79 A 2 Andrews, R.S.: J. Acoust. Soc. Am. **65** (1979) 672.
79 B Bass, F.G., Fuks, J.M.: Wave scattering from statistically rough surfaces, New York: Pergamon Press **1979**.
79 D 1 DeSanto, J.A.: Theoretical methods in ocean acoustics, in: Topics in current physics: Ocean acustics (DeSanto, J.A., ed.), Berlin, Heidelberg, New York: Springer **1979**, pp. 7–77.

79 D 2 DiNapoli, F.R., Deavenport, R.L.: Numerical models of underwater acoustic propagation, in: Topics in current physics: Ocean acoustics (DeSanto, J.A., ed.), Berlin, Heidelberg, New York: Springer **1979**, pp. 79–157.

79 F 1 Fisher, F.H.: J. Acoust. Soc. Am. **65** (1979) 1327.

79 F 2 Flatté, S.M. (ed.): Sound transmission through a fluctuating ocean, Cambridge, New York: Cambridge University Press **1979**.

79 H Hamilton, E.L.: J. Acoust. Soc. Am. **65** (1979) 909.

79 K 1 Kalinowski, A.J.: The Shock and Vibration Digest **11** (1979) 12.

79 K 2 Kaye, G.T., Anderson, V.C.: J. Acoust. Soc. Am. **66** (1979) 842.

79 M 1 Mellen, R.H., Simmons, V.P., Browning, D.G.: J. Acoust. Soc. Am. **65** (1979) 923.

79 M 2 Munk, W., Wunsch, C.: Deep-Sea Res. **26** A (1979) 123.

79 N Neudorfer, M.L.: Offshore Technol. Conf., Dallas, Tx.: 11th Ann. **1979**.

79 P Penrose, J.D., Kaye, G.T.: J. Acoust. Soc. Am. **65** (1979) 374.

79 R Roderick, W.I.: Doppler spectra of bistatic reverberation from the sea surface, New London, Connecticut: U.S. Naval Underwater Systems Center (NUSC), Technical Report No. 6031, **1979**.

79 W Wilson, H.L., Tappert, F.D.: J. Acoust. Soc. Am. **66** (1979) 256.

80 A 1 Akal, T., in: Bottom-interaction ocean acoustics, NATO-Conference Series, Ser. IV: Marine Sciences, Vol. 5, New York and London: Plenum Press **1980**, pp. 557–575.

80 A 2 Anderson, A.L., Hampton, L.D.: J. Acoust. Soc. Am. **67** (1980) 1865.

80 A 3 Andreeva, I.B., Volkova, A.V., Galybin, N.N.: Sov. Phys. Acoust. **26** (1980) 265.

80 H 1 Hamilton, E.L.: J. Acoust. Soc. Am. **68** (1980) 1313.

80 H 2 Hastrup, O.F., Akal, T.: A simple formula to calculate shallow water transmission loss by means of a least-squares surface fit technique, SACLANTCEN SM-139, La Spezia, Italy: SACLANT ASW Research Center, Sept. **1980**.

80 H 3 Hodgkiss, W.S., Anderson, V.C.: J. Acoust. Soc. Am. **67** (1980) 214.

80 J Jensen, F.B., Kuperman, W.A.: J. Acoust. Soc. Am. **67** (1980) 1564.

80 K 1 Kibblewhite, A.C., Hampton, L.D.: J. Acoust. Soc. Am. **67** (1980) 147.

80 K 2 Kung Huang, Clay, C.S.: J. Acoust. Soc. Am. **67** (1980) 795.

80 K 3 Kuperman, W.A., Jensen, F.B. (eds.): Bottom-interacting ocean acoustics, New York: Plenum Press **1980**.

80 L 1 Labianca, F.M.: J. Acoust. Soc. Am. **67** (1980) 1567.

80 L 2 Lauterborn, W. (ed.): Cavitation and inhomogeneities in underwater acoustics, Berlin, Heidelberg, New York: Springer **1980**.

80 L 3 Lovett, J.R.: J. Acoust. Soc. Am. **67** (1980) 338.

80 L 4 Løvik, A., in: Cavitation and inhomogeneities in underwater acoustics (Lauterborn, W., ed.), Springer Series in Electrophysics, Vol. 4, Berlin, Heidelberg, New York: Springer **1980**, p. 211–218.

80 M 1 Medwin, H., in: Cavitation and inhomogeneities in underwater acoustics (Lauterborn, W., ed.), Springer Series in Electrophysics, Vol. 4, Berlin, Heidelberg, New York: Springer **1980**, p. 187–193.

80 M 2 Mellen, R.H., Browning, D.G.: J. Acoust. Soc. Am. **68** (1980) 248.

80 M 3 Mellen, R.H., Simmons, V.P., Browning, D.G.: J. Acoust. Soc. Am. **67** (1980) 341.

80 S Sellschopp, J.: OSTSEE 80. Forschungsfahrt mit WFS PLANET vom 31. August bis 12. September 1980, Fahrtleiterbericht 1980-3 **1980**.

80 T Thiele, R., Schellstede, G.: Forschungsanstalt der Bundeswehr für Wasserschall- und Geophysik (FWG) **1980**, unpublished.

81 C Coppens, A.B.: J. Acoust. Soc. Am. **69** (1981) 862.

81 E Evans, D.L., Watts, D.R.: Wind speed and stress at the sea surface from ambient noise measurements, in: Proceedings of the International Symposium on Acoustic Remote Sensing of the Atmosphere and Oceans, Calgary, Alberta: University Calgary **1981**.

81 K Kuperman, W.A., Jensen, F.B.: Deterministic propagation modelling I: Fundamental principles. NATO Advanced Study Institute on Underwater Acoustics and Signal Processing, Dordrecht, Boston, London: Reidel Publ. Co. **1981**, 125–132.

81 M 1 Mackenzie, K.V.: J. Acoust. Soc. Am. **70** (1981) 801.

81 M 2 Mackenzie, K.V.: J. Acoust. Soc. Am. **70** (1981) 807.

81 M 3 Mellen, R.H., Browning, D.G., Simmons, V.P.: J. Acoust. Soc. Am. **70** (1981) 143.

81 M 4 Mellen, R.H., Browning, D.G., Simmons, V.P.: J. Acoust. Soc. Am. **69** (1981) 1660.

81 M 5 McDaniel, S.T.: J. Acoust. Soc. Am. **70** (1981) 535.

81 N Nichols, R.H.: J. Acoust. Soc. Am. **69** (1981) 974.

81 S 1 Schneider, H.G., in: NATO Advanced Study Institute on Underwater Acoustics and Signal Processing, Dordrecht, Boston, London: Reidel Publ. Co. **1981**, 143–150.

81 S 2 Sellschopp, J.: BORNHOLM 78. Zusammenstellung der Messungen mit dem Thermistor-Schleppkabel, Forschungsanstalt der Bundeswehr für Wasserschall- und Geophysik (FWG), Bericht 1981-19 **1981**.

82 B Behringer, D., Birdsall, T., Brown, M., Cornuelle, B., Heinmiller, R., Knox, R., Metzger, K., Munk, W., Spiesberger, J., Spindel, R., Webb, D., Worcester, P., Wunsch, C.: Nature, London **299** (1981) 121.

82 F 1 Francois, R.E., Garrison, G.R.: J. Acoust. Soc. Am. **72** (1982) 896.

82 F 2 Francois, R.E., Garrison, G.R.: J. Acoust. Soc. Am. **72** (1982) 1879.

82 H 1 Hamilton, E.L., Bachman, R.T.: J. Acoust. Soc. Am. **72** (1982) 1891.

82 H 2 Hamilton, E.L., Bachmann, R.T., Berger, W.H., Johnson, T.C., Mayer, L.A.: J. Sediment. Petrol. **52** (1982) 733.

82 M 1 Mahrt, K.-H.: Personal communication **1982**.

82 M 2 McDaniel, S.T., Gorman, A.D.: J. Geophys. Res. **87**, No. C6 (1982) 4127.

82 M 3 McDaniel, S.T.: J. Acoust. Soc. Am. **72** (1982) 916.

82 N Nelkin, A., Gilmour, G.A.: Development of high resolution side-look sonars, presented at the Acoustical Society of America meeting, Orlando, Florida, 9–12 Nov. **1982**.

82 S 1 Sellschopp, J.: FLEX 76. Zusammenstellung der Messungen mit dem Thermistor-Schleppkabel, Forschungsanstalt der Bundeswehr für Wasserschall- und Geophysik (FWG), Bericht 1982-15 **1982**.

82 S 2 Sellschopp, J.: OSTSEE 82. Forschungsfahrt mit WFS PLANET vom 4. August bis 26. August 1982, Fahrtleiterbericht 1982-5 **1982**.

82 T Thorpe, S.A., Stubbs, A.R., Hall, A.J.: Nature **296** (1982) 636.

82 U Urick, R.J.: Sound propagation in the sea, Peninsula **1982**.

82 Y Young, R.A., Merrill, J.T., Clarke, T.L., Proni, J.R.: Geophys. Res. Letters **9** (1982) 175.

83 D DeFerrari, H.A., Davis, R.I., Nguygen, H., Tusting, R.F., Williams, N.J.: J. Acoust. Soc. Am. **74** (1983) 1448.

83 G Gritsch, H.: Verfahren zur Unterwasserortung mit Schallimpulsen, München: Deutsches Patentamt, Offenlegungsschrift DE 3221013 **1983**.

83 H 1 Hay, A.E.: J. Geophys. Res. **88**, No. C 12 (1983) 7525.

83 H 2 Hsu, C.C., Fisher, F.H.: J. Acoust. Soc. Am. **74** (1983) 564.

83 K Kögler, F.C., Ullrich, J.: Morphologie und tektonische Strukturen im Bereich der Clarion fracture zone des "Sonne seamount", des "Loichi-seamount" und des "Puma-ridge", in: Bundesminister für Forschung und Technologie (BMFT)-Forschungsbericht, Bonn-Bad Godesberg: M 83-001 **1983**.

83 L Lee, D., McDaniel, S.T.: J. Acoust. Soc. Am. **73** (1983) 1441.

83 M 1 McDaniel, S.T., Beebe, J.H., in: Acoustics and the sea bed (Pace, N.G., ed.), Bath, UK: Bath University Press **1983**, pp. 63–70.

83 M 2 McDaniel, S.T., Gorman, A.D.: J. Acoust. Soc. Am. **73** (1983) 1476.

83 M 3 McDaniel, S.T., Gorman, A.D.: J. Acoust. Soc. Am. **74** (1983) 241.

83 M 4 Meißner, R., Theilen, F.: Attenuation of seismic waves in sediments, 11th World Petrolium Congress, London: John Wiley & Sons **1983**.

83 P Pace, N.G. (ed.): Acoustics and the sea bed, Bath, UK: Bath University Press **1983**.

83 S Spiesberger, J.L., Birdsall, T.G., Metzger, K., Knox, R.A., Spofford, C.W., Spindel, R.C.: J. Phys. Oceanogr. **13** (1983) 1836.

83 T Thiele, R., in: Acoustics and the sea bed (Pace, N.G., ed.), Bath, UK: Bath University Press **1983**, pp. 207–213.

83 U Urick, R.J.: Principles of underwater sound, 3rd edition, New York: McGraw-Hill **1983**.

84 B Baggeroer, A.B.: J. Ocean Eng., special issue **1984**, 229.

84 D Dettmann, E.: Forschungsanstalt der Bundeswehr für Wasserschall- und Geophysik (FWG) **1984**, unpublished.

84 G Geyer, D., Wille, P.: J. Acoust. Soc. Am. **75** (1984) 173.

84 H Hirschleber, H.: Seismische Quellen und spezielle Empfänger auf See in Erfassung seismischer Daten, 4. Mintrop Seminar, Bochum: Unikontakt **1984**, 161–200.

84 J Jarvis, F.C.: J. Ocean Eng., special issue **1984**, 253.

84 K Koenigs, P., Monti, J.M., Nützel, B., Herwig, H.: Acoustic scattering from the sea floor, paper presented at 108th meeting of the Acoustical Society of America, Minneapolis, October 12, **1984**.

84 S Schott, F.: Proc. Acoust. current-profiling Symposium, Nov. 1983, Washington, D.C.: NOAA, Natl. Oceanic Atmos. Adm., May **1984**, p. 28–30.

85 B Boehme, H., Chotiros, N.P., Rolleigh, L.D., Pitt, S.P., Garcia, A.L., Goldsberry, T.G., Lamb, R.A.: J. Acoust. Soc. Am. **77** (1985) 962.
85 K Kuperman, W.A., Ferla, M.C.: J. Acoust. Soc. Am. **77** (1985) 2067.
85 N Nghien-Phu, L., Tappert, F.: J. Acoust. Soc. Am. **78** (1985) 642.
85 S 1 Schneider, H.G., Thiele, R., Wille, P.C.: J. Acoust. Soc. Am. **77** (1985) 1409.
85 S 2 Scholz, B., Kroll, W., in: Adaptive methods in underwater acoustics (Urban, H.G., ed.), Dordrecht: Reidel Publ. Co. **1985**, 309–315.
85 S 3 Strasberg, M., in: Adaptive methods in underwater acoustics (Urban, H.G., ed.), Dordrecht: Reidel Publ. Co. **1985**, 125–143.
85 U Urban, H.G. (ed.): NATO Advanced Study Institute on Adaptive Methods in Underwater Acoustics, Dordrecht: Reidel Publ. Co. **1985**.
85 W Wille, P.: NATO Advanced Study Institute on Adaptive Methods in Underwater Acoustics, Lüneburg 1984, Dordrecht, Boston, Lancaster: Reidel Publ. Co. **1985**, ser. C, 151, 13.
86 W 1 Werner, F.: Bedform patterns on continental shelfs, publication in preparation.
86 W 2 Wille, P., Geyer, D.: Measurements of wind dependent acoustic transmission loss in shallow water under breaking wave conditions, 12th International Congress on Acoustics, Halifax **1986** (in prep.).
86 T Thiele, R.: Forschungsanstalt der Bundeswehr für Wasserschall- und Geophysik (FWG), in preparation for publication.

3.3 Optical properties of sea water

3.3.1 Basic facts and equations

3.3.1.1 Definitions and nomenclature

The nomenclature and units follow the recommendations by the IAPSO (International Association of the Physical Sciences of the Ocean) Working Group on Symbols, Units and Nomenclature in Physical Oceanography. In addition the classical American nomenclature and the nomenclature recommended by the IAMAP (International Association of Meteorology and Atmospheric Physics) Radiation Commission in Atmospheric Physics are presented for reasons of comparison in Table 1, see also Højerslev [84 H 1]. In this section spectral densities of radiation quantities (or more simply spectral radiation quantities) are designated by the IAPSO recommended symbol for the respective quantities supplemented by the subscript λ, e.g. L_λ for the spectral density of radiance, i.e. the radiance $L=\int L_\lambda d\lambda$.

Table 1. Fundamental quantities, units, and symbols in physical oceanography. In this section the IAPSO recommended nomenclature is used. For comparison the nomenclature recommended by IAMAP and the classical American nomenclature are represented as well.

Quantity [1])	Units	Symbol		
		IAPSO	IAMAP	Classical American
Radiant energy	J	Q	Q	U
Radiant flux	W	Φ	Φ	P
Radiant intensity	$W sr^{-1}$	I	I	J
Radiance	$W m^{-2} sr^{-1}$	L	L	N
Irradiance	$W m^{-2}$	E	E	H
Radiant exitance	$W m^{-2}$	M	M	W
Downward irradiance	$W m^{-2}$	E_d	$E(-)$, $E\downarrow$	$H(-)$
Upward irradiance	$W m^{-2}$	E_u	$E(+)$, $E\uparrow$	$H(+)$
Vector (net) irradiance	$W m^{-2}$	$\vec{E}$	–	$\vec{H}$
Scalar irradiance	$W m^{-2}$	E_o	–	h
Downward scalar irradiance	$W m^{-2}$	E_{od}	–	$h(-)$
Upward scalar irradiance	$W m^{-2}$	E_{ou}	–	$h(+)$
Spherical irradiance	$W m^{-2}$	E_s	–	–
Irradiance ratio		$R=E_u/E_d$	–	$R=H(+)/H(-)$
		$\bar{\mu}=\vec{E}/E_o$	–	–
Distribution factors		$D_d=E_{od}/E_d$	–	$D-=h-/H-$
		$D_u=E_{ou}/E_u$	–	$D+=h+/H+$
Wavelength	nm	λ	λ	λ
Light absorption coefficient	m^{-1}	a	σ_a	a
Volume light scattering function	$m^{-1} sr^{-1}$	β	γ_s	σ
(Total) light scattering coefficient	m^{-1}	b	σ_s	s
(Beam) light attenuation coefficient	m^{-1}	c	σ_e	α
Forward light scattering coefficient	m^{-1}	b_f	–	f
Backward light scattering coefficient	m^{-1}	b_b	–	b

[1]) The spectral densities of the radiation quantities Q, Φ, ..., E_s are designated by subscript λ. The corresponding units are those of the radiation quantities per unit wavelength.

The analysis of the daylight radiance distribution and the optical quantities derived thereof leads to the following definitions:

Radiant energy Q and radiance L:

Defining the spectral radiant energy $Q_\lambda = \mathrm{d}Q/\mathrm{d}\lambda$ as the radiant energy $\mathrm{d}Q$ in the wavelength interval $\lambda \cdots \lambda + \mathrm{d}\lambda$ divided by $\mathrm{d}\lambda$, the radiant energy $\mathrm{d}^4 Q = \mathrm{d}^3 Q_\lambda \mathrm{d}\lambda$ in the spectral range $\lambda \cdots \lambda + \mathrm{d}\lambda$ transported during a time $\mathrm{d}t$ across an area $\mathrm{d}A$ and in directions confined to a solid angle $\mathrm{d}\omega$ is related to the spectral radiance $L_\lambda = \mathrm{d}L/\mathrm{d}\lambda$ by the expression:

$$\mathrm{d}^3 Q_\lambda \mathrm{d}\lambda = L_\lambda \cos\alpha \, \mathrm{d}\lambda \mathrm{d}A \mathrm{d}\omega \mathrm{d}t \tag{1}$$

in which $\mathrm{d}\lambda$, $\mathrm{d}A$, $\mathrm{d}\omega$, and $\mathrm{d}t$ are all infinitesimal quantities, and α is the angle between the direction considered and the outward normal of $\mathrm{d}A$, Fig. 1.

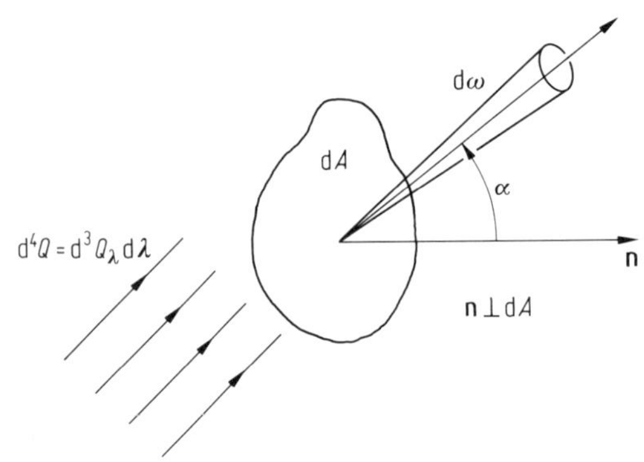

Fig. 1. Visualization of eq. (1).

The spectral radiance L_λ is a function of (i) three space coordinates $\boldsymbol{r}$, (ii) one angle α, which can be expressed in terms of a zenith angle θ and an azimuth angle φ, (iii) wavelength λ, and (iv) time t, i.e. $L_\lambda = L_\lambda(\boldsymbol{r}, \alpha, \lambda, t)$.

In marine optics the three space coordinates are normally given in a three-dimensional, left-handed Cartesian coordinate system as a frame of reference. The (x, y) plane defines a smooth water surface and the z coordinate (oriented downwards) defines the water depth. The vertical plane through origo and the position of the sun defines the azimuth angle $\varphi = 0$. Often the x coordinate is defined for the direction $\varphi = 0$, but it can also follow the direction of a latitude. The zenith angle, i.e. the angle between zenith and the incident radiant energy confined to an element of solid angle $\mathrm{d}\omega$, is termed θ. The above conventions imply that $\mathrm{d}A$ in Fig. 1 is oriented horizontally, and that the unit vector $\boldsymbol{n}$ and the z coordinate both have the same orientation, and $\alpha = \theta$.

From eq. (1) a number of optical quantities can be defined (see also Table 1):

Radiant flux Φ:

The radiant flux Φ is considered to be a flux of radiant energy, i.e. radiant energy per unit time. Introducing the spectral radiant flux $\Phi_\lambda = \mathrm{d}\Phi/\mathrm{d}\lambda$ for the wavelength interval $\lambda \cdots \lambda + \mathrm{d}\lambda$ with radiant flux $\mathrm{d}\Phi$, the definition of Φ applied to eq. (1) leads to the equation:

$$\mathrm{d}^2 \Phi_\lambda \mathrm{d}\lambda = \frac{\mathrm{d}^3 Q_\lambda \mathrm{d}\lambda}{\mathrm{d}t} = L_\lambda \cos\theta \, \mathrm{d}\lambda \mathrm{d}A \mathrm{d}\omega \tag{2}$$

for the radiant flux $\mathrm{d}^3\Phi = \mathrm{d}^2\Phi_\lambda \mathrm{d}\lambda$ in the spectral range $\lambda \cdots \lambda + \mathrm{d}\lambda$ emanating from an area $\mathrm{d}A$ into a solid angle $\mathrm{d}\omega$.

Radiant intensity I:

A given radiating source emits radiant fluxes in various directions at certain radiant intensities. If the radiant flux from such a source becomes increasingly collimated in one direction, the radiant intensity is said to increase in this particular direction. It is therefore natural to define radiant intensity I as radiant flux per unit solid angle of emittance (or acceptance).

Introducing $I_\lambda = \mathrm{d}I/\mathrm{d}\lambda$ the spectral radiant intensity for the wavelength interval $\lambda \cdots \lambda + \mathrm{d}\lambda$ with radiant intensity $\mathrm{d}I$, eq. (2) yields

$$\mathrm{d}I_\lambda \mathrm{d}\lambda = \frac{\mathrm{d}^2\Phi_\lambda \mathrm{d}\lambda}{\mathrm{d}\omega} = L_\lambda \cos\theta \,\mathrm{d}\lambda \mathrm{d}A \tag{3}$$

for the radiant intensity $\mathrm{d}^2 I = \mathrm{d}I_\lambda \mathrm{d}\lambda$ in the spectral range $\lambda \cdots \lambda + \mathrm{d}\lambda$ emitted by an area $\mathrm{d}A$. Note that a perfect parallel light beam is nonexisting according to this definition.

Irradiance (in general) E:

The irradiance E is defined as the radiant flux incident on a surface per unit surface area. Defining the spectral irradiance $E_\lambda = \mathrm{d}E/\mathrm{d}\lambda$ for the wavelength interval $\lambda \cdots \lambda + \mathrm{d}\lambda$ with irradiance $\mathrm{d}E$, eq. (2) gives

$$\mathrm{d}E_\lambda \mathrm{d}\lambda = \frac{\mathrm{d}^2\Phi_\lambda \mathrm{d}\lambda}{\mathrm{d}A} = L_\lambda \cos\theta \,\mathrm{d}\lambda \mathrm{d}\omega \tag{4}$$

for the irradiance $\mathrm{d}^2 E = \mathrm{d}E_\lambda \mathrm{d}\lambda$ in the spectral range $\lambda \cdots \lambda + \mathrm{d}\lambda$ incident from a solid angle $\mathrm{d}\omega$.

Scalar irradiance E_o:

The integral of a radiance distribution at a point over all directions θ, φ around the point:

$$E_{o\lambda} \mathrm{d}\lambda = \mathrm{d}\lambda \int_{4\pi} L_\lambda \mathrm{d}\omega \tag{5}$$

for the scalar irradiance $\mathrm{d}E_o = E_{o\lambda} \mathrm{d}\lambda$ in the spectral range $\lambda \cdots \lambda + \mathrm{d}\lambda$. $E_{o\lambda}$ denotes the spectral scalar irradiance and $\int_{4\pi}$ integration over all solid angles in the upper and lower hemisphere, and $\mathrm{d}\omega = \sin\theta \mathrm{d}\theta \mathrm{d}\varphi$.

Spherical irradiance E_s:

Limit of the ratio of radiant flux onto a spherical surface to the area of the surface as the radius of the sphere tends toward zero with its centre fixed:

$$E_{s\lambda} \mathrm{d}\lambda = \mathrm{d}\lambda \frac{\Phi_\lambda(r)}{4\pi r^2} \tag{6}$$

for the spherical irradiance $\mathrm{d}E_s = E_{s\lambda} \mathrm{d}\lambda$ in the spectral range $\lambda \cdots \lambda + \mathrm{d}\lambda$. $E_{s\lambda}$ is the spectral spherical irradiance and $\Phi_\lambda(r)$ the spectral radiant flux onto the sphere of radius r.

The following relation between E_o and E_s is valid:

$$E_{o\lambda} \mathrm{d}\lambda = 4E_{s\lambda} \mathrm{d}\lambda \tag{7}$$

Downward irradiance E_d:

The radiant flux from the upper hemisphere ($0 \leqq \theta \leqq 90°$, $0 \leqq \varphi \leqq 360°$) on a horizontal infinitesimal area containing the point being considered, divided by the area:

$$E_{d\lambda} \mathrm{d}\lambda = \mathrm{d}\lambda \int_{2\pi} L_\lambda \cos\theta \mathrm{d}\omega \tag{8}$$

for the downward irradiance $\mathrm{d}E_d = E_{d\lambda} \mathrm{d}\lambda$ in the spectral range $\lambda \cdots \lambda + \mathrm{d}\lambda$. $E_{d\lambda}$ denotes the spectral downward irradiance and $\int_{2\pi}$ integration over all solid angles in the upper hemisphere.

Upward irradiance E_u:

The radiant flux from the lower hemisphere ($90° \leqq \theta \leqq 180°$, $0 \leqq \varphi \leqq 360°$) on a horizontal infinitesimal area containing the point being considered, divided by the area:

$$E_{u\lambda} \mathrm{d}\lambda = \mathrm{d}\lambda \int_{-2\pi} L_\lambda |\cos\theta| \mathrm{d}\omega \tag{9}$$

for the upward irradiance $\mathrm{d}E_u = E_{u\lambda} \mathrm{d}\lambda$ in the spectral range $\lambda \cdots \lambda + \mathrm{d}\lambda$. $E_{u\lambda}$ is the spectral upward irradiance, while $\int_{-2\pi}$ implies integration over all solid angles in the lower hemisphere.

Vector irradiance E:

The vector (net) irradiance $dE = E_\lambda d\lambda$ in the spectral range $\lambda \cdots \lambda + d\lambda$ is defined as:

$$E_\lambda d\lambda = d\lambda \int_{4\pi} L_\lambda \cos\theta d\omega = (E_{d\lambda} - E_{u\lambda}) d\lambda \tag{10}$$

with E_λ being the spectral vector irradiance.

Downward scalar irradiance E_{od}:

The downward scalar irradiance $dE_{od} = E_{od\lambda} d\lambda$ in the spectral range $\lambda \cdots \lambda + d\lambda$ is defined as:

$$E_{od\lambda} d\lambda = d\lambda \int_{2\pi} L_\lambda d\omega \tag{11}$$

with $E_{od\lambda}$ being the spectral downward scalar irradiance.

Upward scalar irradiance E_{ou}:

The upward scalar irradiance $dE_{ou} = E_{ou\lambda} d\lambda$ in the spectral range $\lambda \cdots \lambda + d\lambda$ is defined as:

$$E_{ou\lambda} d\lambda = d\lambda \int_{-2\pi} L_\lambda d\omega \tag{12}$$

with $E_{ou\lambda}$ being the spectral upward scalar irradiance.

From eqs. (5), (11) and (12)

$$E_{o\lambda} d\lambda = (E_{od\lambda} + E_{ou\lambda}) d\lambda \tag{13}$$

Note the difference in signs between eqs. (10) and (13).

Quanta irradiance q:

The daylight available for photosynthesis in natural waters (PAR) is confined to the approximate spectral range 350 nm $\leq \lambda \leq$ 700 nm.

The quanta irradiance is normally measured as the number of photons coming from the upper hemisphere and impinging on a horizontal unit area per unit time. The number of photons $N(\lambda_0)d\lambda_0$ per unit time and unit area in the spectral range $d\lambda_0$ can be expressed in terms of the downward irradiance $E_{d\lambda_0} d\lambda_0$:

$$N(\lambda_0) d\lambda_0 = E_{d\lambda_0}(\lambda_0) \lambda_0 h^{-1} v_0^{-1} d\lambda_0 \tag{14}$$

in which h is Planck's constant, v_0 the propagation velocity of light and λ_0 the wavelength, both *in vacuo*. This velocity holds, practically speaking, the same value in air but becomes reduced in sea water according to the relation:

$$v = v_0/n, \tag{15}$$

in which n is the refractive index, being dependent on temperature T, salinity S, pressure p, and wavelength λ according to the approximate form (Austin and Halikas [76 A]):

$$n(T, S, p, \lambda) = n(10\,°\mathrm{C}, 0, 0, 546.1\,\mathrm{nm}) + \frac{\partial n}{\partial T}(T - 10\,°\mathrm{C}) + \frac{\partial n}{\partial S} S + \frac{\partial n}{\partial p} p + \frac{\partial n}{\partial \lambda}(\lambda - 546.1\,\mathrm{nm}) \tag{16}$$

The values for the derivatives in eq. (16) are:

$$\frac{\partial n}{\partial T} = -6.67 \cdot 10^{-5}\,(°\mathrm{C})^{-1}$$

$$\frac{\partial n}{\partial S} = 20.93 \cdot 10^{-2}\,(‰\,\mathrm{S})^{-1}$$

$$\frac{\partial n}{\partial p} = 0.14 \cdot 10^{-9}\,\mathrm{Pa}^{-1}$$

$$\frac{\partial n}{\partial \lambda} = -4.33 \cdot 10^{-5}\,\mathrm{nm}^{-1}$$

and

$$n(10\,°\mathrm{C}, 0, 0, 546.1\,\mathrm{nm}) = 1.33518\,.$$

Practically speaking, the refractive index can be put equal to 4/3 for essentially all natural sea waters and wavelengths. For the change in wavelength between *in vacuo* (or air) and water, respectively, there exists a relation analogous to eq. (15):

$$\lambda = \lambda_0 / n \tag{17}$$

because the frequency for the radiant energy is invariant to changes in refractive indices. Accordingly, the number of photons per unit area and unit time (or any irradiance parameter defined above) will not change discontinuously at any optical interface assuming no reflection: Applying eqs. (8), (14), and (15) the quanta irradiance in water is then defined as:

$$q = \int_{350\,\mathrm{nm}}^{700\,\mathrm{nm}} E_{\mathrm{d}\lambda}(\lambda)\lambda h^{-1} v^{-1} \mathrm{d}\lambda = \int_{350\,\mathrm{nm}}^{700\,\mathrm{nm}} N(\lambda)\mathrm{d}\lambda \tag{18}$$

In photosynthetic studies it is, however, preferable to deal with the number of photons coming from all directions rather than from only the upper hemisphere. Consequently, photosynthetic available daylight should preferably be defined through eq. (5):

$$q_0 = \int_{350\,\mathrm{nm}}^{700\,\mathrm{nm}} E_{\mathrm{o}\lambda}(\lambda)\lambda h^{-1} v^{-1} \mathrm{d}\lambda \tag{19}$$

As a first order approximation the following correction is recommended (Højerslev [78 H]):

$$q_0 = 1.3 q \tag{20}$$

3.3.1.2 Quantities related to radiance and irradiance

The optical properties of sea water (or of any hydrosol) can be divided into two mutually exclusive classes consisting of *inherent* and *apparent* properties. An inherent property is one that, as a physical quantity, has always the same magnitude for a given water mass with a definite composition of dissolved and suspended matter, i.e. variations in the inherent properties are solely caused by concentration changes of dissolved and suspended matter in the water mass. An apparent property is one for which this is not the case.

Often this definition is given in terms of the environmental radiance distribution: An inherent property is one that is independent of changes in the radiance distribution; an apparent property is one for which this is not the case (Preisendorfer [61 P]). Quantities related to irradiance parameters are clearly all apparent properties although they demonstrate quite commonly quasi-inherent behavior.

Apparent properties:

The vertical attenuation coefficient K_L for the radiance L in any direction (θ, φ), at a certain depth z, is the most fundamental apparent property in marine optics, but is not widely determined from field measurements. The definition of K_L reads:

$$K_L = -\frac{1}{L_\lambda \mathrm{d}\lambda} \frac{\partial L_\lambda \mathrm{d}\lambda}{\partial z} \tag{21}$$

Note that K_L depends on the concentration and composition of the suspended and dissolved matter, the position, the depth, the orientation (θ, φ), the wavelength λ, the solar elevation, the cloud cover, the state of the sea, the depth to the bottom and the albedo of the bottom. K_L can be positive, negative, and consequently zero, at certain depths and surface light conditions without violating the principle of conservation of energy, e.g. Lundgren and Højerslev [71 L 1] and Timofeeva [74 T]. However, at great depths in homogeneous waters, K_L becomes eventually a constant for all θ, φ in the so-called asymptotic region, Højerslev and Zaneveld [77 H 6]. In this region all the K-functions defined in eqs. (22)···(26) attain the same constant value as K_L, see eq. (76).

Vertical (diffuse) attenuation coefficients for the irradiances defined in eqs. (5)···(12) and (18) can be defined in the same manner as K_L in eq. (21). The most relevant definitions are listed below:

$$K_{\mathrm{o}} = -\frac{1}{E_{\mathrm{o}\lambda} \mathrm{d}\lambda} \frac{\partial E_{\mathrm{o}\lambda} \mathrm{d}\lambda}{\partial z}, \tag{22}$$

in which K_{o} is the vertical attenuation coefficient for the scalar irradiance in the spectral range $(\lambda, \lambda + \mathrm{d}\lambda)$;

$$K_{\mathrm{d}} = -\frac{1}{E_{\mathrm{d}\lambda} \mathrm{d}\lambda} \frac{\partial E_{\mathrm{d}\lambda} \mathrm{d}\lambda}{\partial z}, \tag{23}$$

Højerslev

in which K_d is the vertical attenuation coefficient for the downward irradiance in the spectral range $(\lambda, \lambda+d\lambda)$;

$$K_u = -\frac{1}{E_{u\lambda}d\lambda}\frac{\partial E_{u\lambda}d\lambda}{\partial z}, \tag{24}$$

in which K_u is the vertical attenuation coefficient for the upward irradiance in the spectral range $(\lambda, \lambda+d\lambda)$;

$$K_E = -\frac{1}{E_{\lambda}d\lambda}\frac{\partial E_{\lambda}d\lambda}{\partial z}, \tag{25}$$

in which K_E is the vertical attenuation coefficient for the vector (net) irradiance in the spectral range $(\lambda, \lambda+d\lambda)$;

$$K_q = -\frac{1}{q}\frac{\partial q}{\partial z}, \tag{26}$$

in which K_q is the vertical attenuation coefficient for the downward quanta irradiance between $\lambda = 350$ nm and $\lambda = 700$ nm.

The depth of the euphotic zone is defined through one of the two relations below:

$$\frac{q(z=0)}{q(z=z(1\%))} = 100, \tag{27}$$

or

$$z_q(1\%) = \frac{\ln 100}{\langle K_q \rangle}, \tag{28}$$

in which $z_q(1\%)$ is the depth of the euphotic zone, i.e. the depth at which the quanta irradiance $q(o)$ just below the surface is reduced by a factor of 100, and

$$\langle K_q \rangle = \frac{1}{z_q(1\%)}\int_0^{z_q(1\%)} K_q(z')dz'. \tag{29}$$

All the vertical attenuation coefficients defined in eqs. (22)···(26) attain always positive values in absorbing and passive (no internal radiating sources) media like sea water. In some *rare* cases with high concentrations of plankton pigments near the surface the sea water cannot be considered passive around $\lambda = 685$ nm due to intense pigment fluorescence. So, in this spectral region and at shallow depths even negative K ($\lambda = 685$ nm) can be observed, Højerslev [75 H]. In homogeneous sea waters K_o, K_d, K_u, and consequently K_E, are almost independent on the surface light conditions (solar elevation) and the depth, i.e. they display quasi-inherent properties. This is, however, not the case for K_q except for turbid homogeneous waters, say for $z(1\%) \leqq 10$ m.

For inhomogeneous sea waters normally horizontally stratified within the relevant length scales, K_o, K_d, K_u, *and* K_q vary only slightly with respect to the solar elevation, e.g. Højerslev [74 H 1, 74 H 2, 82 H 1]. This is the rationale for decomposing any of the above K functions into forms like:

$$K_d = K_w + K_p + K_{ph} + K_y \tag{30}$$

in which the four vertical attenuation coefficient terms $K_w \cdots K_y$ originate from the vertical attenuation due to pure sea water (index w), suspended inorganic matter and detritus (index p), living phytoplankton (index ph), and dissolved organic matter like yellow substance (index y). Such decompositions are not permissible in a strict sense (see the definition of an apparent property), but have proven their applicability at a number of occasions.

Irradiance ratio R:

The irradiance ratio in the spectral range $(\lambda, \lambda+d\lambda)$ is defined as the ratio of the upward to the downward irradiance at a depth in the sea, i.e.:

$$R = \frac{E_{u\lambda}d\lambda}{E_{d\lambda}d\lambda}. \tag{31}$$

Even in stratified sea waters R is slowly varying with depth since

$$\frac{dR}{dz} = R(K_d - K_u) \tag{32}$$

in which $R < 0.1$ and $K_d \simeq K_u$. Changes in the solar elevation have a minor effect on R, being of the order of 10% or less.

Average cosine $\bar{\mu}$:

The average cosine in the spectral range $(\lambda, \lambda+\mathrm{d}\lambda)$ is defined as the ratio of the vector irradiance to the scalar irradiance at a depth in the sea, i.e.:

$$\bar{\mu}=\frac{E_\lambda \mathrm{d}\lambda}{E_{\mathrm{o}\lambda}\mathrm{d}\lambda}=\frac{\mathrm{d}\lambda \int_{4\pi} L_\lambda \cos\theta \,\mathrm{d}\omega}{\mathrm{d}\lambda \int_{4\pi} L_\lambda \mathrm{d}\omega} \equiv \langle\cos\theta\rangle_{4\pi} \quad (33)$$

according to eqs. (5) and (10). The average cosine is of fundamental importance in photosynthetic studies. It is normally equal to 0.75 with an *overall* variation of the order of ± 0.15 for all wavelengths in the visible part of the spectrum and for a very large group of water types, if not them all. The average cosine varies only sligthly with depth and surface light conditions but may change to some extent with wavelengths, e.g. Højerslev [77 H 1, 78 H].

Distribution factors, D_d and D_u:

The downward distribution factor D_d in the spectral range $(\lambda, \lambda+\mathrm{d}\lambda)$ is defined in a reciprocal but similar way as the average cosine, since:

$$D_\mathrm{d}=\frac{E_{\mathrm{od}\lambda}\mathrm{d}\lambda}{E_{\mathrm{d}\lambda}\mathrm{d}\lambda}=\frac{\mathrm{d}\lambda \int_{2\pi} L_\lambda \mathrm{d}\omega}{\mathrm{d}\lambda \int_{2\pi} L_\lambda \cos\theta \,\mathrm{d}\omega} \equiv \langle\cos\theta\rangle_{2\pi}^{-1} \quad (34)$$

The upward distribution factor D_u in the spectral range $(\lambda, \lambda+\mathrm{d}\lambda)$ is defined as:

$$D_\mathrm{u}=\frac{E_{\mathrm{ou}\lambda}\mathrm{d}\lambda}{E_{\mathrm{u}\lambda}\mathrm{d}\lambda}=\frac{\mathrm{d}\lambda \int_{-2\pi} L_\lambda \mathrm{d}\omega}{\mathrm{d}\lambda \int_{-2\pi} L_\lambda |\cos\theta| \mathrm{d}\omega} \equiv \langle|\cos\theta|\rangle_{-2\pi}^{-1}, \quad (35)$$

in which $D_\mathrm{u} \simeq 2.5 \pm 0.3$ for most sea waters. If the radiance in the lower hemisphere is assumed isotropic, i.e. if $L=$constant for $90° \leqq \theta \leqq 180°$ and $0° \leqq \varphi \leqq 360°$, then $D_\mathrm{u}=2$ exactly.

3.3.1.3 Inherent properties

Light absorption coefficient a:

The internal absorptance of an infinitesimally thin layer of the medium (sea water) normal to the beam, divided by the thickness Δr of the layer. Since the absorptance A is defined as the ratio of the radiant flux lost from the beam by means of absorption (transformation of radiant energy into sensible heat, chemical energy, fluorescent energy, etc.) to the incident radiant flux, the definition for a becomes:

$$a=-\frac{\Delta A}{\Delta r}=-\frac{\Delta\Phi}{\Phi\Delta r}. \quad (36)$$

(Total) light scattering coefficient b:

The internal scatterance of an infinitesimally thin layer of the medium (sea water) normal to the beam, divided by the thickness Δr of the layer. Since the scatterance B is defined as the ratio of the radiant flux scattered from the beam (causing no loss in radiant energy but solely changes in direction of photon pathways) to the incident radiant flux, the definition for b becomes:

$$b=-\frac{\Delta B}{\Delta r}=-\frac{\Delta\Phi}{\Phi\Delta r}. \quad (37)$$

(Beam) light attenuation coefficient c:

The internal attenuance of an infinitesimally thin layer of the medium (sea water) normal to the beam, divided by the thickness, Δr of the layer. Since the attenuance C is defined as the ratio of the radiant flux lost from the beam by means of absorption and scattering, to the incident radiant flux, the definition for c becomes:

$$c=-\frac{\Delta C}{\Delta r}=-\frac{\Delta\Phi}{\Phi\Delta r}. \quad (38)$$

For a (nonexisting) parallel and monochromatic beam of infinitesimal cross section, the following relation is valid:

$$a+b=c\,. \tag{39}$$

If the beam is polychromatic instead of monochromatic, part of the radiant flux absorbed at some wavelengths might be re-emitted as a fluorescent radiant flux at other (higher) wavelengths. For such cases, eq. (39) is not true in general.

Volume light scattering function $\beta(\alpha)$:

The radiant intensity (from an infinitesimal volume element $\mathrm{d}V$ in a given direction forming the angle α to the incident beam) per unit of irradiance on the volume and per unit volume, i.e.:

$$\beta(\alpha)=\frac{\mathrm{d}I(\alpha)}{E\cdot\mathrm{d}V}, \tag{40}$$

or, alternatively:

$$\beta(\alpha)=\frac{\mathrm{d}^2\Phi(\alpha)}{\mathrm{d}^2\Phi(\alpha=0)}\cdot(fl\cdot\mathrm{d}\omega)^{-1} \tag{41}$$

in which $\mathrm{d}^2\Phi(\alpha=0)$ is the radiant flux at the volume surface perpendicular to the incident beam, and $\mathrm{d}^2\Phi(\alpha)$ the outgoing radiant flux from $\mathrm{d}V$ in the direction α. Note that $\mathrm{d}^2\Phi(\alpha=0)$ and $\mathrm{d}^2\Phi(\alpha)$ are the radiant fluxes at the volume element. In the real experimental situation $\mathrm{d}V$ is surrounded by sea water, see Fig. 3, and eq. (41) should read:

$$\beta(\alpha)=\frac{\mathrm{d}^2\Phi(\alpha)}{\mathrm{d}^2\Phi(\alpha=0)}(\mathrm{d}l\cdot\mathrm{d}\omega)^{-1}\mathrm{e}^{c(l-l_0+r)} \tag{42}$$

in which c is the light attenuation coefficient of the hydrosol surrounding $\mathrm{d}V$, and $\mathrm{d}\omega$ the solid angle of acceptance for the light sensor.

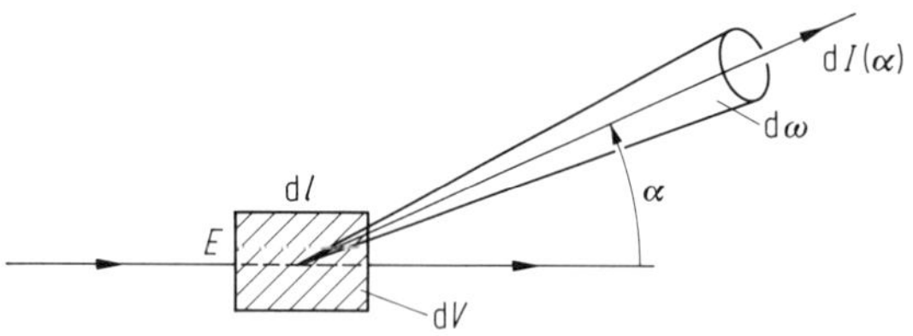

Fig. 2. Visualization of eqs. (40) and (41).

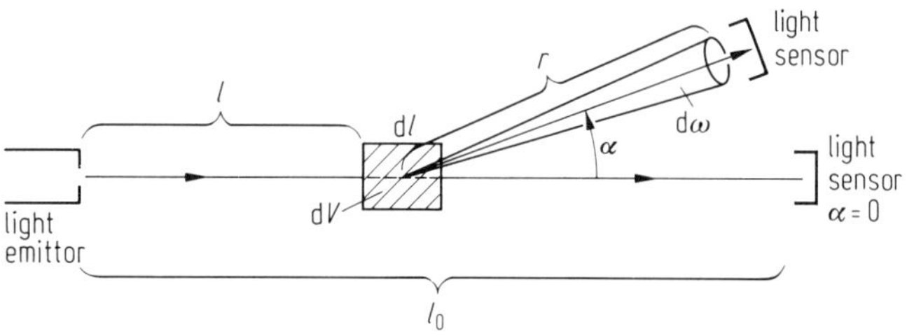

Fig. 3. Visualization of eq. (42).

Forward and backward scattering coefficients, b_f and b_b: The forward scattering coefficient b_f is defined as:

$$b_\mathrm{f}=\int_{2\pi}\beta\,\mathrm{d}\omega\,, \tag{43}$$

and the backward scattering coefficient b_b as:

$$b_\mathrm{b}=\int_{-2\pi}\beta\,\mathrm{d}\omega\,. \tag{44}$$

Since $b=b_\mathrm{f}+b_\mathrm{b}$:

$$b=\int_{4\pi}\beta\,\mathrm{d}\omega\,. \tag{45}$$

It is normally assumed *a priori* that $\beta(\alpha,\varphi_1)=\beta(\alpha,\varphi_2)$ for all φ_1, φ_2, i.e.:

$$b=2\pi\int_0^\pi\beta(\alpha)\sin\alpha\,\mathrm{d}\alpha\,. \tag{46}$$

Combination of eqs. (39) and (45) gives straightforward:

$$a+b=a+\int_{4\pi}\beta \mathrm{d}\omega=c\,. \tag{47}$$

The equation shows that the two independent and most fundamental inherent properties are a and β if *unpolarized radiant energy transfers are considered* (streams of photons). This is the usual assumption in marine optics, but if polarization is considered the light scattering properties of the sea water are completely described through a so-called Mueller scattering matrix, or, alternatively, by the so-called Jones' matrices, consisting of 16 (4 × 4) elements. The Mueller scattering matrix can formally be written as:

$$\begin{Bmatrix} a_1 & b_1 & b_3 & b_5 \\ c_1 & a_2 & b_4 & b_6 \\ c_3 & c_4 & a_4 & b_2 \\ c_5 & c_6 & c_2 & a_4 \end{Bmatrix} \tag{48}$$

The volume light scattering function equals

$$\beta(\alpha)=\tfrac{1}{2}(a_1+a_2+b_1+c_1)\,, \tag{49}$$

and the polarization p produced by the scattering of an unpolarized light beam

$$p(\alpha)=2(a_1-a_2+b_1-c_1)\cdot\beta(\alpha)^{-1}\,. \tag{50}$$

If the sea water, in which matter is suspended, obeys the Mie scattering theory the Mueller scattering matrix reduces to contain four independent coefficients only:

$$\begin{Bmatrix} a_1 & 0 & 0 & 0 \\ 0 & a_2 & 0 & 0 \\ 0 & 0 & a_3 & b_2 \\ 0 & 0 & -b_2 & a_3 \end{Bmatrix} \tag{51}$$

The complete theory can be found in various textbooks and review articles, e.g. Chandrasekhar [50 C], Van de Hulst [57 H], and Morel [73 M 2], and is for this reason omitted here.

3.3.1.4 Equations of radiative transfer

In experimental marine optics in recent years most measurements do not include the polarization of the underwater light field. This is also reflected in the majority of the underwater optical models which are, as a matter of fact, mainly based on irradiance data. For these reasons only the scalar theory for the radiative transfer is considered, i.e. the underwater light field is assumed unpolarized. It will further be assumed that

(i) all radiant flux sources are monochromatic and external, i.e. part of the radiant energy is not transformed into fluorescent energy, and

(ii) the boundaries of the sea extend towards infinity.

First consider that $L=L(x,y,z,\theta,\varphi,\lambda,t)$. There will typically be two classes of time dependences:

1. The time dependence connected with the propagation velocity of light in the sea, i.e. $v=v_0/n$.
2. Variations in the atmospheric light conditions due to changes in solar elevation, cloud cover, etc., and underwater light fluctuations due to waves and whitecaps.

The first class of time dependence is without practical importance but the second one is not.

Correction procedures, such as deck photometers, etc., for the second class of time dependences, having time scales between say one second and twelve hours, have been worked out. Thus for most measuring situations it can simply be assumed that the radiance field is quasi-stationary, i.e. $L=L(x,y,z,\theta,\varphi,\lambda)$, and the classical time-independent equation for radiative transfer becomes:

$$\frac{\partial L(x,y,z,\theta,\varphi)}{\partial x}\sin\theta\cos\varphi+\frac{\partial L(x,y,z,\theta,\varphi)}{\partial y}\sin\theta\sin\varphi+\frac{\partial L(x,y,z,\theta,\varphi)}{\partial z}\cos\theta$$
$$=-c(x,y,z)L(x,y,z,\theta,\varphi)+\int_{4\pi}L(x,y,z,\theta',\varphi')\beta(x,y,z,\alpha)\sin\theta'\mathrm{d}\theta'\mathrm{d}\varphi'\,, \tag{52}$$

in which the specification of λ has been omitted for reasons of convenience, and

$$\cos\alpha = \cos\theta\cos\theta' + \sin\theta\sin\theta'\cos(\varphi - \varphi'). \tag{53}$$

Eq. (52) illustrates clearly the difficulties in measuring L. It has to be measured in at least two vertical points, four horizontal points and in a number of (θ, φ) directions in order to make proper use of eq. (52). There are two different ways to circumvent the problems associated with eq. (52):

1. Assuming a horizontally stratified sea this equation reduces to:

$$\frac{\partial L(z,\theta,\varphi)}{\partial z}\cos\theta = -c(z)L(z,\theta,\varphi) + \int_{4\pi} \beta(z,\alpha)L(z,\theta',\varphi')\sin\theta' \mathrm{d}\theta' \mathrm{d}\varphi. \tag{54}$$

2. Integrating eq. (52) over all solid angles leads to:

$$\frac{\partial}{\partial x}\int_{4\pi} L\sin\theta\cos\varphi \mathrm{d}\omega + \frac{\partial E}{\partial z} = -aE_0. \tag{55}$$

The term

$$\frac{\partial}{\partial y}\int_{4\pi} L\sin\theta\sin\varphi \mathrm{d}\omega$$

vanishes since $L(\varphi) = L(-\varphi)$ due to the definition of $\varphi = 0$. The φ dependence of L can be approximated by an expression of the form:

$$L(\varphi) = A\frac{\varepsilon + 1}{\varepsilon - \cos\varphi}, \qquad \varepsilon > 1, \tag{56}$$

for any point in the sea water. A is a function of L_{max} and L_{min} for any fixed θ (Lundgren and Højerslev [71 L 1]) and ε a constant ($\varepsilon = 1.68$ in the particular case).

Conventional scale analysis for the first and the second term on the left-hand side of eq. (55) results in, when ignoring signs;

$$\frac{\partial}{\partial x}\int_{4\pi} L\sin\theta\cos\varphi \mathrm{d}\omega \Big/ \frac{\partial E}{\partial z} \ll \frac{\partial E_o}{\partial x}\Big/\frac{\partial E}{\partial z} < \frac{\partial E}{\partial x}\Big/\frac{\partial E}{\partial z} \approx 1. \tag{57}$$

The last condition is obtained for a tilt of 45° of the a isosurfaces in the x direction. Accordingly, by using all the conditions in eq. (57), it can safely be assumed that

$$\frac{\partial E}{\partial z} \gg \frac{\partial}{\partial x}\int_{4\pi} L\sin\theta\cos\varphi \mathrm{d}\omega$$

resulting in the now justified and highly useful equation:

$$\frac{\partial E}{\partial z} \cong -aE_o \tag{58}$$

valid in *all* sea waters displaying even the highest inhomogeneities with respect to suspended and dissolved organic matter.

Note that E, E_o, and a depend on position (x, y, z) for the case considered.

For horizontally stratified sea waters,

$$\frac{\partial E}{\partial z} = -aE_0 \tag{59}$$

exactly.

It is sometimes convenient to consider another form of eq. (59):

$$a = K_d\left[1 - R + \frac{1}{K_d}\frac{\mathrm{d}R}{\mathrm{d}z}\right]\frac{E_d}{E_o} \tag{60}$$

which, for most sea waters, can be safely approximated by:

$$a \cong K_d(1 - R)\frac{E_d}{E_o} = K_d\bar{\mu}. \tag{61}$$

An exact expression for a can be derived from eq. (59) and the definitions for K_E and $\bar{\mu}$:

$$a = K_E \bar{\mu}\,. \tag{62}$$

The approximated expression in eq. (61) is seen to be exact if $K_d = K_E$ as normally observed. As stated earlier, K_E (or K_d) and $\bar{\mu}$ are slowly varying functions with respect to surface light conditions and $\bar{\mu}$ is in addition only slightly depth-dependent in general. Accordingly, from the definitions of apparent and inherent properties the vertical variation of a is directly reflected in the vertical variation of K_E (or K_d). This empirical observation is also reflected in the most simple oceanic model for K_d which states:

$$K_d \approx a + b_b \approx a + 0.02b \tag{63}$$

in which $a \approx b$ for visible light.

Assume for the rest of this section that the assumptions leading to eq. (54) are still valid except for one: The assumption of having monochromatic radiant flux sources is now replaced by the assumption of having polychromatic sources (like the sun). This involves that part of the underwater radiant energy may be transformed into fluorescent energy. Consequently, at certain wavelengths and depths there will be internal radiant flux sources as well. This effect, F_*, has to be taken into account together with eq. (54), which, for the actual case, becomes (Højerslev [75 H]):

$$\frac{\partial L_\lambda(z,\theta,\varphi,\lambda_0)}{\partial z}\cos\theta = -c(z,\lambda_0)L_\lambda(z,\theta,\varphi,\lambda_0) + L_*(z,\theta,\varphi,\lambda_0) + F_*(z,\theta,\varphi,\lambda,\lambda_0) \tag{64}$$

where L_* is the (spectral) path function:

$$L_* = \int_{4\pi} \beta(z,\theta,\varphi,\theta',\varphi',\lambda_0)L_\lambda(z,\theta',\varphi',\lambda_0)\sin\theta'\,d\theta'\,d\varphi' \tag{65}$$

for light at wavelength λ and F_* is the (spectral) source function of fluorescent light at wavelength λ_0:

$$F_* = \int_0^\infty f(z,\pi/2,\lambda,\lambda_0)E_{o\lambda}(z,\lambda)\,d\lambda\,. \tag{66}$$

Eq. (66) can be proven from the following considerations: The volume light scattering function $\beta(z,\alpha,\lambda_0)$ is for this case defined by the relation

$$\beta(z,\alpha,\lambda_0) = \frac{dI_{\lambda\,\mathrm{scat.}}(z,\alpha,\lambda_0)\,d\lambda_0}{E_\lambda(z,\lambda_0)\,d\lambda_0\,dV} \tag{67}$$

where $dI_{\lambda\,\mathrm{scat.}}(z,\alpha,\lambda_0)\,d\lambda_0$ is the radiant intensity of the light scattered in the direction α from the incident beam, and $E_\lambda(z,\lambda_0)\,d\lambda_0$ is the irradiance at wavelength λ_0 on the volume element dV at the depth z.

The light scattering coefficient b is found by integration $\beta(z,\alpha,\lambda_0)$ over all solid angles:

$$b(z,\lambda_0) = \int_{4\pi} \beta(z,\alpha,\lambda_0)\,d\omega\,. \tag{68}$$

In case of isotropic scattering, eqs. (68) and (65) give:

$$\beta(z,\alpha,\lambda_0) = \frac{1}{4\pi}b(z,\lambda_0) = \beta(z,\pi/2,\lambda_0) \tag{69}$$

and

$$L_*(z,\theta,\varphi,\lambda_0) = E_{o\lambda}(z,\lambda_0)\beta(z,\pi/2,\lambda_0)\,. \tag{70}$$

Since daylight acts like a broad spectral light source, emission of fluorescent radiant energy takes place in the sea when excited by the shortwave part of the daylight.

A spectral volume scattering function of fluorescent radiant intensity $f(z,\alpha,\lambda,\lambda_0)$ is introduced by the relation:

$$f(z,\alpha,\lambda,\lambda_0)\,d\lambda_0 = \frac{d^2I_{\lambda\,\mathrm{fluor.}}(z,\alpha,\lambda,\lambda_0)\,d\lambda_0}{E_\lambda(z,\lambda)\,d\lambda\,dV} \tag{71}$$

where $d^2I_{\lambda\,\mathrm{fluor.}}(z,\alpha,\lambda,\lambda_0)\,d\lambda_0$ is the radiant intensity of fluorescent light at wavelength λ_0 excited by daylight at wavelength λ and emitted in the direction α from the incident beam. $E_\lambda(z,\lambda)$, $d\lambda$ is the daylight radiant flux at wavelength λ impinging upon the volume element dV at the depth z. The total scattered radiant intensity of fluorescent light at wavelength λ_0 in the direction α from the incident beam is obtained from integration of eq. (71)

over all wavelengths, i.e.:

$$\mathrm{d}I_{\lambda\,\mathrm{fluor.}}(z,\alpha,\lambda_0)=\mathrm{d}V\int_0^\infty E_\lambda(z,\lambda)f(z,\alpha,\lambda,\lambda_0)\mathrm{d}\lambda\,. \tag{72}$$

Normally, fluorescent radiant intensities are scattered isotropically, which means that

$$f(z,\alpha,\lambda,\lambda_0)=f(z,\pi/2,\lambda,\lambda_0)\,. \tag{73}$$

The source function of fluorescent light F_* can now be derived analogous to the derivation of the path function L_* in the case of isotropic scattering, see eqs. (67) and (70):

$$F_*=\int_0^\infty f(z,\pi/2,\lambda,\lambda_0)E_{o\lambda}(z,\lambda)\mathrm{d}\lambda\,. \tag{74}$$

The scalar irradiance can be measured by means of a light absorption meter (Højerslev [75 H]), and the volume scattering function of fluorescent light by aid of a conventional fluorimeter measuring through 90° the excitation spectrum at the emission wavelength λ_0 of a water sample taken at depth z (Parker [68 P]). In practice it suffices to integrate the expression in eq. (74) between $\lambda=295$ nm (lower limit of the incoming daylight radiant energy at the sea surface) and λ_0.

An optically homogeneous water mass acts like a monochromator at great depths because $K_L(\lambda)$ has an absolute minimum.

In the clearest (oligotrophic) waters the wavelength of maximum transmission is around 460 nm for the downward irradiance, thus K_d(460nm) attains its minimum among all the K_d functions. Applying this to eq. (23) it can be seen directly that at great depths in these parts of the ocean only the blue ($\lambda=460$ nm) daylight component will be present.

In turbid (not necessarily eutrophic) waters the wavelength of maximum transmission can even be observed around wavelengths where the pure sea water itself absorbs strongly, i.e. above 600 nm.

But again, these water masses act as monochromators at great optical depths $\int_0^z K_d(z',\lambda)\mathrm{d}z'$.

Note that the water depth z can be small in case of very turbid waters, say less than 10 m. All together, at great depths the time-independent equation of radiative transfer reduces to:

$$\frac{\partial L_\lambda(z,\theta,\varphi,\lambda_0)}{\partial z}\cos\theta=-c(z,\lambda_0)L_\lambda(z,\theta,\varphi,\lambda_0)+L_*(z,\theta,\varphi,\lambda_0) \tag{75}$$

in which the wavelength dependence can be disregarded. Assuming either (i) a and $\beta(\alpha)$ are both independent of depth or (ii) the ratio $a/\beta(\alpha)$ is constant with depth (but dependent on α), Højerslev and Zaneveld [77 H 6] have proven theoretically, without using further assumptions than those contained in eq. (75), the existence of the so-called asymptotic regime, $z\to\infty$, for which:

$$\frac{\partial L}{\partial\varphi}=0 \quad\text{and}\quad -\frac{1}{L(z,\theta)}\frac{\partial L(z,\theta)}{\partial z}=k \tag{76}$$

with k being a constant. So, in the asymptotic regime, the radiance distribution is symmetrical around the vertical and the *shape* of this distribution is invariant to changes with depth. The finding

$$-\frac{1}{L}\frac{\partial L}{\partial z}=k$$

implies that the K functions in eqs. (22)···(26) also attain the constant value k and that

$$\frac{\mathrm{d}R}{\mathrm{d}z}=\frac{\mathrm{d}\bar{\mu}}{\mathrm{d}z}=\frac{\mathrm{d}D_\mathrm{d}}{\mathrm{d}z}=\frac{\mathrm{d}D_\mathrm{u}}{\mathrm{d}z}=0$$

exactly.

The existence of the asymptotic regime implies finally that the *average* of all possible K_L functions, and consequently all the K functions, calculated from the sea surface (or from any finite depth z_0) down to the asymptotic regime equals k, i.e.:

$$\lim_{z\to\infty}\frac{1}{z}\int_0^z K_L(z',\theta,\varphi)\mathrm{d}z'=k$$

or

$$\lim_{z\to\infty} \frac{1}{z-z_0} \int_{z_0}^{z} K_L(z',\theta,\varphi)\mathrm{d}z' = k\,.$$

If the shape of the L distribution and the first derivative of L with respect to depth, $\partial L/\partial z$, are given in the asymptotic regime then the inherent (and apparent) optical properties can be derived, (Zaneveld [74 Z]):

$$c = \lim_{n\to\infty}\left[-\frac{\int_0^\pi \frac{\partial L}{\partial z} P_n(\cos\theta)\cos\theta\sin\theta\,\mathrm{d}\theta}{\int_0^\pi L P_n(\cos\theta)\sin\theta\,\mathrm{d}\theta}\right] \tag{77}$$

$$\beta(\alpha) = \sum_{n=0}^{\infty} \frac{2n+1}{4\pi} P_n(\cos\alpha)\left[c + \frac{\int_0^\pi \frac{\partial L}{\partial z} P_n(\cos\theta)\cos\theta\sin\theta\,\mathrm{d}\theta}{\int_0^\pi L P_n(\cos\theta)\sin\theta\,\mathrm{d}\theta}\right] \tag{78}$$

in which P_n is the Legendre polynomial of the order n. It is possible to integrate eq. (78) over all solid angles in order to find b and then make the subtraction $c-b=a$ in order to get a. However, observing that eq. (62) is strictly valid and,

$$K_E = k = -\frac{1}{L}\frac{\partial L}{\partial z},$$

then by applying eq. (33), it is straightforward to derive a:

$$a = -\frac{\int_0^\pi L\cos\theta\sin\theta\,\mathrm{d}\theta}{\int_0^\pi L\sin\theta\,\mathrm{d}\theta}\,\frac{1}{L}\frac{\partial L}{\partial z}\,. \tag{79}$$

Finally, it is emphasized that all the apparent optical properties in the asymptotic regime become inherent, i.e. they are physical constants which can be given in terms of a and $\beta(\alpha)$.

3.3.2 Optical properties of water

3.3.2.1 Optical properties of pure water and pure sea water

Comprehensive reviews of measurements of the optical properties of pure water and pure sea water performed in the ultraviolet, visible, and infrared part of the spectrum are recently given by Ivanoff [73 I], Smith and Tyler [76 S], Morel and Prieur [77 M], Tam and Patel [79 T] and Smith and Baker [81 S 1]. A few general conclusions can be stated from the above reviews:

1. Preparation of pure (sea) water *standard* batches is extremely difficult. No laboratory has been appointed as being capable to produce such standard batches from any international organization.
2. Numbers for the spectral light attenuation (or absorption) coefficient for pure (sea) waters are unacceptably deviating from one data set to another one in the ultraviolet and visible part of the spectrum, Fig. 4. Only for the red and infrared region the different data sets display a fair agreement.
3. The experimental set-up for measuring the light scattering function for pure (sea) water is, basically speaking, more accessible to optimization compared with the one for measuring the spectral light attenuation (or absorption) coefficient.
4. Estimates of the spectral light attenuation (or absorption) coefficient of pure (sea) water are recommended to be performed from *in-situ* downward irradiance measurements in the clearest natural waters of the world. The different existing irradiance data sets from such waters are consistent throughout the whole spectrum. Moreover, the estimates of the spectral light attenuation (or absorption) coefficient are generally considerably lower than those measured *in-vitro* in the ultraviolet and visible part of the spectrum, Fig. 5.

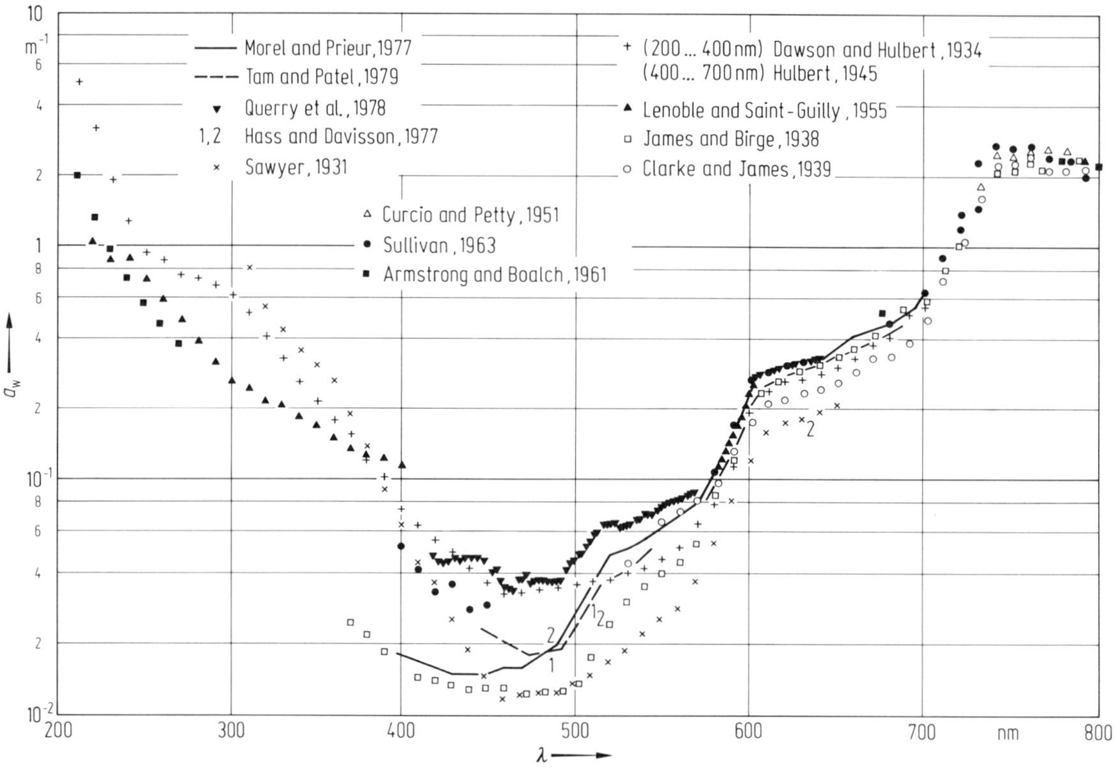

Fig. 4. Light absorption coefficient for pure freshwater a_w vs. wavelength λ as given by various authors (Smith and Baker [81 S 1]).

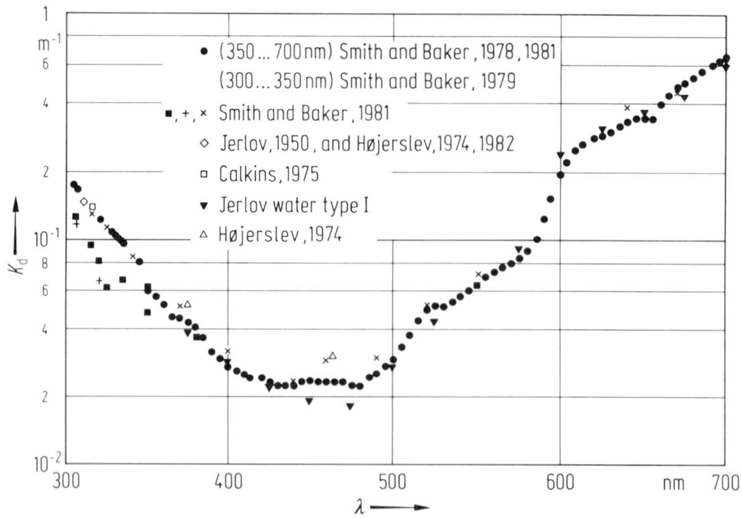

Fig. 5. Vertical attenuation coeffcient for downward irradiance K_d vs. wavelength λ as determined by various authors for clear ocean waters (Smith and Baker [81 S 1].

5. The clearest natural waters of the world are useless for *in-situ* measurements of the volume light scattering function for pure (sea) water due to their pronounced forward light scattering. Accordingly, the spectral volume light scattering function has to be measured *in-vitro*. Spectral measurements of light transmission, $T(\lambda)$, or, alternatively, $c(\lambda)$, are principally performed in a manner defined by eq. (38). Between $T(\lambda)$ and $c(\lambda)$ there exists the following relation:

$$T(\lambda)=\mathrm{e}^{-c(\lambda)r} \tag{80}$$

in which r is the path length traversed by an infinitesimally thin light beam, consequently having no divergence. This condition can never be achieved in the real experimental situation. Therefore, eq. (80) has to be modified into:

$$T(\lambda)=k(\lambda)\mathrm{e}^{-c(\lambda)r} \tag{81}$$

in which $k(\lambda)$ is a calibration constant depending on the geometry, optics, and experimental set-up of the light transmission meter applied (Højerslev [85 H 2]. $k(\lambda)$ is typically of the order of 0.85 ± 0.15. Conventional maximum error analysis applied to eq. (81) give the relative error for c at a fixed wavelength:

$$\frac{\Delta c}{c}=\frac{1}{cr}\left(\frac{\Delta k}{k}+\frac{1}{k}\mathrm{e}^{cr}\Delta T\right) \tag{82}$$

in which ΔT has to be interpreted as the uncertaincy in the measured light transmission due to light sensor stability and the maximum fluctuations around the average radiant flux output of the light transmission meter. Assuming unrealistically $\Delta k=0$ and $k=1$, it is seen that $\Delta c/c$ attains its minimum for $cr=1$. This condition can practically speaking, not be met in the laboratory for measurements of the spectral attenuation coefficient of pure (sea) water in the ultraviolet, blue, and green (in contrast to the red) part of the spectrum because r would have to be 10···40 m in approximately. Typical path lengths are around 3 m, so the differing c values in the before mentioned spectral region can be given at least one explanation, Fig. 4.

The spectral light scattering function for pure (sea) water is most suitably measured in a manner described by eq. (42) and depicted in Fig. 3 for the case where $l+r=l_0$ because the factor multiplying c in eq. (42) vanishes in this case. Conventional maximum error analysis applied to eq. (42) for the case $l+r=l_0$ gives:

$$\frac{\Delta\beta}{\beta}=\Delta T, \tag{83}$$

in which ΔT has to be interpreted as in eq. (82). It should be emphasized that, for the experimental set-up of the light scattering meter, a calibration factor like in case of the light transmission meter has not to be included. The main difficulty for measuring the light scattering properties of pure (sea) water lies clearly in both the purity of the water batch as well as in the geometry and optics of the light scattering meter, which can be directly seen from eq. (42); $\mathrm{d}l$ and $\mathrm{d}\omega$ must be small and well-defined, which is not fulfilled for $\mathrm{d}l$ at small scattering angles. However, in case of pure (sea) water this is not a critical point since the overall variation of $\beta(\alpha)$ equals a factor of 2.

The *in-situ* method for arriving at what is believed the most reliable spectral light absorption coefficients for pure (sea) water makes use of eqs. (58), (60), or (61). The downward (and upward) spectral irradiance measurements are performed in the clearest occuring natural waters of the world (Smith and Baker [81 S 1]). Estimates or models of the spectral average cosine, $\bar{\mu}_{\mathrm{w}}(\lambda)$, must be made for these clear waters in order to calculate the spectral light absorption coefficient for pure (sea) water $a_{\mathrm{w}}(\lambda)$. This procedure is sufficiently accurate for the purpose in question since $\bar{\mu}_{\mathrm{w}}(\lambda)$ is a slowly varying parameter with respect to surface light conditions, wavelength, and depth, see Table 2 and Højerslev [78 H].

Conventional maximum error analysis applied to the eqs. (8), (23), and (61) results in (Højerslev [85 H 2]):

$$\frac{\Delta K_{\mathrm{d}}}{K_{\mathrm{d}}}=\frac{\Delta K_{\mathrm{w}}}{K_{\mathrm{w}}}=\frac{\Delta a_{\mathrm{w}}}{a_{\mathrm{w}}}=4\frac{\mathrm{e}^{2K_{\mathrm{d}}(z_2-z_1)}}{2K_{\mathrm{d}}(z_2-z_1)}\frac{\Delta E_{\mathrm{d}}(z_1)}{E_{\mathrm{d}}(z_1)}+\frac{\Delta z}{z_2-z_1} \tag{84}$$

in which z_1 and z_2 are two depths at which the downward irradiance is measured. Ignoring the last term, meaning that the depth determinations z_1 and z_2 are accurate, $\Delta a_{\mathrm{w}}/a_{\mathrm{w}}$ has its minimum for $2K_{\mathrm{d}}(z_2-z_1)=1$.

This condition can easily be fulfilled in nature at all wavelengths by selecting z_1 and z_2 appropriately. The maximum value for z_2-z_1 equals 30 m around $\lambda=450$ nm. It should finally be noted that the above *in-situ* method is not hampered by contamination problems as is the case for the *in-vitro* method.

The light scattering function for pure (sea) water $\beta_{\mathrm{w}}(\alpha)$ is described through the following equations, Morel [74 M 2]:

$$\beta_{\mathrm{w}}(\alpha)=\beta_{\mathrm{w}}(90^\circ)\left[1+\frac{1-\delta}{1+\delta}\cos^2\alpha\right] \tag{85}$$

and

$$b_w = \frac{8\pi}{3} \beta_w(90°) \frac{2+\delta}{1+\delta} \quad (86)$$

in which the depolarization ratio $\delta = 0.09$.

The wavelength dependence of $\beta_w(\alpha)$ can be expressed in terms of a power law of the form $\lambda^{-4.32}$ (not as λ^{-4} as is the case for the pure Rayleigh scattering). $b_w(S = 35 \cdots 39‰)$ is approximately 30% higher than $b_w(S = 0‰)$.

Table 2. Spectral values of vertical attenuation coefficient for the downward irradiance K_w, light absorption coefficient, a_w, average cosine $\bar{\mu}_w$ for the clearest natural ocean waters, and light scattering coefficient b_w for pure artificial sea water. After Morel [74 M 2], Smith and Baker [81 S 1].

λ nm	K_w m^{-1}	a_w m^{-1}	$\bar{\mu}_w$	b_w m^{-1}	λ nm	K_w m^{-1}	a_w m^{-1}	$\bar{\mu}_w$	b_w m^{-1}
200	3.14	3.07	0.978	0.151	500	0.0271	0.0257	0.948	0.0029
210	2.05	1.99	0.971	0.119	510	0.0370	0.0357	0.965	0.0026
220	1.36	1.31	0.963	0.0995	520	0.0489	0.0477	0.975	0.0024
230	0.968	0.927	0.958	0.0820	530	0.0519	0.0507	0.977	0.0022
240	0.754	0.720	0.955	0.0685	540	0.0568	0.0558	0.982	0.0021
250	0.588	0.559	0.951	0.0575	550	0.0648	0.0638	0.985	0.0019
260	0.481	0.457	0.950	0.0485	560	0.0717	0.0708	0.987	0.0018
270	0.394	0.373	0.947	0.0415	570	0.0807	0.0799	0.990	0.0017
280	0.306	0.288	0.941	0.0353	580	0.109	0.108	0.991	0.0016
290	0.230	0.215	0.935	0.0305	590	0.158	0.157	0.994	0.0015
300	0.154	0.141	0.916	0.0262	600	0.245	0.244	0.996	0.0014
310	0.116	0.105	0.905	0.0229	610	0.290	0.289	0.997	0.0013
320	0.0944	0.0844	0.894	0.0200	620	0.310	0.309	0.997	0.0012
330	0.0765	0.0678	0.886	0.0175	630	0.320	0.319	0.997	0.0011
340	0.0637	0.0561	0.881	0.0153	640	0.330	0.329	0.997	0.0010
350	0.0530	0.0463	0.874	0.0134	650	0.350	0.349	0.997	0.0010
360	0.0439	0.0379	0.863	0.0120	660	0.400	0.400	1	0.0008
370	0.0353	0.0300	0.850	0.0106	670	0.430	0.430	1	0.0008
380	0.0267	0.0220	0.824	0.0094	680	0.450	0.450	1	0.0007
390	0.0233	0.0191	0.820	0.0084	690	0.500	0.500	1	0.0007
400	0.0209	0.0171	0.818	0.0076	700	0.650	0.650	1	0.0007
410	0.0196	0.0162	0.827	0.0068	710	0.834	0.839	1.006	0.0007
420	0.0184	0.0153	0.832	0.0061	720	1.170	1.169	0.999	0.0006
430	0.0172	0.0144	0.837	0.0055	730	1.800	1.799	0.999	0.0006
440	0.0170	0.0145	0.853	0.0049	740	2.380	2.38	1	0.0006
450	0.0168	0.0145	0.863	0.0045	750	2.47	2.47	1	0.0005
460	0.0176	0.0156	0.886	0.0041	760	2.55	2.55	1	0.0005
470	0.0175	0.0156	0.891	0.0037	770	2.51	2.51	1	0.0005
480	0.0194	0.0176	0.907	0.0034	780	2.36	2.36	1	0.0004
490	0.0212	0.0196	0.925	0.0031	790	2.16	2.16	1	0.0004
					800	2.07	2.07	1	0.0004

3.3.2.2 Optical properties of natural sea waters

The optical properties of some coastal and oceanic waters are represented in Table 3. Clear and turbid waters are clearly identified but apart from this the usefulness of such an overview is rather limited. In order to characterize/classify natural waters optically there exist two basically different ways to proceed. The first and original classification scheme is based upon empirical simple relationships between routinely measured optical parameters whereas the second one is based upon the light absorption properties of the sea water concentrations of suspended inorganic matter and detritus, C_p, living phytoplankton, C_{ph}, dissolved organic matter, C_y, and the earlier described optical properties of pure sea water. In neither of the two classification schemes the light scattering function $\beta(\alpha)$ or the light scattering coefficient b are given any pertinent preference due to several reasons of which the most important ones are listed below:

1. The *shape* of the light scattering functions is virtually independent of the water masses from which they have been obtained in the scattering range $3° \leqq \alpha \leqq 170°$ (Figs. 6 and 7 and Kullenberg [74 K]).

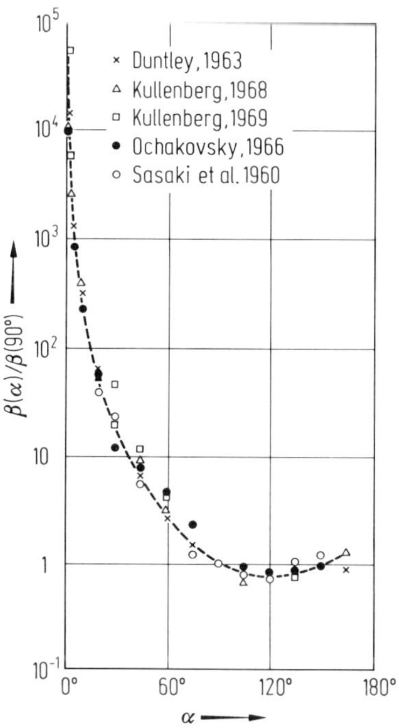

Fig. 6. Ratio of scatterance at α scatterance at 90°, $\beta(\alpha)/\beta(90°)$ vs. scattering angle α (Kullenberg [74 K]). Duntley, 1963, lake; Kullenberg, 1968, Sargasso Sea; Kullenberg, 1969, Baltic; Ochakovsky, 1966, Mediterranean; Sasaki et a., 1960, Pacific.

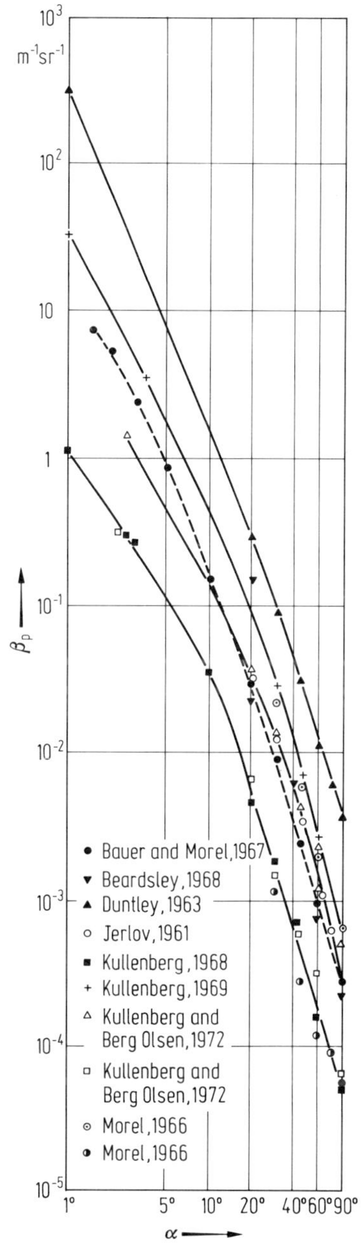

Fig. 7. Particle scattering function β_p vs. scattering angle α (Kullenberg [74 K]). Bauer and Morel, 1967, Mediterranean; Beardsley, 1968, Atlantic; Duntley, 1963, lake surface; Jerlov, 1961, Atlantic surface; Kullenberg, 1968, Sargasso Sea, 10···75 m depth; Kullenberg, 1969, Baltic surface; Kullenberg and Berg Olsen, 1972, Mediterranean, Gibraltar, 10 m depth; Kullenberg and Berg Olsen, 1972, western Mediterranean, 100 m depth; Morel, 1966, Mediterranean, 50 m depth; Morel, 1966, Mediterranean, 2500 m depth.

Table 3. Comparison of inherent properties (Jerlov [76 J]). $c = c_w + b_p + a_p + a_y$ = total attenuation coefficient; c_w = attenuation coefficient for water; b_p = scattering coefficient for particles; a_p = absorption coefficient for particles; a_y = absorption coefficient for yellow substance: Δ = wavelength selectivity.

Region	Wave-length	$c-c_w$ observed	b_p observed	a_p+b_p	a_p	Δa_p	a_p+a_y	$\Delta(a_p+a_y)$	a_y observed	Δa_y	Ref.
	nm	m^{-1}									
Sargasso Sea	440	0.05	0.04				0.01				
Carribean Sea	655	0.06	0.06	0.06	0		0.00				
	440	0.09	0.06				0.03	0.03			Jerlov (1951)
Equator Central Pacific	440	0.09	0.05				0.04				
Romanche Deep	440	0.12	0.07				0.05				
Galapagos	655	0.11	0.07	0.11	0.04		0.04				
	440	0.24	0.08				0.16	0.12			
Off Peru (64 miles)	700	0.39		0.39					0		
	400	0.73		0.64					0.09	0.09	
Galapagos	700	0.16		0.16					0		Burt (1958)
	400	0.25		0.21					0.04	0.04	
Northwestern Galapagos	700	0.07		0.07					0		
	400	0.11		0.08					0.03	0.03	
Continental slope	668	0.05							0		Clarke and James (1939)
	365	0.10		0.08					0.02	0.02	
Bermuda	655	0.10		0.10					0		Ivanoff et al. (1961)
	380	0.20		0.17					0.03	0.03	
Kattegat	655	0.23	0.15	0.23	0.08		0.08		0		
	380	0.54	0.16	0.44	0.27	0.19	0.38	0.30	0.11	0.11	
South Baltic Sea	655	0.27	0.20	0.27	0.07		0.07		0		Jerlov (1955a)
	380	1.15	0.21	0.49	0.28	0.21	0.96	0.89	0.68	0.68	
Bothnian Gulf	655	0.38	0.28	0.38	0.10		0.10		0		
	380	1.72	0.31	0.64	0.33	0.23	1.41	1.31	1.08	1.08	
Skagerrak	655										
	380					0.3		0.4		0.1	Malmberg (1964)
North Atlantic	665								0		
	420								0.03	0.03	

Højerslev

Region	Wave-length	$c-c_w$ observed	b_p observed	a_p+b_p	a_p	Δa_p	a_p+a_y	$\Delta(a_p+a_y)$	a_y observed	Δa_y	Ref.
	nm	m^{-1}									
North Sea	665								0.01		Kalle (1961)
	420								0.10	0.09	
Baltic Sea	665								0.02		
	420								0.33	0.31	
Sargasso Sea	633	0.03	0.02	0.03	0.01		0.01				G. Kullenberg et al. (1970)
	484	0.03			0.01		0.02				
	440	0.05	0.04		0.01		0.01				
	375		0.02								
Western Mediterranean area	655	0.08	0.04	0.08	0.04		0.04		0		Højerslev (1973)
	375	0.06	0.04				0.02				
Villefranche offshore											G. Copin-Montegut et al. (1971)
50 m	250	1.98		0.44					0.98		
1000 m		0.65							0.2		
coastal		3.42		1.65							
Baltic Sea	655				0.15		0.15				Højerslev (1974a)
	375				0.20		0.80		0.6		

2. The number of observations of $\beta(\alpha)$ is limited, especially at small scattering angles, say $\alpha \leqq 3°$, see, for instance, Figs. 8 and 9.

3. There are possibly more observations of b measured by means of b meters integrating between say $3° \leqq \alpha \leqq 180°$, e.g. Højerslev and Larsen [80 H 2], but since

$$0.02 \leqq \int_0^{3°} \beta(\alpha) d\omega \Big/ \int_{3°}^{180°} \beta(\alpha) d\omega \leqq 13.3 \quad (87)$$

assuming the Mie light scattering theory to be applied for $0 \leqq b/c \leqq 1$ and for a Junge particle distribution (Højerslev [85 H 2]), such directly measured b values might be uncertain in some regions, possibly the coastal ones.

4. Measurements of $\beta(45°)$ have earlier been used in optical classification schemes but should be avoided since $\beta(45°)$ does not characterize the light scattering properties of a water mass appropriately, e.g. Table 4 and Kullenberg [74 K].

5. The different natural sea waters in the world contain highly varying concentrations of dissolved organic matter, expressed in terms of the light absorption a_y(375 nm) by a "particle-free" water mass, being clearly uncorrelated with the concentration of suspended matter, expressed in terms of $z_q(1\%)$ (Table 5 and Højerslev [82 H 1]). Since the dissolved organic matter is not scattering the light measurably, in contrast to the suspended matter, the light scattering function is not any *key* parameter in optical water mass classification. In order to avoid confusion at this point it is emphasized that $\beta(\alpha)$ is certainly a main parameter in radiant energy transfers in the ocean, as also stated earlier, as well as a useful auxiliary parameter in optical classification schemes.

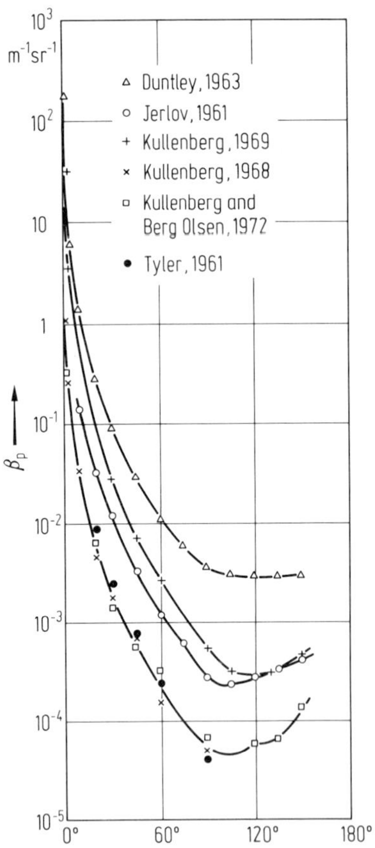

Fig. 8. Particle scattering function β_p observed *in-situ* vs. scattering angle α (Kullenberg, [74 K]). Duntley, 1963, lake; Jerlov, 1961, Atlantic; Kullenberg, 1969, Baltic; Kullenberg, 1968, Sargasso Sea; Kullenberg and Berg Olsen, 1972, Mediterranean; Tyler, 1961, Pacific.

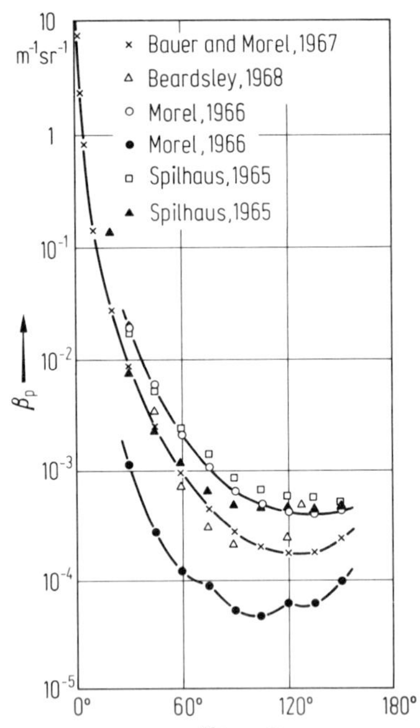

Fig. 9. Particle scattering function β_p observed *in-vitro* vs. scattering angle α (Kullenberg [74 K]). Bauer and Morel, 1967, Mediterranean; Beardsley, 1968, Atlantic; Morel, 1966, Mediterranean, 50 m depth; Morel, 1966, Mediterranean, 250 m depth; Spilhaus, 1965, Atlantic, 50 m depth; Spilhaus, 1965, Atlantic, 700 m depth.

Table 4. Ratio of light scatterance at 45° to total light scattering coefficient, $\beta(45°)/b$. Kullenberg [74 K].

Area	Depth m	Wavelength nm	$\beta(45°)/b$ 10^{-2}	Ref.
Atlantic	surface	465; 625	3.4	Jerlov (1961)
East China Sea	surface	≈450	3.3	Kozlyaninov (1957)
Lake	surface	522	3.3	Duntley (1963)
Mediterranean	surface	545	3.5	Ochakovsky (1966)
Sargasso Sea	10···75	460; 650	3.5	Kullenberg (1968)
Mediterranean	10	633	2.5	Kullenberg and Berg Olsen (1972)
Mediterranean	100	633	3.4	Kullenberg and Berg Olsen (1972)
Baltic Sea	surface	460; 650	2.1	Kullenberg (1969)
Suspension of quartz		–	3.2	Hodkinson (1963)
Suspension of bituminous coal			2.0	Hodkinson (1963)

Table 5. Vertical attenuation coefficients for the downward UV–B irradiance, K_d (310 nm), and the depth of the euphotic zone, z_q (1%), measured in different waters in the marine environment. Højerslev [85 H 1]. a_y: light absorption coefficient for yellow substance.

Area	K_d (310 nm) m^{-1}	z_q (1%) m	a_y (375 nm) m^{-1}
German Bight	5.0	3.5	1.36
German Bight	4.7···5.0	7.6	1.37···1.49
German Bight	0.53···1.77	12	0.07···0.47
German Bight	1.2	25	0.31
Baltic Proper	3.0···3.5	16···20	0.89···1.04
Kattegat	1.2···2.4	17···20	0.30···0.68
Skagerrak	0.59···1.2	20···26	0.12···0.31
Hardangerfjord	0.88···2.9 [1])	20···25	0.32···1.18
North Sea	0.48···0.67 [1])	23···30	0.09···0.15
North Sea	0.37···0.65 [1])	35···50	0.07···0.15
Orkney – Shetland	0.39	16···17	0.04
Shetland – Faroe	0.19	32	≈0
Faroe	0.17···0.19	35···40	≈0
Iceland	0.30 [1])	25···32	0.04
Svalbard	0.20···0.25	50···56	0.01···0.02
Western Greenland	0.17···0.21	46···57	0.01
Western Mediterranean	0.43 [1])	35···40	0.008
Western Mediterranean	0.22 [1])	56···59	0.02
Western Mediterranean	0.16 [1])	70···75	≈0
Gulf of Mexico	0.10···0.12	66···68	≈0
Sargasso Sea	0.15 [2])	112	≈0

[1]) Calculated from irradiance measurements at 372 nm.
[2]) Jerlov's optical classification of water masses (1976).

3.3.3 Optical classifications of sea water

3.3.3.1 General principles of classification

The two basically different optical classification schemes are both based on eq. (61):

$$a \cong \bar{\mu} K_d$$

in which a is an inherent optical property whereas $\bar{\mu}$ and K_d are both apparent optical properties. For fixed wavelengths the almost only existing measurements of $\bar{\mu}$ vs. depth, e.g. Højerslev [73 H, 74 H 1, 75 H, 77 H 2, 78 H], have demonstrated that $\bar{\mu}$ is virtually constant with depth in clear and turbid waters and for any surface light conditions. Consequently, Table 6 shows in essence the spectral variation of $\bar{\mu}$ in clear and turbid waters. It is seen that $\bar{\mu}$ as a first order approximation can be assumed a constant, meaning that $\bar{\mu}$ can be treated as an inherent optical property. A simple optical model from the results of Sauberer and Ruttner [41 S] gives for the surface layers $0.74 \leqq \bar{\mu} \leqq 0.85$ for visible light and all surface light conditions.

Spectral measurements of $\bar{\mu}$ at the same location (Fladen Ground in the North Sea) during and after an intense plankton bloom demonstrate a seasonal variation of the order of $\pm 10\%$ or less, Fig. 10. Accordingly:

$$\bar{\mu} \simeq 0.80\,. \tag{88}$$

Timofeeva ([74 T] and Table 7) has calculated $\bar{\mu}$ in the asymoptotic region as a function of k/c in the range $0 \leqq k/c \leqq 1$ (the lower limit corresponds to a solely scattering medium and the upper one to a solely absorbing medium). Eq. (88) and Table 7 suggest the following first order approximation:

$$k/c \simeq 0.50 \tag{89}$$

on average for visible light in both clear and turbid waters, see also Fig. 11, the measurement of $\bar{\mu}$ in Lake Baikal, and Table 8.

Radiance measurements in very clear and very turbid water (Table 8) reveal that $0.35 \leqq k/c \leqq 0.84$. If these extreme values are applied to Table 7 the following inequality is valid:

$$0.70 \leqq \bar{\mu} \leqq 0.94 \tag{90}$$

which compares reasonably well with the results in Table 6 and Fig. 10.

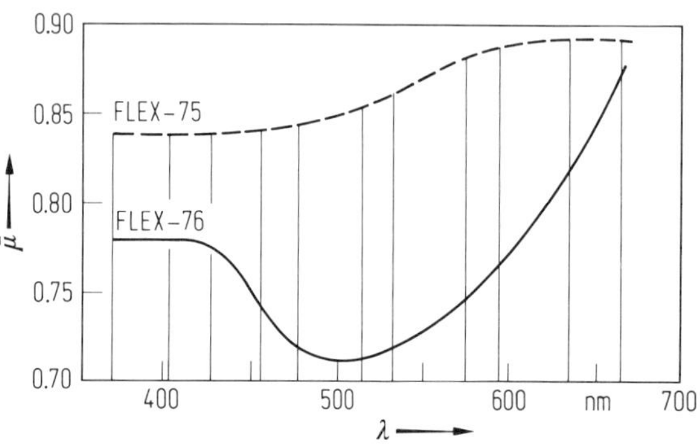

Fig. 10. Spectral average cosine $\bar{\mu} = E/E_0$ from the euphotic zone during FLEX-75 and FLEX-76 at the Fladen Ground in the North Sea. The vertical lines refer to the wavelengths where the measurements were made (Højerslev [82 H 2]).

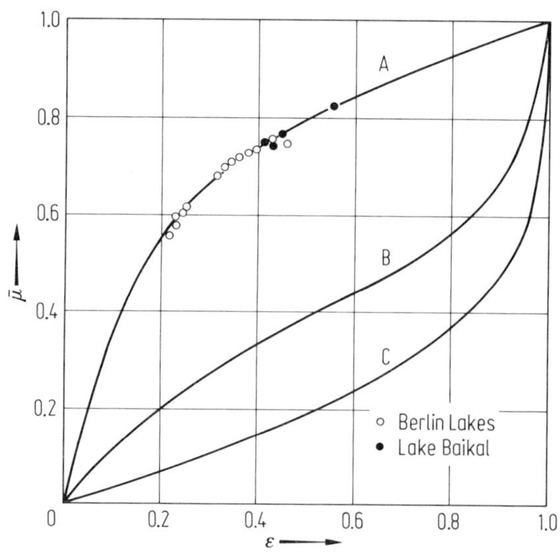

Fig. 11. Dependence of the average cosine $\bar{\mu}$ on $\varepsilon = k/c$ in various media (Timofeeva [74 T]). A: milky medium (Berlin Lakes, Schellenberger [69 S 1], and Lake Baikal); B: medium with latex; C: medium with Rayleigh scattering.

Table 6. Average values of $E_d/E_o \cong \bar{\mu}$ from the euphotic zone. Højerslev [78 H].

Region	Solar elevation [°]	Wave-length nm	E_d/E_o	Author
The Sound	29	372	0.78	Højerslev (1976)
	28	403	0.86	
	27	427	0.83	
	26	455	0.82	
	22	477	0.77	
	20	513	0.70	
	30	532	0.77	
	30	572	0.75	
	32	593	0.81	
	32	633	0.77	
	29	670	0.85	
Fladen Ground FLEX 75	40	525	0.78	Højerslev (1977)
	overcast	403	0.85	
	–	465	0.79	
	–	477	0.84	
	–	532	0.85	
	–	572	0.88	
	–	593	0.88	
	34	465	0.69	
Mediterranean off Gibraltar	68	372	0.88	Højerslev (1973)
	74	427	0.69	
	68	477	0.69	
	70	533	0.73	
	71	572	0.67	
	75	633	0.79	
Off Sardinia	72	372	0.78	Højerslev (1973)
	64	427	0.62	
	60	477	0.73	
	73	533	0.84	
	68	572	0.85	
	74	633	0.93	

Table 7. Dependence of average cosine $\bar{\mu}$ on $\varepsilon(=k/c)$. Timofeeva [74 T].

ε	0	0.01	0.02	0.03	0.04	0.05	0.06	0.07	0.08	0.09
					$\bar{\mu}$					
0.0	0	0.060	0.080	0.117	0.155	0.190	0.223	0.254	0.285	0.313
0.1	0.340	0.366	0.390	0.413	0.434	0.455	0.474	0.492	0.509	0.526
0.2	0.541	0.556	0.570	0.583	0.596	0.608	0.620	0.631	0.641	0.652
0.3	0.661	0.671	0.680	0.688	0.696	0.704	0.712	0.719	0.727	0.734
0.4	0.740	0.747	0.753	0.759	0.765	0.771	0.776	0.782	0.787	0.792
0.5	0.798	0.803	0.807	0.812	0.817	0.821	0.826	0.830	0.835	0.839
0.6	0.843	0.847	0.851	0.856	0.860	0.864	0.867	0.871	0.875	0.879
0.7	0.883	0.887	0.890	0.894	0.898	0.902	0.905	0.909	0.913	0.917
0.8	0.920	0.924	0.928	0.932	0.935	0.939	0.943	0.947	0.951	0.955
0.9	0.958	0.962	0.966	0.970	0.974	0.979	0.983	0.987	0.991	0.996
1.0	1									

Table 8. Optical properties derived from near-asymptotic radiance distributions in the sea. Jerlov [74 J 1].

Region	Depth	Wave-length	Observed		Calculated			Ref.
			k	b	c	a	k/c	
	m	nm	m^{-1}					
Baltic Sea	100	535	0.12	0.25	0.34	0.09	0.35	Jerlov and Nygård (1968)
Sargasso Sea	150	375	0.042	0.030	0.053	0.023	0.79	Lundgren and Højerslev (1971)
	250	425	0.41	0.24	0.48	0.24	0.85	
	400	475	0.27	0.16	0.32	0.16	0.85	
	150	525	0.46	0.23	0.55	0.32	0.84	

Eqs. (61) and (88) imply that K_d is a quasi-inherent optical property for visible light and all depths since K_d can be expressed in terms of a alone (note that b is only implicitely involved through the inequality $0<\bar{\mu}<1$). Many authors have suggested that K_d changes with the solar elevation in the upper surface according to a cosine law but direct measurements, e.g. Højerslev [74 H 1, 74 H 2, 80 H 1], demonstrate almost no such dependence, Figs. 12 and 13 and Table 9. Consequently,

$$a \cong 0.80\, K_d \tag{91}$$

in which K_d is (to a first order) an inherent optical property solely determined by the spatial, temporal, and spectral variations of a. Eq. (91) summarizes the background and forms the basis for the existing two optical classifications.

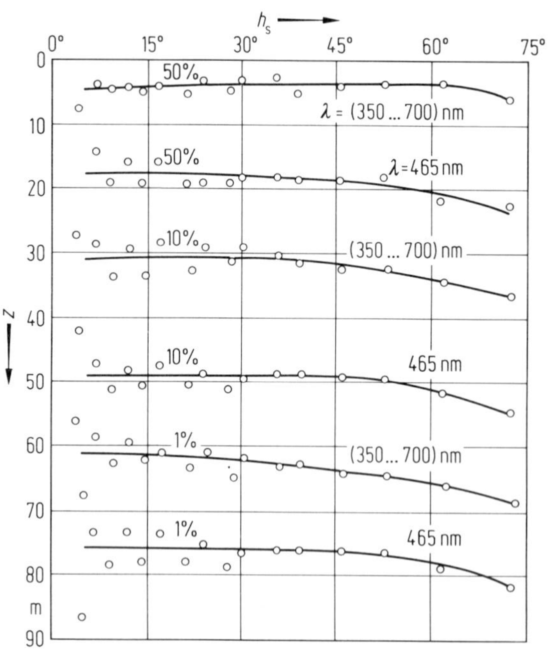

Fig. 12. Depths z of 50%, 10%, and 1% levels at different solar elevations h_s for downward quanta irradiance ($\lambda = 350 \cdots 700$ nm) and for blue downward irradiance ($\lambda = 465$ nm). Clear water off Sardinia. (After Højerslev [74 H 2]).

Table 9. Vertical irradiance attenuation coefficient for the downward irradiance K_d for different solar elevations in the Sargasso Sea. (After Lundgren and Højerslev [71 L 1]).

Depth	Solar elevation	Wavelength [nm]	
m	[°]	375	484
12	25	0.047	0.032
	56	0.043	0.023
37	25	0.041	0.031
	56	0.040	0.022
62	25	0.051	0.033
	56	0.045	0.029
87	25	0.055	0.034
	56	0.051	0.034
125	25	0.054	0.035
	56	0.055	0.036

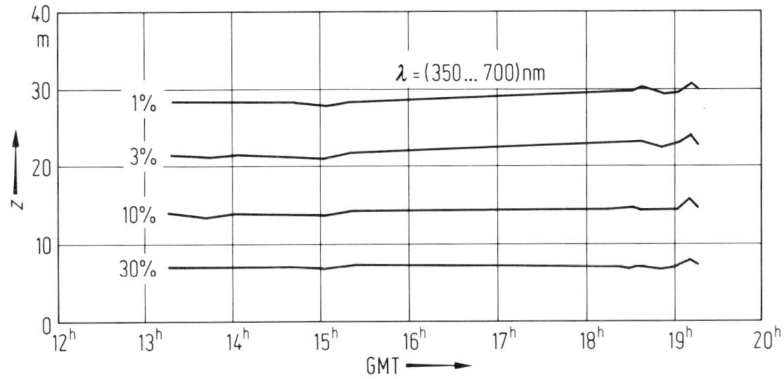

Fig. 13. Depths z of selected percentage levels for the downward quanta irradiance ($\lambda=350\cdots700$ nm) on 26 May, 1976 vs. GMT time at Fladen Ground in the North Sea. *Note:* the depths of the selected levels are virtually constant throughout the day. (Højerslev [80 H 1]).

3.3.3.2 The original optical classification

Jerlov's optical classification of water masses [76 J] is founded on the spectral ($\lambda=310\cdots700$ nm) daylight transmission in the surface layers, expressed in terms of $K_d(\lambda)$, originally at specified solar elevations h_s, for the oceanic water types *I*, *IA*, *IB*, *II*, and *III*, $h_s=90°$, and for the coastal water types, *1*, *3*, *5*, and *9*, $h_s=45°$ (see Fig. 15 and Tables 10 and 11). As explained earlier, the solar elevation constraint can be avoided since $K_d(\lambda)$ is a quasi-inherent optical property. $K_d(\lambda)$ is selected as one appropriate parameter for classification purposes because $K_d(\lambda)$ is easily obtained from vertical profiles of $E_{d\lambda}(\lambda)$ through the relation:

$$K_d(\lambda)=\frac{1}{z_2-z_1}\ln\frac{E_{d\lambda}(\lambda,z_1)}{E_{d\lambda}(\lambda,z_2)}, \quad (92)$$

which is derived from eq. (23). Eq. (92) demonstrates that the two measurements of $E_{d\lambda}(\lambda,z_1)$ and $E_{d\lambda}(\lambda,z_2)$ at the corresponding depths z_1 and z_2, respectively, can be performed by an E_d-meter only measuring in relative units, not in [W m^{-2}] in order to give $K_d(\lambda)$ in the absolute unit [m^{-1}]. Eq. (84) shows that the accuracy of the $K_d(\lambda)$ determinations is satisfactory.

Jerlov's classification [76 J] is based on the idea that a given $K_d(\lambda_0)$ for a given water mass (λ_0 was originally equal to 465 nm) permits the construction of all $K_d(\lambda)$'s between $\lambda=310$ nm and $\lambda=700$ nm, Figs. 14, 15, and 16; Tables 10 and 11.

Fig. 16 depicts that any $K_d(\lambda)$ in the spectral range 310 nm $\leqq\lambda\leqq$ 700 nm for the oceanic water types can be given in terms of K_d(475 nm) through linear relations of the form

$$K_d(\lambda)=\alpha(\lambda)K_d(475\text{ nm})+\gamma(\lambda) \quad (93)$$

in which $\alpha(\lambda)$, $\gamma(\lambda)$ are constants depending on λ and the optical properties of pure sea water at λ and 475 nm, see Table 2 also.

Jerlov's optical classification [76 J] contains the following implicite assumptions:

(1) $$K_d(\lambda)=K_w(\lambda)+K_p(\lambda)+K_{ph}(\lambda)+K_y(\lambda)$$

exactly, see eq. (30).

(2) The concentrations of suspended inorganic matter and detritus (index p), living phytoplankton (index ph) and dissolved organic matter (index y), respectively, are in fixed proportions. Now, point (2) is far from being fulfilled in general. Especially $K_y(\lambda)$ (or the concentration of dissolved organic matter) is not correlated with either $K_p(\lambda)$ or $K_{ph}(\lambda)$ (Figs. 17 and 18, and Table 5). For most open oceanic waters, situated between type *I* and *II* (see Fig. 51), $K_p(\lambda)$ and $K_{ph}(\lambda)$ are believed to be interrelated (Morel [80 M]: "case 1" waters) and, since $K_y(\lambda)$ is small in general, this term is often ignored.

Table 10. Irradiance transmittance in different surface waters. Jerlov [76J].

Water type	Wavelength [nm] 310	350	375	400	425	450	475	500	525	550	575	600	625	650	675	700
Irradiance transmittance [%/m]																
I	86	94	96.3	97.2	97.8	98.1	98.2	97.3	95.8	93.9	91.5	79	74	70	66	57
IA	83	92.5	95	96.3	96.9	97.4	97.5	96.9	95.3	93.5	91.0	78.5	73	69	65	56.5
IB	80	90.5	94	95	95.9	96.5	96.8	95.9	94.7	93	90.5	78	72.5	68.5	64	56
II	69	84	89	91	92.2	93.5	94	93.2	92.7	91.5	89	77	71.5	67	63	54
III	52	73	80	83	85	87.5	89	89	89	88.5	86	74	69	64	60	52
1	17	30	45	60	70	78	84	87	88	88.5	86	74	69	64	60	52
3	9	19	32	46	58	68	75	80	82	83	81	72	67	63	57	49
5	3	10	21	33	46	57	65	70	73	74	72	67	62	58	52	45
7		5	12	21	31	41	49	56	61	63	63	62	58	53	46	40
9		2	5	9	15	21	29	37	46	53	56	55	52	47	40	33
Irradiance transmittance [%/10 m]																
I	22	54	68	76	80	83	83.5	76	65	53	41	9.5	4.7	2.7	1.5	0.4
IA	16	46	60	68	73	77	78	73	62	51	39	9.1	4.5	2.5	1.4	0.3
IB	11	37	52	60	66	70	72	66	58	49	37	8.6	4.3	2.4	1.3	0.3
II	2.5	17	30	38	44	51	54	50	47	41	32	7.4	3.5	1.8	0.9	0.2
III	0.2	4	11	16	20	26	31	32	31	30	23	5.2	2.4	1.2	0.5	0.1
1				0.6	2.7	8.2	18	25	27	30	22	5.0	2.5	1.1	0.6	0.2
3					0.5	2.1	5.5	11	13.5	15	12	3.7	1.8	1.0	0.4	
5						0.4	1.4	2.7	4.5	5	3.7	1.8	0.8	0.4	0.2	
7								0.3	0.7	1.0	1.0	0.8	0.5	0.2		
9										0.2	0.3	0.3	0.2			

Table 11. Vertical irradiance attenuation coefficient for the downward irradiance K_d (0···10 m), in [10^{-2} m^{-1}]. Jerlov [76 J].

Water type	Wavelength [nm]															
	310	350	375	400	425	450	475	500	525	550	575	600	625	650	675	700
I	15	6.2	3.8	2.8	2.2	1.9	1.8	2.7	4.3	6.3	8.9	23.5	30.5	36	42	56
IA	18	7.8	5.2	3.8	3.1	2.6	2.5	3.2	4.8	6.7	9.4	24	31	37	43	57
IB	22	10	6.6	5.1	4.2	3.6	3.3	4.2	5.4	7.2	9.9	24.5	31.5	37.5	43.5	58
II	37	17.5	12.2	9.6	8.1	6.8	6.2	7.0	7.6	8.9	11.5	26	33.5	40	46.5	61
III	65	32	22	18.5	16	13.5	11.6	11.5	11.6	12.0	14.8	29.5	37.5	44.5	52	66
1	180	120	80	51	36	25	17	14	13	12	15	30	37	45	51	65
3	240	170	110	78	54	39	29	22	20	19	21	33	40	46	56	71
5	350	230	160	110	78	56	43	36	31	30	33	40	48	54	65	80
7		300	210	160	120	89	71	58	49	46	46	48	54	63	78	92
9		390	300	240	190	160	123	99	78	63	58	60	65	76	92	110

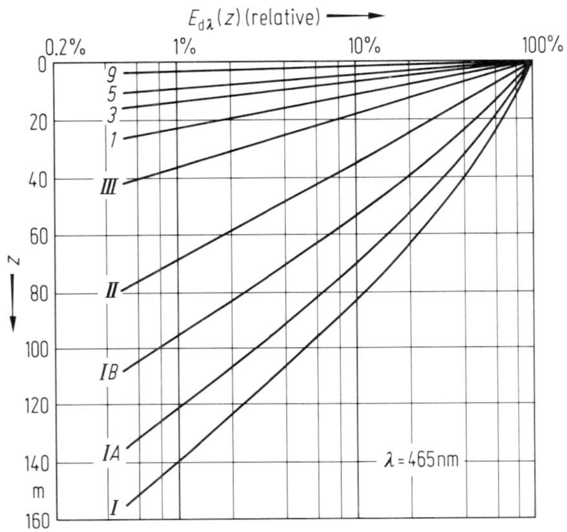

Fig. 14. Depth profiles of downward irradiance, $E_{d\lambda}$(465 nm), in percent of surface irradiance for the oceanic water types *I*, *IA*, *IB*, *II* and *III*, and for the coastal water types *1*, *3*, *5*, and *9* (Jerlov [76 J]).

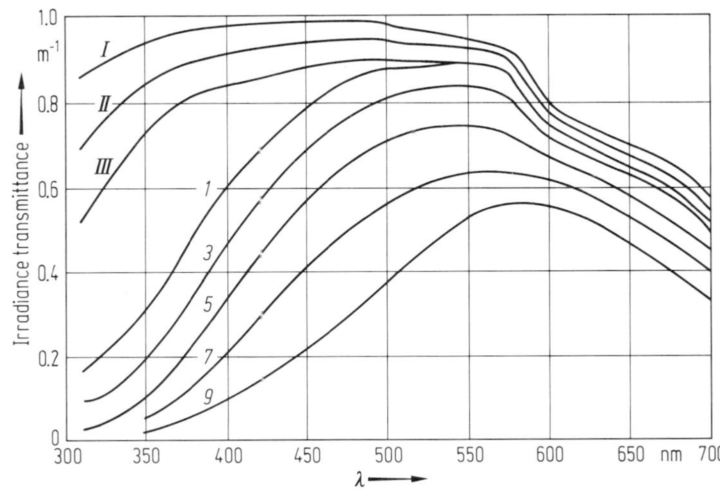

Fig. 15. Transmittance per meter of downward irradiance in the surface layer for optical water types. Oceanic types *I*, *II*, *III* and coastal types *1*, *3*, *5*, *7*, *9* (Jerlov [76 J]).

Højerslev

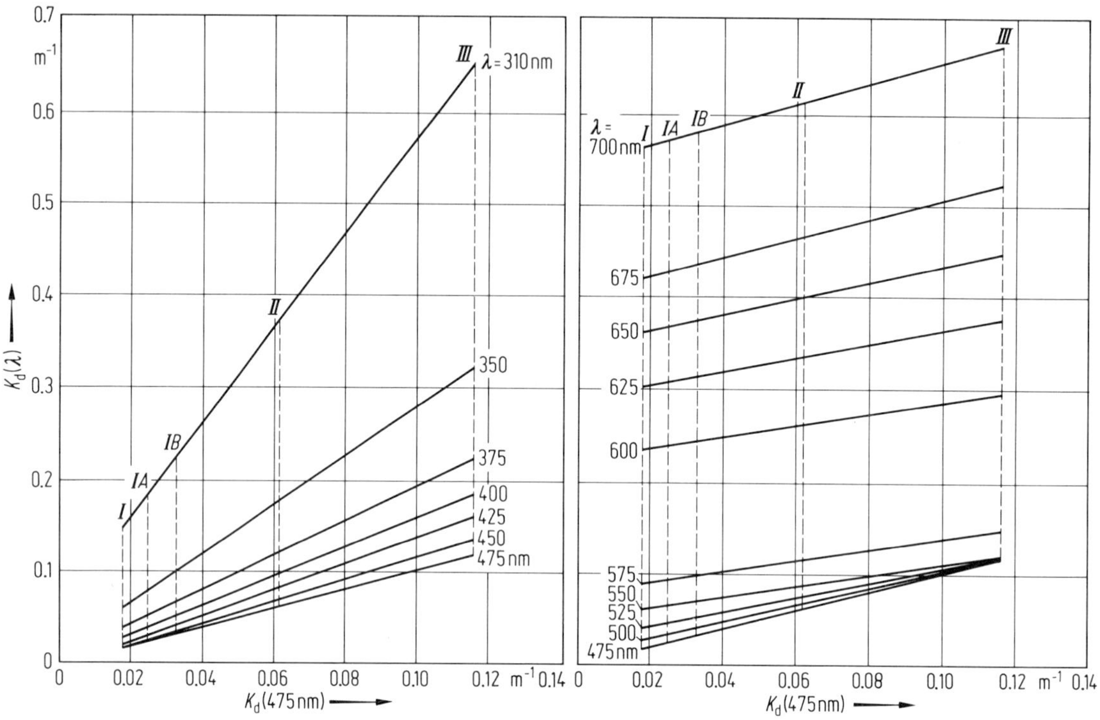

Fig. 16. The attenuation coefficient for the downward irradiance $K_d(\lambda)$ as a function of K_d (475 nm) for different water masses (oceanic types, *I*, *IA*, *IB*, *II*, and *III*) and wavelengths λ (Jerlov [76 J]).

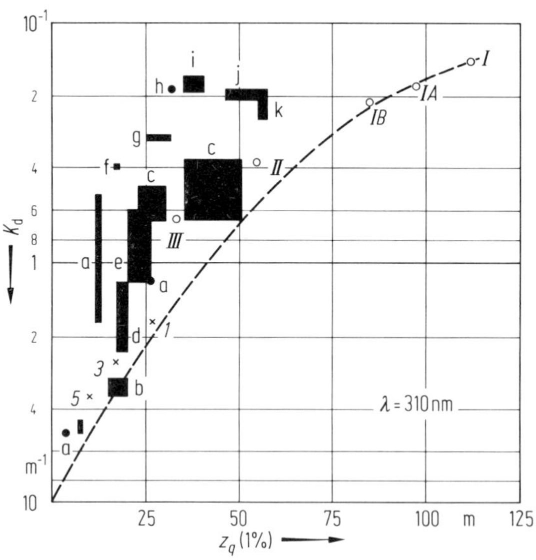

Fig. 17. The vertical attenuation coefficient at $\lambda = 310$ nm, K_d (310 nm), vs. the depth of the euphotic zone, $z_q(1\%)$. (a) The German Bight, (b) The Baltic proper, (c) The Fladen Ground, (d) Kattegat, (e) Skagerrak, (f) Shetland-Orkney Islands, (g) Off South-East Iceland, (h) Shetland-Faroe Islands, (i) Off the Faroe Islands, (j) Off Western Greenland, (k) Off Svalbard. The Roman numbers *I*, *IA*, *IB*, *II*, and *III* (circles) correspond to oceanic water masses and the Arabic numbers *1*, *3*, and *5* (crosses) to coastal water masses according to Jerlov's [76 J] optical classification (Højerslev [82 H 1]).

The concentration of dissolved organic matter, C_y, in [mg/l], can approximately be given in terms of R(310 nm), K_d(310 nm), in [m^{-1}], and $z_q(1\%)$, in [m], through the relation (Højerslev [82 H 1]):

$$C_y = \frac{1-R}{1+4.9R}\left[0.55K_d - \frac{1.21}{z_q(1\%)}\right] - 0.061 \tag{94}$$

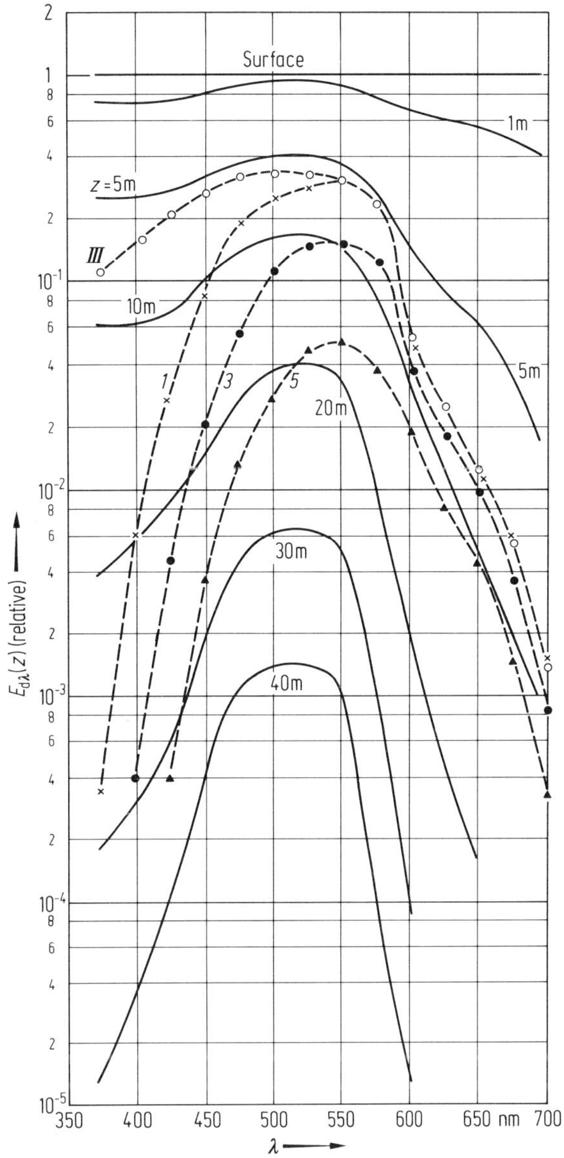

Fig. 18. Measured relative spectral downward irradiance at St. 290, 20 August 1973 $13^h45 \cdots 14^h15$ (solid lines). Included are Jerlov's optical (Jerlov [76 J]), dashed lines, for the water masses *III* (open circles), *I* (crosses), *3* (solid circles), and *5* (triangles), respectively, showing the relative spectral daylight distribution at 10 m depth (Højerslev and Lundgren [77 H 5]).

and the spectral light absorption by yellow substance ("gelbstoff") obeys an exponential law, e.g. Højerslev [82 H 1]:

$$a_y(\lambda) = a_y(\lambda_0)\,e^{-0.014(\lambda-\lambda_0)} \tag{95}$$

in which λ and λ_0 are measured in [nm] and $300\,\text{nm} \leqq \lambda,\ \lambda_0 \leqq 700\,\text{nm}$.

Jerlov's optical classification for the coastal water types *1*, *3*, *5*, and *9* are characterized by a dominating $K_y(\lambda)$ term in the ultraviolet and blue part of the spectrum (see eqs. (30) and (95), and Table 5). These coastal waters are heavily influenced by river discharges into the marine environment. Coastal waters in upwelling regions or around islands are normally not dominated by dissolved organic matter. Accordingly, $K_y(\lambda)$ is small here throughout the visible spectrum (Fig. 18 and Table 5).

In order to circumvent the problems concerning the dissolved organic matter in the sea and still being able to make advantage of Jerlov's optical classification [76 J] there exists a variety of possibilities, see subsections below.

3.3.3.3 Classification of sea water in terms of quanta irradiance

For a large variety of clear and turbid sea waters characterized by high as well as low amounts of dissolved organic matter (Table 12) the following relations can be established (Jerlov [77 J]):

$$z_q(30\%) = 0.50 z_q(10\%) - 0.0023 z_q(10\%)^2 , \tag{96}$$

$$z_q(3\%) = 1.65 z_q(10\%) , \tag{97}$$

$$z_q(1\%) = 2.18 z_q(10\%) \tag{98}$$

in which $z_q(30\%)$, $z_q(10\%)$, $z_q(3\%)$ and $z_q(1\%)$ are the depths in [m] where the downward quanta irradiance ($\lambda = 350 \cdots 700$ nm) is reduced to, respectively, 30%, 10%, 3% and 1% of the surface value. The data set leading to eqs. (96)···(98) is based on average values (Table 12) which are presented in Figs. 19 and 20). Note that eqs. (96)···(98) are almost independent of the solar elevation, e.g. Fig. 13.

Table 12. Stations where downward quanta irradiance ($\lambda = 350 \cdots 700$ nm) observations have been made, Jerlov [77 J].

Station	Region	No. of observations	Ref.
1	Danish waters	37	Højerslev (1974, 1976)
2	North Sea	101	Højerslev (1977, 1982)
3	Off West Africa	16	Kullenberg (personal communication)
4	Off West Africa	22	Morel and Caouménos (1973)
5	Off Iceland	1	Højerslev and Lundgren (1977)
6	California Bay	9	Nygård (1968)
7	West Mediterranean	11	Højerslev (1974)
8	Off Equador, Panama	11	Nygård (1973)
9	Off Equador, Panama	11	Morel (1973)
10	Drake Passage	3	Kullenberg (personal communication)
11	Carribean Sea	7	Nygård (1973)
12	Carribean Sea	7	Morel (1973)
13	Sargasso Sea	1	Nygård (1973)
14	Sargasso Sea	4	Morel (1973)

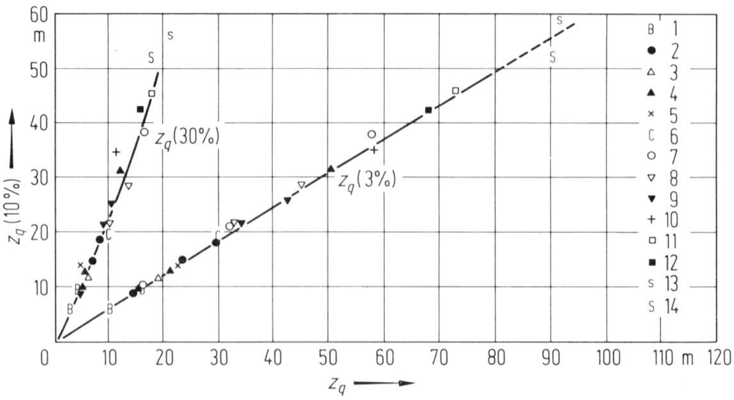

Fig. 19. The 30% level and the 3% level vs. the 10% level for the downward quanta irradiance ($\lambda = 350 \cdots 700$ nm). Mean values for different water masses (Jerlov [77 J]).

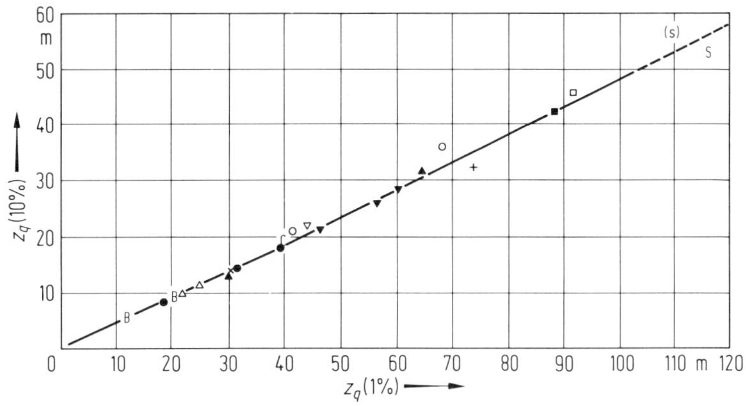

Fig. 20. The 1% level vs. the 10% level for the downward quanta irradiance ($\lambda = 350 \cdots 700$ n). Mean values for different water masses. Symbols as in Fig. 19 (Jerlov [77 J]).

Jerlov's original optical classification [76 J] is based on the spectral values of $K_d(\lambda)$ and fails to work in turbid waters having only small amounts of dissolved organic matter, e.g. Fig. 18. However, if the optical classification scheme is instead based on K_q, or, alternatively, on $z_q(\%)$, then even pronounced variations in the concentration of dissolved organic matter will have only a slight influence on K_q according to eqs. (18), (26), and (95), see also Table 5 and Fig. 17.

The final results (Jerlov [77 J]) are presented in Tables 13 and 14 and Fig. 21. Especially, the coastal waters *1* and *3* comprize now also of waters in which the gelbstoff concentration might be low or negligible.

According to the above optical classification a single measurement of the downward quanta irradiance, like $z_q(10\%)$, is sufficient for the construction of the whole quanta irradiance depth profile within the euphotic zone for the most commonly occurring sea waters.

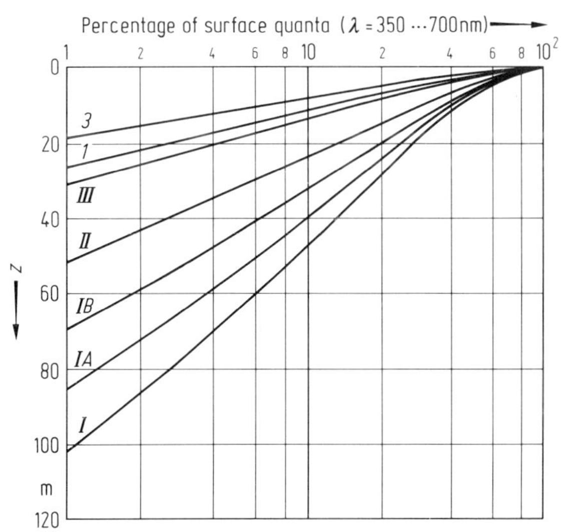

Fig. 21. Depth profiles of downward quanta irradiance ($\lambda = 350 \cdots 700$ nm) as percent of surface value for different optical water types (Jerlov [77 J]).

Højerslev

Table 13. Percentage of downward quanta irradiance ($\lambda = 350 \cdots 700$ nm) at different depths for several optical types of sea water. Jerlov [77 J].

Depth m	Optical types						
	I	*IA*	*IB*	*II*	*III*	*1*	*3*
0	100	100	100	100	100	100	100
2	79	77	75	70	63	54	44
5	64	59	54	48	38	30	21
10	46	42	38	32	17.5	13.3	7.0
15	35.5	32	28	21	8.5	6.0	2.3
20	29	25.5	21	14.2	4.6	3.0	0.73
30	20	16.4	12.0	6.4	1.22	0.65	
40	13.7	10.3	6.7	2.8			
50	9.5	6.6	3.8	1.23			
60	6.5	4.1	2.0				
70	4.3	2.5	1.01				
80	2.9	1.47					
90	1.85						
100	1.14						

Table 14. The vertical attenuation coefficient K_q [m^{-1}] for downward quanta irradiance ($\lambda = 350 \cdots 700$ nm) Jerlov [77 J].

Depth m	Optical types						
	I	*IA*	*IB*	*II*	*III*	*1*	*3*
0··· 2	0.118	0.130	0.144	0.178	0.230	0.310	0.41
2··· 5	0.073	0.089	0.107	0.126	0.169	0.196	0.25
5··· 10	0.065	0.068	0.071	0.099	0.155	0.163	0.22
10··· 15	0.052	0.054	0.061	0.084	0.144	0.154	0.22
15··· 20	0.040	0.042	0.056	0.078	0.123	0.140	0.23
20··· 30	0.039	0.046	0.057	0.080	0.134	0.153	
30··· 40	0.039	0.046	0.058	0.082			
40··· 50	0.037	0.045	0.059	0.082			
50··· 60	0.038	0.048	0.064				
60··· 70	0.040	0.049	0.069				
70··· 80	0.040	0.053					
80··· 90	0.045						
90···100	0.048						

3.3.3.4 Classification of sea water in terms of the color of the sea

The color of the sea is here defined as:

$$F(z=0\,\mathrm{m}) = \frac{L_\lambda(0, 180, \varphi, 447\,\mathrm{nm})}{L_\lambda(0, 180, \varphi, 521\,\mathrm{nm})}, \qquad (99)$$

i.e. as the ratio of the nadir radiances just below the sea surface in the blue ($\lambda = 447$ nm, $\Delta\lambda \approx 15$ nm) and in the green ($\lambda = 521$ nm, $\Delta\lambda \approx 15$ nm) part of the spectrum. Field measurements reveal that F might alternatively be defined as:

$$F(z=0\,\mathrm{m}) = \frac{R(0.447\,\mathrm{nm})}{R(0.521\,\mathrm{nm})} \simeq \frac{E_u(0.447\,\mathrm{nm})}{E_u(0.521\,\mathrm{nm})} \qquad (100)$$

No significant difference between these two definitions has been observed (Højerslev and Jerlov [77 H 4]; Table 15 and Fig. 23).

Table 15. Sea areas where the color index F (areas 1 to 14 and 21 to 29), the downward quanta irradiance q ($\lambda = 350 \cdots 700$ nm) (areas 1 to 29), and the spectral irradiance ratio R (areas 15 to 20) where measured by the author using essentially the same equipment. The numbers 1 to 29 serve to identify the sea areas in Figs. 23 and 24. Højerslev [82 H 2].

Sea area key	Region	No. of observations
1	Danish Belts	10
2	Baltic Sea	14
3	North Sea (FLEX-76) [1]	28
4	North Sea (FLEX-76) [1]	18
5	Off west Africa	9
6	North Sea (FLEX-76) [2]	18
7	North Sea (FLEX-76) [2]	23
8	Gibraltar Strait	2
9	North Sea (FLEX-75) [3]	23
10	North Sea (FLEX-75) [3]	8
11	East of Gibraltar	1
12	East of Sardinia	2
13	Drake Passage	2
14	South of Sardinia	5
15	Baltic Sea	4
16	Gibraltar Strait	1
17	East of Sardinia	3
18	South of Sardinia	1
19	North of Haiti [4]	1
20	Sargasso Sea	1
21	Off the Faroe Islands	30
22	Off Svalbard (NORSEX-79)	5
23	Off the Shetland Islands	2
24	Kattegat	18
25	Off western Greenland	6
26	Skagerrak	15
27	German Bight (MARSEN-79)	26
28	Roskilde Fjord	10
29	Isefjord	10

[1]) Data from the two plankton blooms.
[2]) Data from periods intermediate the two plankton blooms.
[3]) Data from periods after the two plankton blooms.
[4]) Data from Morel (1973).

Table 16. Relation between color index and different quanta percent levels z_q, in [m]. Højerslev [80 H 1].

Color index	Quanta percent levels			
	30%	10%	3%	1%
0.4	4.4	8.9	14.5	19.5
0.5	4.9	10.1	16.5	22.5
0.6	5.4	11.3	18.5	24.5
0.7	5.9	12.5	20.5	27.5
0.8	6.6	13.8	22.5	30
0.9	7.1	15.0	24.5	32.5
1.0	7.6	16.2	26.5	35
1.1	8.1	17.4	28.5	37.5
1.2	8.6	18.6	30.5	40
1.3	9.1	19.8	32	43
1.4	9.6	21.0	34	45.5
1.5	10.0	22.3	36	48
1.6	10.5	23.5	38	50.5
1.7	11.0	24.7	40	53
1.8	11.5	26.0	42	55.5
1.9	12.0	27.2	44	58
2.0	12.4	28.4	46	60.5
2.1	12.9	29.6	48	63
2.2	13.4	30.8	50	65.5
2.3	13.8	32.0	52	68
2.4	14.2	33.2	54	70.5
2.5	14.6	34.5	56	73
2.6	15.0	35.7	57.5	75.5
2.7	15.4	36.9	59.5	78
2.8	15.8	38.1	61.5	80.5
2.9	16.2	39.3	63.5	83
3.0	16.6	40.6	65.5	85.5
3.1	17.0	41.8	67	88
3.2	17.4	43.0	69	90.5
3.3	17.8	44.2	71	93
3.4	18.2	45.4	73	95.5
3.5	18.5	46.7	75	98
3.6	18.8	47.9	77	100.5
3.7	19.1	49.1	79	103
3.8	19.4	50.3	81	105.5
3.9	19.7	51.5	83	108
4.0	20.0	52.7	85	110.5
4.1	20.3	54.0	87	113
4.2	20.6	55.2	89	115.5
4.3	20.9	56.4	90.5	118
4.4	21.2	57.6	92.5	120.5
4.5	21.5	58.9	94.5	123
4.6	21.8	60.1	96.5	125.5

F is only slightly dependent on the surface light conditions. The author has verified this at a number of occasions in both clear and turbid waters as well as during changing cloud conditions and varying solar elevations, e.g. Højerslev [74 H 1], Jerlov [74 J 1], and Fig. 22. This finding is rather striking because it violates any visual experience.

It has been shown empirically that *F* is interrelated to $z_q(10\%)$, or, alternatively, to the depth of the euphotic zone, for what is believed to represent most waters in the marine environment (Table 15, and Figs. 23 and 24). This implies that *F* can be recommended as being one of the suitable parameters for an optical classification of water masses (Højerslev [80 H 1, 82 H 2]), Table 16 and Fig. 25.

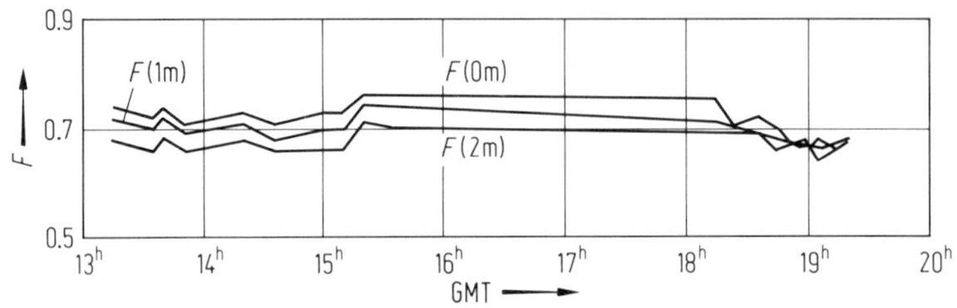

Fig. 22. Color indices *F* obtained on 26 May, 1976, at 0, 1, and 2 m depths, respectively, against GMT time at the Fladen Ground in the North Sea (Højerslev [82 H 2]).

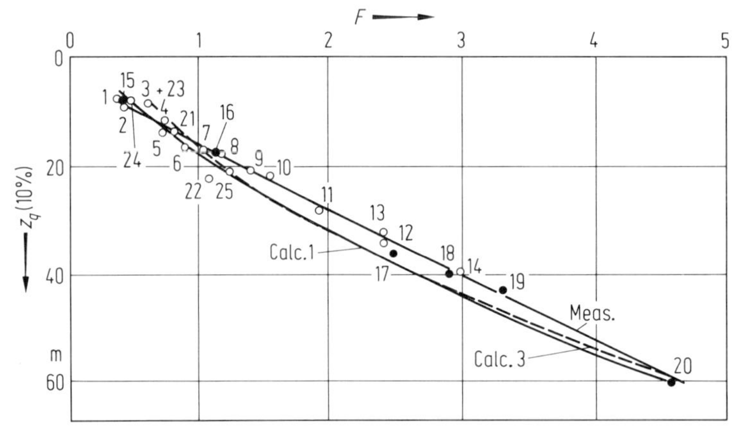

Fig. 23. Color index, *F*, extrapolated to 0 m depth against the depth of the 10% level for the downward quanta irradiance ($\lambda = 350 \cdots 700$ nm), $z_q(10\%)$, for the average experimental case (the straight solid line termed Meas.), and for two theoretical cases called Calc. 1 and Calc. 3. Calc. 1 is depicted as the curved solid line and is obtained from eq. (101) whereas Calc. 3 is depicted as the curved broken line and is obtained from eq. (103). The open circles refer to measurements of the nadir radiances at 447 nm and 521 nm, respectively, and the black dots refer to upward irradiance measurements made at the same wavelengths as in the case of the radiance measurements. The numbers ranging from 1 to 25 refer to the areas of the sea in which the measurements were made (see Table 15 and Højerslev [82 H 2]).

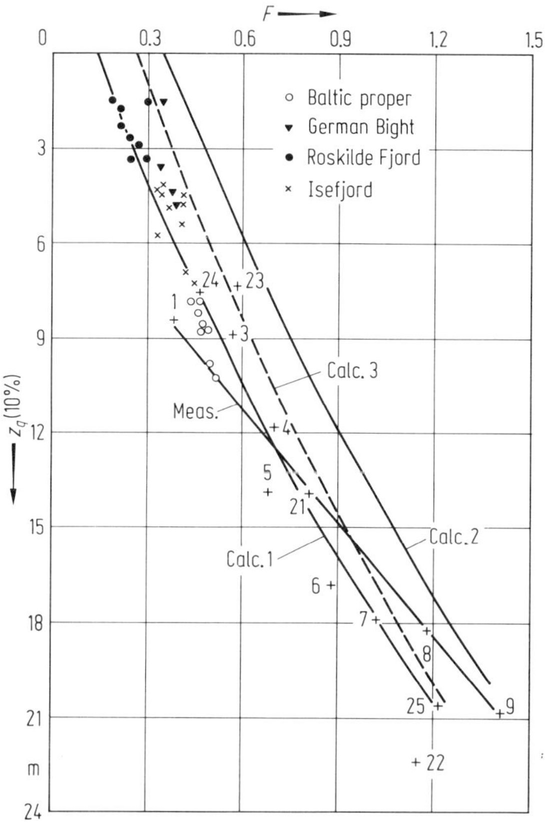

Fig. 24. Color index, F, extrapolated to 0 m depth against the depth of the 10% level for the downward quanta irradiance ($\lambda = 350 \cdots 700$ nm), $z_q(10\%)$, for the average experimental case (the straight solid line termed Meas.), and for three theoretical cases called Calc. 1, Calc. 2, and Calc. 3, respectively. Fig. 24 is an enlargement of the left corner of Fig. 23 so Calc. 1 and Calc. 3 are the same in both figures. Calc. 2 is obtained from eq. (102). The numbers and the symbols refer to the areas of the sea in which the measurements were made (see Table 15 and Højerslev [82 H 2]).

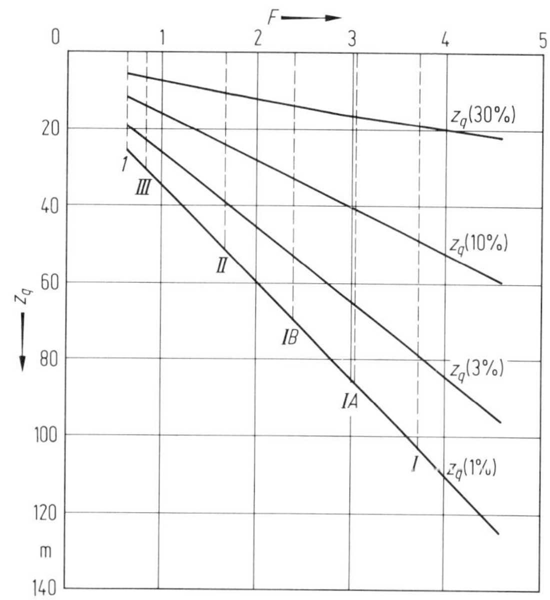

Fig. 25. Empirical relations between the color index obtained at 0 m depth and selected percentage levels for the downward quanta irradiance ($\lambda = 350 \cdots 700$ nm). Included are the optical water classes *I*, *IA*, *IB*, *II*, *III* oceanic, and *1* coastal (Højerslev [80 H 1]).

Three different color/quanta-algorithms have been developed by Højerslev [82 H 2]. The first one, termed Calc. 1 in Figs. 23 and 24, is principally based on Jerlov's optical classification and reads:

$$F_1 = 0.00060 z_q^2(10\%) + 0.038 z_q(10\%) + 0.14\,. \qquad (101)$$

The second one, termed Calc. 2 in Fig. 24, involves as main assumptions that the turbidity of a water mass is inversely proportional to the depth of the euphotic zone and that the backscattering coefficient b_b is independent of wavelength. It reads:

$$F_2 = 0.00050 z_q^2(10\%) + 0.038 z_q(10\%) + 0.35\,. \qquad (102)$$

The third color/quanta-algorithm, termed Calc. 3 in Figs. 23 and 24, includes the optical properties of pure sea water, gelbstoff and suspended matter in an approximate manner (Højerslev [82 H 2]):

$$F_3 = 0.00066 z_q^2(10\%) + 0.029 z_q(10\%) + 0.25 \qquad (103)$$

and

$$F_3 = 0.00014 z_q^2(1\%) + 0.014 z_q(1\%) + 0.25\,. \qquad (104)$$

Note that the above three algorithms are differing only slightly from one another although being derived differently.

3.3.3.5 Classification of sea water in terms of the downward irradiance in a narrow or broad wavelength band

The previous three sections suggest that an optical classification of water masses is feasible from a single measurement of the downward irradiance in a narrow or broad wavelength band. It is evident from Table 5 and Fig. 17 that the appropriate wavelength band must be away from the ultraviolet and blue parts of the spectrum because the highly varying gelbstoff concentrations, even in sea waters displaying the same turbidity, cause a strong absorption in this particular spectral region according to eq. (95). A proper choice of a wavelength band must, heuristically speaking, be situated in the green part of the spectrum. (In the red part of the spectrum the optical properties of the pure sea water itself play a considerable or even a dominating role relative to the optical properties of the sea water mass in question.)

582 daylight spectra of the downward irradiance have been recorded by means of a French irradiance meter in various sea waters. On the basis of this data set, statistical relations between the downward irradiance at different wavelengths (and with different halfwidths) and the downward quanta irradiance ($\lambda = 370 \cdots 700$ nm) are established. They are expressed in the form:

$$q(z)\% = A(z)\,[E_d(\lambda_0)\%]^{B(z)} \tag{105}$$

in which $q(z)\%$ and $E_d(\lambda_0)\%$ are, respectively, the relative downward quanta irradiance and the relative quasi-monochromatic irradiance at wavelength, λ_0 (with 100% just below the surface). The factors and exponents, $A(z)$ and $B(z)$, respectively, are the tabulated depth-dependent parameters obtained from a best fit analysis.

The optimal relations (highest correlation coefficients γ) are found for $\lambda_0 = 500$ nm with halfwidth being set equal to 25 nm or less. Furthermore, a similar but less accurate conversion of downward lux irradiance into downward quanta irradiance is feasible. This enables old marine biological lux data to be transformed back into the recommended downward quanta irradiance, and *vice versa*, implying that Jerlov's [76 J] optical water masses, including those turbid water masses of *low* gelbstoff concentration, can be defined in terms of the downward lux irradiance, Fig. 26.

Independently of the above statistical investigation (Table 17) a Danish data set involving both spectral downward irradiance as well as downward quanta irradiance ($\lambda = 350 \cdots 700$ nm) has been established for different areas and depths in the sea. The measured downward irradiance $E_d(500\text{ nm})\%$ from the Danish data set applied along with the tabulated $A(z)$ and $B(z)$ values in Table 17 in order to calculate the downward quanta irradiances compares favourably with the quanta irradiances measured independently and directly, Fig. 27. The sea waters listed in Fig. 27 represent clear and turbid waters, the latter group having gelbstoff concentrations in varying high and low proportions.

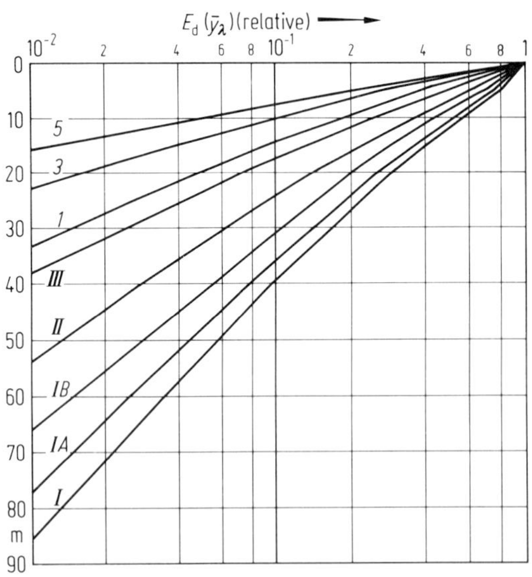

Fig. 26. Depth profiles of the downward lux irradiance, $E_d(\bar{y}_\lambda)$, in percent of surface irradiance for the oceanic types *I*, *IA*, *IB*, *II*, and *III*, and for the coastal types *1*, *3*, and *5* (Højerslev [78 H]).

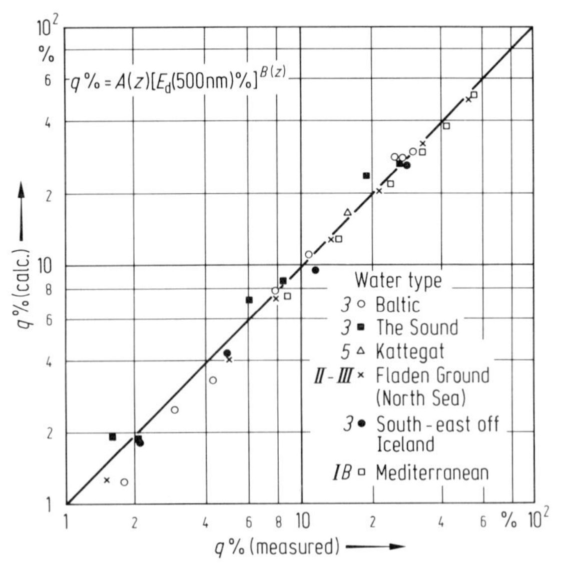

Fig. 27. The relative downward quanta irradiance ($\lambda = 350 \cdots 700$ nm) measured directy vs. the relative downward quanta irradiance calculated from measured values of the relative downward irradiance, E_d (500 nm)%, from eq. (105) and Table 17, respectively [81 H 1].

Through Jerlov's optical classification [76 J] in terms of downward quanta irradiances together with his values for E_d(500 nm)% (Table 10 and Jerlov [78 J]), and $A(z)$ and $B(z)$ in Table 17 it is straightforward to compare measured and calculated (through eq. (105) and Table 17) downward quanta irradiances as depicted in Fig. 28. Since Jerlov's optical data set forms a data set more or less independent of the so-called French and Danish one it seems verified that the parameter E_d(500 nm)% can serve as a highly useful optical classification means for most sea waters (see also Austin and Petzold [81 A]).

Quasi-monochromatic irradiance meters have some advantages compared with wavelength integrating irradiance meters like the quanta meter. The advantages are typically (i) well defined light collecting properties (ii) well defined immersion effect coefficients, e.g. Smith [69 S 2], (iii) well defined instrumental sensitivity (iv) eased calibration procedures and (v) low production costs.

Table 17. Proposed values for $A(z)$ and $B(z)$ for the quasi-monochromatic downward irradiance around $\lambda = 500$ nm, E_d (500 nm), eq. (105), and for the downward lux irradiance, E_d ($\bar{y}_\lambda$). Morel and Højerslev [79 M]. γ: correlation coefficient.

Depth	E_d (500 nm)			E_d ($\bar{y}_\lambda$)		
z m	$A(z)$	$B(z)$	γ	$A(z)$	$B(z)$	γ
5	1.115	0.865	0.967	0.459	1.121	0.967
10	0.725	0.925	0.992	0.378	1.198	0.993
15	0.57	0.965	0.997	0.415	1.216	0.993
20	0.51	0.975	0.997	0.455	1.244	0.994
25	0.475	0.990	0.996	0.515	1.255	0.995
30	0.44	1.000	0.995	0.573	1.275	0.994
40	0.39	1.040	0.994	0.714	1.334	0.995
50	0.34	1.080	0.992	0.968	1.342	0.996
60	0.30	1.180	0.993	1.256	1.391	0.996
80	0.275	1.340	0.994	2.158	1.428	0.996
100	0.34	1.400	0.994	3.342	1.411	0.996

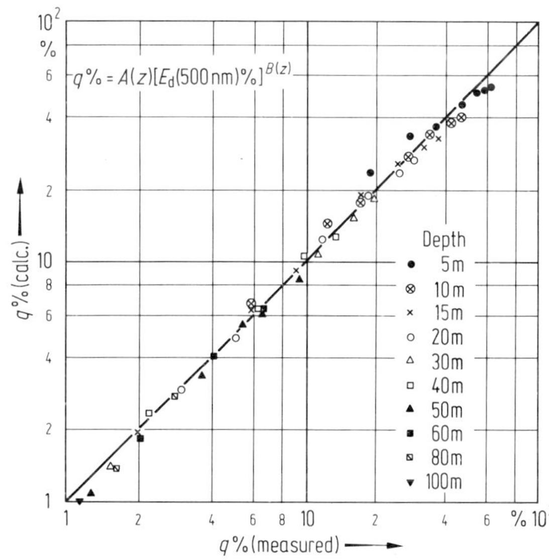

Fig. 28. The relative downward quanta irradiances ($\lambda = 350\cdots700$ nm) taken from Jerlov's [76 J] optical classification of water masses, i.e. *I*, *IA*, *IB*, *II*, *III*, *1*, and *3*, respectively, vs. the relative downward quanta irradiances calculated from the E_d(500 nm)% values for the same optical classes as above from eq. (105) and Table 17, respectively (Jerlov [76 J] and Højerslev [81 H 1]).

Højerslev

3.3.3.6 Classification of sea water in terms of light attenuation and light scattering coefficients

The previous sections suggest strongly that there also exists a possibility for classifying sea waters in terms of the light attenuation coefficient c and the light scattering coefficient b. The observed approximate constancy in the shape of the light scattering function $\beta(\alpha)$ for both clear and turbid waters is the rationale for looking into optical classification schemes in terms of c and b instead of c and $\beta(\alpha)$. Timofeeva [74T] has presented such an optical classification in Tables 7, 18, and 19, and in Figs. 11, 29, and 30. The basic experimental work behind her optical classification is performed *in-vitro* on so-called milky solutions in which the light scattering produced by the suspended matter is far greater than the amount of light scattered by the pure solvent. In case of sea water the analogy would be that $c \gg c_w$ and $b \gg b_w$. This is a highly important observation to make since sea waters always ought to be classified in terms of $c - c_w$ and $b - b_w$, especially in either clear waters or in the red part of the spectrum where c_w is a dominating term (see below). Timofeeva [74T] also assumes that the vertical attenuation coefficient K_d for the downward irradiance is constant throughout the media and equal to its asymptotic value k, see eqs. (76)···(79). This is a first order approximation which in most cases has proven its usefulness *in-situ*. Table 8 shows that $k/c = \varepsilon$ is typically to be found in the range 0.35···0.84 *in-situ*, for which the curvature of $\log E_d$ vs. the optical depth τ is negligible, implying that $K_d \cong k$ as depicted in Fig. 29. Timofeeva's [74T] findings are checked against field measurements, Figs. 11, 30, and 31, and the agreement for the presented examples is satisfactory.

Table 18. Relation between probability of photon-survival $\chi(=b/c)$ and parameter $\varepsilon(=k/c)$. Timofeeva [74T].

ε	0	0.01	0.02	0.03	0.04	0.05	0.06	0.07	0.08	0.09
					χ					
0.0	1	0.999	0.998	0.996	0.994	0.990	0.986	0.982	0.977	0.972
0.1	0.966	0.960	0.953	0.946	0.939	0.932	0.924	0.916	0.908	0.900
0.2	0.892	0.883	0.875	0.866	0.857	0.848	0.839	0.830	0.820	0.811
0.3	0.802	0.792	0.783	0.773	0.763	0.753	0.744	0.734	0.724	0.714
0.4	0.704	0.694	0.684	0.674	0.663	0.653	0.643	0.632	0.622	0.612
0.5	0.601	0.591	0.580	0.570	0.559	0.548	0.538	0.527	0.516	0.505
0.6	0.494	0.483	0.472	0.461	0.450	0.439	0.427	0.416	0.405	0.393
0.7	0.382	0.370	0.359	0.347	0.335	0.324	0.312	0.300	0.288	0.276
0.8	0.264	0.251	0.239	0.227	0.214	0.202	0.189	0.176	0.163	0.150
0.9	0.137	0.124	0.111	0.098	0.084	0.070	0.056	0.043	0.029	0.014
1.0	0									

Table 19. Relation between parameter $\varepsilon(=k/c)$ and probability of photon-survival $\chi(=b/c)$. Timofeeva [74T].

χ	0	0.01	0.02	0.03	0.04	0.05	0.06	0.07	0.08	0.09
					ε					
0.0	1	0.993	0.986	0.979	0.972	0.965	0.957	0.950	0.943	0.936
0.1	0.928	0.921	0.913	0.906	0.898	0.890	0.883	0.875	0.867	0.859
0.2	0.852	0.843	0.835	0.827	0.819	0.811	0.803	0.795	0.786	0.778
0.3	0.770	0.761	0.753	0.745	0.736	0.728	0.719	0.710	0.702	0.693
0.4	0.684	0.675	0.667	0.658	0.649	0.640	0.631	0.622	0.613	0.604
0.5	0.595	0.585	0.576	0.567	0.558	0.548	0.539	0.530	0.520	0.511
0.6	0.501	0.492	0.482	0.472	0.463	0.454	0.443	0.433	0.424	0.414
0.7	0.404	0.394	0.384	0.374	0.364	0.353	0.343	0.333	0.323	0.312
0.8	0.302	0.291	0.280	0.270	0.259	0.248	0.237	0.225	0.214	0.202
0.9	0.190	0.178	0.165	0.152	0.139	0.125	0.110	0.093	0.075	0.052
1.0	0									

Table 20. Observations of the light attenuation coefficient c and of this coefficient related to pure water, $c-c_w$, Jerlov [74 J 1].

c-meter		Region	c [m^{-1}]		$c-c_w$ [m^{-1}]		Ref.
Type	Path-length [m]		380 nm	655 nm	380 nm	655 nm	
Mono-chromator	1	Continental Slope Atlantic Ocean	0.14	0.36			Clarke and James (1939)
Beckman DU	1	Off Peru			0.14	0.06	Burt (1958)
					0.22	0.13	
					0.35	0.21	
					0.60	0.34	
Beckman DU	1	Bermuda waters			0.21	0.12	Visser (1968)
					0.82	0.44	
Transmittance meter	2	Caribbean Sea	0.14	0.36			Jerlov (1951)
		Galapagos	0.27	0.43			
	1	Kattegat	0.58	0.53			Jerlov (1955)
	1	The Sound	0.95	0.56			
	1	Baltic Sea	1.19	0.57			
	1	Bermuda waters	0.24	0.40			Ivanoff et al. (1961)
			0.38	0.48			
			0.45	0.52			
			0.64	0.61			
	1.5	Sargasso Sea	0.06	0.32			Kullenberg et al. (1970)
	1.5	Mediterranean	0.12	0.35			Højerslev (1973)
			0.21	0.38			
			0.26	0.42			

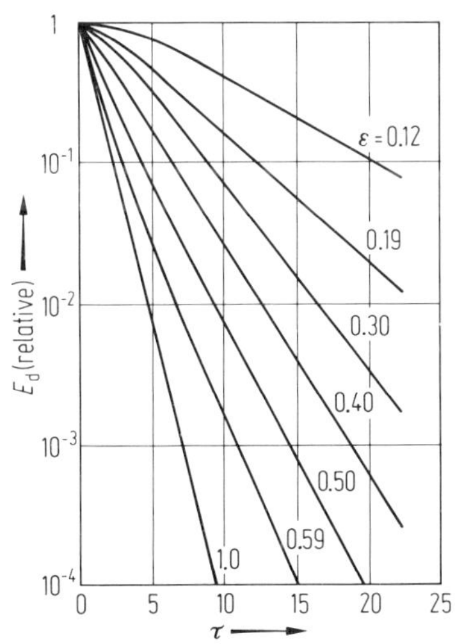

Fig. 29. The change with depth of downward irradiance E_d in milky media with various $\varepsilon = k/c$ (Makarevich and Robilko [69 M]). τ: optical depth.

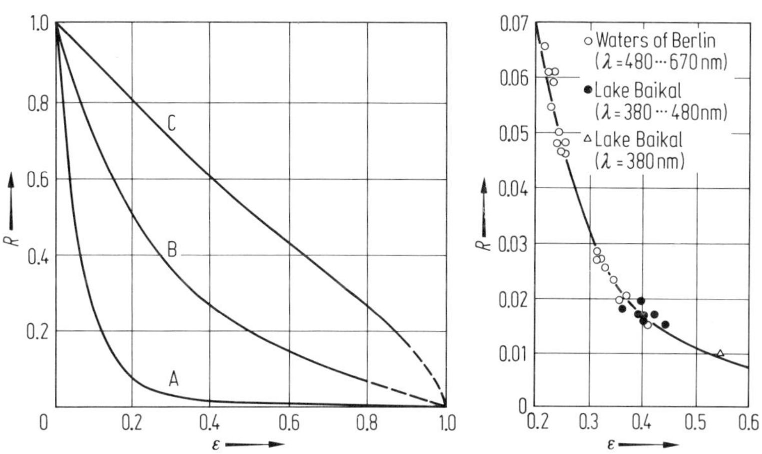

Fig. 30. Dependence of irradiance ratio on $\varepsilon = k/c$ in various media (Timofeeva [71 T]). (A) milky medium; (B) medium with latex; (C) medium with Rayleigh scattering.

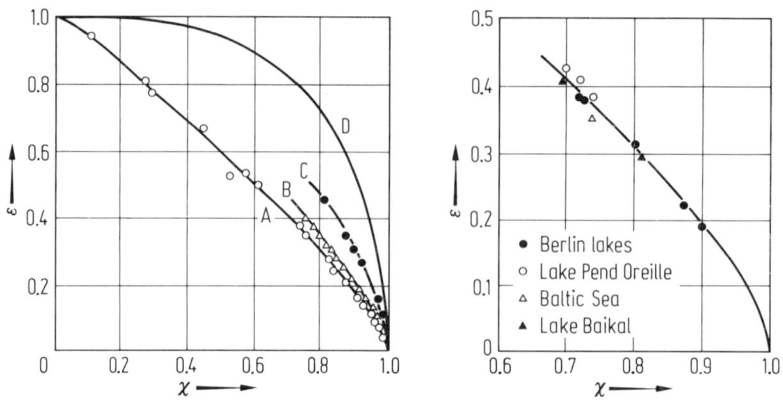

Fig. 31. The dependence of $\varepsilon = k/c$ on photon survival probability $\chi = b/c$ in various media (on the left) and in the sea (on the right); Timofeeva [71 T]. (A) milky medium; (B) milky medium (Herman and Lenoble [64 H]); (C) medium with afcolac (Herman and Lenoble [64 H]); (D) medium with Rayleigh scattering.

For a large group of water masses the following empirical relation is valid, provided no intense plankton blooms, upwelling or resuspension of bottom sediments occurs, e.g. Table 20, Figs. 32 and 33, Højerslev [74 H 1, 82 H 2], and Jerlov [74 J 1]:

$$(c - c_w)_{\lambda = 380\,\mathrm{nm}} - 1.80(c - c_w)_{\lambda = 655\,\mathrm{nm}} = a_y(380\,\mathrm{nm})\,. \tag{106}$$

For all the oceanic water types *I–III*, $a_y(380\,\mathrm{nm})$ is negligible as also suggested in Fig. 32.

Another empirical relation similar to eq. (106) can be given for the same sea waters and constraints as before, e.g. Fig. 34, Højerslev [74 H 1, 77 H 2]:

$$(c - c_w)_{\lambda = 525\,\mathrm{nm}} - 1.15(c - c_w)_{\lambda = 655\,\mathrm{nm}} = a_y(525\,\mathrm{nm}) \simeq 0 \tag{107}$$

in which $a_y(525\,\mathrm{nm})$ can be ignored, practically speaking, for all the water types between *I* oceanic and *3* coastal (Jerlov [76 J]). The point scatter in Figs. 33 and 34 is partly explained by a time delay in the measurements of the three parameters $c(380\,\mathrm{nm})$, $c(525\,\mathrm{nm})$, and $c(655\,\mathrm{nm})$, combined with a stratified water column under the influence of internal waves (Højerslev [77 H 2] and Joseph [55 J 2]).

During the seasonal short and intense plankton blooms in the open sea, eqs. (106) and (107) are not longer valid because the optical properties of living plankton (and their associated chlorophyll-like pigments) become dominating. This results in an increased light attenuation in the blue and red part of the spectrum relative to the one in the green part. In other words: The coefficients 1.80 and 1.15 given in eqs. (106) and (107), respectively, are likely to drop as depicted in Figs. 35 and 36.

For the North Sea from which the data set represented in Figs. 33···36 was obtained, eqs. (106) and (107) change into, respectively:

$$(c - c_w)_{\lambda = 380\,\mathrm{nm}} - 1.12(c - c_w)_{\lambda = 655\,\mathrm{nm}} = a_y(380\,\mathrm{nm}) \tag{108}$$

and

$$(c - c_w)_{\lambda = 525\,\mathrm{nm}} - 0.93(c - c_w)_{\lambda = 655\,\mathrm{nm}} = a_y(525\,\mathrm{nm}) \simeq 0 \tag{109}$$

during the intense plankton bloom optically measured from April, ultimo, to May, medio, 1976. Consequently, a sea water mass regarded as an ecosystem in a mature (balanced) state, i.e. in a steady-state condition, has a tendency to obey eqs. (106) and (107), whereas increased and unbalanced plankton growth gives tendencies towards eqs. (108) and (109), see also Fig. 37 and note the short duration in which the change in inclination of the two regression lines takes place (March to April).

Eq. (106) forms the classical background for measuring the gelbstoff concentration *in-situ* and is sometimes referred to as the Δ EK-method (Joseph [50 J 1] and Jerlov [55 J 1]).

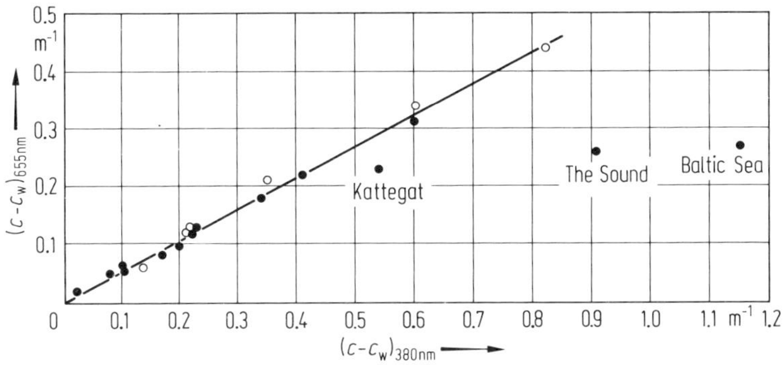

Fig. 32. Relationship between light attenuation coefficients ($c-c_w$) for the red ($\lambda=655$ nm) and for the ultraviolet ($\lambda=380$ nm). Open circles refer to direct measurements of ($c-c_w$). Closed circles refer to in-situ measurements of c subtracted c_w (Jerlov [74 J 1]). c_w: light attenuation coefficient of pure water.

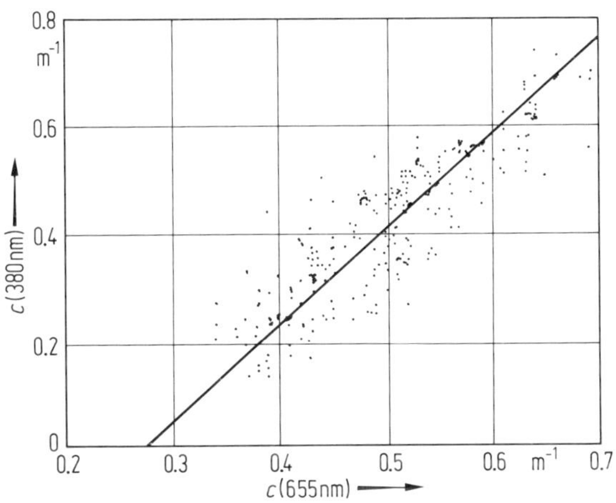

Fig. 33. Light attenuation coefficients in the red part of the spectrum $c(\lambda=655\,\text{nm})$ in relation to light attenuation coefficients in the ultraviolet part $c(\lambda=380\,\text{nm})$. The data are obtained from sea surface to sea bottom during FLEX-75 on the Fladen Ground in the North Sea. Number of points: 672; correlation coefficient, $r=0.943$; slope, $\alpha=1.80$; and intersect with ordinate axis, $q=-0.50$ (Højerslev [82 H 2]).

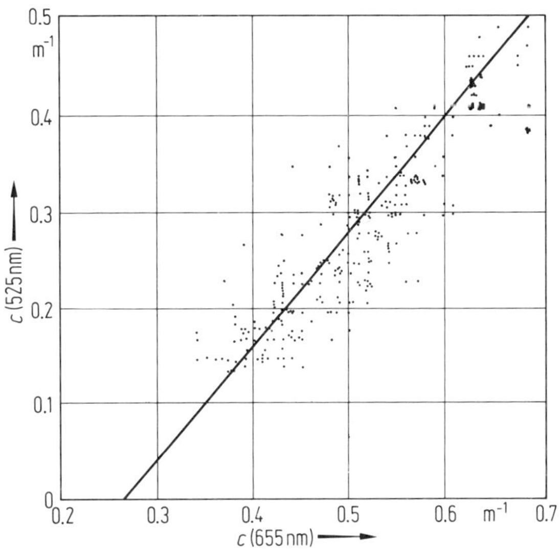

Fig. 34. Light attenuation coefficients in the red part of the spectrum $c(\lambda=655\,\text{nm})$ in relation to light attenuation coefficients in the green part $c(\lambda=525\,\text{nm})$. The data are obtained from sea surface to sea bottom during FLEX-75. Number of points: 651; $r=0.939$; $\alpha=1.13$; and $q=-0.28$ (Højerslev [82 H 2]). See Fig. 33.

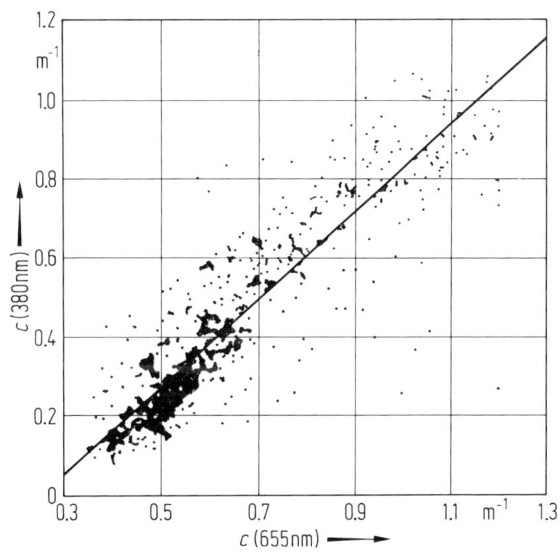

Fig. 35. Light attenuation coefficients in the red part of the spectrum $c(\lambda=655\,\text{nm})$ in relation to light attenuation coefficients in the ultraviolet part $c(\lambda=380\,\text{nm})$. The data are obtained from sea surface to sea bottom during FLEX-76 on the Fladen Ground in the North Sea. Number of points: 1285; $r=0.904$; $\alpha=1.12$; and $q=-0.28$ (Højerslev [82 H 2]). See Fig. 33.

Højerslev

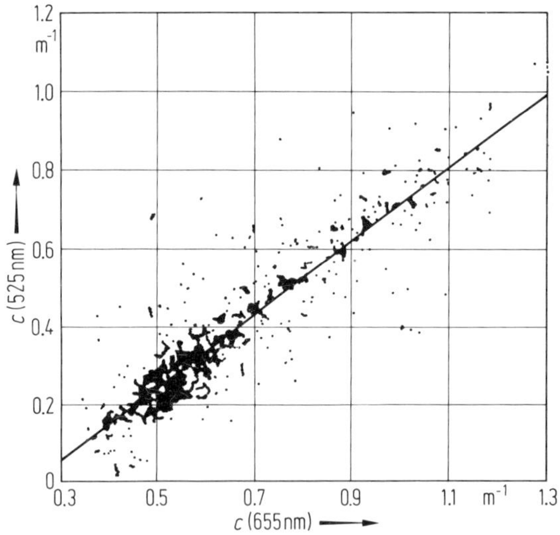

Fig. 36. Light attenuation coefficients in the red part of the spectrum $c(\lambda = 655\ \mathrm{nm})$ in relation to light attenuation coefficients in the green part $c(\lambda = 525\ \mathrm{nm})$. The data are obtained from sea surface to sea bottom during FLEX-76. Number of points: 1285; $r = 0.906$; $\alpha = 0.93$; and $q = -0.23$ (Højerslev [82 H 2]). See Fig. 33.

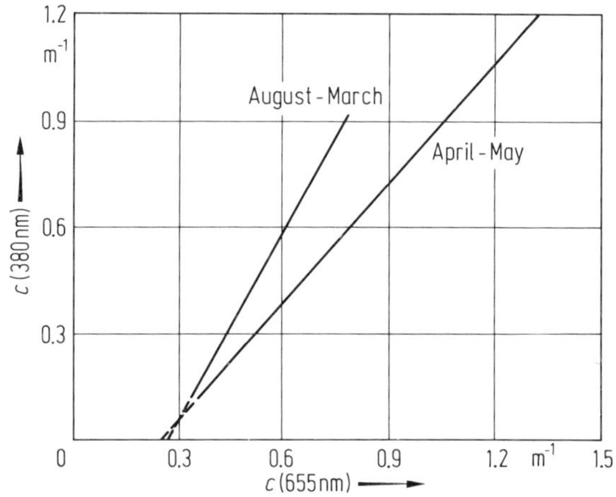

Fig. 37. The light attenuation coefficient $c(\lambda = 655\ \mathrm{nm})$ in relation to the light attenuation coefficient $c(\lambda = 380\ \mathrm{nm})$ measured during FLEX-75 and FLEX-76. (Højerslev [82 H 1]).

The Δ EK-method is not sensitive enough for measuring gelbstoff in the oceans (Højerslev [85 H 1]) but only in coastal waters influenced by river discharges. Thus from the light transmission measurements in the North Sea during August and September, 1975 (FLEX-75), and March, April, May and June, 1976 (FLEX-76), Højerslev [77 H 2, 77 H 3, 82 H 1, 85 H 1], no change in the gelbstoff concentration determined by eq. (106) could be observed. However, far more accurate measurements, in principle daylight irradiance measurements, gave reasons for the same conclusion, Figs. 38, 39, and 40, and Table 5, despite of the occurrencies of two intense plankton blooms.

The empirical findings stated in eqs. (106)···(109) can be interpreted as if the ratio of $(b - b_w)/(c - c_w)$ is normally in a fixed proportion for a given water mass outside the (short) seasons of plankton blooms. (In Timofeeva's [74 T] optical classification schemes $(b - b_w)/(c - c_w) \simeq b/c$). Jerlov [74 J 1] has made two such proposals, Table 21 and Fig. 41, which are suffering from that the $(b - b_w)/(c - c_w)$ ratios seem to be too unprecise in view of optical water mass classification and too impractical in view of other methods. Moreover, the available data set is still insufficient. For the selected group of water types, comprizing of both clear and turbid waters, the two proposals state:

$$\left(\frac{b - b_w}{c - c_w}\right)_{\lambda = 655\ \mathrm{nm}} = 0.70 . \qquad (110)$$

This implies that suspended matter in the mature ecosystem scatters the light about twice as much as it absorbs it in the red part of the spectrum.

$$\left(\frac{b - b_w}{c - c_w}\right)_{\lambda = 380\ \mathrm{nm}} = 0.48 . \qquad (111)$$

Here, in this spectral region, light scattering is about the same as light absorption.

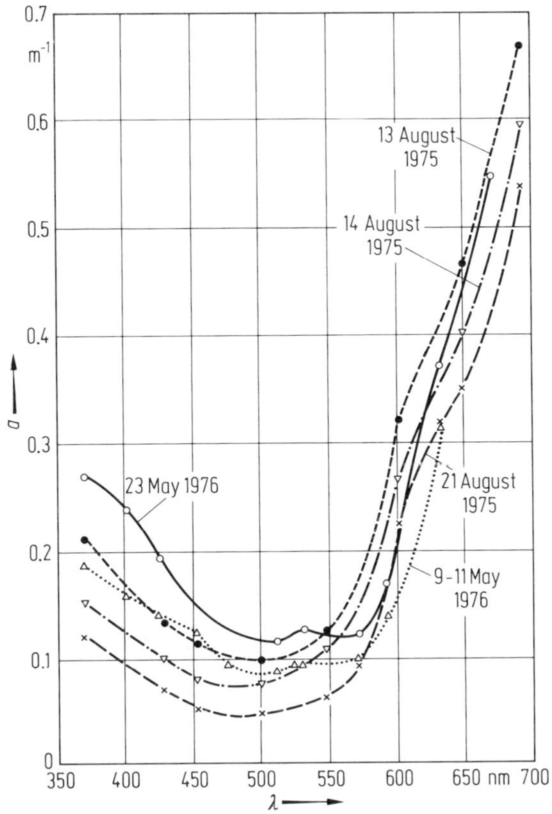

Fig. 38. Spectral light absorption coefficient within the euphotic zone on Fladen Ground before, during, and after the occurrences of two plankton blooms lasting from 20 April to 10 May 1976 and 22 May to 26 May, 1976, respectively. (Højerslev [82 H 2]).

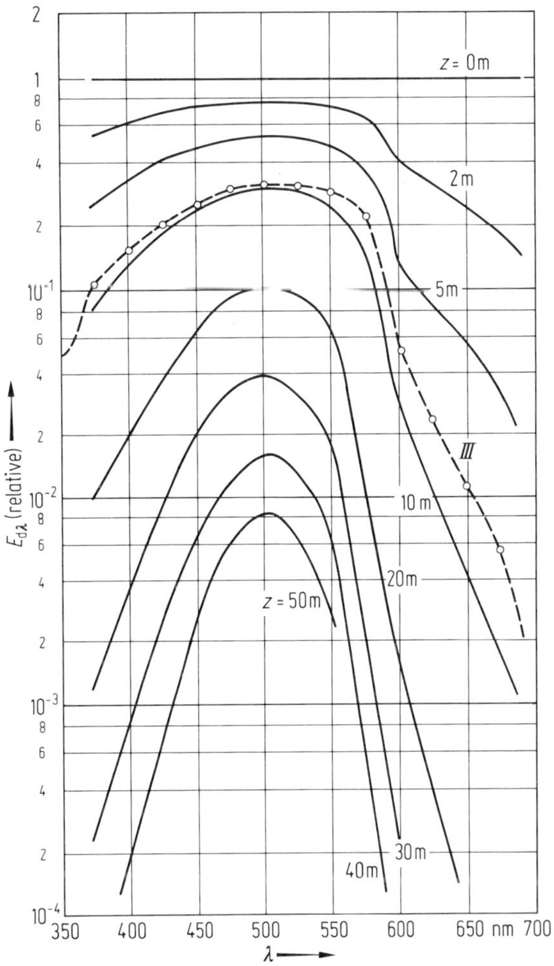

Fig. 39. The relative spectral distribution of the downward daylight irradiance at selected depths on 21 August 1975 at the position 58°46′ N 0°56′ E (solid lines), Højerslev [82 H 2]. The relative spectral distribution for the oceanic optical class *III* at 10 m depth (dashed line), according to Jerlov's optical classification of water masses, is also shown (Jerlov [76 J]).

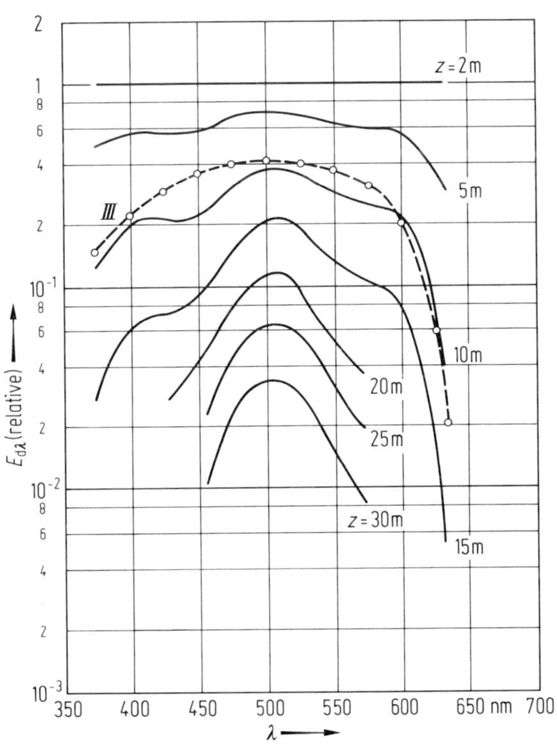

Fig. 40. The relative spectral distributions of the downward daylight irradiance at selected depths (solid lines) at the central position in the so-called FLEX box during the end of first plankton bloom on 10 May 1976 (Højerslev [82 H 2]). The relative spectral distribution for the oceanic optical class *III* at 10 m depth (dashed line), according to Jerlov's optical classification of water masses, is also shown (Jerlov [76 J]).

Table 21. Observations of the light attenuation coefficient in relation to pure water, $c-c_w$, and of the light scattering coefficient for particles b_p. Jerlov [74 J 1].

Region	$c-c_w$ [m^{-1}]		b_p [m^{-1}]	
	380 nm	655 nm	380 nm	655 nm
Sargasso Sea	0.05	0.04	0.03	0.02
	0.06	0.03	0.04	0.02
Caribbean Sea	0.11	0.06	0.06	0.06
Eq. Central Pacific	0.11		0.05	
Romanche Deep	0.14		0.07	
Mediterranean	0.11	0.04	0.04	0.03
	0.16	0.11	0.06	0.07
		0.15		0.11
Galapagos	0.25	0.13	0.09	0.08
Bermuda waters	0.20	0.10	0.10	0.11
	0.25	0.16	0.11	0.12
		0.32		0.23
Kattegat	(0.54)	0.23	0.16	0.15
Baltic Sea	(1.15)	0.27	0.21	0.20
Bothnian Sea	(1.72)	0.38	0.31	0.28

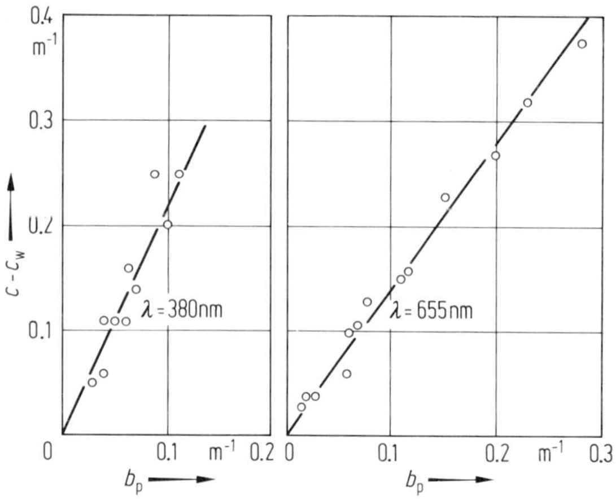

Fig. 41. Light attenuation coefficients ($c-c_w$) as a function of particle scattering coefficients (b_p) for the ultraviolet ($\lambda=380$ nm) and for the red ($\lambda=655$ nm) (Jerlov [74 J 1]). c_w: light attenuation coefficient of pure water.

3.3.3.7 Classification of sea water in terms of light absorption properties of suspended and dissolved matter

The optical properties of sea water are determined by its content of phytoplankton, detritus, sediments, dissolved organic matter, water molecules, ions present, and temperature. For oceanic conditions the concentration of ions (salinity) is virtually invariable and the temperature is confined to limited variations. Thus with respect to optical water mass classification, the contribution from pure sea water to the optical properties of ocean water can be treated as constant blank values, practically speaking, Table 2. The optical properties of plankton, detritus, and sediments are determined through their size distribution, shape, and their relative refractive index given as a complex number in order to include both light scattering and absorption. There is a gradual transition between plankton and detritus, so a division into two such categories might, in certain cases, be questionable from an optical as well as from a biological point of view. In particular, the spectral light absorption by plankton is affected by their healthy state.

The dissolved organic matter (gelbstoff) consists of, chemically speaking, very complex groups which are only partly known. However, optically gelbstoff is simple to deal with. It absorbs the light according to a generally valid exponential law, see eq. (95), Thus by measuring its light absorption at a single wavelength, preferably in the ultraviolet part of the spectrum, the spectral characteristics of a given gelbstoff batch are completely determined.

From spectral irradiance *in-situ* measurements it is possible to arrive at approximate concentrations, typically within ±10%, of phytoplankton pigments, nonchlorophyllous particles and gelbstoff (Prieur and Sathyendranath [81 P] and Table 22). The method is based on the following two equations (Morel and Prieur [75 M]):

$$a(\lambda)=\frac{K_d(\lambda)(1-R(\lambda))\cos j}{0.60+(0.47+2.5R(\lambda))\cos j}, \tag{112}$$

in which j is the sun zenith angle in water, and:

$$a(\lambda)-a_w(\lambda)=Ca_c^*(\lambda)+Pa_p^*(\lambda)+Ya_y^*(\lambda). \tag{113}$$

The light absorption related to each group of substances is expressed as the product of its concentration C, P, or Y and the corresponding specific light absorption a_c^*, a_p^*, or a_y^* [m^{-1} per unit concentration]. Combining eqs. (60) and (112), and assuming $dR(\lambda)/dz=0$, it is found that

$$\bar{\mu}=\frac{(1-R(\lambda))\cos j}{0.60+(0.47+2.5R(\lambda))\cos j}. \tag{114}$$

For the extreme conditions $0 \leqq R(\lambda) \leqq 0.10$ and $0° \leqq j \leqq 48.6°$ (the maximum value for j), the inequality

$$0.55 \leqq \bar{\mu} \leqq 0.93 \tag{115}$$

compares favourably with the experimental findings in Table 6.

Table 22. Spectral values of normalized spectral absorption values of phytoplankton pigments, a_{ph}^*, nonchlorophyllous suspended matter, a_p^*, and gelbstoff, a_y^*, according to Prieur and Sathyendranath [81 P].

λ nm	a_{ph}^*	a_p^*	a_y^*	λ nm	a_{ph}^*	a_p^*	a_y^*
400	0.687	1.892	1.751	555	0.321	0.566	0.200
405	0.781	1.766	1.632	560	0.294	0.565	0.186
410	0.828	1.726	1.522	565	0.273	0.587	0.174
415	0.883	1.635	1.419	570	0.276	0.599	0.162
420	0.913	1.534	1.323	575	0.268	0.569	0.151
425	0.939	1.410	1.234	580	0.291	0.624	0.141
430	0.973	1.274	1.150	585	0.274	0.649	0.131
435	1.001	1.119	1.073	590	0.282	0.718	0.122
440	1.000	1.000	1.000	595	0.249	0.805	0.114
445	0.971	0.914	0.932	600	0.236	0.850	0.106
450	0.944	0.868	0.869	605	0.279	0.913	0.099
455	0.928	0.764	0.811	610	0.252	0.879	0.093
460	0.917	0.691	0.756	615	0.268	0.877	0.086
465	0.902	0.636	0.705	620	0.276	0.875	0.080
470	0.870	0.589	0.657	625	0.299	0.910	0.075
475	0.839	0.567	0.613	630	0.317	0.935	0.070
480	0.798	0.522	0.571	635	0.333	0.992	0.065
485	0.773	0.496	0.533	640	0.334	1.041	0.061
490	0.750	0.483	0.497	645	0.326	1.133	0.057
495	0.717	0.474	0.463	650	0.356	1.227	0.053
500	0.668	0.473	0.432	655	0.389	1.337	0.049
505	0.645	0.480	0.403	660	0.441	1.528	0.046
510	0.618	0.509	0.375	665	0.534	1.542	0.043
515	0.582	0.495	0.350	670	0.595	1.542	0.040
520	0.528	0.488	0.326	675	0.544	1.542	0.037
525	0.504	0.488	0.304	680	0.502	1.542	0.035
530	0.474	0.495	0.284	685	0.420	1.542	0.032
535	0.444	0.505	0.264	690	0.329	1.542	0.030
540	0.416	0.526	0.247	695	0.262	1.542	0.028
545	0.384	0.548	0.230	700	0.215	1.542	0.026
550	0.357	0.548	0.214				

3.3.4 Daylight in the sea

3.3.4.1 Irradiance on the sea surface

The downward irradiance on the sea surface is dependent on a large variety of parameters which are: Latitude, season, and time of the day (solar elevation), optical properties of the aerosols and hydrosols in the atmosphere and the cloud cover, e.g. Robinson [66 R] or Kondratyev [69 K 1]. The spectral range of the downward irradiance on the sea surface is approximately within 295···3000 nm, Fig. 42, and its spectral distribution is virtually independent of solar elevations above 15°, Table 23, and of aerosol concentrations within typical limits.

The spectrum of the downward irradiance is modified by the presence of atmospheric gases and of the Fraunhofer lines whose characteristics are given in Table 24.

These Fraunhofer lines might serve as field calibration means for spectral irradiance meters. Apart from this they play an insignificant role in marine optics with respect to photosynthesis, solar heat input to the sea, etc., as depicted in Fig. 43.

The downward irradiance on the sea surface is often termed the global radiation. It is simply the sum of the direct sun radiation and the diffuse sky radiation.

Fig. 44 demonstrates that the global radiation becomes less diffuse with increasing wavelengths and solar elevations. This fact is sometimes overlooked in albedo calculations. Fig. 45 shows the relative spectral distribution of the direct sun radiation and the diffuse sky radiation. Note the well known strong spectral dependency, $\lambda^{-4.09}$, Morel [80 M] of the diffuse sky radiation (A) and the similar spectral distributions of the global radiation for clear (B) and overcast (C) situations, respectively.

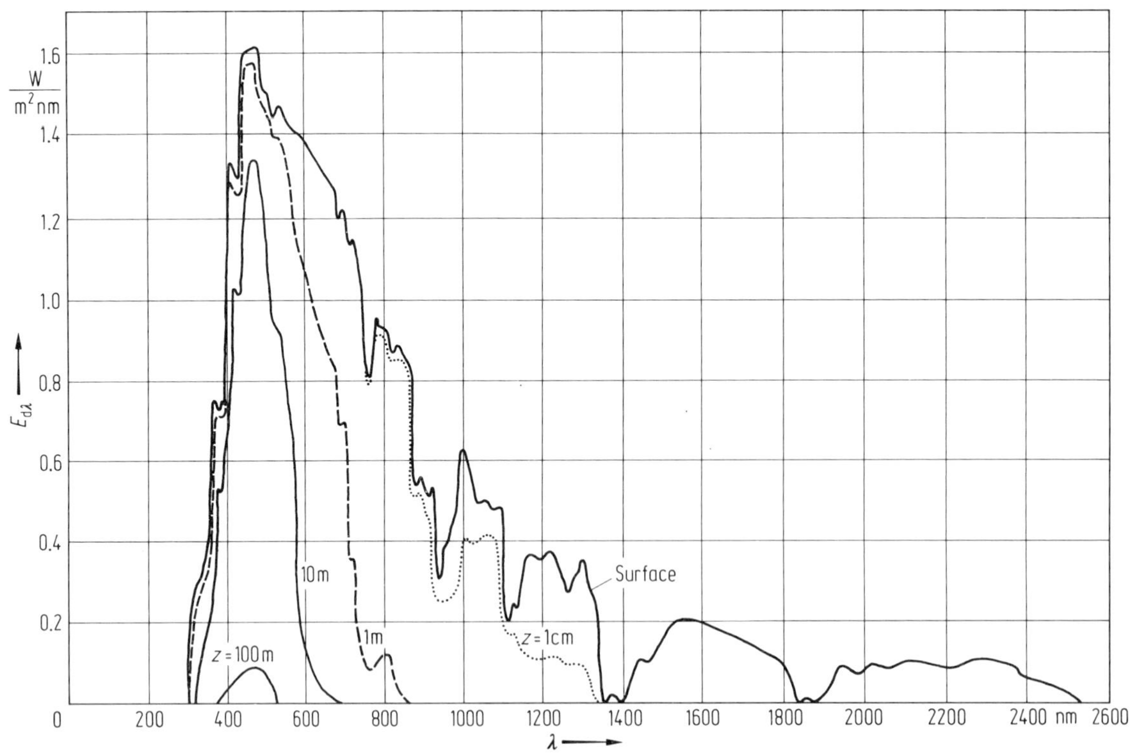

Fig. 42. The complete spectrum of downward daylight irradiance in typical oceanic water at different paths *z* (Jerlov [76 J]).

Table 23. Spectral distribution of irradiance from sun and sky in relative units. After Kimball [24 K].

Wavelength nm	Solar elevation [°] 11.3	30	65
350			190
375			235
400	54	131	307
425	51	153	362
450	48	170	400
475	47	178	417
500	46	179	413
525	44	172	393
550	41	166	377
575	41	165	371
600	41	165	368
625	40	162	254
650	42	160	344
675	42	159	336
700	41	154	318
725	41	145	298
750	38	136	277

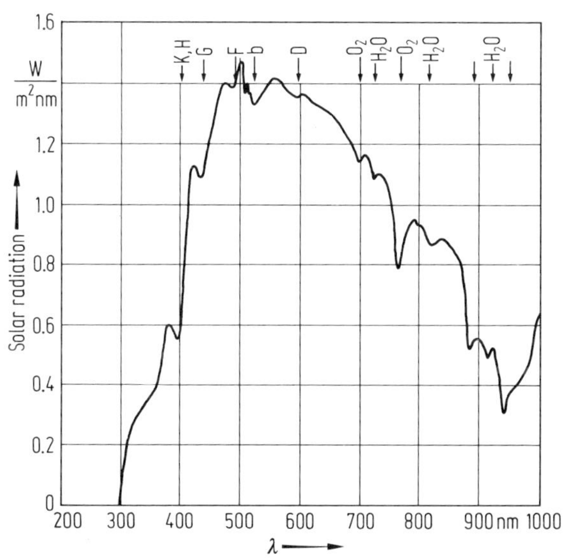

Table 24. Wavelengths of the Fraunhofer lines of the sun's spectrum at 15 °C and 76 cm Hg pressure. (Fabry and Buisson system).

Line	Due to	Wavelength nm
U	Fe	294.79
t	Fe	299.94
T	Fe	302.1067
a	Fe	304.7623
S_1, S_2	Fe	310.0683
	Fe	310.0326
	Fe	309.9943
R	Ca	318.1277
	Ca	317.9343
Q	Fe	328.6773
P	Ti	336.1194
O	Fe	344.1020
N	Fe	358.1210
M	Fe	372.7636
L	Fe	382.0438
K	Ca	393.3684
H	Ca	396.8498
h	H	410.1750
g	Ca	422.6742
G	Fe	430.7914
	Ca	430.7749
G'	H	434.0477
F	H	486.1344
b_4	Fe	516.7510
	Mg	516.7330
b_2	Mg	517.2700
b_1	Mg	518.3621
E_2	Fe	526.9557
D_2	Na	588.9977
D_1	Na	589.5944
C	H	656.2816
B	O	686.9966
A	O	762.1
	O	759.4
Z	–	822.85
Y	–	899.0

Fig. 43. Spectral distribution of direct solar radiation at sea level. Air mass 1.0. Concentration of precipitable water 10 mm. Ozon 3.5 mm. Positions of the most intense Fraunhofer lines and the major absorption bands are indicated (Jerlov [76 J]).

Højerslev

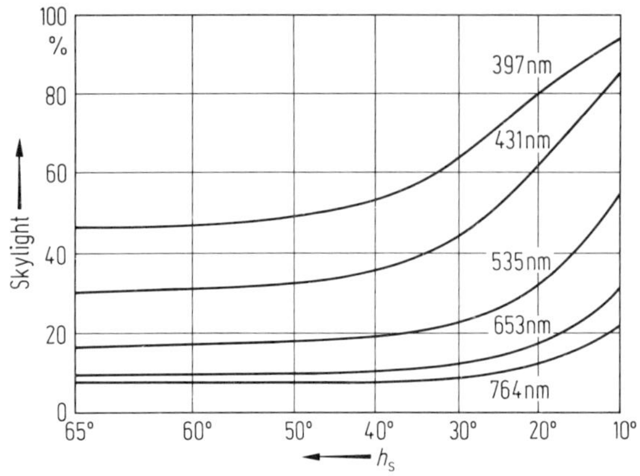

Fig. 44. Percentage of skylight in the global radiation as a function of solar elevation h_s for selected wavelengths between 397 nm and 764 nm. (After Sauberer and Ruttner [41 S]).

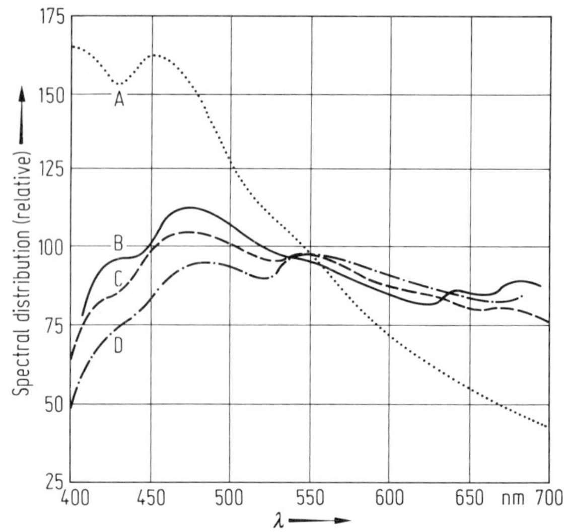

Fig. 45. Average spectral distribution of daylight irradiance (relative units) for (A) zenith sky, (B) sun + sky (horizontal plane), (C) overcast sky, (D) direct sunlight. (After Taylor and Kerr [41 T]).

3.3.4.2 Reflection and albedo at the sea surface

The reflectance for the electric vector of radiation resolved into components parallel and perpendicular to the plane of incidence is given by Fresnel's equations

$$\varrho_{\parallel}=\frac{\tan^2(i-j)}{\tan^2(i+j)}, \tag{116}$$

$$\varrho_{\perp}=\frac{\sin^2(i-j)}{\sin^2(i+j)} \tag{117}$$

in which i is the angle of incidence and j the angle of refraction. The reflected ray is in the plane of incidence and the angle of reflection is equal to i.

Assuming the global radiation to be unpolarized, which is a fair assumption in most fields of marine optics, the light reflection for a flat sea becomes:

$$\varrho=\tfrac{1}{2}(\varrho_{\parallel}+\varrho_{\perp}). \tag{118}$$

At grazing angles of incidence $i=90°$, $j=48.6°$ (termed the critical angle θ_{cr}) defines the so-called Snell circle.

The global radiation, or any of its spectral components, reflected from a flat sea is defined by:

$$r=\frac{\int_{2\pi}\varrho(i)L(z=0+,i,\varphi)\cos i\,d\omega}{\int_{2\pi}L(z=0+,i,\varphi)\cos i\,d\omega} \tag{119}$$

in which $z=0+$ refers to a height just above the sea surface. Note that eq. (119) is only valid for the case of unpolarized light.

The light reflection $\varrho(i)$ can be approximated into:

$$\varrho(i)=0.020+0.28(1-\cos i)^3+0.30(1-\cos i)^6+0.40(1-\cos i)^9 \tag{120}$$

which offers obvious advantages when performing the integration of eq. (119) because the radiance L can be expressed in terms of Legendre polynomials or expressions containing cosine terms, e.g. the cardiodal radiance distribution.

Table 25. Reflectance of radiation against a calm sea surface. Jerlov [76 J].

Angle of incidence i [°]	Reflectance [%]		
	$\varrho_{\parallel}$	$\varrho_{\perp}$	ϱ
0	2.0	2.0	2.0
5	2.0	2.1	2.0
10	1.9	2.1	2.0
15	1.8	2.3	2.0
20	1.7	2.5	2.1
25	1.4	2.7	2.1
30	1.2	3.1	2.1
35	0.9	3.6	2.3
40	0.6	4.3	2.4
45	0.3	5.3	2.8
50	0.1	6.7	3.4
55	0.2	8.6	4.4
60	0.4	11.5	5.9
65	1.7	15.8	8.7
70	4.7	21.9	13.3
75	11.0	31.3	21.2
80	24.0	45.9	34.9
85	49.3	67.4	58.3
90	100	100	100

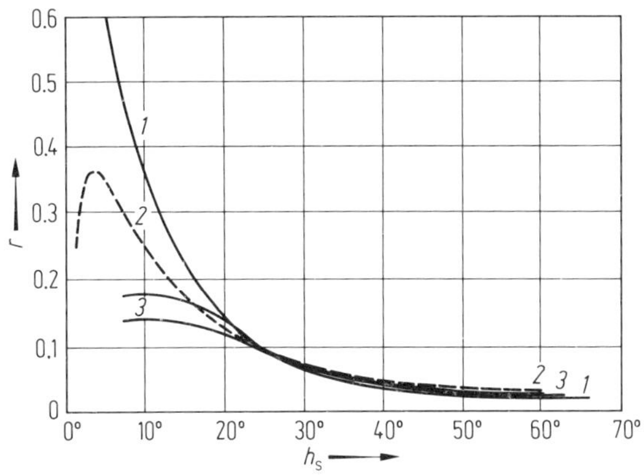

Fig. 46. Curve *1* shows the effect of solar elevation h_s on the reflectance r of a perfectly plane sea surface for the direct solar beam. Curve *2* corresponds to the total solar irradiance (direct plus diffuse) for clear sky and also for a perfectly plane sea surface. Curves *3*, after Cox and Munk (1956), show the effect of h_s, on the reflectance for direct solar beam with a sea state 4, for two different assumptions about multiple reflections produced when the solar elevation is relatively small (Ivanoff [77 I]).

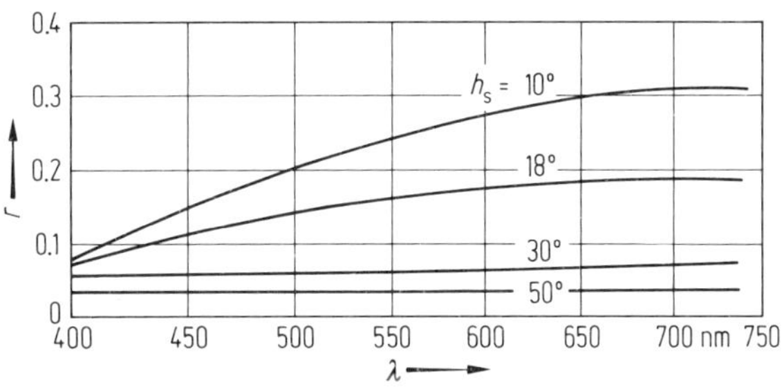

Fig. 47. Reflectance r of global radiation for different solar elevations h_s as a function of wavelengths. Corrected for cosine error of the collector. (After Sauberer and Ruttner [41 S].)

Table 26. Reflectance, in [%], for the diffuse sky radiation, for different assumptions concerning the distribution of radiances in the sky. Ivanoff [77 I].

Sky	Judd (1942)	Burt (1953)	Cox and Munk (1956)	Preisendorfer (1957)	Payne (1972)
			Flat sea		
Clear			10.0		
Cloudy					
Uniform	6.6	6.6	6.6		6.8
Cardioidal Distribution			5.2	5.16	5.2
			Sea state 4		
Clear			7.1···8.8		
Cloudy					
Uniform		5.7	5.0···5.5		5.0
Cardioidal Distribution		4.8	4.3···4.7		4.2

Højerslev

When dealing with the reflectance of global radiation, eq. (119) is sometimes written as

$$r = \varrho_s(h_s)(1 - Q(h_s)) + \langle \varrho_d \rangle Q(h_s) \tag{121}$$

in which the first factor refers to the reflection of the direct sunlight, the sun having the solar elevation h_s. $Q(h_s)$ is the significant ratio of sky radiation to global radiation, e.g. Figs. 44 and 45, and $\langle \varrho_d \rangle$ the average reflection of the diffuse sky radiation, varying between 5.2% and 10%, Table 26.

The spectral variation of $r(h_s)$ in the range 400···750 nm as a function of h_s is depicted in Fig. 47. $r(h_s)$ is virtually independent of all wavelengths 295···3000 nm when h_s exceeds 30°. For any fixed h_s below 30° the spectral dispersion of $r(h_s)$ increases towards the red and infra-red part of the spectrum (compare Figs. 44 and 47).

Surface waves producing no white caps subdue $r(h_s)$ at solar elevations below 20° (Table 26 and Fig. 46). At high wind speeds white caps are produced and the light reflection can easily exceed the one for the case of a flat sea (curve 2 in Fig. 46) at all solar elevations (e.g. Monahan et al. [83 M]). The dependence of $r(h_s)$ on white caps is only partly known.

The albedo of the sea is defined as the ratio of all the radiant energy leaving the sea to that falling on it (per unit area):

$$A = \frac{E_u(z=0+)}{E_d(z=0+)}. \tag{122}$$

Combining this definition with eqs. (31) and (119):

$$A = r + (1 - r_w)\frac{E_u(z=0-)}{E_d(z=0+)} \tag{123}$$

in which $1 - r_w \approx 0.5$ and r_w is the water-air reflectance defined as:

$$r_w = \frac{\int_{2\pi} \varrho(j) L(z=0-, j, \varphi) \cos j \, d\omega}{\int_{2\pi} L(z=0-, j, \varphi) \cos j \, d\omega} \tag{124}$$

in which $z=0-$ refers to just below the sea surface. It follows from simple geometrical considerations, Gershun [39 G 1], that:

$$L(z=0-, j, \varphi) = n^2(1 - \varrho(j))^{-1} L(z=0+, i, \varphi). \tag{125}$$

The angles i and j are interrelated through Snell's law:

$$\frac{\sin i}{\sin j} = n \simeq \tfrac{4}{3} \tag{126}$$

Some experimental values of the albedo during overcast are presented in Table 27. These values are approximately the same as the theoretical values for the reflectance in Table 26. Thus, in case of overcast, the second term in eq. (123) is not crucial.

The reflectance of the global radiation for clear weather conditions attains maximum values at solar elevations around 10° depending on sea state, Fig. 46. This maximum is most pronounced in case of a flat sea. Observations of the albedo at various solar elevations and sea states lead to similar conclusions as depicted in Fig. 48. The existence of such maxima occuring at *low* solar elevations and at clear weather conditions is heuristically understood by inspecting eq. (121), Figs. (44) and (45), and Table 25, respectively:

For a flat sea and $h_s = 90°$, $Q(h_s) \approx 0$ and $r \approx 0.02$. For $h_s = 0°$, $Q(h_s) \approx 1$ and $r \approx 0.06$. At low solar elevations, $1 - Q(h_s) > 0$, $\langle \varrho_d \rangle \approx 0.06$ and $0.20 \leqq \varrho_s(h_s) \leqq 1.00$.

Table 27. Some experimental values of the albedo of the sea (in %) during overcast weather (Ivanoff [77 I]).

Powell and Clarke (1936)	Neiburger (1948)	Burt (1953)	Grichtchenko (1959)	Calathas (1970)	Payne (1972)
6.5	10.5	Flat Sea 9.4	6···11	6.1	5···7.5
		Sea state 4		4.6	

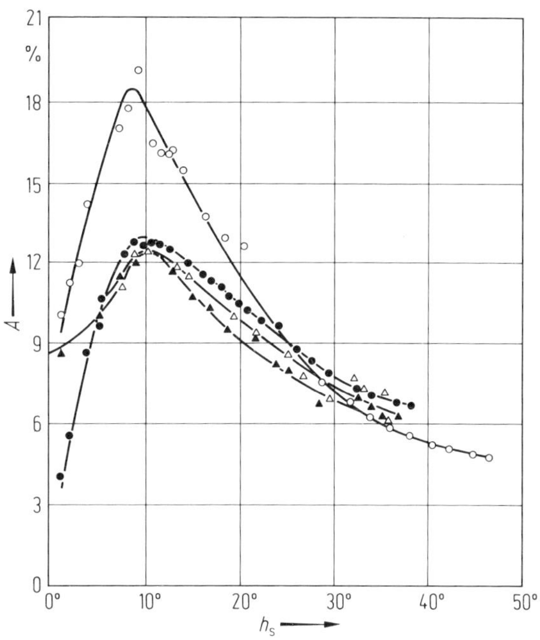

	Wind [m/s]	Sea state	Wave amplitude [m]
○	7	moderately rough	0.5
●	8	rough	1.0
△	15 - 16	very rough	1.5 - 2.0
▲	17 - 20	very rough	1.5 - 2.0

Fig. 48. Variation of the albedo of the sea, A, with solar elevation h_s and roughness of the sea, for clear weather. Measurements made in the Mediterranean Sea from the "Bouée Laboratoire" (42°14′ N, 5°35′ E) by Calathas [70 C 1]. (Ivanoff [77 I]).

3.3.4.3 Irradiance in the sea

The global radiation into the sea becomes eventually absorbed in a manner depending on the optical water type (Jerlov [76 J] and Table 28). Most of the radiant flux becomes transferred into sensible heat. The global radiation used in the photosynthesis by plankton is typically less than 1%.

The results in Table 28 are based upon observations which are difficult to perform routinely at sea because the strong light absorption in the surface layers causes high ambient noise levels in the recordings. This is the rationale for developing models describing the transformation of the global radiation into sensible heat by means of simple input parameters. Højerslev [84 H 2] has presented such a model which only demands one quanta irradiance measurement at one single depth, $z_q(10\%)$, as an input parameter. Alternatively, $z_q(10\%)$ can be replaced by either the color index F, and therebv be remotely determined (Table 16 and Figs. 23···25), or by the monochromatic irradiance around $\lambda = 500$ nm, $E_d(500\,\text{nm})$ (Table 17 and Figs. 27 and 28).

The equation for the quasi-momentaneous heat input into the sea due to the global radiation (300···3000 nm) writes in its most general form for a stagnant sea:

$$\varrho c_p \frac{\partial T}{\partial t} = a E_o \qquad (127)$$

in which ϱ is the density of the sea water, c_p the specific heat capacity at constant pressure, T the temperature of the sea water, a the light absorption coefficient for the global radiation (300···3000 nm), and E_o the scalar irradiance of the global radiation (300···3000 nm). All these parameters are functions of the three space coordinates (x, y, z) and time t. Normally ϱ and c_p are assumed to be fixed constants and eq. (127) is approximated into:

$$\varrho c_p \frac{\partial T}{\partial t} = -\frac{\partial E}{\partial z} \qquad (128)$$

Table 28. Percentage of total downward irradiance ($\lambda = 300 \cdots 3000$ nm) from sun and sky. For oceanic water the solar elevation is 90°; for coastal water 45°. Jerlov [76 J].

Depth m	Oceanic water					Coastal water				
	I	*IA*	*IB*	*II*	*III*	*1*	*3*	*5*	*7*	*9*
0	100	100	100	100	100	100	100	100	100	100
1	44.5	44.1	42.9	42.0	39.4	36.9	33.0	27.8	22.6	17.6
2	38.5	37.9	36.0	34.7	30.3	27.1	22.5	16.4	11.3	7.5
5	30.2	29.0	25.8	23.4	16.8	14.2	9.3	4.6	2.1	1.0
10	22.2	20.8	16.9	14.2	7.6	5.9	2.7	0.69	0.17	0.052
20						1.3	0.29	0.020		
25	13.2	11.1	7.7	4.2	0.97					
50	5.3	3.3	1.8	0.70	0.041	0.022				
75	1.68	0.95	0.42	0.124	0.0018					
100	0.53	0.28	0.10	0.0228						
150	0.056			0.00080						
200	0.0062									

by applying the arguments leading to eq. (58). E stands here for the vector irradiance of the global radiation (300⋯3000 nm). If the sea is supposed to be horizontally stratified T and E are functions of depth z and time t only. Højerslev [85 H 2] has given the approximate expression for E:

$$E(z) = E(z=0-)\,[0.44\,\mathrm{e}^{-\int_0^z K_q \mathrm{d}z} + 0.56\,\mathrm{e}^{-2.35z}] \qquad (129)$$

in which $E(0-)$ is the global radiation (300⋯3000 nm) just below the sea surface, $E(0-) = E(0+)\,(1-r)$, see eq. (121). $E(0-)$ can approximately be given in terms of the downward quanta irradiance (350⋯700 nm) just below the sea surface, $q(0-)$, for which the following relations hold for fairly clear sky situations and a calm sea (Højerslev [80 H 1, 82 H 2] and Fig. 49):

$$q(0-, h_s) = 1200 \sin^{1.4} h_s \cdot 10^{18}\ \text{photons m}^{-2}\,\text{s}^{-1} \qquad (130)$$

and (Jerlov [74 J 2], Bethoux [77 B] and Højerslev [84 H 2]):

$$\frac{q(0-, h_s)}{E(0-, h_s)} = 1.2 \cdot 10^{18}\,\frac{\text{photons m}^{-2}\,\text{s}^{-1}}{\text{W m}^{-2}}. \qquad (131)$$

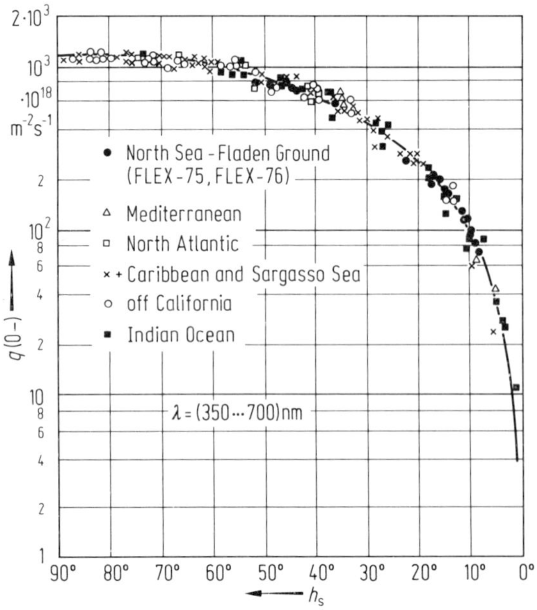

Fig. 49. Downward quanta irradiance ($\lambda = 350 \cdots 700$ nm) just below the sea surface, $q(0-)$, against solar elevation, h_s, on clear days. The data set stems from measurements made in different sea waters. The quanta irradiance just below the sea surface (the 100% level) can be expressed as $q(0-, h_s) = 1200 \sin^{1.4} h_s \cdot 10^{18}$ photons $\text{m}^{-2}\,\text{s}^{-1}$ (Højerslev [80 H 1]).

The reason for replacing $E(0-, h_s)$ by $q(0-, h_s)$ is both the simple empirical findings in eq. (130), the unnecessity of including the reflectance, and, finally, the frequency at which the quanta irradiance is routinely measured at sea.

The depths of the 50%, 30%, 10%, 3%, and 1% levels for the downward quanta irradiance are mutually related to one another through relations which were observed in the course of a year on Fladen Ground (Højerslev [82 H 2]):

$$z_q(50\%) = 0.20 z_q(10\%), \tag{132}$$

$$z_q(30\%) = 0.48 z_q(10\%), \tag{133}$$

$$z_q(3\%) = 1.58 z_q(10\%), \tag{134}$$

$$z_q(1\%) = 2.16 z_q(10\%), \tag{135}$$

being in good agreement with other findings (Jerlov [77 J] and Aas [80 A]). The depths of the above levels were virtually independent of the solar elevation h_s (Fig. 13).

If $z_q(10\%)$ is measured either directly or indirectly by means of the color index or $E_d(500\,\text{nm})$, $E(z)$ in eq. (129) can be obtained at the five depths stated in eqs. (132)···(135) since:

$$0.44\,e^{-\int_0^{z_q(10\%)} K_q dz} \equiv 0.44\,e^{-\ln 10} \equiv 0.44\,e^{-\langle K_q \rangle z_q(10\%)}. \tag{136}$$

The result of this procedure of expressing $E(z)$ in eq. (129) in terms of $q(z)$ is presented in Fig. 50 and compared with the optical classification presented in Table 28 (Jerlov [76 J]). Having $E(z)$, eq. (128) gives straightforward the rate of change in the sea temperature at any depth within the euphotic zone. The present simple model compares favourably with others and the fit to the classification is remarkable for type *IB*, which covers large areas of the oceans (Fig. 51).

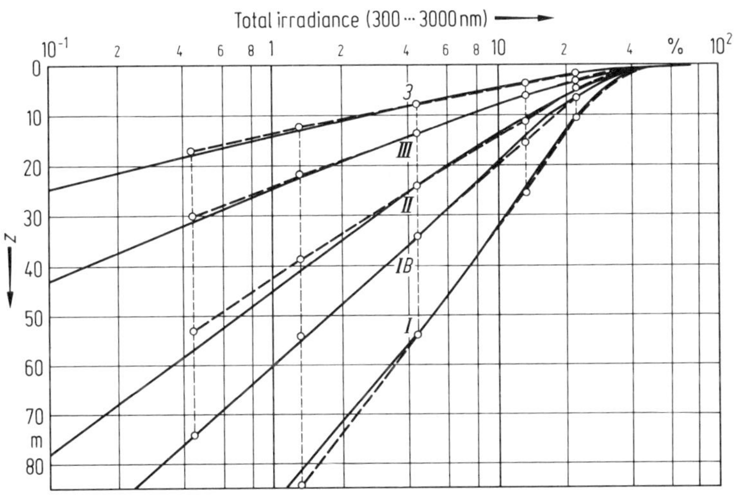

Fig. 50. Depth profiles for the total downward irradiance in percent of surface irradiance for the optical water types *I*, *IB*, *II*, *III* and *3*. Solid curves are from Jerlov [76 J]. Dashed curves are as computed from the solar heat input model applying $z_q(10\%)$ as the only optical input parameter (Højerslev [84 H 2]).

It should be emphasized that $z_q(10\%)$ can be derived from the water leaving nadir radiance at $\lambda = 443$ nm and $\lambda = 520$ nm, e.g. Højerslev [80 H 1, 82 H 2], so the NIMBUS-7, CZCS (Coastal Zone Color Scanner) is one obvious tool in modelling the solar heat input to the marine environment.

The downward spectral irradiance in the range 300···700 nm is a pertinent parameter in marine optics. As seen in subsects. 3.3.2 and 3.3.3, $E_d(\lambda, z)$, or, alternatively, $K_d(\lambda, z)$, is the obvious field parameter if practical points of view are taken into account also. Systematic data collection of $E_d(\lambda, z)$ was initiated by Knudsen [22 K] in Danish sea waters. For some historical information the reader is referred to Sauberer and Ruttner [41 S] and Jerlov [76 J]. In the late 60's a few specialized laboratories developed their own spectral irradiance meters, so this period might be regarded as the beginning of a systematical *global* investigation of $E_d(\lambda, z)$, see, e.g., Figs. 18 and 52···55.

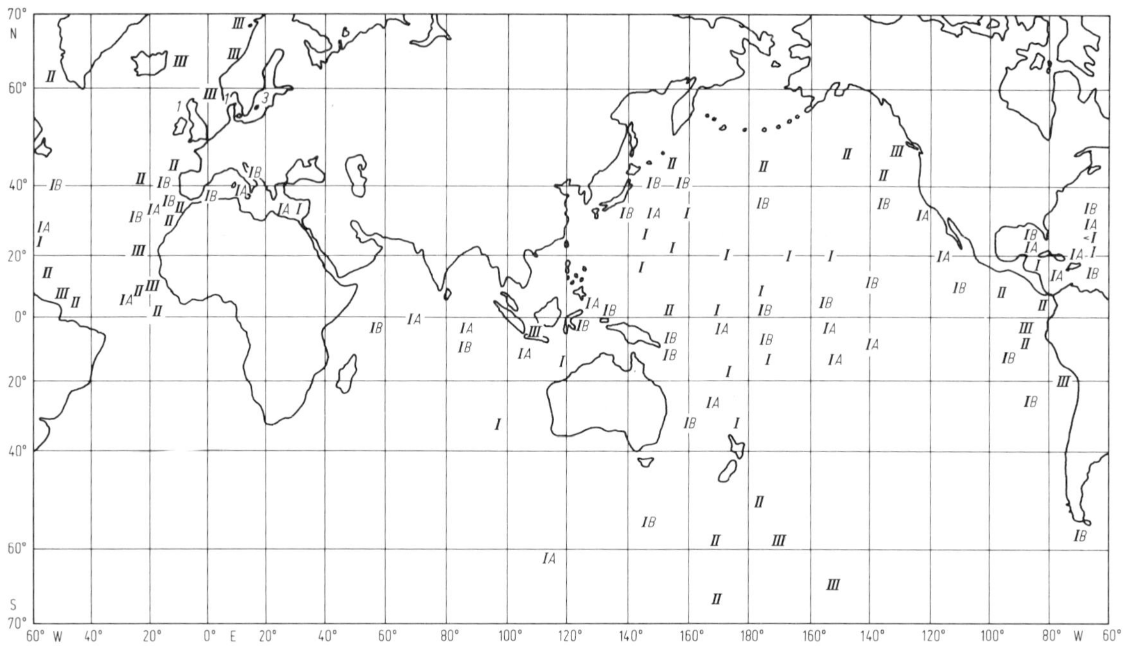

Fig. 51. Regional distribution of optical water types: *I*, *IA*, *IB*, *II*, *III* after (Jerlov [76 J]).

The spectra of the downward irradiance as depicted in Figs. 18 and 52···55 demonstrate some common features, (i) an intermediate maximum in the blue or green part of the spectrum at any depth in the sea, i.e. maximum daylight penetration takes place in this part of the spectrum (where exactly depends on the optical water type, see, e.g., Morel and Caloumenos [74 M 3]), (ii) low irradiance values in the red part of the spectrum for all water types (e.g. Fig. 15), starting quite abruptly around $\lambda = 600$ nm and then decreasing monotoneously with wavelength, (iii) highly varying irradiance values in the blue and ultraviolet part of the spectrum reflecting a highly variable gelbstoff concentration (e.g. eq. (95), Tables 5 and 10, and Figs. 15···18 and 52···55), and (iv) the conformity of the daylight irradiance spectra from a given water mass increases with depth and runs eventually parallel at great optical depth provided optical homogeneity of the sea water in question (Højerslev and Zaneveld [77 H 6]).

The general spectral behavior of E_d holds true for E_u as well because $dR/dz \simeq 0$ due to eq. (32) and exemplified in Fig. 53. Consequently, the spectral behavior of $E = E_d - E_u$ is almost identical with the one for E_d since $dR/dz \simeq 0$ *and* $E_d \gg E_u$. Previously, it has been shown that $\bar{\mu} \equiv E/E_o$ is fairly constant throughout the visible part of the spectrum for most if not all optical water types, see, e.g., Table 6 and eqs. (88) and (90). This implies that the spectral behavior of E_o is almost similar to the one for E (and thereby the one for E_d and E_u, respectively) as depicted in Fig. 56.

Instead of presenting spectra of the downward irradiance as done in Figs. 52···55 it is often more informative to give the spectral distribution of either K_d or a as a function of depth, see eqs. (23), (36), and (112), and the derived approximation $a \simeq \bar{\mu} K_d$. Two such examples are depicted in Figs. 57 and 58. Curve A in Fig. 57 exemplifies an extremely pure (oligotrophic) sea water mass in which the light absorption properties, but not the light scattering properties, of pure (sea) water are dominating, see subsect. 3.3.2 and Fig. 8. Note that K_d has no maximum but only a weakly developed minimum around $\lambda = 450$ nm, see also Table 2. Curves B and D in Fig. 57 are U-shaped from which turbid water masses rich in phytoplankton, detritus, *and* sediments but poor in gelbstoff are identified. Finally, curve C is V-shaped characterizing an eutrophic sea water mass dominated by living phytoplankton. Fig. 58 shows spectral distributions of K_d (or of a) being almost conform with the one represented by curve D in Fig. 57. Such spectral features are typical of a group of coastal waters in which gelbstoff is abundant (all details are given in subsect. 3.3.1.2 and, especially, in subsect. 3.3.3.7).

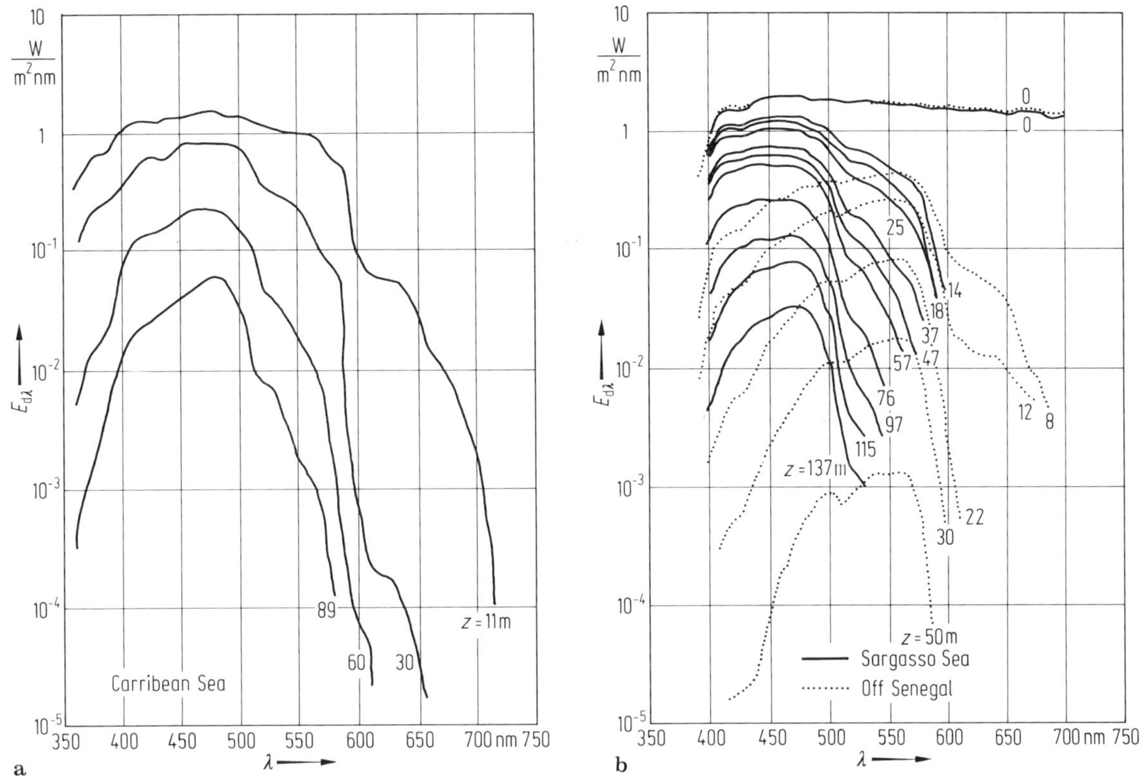

Fig. 52. Spectral distribution of downward irradiance for different depths z. (a) after Smith [73 S]. (b) after Morel and Caloumenos [74 M 3].

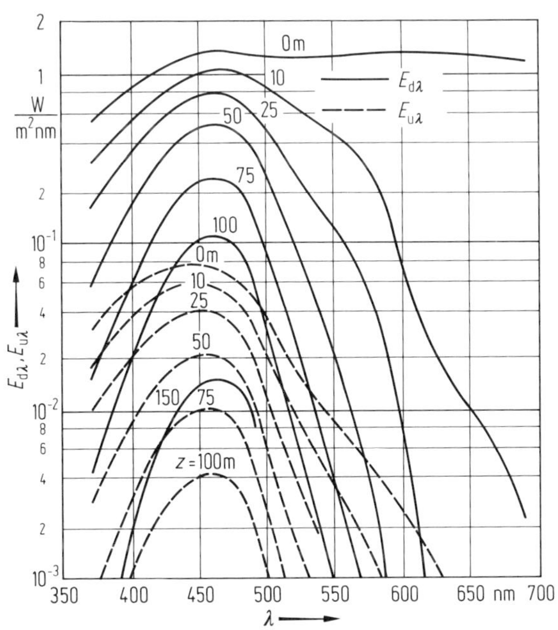

Fig. 53. Comparison between spectral distribution of downward (solid lines) and upward (dashed lines) irradiance for solar elevation of 55···60° and different depths z in the Sargasso Sea (Lundgren and Højerslev [71 L 1]).

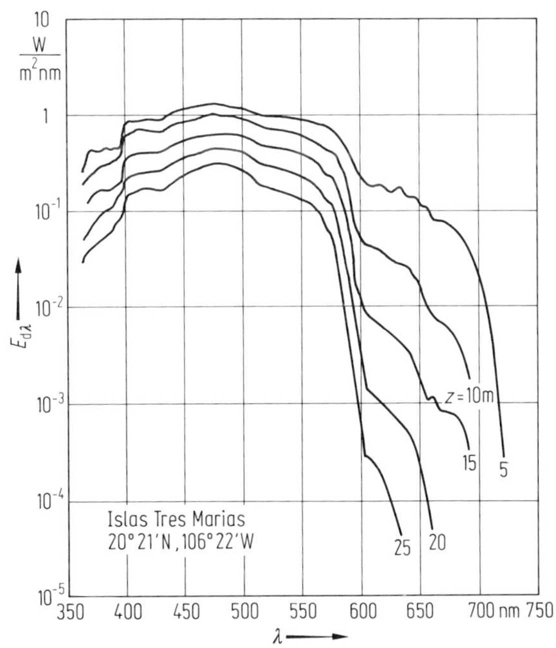

Fig. 54. Spectral distribution of downward irradiance (Tyler and Smith [70 T]).

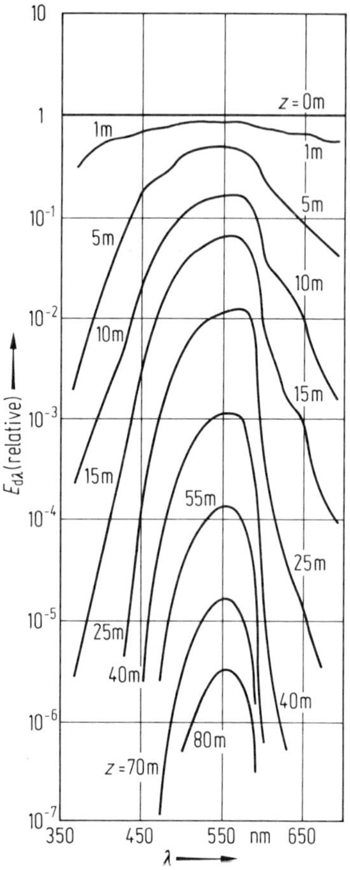

Fig. 55. Baltic proper: $E_{d\lambda}(\lambda, z)$ in normalized values of $E_{d\lambda}(\lambda, 0-)$. Solar elevation: 50° (Højerslev [74 H 1]).

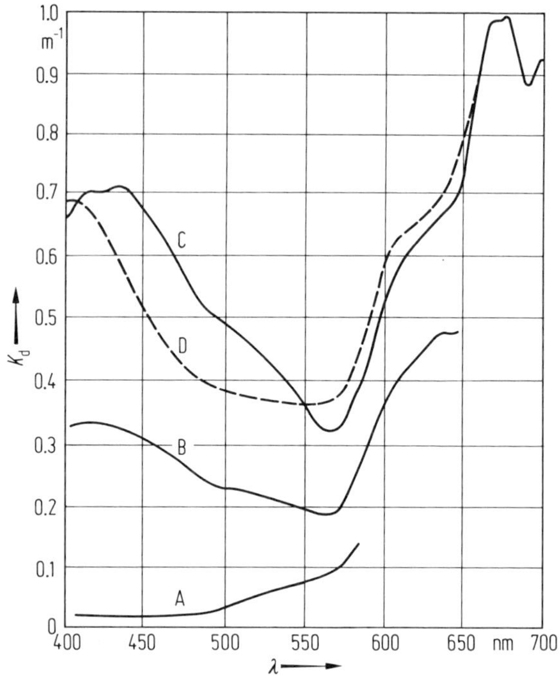

Fig. 57. The vertical attenuation coefficient K_d of downward irradiance for (A) the Sargasso Sea, (B) and (C) the upwelling area off West Africa, and (D) close to the African coast. (After Morel and Prieur [75 M]).

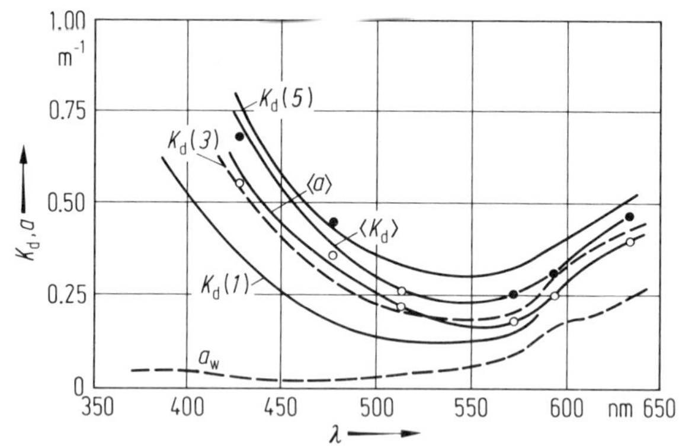

Fig. 58. Average absorption coefficients ⟨a⟩ and average vertical attenuation coefficients ⟨K_d⟩ for the downward irradiance within the euphotic zone as a function of wavelength. Included are the absorption coefficient a_w for pure water and the vertical attenuation coefficients $K_d(1) \cdots K_d(5)$ for the downward irradiance belonging to the coastal types *1*···*5* (Jerlov [76 J] and Højerslev [78 H]).

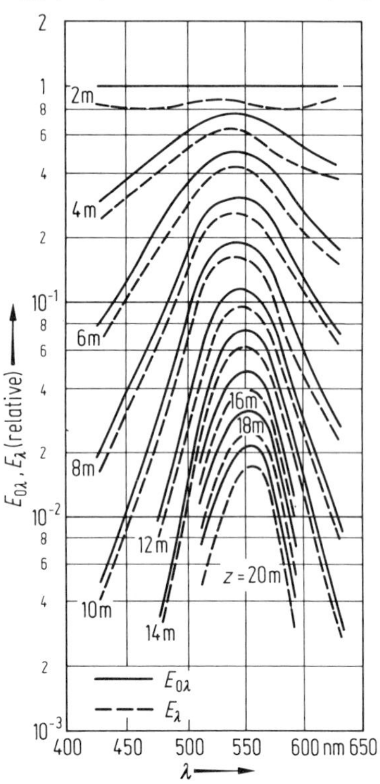

Fig. 56. The spectral distribution of $E_0(z)$ and $E(z)$ in normalized values of $E_0(z=2\,\mathrm{m})$ (Kattegat, Højerslev [76 H]). The daylight spectrum of E and E_0, the vertical distribution of average cosine $\bar{\mu} = E/E_0$ and the vertical distribution of the light absorption coefficient as a function of wavelength are all obtained by aid of a light absorption meter constructed by the author (Højerslev [75 H]).

3.3.4.4 Radiance in the sea

The first systematic measurements of the spectral (scalar) radiance $L_\lambda(\theta, \varphi, z, \lambda)$ were carried out in Swedish sea waters by Jerlov and Liljequist [38 J 2]. They applied the so-called Gershun tube [36 G] for those measurements but other radiance meter set-ups such as the so-called shadowing screen photometer measuring $\int_0^{2\pi} L_\lambda(\theta, \varphi, z, \lambda)\mathrm{d}\varphi$ were applied as well (Petersson [38 P]). The subsequent two decades added essentially only more radiance data from various sea waters and improved instrumental sensitivity through the invention of the photomultiplier (Sasaki et al. [55 S]). In 1960 Tyler presented the first comprehensive radiance data set of $L_\lambda(\theta, \varphi, z)$ for the blue-green part of the spectrum. These data were obtained in a lake called Pend Oreille (Tyler [60 T]).

Radiance measurements at sea have not been performed in a systematic global manner although this parameter is the most fundamental one in marine optics (the reasons for this are given in subsect. 3.3.1.1). There have been attempts to develop radiance meters suitable for routine operations at sea, see, e.g., Lundgren and Højerslev [71 L 1]. Lundgren [71 L 2] and Smith [74 S], but financial restrictions put an end to that (Lundgren and Smith, personal communication). In the decade before 1984 only few radiance field experiments have been made contrasting the considerable theoretical work on the subject, e.g. Kattawar and Plass [72 K 1, 73 K], Raschke [75 R] and Prieur [76 P 2]. Finally extensive experimental work on (vector) radiance *in-vitro* has essentially been reported by Timofeeva [74 T] given results comparable to those obtained *in-situ* at sea.

One example of the radiance distribution in air is depicted in Fig. 59. The maximum value is found in the sun's direction and the minimum value around the anti-sun's direction. Note that the sky light is not uniformly distributed as sometimes assumed in first order light transfer models. Fig. 60 shows schematically the radiative transfer of the global light into the sea. Note especially that the upward radiance L_u outside the Snell circle is totally reflected in case of a flat sea and thus contributing to the air-sea transmitted downward radiances L_s and L_{sky}.

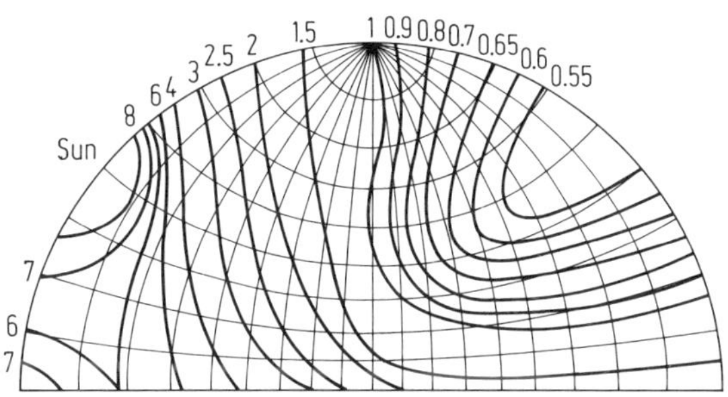

Fig. 59. Daylight radiance distribution at sea surface for a clear sky (normalized at zenith (Kimball [24 K]).

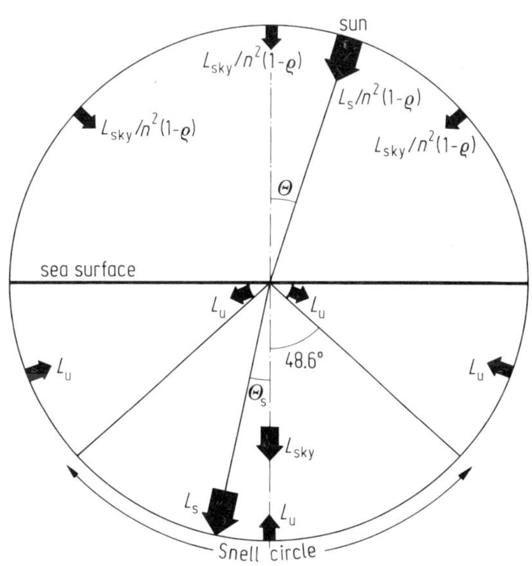

Fig. 60. Simplified surface radiance distribution of daylight (Lundgren and Højerslev [71 L 1]). n: index of refraction of sea water. Radiance from the sun, $L_s/(n^2(1-\varrho))$, incident on the sea surface at zenith angle θ, is transmitted, as radiance L_s, through the surface at angle θ_s determined by Snell's law. L_{sky}: sky-light radiance below the sea surface. Outside the Snell circle the upward radiance L_u is totally reflected.

Some examples of the angular radiance distribution in the vertical plane through the sun are presented in three different ways in Figs. 61···67. The examples given are taken from both clear and turbid waters. In Fig. 65 obtained from a clear water mass the polarization of the (vector) radiance is measured as well using the four Stoke's parameters (Lundgren, personal communication).

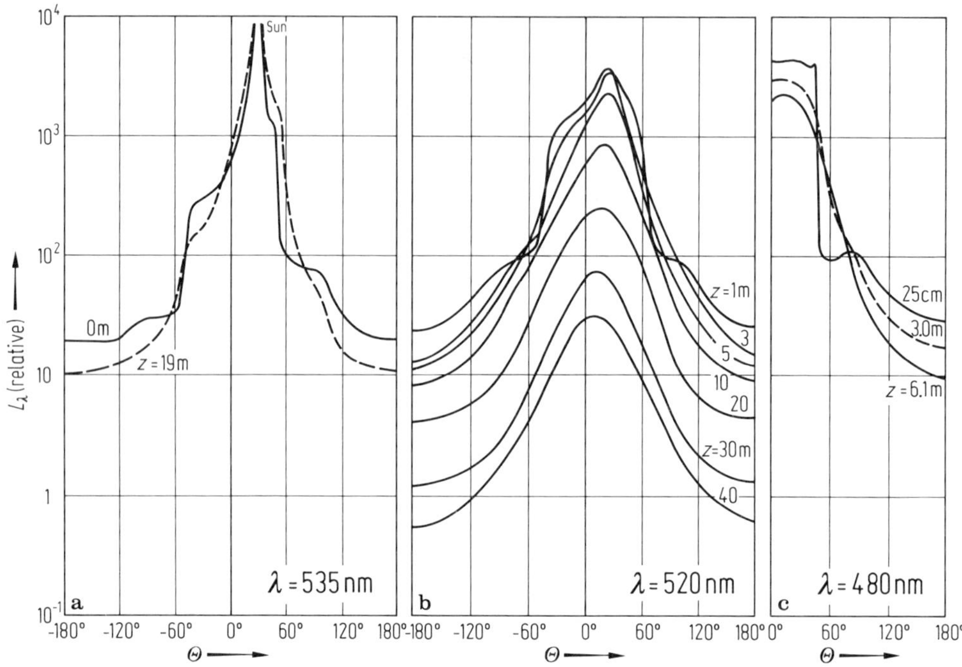

Fig. 61. Spectral radiance distributions L_λ, in relative units, in the vertical plane of the sun for relatively turbid waters and different depths. (a) clear day, Gullmar Fjord, $\lambda = 535$ nm. (After Jerlov and Fukuda [60 J].) (b) hazy weather, off Japan, $\lambda = 520$ nm. (After Sasaki et al. [62 S].) (c) overcast sky, Lake Pend Oreille, $\lambda = 480$ nm. (After Tyler [58 T].) θ: vertical angle.

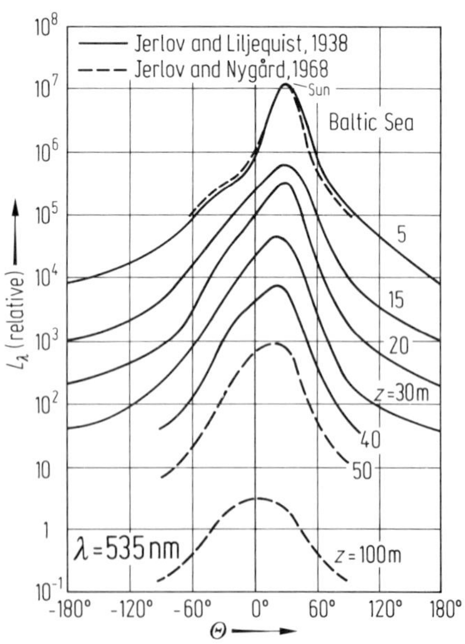

Fig. 62. Spectral radiance distribution L_λ (535 nm), in relative units, in the vertical plane of the sun for different depths (Jerlov and Nygård [68 J]). θ: vertical angle.

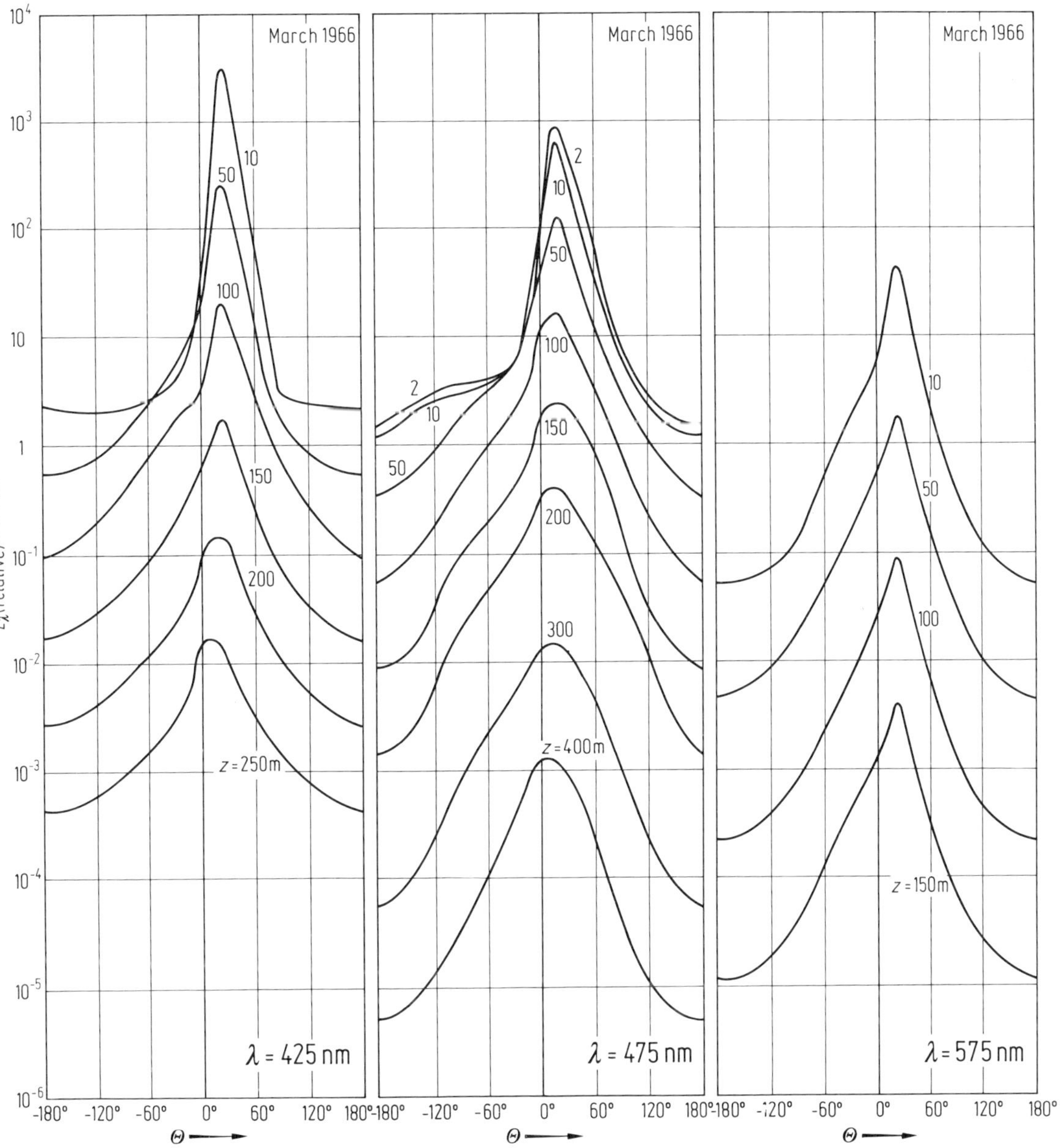

Fig. 63. Relative spectral radiance distributions L_λ in the vertical plane of the sun vs. vertical angle θ for different wavelengths and depths in the Sargasso Sea, 1966. Lundgren and Højerslev [71 L 1]. See also Fig. 64.

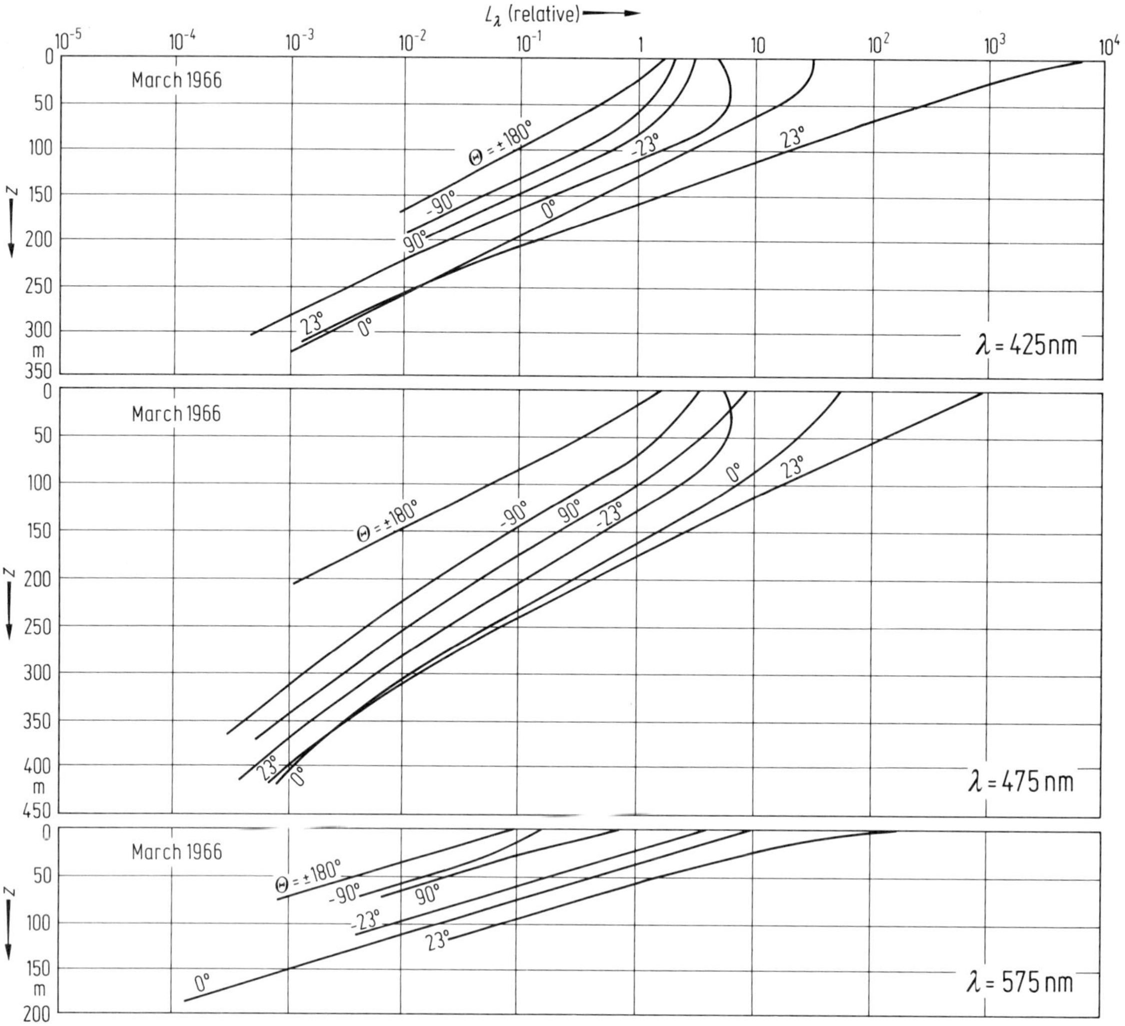

Fig. 64. Relative spectral radiance L_λ in the vertical plane of the sun vs. depth for different wavelengths and vertical angles θ in the Sargasso Sea, 1966. Lundgren and Højerslev [71 L 1]. See also Fig. 63.

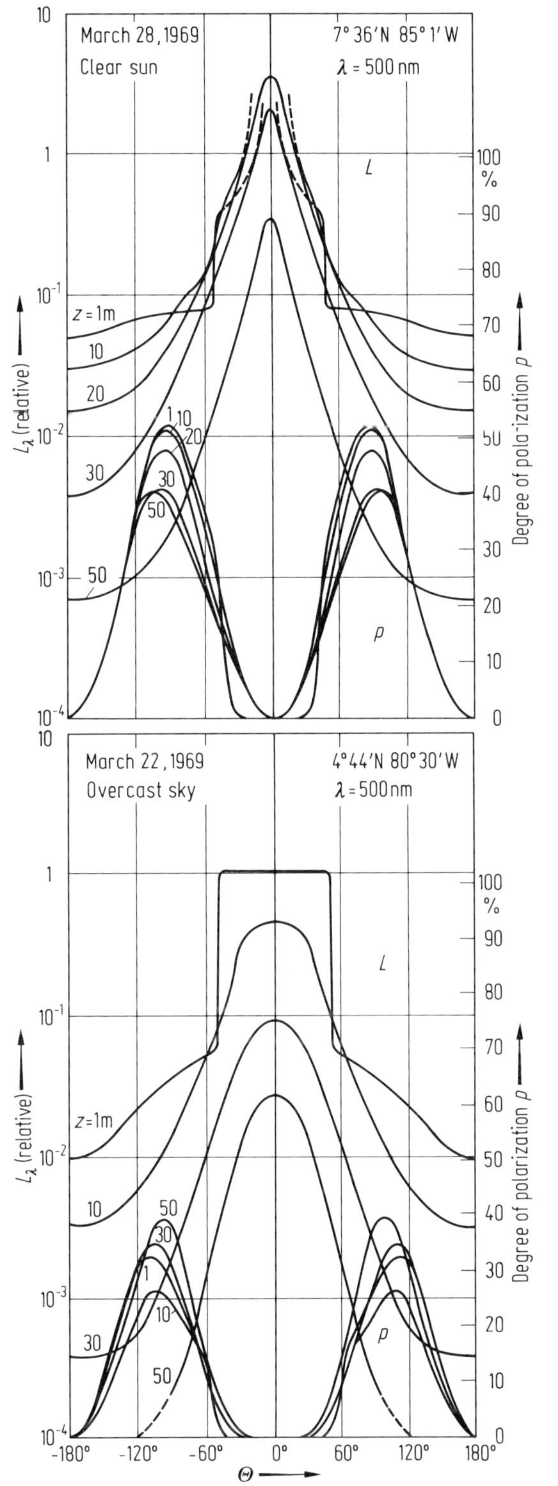

Fig. 65. Distributions of spectral radiance L_λ, in relative units, and degree of polarization as functions of the vertical angle θ measured at various depths in the vertical plane of the sun in Panama Basin during YALOC 69 (Lundgren [71 L 2]).

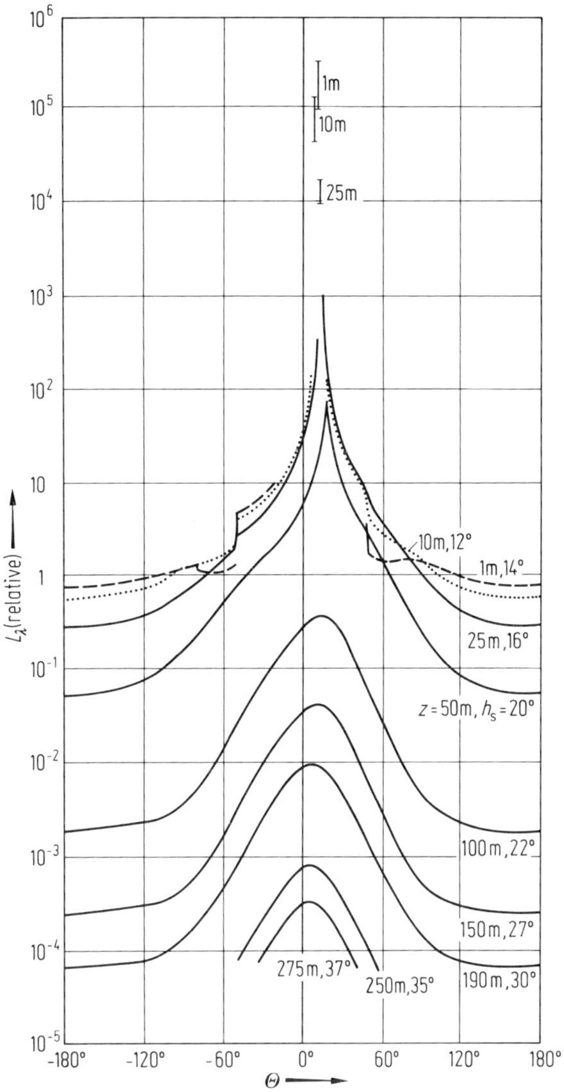

Fig. 66. Change with increase of depth of spectral radiance L_λ in the vertical plane of the sun towards the asymptotic state. Clear water of the Mediterranean. The angular resolution is 1.3° Solar zenith angle is given together with depth. (After Lundgren, personal communication.) θ: vertical angle.

Højerslev

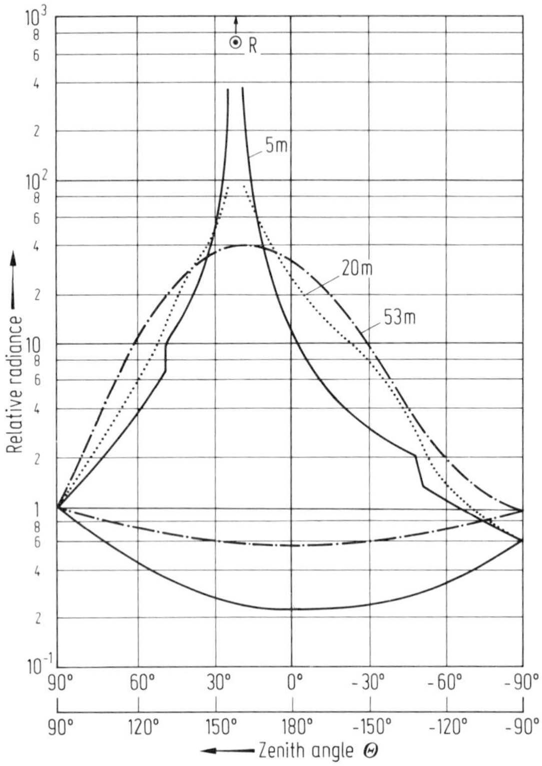

Fig. 67. Relative radiance versus zenith angle for up- and downward radiance in the vertical plane of the sun. The curves have been arbitrarily normalized to unity at $\theta = 90°$ so that they may be readily compared. Other information: zenith angle of the sun 28°, refracted angle of the sun 21°30′, clear skies (less than 0.1 cloud cover), moderate breeze (Beaufort No. 4) with gusts to 25 knots, wind chopped moderate sea (1 to $1\frac{1}{2}$ m waves) in lee of island, instrument suspended 4 m from stern of ship (R/V Ellen B. Scripps), 60 nm bandwidth (FWHM) spectral response approximately centered at the wavelength of maximum transmittance of the water under investigation ($\lambda_{max} \approx 497$ nm obtained using Wratten No. 64 filter), Gulf of California (Fresnel II Cruise–Sta. 2) 16 March 1971, 25°26′ N, 111°08′ W. The upward radiance for 20 m is very nearly the same as the upward radiance for 5 m and is not shown (Smith [74 S]).

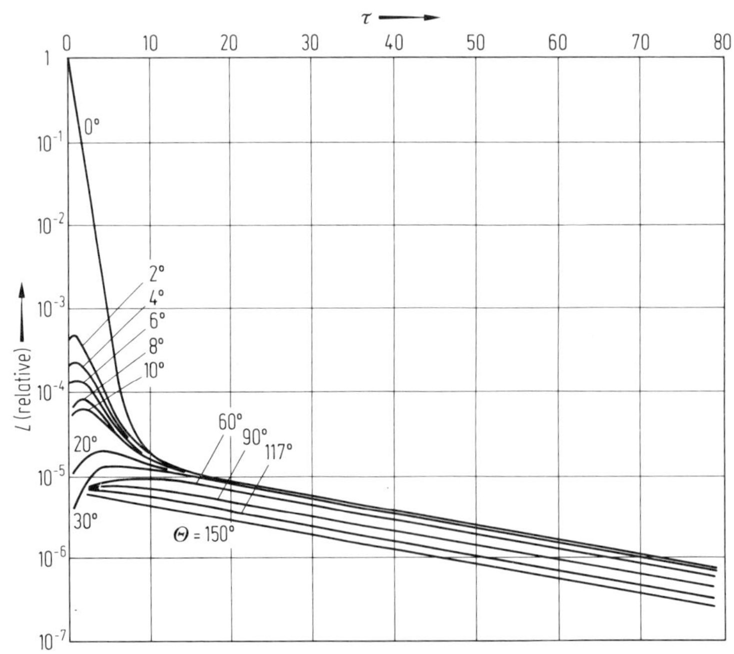

Fig. 68. The change with depth of radiance in various directions, indicated by vertical angle θ, in a milky medium (Timofeeva [74 T]). τ: optical depth.

All the radiance measurements show more similarities than differences although they have been performed by different instrumentations in various waters. In the surface layer the radiance peak is given by L_s, Fig. 60, which, however, becomes rapidly flattened with depth. At greater depth, eventually approaching the asymptotic regime, the radiance peaks at zenith and the radiance becomes symmetrical around the vertical as predicted by theory (Figs. 61b, 62, 63, 65, and 66, and Højerslev and Zaneveld [77 H 6]). The existence of the asymptotic regime implies also that $K_L(\theta, \varphi, z) = k$, i.e. the radiance curves as plotted in Figs. 64 and 68 become eventually parallel with depth provided a homogeneous hydrosol (e.g. Kullenberg et al. [70 K], Lundgren and Højerslev [71 L 1] and Fig. 68).

Table 29 gives the zenith angular radiance distribution in normalized values as a function of k/c (note that L is independent of the azimuth angle φ). It should be emphasized that the numbers in Table 29 stem from *in-vitro* measurements but do agree satisfactorily with *in-situ* measurements as well as with other optical models for deep regime sea water (Prieur and Morel [71 P] and Timofeeva [74 T]).

Table 29. $L(\theta)/L(0^\circ)$ for various zenith angles θ and $\varepsilon(=k/c)$ in the deep regime, $L(\theta) = L(\theta, \varphi)$ [74 T].

	ε										
θ	0	0.1	0.2	0.3	0.4	0.5	0.6	0.7	0.8	0.9	1
					$L(\theta)/L(0^\circ)$						
0	1	1	1	1	1	1	1	1	1	1	1
10	1	0.975	0.965	0.955	0.940	0.935	0.935	0.925	0.925	0.925	0
20	1	0.930	0.865	0.813	0.780	0.760	0.750	0.740	0.740	0.740	0
30	1	0.847	0.715	0.630	0.580	0.540	0.510	0.495	0.490	0.490	0
40	1	0.735	0.545	0.445	0.375	0.325	0.290	0.273	0.255	0.245	0
50	1	0.640	0.393	0.300	0.240	0.185	0.145	0.125	0.110	0.095	0
60	1	0.535	0.290	0.190	0.135	0.095	0.065	0.055	0.045	0.035	0
70	1	0.460	0.212	0.125	0.070	0.047	0.030	0.022	0.017	0.010	0
80	1	0.380	0.143	0.080	0.045	0.030	0.020	0.013	0.011	0.005	0
90	1	0.295	0.100	0.050	0.025	0.015	0.010	0.006	0.003	–	0
100	1	0.240	0.070	0.035	0.013	0.009	0.006	0.003	–	–	0
110	1	0.200	0.056	0.025	0.009	0.007	0.004	0.002	–	–	0
120	1	0.180	0.045	0.020	0.007	0.004	0.003	–	–	–	0
130	1	0.160	0.036	0.015	0.005	0.004	–	–	–	–	0
140	1	0.145	0.028	0.012	0.005	0.003	–	–	–	–	0
150	1	0.135	0.025	0.010	0.004	0.002	–	–	–	–	0
160	1	0.130	0.020	0.009	0.004	0.002	–	–	–	–	0
170	1	0.126	0.016	0.009	0.003	0.002	–	–	–	–	0
180	1	0.120	0.015	0.008	0.003	0.002	–	–	–	–	0

In the surface layers the Snell circle defined by the critical angle $\approx 48^\circ 6$, Fig. 60, can be observed through a sudden change in L around $48^\circ 6$, Figs. 61 and 65⋯67. These radiance measurements have all been performed in many zenith angle steps in contrast to those in Figs. 62 and 63. Finally, it is demonstrated that $K_L(\theta, \varphi, z)$ may be positive, negative, and thus zero in the surface layers depending on the surface light conditions, $\theta \leqq 90^\circ$ and k/c, e.g. Figs. 64 and 68, Lauscher [55 L], or Jerlov [76 J]. The observed radiance maximum at some small optical depths is favoured by a clear sky, intermediate solar elevations, and small k/c ranges, i.e. when the light scattering is dominating the light absorption.

Note the absence of the radiance maximum for $\lambda = 525$ nm in Fig. 64 where k/c is relatively high for the Sargasso Sea.

Till now, only the zenith angular distribution of the radiance in the vertical plane through the sun has been reported upon. A first order approximation of the azimuth angular distribution of the radiance has been suggested by Lundgren and Højerslev [71 L 1]. It is based on spectral field results from the Sargasso Sea so that a considerable range of k/c is covered.

One such example is depicted in Fig. 69 and the approximate expression

$$L(\theta_0, \varphi) = L(\theta_0, -\varphi)$$

reads:

$$L(\theta_0, 0^\circ) = L(\theta_0, 180^\circ)\frac{\mathrm{e}+1}{\mathrm{e}-\cos\varphi}$$

in which θ_0 is any fixed zenith angle between 0° and 180°,

$$L(\theta_0, 0°) = L_{max}(\theta_0), \quad L(\theta_0, 180°) = L_{min}(\theta_0),$$

and e a constant depending on k/c.

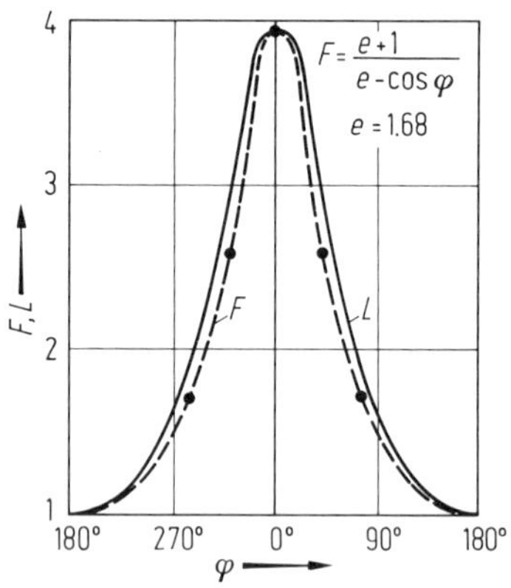

Fig. 69. Comparison between a measured radiance distribution $L(\varphi)$ for various constant zenith angles and the approximative distribution $F(\varphi)$. (Lundgren and Højerslev [71 L 1]). φ: azimuth angle.

3.3.5 Visibility in the sea

3.3.5.1 Basic definitions

The oldest field in marine optics concerns the visibility of various sea waters (e.g. Krümmel [1886 K] or Sauberer and Ruttner [41 S] for the historical background). The first appearance of a rigorous visibility theory is commonly attributed to the work by Duntley and his collaborators, [52 D], and [63 D].

Le Grand [39 G 2] and Sauberer and Ruttner [41 S] have contributed as well – among many others. The theory by Duntley et al., [52 D] and [63 D] suffers from the unrealistic assumption that $K_L(\theta, \varphi, z)$ has to be a *positive* constant for all θ, φ, z. This is especially not the case in surface layers (e.g. Figs. 64 and 68). In the following a more general and also simpler visibility theory is presented, being in good accordance with experimental findings, Højerslev [77 H 1, 78 H], exemplified by Secchi disc results (Højerslev [77 H 1]).

Theory:

A horizontally stratified infinite sea water is assumed for which the equation of radiative transfer of the light field reads:

$$\frac{\partial L(\theta, \varphi, z)}{\partial r} = -c(z)L(\theta, \varphi, z) + L_*(\theta, \varphi, z) \tag{137}$$

in which δr is a pathway in the direction of radiance propagation and $L_*(\theta, \varphi, z)$ is the path function defined by:

$$L_*(\theta, \varphi, z) = \int_{4\pi} \beta(\theta, \varphi, \theta', \varphi', z) L(\theta', \varphi', z) \sin\theta' \, d\theta' d\varphi'. \tag{138}$$

Consider now a target in the sea of such a size that the (diffuse) radiance impinging *on* the target will not be affected by the introduction of the target.

For this case the equation of radiative transfer from the target and into the surroundings becomes:

$$\frac{\partial L_t(\theta, \varphi, z)}{\partial r} = -c(z)L_t(\theta, \varphi, z) + L_*(\theta, \varphi, z) \tag{139}$$

in which $L_t(\theta, \varphi, z)$ is the target leaving radiance. Subtraction of eq. (137) from eq. (139) and subsequent integration over the pathway r leads to:

$$L_t(\theta, \varphi, r) - L(\theta, \varphi, r) = (L_t(\theta, \varphi, 0) - L(\theta, \varphi, 0))\, e^{-\int_0^r c(r) dr} \tag{140}$$

in which $\int_0^r c dr$ equals the optical depth along the pathway r from the target. This proves that the radiance differences between the target and its background follow the attenuation of an infinitesimally narrow light beam.

The definitions of the inherent contrast $C(r=0)$ and the apparent contrast $C(r)$ are, respectively:

$$C(r=0)=\frac{L_t(\theta,\varphi,0)-L(\theta,\varphi,0)}{L(\theta,\varphi,0)} \tag{141}$$

$$C(r=r)=\frac{L_t(\theta,\varphi,r)-L(\theta,\varphi,r)}{L(\theta,\varphi,r)}. \tag{142}$$

Moreover, the definition of $K_L(\theta,\varphi,z)$ gives upon integration along the pathway r:

$$L(\theta,\varphi,r)=L(\theta,\varphi,0)\,\mathrm{e}^{-\int_0^r K_L(\theta,\varphi,r)\cos\theta\,\mathrm{d}r} \tag{143}$$

resembling the expression in eq. (140).

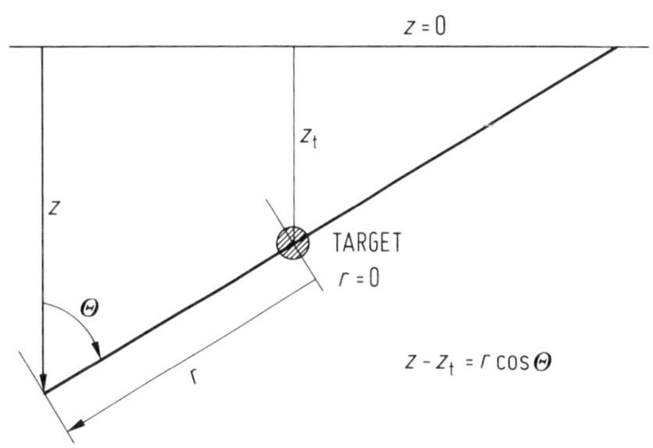

Fig. 70. A suitable coordinate system for visibility calculations.

Division of eq. (140) by $L(\theta,\varphi,r)$ on its left side, application of eq. (143) to its right side, and introduction of the definitions in eqs. (141) and (142) result in the fundamental equation of contrast transfer for stratified sea waters:

$$\frac{C(\theta,\varphi,z)}{C(\theta,\varphi,z_t)}=\mathrm{e}^{\int_0^r(-c(r)+K_L(\theta,\varphi,r)\cos\theta)\mathrm{d}r}. \tag{144}$$

For the visibility in any horizontal direction eq. (144) reduces to

$$\frac{C(90^\circ,\varphi,r)}{C(90^\circ,\varphi,0)}=\mathrm{e}^{-c(z_t)r}. \tag{145}$$

The vertical visibility defined from the sea surface and downwards, i.e. the case of a Secchi disc, becomes:

$$\frac{C(180^\circ,0^\circ,0)}{C(180^\circ,0^\circ,D)}=\mathrm{e}^{\int_0^D(c(z)+K_L(180^\circ,0,z)\mathrm{d}z} \tag{146}$$

in which D is the so-called Secchi disc depth.

3.3.5.2 The Secchi disc

The Secchi disc was systematically applied in the optical oceanography since 1865 as a means to determine the transparency of the sea (for historical details, see Sauberer and Ruttner [41 S] and Collier, Finlayson and Cake [68 C]). Later the Secchi disc was adopted by marine biologists in order to measure the depth of the euphotic zone defined as the depth at which 1% of the surface quanta irradiance is found. This depth $z_q(1\%)$ depends on the inherent optical properties of the sea water, the state of the sea, and the surface light conditions. However, for much practical biological work the depth of the euphotic zone might be considered constant for a given water mass (Højerslev [74 H 1, 74 H 2]) independent of the surface light conditions.

The relationship between the Secchi disc depth D, i.e. the depth where the Secchi disc disappears from sight, and $z_q(1\%)$ in its most simple form can be written:

$$z_q(1\%)=AD \tag{147}$$

where A is a dimensionless constant. Quite often another relationship involving the average vertical attenuation coefficient for the downward quanta irradiance $\langle K_q \rangle$ and D is given in the oceanographic literature namely

$$\langle K_q \rangle = B/D \tag{148}$$

where B is another dimensionless constant equal to $\ln 100/A = 4.6/A$.

The Secchi disc method as practiced is simple. The procedure is as follows:

1) Measure the depth where the Secchi disc disappears.
This depth will be D.
2) Calculate $z_q(1\%)$ for a chosen value of the constant A. The choice of A is virtually only dependent on the water mass and the surface light conditions.
3) Assume a $K_q(z)$ independent of depth, $K_q(z) = \langle K_q \rangle$.
4) Calculate the downward quanta irradiance from the relation

$$q(z) = \exp(-\langle K_q \rangle z).$$

The method is seemingly arbitrary: the human eye is involved, a value for A has to be decided upon, and, finally, a constant $K_q(z)$ has to be assumed which is normally not the case. Now, the latter objection has been overcome by Jerlov [77 J] since he has established functional relationships between $z_q(1\%)$ and $z_q(3\%)$, $z_q(10\%)$, and $z_q(30\%)$, i.e. the depths of the 1%, 3%, 10%, and 30% levels for the downward quanta irradiance, respectively. Accordingly, knowing $z_q(1\%)$ from a Secchi disc measurement the variation of $K_q(z)$ with depth can be taken into account unless the water mass is highly stratified which might be the case in upwelling areas, in plankton blooms, where fresh water run-out encounters sea water, etc.

The crucial point is, however, the choice of the factors B and A in eqs. (147) and (148). Poole and Atkins [29 P] show from data gathered in the English Channel that $B = 1.7$. Clarke [41 C] found the same value in waters east of the Caribbean region and south of the Bermuda region. Graham [66 G] found in the North Eastern Pacific by a somewhat different approach that B varies between 0.3 and 1.6. As stated by Steeman Nielsen [75 S]: "It is clear that a constant value for the factor B cannot be correct for all water types". He gives a value for B equal to 1.85 valid in Danish coastal waters.

The Secchi disc theory is given by eq. (146) for the case of great water depths. This condition is, however, not crucial since a water depth of the order of three times the Secchi depth is a sufficient constraint (Højerslev [80 H 1]). For reasons of simplicity the disappearance of the Secchi disc is assumed measured from just below the water surface (details are given by Højerslev [77 H 1, 78 H, 82 H 2]). Moreover, eq. (146) is modified into the conventional Secchi disc equation:

$$(c + K_u)D = \ln \frac{C(180^\circ, 0^\circ, 0)}{C(180^\circ, 0^\circ, D)} \tag{149}$$

for *photopic* vision. K_u, the upward attenuation coefficient for lux irradiance, and c, the attenuation coefficient for lux, have to be interpreted as *average* values along the vertical distance D. $K_u(\text{lux}) \simeq K_d(\text{lux})$ because the irradiance ratio $R(\text{lux}) \simeq$ constant as shown earlier.

$$C(180^\circ, 0, D) = 0.0066 \tag{150}$$

according to Blackwell [46 B]. Optical and independent measurements by Højerslev [86 H] of $c(\text{lux})$, $K_u(\text{lux})$, $D(\text{lux})$, and $C(180^\circ, 0, 0)$ for photopic vision resulted in $C(180^\circ, 0, D) = 0.0070$, thus confirming eq. (150) and the validity of eq. (149). Von Helmholtz: $C(180^\circ, 0^\circ, 0) = 0.0075$; e.g. [41 S]. Assuming the Secchi disc to be a perfect Lambert emitter with reflectance ϱ and the upwelling radiance to be uniformly distributed, $C(180^\circ, 0^\circ, 0)$ becomes:

$$C(180^\circ, 0^\circ, 0) = \frac{\varrho E_d - E_u}{E_u} \simeq \frac{\varrho}{R}. \tag{151}$$

Inserting eqs. (150) and (151) into eq. (149), applying the assumption $K_u \simeq K_d$, and omitting the *lux concept* for matters of convenience only, it is found that

$$(c + K_d)D = 4.80 - \ln R, \tag{152}$$

in which ϱ is put equal to 0.8. (If $\varrho = 0.9$ the number reads 4.91 instead of 4.80 and for $\varrho = 0.7$ it reads 4.66, i.e. the reflectance ϱ of the Secchi disc is not crucial.)

The diameter of the Secchi disc is normally around 30 cm. The size of the disc has a slight effect when measuring D but *not* in such a way that a greater diameter automatically involves a greater D. It is true that a decrease in the angular subtense of the Secchi disc leads to a reduced contrast vision (Blackwell [46 B]) but at the same time the assumption stating that the disc must not disturb the surrounding light field should not be violated. As a matter of fact it has been found from own experiments that great Secchi discs used in turbid waters may give a lesser D than smaller discs.

Equation (152) contains 3 unknowns, i.e. c, K_d, and R.

According to all published work concerning the Secchi disc it is clearly expressed that there are no ways by which the terms c and K_d can be found separately from downward Secchi disc observations (an academic but highly impractical method where both upward and downward observations are involved allows the above-mentioned separation (Preisendorfer [76 P 1]). It is, however, not impossible to separate the term $(c+K_d)$ within reasonable limit of errors into c and K_d from a single Secchi disc measurement, which is shown both theoretically and experimentally below. The optical model given by Timofeeva [74 T] has been applied. This gives the irradiance ratio R and the vertical attenuation coefficient K_d expressed in terms of the light attenuation coefficient c as a function of the photon survival probability b/c, see Table 30. This model allows the left side of eq. (152) to be written in terms of c and D so that a calculation of Dc as a function of b/c is now possible. The result of this computation is given in Table 30. All details are given by Højerslev [77 H 1, 78 H, 82 H 2] including Secchi disc measurements performed also *above* the sea surface at various surface light conditions. It has been concluded that $Dc \simeq 6$ for daylight conditions ensuring photopic vision as well as for a large variety of sea waters, see e.g. Fig. 71.

Table 30. The Secchi disc parameters according to eq. (152), Timofeeva [71 T, 74 T], as a function of the photon survival probability b/c.

b/c	K_d/c	R	Dc
0.1	0.93	0.0014	5.89
0.2	0.85	0.0021	5.93
0.3	0.77	0.0032	5.96
0.4	0.68	0.0050	6.01
0.5	0.60	0.0075	6.06
0.6	0.50	0.0123	6.13
0.7	0.40	0.0203	6.21
0.8	0.30	0.0334	6.31
0.9	0.19	0.0580	6.43

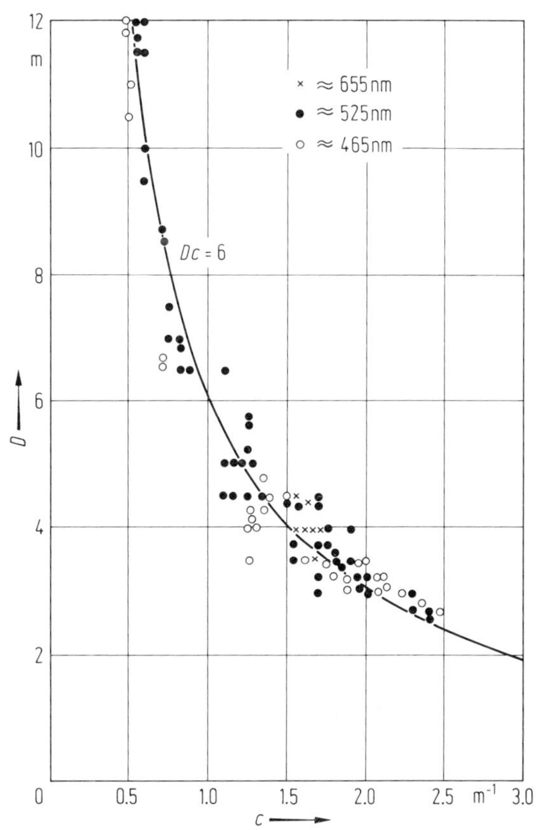

Fig. 71. Spectral Secchi disc depths D vs. spectral attenuation coefficients c. Full circles refer to measurements of the parameters D, c in the green part of the spectrum $\approx$525 nm by using VG 9 glass filters (Schott und Gen.) in both the light transmission meter and in the Secchi disc measurements. Crosses refer to measurements in the same way as above of D, c in the red part of the spectrum $\approx$655 nm by using RG 1 glass filters and, finally, open circles refer to measurements of D, c in the blue part of the spectrum $\approx$465 nm by using a glass filter combination of GG 5 and BG 12. The line represents the relation $Dc = 6$. All the points are obtained in Danish waters at different surface light conditions. Three different Secchi discs have been applied with diameters around 20, 30, and 100 cm, respectively. Sometimes observations were made from deck; at other occasions they were made just below the water surface by means of a water viewer. No significant difference in the observed values from one type of a measurement to another one could be observed (Højerslev [77 H 1]).

According to Table 30:

$$Dc = 6.10 \qquad (153)$$

on average, so eq. (152) can be approximated into:

$$K_d D = -1.30 - \ln R \qquad (154)$$

and since (Timofeeva [71 T])

$$R \simeq 0.15\,e^{-5K/c} \qquad (155)$$

$$K_d D = 3.3 \qquad (156)$$

for photopic vision from just below the sea surface. K_d(lux) can be given in terms of K_q or, alternatively, the 10% level for the downward lux irradiance can be approximated into the 10% level for the downward quanta irradiance (Tables 17, 31, and 32, and Figs. 72 and 73). It should be emphasized that the findings presented in Tables 31 and 32 as well as in Figs. 72 and 73 refer to Secchi disc observations made in calm stratified sea waters *above* the sea surface at different surface light levels. By comparing eqs. (148), (154) and (156) it is seen that $B \simeq -1.30 - \ln R$. This implies that R(lux) is typically of the order of 0.01 which is actually the case for most regions in the sea. B-values of the order of 1 are thus very exceptional if not wrong.

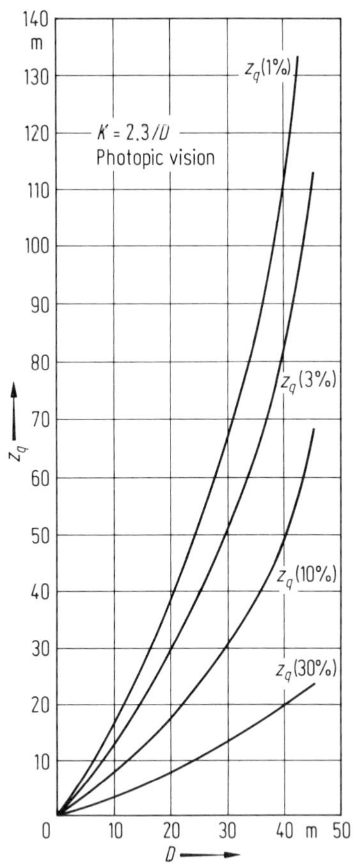

Fig. 72. Secchi disc depths D vs. the 30%, 10%, and 1% light level z_q for the downward quanta irradiance ($\lambda = 350 \cdots 700$ nm) at intermediate light levels and in turbid waters (Højerslev [77 H 1]). K: vertical attenuation coefficient for the downward lux irradiance.

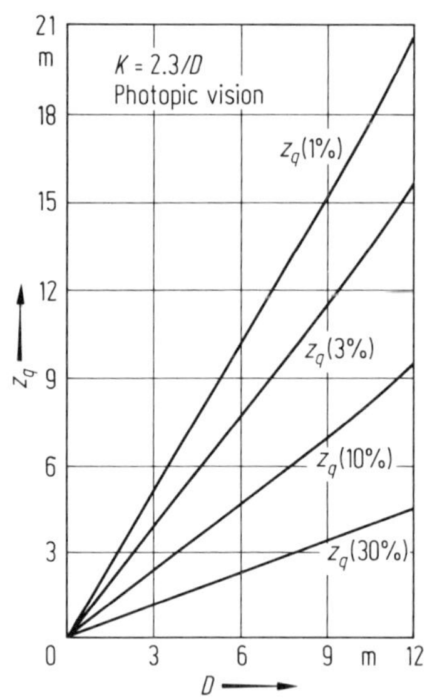

Fig. 73. Secchi disc depths D vs. the 30%, 10%, 3%, and 1% light levels for the downward quanta irradiance ($\lambda = 350 \cdots 700$ nm) at intermediate daylight levels (Table 32; Højerslev [77 H 1]). K: vertical attenuation coefficient for the downward lux irradiance.

Table 31. The depths of selected quanta levels z_q, in [m], as a function of the Secchi disc depth D, in [m], for an intermediate to a high sun and blue sky. Included is the depth z_q (10%), in [m], of the 10%-level for the downward lux irradiance vs. D. Højerslev [77 H 1]. Assumption: $K(\bar{y}_\lambda)=3/D(\bar{y}_\lambda)$ between the 100%-level and the 10%-level for the downward lux irradiance $\equiv E_d(\bar{y}_\lambda)$.

D	z_q (30%)	z_q (10%)	z_q (3%)	z_q (1%)	Lux irradiance z_q (10%)
5	1.48	3.00	4.95	6.53	3.83
6	1.77	3.59	5.92	7.82	4.60
8	2.34	4.79	7.90	10.43	6.13
10	2.89	5.95	9.82	12.96	7.67
12	3.48	7.19	11.86	15.66	9.20
14	4.03	8.39	13.84	18.27	10.74
16	4.65	9.73	16.05	21.19	12.26
18	5.20	10.96	18.08	23.87	13.81
20	6.01	12.78	21.09	27.83	15.33
22	6.58	14.07	23.22	30.64	16.89
25	7.76	16.82	27.75	36.63	19.17
30	9.61	21.30	35.15	46.39	23.00
35	11.55	26.28	43.36	57.24	26.81
40	13.63	31.95	52.72	69.59	30.67
45	16.07	39.20	64.68	85.38	34.50
50	18.19	46.18	76.20	100.58	38.33

Table 32. The depths of selected quanta levels z_q, in [m], as a function of the Secchi disc depth D, in [m], for a low sun or overcast. Included is the depth z_q (10%), in [m], of the 10%-level for the downward lux irradiance vs. D. Højerslev [77 H 1]. Assumption: $K(\bar{y}_\lambda)=2.3/D(\bar{y}_\lambda)$ between the 100%-level and the 10%-level for the downward lux irradiance $\equiv E_d(\bar{y}_\lambda)$. Note: The depth of the 10%-level for $E_d(\bar{y}_\lambda)=D$.

D	z_q (30%)	z_q (10%)	z_q (3%)	z_q (1%)	Lux irradiance z (10%)
5	1.92	3.90	6.44	8.49	5
6	2.29	4.69	7.74	10.21	6
8	3.04	6.25	10.31	13.61	8
10	3.76	7.81	12.89	17.01	10
12	4.55	9.52	15.71	20.73	12
14	5.44	11.48	18.94	25.00	14
16	6.26	13.33	21.99	29.03	16
18	7.21	15.52	25.61	33.80	18
20	8.20	17.86	29.47	38.90	20
22	9.08	20.00	33.00	43.56	22
25	10.60	23.81	39.29	51.86	25
30	13.26	30.93	51.03	67.37	30
35	16.25	39.77	65.62	86.62	35
40	19.08	49.38	81.48	107.55	40

Finally, it has been found that a waveheight amounting 1 m along the Secchi disc string reduces the Secchi disc depth D observed by means of a water viewer by about 40% (e.g. Højerslev and Lundgren [77 H 5]).

Højerslev

3.3.6 Passive remote sensing

The possibility of using remote sensing, defined as a measurement in which the sensor is not in direct contact with the object to be measured, for determining the sea water content of suspended and dissolved matter has been recognized for some years (e.g. Clarke et al. [70 C 2]). However, until the launch of the NIMBUS-7, CZCS (Coastal Zone Color Scanner) satellite in 1978, remote sensing experiments were performed on small scales only.

The color of the sea is defined as either

(i) the spectral distribution of the upward nadir radiance just below the sea surface, or

(ii) the spectral distribution of the irradiance reflectance just below the sea surface, or, finally,

(iii) in terms of a color index (see eqs. (99) and (100) and Table 33). As seen earlier (Table 16) the depth of the euphotic zone, $z_q(1\%)$ in [m], can be given in terms of a color index:

$$z_q(1\%) = -65.4 + \sqrt{3200 + 7692 L_\lambda(443\,\text{nm})/L_\lambda(520\,\text{nm})} \tag{157}$$

(Højerslev [81 H]) or

$$z_q(1\%) = -65.4 + \sqrt{3200 + 6923 (L_\lambda(443\,\text{nm})/L_\lambda(550\,\text{nm}))^{0.742}} \tag{158}$$

(Austin and Petzold [81 A] and Højerslev [81 H 2, 85 H 2]).

The upwelling spectral radiance signal recieved by a remote sensor, L_z, can be expressed as (Fig. 74):

$$L_z = \left[\frac{T_w}{n^2} \cdot L_\lambda + \varrho L_s\right] \cdot T_a + L^* \tag{159}$$

in which n is the refractive index of sea water, T_w and T_a the transmission values of the nadir radiance through the sea surface and through the air column between the sea surface and the remote sensor, respectively, ϱ the sea surface reflectance (Table 25), L^* the (spectral) path function in air (compare with eq. (65)), L_s the downwelling spectral sky radiance, and L_λ the color of the sea according to one of the previous definitions.

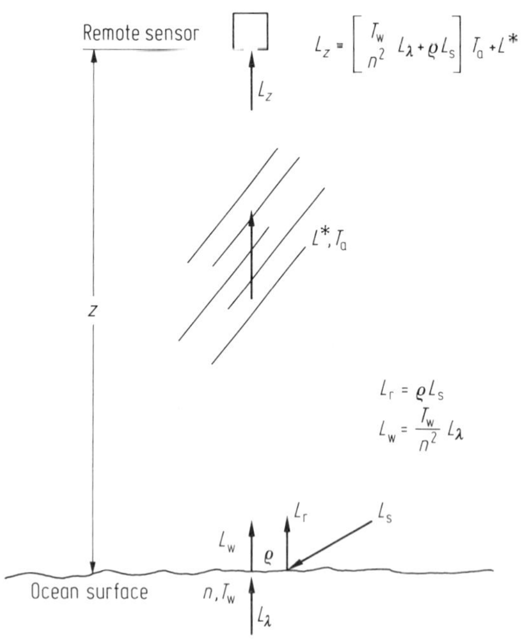

Fig. 74. Component parts of the upward radiance signal L_z. n denotes the refractive index of sea water, T_w and T_a the transmittance of the nadir radiance through, respectively, the sea surface and the air between sea surface and remote sensor, ϱ the reflectance of the sea surface, L_s the downward (spectral) sky radiance, and L^* the (spectral) path function in air (Austin [74 A]).

The far greatest term in eq. (159) is L^* increasing towards the blue part of the spectrum, Fig. 75. Consequently, remote sensing measurements of L_λ in this spectral region are hampered by a considerable background noise level. The two terms $(T_w/n^2)L_\lambda$ and ϱL_s in eq. (159) are typically of the same order of magnitude and T_w/n^2 is, practically speaking, independent of wavelength in the visible part of the spectrum, see subsect. 3.3.1 and Table 25. This means that the color of the sea is invariant across a calm sea surface.

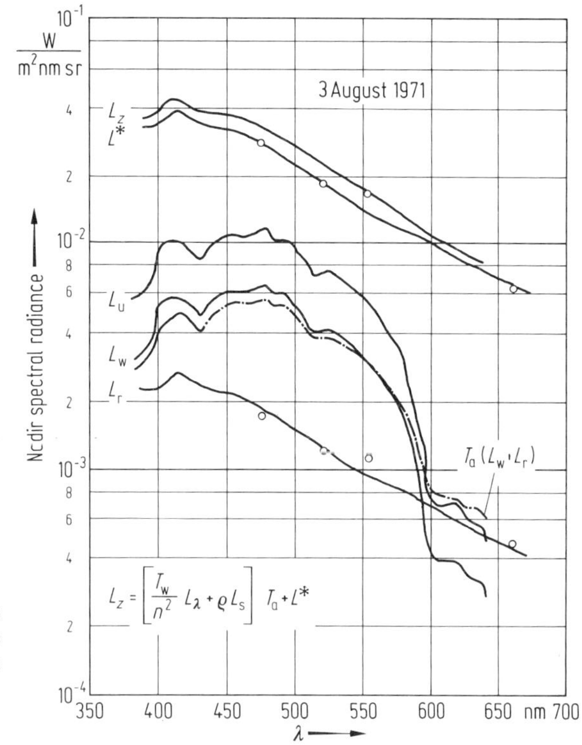

Fig. 75. Computed apparent spectral radiance of the ocean (and its components) as observed above the atmosphere. Blue-green water (Austin [74 A]). For the components of the spectral radiance, see Fig. 74.

In the following, the in-water nadir spectral radiance L_λ, or $R(\lambda)$, will be considered only. The spectrum of L_λ depends on the concentration, nature and vertical distribution of the suspended and dissolved matter, and the optical properties of pure sea water. Conventionally, sea water is said to contain chlorophyll-like phytoplankton pigments, nonchlorophyllous suspended matter and gelbstoff, see Table 22 and Figs. 76 and 77.

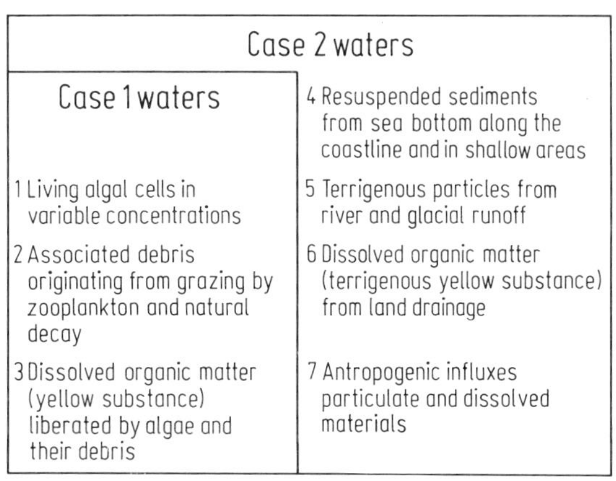

Fig. 76. Schematic presentation of case 1 and case 2 waters (Santhyendranath and Morel [83 S]). Note: The optical water types *I*, *IA*, *IB*, *II*, and *III* according to Jerlov's [76 J] optical classification are typically case 1 waters.

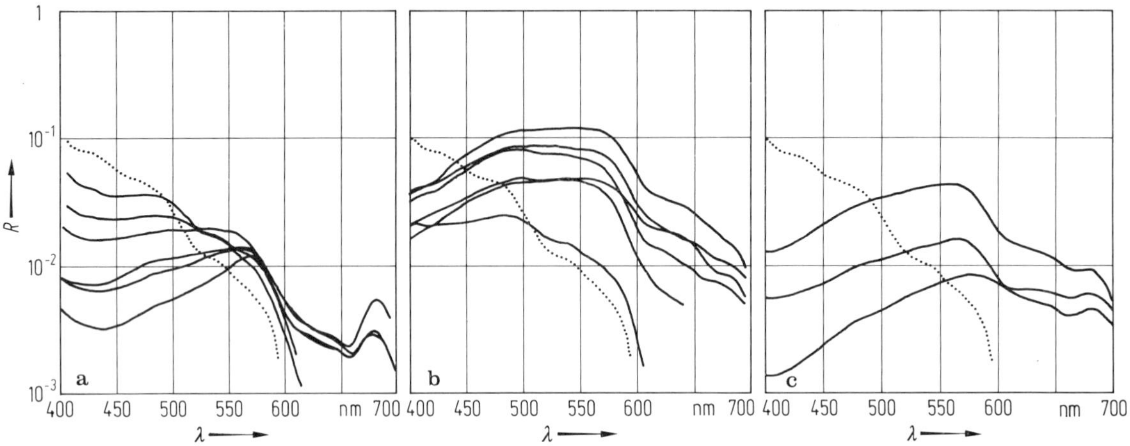

Fig. 77. Some observed spectra of irradiance ratio R. The spectra presented here have been selected to represent the three types of spectra described in the text. In a, b, and c, the spectrum of a clear deep blue sea with little suspended or dissolved organic material, is presented for comparison with the other spectra. These spectra were observed in the course of cruises DISCOVERER, CINECA 5, FOSBER, and C FOX, in different marine regions (Atlantic, Mediterranean, and Pacific). Different combinations of these three types also occur in nature. a: Case 1 waters, with different quantities of phytoplankton. b: Suspended-sediment-dominated case 2 waters. c: Yellow-substance-dominated case 2 waters. Dotted line: Clear ocean water (Sathyendranath and Morel [83 S]).

Pigment algorithms:

The pigment algorithms for the retrieval of surface chlorophyll a and phaeophytin a are based on sea truth data from various sea waters, Table 33. They display considerable discrepancies for coastal waters but are in acceptable agreement for blue oligotrophic waters (Fig. 78). This has led to the concept of case 1 and case 2 waters (Morel [80 M]) which can be defined as follows (see also Fig. 76): The surface pigment concentration for case 1 waters can be given in terms of a single algorithm, see Table 33, whereas this is not so for case 2 waters as depicted in Fig. 79.

The pigment algorithm by Højerslev [81 H 1]:

Surface concentration of $M = \text{Chl.}\, a + \text{phaeo.}\, a$, M in [mg m^{-3}],

$$M = 54.5(1 + 2.57 L_\lambda(447\,\text{nm})/L_\lambda(521\,\text{nm}))^{-3.40} \qquad (160)$$

was tested in the blue and oligotrophic waters of the Gulf of Mexico in 1982 (Table 34) and Højerslev [85 H 2]. The other algorithms presented in Table 33 would lead to almost similar pigment retrievals.

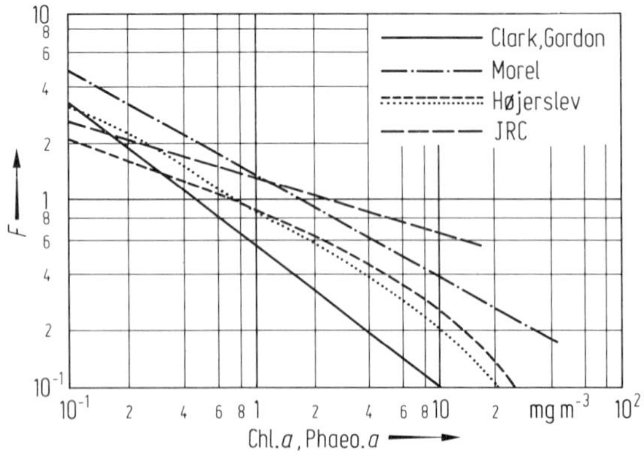

Fig. 78. Some suggested algorithms for sea color (color index F) and surface chlorophyll a and phaeophytin a for marine waters with small amounts of yellow substance. The algorithm by Morel (1980) is based on the spectral irradiance ratio E_u/E_d just below the sea surface whereas the remaining three categories of algorithms are based on the spectral nadir radiance at the sea surface. The algorithm Clark, Gordon can be found in a paper by Gordon and Clark (1980), the two algorithms presented by the author can be found elsewhere (Højerslev (1980, 1981)), and the algorithm termed JRC is developed at the Joint Research Centre at Ispra, Italy (B. Sturm, personal communication). Højerslev [81 H 2] and Table 33.

Table 33. List of algorithms of the type $M = A[CL_\lambda(180°, 0, \lambda_1)/L_\lambda(180°, 0, \lambda_2) + D]^B$ or $M = A[R(\lambda_1)/R(\lambda_2)]^B$ proposed by different authors. N is the number of observations used in the regression analysis and r^2 is the square of the correlation coefficient. (After Santhyendranath and Morel [83 S]).

r_{ij}	A	B	N	r^2	C	D	Ref.
M = Chl. a + Phaeo. a [mg m^{-3}]							
L_{443}/L_{550}	0.776	−1.329	55	0.91	1	0	Clark (1981)
L_{443}/L_{520}	0.551	−1.806	55	0.87	1	0	
L_{520}/L_{550}	1.694	−4.449	55	0.91	1	0	
L_{520}/L_{670}	43.85	−1.372	55	0.88	1	0	
L_{440}/L_{550}	0.54	−1.13	7	0.96	1	0	Clark et al. (1980)
L_{440}/L_{550}	0.505	−1.269	21	0.978	1	0	Gordon and Clark (1980)
L_{440}/L_{520}	0.415	−1.795	21	0.941	1	0	
L_{520}/L_{550}	0.843	−3.975	21	0.941	1	0	
R_{440}/R_{560}	1.92	−1.80	67	0.97	1	0	Morel (1980)
L_{443}/L_{550}	0.783	−2.12		0.94	1	0	Smith and Wilson (1981)
L_{443}/L_{520}	0.483	−3.08		0.88	1	0	
L_{520}/L_{550}	2.009	−5.93		0.95	1	0	
L_{443}/L_{550}	2.45	−3.89	6	0.61	1	0	Sturm (1980)
L_{443}/L_{550}	1.13	−1.71	454		1	0	Gordon et al. (1982)
L_{443}/L_{550}	1.216	−2.589			1	0	Smith and Baker (1982)
L_{447}/L_{521}	54.5	−3.40			2.57	1	Højerslev (1981)
L_{447}/L_{550}	232	−4.54			2.31	1	Højerslev (1981), Austin and Petzold (1981)
M = Total suspended particles (g m^{-3})							
L_{440}/L_{550}	0.4	−0.88	9	0.92	1	0	Clark et al. (1980)
L_{440}/L_{520}	0.33	−1.09	9	0.94	1	0	
L_{520}/L_{550}	0.76	−4.38	9	0.77	1	0	
L_{443}/L_{550}	0.24	−0.98		0.86	1	0	Clark and Baker (personal communication)
L_{520}/L_{550}	0.45	−3.30		0.86	1	0	
L_{520}/L_{670}	5.30	−1.04		0.85	1	0	

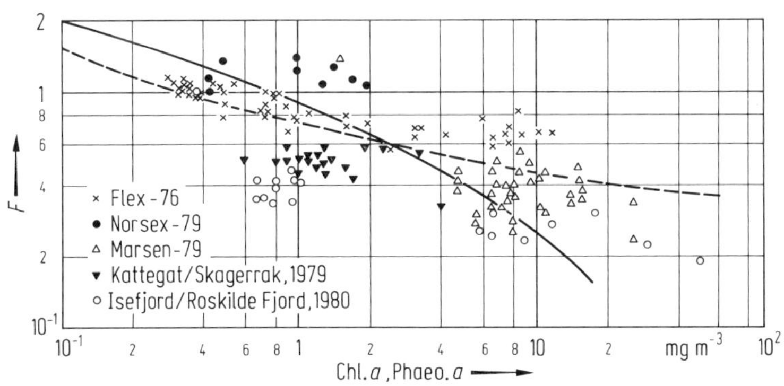

Fig. 79. Color index F, extrapolated to 0 m depth in relation to surface concentration of chlorophyll a and phaeophytin a (Chl. a + Phaeo. a). FLEX-76 (pigment data from A. Weber, Hamburg, personal communication); NORSEX-79: off Svalbard (pigment data from B. Heimdal, Bergen, personal communication); MARSEN-79 (pigment data from R. Doerffer, Geesthacht, personal communication); Kattegat/Skagerrak, 1979 (pigment data from A.M. Larsson, Gothenburg, personal communication); Isefjord/Roskilde Fjord, 1980) pigment data from B. Møller, Copenhagen, personal communication). The solid curve corresponds to the algorithm previously presented by Højerslev (1981) and the stippled curve is based on an algorithm, derived by Højerslev (1982) as well as from the biological findings of Lorenzen (1970). Note: The sea waters in the central part of the North Sea (FLEX-76) and those off Svalbard (NORSEX-79) belong to the case 1 waters (Højerslev [82 H 2]).

Table 34. Surface values of chlorophyll *a*, phaeophytin *a*, and the color index *F*. The pigment values in the second and third column were measured using standard laboratory techniques, i.e. fluorescence and acid (J. Yoder, personal communication). The chlorophyll *a* values in the fourth column were obtained by means of an *in-situ* fluorescence meter produced by Hundahl and Holck [80 H 3]. These values are typically higher than those in the second column, indicating that the phaeopigments may be partly fluorescent. The values in the fifth column are simply the sum of those in the third and the fourth columns. The derived surface pigment concentrations are obtained through the color index algorithm in eq. (160). Højerslev [85 H 2].

Station no.	Yoder		Højerslev		Derived	Color index
	Chl. *a*. $mg\,m^{-3}$	Phaeo. *a*. $mg\,m^{-3}$	Chl. *a*. $mg\,m^{-3}$	Chl. *a*. + phaeo. *a*. $mg\,m^{-3}$	Chl. *a*. + phaeo. *a*. $mg\,m^{-3}$	*F*
3	0.07	0.06	0.13	0.24	0.15	1.82
5	0.06	0.04	0.08	0.13	0.14	1.88
7	0.07	0.05	0.08	0.14	0.14	1.87
10	0.05	0.05	0.11	0.22	0.13	1.92
24	0.06	0.05	0.06	0.11	0.14···0.17	1.74···1.84
26	0.06	0.04	0.10	0.17	0.14···0.16	1.77···1.87
50	–	–	–	–	0.13	1.90
62	0.07	0.05	0.08	0.14	0.15···0.16	1.81···1.86
72	–	–	–	–	0.15	1.82
82	–	–	–	–	0.15···0.16	1.78···1.80
87	0.07	0.05	0.09	0.15	0.14	1.88
89	–	–	0.10	–	0.13···0.15	1.80···1.91
106	0.08	0.05	0.14	0.21	0.21	1.61
110	0.43	0.26	0.09	0.14	0.19	1.67

Penetration depth and vertical distribution of chlorophyll-like pigments:

The penetration depth z_{90}, i.e. the depth down to which the chlorophyll-like pigments are contributing with about 90% to the in-water surface nadir radiance L_λ, is defined as:

$$z_{90} = \frac{1}{K} \qquad (161)$$

in which K is the minimum irradiance vertical attenuation coefficient (Gordon and McCluney [76 G]). At the depth z_{90} the downward irradiance for the most penetrating daylight is reduced from the surface value by approximately 60% or 1/e. The depth z_{90} corresponds roughly to the one half of a white Secchi disc depth or to the depth of disappearance of a black Secchi disc (Højerslev [77 H 1]).

Assuming $K_d = K_u = K$, i.e. R constant with depth, and independent of solar elevation, it is straightforward to give an alternative definition for z_{90}:

$$(1 - e^{-2Kz_{90}}) = 0.90 \qquad (162)$$

stating that below z_{90} the signal contribution to L_λ is around 10% for the most penetrating daylight. Figs. 80 and 81 show two examples of chlorophyll maxima which would not be observed by means of remote sensing techniques (Højerslev [85 H 2]).

At mid-latitudes the surface light conditions cause normally no photoinhibition on the phytoplankton growth so often no intermediate chlorophyll maxima are observed (Figs. 82 and 83). The downward extent of the plankton bloom depicted is *not* controlled by the daylight field but instead by the presence of a pycnocline at a depth approximately equal to $z_q(0.01\%)$.

The phytoplankton movements between the sea surface and the pycnocline are not hampered by any density stratification but are instead governed by the turbulence field in this upper water layer.

This may lead to a pronounced stratification of the biomass as can be observed in Figs. 82 and 83.

A stratified water column of chlorophyll-like pigments influences the water leaving nadir radiance L_λ in a manner depending on the vertical distribution of these pigments. Gordon and Clark [80 G 1] have shown that the equivalent pigment concentration $\langle M \rangle$ in a stratified sea, effecting L_λ in the same manner as if the sea was not stratified, can be expressed by:

$$\langle M \rangle = \frac{\int_0^{z_{90}} M(z) f(z) \mathrm{d}z}{\int_0^{z_{90}} f(z) \mathrm{d}z} \tag{163}$$

in which

$$f(z) = \exp\left[-\int_0^{z_{90}} 2K(z') \mathrm{d}z' \right]$$

and $M(z)$ is the pigment concentration at the depth z. Making the same assumptions as those leading to eq. (162) the expression for $\langle M \rangle$ is straightforward to derive.

At shallow water depth the bottom albedo might contribute to L_λ (e.g. Joseph [50 J 2], Plass and Kattawar [72 P 2], and McCluney [74 M 1]). From a distinctly different approach, Højerslev [77 H 1] has found that a bottom at a depth greater than 3 times the Secchi disc depth D, corresponding to an optical depth $\int_0^z c(z')\mathrm{d}z' \geqq 2$, has only a negligible influence on L_λ. For shallow waters, which are normally turbid and case 2 waters, the reflectance from the bottom can be ignored for water depths typically greater than, say, 20 m.

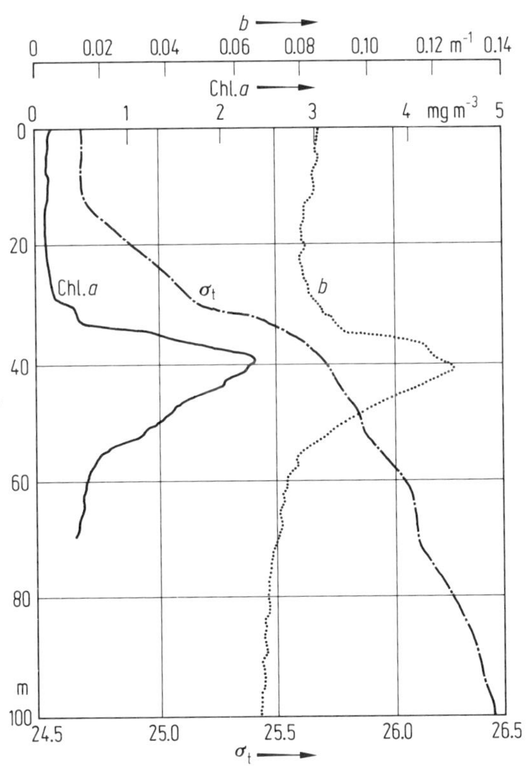

Fig. 80. The vertical distribution of chlorophyll *a*, light scattering coefficient *b*, and sigma-t, σ_t, in the Gulf of Mexico off Southwest Florida, April 1982 (Højerslev [85 H 2]). Note that the maxima of both Chl. *a* and *b* (suspended matter) are developed below the pycnocline because they are probably caused by the strong surface light conditions.

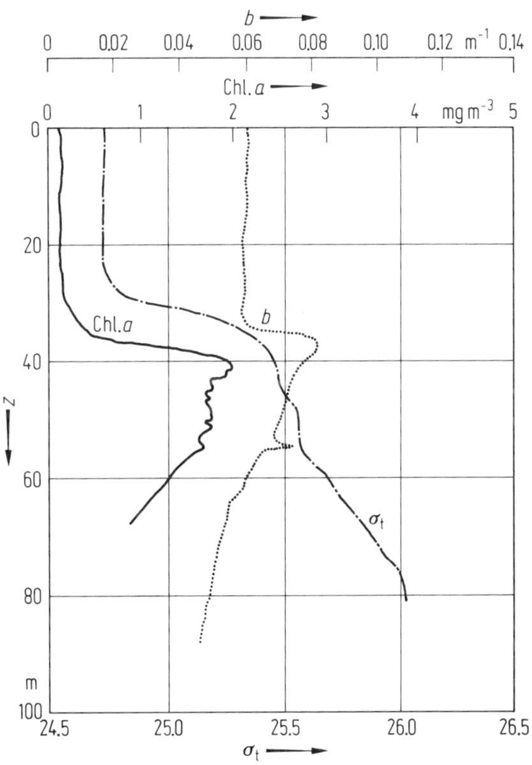

Fig. 81. The vertical distribution of chlorophyll *a*, light scattering coefficient *b*, and sigma-t σ_t, in the Gulf of Mexico off Southwest Florida, April 1982 (Højerslev [85 H 2]). Note the vertical extent of the Chl. *a* and *b* maxima below the pycnocline. The difference in the width of the two Chl. *a* maxima in Figs. 80 and 81 is most likely controlled by the dynamical features in the regime.

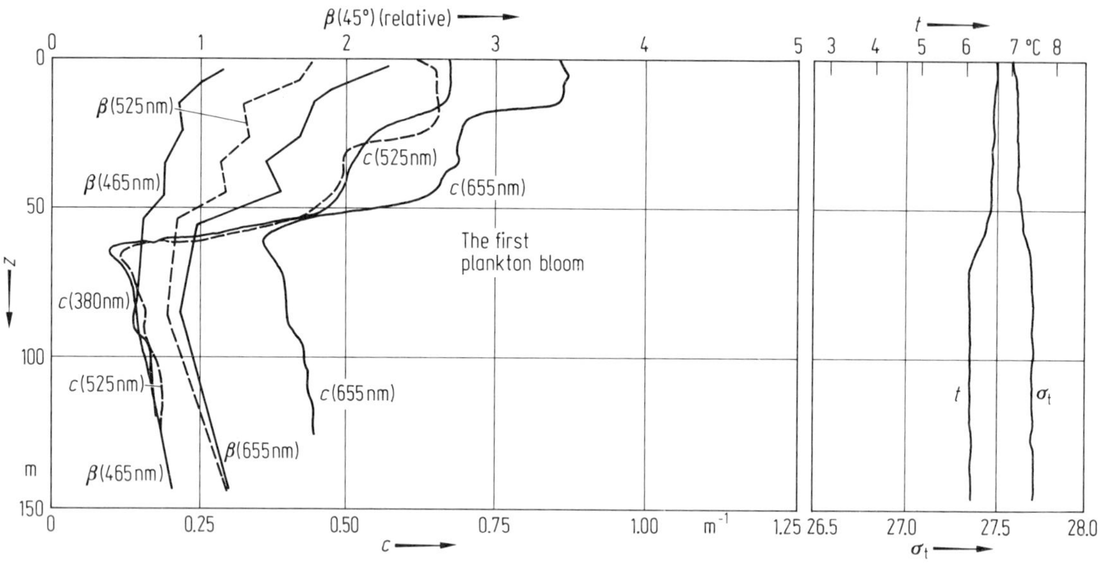

Fig. 82. Spectral light attenuation coefficients, c(380 nm), c(525 nm), and c(655 nm), relative spectral light scattering through the fixed angle of 45°, β(465 nm), β(525 nm), and β(655 nm), sigma-t, σ_t, and temperature t vs. depth on the Fladen Ground in the North Sea on April 27, 1976. The optical parameters are all a measure of the biomass consisting of mainly living phytoplankton (Weber, personal communication). Note that the lower limit of the plankton bloom coincides with the pycnocline around which the surface quanta irradiance ($\lambda = 350 \cdots 700$ nm) is reduced to about a factor of 10^{-4} (Højerslev [77 H 3]).

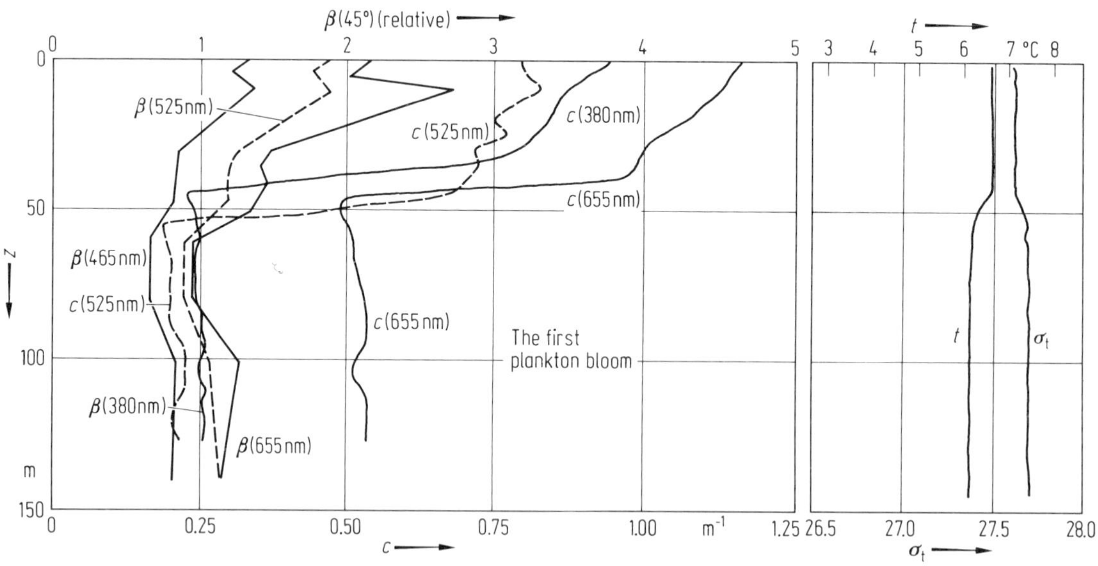

Fig. 83. Spectral light attenuation coefficients c(380 nm), c(525 nm), and c(655 nm), relative spectral light scattering through the fixed angle of 45° β(465 nm), β(525 nm), and β(655 nm), sigma-t, σ_t, and temperature t vs. depth on the Fladen Ground in the North Sea on April 29, 1976. Note the rapid growth of the plankton bloom within two days only (compare with Fig. 82) and the abrupt change in the vertical biomass distribution governed by the now shallower pycnocline (Højerslev [77 H 3]).

3.3.7 References for 3.3

1886 K Krümmel, O.: Der Ozean, Leipzig und Prag **1886**, 242 pp.
22 K Knudsen, M.: Cons. Perm. Int. Explor. Mer Publ. Circ. **76** (1922) 16 pp.
24 K Kimball, H.H.: Mon. Weather Rev. **52** (1924) 475.
29 P Poole, H.H., Atkins, W.R.G.: J. Mar. Biol. Ass. U.K. **16** (1929) 297.
31 S Sawyer, W.R.: Contrib. Can. Biol. Fish., New Ser. **7** (1931) 74.
34 D Dawson, L.H., Hulbert, E.O.: J. Opt. Soc. Am. **21** (1934) 175.
36 G Gershun, A.: Izv. Akad. Nauk SSSR, Otd. Mat. Estestv. Nauk **1936**, p. 417–430.
36 P Powell, W.M., Clarke, G.L.: J. Opt. Soc. Am. **26** (1936) 111.
38 J 1 James, H.R., Birge, F.A.: Trans. Wis. Acad. Sci. **31** (1938) 1.
38 J 2 Jerlov, (Johnson), N.G., Liljequist, G.: Sven. Hydrogr. Biol. Komm. Skr., Ny Ser. Hydrogr. **14** (1938) 15 pp.
38 P Petersson, H.: Rapport et Process-Verbaux des Reunions du Conseil Permanent Int. Explor. Mer **V 108 (Pt 2)** (1938) 9.
39 C Clarke, G.L., James, H.R.: J. Opt. Soc. Am. **29** (1939) 43.
39 G 1 Gershun, A.: J. Math. Phys. **18** (1939) 51.
39 G 2 Le Grand, Y.: Ann. Inst. Oceanogr. Paris **XIX** (1939) 393.
41 C Clarke, G.L.: J. Mar. Res. Vol. **IV**, No. **3** (1941) 221.
41 S Sauberer, F., Ruttner, F.: Die Strahlungsverhältnisse der Binnengewässer, Leipzig: Akademischer Verlag **194**, 240 pp.
41 T Taylor, A.H., Kerr, G.P.: J. Opt. Soc. Am. **31** (1941) 3.
45 H Hulbert, E.O.: J. Opt. Soc. Am. **35** (1945) 698.
46 B Blackwell, H.R.: J. Opt. Soc. Am. **36** (1946) 624.
48 N Neiburger, M.: Trans. Am. Geophys. Union **29** (1948) 647.
50 C Chandrasekhar, S.: Radiative transfer, London: Oxford University Press **1950**, 393 pp.
50 J 1 Joseph, J.: Dtsch. Hydrogr. Z. **3** (1950) 69.
50 J 2 Joseph, J.: Dtsch. Hydrogr. Z. **3** (1950) 324.
51 C Curcio, J.A., Petty, C.C.: J. Opt. Soc. Am. **41** (1951) 302.
51 J Jerlov, N.G.: Optical studies of ocean water, Rep. Swedish Deep-Sea Exped. 1947–1948, **3** (1951) 1.
52 D Duntley, S.Q.: The visibility of submerged subjects. Visibility Lab., Cambridge, Mass.: Mass. Inst. Technology **1952**, 74 pp.
53 B Burt, W.V.: Trans. Am. Geophys. Union **34** (1953) 199.
55 J 1 Jerlov, N.G.: Medd. Oceanogr. Inst. Gothenburg **25** (1955) 19 pp.
55 J 2 Joseph, J.: Proc. UNESCO Symp. Phys. Oceanogr., Tokyo **1955**, p. 59.
55 L Lauscher, F., in: Handbuch der Geophysik (Linke, F., Möller, F., eds.), Berlin: Springer **1955**, p. 723.
55 S Sasaki, T., Okami, N., Watanabe, S., Oshiba, G.: J. Sci. Res. Inst. Tokyo **49** (1955) 103.
57 H Van de Hulst, H.C.: Light scattering by small particles, New York: Wiley-Interscience **1957**, 470 pp.
57 K Kozlyaninov, M.V.: New instrument for measuring the optical properties of sea water. Tr. Inst. Okeanol., Akad. Nauk SSSR **25** (1957) 134 pp.
58 B Burt, W.V.: Deep-Sea Res. **5** (1958) 51.
58 T Tyler, J.: J. Mar. Res. **16** (1958) 96.
59 G Grichtchenko, D.L.: Tr. Gl. Geofiz. Obs. A.J. Boejkova **80** (1959) 32.
60 J Jerlov, N.G., Fukuda, M.: Tellus **12** (1960) 348.
60 T Tyler, J.: Bull. Scripps Inst. Oceanogr., University of California **7** (1960) 363.
61 A Armstrong, F.A.J., Boalch, G.T.: Union Géod. Géophys. Int. Monogr. **10** (1961) 63.
61 I Ivanoff, A., Jerlov, N.G., Waterman, T.H.: Limnol. Oceanogr. **6** (1961) 129.
61 J Jerlov, N.G.: Medd. Oceanogr. Inst. Gothenburg, Ser. B, **8** (1961) 40 pp.
61 K Kalle, K.: Union Géod. Géophys. Int. Monogr. **10** (1961) 59.
61 P Preisendorfer, R.W.: Union Géod. Géophys. Int. Monogr. **10** (1961) 11.
61 T Tyler, J.: Limnol. Oceanogr. **6** (1961) 451.
62 S Sasaki, T., Watanabe, S., Oshiba, G., Okami, N., Kajihara, M.: Bull. Jpn. Soc. Sci. Fish. **28** (1962) 489.
63 D Duntley, S.Q.: J. Opt. Soc. Am. **53** (1963) 15.
63 H Hodkinson, J.R., in: I.C.E.S. Electromagnetic Scattering (Kerker, M., ed.), London: Pergamon Press **5** (1963) 87.
63 S Sullivan, S.A.: J. Opt. Soc. Am. **53** (1963) 962.

64 H Herman, M., Lenoble, J.: Rev. Opt. **11** (1964) 552.
65 S Spilhaus, A.F.: Observations of Light in Sea Water. Office Naval Res., Washington, D.C., Cambridge, Mass.: Dept. Geol. Geophys., Mass. Inst. Technol. **1965**, 242 pp.
66 G Graham, J.J.: Limnol. Oceanogr. **11** (1966) 184.
66 M Morel, A.: J. Chim. Phys. **10** (1966) 1359.
66 O Ochakovsky, Yu.E.: U.S. Dept. Comm., Joint. Publ. Res. Ser., Rep. **36** (816) (1966) 98.
66 R Robinson, N.: Solar radiation, Amsterdam: Elsevier Scientific Publ. Comp. **1966**, 347 pp.
67 B Bauer, O., Morel, A.: Ann. Geophys. **23** (1967) 109.
68 B Beardsley, G.F., jr.: J. Opt. Soc. Am. **58** (1) (1968) 52.
68 C Collier, A., Finlayson, G.M., Cake, E.W.: Limnol. Oceanogr. **13** (1968) 391.
68 J Jerlov, N.G., Nygård, K.: Rep. Inst. Phys. Oceanogr., Univ. Copenhagen **1** (1968) 7 pp.
68 K Kullenberg, G.: Deep-Sea Res. **15** (1968) 423.
68 N Nygård, K.: Rep. Inst. Phys. Oceanogr., Univ. Copenhagen **2** (1968) 12 pp.
68 P Parker, A.C.: Photoluminescence of solutions, Amsterdam: Elsevier Scientific Publ. Comp. **1968**, 544 pp.
69 J Jerlov, N.G., Nygård, K.: Rep. Inst. Phys. Oceanogr., Univ. Copenhagen **10** (1969) 29 pp.
69 K 1 Kondratyev, K.V.: Radiation in the Atmosphere, London: Academic Press **1969**, 912 pp.
69 K 2 Kullenberg, G.: Rep. Inst. Phys. Oceanogr., Univ. Copenhagen **5** (1969) 12 pp.
69 M Makarevich, S.A., Robilko, A., in: Optike rasseivayushchikh sred. (Ivanov, A.P., ed.), Minsk: Izd. Nauka i tekhnika **1969**, 175.
69 S 1 Schellenberger, G.: Acta Hydrophys. **14** (1969) 112 pp.
69 S 2 Smith, R.C.: J. Mar. Res. **27** (1969) 341.
70 C 1 Calathas, J.: Contribution à l'etude de la reflexion de la lumière du jour à la surface de la mer. Thèse 3ème cycle, Univ. Paris, **VI** (1970).
70 C 2 Clarke, G.L., Ewing, G.C., Lorenzen, C.J.: Science **167** (1970) 1119.
70 K Kullenberg, G., Lundgren, B., Malmberg, Sv.Å., Nygård, K., Højerslev, N.K.: Rep. Inst. Phys. Oceanogr., Univ. Copenhagen **11** (1970) 18 pp.
70 L Lorenzen, C.J.: Limnol. Oceanogr. **15** (1970) 479.
70 T Tyler, J., Smith, R.C.: Measurements of spectral irradiance under water, Ocean Series **1**, New York: Gordon and Breach **1970**, 103 pp.
71 C Copin-Montegut, G., Ivanoff, A., Saliot, A.: C.R. Acad. Sci., Sér. B **272** (1971) 1453.
71 L 1 Lundgren, B., Højerslev, N.K.: Rep. Inst. Phys. Oceanogr., Univ. Copenhagen **14** (1971) 44 pp.
71 L 2 Lundgren, B.: Rep. Inst. Phys. Oceanogr., Univ. Copenhagen **17** (1971) 34 pp.
71 P Prieur, L., Morel, A.: Cah. Oceanogr. **23** (1971) 35.
71 T Timofeeva, V.A.: Izv. Atmos. Oceanic Phys. **7** (1971) 467.
72 K 1 Kattawar, G.W., Plass, G.N.: J. Phys. Oceanogr. **2** (1972) 146.
72 K 2 Kullenberg, G., Berg Olsen, N.: Rep. Inst. Phys. Oceanogr., Univ. Copenhagen **19** (1972) 15 pp.
72 P 1 Payne, R.E.: J. Atmos. Sci. **29** (1972) 959.
72 P 2 Plass, G.N., Kattawar, G.W.: J. Phys. Oceanogr. **2** (1972) 139.
73 H Højerslev, N.K.: Rep. Inst. Phys. Oceanogr., Univ. Copenhagen **21** (1973) 70 pp.
73 I Ivanoff, A.: Facteurs physiques, chimiques et biologiques affectant la propagation de la lumière dans les eaux de mer, in: Optics of the sea (interface and in-water transmission and imaging)-AGARD Lect. Ser., No. **61** (1973) 45 pp.
73 K Kattawar, G.W., Plass, G.N.: J. Quant. Spectrosc. Radiat. Transfer **13** (1973) 1065.
73 M 1 Morel, A.: Data Rep. SCOR Discoverer Exp. May 1970-I, SIO ref. 73-16, **1973.**
73 M 2 Morel, A.: Diffusion de la lumière par les eaux de mer. Résultats expérimentaux et approche théorique, in: Optics of the Sea (interface and in-water transmission and imaging)-AGARD Lect. Ser., No. **61** (1973) 76 pp.
73 M 3 Morel, A., Caloumenos, L.: Mesure d'éclairments sous-marin, flux de photons et analyse spectrale (Campagnes Harmattan et Cineca II, 3e Partie). Présentations des results. Rap. No. 11, Villefranche-sur-Mer: Lab. d. Oceanogr. Phys. **1973**, 180 pp.
73 N Nygård, K.: Data Rep. SCOR Discoverer Exp. May 1970-I, SIO ref. 73-16, **1973**
73 S Smith, R.C.: Scripps spectroradiometer data. Data Rep. Discoverer Exp. May 1970-II, SIO ref. 73-16, **1973.**
74 A Austin, R.W., in: Optical Aspects of Oceanography (Jerlov, N.G., Steemann Nielsen, E., eds.), New York: Academic Press **1974**, 317.
74 H 1 Højerslev, N.K.: Rep. Inst. Phys. Oceanogr., Univ. Copenhagen **23** (1974) 89 pp.

74 H 2 Højerslev, N.K.: Rep. Inst. Phys. Oceanogr., Univ. Copenhagen **26** (1974) 38 pp.
74 J 1 Jerlov, N.G., in: Optical Aspects of Oceanography (Jerlov, N.G., Steemann Nielsen, E., eds.), New York: Academic Press **1974**, 77.
74 J 2 Jerlov, N.G.: Rep. Inst. Phys. Oceanogr., Univ. Copenhagen **24** (1974) 5 pp.
74 K Kullenberg, G., in: Optical Aspects of Oceanography (Jerlov, N., Steemann Nielsen, E., eds.), New York: Academic Press **1974**, 25.
74 M 1 McCluney, W.R.: Estimation of Sunlight Penetration in the Sea for Remote Sensing, Goddard Space Flight Center, X-913-74-113, **1974**, 17 pp.
74 M 2 Morel, A., in: Optical Aspects of Oceanography (Jerlov, N., Steemann Nielsen, Ed., eds.), New York: Academic Press **1974**, 1.
74 M 3 Morel, A., Calouménos, L.: Tethys **6** (1974) 93.
74 S Smith, R.C., in: Optical Aspects of Oceanography (Jerlov, N., Steemann Nielsen, E., eds.), New York: Academic Press **1974**, 95.
74 T Timofeeva, A., in: Optical Aspects of Oceanography (Jerlov, N., Steemann Nielsen, E., eds.), New York: Academic Press **1974**, 177.
74 Z Zaneveld, J.R.V., in: Optical Aspects of Oceanography (Jerlov, N., Steemann Nielsen, E., eds.), New York: Academic Press **1974**, 121.
75 G Gordon, H.R., McCluney, W.R.: Appl. Optics **14** (1975) 413.
75 H Højerslev, N.K.: Limnol. Oceanogr. **20** (6) (1975) 1024.
75 M Morel, A., Prieur, L.: Rapp. Cent. Rech. Océanogr. Villefranche-sur-Mer **17** (1975) 17.
75 R Raschke, E.: Deep-Sea Res. **22** (1975) 659.
75 S Steemann Nielsen, E.: Marine photosynthesis, with special emphasis on the ecological aspects, Amsterdam: Elsevier Scientific Publ. Comp. **1975**, 123 pp.
76 A Austin, R.W., Halikas, G.: The index of refraction of seawater, San Diego: Visibility Lab. SIO ref. **76-1** (1976) 121 pp.
76 H Højerslev, N.K., in: Bæltprojektet, Fysiske undersøgelser (Nielsen, A., ed.), Miljøstyrelsen **1976**, 101.
76 J Jerlov, N.G.: Marine Optics, 2nd ed., Amsterdam: Elsevier Scientific Publ. Comp. **1976**, 203 pp.
76 L Lundgren, B.: Rep. Inst. Phys. Oceanogr., Univ. Copenhagen **30** (1976) 38 pp.
76 P 1 Preisendorfer, R.W.: Hydrologic Optics, U.S. Dept. Comm., NOAA-Environ. Res. Lab. Vol. **I** (1976) 218 pp.
76 P 2 Prieur, L.: Transfert radiatif dans les eaux de mer. Application a la determination de parametres optiques caracterisant leur teneur en substances dissoute et leur conteny en particules. Thése de doctorat d'Etat ès Sciences Physiques, No. d'Enregistrement au C.N.R.S., A.O. 12187, **1976**, 300 pp.
76 S Smith, R.C., Tyler, J.: Photochem. Photobiol. Rev. Vol. **I** (1976) 117.
77 B Bethoux, J.P.: Contribution a l'etude termique de la Mer Mediterranée. Thése de doctorate d'Etat ès Sciences Physiques, No. d'Enregistrement au C.N.R.S., A.O., 24 Mars **1977**, 192 pp.
77 H 1 Højerslev, N.K.: Spectral Daylight Irradiance and Light Transmittance in Natural Waters Measured by Means of a Secchi Disc only. ICES-C.M. 77, Paper C, **42** (1977) 7 pp.
77 H 2 Højerslev, N.K.: Rep. Inst. Phys. Oceanogr., Univ. Copenhagen **32** (1977) 39 pp.
77 H 3 Højerslev, N.K.: Optical measurements at Fladen Ground in March-June 1976 "Meteor-FLEX-76", in FLEX Atlas, SFB 94, Hamburg **1977**, 12 pp.
77 H 4 Højerslev, N.K., Jerlov, N.G.: Rep. Inst. Phys. Oceanogr., Univ. Copenhagen **35** (1977) 12 pp.
77 H 5 Højerslev, N.K., Lundgren, B.: Rep. Inst. Phys. Oceanogr., Univ. Copenhagen **33** (1977) 63 pp.
77 H 6 Højerslev, N.K., Zaneveld, J.R.V.: Rep. Inst. Phys. Oceanogr., Univ. Copenhagen **34** (1977) 16 pp.
77 I Ivanoff, A., in: Modelling and Prediction of the Upper Layers of the Ocean (Kraus, E.B., ed.), Oxford, New York: Pergamon Press **1977**, 47.
77 J Jerlov, N.G.: J. Cons. Int. Explor. Mer **37** (1977) 281.
77 M Morel, A., Prieur, L.: Limnol. Oceanogr. **22** (1977) 709.
78 H Højerslev, N.K.: J. Cons. Int. Explor. Mer **38** (1978) 131.
78 J Jerlov, N.G.: Rep. Inst. Phys. Oceanogr., Univ. Copenhagen **36** (1978) 46 pp.
78 Q Querry, M.R., Cary, P.G., Waring, R.C.: Appl. Opt. **17** (1978) 3587.
79 M Morel, A., Højerslev, N.K.: IAPSO, Radiant energy in the sea, IUGG Canberra **1979**, 77.
79 T Tam, A.C., Patel, C.K.N.: Appl. Opt. **18** (1979) 3348.
80 A Aas, E.: Rep. Inst. Phys. Oceanogr., Univ. Copenhagen **42** (1980) 11.
80 C Clark, D.K., Baker, E.T., Strong, A.E.: Boundary-Layer Meteorol. **18** (1980) 287.
80 G 1 Gordon, H.R., Clark, D.K.: Appl. Opt. **19** (1980) 3428.
80 G 2 Gordon, H.R., Clark, D.K.: Boundary-Layer Meteorol. **18** (1980) 299.
80 H 1 Højerslev, N.K.: Boundary-Layer Meteorol. **18** (1980) 203.

80 H 2 Højerslev, N.K., Larsen, K.: Rep. Inst. Phys. Oceanogr., Univ. Copenhagen **42** (1980) 155.

80 H 3 Hundahl, H., Holck, J.: Rep. Inst. Phys. Oceanogr., Univ. Copenhagen **42** (1980) 145.

80 M Morel, A.: Boundary-Layer Meteorol. **18** (1980) 177.

80 S Sturm, B.: XXVII Rassegna Internatzionale Elettronica Nucleare ed Aerospeziale, Rome, March 7–16, **1980**.

81 A Austin, R.W., Petzold, T.J., in: Oceanography from space (Gower, J.F.R., ed.), New York: Plenum Press **1981**, 239.

81 C Clark, D.K., in: Oceanography from space (Gower, J.F.R., ed.), New York: Plenum Press **1981**, 227.

81 H 1 Højerslev, N.K.: Optical properties of ocean waters; Int. Assoc. Met. Atm. Phys., Radiation Comm., IAMAP rep. on radiation transfer in the oceans and remote sensing of ocean properties, Hamburg **1981**, p. 3–9.

81 H 2 Højerslev, N.K., in: Oceanography from space (Gowers, J.F.R., ed.), New York: Plenum Press **1981**, 347.

81 P Prieur, L., Sathyendranath, G.: Limnol. Oceanogr. **26** (1981) 671.

81 S 1 Smith, R.C., Baker, K.S.: Appl. Opt. **20** (1981) 177.

81 S 2 Smith, R.C., Wilson, W.H., in: Oceanography from space (Gower, J.F.R., ed.), New York: Plenum Press **1981**, 281.

82 H 1 Højerslev, N.K., in: The role of solar ultraviolet radiation in marine ecosystems (Calkins, J., ed.), New York: Plenum Press **1982**, 263.

82 H 2 Højerslev, N.K.: J. Cons. Int. Explor. Mer **40** (1982) 272.

82 S Smith, R.C., Baker, K.S.: J. Mar. Biol. **66** (1982) 269.

83 M Monahan, E.C., Spillane, M.C., Bowyer, P.A., Doyle, D.M., Stabeno, P.J.: Rep. Univ. College, Galway **5** (1983) 92 pp.

83 S Sathyendranath, S., Morel, A., in: Remote Sensing Applications in Marine Science and Technology (Cracknell, A.P., ed.), Dordrecht, Boston, London: Reidel Publ. Comp. **1983**, 323.

84 G Gordon, H.R., Morel, A., in: Lecture notes in coastal and estuarine studies series, New York: Springer **1984**, 114 pp.

84 H 1 Højerslev, N.K.: Rep. Inst. Phys. Oceanogr., Univ. Copenhagen **47** (1984) 32 pp.

84 H 2 Højerslev, N.K.: The solar heat input into the upper marine environment, in: Large-scale oceanographic experiments and satellites (Gauthier, C., Fieux, M., eds.), Dordrecht, Boston, London: Reidel Publ. Comp. **1984**, pp. 247–256.

85 H 1 Højerslev, N.K.: Occurrence and optical effects of gelbstoff, in press.

85 H 2 Højerslev, N.K.: Bio-optical measurements in the Southwest Florida Shelf ecosystem, J. Cons. Int. Explor. Mer **42** (1985) 65.

86 H Højerslev, N.K.: unpublished.

3.4 Chemical properties of sea water

3.4.0 List of symbols and abbreviations

A	total mass of an element in the oceans
C	concentration of a constituent
C_{35}	concentration of a constituent at $S=35$
Cl	chlorinity
K_{15}	electrical conductivity ratio of sea water and KCl solution with KCl mass fraction of $32.4356 \cdot 10^{-3}$, at $T=15\,°C$ and $p=101325$ Pa
L	solubility of a gas in sea water from air saturated with water, at $p=101325$ Pa, in [cm^3 (STP) dm^{-3}], [mol kg^{-1}]
p	pressure, in [Pa]
S	practical salinity; salinity according to old salinity scales, in [‰]
S_A	absolute salinity
T	temperature, in [°C], [K]
t	time, in [a]
t_r	residence time of an element in the oceans, in [a]
DOC	dissolved organic C
STP	standard temperature (0 °C) and pressure (1 standard atmosphere ≙ 101325 Pa)

3.4.1 General remarks

Sea water contains a great number of inorganic and organic substances. The bulk of these substances is dissolved in the water, the rest is suspended as particles. By convenience dissolved substances are defined as those which pass a 0.5 μm filter. This division implies that most of the colloidal forms are included in the dissolved matter.

The suspended particulate matter in sea water is very variable in composition. Its inorganic fraction consists mainly of minerals such as feldspar, clays and quartz, or of siliceous and calcareous components of living or dead organisms. The organic particles are mainly living organisms and their decay products. The waters of the open ocean contain very little suspended particulate matter.

The tables of this section refer to substances which are dissolved in the sea water. Particulate matter is normally excluded.

A lot of substances enter the sea as contaminants or pollutants, mobilized or produced by man. These substances are mentioned here only marginally. For recent textbooks about the pollution of the marine environment, see, for instance, [72 D, 72 R, 75 G, 76 G 1, 76 J, 81 G].

3.4.2 Chlorinity, chlorosity, salinity

(cf. sect. 3.1) [79 I, 81 U 1, 81 U 2]

The chlorinity of sea water was defined in 1902 as the mass, in [g], of Cl equivalent to the mass of halogens contained in 1 kg of sea water. In order to make the chlorinity independent of changes in the accepted atomic masses of Cl and Ag (used for titration), it was later (1937) redefined:

The *chlorinity* of a sample of sea water represents 0.3285234 times the ratio of the mass of pure reference Ag ("Atomgewichtssilber") necessary to precipitate the halogens contained in the sample to the mass of this sample.

The mean chlorinity of all seas is close to $19.22 \cdot 10^{-3}$.

Chlorosity is defined according to chlorinity, but is related to 1 dm^3 of sea water at 20 °C, and not to 1 kg.

The salinity of sea water was defined in 1902 as the mass, in [g], of dissolved inorganic matter in 1 kg of sea water after all bromide and iodide have been replaced by the equivalent amount of chloride, and all carbonate converted to oxide.

The original relationship between chlorinity Cl and salinity S was

$$S = 0.030 + 1.8050\,Cl, \tag{1}$$

with both S and Cl in [‰].

The adoption of precise conductometric methods for the determination of the salinity of sea water lead to a new definition of salinity in terms of (relative) electrical conductivity in 1966, see sect. 3.1, eqs. (2)···(4). This definition was based on an extensive survey of the conductivity – chlorinity relationship of water samples from all parts of the oceans. Chlorinity was related to this newly defined salinity by the equation

$$S = 1.80655\,Cl. \tag{2}$$

This equation gives results identical to eq. (1) at a salinity of $35.00 \cdot 10^{-3}$.

However, even this definition of salinity appeared unsatisfactory for several reasons. So, salinity was again redefined in 1978. The new concept distinguishes between two different quantities:

1. *Absolute salinity*, symbol S_A, is defined as the ratio of mass of dissolved material in sea water to the mass of sea water.

2. The *practical salinity*, symbol S, is defined in terms of the ratio K_{15} of the electrical conductivity of a sea water sample at the temperature of 15 °C and the pressure of 1 standard atmosphere (101 325 Pa) to that of a KCl solution in which the mass fraction of KCl is $32.4356 \cdot 10^{-3}$, at the same temperature and pressure. For details, see sect. 3.1, eq. (5). The K_{15} value exactly equal to 1 corresponds, by definition, to a practical salinity exactly equal to 35.

The standard KCl solution has the same conductivity at 15 °C and 1 standard atmosphere (101 325 Pa) as sea water from the North Atlantic of chlorinity $19.374 \cdot 10^{-3}$ (equivalent to a salinity of $35.000 \cdot 10^{-3}$ on the old salinity scale) and thus provides continuity with previous salinity scales.

It should be noted that the practical salinity values are 1000 times the values of the salinity of the same samples of sea water obtained on the previous scales. For instance, a sample of sea water having a salinity of 0.03500 (= 35.00‰) on the old scale will have a practical salinity of 35.00. The absolute salinity S_A of this sample would be about $35.15 \cdot 10^{-3}$.

Chlorinity is henceforth to be regarded as a separate, independent variable in describing the properties of sea water.

The new definitions of absolute salinity and practical salinity, 1978, are to be used by all oceanographers from 1 January 1982 onward. The new tables [81 U 2] are intended to replace the old ones (1966, 1971).

In the open oceans, the practical salinity of the surface water varies between 32 and 38. Lower salinities are found near the coasts because of the discharge of rivers. Landlocked seas show the greatest range of salinities from 1 in the northern Baltic Sea to 40 in the Mediterranean Sea and 41 in the Red Sea. The mean salinity of all sea waters is about 34.72.

3.4.3 Chemical composition of sea water

[71 R, 75 D, 75 R]

For geochemical reasons we can assume that all the elements found on the continents are also contained in the sea. At present time, about 80 elements have been identified or determined in sea water. With the development of more sensitive analytical methods, it is probable that the other naturally occurring elements will soon be detected.

The concentrations of the elements in sea water differ considerably, ranging from 20 g dm^{-3} for chlorine to less than 1 ng dm^{-3} for some rare elements.

The chemical species of the elements in sea water are more or less complicated. Only the noble gases exist in the elemental form. All the other elements are present as inorganic molecules, simple or complex ions, undissociated acids or hydroxides, ion pairs, organic molecules, etc. [75 S 2].

Table 1 contains the average abundances of the elements dissolved in ocean water and the probable main dissolved chemical species.

For practical reasons it is appropriate to treat the major constituents, the minor constituents, the dissolved atmospheric gases, and the organic compounds separately.

3.4.4 Residence times of elements in the oceans

[71 G]

The most useful approach for the description of the reactivity of an element in the oceanic chemical system is the residence time. This is defined as the average time an element remains in ocean water between its introduction into the water (mainly by rivers) and its removal (mainly by incorporation into the sediments). The residence time t_r can be calculated from

$$t_r = A(\mathrm{d}A/\mathrm{d}t)^{-1}, \qquad (4)$$

where A is the total mass of the element dissolved in the oceans, and $\mathrm{d}A/\mathrm{d}t$ is the mass introduced to or removed from the oceans per time.

Elements which are reactive in sea water, such as Al, Fe, Th, and Pb, generally have short residence times, while elements with low geochemical reactivity like Na, K, Mg, Cl, Br, and B have long residence times.

A classification of elements according to their residence times is given in Table 2 which is based upon river input data.

Table 2. Elements classified according to their residence times t_r in the oceans, mainly based on [71 G].

t_r [a]	Elements							
$10^2 \cdots 3\cdot10^2$	Al	Fe	Th					
$3\cdot10^2 \cdots 10^3$	La	Pb						
$10^3 \cdots 3\cdot10^3$								
$3\cdot10^3 \cdots 10^4$	Cr	Mn	Ga	Sb				
$10^4 \cdots 3\cdot10^4$	Si	Ti	Cu	Zn	Se			
$3\cdot10^4 \cdots 10^5$	Sc	V	Co	Ni	As	Ag	Ba	Hg
$10^5 \cdots 3\cdot10^5$	P	W	Au					
$3\cdot10^5 \cdots 10^6$	F	Mo	J	Cs				
$10^6 \cdots 3\cdot10^6$	Li	Ca	U					
$3\cdot10^6 \cdots 10^7$	K	Rb	Sr					
$10^7 \cdots 3\cdot10^7$	B	Mg	S					
$3\cdot10^7 \cdots 10^8$	Na	Cl	Br					

In the concept of the residence time a steady-state system is assumed in which the rate of addition of an element to the sea water is compensated by the rate of removal. A further assumption is that complete mixing of the input takes place in times that are short compared to the residence times. This cannot be true for the reactive elements, because the deep water circulation of the oceans requires about $2\cdot10^3$ years. Furthermore, many data for A and especially for $\mathrm{d}A/\mathrm{d}t$ are not at all certain. From this it is evident that numbers for residence times should be used with caution. For the same reason the relatively broad classification in Table 2 seems to be justified.

It could be expected that the concentration of an element in the oceans would be regulated by the solubility of the least soluble compound which it forms in sea water. However, this seems to be true only for a small number of elements, e.g. La which is precipitated as $LaPO_4$. For the other elements the observed concentrations in sea water are much lower than those calculated from solubility equilibria. In other words, most of the elements are undersaturated in the oceans, partly by some orders of magnitude. This fact indicates that the removal of many elements from sea water is accomplished by other mechanisms, such as precipitation or co-precipitation with the hard parts of organisms (e.g. the shells of diatoms), with mineral particles (e.g. alumo-silicates), or with particulate iron or manganese hydroxides.

Weichart

3.4.5 Solubility of gases in sea water

[75 K, 75 S 1]

The solubility of a gas in sea water decreases with increasing temperature and salinity and is directly proportional to its partial pressure in the gas phase. At equilibrium, the partial pressure in the sea water is equal to the partial pressure in the atmosphere.

At the surface of the oceans, the gases are equilibrated between the atmosphere and the water according to their solubilities. In the case of total CO_2 two solubilities are involved. One is the normal (physical) solubility of molecular CO_2 which follows the rules mentioned above. The other is a "chemical solubility". The concentration of the chemically bound CO_2 (HCO_3^- and CO_3^{2-}) is by more than two orders of magnitude greater than that of the molecular CO_2 (cf. subsect. 3.4.3.3).

When the sea water loses contact with the atmosphere and moves downwards, it carries with it a near equilibrium concentration of each atmospheric gas. For most of the gases, except O_2 and CO_2, these concentrations are preserved almost unchanged as long as the water stays in the deep (cf. subsect. 3.4.3.3).

For practical oceanographic purposes the solubilities of atmospheric gases in sea water are generally defined in terms of the volumes, in [cm^3], of the gases at standard temperature and pressure (STP), i.e. at $T = 0\,°C$ and $p = 101\,325$ Pa, which will be dissolved by 1 dm^3 of the water from air saturated with water vapour, under a total pressure of 1 standard atmosphere (101 325 Pa).

For comparative purposes oceanographers often express their results in terms of percentage saturation, which is calculated from the observed concentration of the gas concerned and its solubility in sea water of the in-situ temperature and salinity.

The solubility of gases in sea water can be expressed by the equation:

$$\ln L = A_1 + A_2(100/T) + A_3 \ln(T/100) + A_4(T/100) + S[B_1 + B_2(T/100) + B_3(T/100)^2]\,. \tag{5}$$

Herein is

L the solubility of the gas concerned from air saturated with water, at a total pressure of 1 standard atmosphere (101 325 Pa),
T the absolute temperature, in [K], and
S the practical salinity.

In Table 3 the solubility of O_2 in sea water is listed for different temperatures and salinities. The data were calculated from eq. (5) using the following constants, which yield the solubility in [cm^3 (STP) dm^{-3}]

$A_1 = -173.4292$ $B_1 = -0.033096$
$A_2 = 249.6339$ $B_2 = 0.014259$
$A_3 = 143.3483$ $B_3 = -0.0017000$.
$A_4 = -21.8492$

The solubility of N_2 in sea water in [cm^3 (STP) dm^{-3}] can be calculated from eq. (5) substituting the following constants:

$A_1 = -172.4965$ $B_1 = -0.049781$
$A_2 = 248.4262$ $B_2 = 0.025018$
$A_3 = 143.0738$ $B_3 = -0.003486$.
$A_4 = -21.7120$

A set of data, in [cm^3 (STP) dm^{-3}], for the solubility of N_2 at different temperatures and salinities is given in Table 4.

The solubilities of N_2, O_2, and the noble gases in sea water at a salinity of 35 for selected temperatures can be obtained from Table 5, in [μmol kg^{-1}] or [nmol kg^{-1}].

For Table 3, see next page.

Table 4. Solubility L of N_2 in sea water from air saturated with water, at a total pressure of 1 standard atmosphere (101325 Pa), based on [69 M, 70 W].

T [°C]	Salinity 0	10	20	30	34	35	36	38	40
	L [cm^3 (STP) dm^{-3}]								
0	18.42	17.10	15.87	14.73	14.30	14.19	14.09	13.88	13.67
1	17.95	16.67	15.48	14.38	13.96	13.86	13.75	13.55	13.35
2	17.50	16.26	15.11	14.04	13.64	13.54	13.44	13.24	13.05
3	17.07	15.87	14.75	13.72	13.32	13.23	13.13	12.94	12.76
4	16.65	15.49	14.41	13.41	13.03	12.93	12.84	12.66	12.47
5	16.26	15.13	14.09	13.11	12.74	12.65	12.56	12.38	12.21
6	15.88	14.79	13.77	12.83	12.47	12.38	12.29	12.12	11.95
8	15.16	14.14	13.18	12.29	11.95	11.87	11.79	11.62	11.46
10	14.51	13.54	12.64	11.80	11.48	11.40	11.32	11.17	11.01
12	13.90	12.99	12.14	11.34	11.04	10.96	10.89	10.74	10.60
14	13.34	12.48	11.67	10.92	10.63	10.56	10.49	10.35	10.21
16	12.83	12.01	11.24	10.53	10.25	10.19	10.12	9.99	9.68
18	12.35	11.57	10.84	10.16	9.90	9.84	9.77	9.65	9.52
20	11.90	11.16	10.47	9.82	9.57	9.51	9.45	9.33	9.21
22	11.48	10.78	10.12	9.50	9.26	9.21	9.15	9.03	8.92
24	11.09	10.42	9.79	9.20	8.89	8.92	8.87	8.76	8.65
26	10.73	10.09	9.49	8.92	8.71	8.65	8.60	8.50	8.39
28	10.38	9.77	9.20	8.66	8.45	8.40	8.35	8.25	8.15
30	10.06	9.48	8.93	8.41	8.21	8.16	8.12	8.02	7.92

Table 5. Solubility L of N_2, O_2, and noble gases in sea water from air saturated with water, at a total pressure of 1 standard atmosphere (101325 Pa), and a salinity of 35, based on [75 K].

T [°C]	L [µmol kg^{-1}]			L [nmol kg^{-1}]			
	N_2	O_2	Ar	Ne	Kr	He	Xe
0	616.4	349.5	16.98	7.88	4.1	1.77	0.66
5	549.6	308.1	15.01	7.55	3.6	1.73	0.56
10	495.6	274.8	13.42	7.26	3.1	1.70	0.46
15	451.3	247.7	12.11	7.00	2.7	1.68	0.39
20	414.4	225.2	11.03	6.77	2.4	1.66	0.33
25	383.4	206.3	10.11	6.56	2.2	1.65	0.29
30	356.8	190.3	9.33	6.36	2.0	1.64	0.25

Table 3. Solubility L of O_2 in sea water from air saturated with water, at a total pressure of 1 standard atmosphere (101325 Pa) [73 U].

T [°C]	Salinity 0	2	4	6	8	10	12	14	16	18	20	22	24	26	28	30	32	34	35	36	38	40
	L [cm^3 (STP) dm^{-3}]																					
0	10.22	10.08	9.94	9.81	9.67	9.54	9.41	9.29	9.16	9.04	8.91	8.79	8.67	8.56	8.44	8.32	8.21	8.10	8.05	7.99	7.88	7.77
1	9.94	9.80	9.67	9.54	9.41	9.28	9.16	9.04	8.91	8.79	8.68	8.56	8.44	8.33	8.22	8.11	8.00	7.89	7.84	7.78	7.68	7.58
2	9.67	9.54	9.41	9.28	9.16	9.04	8.92	8.80	8.68	8.56	8.45	8.34	8.22	8.11	8.01	7.90	7.79	7.69	7.64	7.59	7.48	7.38
3	9.41	9.28	9.16	9.04	8.92	8.80	8.68	8.57	8.45	8.34	8.23	8.12	8.01	7.91	7.80	7.70	7.60	7.50	7.45	7.40	7.30	7.20
4	9.16	9.04	8.92	8.81	8.69	8.57	8.46	8.35	8.24	8.13	8.02	7.92	7.81	7.71	7.61	7.51	7.41	7.31	7.26	7.22	7.12	7.03
5	8.93	8.81	8.70	8.58	8.47	8.36	8.25	8.14	8.03	7.93	7.83	7.72	7.62	7.52	7.42	7.33	7.23	7.14	7.09	7.04	6.95	6.86
6	8.70	8.59	8.48	8.37	8.26	8.15	8.05	7.94	7.84	7.74	7.64	7.54	7.44	7.34	7.25	7.15	7.06	6.97	6.92	6.88	6.79	6.70
7	8.49	8.38	8.27	8.16	8.06	7.95	7.85	7.75	7.65	7.55	7.45	7.36	7.26	7.17	7.08	6.98	6.89	6.81	6.76	6.72	6.63	6.55
8	8.28	8.17	8.07	7.97	7.86	7.76	7.66	7.57	7.47	7.37	7.28	7.19	7.09	7.00	6.91	6.82	6.74	6.65	6.61	6.57	6.48	6.40
9	8.08	7.98	7.88	7.78	7.68	7.58	7.48	7.39	7.30	7.20	7.11	7.02	6.93	6.84	6.76	6.67	6.59	6.50	6.46	6.42	6.34	6.26
10	7.89	7.79	7.69	7.60	7.50	7.41	7.31	7.22	7.13	7.04	6.95	6.86	6.78	6.69	6.61	6.52	6.44	6.36	6.32	6.28	6.20	6.12
11	7.71	7.61	7.52	7.42	7.33	7.24	7.15	7.06	6.97	6.88	6.80	6.71	6.63	6.54	6.46	6.38	6.30	6.22	6.18	6.14	6.07	5.99
12	7.53	7.44	7.35	7.26	7.17	7.08	6.99	6.90	6.82	6.73	6.65	6.56	6.48	6.40	6.32	6.24	6.17	6.09	6.05	6.01	5.94	5.87
13	7.37	7.27	7.18	7.10	7.01	6.92	6.84	6.75	6.67	6.59	6.50	6.42	6.34	6.27	6.19	6.11	6.04	5.96	5.93	5.89	5.82	5.74
14	7.20	7.12	7.03	6.94	6.86	6.77	6.69	6.61	6.53	6.45	6.37	6.29	6.21	6.14	6.06	5.99	5.91	5.84	5.80	5.77	5.70	5.63
15	7.05	6.96	6.88	6.79	6.71	6.63	6.55	6.47	6.39	6.31	6.24	6.16	6.08	6.01	5.94	5.87	5.79	5.72	5.69	5.65	5.58	5.52
16	6.90	6.81	6.73	6.65	6.57	6.49	6.41	6.34	6.26	6.18	6.11	6.03	5.96	5.89	5.82	5.75	5.68	5.61	5.58	5.54	5.48	5.41
17	6.75	6.67	6.59	6.51	6.44	6.36	6.28	6.21	6.13	6.06	5.99	5.91	5.84	5.77	5.70	5.64	5.57	5.50	5.47	5.43	5.37	5.31
18	6.61	6.54	6.46	6.38	6.31	6.23	6.16	6.08	6.01	5.94	5.87	5.80	5.73	5.66	5.59	5.53	5.46	5.40	5.36	5.33	5.27	5.21
19	6.48	6.40	6.33	6.25	6.18	6.11	6.03	5.96	5.89	5.82	5.75	5.69	5.62	5.55	5.49	5.42	5.36	5.29	5.26	5.23	5.17	5.11
20	6.35	6.28	6.20	6.13	6.06	5.99	5.92	5.85	5.78	5.71	5.64	5.58	5.51	5.45	5.38	5.32	5.26	5.20	5.17	5.14	5.07	5.02
21	6.23	6.15	6.08	6.01	5.94	5.87	5.80	5.74	5.67	5.60	5.54	5.47	5.41	5.35	5.28	5.22	5.16	5.10	5.07	5.04	4.98	4.93
22	6.11	6.04	5.97	5.90	5.83	5.76	5.69	5.63	5.56	5.50	5.44	5.37	5.31	5.25	5.19	5.13	5.07	5.01	4.98	4.95	4.89	4.84
23	5.99	5.92	5.85	5.79	5.72	5.65	5.59	5.52	5.46	5.40	5.34	5.28	5.21	5.15	5.10	5.04	4.98	4.92	4.89	4.87	4.81	4.75
24	5.88	5.81	5.74	5.68	5.61	5.55	5.49	5.42	5.36	5.30	5.24	5.18	5.12	5.06	5.01	4.95	4.89	4.84	4.81	4.78	4.73	4.67
25	5.77	5.70	5.64	5.58	5.51	5.45	5.39	5.33	5.27	5.21	5.15	5.09	5.03	4.98	4.92	4.86	4.81	4.75	4.73	4.70	4.65	4.59
26	5.66	5.60	5.54	5.48	5.41	5.35	5.29	5.23	5.17	5.12	5.06	5.00	4.95	4.89	4.83	4.78	4.73	4.67	4.65	4.62	4.57	4.52
27	5.56	5.50	5.44	5.38	5.32	5.26	5.20	5.14	5.08	5.03	4.97	4.92	4.86	4.81	4.75	4.70	4.65	4.60	4.57	4.54	4.49	4.44
28	5.46	5.40	5.34	5.28	5.23	5.17	5.11	5.05	5.00	4.94	4.89	4.83	4.78	4.73	4.67	4.62	4.57	4.52	4.50	4.47	4.42	4.37
29	5.37	5.31	5.25	5.19	5.14	5.08	5.02	4.97	4.91	4.86	4.81	4.75	4.70	4.65	4.60	4.55	4.50	4.45	4.42	4.40	4.35	4.30
30	5.28	5.22	5.16	5.10	5.05	4.99	4.94	4.89	4.83	4.78	4.73	4.68	4.62	4.57	4.52	4.47	4.43	4.38	4.35	4.33	4.28	4.24

Table 3 (continued)

T [°C]	Salinity 0	2	4	6	8	10	12	14	16	18	20	22	24	26	28	30	32	34	35	36	38	40
	L [cm^3 (STP) dm^{-3}]																					
31	5.19	5.13	5.07	5.02	4.96	4.91	4.86	4.80	4.75	4.70	4.65	4.60	4.55	4.50	4.45	4.40	4.36	4.31	4.28	4.26	4.22	4.17
32	5.10	5.04	4.99	4.94	4.88	4.83	4.78	4.73	4.68	4.63	4.58	4.53	4.48	4.43	4.38	4.33	4.29	4.24	4.22	4.20	4.15	4.11
33	5.01	4.96	4.91	4.86	4.80	4.75	4.70	4.65	4.60	4.55	4.50	4.46	4.41	4.36	4.31	4.27	4.22	4.18	4.15	4.13	4.09	4.04
34	4.93	4.88	4.83	4.78	4.73	4.68	4.63	4.58	4.53	4.48	4.43	4.39	4.34	4.29	4.25	4.20	4.16	4.11	4.09	4.07	4.03	3.98
35	4.85	4.80	4.75	4.70	4.65	4.60	4.55	4.51	4.46	4.41	4.36	4.32	4.27	4.23	4.18	4.14	4.10	4.05	4.03	4.01	3.97	3.93

Table 6. Preparation of artificial sea water ($S = 35.0$) according to Kalle [65 K] (corrected).

Solution A		Solution B	
NaCl	239.0 g	$Na_2SO_4 \cdot 10\,H_2O$	90.6 g
$MgCl_2 \cdot 6\,H_2O$	108.3 g	$NaHCO_3$	2.00 g
$CaCl_2$ sicc.	11.5 g	H_3BO_3	0.27 g
KCl	6.82 g	NaF	0.03 g
KBr	0.99 g		
$SrCl_2 \cdot 6\,H_2O$	0.40 g		
Aqua dest.	8560 cm^3	Aqua dest.	1000 cm^3

Solution B is poured into solution A under stirring. A small amount of precipitate is allowed to settle for one day. The supernatant solution is filtered through a paper filter (small pore size, e.g. Schleicher & Schüll Nr. 589).

Table 7. Preparation of artificial sea water ($S = 35.0$) according to Graßhoff [76 G 2], based on [67 K].

Solution 1		Solution 2	
NaCl	23.9 g	$MgCl_2 \cdot 6\,H_2O$	10.8 g
Na_2SO_4	4.0 g	$CaCl_2 \cdot 2\,H_2O$	1.5 g
KCl	0.7 g	$SrCl_2 \cdot 6\,H_2O$	0.025 g
$NaHCO_3$	0.2 g		
KBr	0.1 g		
H_3BO_3	0.030 g		
NaF	0.003 g		
Aqua dest.	500 cm^3	Aqua dest.	455 cm^3

While stirring, solution 2 is added to solution 1. The exact salinity is determined from electrical conductivity.

3.4.6 Artificial sea water

[40 L, 65 K, 76 G 2]

In many chemical and biological investigations artificial sea water of a known composition is needed. Several recipes are in use. Most of them include all the major constituents. The recipes of Kalle [65 K] and of Graßhoff [76 G 2] are given in the Tables 6 and 7 on the previous page.

In most chemical work a synthetic sea water containing only the two main cations and the three main anions serves its purpose. According to Graßhoff [76 G 2] it is prepared by dissolving 32 g NaCl, 14 g $MgSO_4 \cdot 7H_2O$ and 0.2 g $NaHCO_3$ in distilled water, diluting to one liter. The salinity of this water is 34.2.

All the reagents used should be at least of analytical grade. However, even these reagents always contain traces of impurities. Many of them (e.g. Fe, Mn, Cu, Ni, Co, Zn, phosphate and arsenate) can be stripped by a co-precipitation technique: To about one liter of artificial sea water some drops of 1 n NaOH are added, until a slight stable precipitate of magnesium hydroxide is formed. The solution is boiled and left over night for settling of the precipitate. Then the supernatant clear solution is transferred to a clean bottle and the pH is readjusted to 8 by adding some drops of diluted HCl.

3.4.7 References for 3.4

40 L Lyman, J., Fleming, R.: J. Mar. Res. **3** (1940) 134.
65 K Kalle, K., in: Allgemeine Meereskunde (Dietrich, G., Kalle, K., eds.) 43, Berlin: Gebrüder Borntraeger **1965**.
67 K Kester, D.R., Duedall, U.W., Connore, D.N., Pytkowicz, R.M.: Limnol. Oceanogr. **12** (1967) 176.
69 M Murray, C.N., Riley, J.P.: Deep-Sea Res. **16** (1969) 297.
70 W Weiss, R.F.: Deep-Sea Res. **17** (1970) 721.
71 G Goldberg, E.D., Broecker, W.S., Gross, M.G., Turekian, K.K., in: Radioactivity in the Marine Environment, 137, Washington: National Acad. Sciences **1971**.
71 R Riley, J.P., Chester, R.: Introduction to Marine Chemistry, London: Academic Press **1971**.
72 D Dyrssen, D., Jagner, D. (eds.): The Changing Chemistry of the Oceans, Stockholm: Almquist and Wiksell; New York: Wiley and Sons **1972**.
72 R Ruivo, M. (ed.): Marine Pollution and Sea Life, Rome, FAO, West Byfleet, Fishing New Ltd. **1972**.
73 U UNESCO: International Oceanographic Tables, Paris and Wormley (England), Vol. 2 **1973**.
74 B Broecker, W.S.: Chemical Oceanography, New York: Harcourt Brace Jovanovich **1974**.
75 B Brewer, P.G., in: Chemical Oceanography (Riley, J.P., Skirrow, G., eds.), I, 415, London: Academic Press **1975**.
75 D Dietrich, G., Kalle, K., Krauss, W., Siedler, G.: Allgemeine Meereskunde, Berlin, Stuttgart: Gebrüder Borntraeger **1975**.
75 G Goldberg, E.D., in: Chemical Oceanography (Riley, J.P., Skirrow, G., eds.), III, 39, London: Academic Press **1975**.
75 K Kester, D.R., in: Chemical Oceanography (Riley, J.P., Skirrow, G., eds.), I, 497, London: Academic Press **1975**.
75 R Riley, J.P., in: Chemical Oceanography (Riley, J.P., Skirrow, G., eds.), III, 193, London: Academic Press **1975**.
75 S 1 Sirrow, G., in: Chemical Oceanography (Riley, J.P., Skirrow, G., eds.), II, 1, London: Academic Press **1975**.
75 S 2 Stumm, W., Brauner, P.A., in: Chemical Oceanography (Riley, J.P., Skirrow, G., eds.), I, 173, London: Academic Press **1975**.
75 W 1 Williams, P.J., in: Chemical Oceanography (Riley, J.P., Skirrow, G., eds.), II, 301, London: Academic Press **1975**.
75 W 2 Wilson, T.R.S., in: Chemical Oceanography (Riley, J.P., Skirrow, G., eds.), I, 365, London: Academic Press **1975**.
76 G 1 Goldberg, E.D.: The Health of the Oceans, Paris: UNESCO Press **1976**.
76 G 2 Graßhoff, K. (ed.): Methods of Seawater Analysis, Weinheim: Verlag Chemie **1976**.
76 J Johnston, R.J. (ed.): Marine Pollution, London: Academic Press **1976**.
79 I IAPSO: SUN-Report, Publication Scientifique Nr. 31, Paris: IUGG Publications Office **1979**.
81 D Duursma, E.K., Dawson, R. (eds.): Marine organic chemistry, Amsterdam: Elsevier **1981**.
81 G Gerlach, S.A.: Marine Pollution, Diagnosis and Therapy, Berlin, Heidelberg, New York: Springer **1981**.
81 U 1 UNESCO: Background papers and supporting data on the Practical Salinity Scale 1978. UNESCO Tech. Pap. Mar. Sci. Nr. 37, Paris: UNESCO **1981**.
81 U 2 UNESCO: International oceanographic tables. UNESCO Tech. Pap. Mar. Sci. Nr. 39, III, Paris: UNESCO **1981**.
81 W Whitfield, M., Turner, D.R., Dickson, A.G., in: River Inputs to Ocean Systems, 132, New York: United Nations **1981**.
83 W Wong, C.S., Boyle, E., Bruland, K.W., Burton, J.D., Goldberg, E.D.: Trace Metals in Sea Water, Nato Conference Series, New York and London: Plenum Press **1983**.

Weichart

3.4.3.1 Major constituents, alkalinity, pH

[75 W 2]

The major constituents of sea water have been defined as those which make a significant contribution to salinity. This means that all constituents present at concentrations higher than 1 mg dm^{-3} belong to this group, namely Cl^-, Na^+, SO_4^{2-}, Mg^{2+}, Ca^{2+}, K^+, HCO_3^-, Br^-, Sr^{2+}, H_3BO_3, and F^-. $Si(OH)_4$, although present at 2 mg dm^{-3}, is conventionally excluded, because, as a consequence of its biological activity, its concentration is subject to large variations. In Table 1 the major constituents are given in bold types.

In the oceans the concentrations of the major constituents are almost constant. In adjacent seas like the Mediterranean, the North Sea, etc., their ratios are almost constant, and their concentrations nearly proportional to salinity. So, the major constituents are usually said to be conservative.

For the practical use, the concentration of a major constituent in a sea water sample can be calculated from the equation

$$C \approx \frac{S}{35} C_{35}. \qquad (3)$$

Herein is

C the concentration of the constituent in the sample,
C_{35} the concentration of the constituent at $S=35$, to be taken from Table 1,
S the practical salinity of the sample.

Under typical sea conditions, most of the major constituents (e.g. Cl^-, Na^+, SO_4^{2-}, K^+) follow eq. (3) very well (within the limits of analytical error), while others show greater deviations (for HCO_3^- up to $\pm 10\%$).

Under some atypical conditions, considerable irregularities may be found. Such conditions can occur in estuaries and fjords, in land-locked seas and anoxic basins, in areas with submarine volcanism or with admixture of geological brines, etc.

While the major cations do all belong to strong bases, some of the major anions, namely HCO_3^-, CO_3^{2-}, and $B(OH)_4^-$, belong to weak acids. This fact is responsible for the buffer capacity, the alkalinity, and the slightly alkaline reaction (pH) of sea water.

The alkalinity is the sum of the anions of H_2CO_3 and H_3BO_3. It is expressed in equivalent per dm^3 or in equivalent per kg. It corresponds to the amount of strong univalent acid, in [mol], required to neutralize, respectively, 1 dm^3 or 1 kg of sea water. In the open oceans the alkalinity is almost proportional to salinity. At a salinity of 35 (like in Table 1), the alkalinity is close to $2.45 \cdot 10^{-3}$ equivalent per dm^3.

In the open ocean the pH of sea water is relatively constant and only rarely falls outside the range 7.8 to 8.4. More extreme values may be found in restricted locations such as estuaries, land-locked basins, etc.

3.4.3.2 Minor constituents

[75 B]

Unlike the major constituents, most of the minor constituents (see Table 1) are not conservative in sea water. This is partly because of their greater geochemical or biological reactivity, and partly because processes like biological uptake or input from land will produce much greater relative effects on constituents occurring at low concentrations.

Variations in the concentrations of minor constituents may be caused by natural processes or by human activity (contamination, pollution). In detail, we have to take into account: uptake or release by organisms (e.g. of phosphate, nitrate, silicate, and some trace metals) which normally result in a vertical transport from the surface into deeper layers or to the bottom, adsorption onto inorganic particles in the water and subsequent downwards transport, run-off from land, fallout or rainout from the atmosphere, introduction from the sea floor (e.g. by submarine volcanoes, or geological brines), etc. For some of the minor constituents the variations may reach one or two, for the micronutrients (phosphate, nitrate, silicate) even more than two orders of magnitude.

In addition, some of the data on the minor constituents listed in Table 1 are unreliable owing to considerable uncertainties in the analytical procedures. Furthermore, only few of the published analyses are those of waters from below the thermocline which would be more representative of the oceans as a whole than surface waters.

Thus, it is evident that for a number of minor constituents the data in Table 1 have to be used with caution.

Weichart

Table 1. Average abundances of chemical elements dissolved in ocean water ($S = 35.00$, $T \approx 10\,°C$) and probable main dissolved chemical species in oxic sea water (mainly according to [75 B, 75 R, 75 W 2, 81 W, 83 W] with additions and corrections). Elements are listed in the order of increasing atomic number. Major constituents in bold face.

Element		Concentration C_{35} $\mu g\,dm^{-3}$	Probable main species
Hydrogen	**H**	$1.10 \cdot 10^8$	H_2O
Helium	He	$7.0 \cdot 10^{-3}$	He (gas)
Lithium	Li	$1.8 \cdot 10^2$	Li^+
Beryllium	Be	$5 \cdot 10^{-3}$	$Be(OH)^+$
Boron	**B**	$4.6 \cdot 10^3$	H_3BO_3, $B(OH)_4^-$
Carbon	**C**	$2.8 \cdot 10^4$	HCO_3^-, $MgCO_3$
		$1 \cdot 10^3$	organic compounds
		$2 \cdot 10^2$	CO_2 (gas)
Nitrogen	N	$1.4 \cdot 10^4$	N_2 (gas)
		$5.0 \cdot 10^2$	NO_3^-, NO_2^-, organic compounds (under anoxic conditions: NH_4^+)
Oxygen	**O**	$8.80 \cdot 10^8$	H_2O
		$6 \cdot 10^3$	O_2 (gas)
Fluorine	**F**	$1.3 \cdot 10^3$	F^-, MgF^+
Neon	Ne	0.12	Ne (gas)
Sodium	**Na**	$1.106 \cdot 10^7$	Na^+
Magensium	**Mg**	$1.323 \cdot 10^6$	Mg^{2+}, $MgSO_4$
Aluminum	Al	2	$Al(OH)_4^-$
Silicon	Si	$2 \cdot 10^3$	$Si(OH)_4$
Phosphorus	P	60	HPO_4^{2-}, $MgPO_4^-$, organic compounds
Sulphur	**S**	$9.29 \cdot 10^5$	SO_4^{2-}, $NaSO_4^-$, $MgSO_4$ (under anoxic conditions also HS^-)
Chlorine	**Cl**	$1.987 \cdot 10^7$	Cl^-
Argon	Ar	$4.5 \cdot 10^2$	Ar (gas)
Potassium	**K**	$4.10 \cdot 10^5$	K^+
Calcium	**Ca**	$4.23 \cdot 10^5$	Ca^{2+}, $CaSO_4$
Scandium	Sc	$6 \cdot 10^{-4}$	$Sc(OH)_3$
Titanium	Ti	1	$Ti(OH)_4$
Vanadium	V	2	$H_2VO_4^-$, HVO_4^{2-}
Chromium	Cr	0.3	$Cr(OH)_3$, CrO_4^{2-}
Manganese	Mn	0.1	Mn^{2+}, $MnCl^+$
Iron	Fe	1	$Fe(OH)_2^+$, $Fe(OH)_3$
Cobalt	Co	$5 \cdot 10^{-2}$	Co^{2+}, chloro complexes
Nickel	Ni	1	Ni^{2+}, chloro complexes
Copper	Cu	0.5	$CuCO_3$, Cu^{2+}, hydroxy and chloro complexes
Zink	Zn	2	Zn^{2+}, chloro and hydroxy complexes
Gallium	Ga	$3 \cdot 10^{-2}$	$Ga(OH)_4^-$
Germanium	Ge	$5 \cdot 10^{-2}$	$Ge(OH)_4$
Arsenic	As	3	$HAsO_4^{2-}$, $H_2AsO_4^-$
Selenium	Se	0.1	SeO_3^{2-}
Bromine	**Br**	$6.9 \cdot 10^4$	Br^-
Krypton	Kr	0.2	Kr (gas)
Rubidium	Rb	$1.2 \cdot 10^2$	Rb^+
Strontium	**Sr**	$8.2 \cdot 10^3$	Sr^{2+}, $SrSO_4$
Yttrium	Y	$1 \cdot 10^{-3}$	$Y(OH)_3$, carbonate complexes
Zirconium	Zr	$3 \cdot 10^{-2}$	$Zr(OH)_4$
Niobium	Nb	$1 \cdot 10^{-2}$	

Table 1 (continued)

Element		Concentration C_{35} µg dm^{-3}	Probable main species
Molybdenum	Mo	10	MoO_4^{2-}
Technetium	Tc		
Ruthenium	Ru		
Rhodium	Rh		
Palladium	Pd		
Silver	Ag	$4 \cdot 10^{-2}$	$AgCl_2^-$, $AgCl_3^{2-}$
Cadmium	Cd	$4 \cdot 10^{-2}$	$CdCl_2$, $CdCl^+$
Indium	In	$1 \cdot 10^{-4}$	$In(OH)_2^+$
Tin	Sn	$1 \cdot 10^{-2}$	hydroxy complexes
Antimony	Sb	0.2	hydroxy complexes
Tellurium	Te		$HTeO_3^-$
Iodine	I	60	IO_3^-, I^-
Xenon	Xe	$5 \cdot 10^{-2}$	Xe (gas)
Caesium	Cs	0.4	Cs^+
Barium	Ba	20	Ba^{2+}, $BaSO_4$
Lanthanum	La	$3 \cdot 10^{-3}$	La^{3+} and hydroxy complexes
Cerium	Ce	$2 \cdot 10^{-3}$	Ce^{3+} and hydroxy complexes
Praseodymium	Pr	$6 \cdot 10^{-4}$	Pr^{3+} and hydroxy complexes
Neodymium	Nd	$3 \cdot 10^{-3}$	Nd^{3+} and hydroxy complexes
Promethium	Pm		Pm^{3+} and hydroxy complexes
Samarium	Sm	$5 \cdot 10^{-4}$	Sm^{3+} and hydroxy complexes
Europium	Eu	$1 \cdot 10^{-4}$	Eu^{3+} and hydroxy complexes
Gadolinium	Gd	$1 \cdot 10^{-3}$	Gd^{3+} and hydroxy complexes
Terbium	Tb	$1 \cdot 10^{-4}$	Tb^{3+} and hydroxy complexes
Dysprosium	Dy	$1 \cdot 10^{-3}$	Dy^{3+} and hydroxy complexes
Holmium	Ho	$2 \cdot 10^{-4}$	Ho^{3+} and hydroxy complexes
Erbium	Er	$1 \cdot 10^{-3}$	Er^{3+} and hydroxy complexes
Thulium	Tm	$2 \cdot 10^{-4}$	Tm^{3+} and hydroxy complexes
Ytterbium	Yb	$1 \cdot 10^{-3}$	Yb^{3+} and hydroxy complexes
Lutetium	Lu	$2 \cdot 10^{-4}$	hydroxy and carbonate complexes
Hafnium	Hf	$7 \cdot 10^{-3}$	hydroxy complexes
Tantalum	Ta	$2 \cdot 10^{-3}$	
Tungsten	W	0, 1	WO_4^{2-}
Rhenium	Re	$4 \cdot 10^{-3}$	ReO_4^-
Osmium	Os		
Iridium	Ir		
Platinum	Pt		
Gold	Au	$4 \cdot 10^{-3}$	$AuCl_2^-$
Mercury	Hg	$5 \cdot 10^{-3}$	$HgCl_4^{2-}$, $HgCl_2$
Thallium	Tl	$1 \cdot 10^{-2}$	Tl^+, hydroxy complexes of Tl^{3+}(?)
Lead	Pb	$5 \cdot 10^{-3}$	carbonate, chloro and hydroxy complexes
Bismuth	Bi	$2 \cdot 10^{-2}$	BiO^+, $Bi(OH)_2^+$
Polonium	Po		PoO_3^{2-}
Astatine	At		
Radon	Rn	$6 \cdot 10^{-13}$	Rn (gas)
Francium	Fr		
Radium	Ra	$7 \cdot 10^{-8}$	Ra^{2+}
Actinium	Ac		
Thorium	Th	$1 \cdot 10^{-4}$	$Th(OH)_4$
Protactinium	Pa		
Uranium	U	3.2	$UO_2(CO_3)_3^{4-}$

Weichart

3.4.3.3 Dissolved atmospheric gases

[75 K, 75 S 1]

The noble gases behave in the sea in a conservative manner. The same is true for N_2, although it can be converted to and from fixed N by some organisms. The concentrations of these gases in the sea (see Table 1) are mainly determined by their atmospheric content and their solubility.

On the other hand, O_2 and CO_2 are heavily involved with the biological and chemical processes taking place within the sea. O_2 can vary within wide limits. In cold surface waters with high primary production the oxygen concentration can rise up to almost $10\,cm^3\,dm^{-3}$, for unit, see subsect. 3.4.5. Below the mixed surface layer it may decrease to zero, e.g. in some intermediate and deep waters of the Pacific, and in the deep waters of the Baltic Sea. The variations of CO_2 are very much moderated by the huge reservoirs of HCO_3^- and CO_3^{2-}, which make up 99% of the total carbon dioxide in the sea (cf. Table 1). The concentrations of total CO_2 in the oceans vary between $2.0\,mmol\,dm^{-3}$ in warm surface water, $2.25\,mmol\,dm^{-3}$ in the deep Atlantic, and $2.5\,mmol\,dm^{-3}$ in the deep Pacific. The corresponding concentrations of molecular CO_2 are 0.01, 0.015, and $0.02\,mmol\,dm^{-3}$ [74 B].

The solubilities of O_2, N_2, and noble gases in sea water are listed in Tables 3···5 (see subsect. 3.4.5).

Other gases such as CO, CH_4, and NO_2 are not considered here, because those are normally treated in the context of marine pollution.

3.4.3.4 Organic compounds

[75 W 1, 81 D]

The organic matter in the sea is customarily divided into two categories: dissolved and particulate (cf. subsect. 3.4.1). The concentration of dissolved organic matter usually exceeds the particulate organic fraction by a factor of 10···20. Only 20% of the particulate matter, on average, consists of living cells.

Nearly all the organic matter has originally been produced by organisms. Therefore, the organic compounds in the sea may embrace the whole range of their metabolic and decay products.

Man produces great amounts of organic chemicals which partly enter the sea as contaminants. Serious pollution problems can be caused especailly by such chemicals which were not present in the oceans so far (e.g. chlorinated hydrocarbons). However, as already explained in subsect. 3.4.1, we shall not consider aspects of marine pollution in this context.

The total concentration of dissolved organic carbon (DOC) in ocean waters usually ranges between 0.3 and $3\,mg\,dm^{-3}$, the average lying close to $1\,mg\,dm^{-3}$ (cf. Table 1).

Between the surface and a few hundred meters depth the concentration of DOC is generally a little higher than in the deeper water. The actual concentration depends on the balance between the rate at which dissolved organic compounds are formed by excretion, after-death release, autolysis, etc., and the rate at which they are removed by hydrolysis, biochemical breakdown, etc.

The major fraction of the dissolved organic carbon in the sea consists of a complex material which is very resistant to biochemical attack. The mean age (cf. subsect. 3.4.4) of this fraction is about 2000 years. As a result of polycondensation and polymerisation, its molecular mass is relatively high.

The minor fraction of the dissolved organic carbon is composed of labile compounds which comprise all sorts of anabolic and katabolic products, such as sugars, amino acids, nucleic acids, carboxylic acids, vitamins, enzymes, hydrocarbons, phenols, ketones, aldehydes, metal complexing agents, urea, etc.

The average concentrations of fatty acids, free amino acids, and free sugars in sea water are about $10\,\mu g\,dm^{-3}$ C for each group. The individual compounds of these groups usually fall in the range $0.1\cdots10\,\mu g\,dm^{-3}$. The average concentrations of combined amino acids lie near $50\,\mu g\,dm^{-3}$ C, those of combined sugars near $200\,\mu g\,dm^{-3}$ C.

In the world oceans the turn-over of the labile compounds is about $4\cdot10^{13}\,kg\,a^{-1}$ C, the turn-over of the stable fraction about $1\cdot10^{12}\,kg\,a^{-1}$ C, corresponding to roughly $1\,\mu g\,dm^{-3}\,a^{-1}$ C.

In the upper layer of the oceans down to a depth of a few hundred meters the labile fraction plays an important ecological role, mainly as a source of energy for other marine organisms. Its actual concentration is usually much smaller than that of the stable fraction. Even in the surface layer of highly productive areas the dissolved labile compounds account for less than half of the total dissolved organic carbon.

In waters deeper than a few hundred meters practically all the dissolved organic material consists of the stable compounds which are (at least actually) resistant to bacterial attack.

In Table 1 organic substances are mentioned merely under C, P, and N. In fact, about 10% of the phosphorus and the nitrogen in the oceans are present in the organic form. On the other hand, there are also some heavy metals (e.g. Cu, Ni, and Zn) which are partly complexed or chelated by organics dissolved in the sea water. These organo-metallic species are not included in Table 1, because there exists little reliable knowledge about them.

Weichart